ELEMENTS
OF ENGINEERING
ELECTROMAGNETICS

FIFTH EDITION

ELEMENTS OF ENGINEERING ELECTROMAGNETICS

FIFTH EDITION

Nannapaneni Narayana Rao

Professor of Electrical and Computer Engineering
University of Illinois at Urbana–Champaign

PRENTICE HALL
UPPER SADDLE RIVER, NEW JERSEY 07458

Library of Congress Cataloging-in-Publication Data

Narayana Rao. Nannapaneni;
   Elements of engineering electromagnetics / Nannapaneni Narayana
Rao.—5th ed.
     p. cm.
   Includes bibliographical references and index.
   ISBN 0-13-013201-2
   1. Electromagnetic theory.   I. Title.
QC670.N3 2000   98-52162
621.3'01—dc21   CIP

Associate editor: ALICE DWORKIN
Production supervisor: ANN MARIE KALAJIAN
Editor-in-chief: MARCIA HORTON
Assistant VP of production and manufacturing: DAVID W. RICCARDI
Managing editor: EILEEN CLARK
Manufacturing buyer: PAT BROWN
Creative director: JAYNE CONTI
Cover design: BRUCE KENSELAAR
Copy editor: PETER ZURITA
Editorial assistant: DAN DE PASQUALE

© 2000 by Prentice-Hall, Inc.
Upper Saddle River, New Jersey 07458

10 9 8 7 6 5 4 3 2

ISBN 0-13-013201-2

Prentice-Hall International (UK) Limited, *London*
Prentice-Hall of Australia Pty. Limited, *Sydney*
Prentice-Hall Canada Inc., *Toronto*
Prentice-Hall Hispanoamericana, S.A., *Mexico*
Prentice-Hall of India Private Limited, *New Delhi*
Prentice-Hall of Japan, Inc., *Tokyo*
Prentice-Hall (Singapore) Pte. Ltd, *Singapore*
Editora Prentice-Hall do Brasil, Ltda., *Rio de Janeiro*

Om Shri Ganeshaya Namaha

In deep appreciation of
the profound influence on my professional career
by the late Edward C. Jordan (1910–1991),
renowned author of the famous text book,
"Electromagnetic Waves and Radiating Systems,"
first published in 1950 by Prentice Hall,
and revered head of my department
for 25 years (1954–1979).

Contents

Preface

The introductory textbooks on engineering electromagnetics can be classified broadly into three categories:

Electro-magnetics textbooks and this edition

1. **One-semester** textbooks based on the traditional approach, covering essentially electrostatics and magnetostatics and culminating in Maxwell's equations with some discussion of their applications.
2. **Two-semester** textbooks with the first half or more covering electrostatics and magnetostatics as in category 1 and the remainder devoted to topics associated with electromagnetic waves.
3. **One-** or two-semester textbooks that have deviated from the traditional approach with the degree and nature of deviation depending on the author.

The bulk of the textbooks fall into categories 1 and 2, and only a small minority, including this book, belong to category 3. The deviation from the traditional approach originated with the first edition, a one-semester text with the basic material built up on time-varying fields and their engineering applications so as to enhance its utility for the one-semester student of engineering electromagnetics, while enabling the student to take further (elective) courses in electromagnetics to learn many of the same field concepts and mathematical tools provided by the traditional treatment. In preparing the second edition, a major revision of the first edition was undertaken by expanding the text for one- or two-semester usage to provide flexibility, while preserving the basic philosophy of the first edition, arising from the assertion that as a prerequisite to the first EE course in fields, most schools have an engineering physics course in which the students are exposed to the historical treatment of electricity and magnetism. Subsequent editions have further enhanced the usage by incorporating changes and adding material to satisfy the prerequisite needs pertinent to emerging technologies. For example, the substantial changes leading to the fourth edition have been prompted by the increasing need for coverage of material at the introductory level for application beyond the microwave region into the optical regime of the electromagnetic spectrum with the advent of the era of photonics over-lapping with that of electronics.

Electro-magnetic technologies

In preparing this fifth edition, I have carried the deviation from the tradi-tional approach even further as we are nearing the end of the twentieth century and approaching the dawn of the twenty-first century. To put this in proper perspective, reference is made to the article by Anthony DeMaria,[1] which de-scribes how in the family tree of major technologies that are based on Maxwell's equations, nineteenth century electrically and magnetically based technologies (electromechanics) may be viewed as the grandparents, twentieth century electronics hardware and software as the parents, and the subgroups of the twenty-first century photonics technology as the children, as illustrated in the following diagram (© 1991 IEEE):

Thus, in this edition, I have reorganized the material of the fourth edition and added topics to associate chapters or parts of chapters with the blocks in the family of electromagnetic technologies. To do this, I have divided the book into two parts. Part I, entitled "Fundamental Elements," comprises the first five chapters:

1. Vectors and Fields
2. Fields and materials
3. Maxwell's Equations in Integral form and Boundary Conditions

---

[1] Anthony J. DeMaria, "Maxwell's Children Light the Way," *IEEE Circuits and Devices Magazine,* vol. 7, No. 2, March 1991, pp. 36–43.

4. Maxwell's Equations in Differential Form, Potential Functions, and Energy Storage
5. Uniform Plane Waves and Power Flow in an Electromagnetic Field

These chapters are the same as Chapters 1, 2, 3, 4, and 6, respectively, of the fourth edition. Note that the chapter on uniform plane waves is considered to be fundamental (instead of an application of Maxwell's equations) and immediately follows the chapters on Maxwell's equations. Part II, entitled "Applied Elements," comprises the remaining seven chapters:

6. Field and Line Essentials for Digital Electronics
7. Transmission Lines for Communications
8. Topics in Electric- and Magnetic-Field Systems
9. Guided Wave Principles for Electronics and Optoelectronics
10. Several Topics for Electronics and Photonics
11. Principles of Radiation and Antennas
12. Topics in Numerical Electromagnetics

These chapters constitute a reorganization of the material of Chapters 5, 7, 8, 9, 10, and 11 of the fourth edition with a few additional topics included to provide the coverage to achieve the stated goal, as evident from the titles of the chapters.

Thus, some of the salient features of the thread of development of the material, evident from a reading of the table of contents, are the following:

Thread of development of material

1. Introduces basic concepts of vectors and fields for static as well as time-varying cases at the outset and vector calculus concepts later as needed.
2. Discusses materials following the presentation of electric and magnetic-field concepts prior to the study of Maxwell's equations.
3. Introduces collectively Maxwell's equations for time-varying fields, first in integral form and then in differential form.
4. Considers boundary conditions following Maxwell's equations in integral form, and potential functions and associated equations following Maxwell's equations in differential form.
5. Obtains uniform plane wave solutions by considering the infinite plane current sheet source first in free space and then in a material medium.
6. Provides continuous coverage of field and line topics essential for digital electronics, beginning with p–n junction and circuit elements and progressing through transmission-line time-domain analysis leading to interconnections between logic gates and culminating in crosstalk on transmission lines.
7. Presents the sinusoidal steady-state analysis of transmission lines comprising the topics of standing waves, resonance, power transfer, and matching with emphasis on computer and graphical solutions.
8. Introduces quasistatic approximations of Maxwell's equations as the basis for electric- and magnetic-field systems, important for the study of electromechanics, and considers other topics pertinent to electromechanical systems.

9. Develops principles of guided waves for both electronics and optoelectronics by confining the treatment to one-dimensional waveguides comprising parallel-plate metallic waveguides followed by dielectric slab waveguides.

10. Devotes a chapter to several topics pertinent to electronics and photonics, including two-dimensional metallic waveguides and optical fibers, pulse broadening due to dispersion, interference and diffraction, and wave propagation in an anisotropic medium.

11. Introduces radiation by obtaining the complete field solution to the Hertzian dipole field through the magnetic vector potential, and then develops the basic concepts of antennas.

12. Devotes a chapter to numerical methods comprising the finite-difference method, the method of moments, the finite-element method, and the finite-difference time-domain method, including their formulation and application to the solution of electromagnetic field problems.

While the entire book can be used for a two-semester course sequence, this edition incorporates flexibility to facilitate its adoption for a one-semester course of content depending on the background preparation for the course and the needs of the curriculum, as in the case of the second, third, and fourth editions. Specifically, this edition addresses the needs of a first course required for electrical engineering students as well as computer engineering students, and more useful than the traditional treatment, followed by one or more required or elective courses for electrical engineering students, that build up on the first course. Chapters 3 through 6 constitute the core material for this course, with Chapter 2 serving as a review or an enhancement of what the students already learn in the sophomore physics course on electricity and magnetism.

Teaching and learning aids   As in the previous editions, a number of teaching and learning aids are employed: (1) examples distributed throughout the text, (2) discussions of practical applications of field concepts and phenomena interspersed among presentations of basic subject matter, (3) descriptions of brief experimental demonstrations suitable for presentation in the classroom, (4) summary of material and review questions (**Q**) for each chapter, (5) inclusion of drill problems (**D**) at the end of each section, (6) marginal notes, and (7) keywords (**K**) at the end of each section. I have omitted the PC program listings in BASIC but retained the details pertinent to the computations and the results. New to this edition are review problems (**R**) at the end of each chapter, following the homework problems (**P**). For the book, there are a total of 105 Examples, 162 **D** Problems, 426 **P** Problems, 415 **Q** Questions, and 80 **R** Problems. Answers are provided for 40 % of the **P** Problems.

Acknowledgment   I wish to express my appreciation to the many colleagues at the University of Illinois at Urbana–Champaign who have taught from the previous editions of the book during the period from 1977 to 1998. Listed in alphabetical order are J. Bernhard, M.T. Birand, D.J. Brady, A. Cangellaris, K.Y. Cheng, W.C. Chew, S.L. Chuang, D.H. Cooper, K. Coperich, C. Daft, T.A. DeTemple, M. Feng, S.J. Franke, S. Fridman, L.A. Frizzell, S.D. Gedney, R. Gilbert, S. Gnanalingam, Y. Gu, K.C. Hsieh, S. Iyer, J. Joseph, K. Kim, P.W. Klock, J. Kolodzey, E. Kudeki,

J.P. Leburton, S.W. Lee, C.H. Liu, C. Lu, R.L. Magin, K. Mahadevan, P.E. Mast, E.A. Mechtly, H. Merkelo, E. Michielssen, K.L. Miller, R. Mittra, H. Morkoc, G.C. Papen, A.F. Peterson, A. Rahhal-Arabi, P.L. Ransom, U. Ravaioli, J. Schutt-Aine, C.F. Sechrist, Jr., L.G. Smith, J. Song, A. Steinbach, G. Swenson, D.R. Tanner, R.J. Turnbull, R. Wagner, D. Weile, A.W. Wernik, K. Whites, K.C. Yeh, and J. Zhang. Thanks are also due to the numerous users at other schools. I am grateful to Phillip Christie of the University of Delaware, Robert Nevels of Texas A&M University, and Philip Serafim of Northeastern University, for serving as reviewers of the manuscript for this edition. Many individuals in the department have provided support over the years. Sheryle Carpenter and Laurie Fisher performed the office duties in an admirable manner that ensured the smooth functioning of the office at all times during my tenure as Associate Head of the department since 1987. The typing of the new portions of the manuscript was done by Kelly Voyles in a prompt and skillful manner. As always, I am deeply indebted to my wife Sarojini for her continued understanding and patience.

About the dedication

The first part of the dedication offering Salutations to Lord Ganesha was inspired in part by my visit in January 1999 to the Bandung Institute of Technology, Bandung, Indonesia, where the Image of Ganesha adorns the entrance to the Institute on Jalan Ganesha. With reference to the second part of the dedication pertaining to the profound influence on my professional career by the late E.C. Jordan, I also wish to express my appreciation for the opportunities provided to me by my department heads G.W. Swenson, Jr., E.W. Ernst, T.N. Trick, and S.M. Kang, following his tenure.

N. Narayana Rao

Urbana, Illinois
February 1999

About the Author

Nannapaneni Narayana Rao was born in Kakumanu, Guntur District, Andhra Pradesh, India. Prior to coming to the United States in 1958, he received a B. Sc. degree in physics from the University of Madras in 1952 and a Diploma in Electronics from the Madras Institute of Technology in 1955. In the United States, he attended the University of Washington, receiving M.S. and Ph.D. degrees in Electrical Engineering in 1960 and 1965, respectively. In 1965, he joined the faculty of the Department of Electrical Engineering, now the Department of Electrical and Computer Engineering, of the University of Illinois at Urbana–Champaign, where he carried out research in the general area of ionospheric propagation and authored the undergraduate textbook *Basic Electromagnetics with Applications* (Prentice Hall, 1972), prior to the four previous editions of this book (1977, 1987, 1991, and 1994).

Professor Rao has received numerous awards and honors for his teaching and curricular activities. These include the first Award in Engineering in 1983 from the Telugu Association of North America, an association of Telugu-speaking people of origin in the State of Andhra Pradesh, India, with the citation, "Dedicated teacher and outstanding contributor to electromagnetics"; a plaque of highest appreciation from the Faculty of Technology, University of Indonesia in Jakarta, for curriculum development in 1985–1986; the Campus Undergraduate Instructional Awards in 1982 and 1988, the Everitt Award for Teaching Excellence from the College of Engineering in 1987, the Campus Award for Teaching Excellence and the first Oakley Award for Innovation in Instruction in 1989, and the Halliburton Award for Engineering Education Leadership from the College of Engineering in 1991, all at the University of Illinois at Urbana–Champaign; election to the Fellow grade of the Institute of Electrical and Electronics Engineers (IEEE) in 1989 for contributions to electrical engineering education and ionospheric propagation; the AT&T Foundation Award for Excellence in Instruction of Engineering Students from the Illinois–Indiana Section of the American Society for Engineering Education

(ASEE) in 1991; the ASEE Centennial Certificate in 1993 for exceptional contribution to the ASEE and the profession of engineering; and the IEEE Technical Field Award in Undergraduate Teaching in 1994 with the citation, "For inspirational teaching of undergraduate students and the development of innovative instructional materials for teaching courses in electromagnetics."

PART ONE

FUNDAMENTAL ELEMENTS

A. BASIC FIELD CONCEPTS

1. VECTORS AND FIELDS
2. FIELDS AND MATERIALS

B. MAXWELL'S EQUATIONS AND UNIFORM PLANE WAVES

3. MAXWELL'S EQUATIONS IN INTEGRAL FORM AND BOUNDARY CONDITIONS
4. MAXWELL'S EQUATIONS IN DIFFERENTIAL FORM, POTENTIAL FUNCTIONS, AND ENERGY STORAGE
5. UNIFORM PLANE WAVES AND POWER FLOW IN AN ELECTROMAGNETIC FIELD

A. BASIC FIELD CONCEPTS

Vectors and Fields

Electromagnetics deals with the study of electric and magnetic fields. It is at once apparent that we need to familiarize ourselves with the concept of a field, and in particular with electric and magnetic fields. These fields are vector quantities and their behavior is governed by a set of laws known as Maxwell's equations. The mathematical formulation of Maxwell's equations and their subsequent application in our study of the elements of engineering electromagnetics require that we first learn the basic rules pertinent to mathematical manipulations involving vector quantities. With this goal in mind, we devote this chapter to vectors and fields.

We first study certain simple rules of vector algebra without the implication of a coordinate system and then introduce the Cartesian, cylindrical, and spherical coordinate systems. After learning the vector algebraic rules, we turn our attention to a discussion of scalar and vector fields, static as well as time-varying, by means of some familiar examples. We devote particular attention to sinusoidally time-varying fields, scalar as well as vector, and to the phasor technique of dealing with sinusoidally time-varying quantities. With this general introduction to vectors and fields, in Chapter 2, we study the concepts of electric and magnetic fields, from considerations of the experimental laws of Coulomb and Ampère.

1.1 VECTOR ALGEBRA

In the study of elementary physics, we come across several quantities such as mass, temperature, velocity, acceleration, force, and charge. Some of these quantities have associated with them not only a magnitude but also a direction in space, whereas others are characterized by magnitude only. The former class of quantities are known as *vectors* and the latter class of quantities are known as

Vectors versus scalars

3

scalars. Mass, temperature, and charge are scalars, whereas velocity, acceleration, and force are vectors. Other examples are voltage and current for scalars
and electric and magnetic fields for vectors.

Vector quantities are represented by symbols in boldface roman type (e.g.,
A), to distinguish them from scalar quantities, which are represented by symbols
in lightface italic type (e.g., *A*). Graphically, a vector, say, **A**, is represented by a
straight line with an arrowhead pointing in the direction of **A** and having a
length proportional to the magnitude of **A**, denoted $|\mathbf{A}|$ or simply *A*. Figure 1.1
shows four vectors drawn to the same scale. If the top of the page represents
north, then vectors **A** and **B** are directed eastward, with the magnitude of **B**
being twice that of **A**. Vector **C** is directed toward the northeast and has a magnitude three times that of **A**. Vector **D** is directed toward the southwest and has
a magnitude equal to that of **C**. Since **C** and **D** are equal in magnitude but opposite in direction, one is the negative of the other.

*Unit vector
defined*

Since a vector may have in general an arbitrary orientation in three dimensions, we need to define a set of three reference directions at each and every
point in space in terms of which we can describe vectors drawn at that point. It
is convenient to choose these three reference directions to be mutually orthogonal, as, for example, east, north, and upward, or the three contiguous edges of a
rectangular room. Thus, let us consider three mutually orthogonal reference directions and direct *unit vectors* along the three directions as shown, for example,
in Fig. 1.2(a). A unit vector has magnitude *unity*. We shall represent a unit vector by the symbol **a** and use a subscript to denote its direction. We shall denote
the three directions by subscripts 1, 2, and 3. We note that for a fixed orientation
of $\mathbf{a}_1$, two combinations are possible for the orientations of $\mathbf{a}_2$ and $\mathbf{a}_3$, as shown in
Figs. 1.2(a) and (b). If we take a right-hand screw and turn it from $\mathbf{a}_1$ to $\mathbf{a}_2$
through the 90° angle, it progresses in the direction of $\mathbf{a}_3$ in Fig. 1.2(a) but opposite to the direction of $\mathbf{a}_3$ in Fig. 1.2(b). Alternatively, a left-hand screw when
turned from $\mathbf{a}_1$ to $\mathbf{a}_2$ in Fig. 1.2(b) will progress in the direction of $\mathbf{a}_3$. Hence the
set of unit vectors in Fig. 1.2(a) corresponds to a right-handed system, whereas
the set in Fig. 1.2(b) corresponds to a left-handed system. We shall work consistently with the right-handed system.

A vector of magnitude different from unity along any of the reference
directions can be represented in terms of the unit vector along that direction.

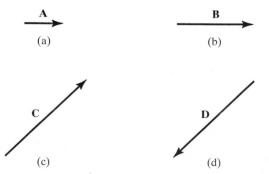

FIGURE 1.1   Graphical representation
of vectors.

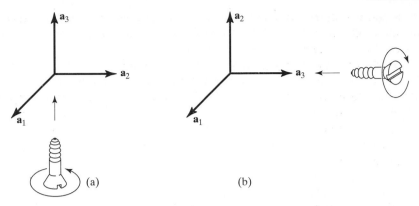

FIGURE 1.2 (a) Set of three orthogonal unit vectors in a right-handed system. (b) Set of three orthogonal unit vectors in a left-handed system.

Thus $4\mathbf{a}_1$ represents a vector of magnitude 4 units in the direction of $\mathbf{a}_1$, $6\mathbf{a}_2$ represents a vector of magnitude 6 units in the direction of $\mathbf{a}_2$, and $-2\mathbf{a}_3$ represents a vector of magnitude 2 units in the direction opposite to that of $\mathbf{a}_3$, as shown in Fig. 1.3. Two vectors are added by placing the beginning of the second vector at the tip of the first vector and then drawing the sum vector from the beginning of the first vector to the tip of the second vector. Thus, to add $4\mathbf{a}_1$ and $6\mathbf{a}_2$, we simply slide $6\mathbf{a}_2$ without changing its direction until its beginning coincides with the tip of $4\mathbf{a}_1$ and then draw the vector $(4\mathbf{a}_1 + 6\mathbf{a}_2)$ from the beginning of $4\mathbf{a}_1$ to the tip of $6\mathbf{a}_2$, as shown in Fig. 1.3. To see this, imagine that on the floor of an empty rectangular room, you are going from one corner to the opposite corner. Then to reach the destination, you can first walk along one edge and then along the second edge. Alternatively, you can go straight to the destination along the diagonal. By adding $-2\mathbf{a}_3$ to the vector $(4\mathbf{a}_1 + 6\mathbf{a}_2)$ in a similar manner, we obtain the vector $(4\mathbf{a}_1 + 6\mathbf{a}_2 - 2\mathbf{a}_3)$, as shown in Fig. 1.3. We note that the magnitude of $(4\mathbf{a}_1 + 6\mathbf{a}_2)$ is $\sqrt{4^2 + 6^2}$ or 7.211 and that the magnitude of $(4\mathbf{a}_1 + 6\mathbf{a}_2 - 2\mathbf{a}_3)$ is $\sqrt{4^2 + 6^2 + 2^2}$, or 7.483. Conversely to the foregoing discussion, a vector $\mathbf{A}$ at a given point is simply the superposition of three vectors $A_1\mathbf{a}_1$, $A_2\mathbf{a}_2$, and $A_3\mathbf{a}_3$, which are the projections of $\mathbf{A}$ onto the reference directions at that point. $A_1, A_2,$

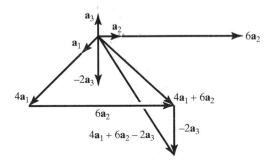

FIGURE 1.3 Graphical addition of vectors.

and A_3 are known as the *components* of **A** along the 1, 2, and 3 directions, respectively. Thus

$$\boxed{\mathbf{A} = A_1\mathbf{a}_1 + A_2\mathbf{a}_2 + A_3\mathbf{a}_3} \tag{1.1}$$

We now consider three vectors, **A**, **B**, and **C**, given by

$$\mathbf{A} = A_1\mathbf{a}_1 + A_2\mathbf{a}_2 + A_3\mathbf{a}_3 \tag{1.2a}$$

$$\mathbf{B} = B_1\mathbf{a}_1 + B_2\mathbf{a}_2 + B_3\mathbf{a}_3 \tag{1.2b}$$

$$\mathbf{C} = C_1\mathbf{a}_1 + C_2\mathbf{a}_2 + C_3\mathbf{a}_3 \tag{1.2c}$$

at a point and discuss several algebraic operations involving vectors as follows.

Vector addition and subtraction.   Since a given pair of like components of two vectors are parallel, addition of two vectors consists simply of adding the three pairs of like components of the vectors. Thus,

$$\begin{aligned}
\mathbf{A} + \mathbf{B} &= (A_1\mathbf{a}_1 + A_2\mathbf{a}_2 + A_3\mathbf{a}_3) + (B_1\mathbf{a}_1 + B_2\mathbf{a}_2 + B_3\mathbf{a}_3) \\
&= (A_1 + B_1)\mathbf{a}_1 + (A_2 + B_2)\mathbf{a}_2 + (A_3 + B_3)\mathbf{a}_3
\end{aligned} \tag{1.3}$$

Vector subtraction is a special case of addition. Thus,

$$\begin{aligned}
\mathbf{B} - \mathbf{C} &= \mathbf{B} + (-\mathbf{C}) \\
&= (B_1\mathbf{a}_1 + B_2\mathbf{a}_2 + B_3\mathbf{a}_3) + (-C_1\mathbf{a}_1 - C_2\mathbf{a}_2 - C_3\mathbf{a}_3) \\
&= (B_1 - C_1)\mathbf{a}_1 + (B_2 - C_2)\mathbf{a}_2 + (B_3 - C_3)\mathbf{a}_3
\end{aligned} \tag{1.4}$$

Multiplication and division by a scalar.   Multiplication of a vector **A** by a scalar m is the same as repeated addition of the vector. Thus,

$$m\mathbf{A} = m(A_1\mathbf{a}_1 + A_2\mathbf{a}_2 + A_3\mathbf{a}_3) = mA_1\mathbf{a}_1 + mA_2\mathbf{a}_2 + mA_3\mathbf{a}_3 \tag{1.5}$$

Division by a scalar is a special case of multiplication by a scalar. Thus,

$$\frac{\mathbf{B}}{n} = \frac{1}{n}(\mathbf{B}) = \frac{B_1}{n}\mathbf{a}_1 + \frac{B_2}{n}\mathbf{a}_2 + \frac{B_3}{n}\mathbf{a}_3 \tag{1.6}$$

Magnitude of a vector.   From the construction of Fig. 1.3 and the associated discussion, we have

$$\boxed{|\mathbf{A}| = |A_1\mathbf{a}_1 + A_2\mathbf{a}_2 + A_3\mathbf{a}_3| = \sqrt{A_1^2 + A_2^2 + A_3^2}} \tag{1.7}$$

Unit vector along A.   The unit vector $\mathbf{a}_A$ has a magnitude equal to unity, but its direction is the same as that of **A**. Hence,

$$\boxed{\mathbf{a}_A = \frac{\mathbf{A}}{|\mathbf{A}|} = \frac{A_1}{|\mathbf{A}|}\mathbf{a}_1 + \frac{A_2}{|\mathbf{A}|}\mathbf{a}_2 + \frac{A_3}{|\mathbf{A}|}\mathbf{a}_3} \tag{1.8}$$

Scalar or dot product of two vectors.   The scalar or dot product of two   *Dot product*
vectors **A** and **B** is a scalar quantity equal to the product of the magnitudes of **A**
and **B** and the cosine of the angle between **A** and **B**. It is represented by a boldface dot between **A** and **B**. Thus, if α is the angle between **A** and **B**, then

$$\boxed{\mathbf{A} \cdot \mathbf{B} = |\mathbf{A}||\mathbf{B}| \cos \alpha = AB \cos \alpha} \qquad (1.9)$$

For the unit vectors $\mathbf{a}_1, \mathbf{a}_2, \mathbf{a}_3$, we have

$$\mathbf{a}_1 \cdot \mathbf{a}_1 = 1 \qquad \mathbf{a}_1 \cdot \mathbf{a}_2 = 0 \qquad \mathbf{a}_1 \cdot \mathbf{a}_3 = 0 \qquad (1.10a)$$

$$\mathbf{a}_2 \cdot \mathbf{a}_1 = 0 \qquad \mathbf{a}_2 \cdot \mathbf{a}_2 = 1 \qquad \mathbf{a}_2 \cdot \mathbf{a}_3 = 0 \qquad (1.10b)$$

$$\mathbf{a}_3 \cdot \mathbf{a}_1 = 0 \qquad \mathbf{a}_3 \cdot \mathbf{a}_2 = 0 \qquad \mathbf{a}_3 \cdot \mathbf{a}_3 = 1 \qquad (1.10c)$$

By noting that $\mathbf{A} \cdot \mathbf{B} = A(B \cos \alpha) = B(A \cos \alpha)$, we observe that the dot
product operation consists of multiplying the magnitude of one vector by the
scalar obtained by projecting the second vector onto the first vector, as shown in
Figs. 1.4(a) and (b). The dot product operation is commutative since

$$\mathbf{B} \cdot \mathbf{A} = BA \cos \alpha = AB \cos \alpha = \mathbf{A} \cdot \mathbf{B} \qquad (1.11)$$

The distributive property also holds for the dot product, as can be seen from the
construction of Fig. 1.4(c), which illustrates that the projection of **B** + **C** onto **A**
is equal to the sum of the projections of **B** and **C** onto **A**. Thus,

$$\mathbf{A} \cdot (\mathbf{B} + \mathbf{C}) = \mathbf{A} \cdot \mathbf{B} + \mathbf{A} \cdot \mathbf{C} \qquad (1.12)$$

Using this property, we have

$$\begin{aligned}
\mathbf{A} \cdot \mathbf{B} = {} & (A_1\mathbf{a}_1 + A_2\mathbf{a}_2 + A_3\mathbf{a}_3) \cdot (B_1\mathbf{a}_1 + B_2\mathbf{a}_2 + B_3\mathbf{a}_3) \\
= {} & A_1\mathbf{a}_1 \cdot B_1\mathbf{a}_1 + A_1\mathbf{a}_1 \cdot B_2\mathbf{a}_2 + A_1\mathbf{a}_1 \cdot B_3\mathbf{a}_3 \\
& + A_2\mathbf{a}_2 \cdot B_1\mathbf{a}_1 + A_2\mathbf{a}_2 \cdot B_2\mathbf{a}_2 + A_2\mathbf{a}_2 \cdot B_3\mathbf{a}_3 \\
& + A_3\mathbf{a}_3 \cdot B_1\mathbf{a}_1 + A_3\mathbf{a}_3 \cdot B_2\mathbf{a}_2 + A_3\mathbf{a}_3 \cdot B_3\mathbf{a}_3
\end{aligned}$$

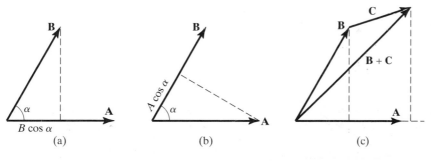

FIGURE 1.4   (a) and (b) For showing that the dot product of two vectors **A** and **B** is the product of the magnitude of one vector and the projection of the second vector onto the first
vector. (c) For proving the distributive property of the dot product operation.

Then using the relationships (1.10a)–(1.10c), we obtain

$$\mathbf{A} \cdot \mathbf{B} = A_1B_1 + A_2B_2 + A_3B_3 \qquad (1.13)$$

Thus, the dot product of two vectors is the sum of the products of the like components of the two vectors.

Finding angle between two vectors   From (1.9) and (1.13), we note that the angle between the vectors **A** and **B** is given by

$$\alpha = \cos^{-1}\frac{\mathbf{A} \cdot \mathbf{B}}{AB} = \cos^{-1}\frac{A_1B_1 + A_2B_2 + A_3B_3}{AB} \qquad (1.14)$$

Thus, the dot product operation is useful for finding the angle between two vectors. In particular, the two vectors are perpendicular if $\mathbf{A} \cdot \mathbf{B} = A_1B_1 + A_2B_2 + A_3B_3 = 0$.

Cross product   **Vector or cross product of two vectors.**   The vector or cross product of two vectors **A** and **B** is a vector quantity whose magnitude is equal to the product of the magnitudes of **A** and **B** and the sine of the smaller angle α between **A** and **B** and whose direction is normal to the plane containing **A** and **B** and toward the side of advance of a right-hand screw as it is turned from **A** to **B** through the angle α, as shown in Fig. 1.5. It is represented by a boldface cross between **A** and **B**. Thus, if $\mathbf{a}_N$ is the unit vector in the direction of advance of the right-hand screw, then

$$\mathbf{A} \times \mathbf{B} = |\mathbf{A}||\mathbf{B}| \sin \alpha\, \mathbf{a}_N = AB \sin \alpha\, \mathbf{a}_N \qquad (1.15)$$

For the unit vectors $\mathbf{a}_1, \mathbf{a}_2, \mathbf{a}_3$, we have

$$\mathbf{a}_1 \times \mathbf{a}_1 = 0 \qquad \mathbf{a}_1 \times \mathbf{a}_2 = \mathbf{a}_3 \qquad \mathbf{a}_1 \times \mathbf{a}_3 = -\mathbf{a}_2 \qquad (1.16a)$$

$$\mathbf{a}_2 \times \mathbf{a}_1 = -\mathbf{a}_3 \qquad \mathbf{a}_2 \times \mathbf{a}_2 = 0 \qquad \mathbf{a}_2 \times \mathbf{a}_3 = \mathbf{a}_1 \qquad (1.16b)$$

$$\mathbf{a}_3 \times \mathbf{a}_1 = \mathbf{a}_2 \qquad \mathbf{a}_3 \times \mathbf{a}_2 = -\mathbf{a}_1 \qquad \mathbf{a}_3 \times \mathbf{a}_3 = 0 \qquad (1.16c)$$

Note that the cross product of two identical unit vectors is the null vector **0**, that is, a vector whose components are all zero. If we arrange the unit vectors in the manner $\mathbf{a}_1\mathbf{a}_2\mathbf{a}_3\mathbf{a}_1\mathbf{a}_2$, then going to the right, the cross product of any two successive unit vectors is the following unit vector, whereas going to the left, the cross

FIGURE 1.5   Cross product operation $\mathbf{A} \times \mathbf{B}$.

product of any two successive unit vectors is the negative of the following unit vector.

The cross product operation is not commutative since

$$\mathbf{B} \times \mathbf{A} = |\mathbf{B}||\mathbf{A}| \sin \alpha \, (-\mathbf{a}_N) = -AB \sin \alpha \, \mathbf{a}_N = -\mathbf{A} \times \mathbf{B} \qquad (1.17)$$

The distributive property holds for the cross product (we shall prove this later in this section) so that

$$\mathbf{A} \times (\mathbf{B} + \mathbf{C}) = \mathbf{A} \times \mathbf{B} + \mathbf{A} \times \mathbf{C} \qquad (1.18)$$

Using this property and the relationships (1.16a)–(1.16c), we obtain

$$
\begin{aligned}
\mathbf{A} \times \mathbf{B} &= (A_1\mathbf{a}_1 + A_2\mathbf{a}_2 + A_3\mathbf{a}_3) \times (B_1\mathbf{a}_1 + B_2\mathbf{a}_2 + B_3\mathbf{a}_3) \\
&= A_1\mathbf{a}_1 \times B_1\mathbf{a}_1 + A_1\mathbf{a}_1 \times B_2\mathbf{a}_2 + A_1\mathbf{a}_1 \times B_3\mathbf{a}_3 \\
&\quad + A_2\mathbf{a}_2 \times B_1\mathbf{a}_1 + A_2\mathbf{a}_2 \times B_2\mathbf{a}_2 + A_2\mathbf{a}_2 \times B_3\mathbf{a}_3 \\
&\quad + A_3\mathbf{a}_3 \times B_1\mathbf{a}_1 + A_3\mathbf{a}_3 \times B_2\mathbf{a}_2 + A_3\mathbf{a}_3 \times B_3\mathbf{a}_3 \\
&= A_1B_2\mathbf{a}_3 - A_1B_3\mathbf{a}_2 - A_2B_1\mathbf{a}_3 + A_2B_3\mathbf{a}_1 \\
&\quad + A_3B_1\mathbf{a}_2 - A_3B_2\mathbf{a}_1 \\
&= (A_2B_3 - A_3B_2)\mathbf{a}_1 + (A_3B_1 - A_1B_3)\mathbf{a}_2 \\
&\quad + (A_1B_2 - A_2B_1)\mathbf{a}_3
\end{aligned}
$$

This can be expressed in determinant form in the manner

$$\mathbf{A} \times \mathbf{B} = \begin{vmatrix} \mathbf{a}_1 & \mathbf{a}_2 & \mathbf{a}_3 \\ A_1 & A_2 & A_3 \\ B_1 & B_2 & B_3 \end{vmatrix} \qquad (1.19)$$

Finding unit vector normal to two vectors

The cross product operation is useful for obtaining the unit vector normal to two given vectors at a point. This can be seen by rearranging (1.15) in the manner

$$\mathbf{a}_N = \frac{\mathbf{A} \times \mathbf{B}}{AB \sin \alpha} = \frac{\mathbf{A} \times \mathbf{B}}{|\mathbf{A} \times \mathbf{B}|} \qquad (1.20)$$

Triple cross product.   A triple cross product involves three vectors in two cross product operations. Caution must be exercised in evaluating a triple cross product since the order of evaluation is important; that is, $\mathbf{A} \times (\mathbf{B} \times \mathbf{C})$ is not in general equal to $(\mathbf{A} \times \mathbf{B}) \times \mathbf{C}$. This can be illustrated by means of a simple example involving unit vectors. Thus, if $\mathbf{A} = \mathbf{a}_1$, $\mathbf{B} = \mathbf{a}_1$, and $\mathbf{C} = \mathbf{a}_2$, then

$$\mathbf{A} \times (\mathbf{B} \times \mathbf{C}) = \mathbf{a}_1 \times (\mathbf{a}_1 \times \mathbf{a}_2) = \mathbf{a}_1 \times \mathbf{a}_3 = -\mathbf{a}_2$$

whereas

$$(\mathbf{A} \times \mathbf{B}) \times \mathbf{C} = (\mathbf{a}_1 \times \mathbf{a}_1) \times \mathbf{a}_2 = 0 \times \mathbf{a}_2 = 0$$

Scalar triple product.   The scalar triple product involves three vectors in a dot product operation and a cross product operation as, for example, $\mathbf{A} \cdot \mathbf{B} \times \mathbf{C}$.

It is not necessary to include parentheses since this quantity can be evaluated in only one manner, that is, by evaluating $\mathbf{B} \times \mathbf{C}$ first and then dotting the resulting vector with $\mathbf{A}$. It is meaningless to try to evaluate the dot product first since it results in a scalar quantity, and hence we cannot proceed any further. From (1.19), we have

$$\mathbf{A} \cdot \mathbf{B} \times \mathbf{C} = (A_1\mathbf{a}_1 + A_2\mathbf{a}_2 + A_3\mathbf{a}_3) \cdot \begin{vmatrix} \mathbf{a}_1 & \mathbf{a}_2 & \mathbf{a}_3 \\ B_1 & B_2 & B_3 \\ C_1 & C_2 & C_3 \end{vmatrix}$$

From (1.13), we then have

$$\mathbf{A} \cdot \mathbf{B} \times \mathbf{C} = \begin{vmatrix} A_1 & A_2 & A_3 \\ B_1 & B_2 & B_3 \\ C_1 & C_2 & C_3 \end{vmatrix} \tag{1.21}$$

Since the value of the determinant on the right side of (1.21) remains unchanged if the rows are interchanged in a cyclical manner,

$$\mathbf{A} \cdot \mathbf{B} \times \mathbf{C} = \mathbf{B} \cdot \mathbf{C} \times \mathbf{A} = \mathbf{C} \cdot \mathbf{A} \times \mathbf{B} \tag{1.22}$$

The scalar triple product has the geometrical meaning that its absolute value is the volume of the parallelepiped having the three vectors as three of its contiguous edges, as will be shown in Section 1.2.

We shall now show that the distributive law holds for the cross product operation by using (1.22). Thus, let us consider $\mathbf{A} \times (\mathbf{B} + \mathbf{C})$. Then, if $\mathbf{D}$ is any arbitrary vector, we have

$$\mathbf{D} \cdot \mathbf{A} \times (\mathbf{B} + \mathbf{C}) = (\mathbf{B} + \mathbf{C}) \cdot (\mathbf{D} \times \mathbf{A}) = \mathbf{B} \cdot (\mathbf{D} \times \mathbf{A}) + \mathbf{C} \cdot (\mathbf{D} \times \mathbf{A})$$
$$= \mathbf{D} \cdot \mathbf{A} \times \mathbf{B} + \mathbf{D} \cdot \mathbf{A} \times \mathbf{C} = \mathbf{D} \cdot (\mathbf{A} \times \mathbf{B} + \mathbf{A} \times \mathbf{C})$$

where we have used the distributive property of the dot product operation. Since this equality holds for any $\mathbf{D}$, it follows that

$$\mathbf{A} \times (\mathbf{B} + \mathbf{C}) = \mathbf{A} \times \mathbf{B} + \mathbf{A} \times \mathbf{C}$$

Example 1.1

Given three vectors

$$\mathbf{A} = \mathbf{a}_1 + \mathbf{a}_2$$
$$\mathbf{B} = \mathbf{a}_1 + 2\mathbf{a}_2 - 2\mathbf{a}_3$$
$$\mathbf{C} = \mathbf{a}_2 + 2\mathbf{a}_3$$

let us carry out several of the vector algebraic operations:

a. $\mathbf{A} + \mathbf{B} = (\mathbf{a}_1 + \mathbf{a}_2) + (\mathbf{a}_1 + 2\mathbf{a}_2 - 2\mathbf{a}_3) = 2\mathbf{a}_1 + 3\mathbf{a}_2 - 2\mathbf{a}_3$

b. $\mathbf{B} - \mathbf{C} = (\mathbf{a}_1 + 2\mathbf{a}_2 - 2\mathbf{a}_3) - (\mathbf{a}_2 + 2\mathbf{a}_3) = \mathbf{a}_1 + \mathbf{a}_2 - 4\mathbf{a}_3$

c. $4\mathbf{C} = 4(\mathbf{a}_2 + 2\mathbf{a}_3) = 4\mathbf{a}_2 + 8\mathbf{a}_3$

d. $|\mathbf{B}| = |\mathbf{a}_1 + 2\mathbf{a}_2 - 2\mathbf{a}_3| = \sqrt{(1)^2 + (2)^2 + (-2)^2} = 3$

e. $\mathbf{i}_B = \dfrac{\mathbf{B}}{|\mathbf{B}|} = \dfrac{\mathbf{a}_1 + 2\mathbf{a}_2 - 2\mathbf{a}_3}{3} = \dfrac{1}{3}\mathbf{a}_1 + \dfrac{2}{3}\mathbf{a}_2 - \dfrac{2}{3}\mathbf{a}_3$

f. $\mathbf{A} \cdot \mathbf{B} = (\mathbf{a}_1 + \mathbf{a}_2) \cdot (\mathbf{a}_1 + 2\mathbf{a}_2 - 2\mathbf{a}_3) = (1)(1) + (1)(2) + (0)(-2) = 3$

g. Angle between $\mathbf{A}$ and $\mathbf{B} = \cos^{-1}\dfrac{\mathbf{A} \cdot \mathbf{B}}{AB} = \cos^{-1}\dfrac{3}{(\sqrt{2})(3)} = 45°$

h. $\mathbf{A} \times \mathbf{B} = \begin{vmatrix} \mathbf{a}_1 & \mathbf{a}_2 & \mathbf{a}_3 \\ 1 & 1 & 0 \\ 1 & 2 & -2 \end{vmatrix} = (-2 - 0)\mathbf{a}_1 + (0 + 2)\mathbf{a}_2 + (2 - 1)\mathbf{a}_3$

$$= -2\mathbf{a}_1 + 2\mathbf{a}_2 + \mathbf{a}_3$$

i. Unit vector normal to $\mathbf{A}$ and $\mathbf{B} = \dfrac{\mathbf{A} \times \mathbf{B}}{|\mathbf{A} \times \mathbf{B}|} = -\dfrac{2}{3}\mathbf{a}_1 + \dfrac{2}{3}\mathbf{a}_2 + \dfrac{1}{3}\mathbf{a}_3$

j. $(\mathbf{A} \times \mathbf{B}) \times \mathbf{C} = \begin{vmatrix} \mathbf{a}_1 & \mathbf{a}_2 & \mathbf{a}_3 \\ -2 & 2 & 1 \\ 0 & 1 & 2 \end{vmatrix} = 3\mathbf{a}_1 + 4\mathbf{a}_2 - 2\mathbf{a}_3$

k. $\mathbf{A} \cdot \mathbf{B} \times \mathbf{C} = \begin{vmatrix} 1 & 1 & 0 \\ 1 & 2 & -2 \\ 0 & 1 & 2 \end{vmatrix} = (1)(6) + (1)(-2) + (0)(1) = 4$

K1.1. Scalars; Vectors; Unit vectors; Right-handed system; Components of a vector; Vector addition; Multiplication of vector by a scalar; Magnitude of a vector; Dot product; Cross product; Triple cross product; Scalar triple product.

D1.1. Vector $\mathbf{A}$ has a magnitude of 4 units and is directed toward north. Vector $\mathbf{B}$ has a magnitude of 3 units and is directed toward east. Vector $\mathbf{C}$ has a magnitude of 4 units and is directed 30° toward south from west. Find the following: **(a)** $\mathbf{A} + \mathbf{C}$; **(b)** $|\mathbf{A} - \mathbf{B}|$; **(c)** $3\mathbf{A} + 4\mathbf{B} + 3\mathbf{C}$; **(d)** $\mathbf{B} \cdot (\mathbf{A} - \mathbf{C})$; and **(e)** $\mathbf{A} \times (\mathbf{B} \times \mathbf{C})$.
Ans. **(a)** 4 units directed 60° west of north; **(b)** 5; **(c)** 6.212 units directed 15° east of north; **(d)** 10.392; **(e)** 24 units directed westward.

D1.2. Given three vectors

$$\mathbf{A} = 3\mathbf{a}_1 + 2\mathbf{a}_2 + \mathbf{a}_3$$
$$\mathbf{B} = \mathbf{a}_1 + \mathbf{a}_2 - \mathbf{a}_3$$
$$\mathbf{C} = \mathbf{a}_1 + 2\mathbf{a}_2 + 3\mathbf{a}_3$$

Find the following: **(a)** $|\mathbf{A} + \mathbf{B} - 4\mathbf{C}|$; **(b)** unit vector along $(\mathbf{A} + 2\mathbf{B} - \mathbf{C})$; **(c)** $\mathbf{A} \cdot \mathbf{C}$; **(d)** $\mathbf{B} \times \mathbf{C}$; and **(e)** $\mathbf{A} \cdot \mathbf{B} \times \mathbf{C}$.
Ans. **(a)** 13; **(b)** $(2\mathbf{a}_1 + \mathbf{a}_2 - 2\mathbf{a}_3)/3$; **(c)** 10; **(d)** $5\mathbf{a}_1 - 4\mathbf{a}_2 + \mathbf{a}_3$; **(e)** 8.

D1.3. Three vectors $\mathbf{A}, \mathbf{B}$, and $\mathbf{C}$ are given by

$$\mathbf{A} = \mathbf{a}_1 + 2\mathbf{a}_2 + 2\mathbf{a}_3$$
$$\mathbf{B} = 2\mathbf{a}_1 + \mathbf{a}_2 - 2\mathbf{a}_3$$
$$\mathbf{C} = \mathbf{a}_1 - \mathbf{a}_2 + \mathbf{a}_3$$

Find the following: **(a)** $\mathbf{A} \times (\mathbf{B} \times \mathbf{C})$; **(b)** $\mathbf{B} \times (\mathbf{C} \times \mathbf{A})$; and **(c)** $\mathbf{C} \times (\mathbf{A} \times \mathbf{B})$.
Ans. **(a)** $2\mathbf{a}_1 + \mathbf{a}_2 - 2\mathbf{a}_3$; **(b)** $\mathbf{a}_1 + 2\mathbf{a}_2 + 2\mathbf{a}_3$; **(c)** $-3\mathbf{a}_1 - 3\mathbf{a}_2$.

1.2   CARTESIAN COORDINATE SYSTEM

In the preceding section, we introduced the technique of expressing a vector at a point in space in terms of its component vectors along a set of three mutually orthogonal directions defined by three mutually orthogonal unit vectors at that point. Now to relate vectors at one point in space to vectors at another point in space, we must define the set of three reference directions at each and every point in space. To do this in a systematic manner, we need to use a coordinate system. Although there are several different coordinate systems, we shall be concerned with only three of those, namely, the Cartesian, cylindrical, and spherical coordinate systems. The *Cartesian coordinate system,* also known as the *rectangular coordinate system,* is the simplest of the three since it permits the geometry to be simple, yet sufficient to study many of the elements of engineering electromagnetics. We introduce the Cartesian coordinate system in this section and devote the next section to the cylindrical and spherical coordinate systems.

   The Cartesian coordinate system is defined by a set of three mutually orthogonal planes, as shown in Fig. 1.6(a). The point at which the three planes intersect is known as the origin O. The origin is the reference point relative to which we locate any other point in space. Each pair of planes intersects in a straight line. Hence, the three planes define a set of three straight lines that form the coordinate axes. These coordinate axes are denoted as the x-, y-, and z-axes. Values of x, y, and z are measured from the origin; hence, the coordinates of the origin are $(0, 0, 0)$, that is, $x = 0$, $y = 0$, and $z = 0$. Directions in which values of x, y, and z increase along the respective coordinate axes are indicated by arrowheads. The same set of three directions is used to erect a set of three unit vectors, denoted $\mathbf{a}_x$, $\mathbf{a}_y$, and $\mathbf{a}_z$, as shown in Fig. 1.6(a), for the purpose of describing

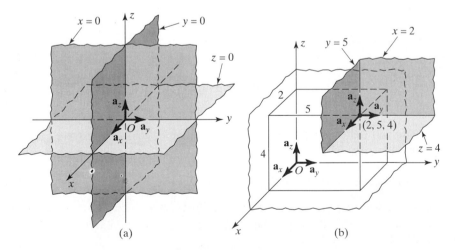

FIGURE 1.6   Cartesian coordinate system. (a) The three orthogonal planes defining the coordinate system. (b) The unit vectors in the Cartesian coordinate system are uniform.

vectors drawn at the origin. Note that the positive x-, y-, and z-directions are chosen such that they form a right-handed system, that is, a system for which $\mathbf{a}_x \times \mathbf{a}_y = \mathbf{a}_z$.

On one of the three planes, namely, the yz-plane, the value of x is constant and equal to zero, its value at the origin, since movement on this plane does not require any movement in the x-direction. Similarly, on the zx-plane, the value of y is constant and equal to zero, and on the xy-plane, the value of z is constant and equal to zero. Any point other than the origin is now given by the intersection of three planes

$$
\begin{array}{l}
x = \text{constant} \\
y = \text{constant} \\
z = \text{constant}
\end{array}
\tag{1.23}
$$

obtained by incrementing the values of the coordinates by appropriate amounts. For example, by displacing the $x = 0$ plane by 2 units in the positive x-direction, the $y = 0$ plane by 5 units in the positive y-direction, and the $z = 0$ plane by 4 units in the positive z-direction, we obtain the planes $x = 2$, $y = 5$, and $z = 4$, respectively, which intersect at point $(2, 5, 4)$, as shown in Fig. 1.6(b). The intersections of pairs of these planes define three straight lines along which we can erect the unit vectors $\mathbf{a}_x$, $\mathbf{a}_y$, and $\mathbf{a}_z$ toward the directions of increasing values of $x, y,$ and z, respectively, for the purpose of describing vectors drawn at that point. These unit vectors are parallel to the corresponding unit vectors drawn at the origin, as can be seen from Fig. 1.6(b). The same is true for any point in space in the Cartesian coordinate system. Thus, each one of the three unit vectors in the Cartesian coordinate system has the same direction at all points, and hence it is uniform. This behavior does not, however, hold for all unit vectors in the cylindrical and spherical coordinate systems, as we shall see in the next section.

Expression for vector joining two points

It is now a simple matter to apply what we have learned in Section 1.1 to vectors in Cartesian coordinates. All we need to do is to replace the subscripts 1, 2, and 3 for the unit vectors and the components along the unit vectors by the subscripts $x, y,$ and z, respectively, and also utilize the property that $\mathbf{a}_x, \mathbf{a}_y,$ and $\mathbf{a}_z$ are uniform vectors. Thus, let us, for example, obtain the expression for the vector $\mathbf{R}_{12}$ drawn from point $P_1(x_1, y_1, z_1)$ to point $P_2(x_2, y_2, z_2)$, as shown in Fig. 1.7. To do this, we note that the position vector $\mathbf{r}_1$ drawn from the origin to the point P_1 is given by

$$
\mathbf{r}_1 = x_1\mathbf{a}_x + y_1\mathbf{a}_y + z_1\mathbf{a}_z
\tag{1.24a}
$$

The *position vector* is so called because it defines the position of the point in space relative to the origin. Similarly, the position vector $\mathbf{r}_2$ drawn from the origin to the point P_2 is given by

$$
\mathbf{r}_2 = x_2\mathbf{a}_x + y_2\mathbf{a}_y + z_2\mathbf{a}_z
\tag{1.24b}
$$

FIGURE 1.7   For obtaining the expression for the vector $\mathbf{R}_{12}$ from $P_1(x_1, y_1, z_1)$ to $P_2(x_2, y_2, z_2)$.

Since, from the rule for vector addition, $\mathbf{r}_1 + \mathbf{R}_{12} = \mathbf{r}_2$, we obtain the vector $\mathbf{R}_{12}$ to be

$$\boxed{\begin{aligned} \mathbf{R}_{12} &= \mathbf{r}_2 - \mathbf{r}_1 \\ &= (x_2 - x_1)\mathbf{a}_x + (y_2 - y_1)\mathbf{a}_y + (z_2 - z_1)\mathbf{a}_z \end{aligned}} \tag{1.25}$$

Thus, to find the components of the vector drawn from one point to another in the Cartesian coordinate system, we simply subtract the coordinates of the initial point from the corresponding coordinates of the final point. These components are just the distances one has to travel along the x-, y-, and z-directions, respectively, if one chooses to go from P_1 to P_2 by traveling parallel to the coordinate axes instead of traveling along the direct straight-line path.

Proceeding further, we can obtain the unit vector along the line drawn from P_1 to P_2 to be

$$\boxed{\mathbf{a}_{12} = \frac{\mathbf{R}_{12}}{R_{12}} = \frac{(x_2 - x_1)\mathbf{a}_x + (y_2 - y_1)\mathbf{a}_y + (z_2 - z_1)\mathbf{a}_z}{[(x_2 - x_1)^2 + (y_2 - y_1)^2 + (z_2 - z_1)^2]^{1/2}}} \tag{1.26}$$

For a numerical example, if P_1 is $(1, -2, 0)$ and P_2 is $(4, 2, 5)$, then

$$\mathbf{R}_{12} = 3\mathbf{a}_x + 4\mathbf{a}_y + 5\mathbf{a}_z$$

$$\mathbf{a}_{12} = \frac{1}{5\sqrt{2}} (3\mathbf{a}_x + 4\mathbf{a}_y + 5\mathbf{a}_z)$$

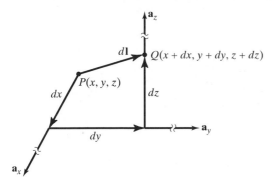

FIGURE 1.8 Differential length vector $d\mathbf{l}$.

In our study of electromagnetic fields, we have to work with line, surface, and volume integrals. These involve differential lengths, surfaces, and volumes, obtained by incrementing the coordinates by infinitesimal amounts. Since in the Cartesian coordinate system, the three coordinates represent lengths, the differential length elements obtained by incrementing one coordinate at a time, keeping the other two coordinates constant, are $dx\ \mathbf{a}_x$, $dy\ \mathbf{a}_y$, and $dz\ \mathbf{a}_z$ for the x-, y-, and z-coordinates, respectively.

Differential length vector. The differential length vector $d\mathbf{l}$ is the vector drawn from a point $P(x, y, z)$ to a neighboring point $Q(x + dx, y + dy, z + dz)$ obtained by incrementing the coordinates of P by infinitesimal amounts. Thus, it is the vector sum of the three differential length elements, as shown in Fig. 1.8, and given by

$$d\mathbf{l} = dx\ \mathbf{a}_x + dy\ \mathbf{a}_y + dz\ \mathbf{a}_z \qquad (1.27)$$

The differential lengths dx, dy, and dz in (1.27) are, however, not independent of each other since in the evaluation of line integrals, the integration is performed along a specified path on which the points P and Q lie. We shall illustrate this by means of an example.

Example 1.2

Let us consider the curve $x = y = z^2$ and obtain the expression for the differential length vector $d\mathbf{l}$ along the curve at the point $(1, 1, 1)$ and having the projection dz on the z-axis.

Finding differential length vector along a curve

The geometry pertinent to the problem is shown in Fig. 1.9. From elementary calculus, we know that for $x = y = z^2$, $dx = dy = 2z\ dz$. In particular, at the point $(1, 1, 1)$, $dx = dy = 2\ dz$. Thus,

$$\begin{aligned}
d\mathbf{l} &= dx\ \mathbf{a}_x + dy\ \mathbf{a}_y + dz\ \mathbf{a}_z \\
&= 2\ dz\ \mathbf{a}_x + 2\ dz\ \mathbf{a}_y + dz\ \mathbf{a}_z \\
&= (2\mathbf{a}_x + 2\mathbf{a}_y + \mathbf{a}_z)\ dz
\end{aligned}$$

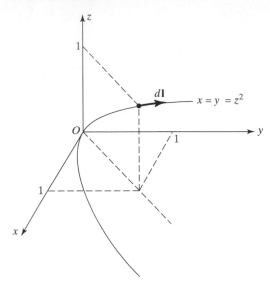

FIGURE 1.9   For finding the differential length vector along the curve $x = y = z^2$.

Note that the z-component of the $d\mathbf{l}$ vector found is dz, thereby satisfying the requirement of projection dz on the z-axis specified in the problem.

Differential length vectors are useful for finding the unit vector normal to a surface at a point on that surface. This is done by considering two differential length vectors at the point under consideration and tangential to two curves on the surface and then using (1.20). Thus, with reference to Fig. 1.10, we have

$$\mathbf{a}_n = \frac{d\mathbf{l}_1 \times d\mathbf{l}_2}{|d\mathbf{l}_1 \times d\mathbf{l}_2|} \tag{1.28}$$

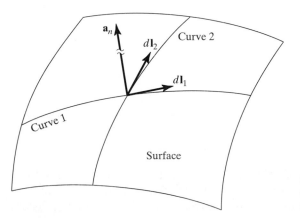

FIGURE 1.10   Finding the unit vector normal to a surface by using differential length vectors.

Let us consider an example.

Example 1.3

Find the unit vector normal to the surface $2x^2 + y^2 = 6$ at the point $(1, 2, 0)$.

Finding unit normal vector at a point on a surface

    With reference to the construction shown in Fig. 1.11, we consider two differential length vectors $d\mathbf{l}_1$ and $d\mathbf{l}_2$ at the point $(1, 2, 0)$. The vector $d\mathbf{l}_1$ is along the straight line $x = 1$, $y = 2$, whereas the vector $d\mathbf{l}_2$ is tangential to the curve $2x^2 + y^2 = 6$, $z = 0$. For $x = 1$ and $y = 2$, $dx = dy = 0$. Hence,

$$d\mathbf{l}_1 = dz\, \mathbf{a}_z$$

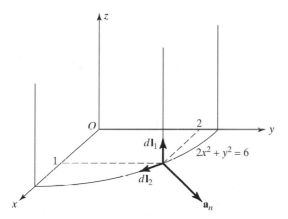

FIGURE 1.11   Example of finding the unit vector normal to a surface.

For $2x^2 + y^2 = 6$ and $z = 0$, $4x\,dx + 2y\,dy = 0$ and $dz = 0$. Specifically, at the point $(1, 2, 0)$, $dy = -dx$ and $dz = 0$. Hence,

$$d\mathbf{l}_2 = dx\,\mathbf{a}_x - dx\,\mathbf{a}_y = dx\,(\mathbf{a}_x - \mathbf{a}_y)$$

The unit normal vector is then given by

$$\mathbf{a}_n = \frac{dz\,\mathbf{a}_z \times dx\,(\mathbf{a}_x - \mathbf{a}_y)}{|dz\,\mathbf{a}_z \times dx\,(\mathbf{a}_x - \mathbf{a}_y)|}$$

$$= \frac{1}{\sqrt{2}}\,(\mathbf{a}_x + \mathbf{a}_y)$$

Differential surface vector.   Two differential length vectors $d\mathbf{l}_1$ and $d\mathbf{l}_2$ originating at a point define a differential surface whose area dS is that of the parallelogram having $d\mathbf{l}_1$ and $d\mathbf{l}_2$ as two of its adjacent sides, as shown in Fig. 1.12(a). From simple geometry and the definition of the cross product of two vectors, it can be seen that

$$dS = dl_1\,dl_2 \sin\alpha = |d\mathbf{l}_1 \times d\mathbf{l}_2| \tag{1.29}$$

In the evaluation of surface integrals, it is convenient to define a differential surface vector $d\mathbf{S}$ whose magnitude is the area dS and whose direction is normal to the differential surface. Thus, recognizing that the normal vector can be directed

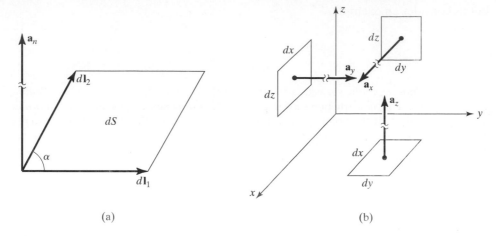

FIGURE 1.12   (a) Illustrating the differential surface vector concept. (b) Differential surface vectors in the Cartesian coordinate system.

to either side of a surface, we can write

$$d\mathbf{S} = \pm dS\, \mathbf{a}_n = \pm |d\mathbf{l}_1 \times d\mathbf{l}_2|\mathbf{a}_n$$

or

$$d\mathbf{S} = \pm d\mathbf{l}_1 \times d\mathbf{l}_2 \tag{1.30}$$

Applying (1.30) to pairs of three differential length elements $dx\, \mathbf{a}_x$, $dy\, \mathbf{a}_y$, and $dz\, \mathbf{a}_z$, we obtain the differential surface vectors

$$\pm dy\, \mathbf{a}_y \times dz\, \mathbf{a}_z = \pm dy\, dz\, \mathbf{a}_x \tag{1.31a}$$

$$\pm dz\, \mathbf{a}_z \times dx\, \mathbf{a}_x = \pm dz\, dx\, \mathbf{a}_y \tag{1.31b}$$

$$\pm dx\, \mathbf{a}_x \times dy\, \mathbf{a}_y = \pm dx\, dy\, \mathbf{a}_z \tag{1.31c}$$

associated with the planes $x =$ constant, $y =$ constant, and $z =$ constant, respectively. These are shown in Fig. 1.12(b) for the plus signs in (1.31a)–(1.31c).

Differential volume.   Three differential length vectors $d\mathbf{l}_1$, $d\mathbf{l}_2$, and $d\mathbf{l}_3$ originating at a point define a differential volume dv which is that of the parallelepiped having $d\mathbf{l}_1$, $d\mathbf{l}_2$, and $d\mathbf{l}_3$ as three of its contiguous edges, as shown in Fig. 1.13(a). From simple geometry and the definitions of cross and dot products, it can be seen that

$$dv = \text{area of the base of the parallelepiped} \times \text{height of the parallelepiped}$$

$$= |d\mathbf{l}_1 \times d\mathbf{l}_2||d\mathbf{l}_3 \cdot \mathbf{a}_n|$$

$$= |d\mathbf{l}_1 \times d\mathbf{l}_2| \frac{|d\mathbf{l}_3 \cdot d\mathbf{l}_1 \times d\mathbf{l}_2|}{|d\mathbf{l}_1 \times d\mathbf{l}_2|}$$

$$= |d\mathbf{l}_3 \cdot d\mathbf{l}_1 \times d\mathbf{l}_2|$$

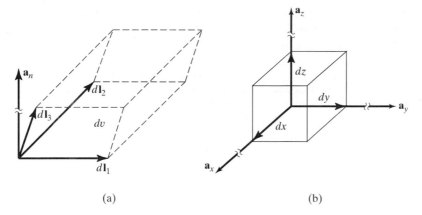

FIGURE 1.13   (a) Parallelepiped defined by three differential length vectors originating at a point. (b) Differential volume in the Cartesian coordinate system.

or

$$dv = \left| d\mathbf{l}_1 \cdot d\mathbf{l}_2 \times d\mathbf{l}_3 \right| \tag{1.32}$$

Thus, the scalar triple product of three vectors originating from a point has the meaning that its absolute value is the volume of the parallelepiped having the three vectors as three of its contiguous edges.

For the three differential length elements $dx\ \mathbf{a}_x$, $dy\ \mathbf{a}_y$, and $dz\ \mathbf{a}_z$ associated with the Cartesian coordinate system, we obtain the differential volume to be

$$dv = dx\,dy\,dz \tag{1.33}$$

which is that of the rectangular parallelepiped shown in Fig. 1.13(b).

We shall conclude this section with a brief review of some elementary analytic geometrical details that will be useful in our study of electromagnetics. An arbitrary surface is defined by an equation of the form

$$f(x, y, z) = 0 \tag{1.34}$$

In particular, the equation for a plane surface making intercepts a, b, and c on the x-, y-, and z-axes, respectively, is given by

$$\frac{x}{a} + \frac{y}{b} + \frac{z}{c} - 1 = 0 \tag{1.35}$$

Since a curve is the intersection of two surfaces, an arbitrary curve is defined by a pair of equations

$$f(x, y, z) = 0 \quad \text{and} \quad g(x, y, z) = 0 \tag{1.36}$$

Alternatively, a curve is specified by a set of three parametric equations

$$x = x(t), \quad y = y(t), \quad z = z(t) \tag{1.37}$$

where t is an independent parameter. For example, a straight line passing through the origin and making equal angles with the positive x-, y-, and z-axes is given by the pair of equations $y = x$ and $z = x$, or by the set of three parametric equations $x = t$, $y = t$, and $z = t$.

K1.2. Cartesian or rectangular coordinate system; Orthogonal surfaces; Unit vectors; Position vector; Vector joining two points; Differential length vector; Differential surface vector; Differential volume.

D1.4. Three points P_1, P_2, and P_3 are given by $(1, -2, 2)$, $(3, 1, 0)$, and $(5, 2, -2)$, respectively. Obtain the following: **(a)** the vector drawn from P_1 to P_2; **(b)** the straight-line distance from P_2 to P_3; and **(c)** the unit vector along the line from P_1 to P_3.
Ans.   **(a)** $(2\mathbf{a}_x + 3\mathbf{a}_y - 2\mathbf{a}_z)$;   **(b)** 3;   **(c)** $(\mathbf{a}_x + \mathbf{a}_y - \mathbf{a}_z)/\sqrt{3}$.

D1.5. For each of the following straight lines, find the differential length vector along the line and having the projection dz on the z-axis: **(a)** $x = 3$, $y = -4$; **(b)** $x + y = 0$, $y + z = 1$; and **(c)** the line passing through the points $(2, 0, 0)$ and $(0, 0, 1)$.
Ans.   **(a)** $dz\, \mathbf{a}_z$;   **(b)** $(\mathbf{a}_x - \mathbf{a}_y + \mathbf{a}_z)\, dz$;   **(c)** $(-2\mathbf{a}_x + \mathbf{a}_z)\, dz$.

D1.6. For each of the following pairs of points, obtain the equation for the straight line passing through the points: **(a)** $(1, 2, 0)$ and $(3, 4, 0)$; **(b)** $(0, 0, 0)$ and $(2, 2, -1)$; and **(c)** $(1, 1, 1)$ and $(3, -2, 4)$.
Ans.   **(a)** $y = x + 1$, $z = 0$;   **(b)** $x = y = -2z$;   **(c)** $3x + 2y = 5$, $3x - 2z = 1$.

1.3   CYLINDRICAL AND SPHERICAL COORDINATE SYSTEMS

Cylindrical coordinate system

In the preceding section, we learned that the Cartesian coordinate system is defined by a set of three mutually orthogonal surfaces, all of which are planes. The cylindrical and spherical coordinate systems also involve sets of three mutually orthogonal surfaces. For the cylindrical coordinate system, the three surfaces are a cylinder and two planes, as shown in Fig. 1.14(a). One of these planes is the same as the $z =$ constant plane in the Cartesian coordinate system. The second plane contains the z-axis and makes an angle ϕ with a reference plane, conveniently chosen to be the xz-plane of the Cartesian coordinate system. This plane is therefore defined by $\phi =$ constant. The cylindrical surface has the z-axis as its axis. Since the radial distance r from the z-axis to points on the cylindrical surface is a constant, this surface is defined by $r =$ constant. Thus, the three orthogonal surfaces defining the cylindrical coordinates of a point are

$$\boxed{\begin{aligned} r &= \text{constant} \\ \phi &= \text{constant} \\ z &= \text{constant} \end{aligned}} \tag{1.38}$$

Only two of these coordinates (r and z) are distances; the third coordinate (ϕ) is an angle. We note that the entire space is spanned by varying r from 0 to ∞, ϕ from 0 to 2π, and z from $-\infty$ to ∞.

The origin is given by $r = 0$, $\phi = 0$, and $z = 0$. Any other point in space is given by the intersection of three mutually orthogonal surfaces obtained by

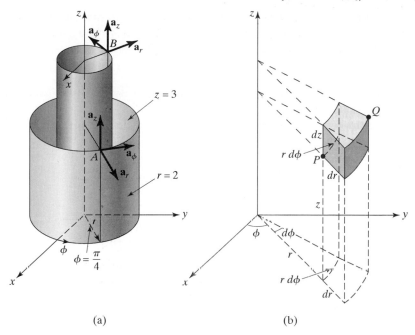

(a)

(b)

FIGURE 1.14 Cylindrical coordinate system. (a) Orthogonal surfaces and unit vectors.
(b) Differential volume formed by incrementing the coordinates.

incrementing the coordinates by appropriate amounts. For example, the
intersection of the three surfaces $r = 2$, $\phi = \pi/4$, and $z = 3$ defines the point
$A(2, \pi/4, 3)$, as shown in Fig. 1.14(a). These three orthogonal surfaces define
three curves that are mutually perpendicular. Two of these are straight lines and
the third is a circle. We draw unit vectors, $\mathbf{a}_r$, $\mathbf{a}_\phi$, and $\mathbf{a}_z$ tangential to these curves
at the point A and directed toward increasing values of r, ϕ, and z, respectively.
These three unit vectors form a set of mutually orthogonal unit vectors in terms
of which vectors drawn at A can be described. In a similar manner, we can draw
unit vectors at any other point in the cylindrical coordinate system, as shown, for
example, for point $B(1, 3\pi/4, 5)$ in Fig. 1.14(a). It can now be seen that the unit
vectors $\mathbf{a}_r$ and $\mathbf{a}_\phi$ at point B are not parallel to the corresponding unit vectors at
point A. Thus, unlike in the Cartesian coordinate system, the unit vectors $\mathbf{a}_r$ and
$\mathbf{a}_\phi$ in the cylindrical coordinate system do not have the same directions everywhere; that is, they are not uniform. Only the unit vector $\mathbf{a}_z$, which is the same as
in the Cartesian coordinate system, is uniform. Finally, we note that for the
choice of ϕ as in Fig. 1.14(a), that is, increasing from the positive x-axis toward
the positive y-axis, the coordinate system is right-handed, that is, $\mathbf{a}_r \times \mathbf{a}_\phi = \mathbf{a}_z$.

To obtain expressions for the differential lengths, surfaces, and volumes in
the cylindrical coordinate system, we now consider two points $P(r, \phi, z)$ and
$Q(r + dr, \phi + d\phi, z + dz)$, where Q is obtained by incrementing infinitesimally each coordinate from its value at P, as shown in Fig. 1.14(b). The three
orthogonal surfaces intersecting at P, and the three orthogonal surfaces
intersecting at Q define a box that can be considered to be rectangular since

dr, $d\phi$, and dz are infinitesimally small. The three differential length elements forming the contiguous sides of this box are dr, $\mathbf{a}_r$, $r\, d\phi\, \mathbf{a}_\phi$, and $dz\, \mathbf{a}_z$. The differential length vector $d\mathbf{l}$ from P to Q is thus given by

$$d\mathbf{l} = dr\, \mathbf{a}_r + r\, d\phi\, \mathbf{a}_\phi + dz\, \mathbf{a}_z \qquad (1.39)$$

The differential surface vectors defined by pairs of the differential length elements are

$$\pm r\, d\phi\, \mathbf{a}_\phi \times dz\, \mathbf{a}_z = \pm r\, d\phi\, dz\, \mathbf{a}_r \qquad (1.40a)$$

$$\pm dz\, \mathbf{a}_z \times dr\, \mathbf{a}_r = \pm dr\, dz\, \mathbf{a}_\phi \qquad (1.40b)$$

$$\pm dr\, \mathbf{a}_r \times r\, d\phi\, \mathbf{a}_\phi = \pm r\, dr\, d\phi\, \mathbf{a}_z \qquad (1.40c)$$

These are associated with the $r = $ constant, $\phi = $ constant, and $z = $ constant surfaces, respectively. Finally, the differential volume dv formed by the three differential lengths is simply the volume of the box; that is,

$$dv = (dr)(r\, d\phi)(dz) = r\, dr\, d\phi\, dz \qquad (1.41)$$

Spherical coordinate system   For the spherical coordinate system, the three mutually orthogonal surfaces are a sphere, a cone, and a plane, as shown in Fig. 1.15(a). The plane is the same as the $\phi = $ constant plane in the cylindrical coordinate system. The sphere has the origin as its center. Since the radial distance r from the origin to points on the spherical surface is a constant, this surface is defined by $r = $ constant. The spherical coordinate r should not be confused with the cylindrical coordinate r. When these two coordinates appear in the same expression, we shall use the subscripts c and s to distinguish between cylindrical and spherical. The cone has its vertex at the origin and its surface is symmetrical about the z-axis. Since the angle θ is the angle that the conical surface makes with the z-axis, this surface is defined by $\theta = $ constant. Thus, the three orthogonal surfaces defining the spherical coordinates of a point are

$$\begin{aligned} r &= \text{constant} \\ \theta &= \text{constant} \\ \phi &= \text{constant} \end{aligned} \qquad (1.42)$$

Only one of these coordinates (r) is distance; the other two coordinates (θ and ϕ) are angles. We note that the entire space is spanned by varying r from 0 to ∞, θ from 0 to π, and ϕ from 0 to 2π.

The origin is given by $r = 0$, $\theta = 0$, and $\phi = 0$. Any other point in space is given by the intersection of three mutually orthogonal surfaces obtained by in-

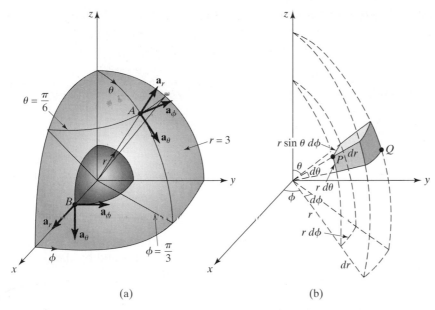

FIGURE 1.15 Spherical coordinate system. (a) Orthogonal surfaces and unit vectors. (b) Differential volume formed by incrementing the coordinates.

crementing the coordinates by appropriate amounts. For example, the intersection of the three surfaces $r = 3$, $\theta = \pi/6$, and $\phi = \pi/3$ defines the point $A(3, \pi/6, \pi/3)$, as shown in Fig. 1.15(a). These three orthogonal surfaces define three curves that are mutually perpendicular. One of these is a straight line and the other two are circles. We draw unit vectors $\mathbf{a}_r$, $\mathbf{a}_\theta$, and $\mathbf{a}_\phi$ tangential to these curves at point A and directed toward increasing values of r, θ, and ϕ, respectively. These three unit vectors form a set of mutually orthogonal unit vectors in terms of which vectors drawn at A can be described. In a similar manner, we can draw unit vectors at any other point in the spherical coordinate system, as shown, for example, for point $B(1, \pi/2, 0)$ in Fig. 1.15(a). It can now be seen that these unit vectors at point B are not parallel to the corresponding unit vectors at point A. Thus, in the spherical coordinate system all three unit vectors $\mathbf{a}_r$, $\mathbf{a}_\theta$, and $\mathbf{a}_\phi$ do not have the same directions everywhere; that is, they are not uniform. Finally, we note that for the choice of θ as in Fig. 1.15(a), that is, increasing from the positive z-axis toward the xy-plane, the coordinate system is right-handed, that is, $\mathbf{a}_r \times \mathbf{a}_\theta = \mathbf{a}_\phi$.

To obtain expressions for the differential lengths, surfaces, and volume in the spherical coordinate system, we now consider two points $P(r, \theta, \phi)$ and $Q(r + dr, \theta + d\theta, \phi + d\phi)$, where Q is obtained by incrementing infinitesimally each coordinate from its value at P, as shown in Fig. 1.15(b). The three orthogonal surfaces intersecting at P and the three orthogonal surfaces intersecting at Q define a box that can be considred to be rectangular since dr, $d\theta$ and $d\phi$ are infinitesimally small. The three differential length elements forming the contiguous sides of this box are $dr\,\mathbf{a}_r$, $r\,d\theta\,\mathbf{a}_\theta$, and $r\sin\theta\,d\phi\,\mathbf{a}_\phi$. The differential length vector $d\mathbf{l}$ from P to Q is thus given by

$$dl = dr\,\mathbf{a}_r + r\,d\theta\,\mathbf{a}_\theta + r\sin\theta\,d\phi\,\mathbf{a}_\phi \tag{1.43}$$

The differential surface vectors defined by pairs of the differential length elements are

$$\pm r\,d\theta\,\mathbf{a}_\theta \times r\sin\theta\,d\phi\,\mathbf{a}_\phi = \pm r^2\sin\theta\,d\theta\,d\phi\,\mathbf{a}_r \tag{1.44a}$$

$$\pm r\sin\theta\,d\phi\,\mathbf{a}_\phi \times dr\,\mathbf{a}_r = \pm r\sin\theta\,dr\,d\phi\,\mathbf{a}_\theta \tag{1.44b}$$

$$\pm dr\,\mathbf{a}_r \times r\,d\theta\,\mathbf{a}_\theta = \pm r\,dr\,d\theta\,\mathbf{a}_\phi \tag{1.44c}$$

These are associated with the $r = $ constant, $\theta = $ constant, and $\phi = $ constant surfaces, respectively. Finally, the differential volume dv formed by the three differential lengths is simply the volume of the box, that is,

$$dv = (dr)(r\,d\theta)(r\sin\theta\,d\phi) = r^2\sin\theta\,dr\,d\theta\,d\phi \tag{1.45}$$

Conversions between coordinate systems   In the study of electromagnetics, it is useful to be able to convert the coordinates of a point and vectors drawn at a point from one coordinate system to another, particularly from the cylindrical system to the Cartesian system and vice versa, and from the spherical system to the Cartesian system and vice versa. To derive first the relationships for the conversion of the coordinates, let us consider Fig. 1.16(a), which illustrates the geometry pertinent to the coordinates of a point P in the three different coordinate systems. Thus, from simple geometrical considerations, we have

$$x = r_c \cos\phi \qquad y = r_c \sin\phi \qquad z = z \tag{1.46a}$$

$$x = r_s \sin\theta \cos\phi \qquad y = r_s \sin\theta \sin\phi \qquad z = r_s \cos\theta \tag{1.46b}$$

Conversely, we have

$$r_c = \sqrt{x^2 + y^2} \qquad \phi = \tan^{-1}\frac{y}{x} \qquad z = z \tag{1.47a}$$

$$r_s = \sqrt{x^2 + y^2 + z^2} \qquad \theta = \tan^{-1}\frac{\sqrt{x^2 + y^2}}{z} \qquad \phi = \tan^{-1}\frac{y}{x} \tag{1.47b}$$

Relationships (1.46a) and (1.47a) correspond to conversion from cylindrical coordinates to Cartesian coordinates, and vice versa. Relationships (1.46b) and

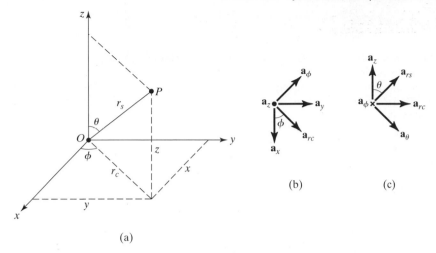

(a)

(b)

(c)

FIGURE 1.16 (a) Converting coordinates of a point from one coordinate system to another. (b) and (c) Expressing unit vectors in cylindrical and spherical coordinate systems, respectively, in terms of unit vectors in the Cartesian coordinate system.

(1.47b) correspond to conversion from spherical coordinates to Cartesian coordinates, and vice versa. It should be noted that in computing ϕ from y and x, consideration should be given to the quadrant of the xy-plane in which the projection of the point P onto the xy-plane lies.

Considering next the conversion of vectors from one coordinate system to another, we note that to do this, we need to express each of the unit vectors of the first coordinate system in terms of its components along the unit vectors in the second coordinate system. From the definition of the dot product of two vectors, the component of a unit vector along another unit vector, that is, the cosine of the angle between the unit vectors, is simply the dot product of the two unit vectors. Thus, considering the sets of unit vectors in the cylindrical and Cartesian coordinate systems, we have with the aid of Fig. 1.16(b),

$$\mathbf{a}_{rc} \cdot \mathbf{a}_x = \cos \phi \qquad \mathbf{a}_{rc} \cdot \mathbf{a}_y = \sin \phi \qquad \mathbf{a}_{rc} \cdot \mathbf{a}_z = 0 \qquad (1.48a)$$

$$\mathbf{a}_\phi \cdot \mathbf{a}_x = -\sin \phi \qquad \mathbf{a}_\phi \cdot \mathbf{a}_y = \cos \phi \qquad \mathbf{a}_\phi \cdot \mathbf{a}_z = 0 \qquad (1.48b)$$

$$\mathbf{a}_z \cdot \mathbf{a}_x = 0 \qquad \mathbf{a}_z \cdot \mathbf{a}_y = 0 \qquad \mathbf{a}_z \cdot \mathbf{a}_z = 1 \qquad (1.48c)$$

Similarly, for the sets of unit vectors in the spherical and Cartesian coordinate systems, we obtain with the aid of Fig. 1.16(b) and (c),

$$\mathbf{a}_{rs} \cdot \mathbf{a}_x = \sin \theta \cos \phi \qquad \mathbf{a}_{rs} \cdot \mathbf{a}_y = \sin \theta \sin \phi \qquad \mathbf{a}_{rs} \cdot \mathbf{a}_z = \cos \theta \qquad (1.49a)$$

$$\mathbf{a}_\theta \cdot \mathbf{a}_x = \cos \theta \cos \phi \qquad \mathbf{a}_\theta \cdot \mathbf{a}_y = \cos \theta \sin \phi \qquad \mathbf{a}_\theta \cdot \mathbf{a}_z = -\sin \theta \qquad (1.49b)$$

$$\mathbf{a}_\phi \cdot \mathbf{a}_x = -\sin \phi \qquad \mathbf{a}_\phi \cdot \mathbf{a}_y = \cos \phi \qquad \mathbf{a}_\phi \cdot \mathbf{a}_z = 0 \qquad (1.49c)$$

We shall now illustrate the use of these relationships by means of an example.

Example 1.4

Let us consider the vector $3\mathbf{a}_x + 4\mathbf{a}_y + 5\mathbf{a}_z$ at the point $(3, 4, 5)$ and convert it to one in spherical coordinates.

First, from the relationships (1.47b), we obtain the spherical coordinates of the point $(3, 4, 5)$ to be

$$r_s = \sqrt{3^2 + 4^2 + 5^2} = 5\sqrt{2}$$

$$\theta = \tan^{-1} \frac{\sqrt{3^2 + 4^2}}{5} = \tan^{-1} 1 = 45°$$

$$\phi = \tan^{-1} \tfrac{4}{3} = 53.13°$$

Then, noting from the relationships (1.49) that

$$
\begin{bmatrix} \mathbf{a}_x \\ \mathbf{a}_y \\ \mathbf{a}_z \end{bmatrix}
=
\begin{bmatrix}
(\mathbf{a}_x \cdot \mathbf{a}_{rs}) & (\mathbf{a}_x \cdot \mathbf{a}_\theta) & (\mathbf{a}_x \cdot \mathbf{a}_\phi) \\
(\mathbf{a}_y \cdot \mathbf{a}_{rs}) & (\mathbf{a}_y \cdot \mathbf{a}_\theta) & (\mathbf{a}_y \cdot \mathbf{a}_\phi) \\
(\mathbf{a}_z \cdot \mathbf{a}_{rs}) & (\mathbf{a}_z \cdot \mathbf{a}_\theta) & (\mathbf{a}_z \cdot \mathbf{a}_\phi)
\end{bmatrix}
\begin{bmatrix} \mathbf{a}_{rs} \\ \mathbf{a}_\theta \\ \mathbf{a}_\phi \end{bmatrix}
$$

$$
=
\begin{bmatrix}
\sin\theta\cos\phi & \cos\theta\cos\phi & -\sin\phi \\
\sin\theta\sin\phi & \cos\theta\sin\phi & \cos\phi \\
\cos\theta & -\sin\theta & 0
\end{bmatrix}
\begin{bmatrix} \mathbf{a}_{rs} \\ \mathbf{a}_\theta \\ \mathbf{a}_\phi \end{bmatrix}
$$

we obtain at the point under consideration

$$
\begin{bmatrix} \mathbf{a}_x \\ \mathbf{a}_y \\ \mathbf{a}_z \end{bmatrix}
=
\begin{bmatrix}
0.3\sqrt{2} & 0.3\sqrt{2} & -0.8 \\
0.4\sqrt{2} & 0.4\sqrt{2} & 0.6 \\
0.5\sqrt{2} & -0.5\sqrt{2} & 0
\end{bmatrix}
\begin{bmatrix} \mathbf{a}_{rs} \\ \mathbf{a}_\theta \\ \mathbf{a}_\phi \end{bmatrix}
$$

$$
\begin{aligned}
3\mathbf{a}_x + 4\mathbf{a}_y + 5\mathbf{a}_z &= 3(0.3\sqrt{2}\,\mathbf{a}_{rs} + 0.3\sqrt{2}\,\mathbf{a}_\theta - 0.8\mathbf{a}_\phi) \\
&\quad + 4(0.4\sqrt{2}\,\mathbf{a}_{rs} + 0.4\sqrt{2}\,\mathbf{a}_\theta + 0.6\mathbf{a}_\phi) \\
&\quad + 5(0.5\sqrt{2}\,\mathbf{a}_{rs} - 0.5\sqrt{2}\,\mathbf{a}_\theta) \\
&= 5\sqrt{2}\,\mathbf{a}_{rs}
\end{aligned}
$$

This result is to be expected since the given vector has components equal to the coordinates of the point at which it is specified. Hence, its magnitude is equal to the distance of the point from the origin, that is, the spherical coordinate r_s of the point, and its direction is along the line drawn from the origin to the point, that is, along the unit vector $\mathbf{a}_{rs}$ at that point. In fact, the given vector is a particular case of the position vector $x\mathbf{a}_x + y\mathbf{a}_y + z\mathbf{a}_z = r_s\mathbf{a}_{rs}$, which is the vector drawn from the origin to the point (x, y, z).

K1.3. Cylindrical coordinate system; Orthogonal surfaces; Unit vectors; Differential lengths, surfaces, and volume; Spherical coordinate system; Orthogonal surfaces; Unit vectors; Differential lengths, surfaces, and volume; Conversions between coordinate systems.

D1.7. Convert into Cartesian coordinates each of the following points: **(a)** $(2, 5\pi/6, 3)$ in cylindrical coordinates; **(b)** $(4, 4\pi/3, -1)$ in cylindrical coordinates; **(c)** $(4, 2\pi/3, \pi/6)$ in spherical coordinates; and **(d)** $(\sqrt{8}, \pi/4, \pi/3)$ in spherical coordinates.
Ans. **(a)** $(-\sqrt{3}, 1, 3)$; **(b)** $(-2, -2\sqrt{3}, -1)$; **(c)** $(3, \sqrt{3}, -2)$; **(d)** $(1, \sqrt{3}, 2)$.

D1.8. Convert into cylindrical coordinates the following points specified in Cartesian coordinates: **(a)** $(-2, 0, 1)$; **(b)** $(1, -\sqrt{3}, -1)$; and **(c)** $(-\sqrt{2}, -\sqrt{2}, 3)$.

Ans.   **(a)** $(2, \pi, 1)$;   **(b)** $(2, 5\pi/3, -1)$;   **(c)** $(2, 5\pi/4, 3)$.

D1.9. Convert into spherical coordinates the following points specified in Cartesian coordinates: **(a)** $(0, -2, 0)$; **(b)** $(-3, \sqrt{3}, 2)$; and **(c)** $(-\sqrt{2}, 0, -\sqrt{2})$.

Ans.   **(a)** $(2, \pi/2, 3\pi/2)$;   **(b)** $(4, \pi/3, 5\pi/6)$;   **(c)** $(2, 3\pi/4, \pi)$.

1.4   SCALAR AND VECTOR FIELDS

Before we take up the task of studying electromagnetic fields, we must understand what is meant by a *field*. A field is associated with a region in space, and we say that a field exists in the region if there is a physical phenomenon associated with points in that region. For example, in everyday life we are familiar with the earth's gravitational field. We do not "see" the field in the same manner as we see light rays, but we know of its existence in the sense that objects are acted upon by the gravitational force of Earth. In a broader context, we can talk of the field of any physical quantity as being a description, mathematical or graphical, of how the quantity varies from one point to another in the region of the field and with time. We can talk of scalar or vector fields depending on whether the quantity of interest is a scalar or a vector. We can talk of static or time-varying fields depending on whether the quantity of interest is independent of time or changing with it.

We shall begin our discussion of fields with some simple examples of scalar fields. Thus let us consider the case of the conical pyramid shown in Fig. 1.17(a). A description of the height of the pyramidal surface versus position on its base is an example of a scalar field involving two variables. Choosing the origin to be the projection of the vertex of the cone onto the base and setting up an xy-coordinate system to locate points on the base, we obtain the height field as a function of x and y to be   *Scalar fields*

$$h(x, y) = 6 - 2\sqrt{x^2 + y^2} \tag{1.50}$$

Although we are able to depict the height variation of points on the conical surface graphically by using the third coordinate for h, we will have to be content with the visualization of the height field by a set of constant-height contours on the xy-plane if only two coordinates were available, as in the case of a two-dimensional space. For the field under consideration, the constant-height contours are circles in the xy-plane centered at the origin and equally spaced for equal increments of the height value, as shown in Fig. 1.17(a).

For an example of a scalar field in three dimensions, let us consider a rectangular room and the distance field of points in the room from one corner of the room, as shown in Fig. 1.17(b). For convenience, we choose this corner to be the origin O and set up a Cartesian coordinate system with the three contiguous edges meeting at that point as the coordinate axes. Each point in the room is defined by a set of values for the three coordinates x, y, and z. The distance r from the origin to that point is $\sqrt{x^2 + y^2 + z^2}$. Thus, the distance field of points in the room from the origin is given by

$$r(x, y, z) = \sqrt{x^2 + y^2 + z^2} \tag{1.51}$$

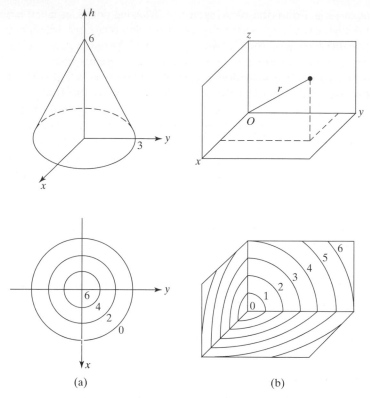

(a)   (b)

FIGURE 1.17   (a) Conical pyramid lying above the xy-plane and a set of constant-height contours for the conical surface. (b) Rectangular room and a set of constant-distance surfaces depicting the distance field of points in the room from one corner of the room.

Since the three coordinates are already used up for defining the points in the field region, we have to visualize the distance field by means of a set of constant-distance surfaces. A constant-distance surface is a surface for which points on it correspond to a particular constant value of r. For the case under consideration, the constant-distance surfaces are spherical surfaces centered at the origin and are equally spaced for equal increments in the value of the distance, as shown in Fig. 1.17(b).

The fields we have discussed thus far are static fields. A simple example of a time-varying scalar field is provided by the temperature field associated with points in a room, especially when it is being heated or cooled. Just as in the case of the distance field of Fig. 1.17(b), we set up a three-dimensional coordinate system and to each set of three coordinates corresponding to the location of a point in the room, we assign a number to represent the temperature T at that point. Since the temperature at that point, however, varies with time t, this number is a function of time. Thus, we describe mathematically the time-varying temperature field in the room by a function $T(x, y, z, t)$. For any given instant of time, we can visualize a set of constant-temperature or isothermal surfaces corresponding to particular values of T as representing the temperature field for that value of time. For a different instant of time, we will have a different set of isothermal surfaces for the same values of T. Thus, we can visualize the time-varying temperature field in the room by a set of isothermal surfaces continuously changing their shapes as though in a motion picture.

The foregoing discussion of scalar fields may now be extended to vector *Vector fields* fields by recalling that a vector quantity has associated with it a direction in space in addition to magnitude. Hence, to describe a vector field, we attribute to each point in the field region a vector that represents the magnitude and direction of the physical quantity under consideration at that point. Since a vector at a given point can be expressed as the sum of its components along the set of unit vectors at that point, a mathematical description of the vector field involves simply the descriptions of the three component scalar fields. Thus, for a vector field **F** in the Cartesian coordinate system, we have

$$\mathbf{F}(x, y, z, t) = F_x(x, y, z, t)\mathbf{a}_x + F_y(x, y, z, t)\mathbf{a}_y + F_z(x, y, z, t)\mathbf{a}_z \qquad (1.52)$$

Similar expressions in cylindrical and spherical coordinate systems are as follows:

$$\mathbf{F}(r, \phi, z, t) = F_r(r, \phi, z, t)\mathbf{a}_r + F_\phi(r, \phi, z, t)\mathbf{a}_\phi + F_z(r, \phi, z, t)\mathbf{a}_z \qquad (1.53a)$$

$$\mathbf{F}(r, \theta, \phi, t) = F_r(r, \theta, \phi, t)\mathbf{a}_r + F_\theta(r, \theta, \phi, t)\mathbf{a}_\theta + F_\phi(r, \theta, \phi, t)\mathbf{a}_\phi \qquad (1.53b)$$

We should however recall that the unit vectors $\mathbf{a}_r$ and $\mathbf{a}_\phi$ in (1.53a) and all three unit vectors in (1.53b) are themselves nonuniform but known functions of the coordinates.

A vector field is described by a set of *direction lines,* also known as *stream* *Finding* *lines* and *flux lines.* A direction line is a curve constructed such that the field is *equations for* tangential to the curve for all points on the curve. To find the equations for the *direction lines* direction lines for a specified vector field **F**, we consider the differential length *of a vector* vector $d\mathbf{l}$ tangential to the curve. Then since **F** and $d\mathbf{l}$ are parallel, their compo- *field* nents must be in the same ratio. Thus, in the Cartesian coordinate system, we obtain the differential equation

$$\frac{dx}{F_x} = \frac{dy}{F_y} = \frac{dz}{F_z} \qquad (1.54)$$

which upon integration gives the required algebraic equation. Similar expressions in cylindrical and spherical coordinate systems are as follows:

$$\frac{dr}{F_r} = \frac{r\,d\phi}{F_\phi} = \frac{dz}{F_z} \qquad (1.55)$$

$$\frac{dr}{F_r} = \frac{r\,d\theta}{F_\theta} = \frac{r\sin\theta\,d\phi}{F_\phi} \qquad (1.56)$$

We shall illustrate the concept of direction lines and the use of (1.54)–(1.56) to obtain the equations for the direction lines by means of an example.

Example 1.5

Consider a circular disk of radius a rotating with constant angular velocity ω about an axis normal to the disk and passing through its center. We wish to describe the linear velocity vector field associated with points on the rotating disk.

We choose the center of the disk to be the origin and set up a two-dimensional coordinate system, as shown in Fig. 1.18(a). Note that we have a choice of the coordinates (x, y) or the coordinates (r, ϕ). We know that the magnitude of the linear velocity of a point on the disk is then equal to the product of the angular velocity ω and the radial distance r of the point from the center of the disk. The direction of the linear velocity is tangential to the circle drawn through that point and concentric with the disk. Hence, we may depict the linear velocity field by drawing at several points on the disk vectors that are tangential to the circles concentric with the disk and passing through those points, and whose lengths are proportional to the radii of the circles, as shown in Fig. 1.18(b), where the points are carefully selected to reveal the circular symmetry of the field with respect to the center of the disk. We find, however, that this method of representation of the vector field results in a congested sketch of vectors. Hence, we may simplify the sketch by omitting the vectors and simply placing arrowheads along the circles, thereby obtaining a set of direction lines. We note that for the field under consideration, the

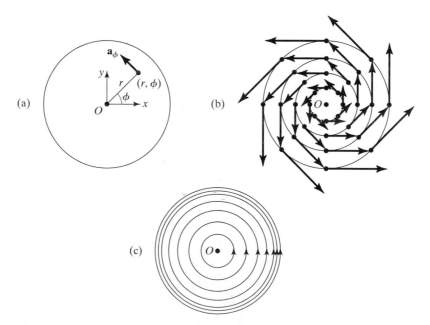

FIGURE 1.18   (a) Rotating disk. (b) Linear velocity vector field associated with points on the rotating disk. (c) Same as (b) except that the vectors are omitted and the density of direction lines is used to indicate the magnitude variation.

direction lines are also contours of constant magnitude of the velocity, and hence by increasing the density of the direction lines as r increases, we can indicate the magnitude variation, as shown in Fig. 1.18(c).

For this simple example, we have been able to obtain the direction lines without resorting to the use of mathematics. We shall now consider the mathematical determination of the direction lines and show that the same result is obtained. To do this, we note that the linear velocity vector field is given by

$$\mathbf{v}(r, \phi) = \omega r \mathbf{a}_\phi$$

Then, considering that the geometry associated with the problem is two-dimensional and using (1.55), we have

$$\frac{dr}{0} = \frac{r \, d\phi}{\omega r}$$

or

$$dr = 0$$
$$r = \text{constant}$$

which represents circles centered at the origin, as in Fig. 1.18(c).

If we wish to obtain the equations for the direction lines using Cartesian coordinates, we first write

$$\mathbf{v}(x, y) = \omega r(\mathbf{a}_\phi \cdot \mathbf{a}_x)\mathbf{a}_x + \omega r(\mathbf{a}_\phi \cdot \mathbf{a}_y)\mathbf{a}_y$$
$$= \omega r(-\sin \phi \, \mathbf{a}_x + \cos \phi \, \mathbf{a}_y)$$
$$= \omega(-y\mathbf{a}_x + x\mathbf{a}_y)$$

Then from (1.54), we have

$$\frac{dx}{-y} = \frac{dy}{x}$$

or

$$x \, dx + y \, dy = 0$$
$$x^2 + y^2 = \text{constant}$$

which again represents circles centered at the origin.

K1.4. Field; Static field, Time-varying field; Scalar field; Constant magnitude contours and surfaces; Vector field; Direction lines.

D1.10. The time-varying temperature field in a certain region of space is given by

$$T(x, y, z, t) = T_0\{[x(1 + \sin \pi t)]^2 + [2y(1 - \cos \pi t)]^2 + 4z^2]\}$$

where T_0 is a constant. Find the shapes of the constant-temperature surfaces for each of the following values of t: **(a)** $t = 0$; **(b)** $t = 0.5$ s; and **(c)** $t = 1$ s. *Ans.* **(a)** elliptic cylinders; **(b)** spheres; **(c)** ellipsoids.

D1.11. For the vector field $\mathbf{F} = (3x - y)\mathbf{a}_x + (x + z)\mathbf{a}_y + (2y - z)\mathbf{a}_z$, find the following: **(a)** the magnitude of $\mathbf{F}$ and the unit vector along $\mathbf{F}$ at the point $(1, 1, 0)$; **(b)** the point at which the magnitude of $\mathbf{F}$ is 3 and the direction of $\mathbf{F}$ is along the unit vector $\frac{1}{3}(2\mathbf{a}_x + 2\mathbf{a}_y + \mathbf{a}_z)$; and **(c)** the point at which the magnitude of $\mathbf{F}$ is 3 and the direction of $\mathbf{F}$ is along the unit vector $\mathbf{a}_z$.

Ans.   **(a)** $3, \frac{1}{3}(2\mathbf{a}_x + \mathbf{a}_y + 2\mathbf{a}_z)$;   **(b)** $(1, 1, 1)$ **(c)** $(0.6, 1.8, -0.6)$.

D1.12.   A vector field is given in cylindrical coordinates by

$$\mathbf{F} = \frac{1}{r^2}\left(\cos\phi\, \mathbf{a}_r + \sin\phi\, \mathbf{a}_\phi\right)$$

Express the vector $\mathbf{F}$ in Cartesian coordinates at each of the following points specified in Cartesian coordinates: **(a)** $(1, 0, 0)$; **(b)** $(1, -1, -3)$; and **(c)** $(1, \sqrt{3}, -4)$.

Ans.   **(a)** $\mathbf{a}_x$;   **(b)** $-\frac{1}{2}\mathbf{a}_y$;   **(c)** $\frac{1}{8}(-\mathbf{a}_x + \sqrt{3}\,\mathbf{a}_y)$.

1.5   SINUSOIDALLY TIME-VARYING FIELDS

In our study of electromagnetics, we will be particularly interested in fields that vary sinusoidally with time. Sinusoidally time-varying fields are important because of their natural occurrence and ease of generation. For example, when we speak, we emit sine waves; when we tune our radio dial to a broadcast station, we receive sine waves; and so on. Also, any function for which the time variation is arbitrary can be expressed in terms of sinusoidally time-varying functions having a discrete or continuous spectrum of frequencies, depending on whether the function is periodic or aperiodic. Thus, if the response of a system to a sinusoidal excitation is known, its response for a nonsinusoidal excitation can be found.

Let us first consider a scalar sinusoidal function of time. Such a function is given by an expression of the form $A\cos(\omega t + \phi)$. In this expression, A is the amplitude of the sinusoidal variation and $(\omega t + \phi)$ is the phase. In particular, the phase of the function at $t = 0$ is ϕ. The quantity $\omega = 2\pi f = 2\pi/T$ is the rate of change of phase with time and is known as the radian frequency, having the units radians per second. The quantity $f = \omega/2\pi$ is the number of times the phase changes by 2π radians in 1 second and is known as the linear frequency or simply frequency, having the units hertz, abbreviated Hz. The quantity $T = 1/f = 2\pi/\omega$ is the period, that is, the time interval in which the phase changes by 2π radians. A plot of the function versus t shown in Fig. 1.19 illustrates how the function changes periodically between positive and negative values. Note that the value of the function at $t = 0$ is $A\cos\phi$ and that a positive maximum occurs for $\omega t + \phi = 0$ or $t = -\phi/\omega$. For a numerical example, let us consider a sinusoidal function having the period 10^{-3} s, amplitude 5 units, and a positive maximum at $t = 2 \times 10^{-4}$ s. Then the expression for the function is given by $5\cos(2\pi \times 10^3 t - 0.4\pi)$.

If we now have a sinusoidally time-varying scalar field, we can visualize the field quantity varying sinusoidally with time at each point in the field region with the amplitude and phase governed by the spatial dependence of the field quantity. Thus, for example, the field $Ae^{-\alpha z}\cos(\omega t - \beta z)$, where A, α, and β are positive constants, is characterized by sinusoidal time variations with amplitude decreasing exponentially with z and the phase at any given time decreasing linearly with z.

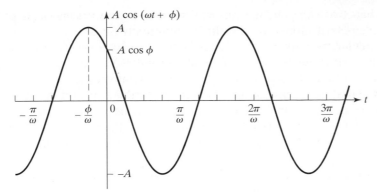

FIGURE 1.19   Sinusoidally time-varying scalar function $A \cos(\omega t + \phi)$.

For a sinusoidally time-varying vector field, the behavior of each component of the field may be visualized in the manner just discussed. If we now fix our attention on a particular point in the field region, we can visualize the sinusoidal variation with time of a particular component at that point by a vector changing its magnitude and direction as shown, for example, for the x-component in Fig. 1.20(a). Since the tip of the vector simply moves back and forth along a line, which in this case is parallel to the x-axis, the component vector is said to be *linearly polarized* in the x-direction. Similarly, the sinusoidal variation with time of the y-component of the field can be visualized by a vector changing its magnitude and direction, as shown in Fig. 1.20(b), not necessarily with the same amplitude and phase as those of the x-component. Since the tip of the vector moves back and forth parallel to the y-axis, the y-component is said to be linearly polarized in the y-direction. In the same manner, the z-component is linearly polarized in the z-direction.

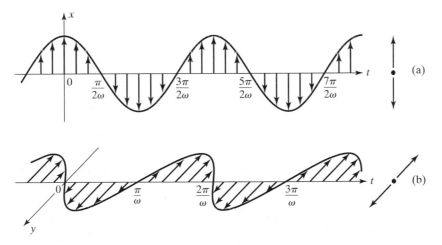

FIGURE 1.20   (a) Time variation of a linearly polarized vector in the x-direction. (b) Time variation of a linearly polarized vector in the y-direction.

Polarization of sinusoidally time-varying fields

When two component sinusoidally time-varying vectors at a point are added, the polarization of the resulting sinusoidally time-varying vector can be linear, circular, or elliptical, that is, the tip of the vector can describe a straight line, a circle, or an ellipse with time, as shown in Fig. 1.21, depending on the relative amplitudes and phase angles of the component vectors.

Note that in the case of linear polarization, the direction of the vector remains along a straight line, but its magnitude changes with time. For circular polarization, the magnitude remains constant, but its direction changes with time. Elliptical polarization is characterized by both magnitude and direction of the vector changing with time.

Linear polarization

If two component sinusoidally time-varying vectors have arbitrary amplitudes but are in phase or in phase opposition as, for example,

$$\mathbf{F}_1 = F_1 \cos (\omega t + \phi) \, \mathbf{a}_x \tag{1.57a}$$

$$\mathbf{F}_2 = \pm F_2 \cos (\omega t + \phi) \, \mathbf{a}_y \tag{1.57b}$$

then the sum vector $\mathbf{F} = \mathbf{F}_1 + \mathbf{F}_2$ is linearly polarized in a direction making an angle

$$\alpha = \tan^{-1} \frac{F_y}{F_x} = \pm \tan^{-1} \frac{F_2}{F_1} \tag{1.58}$$

with the x-direction, as shown in the series of sketches in Fig. 1.22 for the in-phase case illustrating the time history of the magnitude and direction of $\mathbf{F}$ over an interval of one period. The reasoning can be extended to two (or more) linearly polarized vectors not necessarily along the coordinate axes but all of them

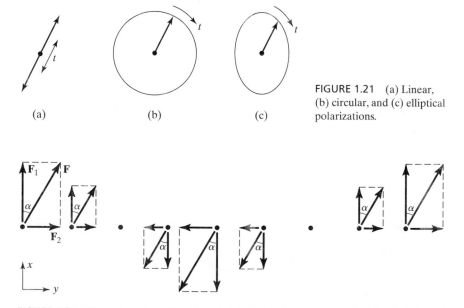

(a)   (b)   (c)

FIGURE 1.21   (a) Linear, (b) circular, and (c) elliptical polarizations.

FIGURE 1.22   The sum vector of two linearly polarized vectors in phase is a linearly polarized vector.

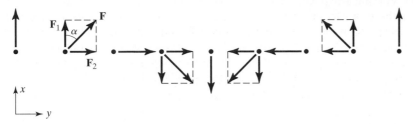

FIGURE 1.23   Circular polarization.

in phase. Thus, the sum vector of any number of linearly polarized vectors having different directions and amplitudes but in phase is also a linearly polarized vector.

If two component sinusoidally time-varying vectors have equal amplitudes, differ in direction by 90°, and differ in phase by $\pi/2$, as, for example, *Circular polarization*

$$\mathbf{F}_1 = F_0 \cos (\omega t + \psi) \, \mathbf{a}_x \tag{1.59a}$$

$$\mathbf{F}_2 = F_0 \sin (\omega t + \phi) \, \mathbf{a}_y \tag{1.59b}$$

then, to determine the *polarization* of the sum vector $\mathbf{F} = \mathbf{F}_1 + \mathbf{F}_2$, we note that the magnitude of $\mathbf{F}$ is given by

$$|\mathbf{F}| = |F_0 \cos (\omega t + \phi) \, \mathbf{a}_x + F_0 \sin (\omega t + \phi) \, \mathbf{a}_y| = F_0 \tag{1.60}$$

and that the angle α which $\mathbf{F}$ makes with $\mathbf{a}_x$ is given by

$$\alpha = \tan^{-1} \frac{F_y}{F_x} = \tan^{-1} \frac{F_0 \sin (\omega t + \phi)}{F_0 \cos (\omega t + \phi)} = \omega t + \phi \tag{1.61}$$

Thus, the sum vector rotates with constant magnitude F_0 and at a rate of ω rad/s, so that its tip describes a circle. The sum vector is then said to be *circularly polarized.* The series of sketches in Fig. 1.23 illustrates the time history of the magnitude and direction of $\mathbf{F}$ over an interval of one period.

The reasoning can be generalized to two linearly polarized vectors not necessarily along the coordinate axes. Thus, if two linearly polarized vectors satisfy the three conditions of (1) equal amplitudes, (2) perpendicularity in direction, and (3) phase difference of 90°, then their sum vector is circularly polarized. We shall illustrate this by means of an example.

Example 1.6

Given two vectors

$$\mathbf{F}_1 = (3\mathbf{a}_x - 4\mathbf{a}_z) \cos 2\pi \times 10^6 t$$
$$\mathbf{F}_2 = 5\mathbf{a}_y \sin 2\pi \times 10^6 t$$

at a point. Note that the vector $\mathbf{F}_1$ consists of two components (x and z) that are in phase opposition. Hence, it is linearly polarized, but along the direction of $(3\mathbf{a}_x - 4\mathbf{a}_z)$. The vector $\mathbf{F}_2$ is linearly polarized along the y-direction. We wish to determine the polarization of the vector $\mathbf{F} = \mathbf{F}_1 + \mathbf{F}_2$.

Since the two linearly polarized vectors $\mathbf{F}_1$ and $\mathbf{F}_2$ are not in phase, we rule out the possibility of $\mathbf{F}$ being linearly polarized. In fact, since $\mathbf{F}_1$ varies with time in a cosine manner, whereas $\mathbf{F}_2$ varies in a sine manner, we note that $\mathbf{F}_1$ and $\mathbf{F}_2$ differ in phase by 90°. The amplitude of $\mathbf{F}_1$ is $\sqrt{3^2 + (-4)^2}$, or 5, which is equal to that of $\mathbf{F}_2$. Also,

$$\mathbf{F}_1 \cdot \mathbf{F}_2 = (3\mathbf{a}_x - 4\mathbf{a}_z) \cdot 5\mathbf{a}_y = 0$$

so that $\mathbf{F}_1$ and $\mathbf{F}_2$ are perpendicular. Thus, $\mathbf{F}_1$ and $\mathbf{F}_2$ satisfy all three conditions for the sum of two linearly polarized vectors to be circularly polarized. Therefore, $\mathbf{F}$ is circularly polarized.

Alternatively, we observe that

$$
\begin{aligned}
|\mathbf{F}| &= |\mathbf{F}_1 + \mathbf{F}_2| \\
&= |3 \cos 2\pi \times 10^6 t\, \mathbf{a}_x + 5 \sin 2\pi \times 10^6 t\, \mathbf{a}_y - 4 \cos 2\pi \times 10^6 t\, \mathbf{a}_z| \\
&= (25 \cos^2 2\pi \times 10^6 t + 25 \sin^2 2\pi \times 10^6 t)^{1/2} \\
&= \sqrt{25} = 5, \text{ a constant with time}
\end{aligned}
$$

Hence, $\mathbf{F}$ is circularly polarized.

Elliptical polarization   For the general case in which the conditions for the sum vector to be linearly polarized or circularly polarized are not satisfied, the sum vector is *elliptically polarized;* that is, its tip describes an ellipse. Thus, linear and circular polarizations are special cases of elliptical polarization. For example, the ellipse described by the tip of the vector resulting from the superposition of two sinusoidally time-varying, orthogonal component vectors $\mathbf{F}_1 = A \cos(\omega t + \phi)\, \mathbf{a}_x$ and $\mathbf{F}_2 = B \cos(\omega t + \theta)\, \mathbf{a}_y$ of the same frequency, for values of $A = 40$, $B = 60$, $\phi = 60°$, and $\theta = 105°$, is shown in Fig. 1.24, where the component vectors and the resultant vector are also shown for one value of time, and the interval between the dots is one-hundredth of the period $2\pi/\omega$.

Relevance of polarization in reception of radio waves   An example in which polarization is relevant is in the reception of radio waves. Here when we talk about polarization, we generally refer to the direction of the electric field of the wave. If the incoming signal is linearly polarized, then

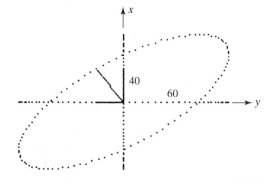

FIGURE 1.24   Ellipse traced by the tip of the vector $40 \cos(\omega t + 60°)\, \mathbf{a}_x + 60 \cos(\omega t + 105°)\, \mathbf{a}_y$.

for *maximum* voltage to be induced in a *linear* receiving antenna, the antenna must be oriented parallel to the direction of polarization of the signal. Any other orientation of the antenna will result in a smaller induced voltage since the antenna "sees" only that component of the electric field parallel to itself. In particular, if the antenna is in the plane perpendicular to the direction of polarization of the incoming signal, *no* voltage is induced. On the other hand, if the incoming signal is circularly or elliptically polarized, a voltage is induced in the antenna except for one orientation that is along the line perpendicular to the plane of the circle or the ellipse.

K1.5. Sinusoidal function; Sinusoidally time-varying scalar field; Sinusoidally time-varying vector field; Polarization of sinusoidally time-varying vector field; Linear polarization; Circular polarization; Elliptical polarization

D1.13. Write the expression for each of the sinusoidal functions of time having the specified characteristics: **(a)** period equal to 1 s, amplitude equal to 10 units, and a positive maximum occurring at $t = 0$; **(b)** frequency equal to 10^6 Hz, value at $t = 0$ equal to 5 units and a positive maximum occurring at $t = \frac{1}{6} \times 10^{-6}$ s; and **(c)** period equal to 2 s, amplitude equal to 5 units, and a negative maximum occurring at $\frac{3}{4}$ s.

Ans. **(a)** $10 \cos 2\pi t$; **(b)** $10 \cos (2\pi \times 10^6 t - \pi/3)$; **(c)** $5 \cos (\pi t + \pi/4)$.

D1.14. Two sinusoidally time-varying vector fields are given by

$$\mathbf{F}_1 = F_0 \cos (2\pi \times 10^8 t - 2\pi z) \, \mathbf{a}_x$$

$$\mathbf{F}_2 = F_0 \cos (2\pi \times 10^8 t - 3\pi z) \, \mathbf{a}_y$$

Find the polarization of $\mathbf{F}_1 + \mathbf{F}_2$ at each of the following points: **(a)** $(3, 4, 0)$; **(b)** $(3, -2, 0.5)$; **(c)** $(-2, 1, 1)$; and **(d)** $(-1, -3, 0.2)$.

Ans. **(a)** Linear; **(b)** circular; **(c)** linear; **(d)** elliptical.

D1.15. A sinusoidally time-varying vector field is given at a point by $\mathbf{F} = 1 \cos (\omega t + 60°) \, \mathbf{a}_x + 1 \cos (\omega t + \alpha) \, \mathbf{a}_y$. Find the value(s) of α between 0° and 360° for each of the following cases: **(a)** $\mathbf{F}$ is linearly polarized along a line lying in the second and fourth quadrants; **(b)** $\mathbf{F}$ is circularly polarized with the sense of rotation from the $+x$-direction toward the $+y$-direction with time; and **(c)** $\mathbf{F}$ is circularly polarized with the sense of rotation from the $+y$-direction toward the $+x$-direction with time.

Ans. **(a)** 240°; **(b)** 330°; **(c)** 150°.

1.6 COMPLEX NUMBERS AND PHASOR TECHNIQUE

In this section, we discuss a mathematical technique known as the phasor technique pertinent to operations involving sinusoidally time-varying quantities. The technique simplifies the solution of a differential equation in which the steady-state response for a sinusoidally time-varying excitation is to be determined, by reducing the differential equation to an algebraic equation involving phasors. A phasor is a complex number or a complex variable. We first review complex numbers and associated operations.

A complex number has two parts: a real part and an imaginary part. Imaginary numbers are square roots of negative real numbers. To introduce the concept of an imaginary number, we define

$$\boxed{\sqrt{-1} = j} \tag{1.62a}$$

or

$$\boxed{(\pm j)^2 = -1} \tag{1.62b}$$

Thus, $j5$ is the positive square root of -25, $-j10$ is the negative square root of -100, and so on. A complex number is written in the form $a + jb$, where a is the real part and b is the imaginary part. Examples are

$$3 + j4 \qquad -4 + j1 \qquad -2 - j2 \qquad 2 - j3$$

Rectangular form

A complex number is represented graphically in a complex plane by using two orthogonal axes, corresponding to the real and imaginary parts, as shown in Fig. 1.25, in which are plotted the numbers just listed. Since the set of orthogonal axes resembles the rectangular coordinate axes, the representation $(a + jb)$ is known as the rectangular form.

Exponential and polar forms

An alternative form of representation of a complex number is the exponential form $Ae^{j\phi}$, where A is the magnitude and ϕ is the phase angle. To convert from one form to another, we first recall that

$$e^x = 1 + x + \frac{x^2}{2!} + \frac{x^3}{3!} + \cdots \tag{1.63}$$

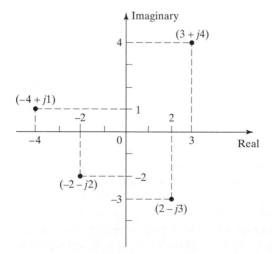

FIGURE 1.25   Graphical representation of complex numbers in rectangular form.

Substituting $x = j\phi$, we have

$$e^{j\phi} = 1 + j\phi + \frac{(j\phi)^2}{2!} + \frac{(j\phi)^3}{3!} + \cdots$$

$$= 1 + j\phi - \frac{\phi^2}{2!} - j\frac{\phi^3}{3!} + \cdots \qquad (1.64)$$

$$= \left(1 - \frac{\phi^2}{2!} + \cdots\right) + j\left(\phi - \frac{\phi^3}{3!} + \cdots\right)$$

$$= \cos\phi + j\sin\phi$$

This is the so-called *Euler's identity*. Thus,

$$\boxed{\begin{array}{c} Ae^{j\phi} = A(\cos\phi + j\sin\phi) \\ = A\cos\phi + jA\sin\phi \end{array}} \qquad (1.65)$$

Now, equating the two forms of the complex numbers, we have

$$a + jb = A\cos\phi + jA\sin\phi$$

or

$$\boxed{a = A\cos\phi} \qquad (1.66a)$$

$$\boxed{b = A\sin\phi} \qquad (1.66b)$$

These expressions enable us to convert from exponential form to rectangular form. To convert from rectangular form to exponential form, we note that

$$a^2 + b^2 = A^2$$

$$\cos\phi = \frac{a}{A} \qquad \sin\phi = \frac{b}{A} \qquad \tan\phi = \frac{b}{a}$$

Thus,

$$\boxed{\begin{array}{l} A = \sqrt{a^2 + b^2} \qquad\qquad\qquad (1.67a) \\[2mm] \phi = \tan^{-1}\frac{b}{a} \qquad\qquad\qquad (1.67b) \end{array}}$$

Note that in the determination of ϕ, the signs of $\cos\phi$ and $\sin\phi$ should be considered to see if it is necessary to add π to the angle obtained by taking the inverse tangent of b/a.

In terms of graphical representation, A is simply the distance from the origin of the complex plane to the point under consideration, and ϕ is the angle measured counterclockwise from the positive real axis ($\phi = 0$) to the line drawn from the origin to the complex number, as shown in Fig. 1.26. Since this representation is akin to the polar coordinate representation of points in two-dimensional space, the complex number is also written as $A\underline{/\phi}$, the polar form.

Turning now to Euler's identity, we see that for $\phi = \pm\pi/2$, $Ae^{\pm j\pi/2} = A \cos \pi/2 \pm jA \sin \pi/2 = \pm jA$. Thus, purely imaginary numbers correspond to $\phi = \pm\pi/2$. This justifies why the vertical axis, which is orthogonal to the real (horizontal) axis, is the imaginary axis.

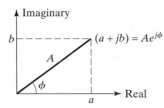

Conversion from rectangular to exponential or polar form

The complex numbers in rectangular form plotted in Fig. 1.25 may now be converted to exponential form (or polar form):

$$3 + j4 = \sqrt{3^2 + 4^2}\, e^{j\tan^{-1}(4/3)} = 5e^{j0.295\pi} = 5\underline{/53.13°}$$

$$-4 + j1 = \sqrt{4^2 + 1^2}\, e^{j[\tan^{-1}(-1/4) + \pi]} = 4.12e^{j0.922\pi} = 4.12\underline{/165.96°}$$

$$-2 - j2 = \sqrt{2^2 + 2^2}\, e^{j[\tan^{-1}(1) + \pi]} = 2.83e^{j1.25\pi} = 2.83\underline{/225°}$$

$$2 - j3 = \sqrt{2^2 + 3^2}\, e^{j\tan^{-1}(-3/2)} = 3.61e^{-j0.313\pi} = 3.61\underline{/-56.31°}.$$

These are shown plotted in Fig. 1.27. It can be noted that in converting from rectangular to exponential (or polar) form, the angle ϕ can be correctly determined if the number is first plotted in the complex plane to see in which quadrant it lies. Also note that angles traversed in the clockwise sense from the positive real axis are negative angles. Furthermore, adding or subtracting an integer multiple of 2π to the angle does not change the complex number.

FIGURE 1.26   Graphical representation of a complex number in exponential form or polar form.

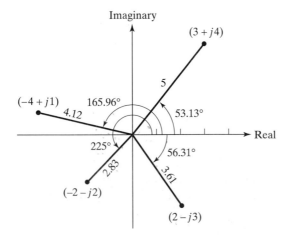

FIGURE 1.27   Polar form representation of the complex numbers of Fig. 1.25.

Arithmetic of complex numbers

Complex numbers are added (or subtracted) by simply adding (or subtracting) their real and imaginary parts separately as follows:

$$(3 + j4) + (2 - j3) = 5 + j1$$
$$(2 - j3) - (-4 + j1) = 6 - j4$$

Graphically, this procedure is identical to the parallelogram law of addition (or subtraction) of two vectors.

Two complex numbers are multiplied by multiplying each part of one complex number by each part of the second complex number and adding the four products according to the rule of addition as follows:

$$(3 + j4)(2 - j3) = 6 - j9 + j8 - j^2(12)$$
$$= 6 - j9 + j8 + 12$$
$$= 18 - j1$$

Two complex numbers whose real parts are equal but whose imaginary parts are the negative of each other are known as complex conjugates. Thus, $(a - jb)$ is the complex conjugate of $(a + jb)$, and vice versa. The product of two complex conjugates is a real number:

$$(a + jb)(a - jb) = a^2 - jab + jba + b^2 = a^2 + b^2 \qquad (1.68)$$

This property is used in division of one complex number by another by multiplying both the numerator and the denominator by the complex conjugate of the denominator and then performing the division by real number. For example,

$$\frac{3 + j4}{2 - j3} = \frac{(3 + j4)(2 + j3)}{(2 - j3)(2 + j3)} = \frac{-6 + j17}{13} = -0.46 + j1.31$$

The exponential form is particularly useful for multiplication, division, and other operations, such as raising to the power, since the rules associated with exponential functions are applicable. Thus,

$$(A_1 e^{j\phi_1})(A_2 e^{j\phi_2}) = A_1 A_2 e^{j(\phi_1 + \phi_2)} \qquad (1.69a)$$

$$\frac{A_1 e^{j\phi_1}}{A_2 e^{j\phi_2}} = \frac{A_1}{A_2} e^{j(\phi_1 - \phi_2)} \qquad (1.69b)$$

$$(A e^{j\phi})^n = A^n e^{jn\phi} \qquad (1.69c)$$

Let us consider some numerical examples:

a. $(5e^{j0.295\pi})(3.61e^{-j0.313\pi}) = 18.05e^{-j0.018\pi}$

b. $\dfrac{5e^{j0.295\pi}}{3.61e^{-j0.313\pi}} = 1.39e^{j0.608\pi}$

c. $(2.83e^{j1.25\pi})^4 = 64.14e^{j5\pi} = 64.14e^{j\pi}$

d. $\sqrt{4.12e^{j0.922\pi}} = [4.12e^{j(0.922\pi + 2k\pi)}]^{1/2}, \qquad k = 0, 1, 2, \ldots$
$$= \sqrt{4.12}\, e^{j(0.461\pi + k\pi)}, \qquad k = 0, 1$$
$$= 2.03e^{j0.461\pi}, \text{ or } 2.03e^{j1.461\pi}$$

Note that in evaluating the square roots, although k can assume an infinite number of integer values, only the first two need to be considered since the numbers repeat themselves for higher values of integers. Similar considerations apply for cube roots, and so on.

Phasor defined

Having reviewed complex numbers, we are now ready to discuss the phasor technique. The basis behind the phasor technique lies in the fact that since

$$\boxed{Ae^{jx} = A \cos x + jA \sin x} \tag{1.70}$$

we can write

$$\boxed{A \cos x = \mathrm{Re}[Ae^{jx}]} \tag{1.71}$$

where Re stands for "real part of." In particular, if $x = \omega t + \theta$, then we have

$$\begin{aligned}
A \cos (\omega t + \theta) &= \mathrm{Re}[Ae^{j(\omega t + \theta)}] \\
&= \mathrm{Re}[Ae^{j\theta}e^{j\omega t}] \\
&= \mathrm{Re}[\overline{A}e^{j\omega t}]
\end{aligned} \tag{1.72}$$

where $\overline{A} = Ae^{j\theta}$ is known as the phasor (the overbar denotes that $\overline{A}$ is complex) corresponding to $A \cos (\omega t + \theta)$. Thus, the phasor corresponding to a cosinusoidally time-varying function is a complex number having magnitude same as the amplitude of the cosine function and phase angle equal to the phase of the cosine function for $t = 0$. To find the phasor corresponding to a sine function, we first convert it into a cosine function and proceed as in (1.72). Thus,

$$\begin{aligned}
B \sin (\omega t + \phi) &= B \cos (\omega t + \phi - \pi/2) \\
&= \mathrm{Re}[Be^{j(\omega t + \phi - \pi/2)}] \\
&= \mathrm{Re}[Be^{j(\phi - \pi/2)}e^{j\omega t}]
\end{aligned} \tag{1.73}$$

Hence, the phasor corresponding to $B \sin (\omega t + \phi)$ is $Be^{j(\phi - \pi/2)}$, or $Be^{j\phi}e^{-j\pi/2}$, or $-jBe^{j\phi}$.

Addition of two sine functions

Let us now consider the addition of two sinusoidally time-varying functions (of the same frequency), for example, $5 \cos \omega t$ and $10 \sin (\omega t - 30°)$, by using the phasor technique. To do this, we proceed as follows:

$$\begin{aligned}
5 \cos \omega t + 10 \sin (\omega t - 30°) &= 5 \cos \omega t + 10 \cos (\omega t - 120°) \\
&= \mathrm{Re}[5e^{j\omega t}] + \mathrm{Re}[10e^{j(\omega t - 2\pi/3)}] \\
&= \mathrm{Re}[5e^{j0}e^{j\omega t}] + \mathrm{Re}[10e^{-j2\pi/3}e^{j\omega t}] \\
&= \mathrm{Re}[5e^{j0}e^{j\omega t} + 10e^{-j2\pi/3}e^{j\omega t}] \\
&= \mathrm{Re}[(5e^{j0} + 10e^{-j2\pi/3})e^{j\omega t}] \\
&= \mathrm{Re}\{[(5 + j0) + (-5 - j8.66)]e^{j\omega t}\} \\
&= \mathrm{Re}[(0 - j8.66)e^{j\omega t}] \\
&= \mathrm{Re}[8.66e^{-j\pi/2}e^{j\omega t}] \\
&= \mathrm{Re}[8.66e^{j(\omega t - \pi/2)}] \\
&= 8.66 \cos (\omega t - 90°)
\end{aligned} \tag{1.74}$$

In practice, we need not write all the steps just shown. First, we express all functions in their cosine forms and then recognize the phasor corresponding to each function. For the foregoing example, the complex numbers $5e^{j0}$ and $10e^{-j2\pi/3}$ are the phasors corresponding to 5 cos ωt and 10 sin $(\omega t - 30°)$, respectively. Then we add the phasors and from the sum phasor write the required cosine function as one having the amplitude the same as the magnitude of the sum phasor and the argument equal to ωt plus the phase angle of the sum phasor. Thus, the steps involved are as shown in the block diagram of Fig. 1.28.

We shall now discuss the solution of a differential equation for sinusoidal steady-state response by using the phasor technique. To do this, let us consider the problem of finding the steady-state solution for the current $I(t)$ in the simple RL series circuit driven by the voltage source $V(t) = V_m \cos(\omega t + \phi)$, as shown in Fig. 1.29. From Kirchhoff's voltage law, we then have

Solution of differential equation

$$RI(t) + L\frac{dI(t)}{dt} = V_m \cos(\omega t + \phi) \qquad (1.75)$$

We know that the steady-state solution for the current must also be a cosine function of time having the same frequency as that of the voltage source but not necessarily in phase with it. Hence, let us assume

$$I(t) = I_m \cos(\omega t + 0) \qquad (1.76)$$

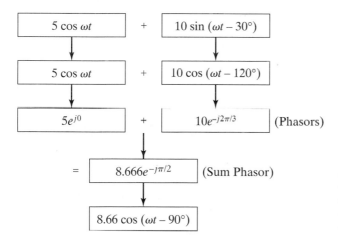

FIGURE 1.28   Block diagram of steps involved in the application of the phasor technique to the addition of two sinusoidally time-varying functions.

FIGURE 1.29 RL series circuit.

The problem now consists of finding I_m and θ.

Using the phasor concept, we write

$$
\begin{aligned}
V_m \cos(\omega t + \phi) &= \text{Re}[V_m e^{j(\omega t + \phi)}] \\
&= \text{Re}[V_m e^{j\phi} e^{j\omega t}] \\
&= \text{Re}[\bar{V} e^{j\omega t}]
\end{aligned}
\tag{1.77a}
$$

$$
\begin{aligned}
I_m \cos(\omega t + \theta) &= \text{Re}[I_m e^{j(\omega t + \theta)}] \\
&= \text{Re}[I_m e^{j\theta} e^{j\omega t}] \\
&= \text{Re}[\bar{I} e^{j\omega t}]
\end{aligned}
\tag{1.77b}
$$

where $\bar{V} = V_m e^{j\phi}$ and $\bar{I} = I_m e^{j\theta}$ are the phasors corresponding to $V(t) = V_m \cos(\omega t + \phi)$ and $I(t) = I_m \cos(\omega t + \theta)$, respectively. Substituting these into the differential equation, we have

$$
R\{\text{Re}[\bar{I} e^{j\omega t}]\} + L \frac{d}{dt}\{\text{Re}[\bar{I} e^{j\omega t}]\} = \text{Re}[\bar{V} e^{j\omega t}]
\tag{1.78}
$$

Since R and L are constants, and since d/dt and Re can be interchanged, we can simplify this equation in accordance with the following steps:

$$
\text{Re}[R\bar{I} e^{j\omega t}] + \text{Re}\left[L \frac{d}{dt}(\bar{I} e^{j\omega t}) \right] = \text{Re}[\bar{V} e^{j\omega t}]
$$

$$
\text{Re}[R\bar{I} e^{j\omega t}] + \text{Re}[j\omega L\bar{I} e^{j\omega t}] = \text{Re}[\bar{V} e^{j\omega t}]
\tag{1.79}
$$

$$
\text{Re}[(R\bar{I} + j\omega L\bar{I}) e^{j\omega t}] = \text{Re}[\bar{V} e^{j\omega t}]
$$

Let us now consider two values of ωt, say, $\omega t = 0$ and $\omega t = \pi/2$. For $\omega t = 0$, we obtain

$$
\text{Re}(R\bar{I} + j\omega L\bar{I}) = \text{Re}(\bar{V})
\tag{1.80}
$$

For $\omega t = \pi/2$, we obtain

$$
\text{Re}[j(R\bar{I} + j\omega L\bar{I})] = \text{Re}[j\bar{V}]
$$

or

$$
\text{Im}(R\bar{I} + j\omega L\bar{I}) = \text{Im}(\bar{V})
\tag{1.81}
$$

where Im stands for "imaginary part of." Now, since the real parts as well as the imaginary parts of $(R\bar{I} + j\omega L\bar{I})$ and $\bar{V}$ are equal, it follows that the two complex numbers are equal. Thus,

$$
\boxed{R\bar{I} + j\omega L\bar{I} = \bar{V}}
\tag{1.82}
$$

By solving this equation, we obtain $\bar{I}$ and hence I_m and θ. Note that by using the phasor technique, we have reduced the problem of solving the differential equation (1.75) to one of solving the phasor (algebraic) equation (1.82). In fact, the phasor equation can be written directly from the differential equation without the necessity of the intermediate steps by recognizing that the time functions $I(t)$ and $V(t)$ are replaced by their phasors $\bar{I}$ and $\bar{V}$, respectively, and d/dt is replaced by $j\omega$. We have here included the intermediate steps merely to illustrate the basis behind the phasor technique. We shall now consider an example.

Example 1.7

For the circuit of Fig. 1.29, let us assume that $R = 1 \ \Omega$, $L = 10^{-3}$ H, and $V(t) = 10 \cos{(1000t + 30°)}$ V and obtain the steady-state solution for $I(t)$.

The differential equation for $I(t)$ is given by

$$I + 10^{-3}\frac{dI}{dt} = 10 \cos{(1000t + 30°)}$$

Replacing the current and voltage by their phasors $\bar{I}$ and $10e^{j\pi/6}$, respectively, and d/dt by $j\omega = j1000$, we obtain the phasor equation

$$\bar{I} + 10^{-3}(j1000\bar{I}) = 10e^{j\pi/6}$$

or

$$\bar{I}(1 + j1) = 10e^{j\pi/6}$$

$$\bar{I} = \frac{10e^{j\pi/6}}{1 + j1} = \frac{10e^{j\pi/6}}{\sqrt{2} \, e^{j\pi/4}}$$

$$= 7.07e^{-j\pi/12}$$

Having determined the value of $\bar{I}$, we now find the required solution to be

$$I(t) = \text{Re}[\bar{I}e^{j\omega t}]$$

$$= \text{Re}[7.07e^{-j\pi/12}e^{j1000t}]$$

$$= 7.07 \cos{(1000t - 15°)} \text{ A}$$

K1.6. Complex number; Rectangular form, exponential form, polar form, and conversions; Complex number arithmetic; Phasor; Phasor technique.

D1.16. Express the following complex numbers in exponential form: **(a)** $5 + j12$; **(b)** $-24 - j7$; and **(c)** $-1 + j\sqrt{3}$.
Ans. **(a)** $13e^{j0.374\pi}$; **(b)** $25e^{j1.09\pi}$; **(c)** $2e^{j0.667\pi}$.

D1.17. Find: **(a)** $5e^{j0.205\pi} + (1 + j2)$ in exponential form; **(b)** $(-\sqrt{3} + j1)^5$ in rectangular form; and **(c)** $(3 + j1)(1 - j3)/(4 + j3)$ in polar form.
Ans. **(a)** $7.07e^{j0.25\pi}$; **(b)** $(16\sqrt{3} + j16)$; **(c)** $2\underline{/-90°}$.

D1.18. Using the phasor technique, express each of the following functions as a single cosinusoidal function of time: **(a)** $4 \cos{(\omega t + 30°)} - 3 \cos{(\omega t + 120°)}$; **(b)** $5 \sin{(\omega t + 60°)} - 5 \cos{(\omega t + 30°)} - 3 \sin{\omega t}$.
Ans. **(a)** $5 \cos{(\omega t - 6.87°)}$; **(b)** $2 \cos{(\omega t - 90°)}$.

1.7 SUMMARY

We first learned in this chapter several rules of vector algebra that are necessary for our study of the elements of engineering electromagnetics by considering vectors expressed in terms of their components along three mutually orthogonal directions. To carry out the manipulations involving vectors at different points in space in a systematic manner, we then introduced the Cartesian coordinate system and discussed the application of the vector algebraic rules to vectors in the Cartesian coordinate system. To summarize these rules, we consider three vectors

$$\mathbf{A} = A_x\mathbf{a}_x + A_y\mathbf{a}_y + A_z\mathbf{a}_z$$

$$\mathbf{B} = B_x\mathbf{a}_x + B_y\mathbf{a}_y + B_z\mathbf{a}_z$$

$$\mathbf{C} = C_x\mathbf{a}_x + C_y\mathbf{a}_y + C_z\mathbf{a}_z$$

in a right-handed Cartesian coordinate system, that is, with $\mathbf{a}_x \times \mathbf{a}_y = \mathbf{a}_z$. We then have

$$\mathbf{A} + \mathbf{B} = (A_x + B_x)\mathbf{a}_x + (A_y + B_y)\mathbf{a}_y + (A_z + B_z)\mathbf{a}_z$$

$$\mathbf{B} - \mathbf{C} = (B_x - C_x)\mathbf{a}_x + (B_y - C_y)\mathbf{a}_y + (B_z - C_z)\mathbf{a}_z$$

$$m\mathbf{A} = mA_x\mathbf{a}_x + mA_y\mathbf{a}_y + mA_z\mathbf{a}_z$$

$$\frac{\mathbf{B}}{n} = \frac{B_x}{n}\mathbf{a}_x + \frac{B_y}{n}\mathbf{a}_y + \frac{B_z}{n}\mathbf{a}_z$$

$$|\mathbf{A}| = \sqrt{A_x^2 + A_y^2 + A_z^2}$$

$$\mathbf{a}_A = \frac{A_x}{\sqrt{A_x^2 + A_y^2 + A_z^2}}\mathbf{a}_x + \frac{A_y}{\sqrt{A_x^2 + A_y^2 + A_z^2}}\mathbf{a}_y + \frac{A_z}{\sqrt{A_x^2 + A_y^2 + A_z^2}}\mathbf{a}_z$$

$$\mathbf{A} \cdot \mathbf{B} = A_xB_x + A_yB_y + A_zB_z$$

$$\mathbf{A} \times \mathbf{B} = \begin{vmatrix} \mathbf{a}_x & \mathbf{a}_y & \mathbf{a}_z \\ A_x & A_y & A_z \\ B_x & B_y & B_z \end{vmatrix}$$

$$\mathbf{A} \cdot \mathbf{B} \times \mathbf{C} = \begin{vmatrix} A_x & A_y & A_z \\ B_x & B_y & B_z \\ C_x & C_y & C_z \end{vmatrix}$$

Other useful expressions are

$$d\mathbf{l} = dx\,\mathbf{a}_x + dy\,\mathbf{a}_y + dz\,\mathbf{a}_z$$

$$d\mathbf{S} = \pm dy\,dz\,\mathbf{a}_x, \pm dz\,dx\,\mathbf{a}_y, \pm dx\,dy\,\mathbf{a}_z$$

$$dv = dx\,dy\,dz$$

We then discussed the cylindrical and spherical coordinate systems, and conversions between these coordinate systems and the Cartesian coordinate system. Relationships for carrying out the coordinate conversions are as follows:

CYLINDRICAL TO CARTESIAN, AND VICE VERSA

$$x = r \cos \phi \qquad y = r \sin \phi \qquad z = z$$

$$r = \sqrt{x^2 + y^2} \qquad \phi = \tan^{-1} \frac{y}{x} \qquad z = z$$

SPHERICAL TO CARTESIAN, AND VICE VERSA

$$x = r \sin \theta \cos \phi \qquad y = r \sin \theta \sin \phi \qquad z = r \cos \theta$$

$$r = \sqrt{x^2 + y^2 + z^2} \qquad \theta = \tan^{-1} \frac{\sqrt{x^2 + y^2}}{z} \qquad \phi = \tan^{-1} \frac{y}{x}$$

Other useful expressions are as follows:

CYLINDRICAL

$$dl = dr \, \mathbf{a}_r + r \, d\phi \, \mathbf{a}_\phi + dz \, \mathbf{a}_z$$
$$dS = \pm r \, d\phi \, dz \, \mathbf{a}_r, \ \pm dr \, dz \, \mathbf{a}_\phi, \ \pm r \, dr \, d\phi \, \mathbf{a}_z$$
$$dv = r \, dr \, d\phi \, dz$$

SPHERICAL

$$dl = dr \, \mathbf{a}_r + r \, d\theta \, \mathbf{a}_\theta + r \sin \theta \, d\phi \, \mathbf{a}_\phi$$
$$dS = \pm r^2 \sin \theta \, d\theta \, d\phi \, \mathbf{a}_r, \ \pm r \sin \theta \, dr \, d\phi \, \mathbf{a}_\theta, \ \pm r \, dr \, d\theta \, \mathbf{a}_\phi$$
$$dv = r^2 \sin \theta \, dr \, d\theta \, d\phi$$

Next we discussed the concepts of scalar and vector fields, static and time-varying, by means of some simple examples such as the height of points on a conical surface above its base, the temperature field of points in a room, and the velocity vector field associated with points on a disk rotating about its center. We learned about the visualization of fields by means of constant-magnitude contours or surfaces and in addition by means of direction lines in the case of vector fields. We also discussed the mathematical technique of obtaining the equations for the direction lines of a vector field.

Particular attention was then devoted to sinusoidally time-varying fields, in view of their importance in our study of electromagnetics. Polarization of sinusoidally time-varying vector fields was then considered. In the general case, the polarization is elliptical; that is, the tip of the vector at a point describes an ellipse with time. Linear and circular polarizations are special cases.

Finally, a review of complex numbers and associated operations was presented as a prelude to the discussion of phasor technique. The phasor technique was illustrated by considering two examples: (1) addition of two sinusoidal functions of time and (2) solution of a differential equation for the steady-state response due to a sinusoidal excitation.

REVIEW QUESTIONS

Q1.1. Give some examples of scalars.

Q1.2. Give some examples of vectors.

Q1.3. Is it necessary for the reference vectors $\mathbf{a}_1$, $\mathbf{a}_2$, and $\mathbf{a}_3$ to be an orthogonal set?

Q1.4. State whether $\mathbf{a}_1$, $\mathbf{a}_2$, and $\mathbf{a}_3$ directed westward, northward, and downward, respectively, is a right-handed or a left-handed set.

Q1.5. State all conditions for which $\mathbf{A} \cdot \mathbf{B} = 0$.

Q1.6. State all conditions for which $\mathbf{A} \times \mathbf{B} = \mathbf{0}$.

Q1.7. What is the significance of $\mathbf{A} \cdot \mathbf{B} \times \mathbf{C} = 0$?

Q1.8. What is the significance of $\mathbf{A} \times (\mathbf{B} \times \mathbf{C}) = \mathbf{0}$?

Q1.9. What is the particular advantageous characteristic associated with the unit vectors in the Cartesian coordinate system?

Q1.10. What is the position vector?

Q1.11. What is the total distance around the circumference of a circle of radius 1 m? What is the total vector distance around the circle?

Q1.12. Discuss the application of differential length vectors to find a unit vector normal to a surface at a point on the surface.

Q1.13. Discuss the concept of a differential surface vector.

Q1.14. What is the total surface area of a cube of sides 1 m? Assuming the normals to the surfaces to be directed outward of the cubical volume, what is the total vector surface area of the cube?

Q1.15. Describe the three orthogonal surfaces that define the cylindrical coordinates of a point.

Q1.16. Which of the unit vectors in the cylindrical coordinate system are not uniform? Explain.

Q1.17. Discuss the conversion from the cylindrical coordinates of a point to its Cartesian coordinates, and vice versa.

Q1.18. Describe the three orthogonal surfaces that define the spherical coordinates of a point.

Q1.19. Discuss the nonuniformity of the unit vectors in the spherical coordinate system.

Q1.20. Discuss the conversion from the spherical coordinates of a point to its Cartesian coordinates, and vice versa.

Q1.21. Describe briefly your concept of a scalar field and illustrate with an example.

Q1.22. Describe briefly your concept of a vector field and illustrate with an example.

Q1.23. How do you depict pictorially the gravitational field of Earth?

Q1.24. Discuss the procedure for obtaining the equations for the direction lines of a vector field.

Q1.25. Discuss the parameters associated with a sinusoidal function of time.

Q1.26. A sinusoidally time-varying vector is expressed in terms of its components along the x-, y-, and z-axes. What is the polarization of each of the components?

Q1.27. What are the conditions for the sum of two linearly polarized sinusoidally time-varying vectors to be circularly polarized?

Q1.28. What is the polarization for the general case of the sum of two sinusoidally time-varying linearly polarized vectors having arbitrary amplitudes, phase angles, and directions?

Q1.29. Considering the second hand on your analog watch to be a vector, state its polarization. What is the frequency?

Q1.30. Discuss the relevance of polarization in the reception of radio waves.

Q1.31. Discuss the conversion of a complex number from rectangular form to exponential (or polar) form.

Q1.32. How do you perform division of one complex number by another using their rectangular forms?

Q1.33. What is a phasor?

Q1.34. Is there any relationship between a phasor and a vector? Explain.

Q1.35. Describe the phasor technique of adding two sinusoidal functions of time.

Q1.36. Describe the phasor technique of solving a differential equation for the sinusoidal steady-state solution.

PROBLEMS

Section 1.1

P1.1. A bug starts at a point and travels 1 m northward, s m eastward, s^2 m southward, s^3 m westward, and so on, where $s < 1$, making a 90°-turn to the right and traveling in the new direction s times the distance traveled in the previous direction. Find the value of s for each of the following cases: **(a)** the total distance traveled by the bug is 1.5 m; **(b)** the straight-line distance from the initial position to the final position of the bug is 0.8 m; and **(c)** the final position of the bug relative to its initial position is 30° east of north.

P1.2. Three vectors $\mathbf{A}, \mathbf{B}$, and $\mathbf{C}$ satisfy the equations

$$\mathbf{A} + \mathbf{B} - \mathbf{C} = 2\mathbf{a}_1 + \mathbf{a}_2$$
$$\mathbf{A} + 2\mathbf{B} + 3\mathbf{C} = -2\mathbf{a}_1 + 5\mathbf{a}_2 + 5\mathbf{a}_3$$
$$2\mathbf{A} - \mathbf{B} + \mathbf{C} = \mathbf{a}_1 + 5\mathbf{a}_2$$

By writing a matrix equation for the 3×3 matrix

$$\begin{bmatrix} A_1 & A_2 & A_3 \\ B_1 & B_2 & B_3 \\ C_1 & C_2 & C_3 \end{bmatrix}$$

and solving it, obtain the vectors $\mathbf{A}, \mathbf{B}$, and $\mathbf{C}$.

P1.3. Two vectors $\mathbf{A}$ and $\mathbf{B}$ originate from a common point. **(a)** If $\mathbf{C} = \mathbf{B} - \mathbf{A}$ comprises the third side of the triangle, obtain using $\mathbf{C} \cdot \mathbf{C} = (\mathbf{B} - \mathbf{A}) \cdot (\mathbf{B} - \mathbf{A})$ the law of cosines relating C to A, B, and the angle α between $\mathbf{A}$ and $\mathbf{B}$. **(b)** Find the expression for the distance from the common point to the side $\mathbf{C}$, in terms of $\mathbf{A}$ and $\mathbf{B}$ only.

P1.4. Four vectors drawn from a common point are given as follows:

$$\mathbf{A} = 2\mathbf{a}_1 - m\mathbf{a}_2 - \mathbf{a}_3$$
$$\mathbf{B} = m\mathbf{a}_1 + \mathbf{a}_2 - 2\mathbf{a}_3$$
$$\mathbf{C} = \mathbf{a}_1 + m\mathbf{a}_2 + 2\mathbf{a}_3$$
$$\mathbf{D} = m^2\mathbf{a}_1 + m\mathbf{a}_2 + \mathbf{a}_3$$

Find the value(s) of m for each of the following cases: **(a)** **A** is perpendicular to **B**; **(b)** **B** is parallel to **C**; **(c)** **A**, **B**, and **C** lie in the same plane; and **(d)** **D** is perpendicular to both **A** and **B**.

P1.5. Show that the tips of three vectors **A**, **B**, and **C** originating from a common point lie along a straight line if $\mathbf{A} \times \mathbf{B} + \mathbf{B} \times \mathbf{C} + \mathbf{C} \times \mathbf{A} = \mathbf{0}$. Provide a geometric interpretation for this result.

P1.6. Show that the tips of four vectors **A**, **B**, **C**, and **D** originating from a common point lie in a plane if $(\mathbf{A} - \mathbf{B}) \cdot (\mathbf{A} - \mathbf{C}) \times (\mathbf{A} - \mathbf{D}) = 0$. Then determine if the tips of $\mathbf{A} = \mathbf{a}_1, \mathbf{B} = 2\mathbf{a}_2, \mathbf{C} = 2\mathbf{a}_3$, and $\mathbf{D} = \mathbf{a}_1 + 2\mathbf{a}_2 - 2\mathbf{a}_3$ lie in a plane.

P1.7. **(a)** Show that

$$\mathbf{A} \times (\mathbf{B} \times \mathbf{C}) = (\mathbf{A} \cdot \mathbf{C})\mathbf{B} - (\mathbf{A} \cdot \mathbf{B})\mathbf{C}$$

(b) Using the result of part (a), show the following:

(i) $\mathbf{A} \times (\mathbf{B} \times \mathbf{C}) + \mathbf{B} \times (\mathbf{C} \times \mathbf{A}) + \mathbf{C} \times (\mathbf{A} \times \mathbf{B}) = \mathbf{0}$
(ii) $(\mathbf{A} \times \mathbf{B}) \cdot (\mathbf{B} \times \mathbf{C}) \times (\mathbf{C} \times \mathbf{A}) = (\mathbf{A} \times \mathbf{B} \cdot \mathbf{C})^2$

Section 1.2

P1.8. Three points are given by $A(12, 0, 0)$, $B(0, 15, 0)$, and $C(0, 0, -20)$. Find the following: **(a)** the distance from B to C; **(b)** the component of the vector from A to C along the vector from B to C; and **(c)** the perpendicular distance from A to the line through B and C.

P1.9. Consider four points (x_1, y_1, z_1), (x_2, y_2, z_2), (x_3, y_3, z_3), and (x_4, y_4, z_4). Show that the center point (x_0, y_0, z_0) of the sphere passing through these points is given by the solution of the equation

$$2\begin{bmatrix} x_2 - x_1 & y_2 - y_1 & z_2 - z_1 \\ x_3 - x_1 & y_3 - y_1 & z_3 - z_1 \\ x_4 - x_1 & y_4 - y_1 & z_4 - z_1 \end{bmatrix}\begin{bmatrix} x_0 \\ y_0 \\ z_0 \end{bmatrix} = \begin{bmatrix} (x_2^2 + y_2^2 + z_2^2) - (x_1^2 + y_1^2 + z_1^2) \\ (x_3^2 + y_3^2 + z_3^2) - (x_1^2 + y_1^2 + z_1^2) \\ (x_4^2 + y_4^2 + z_4^2) - (x_1^2 + y_1^2 + z_1^2) \end{bmatrix}$$

Then find the center point of the sphere and its radius if the four points are $(1, 1, 4)$, $(3, 3, 2)$, $(2, 3, 3)$, and $(3, 2, 3)$.

P1.10. **(a)** Two vectors **A** and **B** originate from a common point $P(x_1, y_1, z_1)$. Show that the equation for the plane in which the two vectors lie is given by

$$\mathbf{A} \times \mathbf{B} \cdot (\mathbf{r} - \mathbf{r}_1) = 0$$

where $\mathbf{r} = x\mathbf{a}_x + y\mathbf{a}_y + z\mathbf{a}_z$ is the position vector and $\mathbf{r}_1 = x_1\mathbf{a}_x + y_1\mathbf{a}_y + z_1\mathbf{a}_z$ is the vector from the origin to the point P.

(b) Using the result of part (a), find the equation for the plane containing the points $(1, 1, 2)$, $(2, 2, 0)$, and $(3, 0, 1)$.

P1.11. Find the expression for the differential length vector tangential to the curve $x + y = 2, y = z^2$ at an arbitrary point on the curve and having the projection dz on the z-axis. Then obtain the differential length vectors tangential to the curve at the points **(a)** $(2, 0, 0)$, **(b)** $(1, 1, 1)$, and **(c)** $(-2, 4, 2)$.

P1.12. Find the expression for the unit vector normal to the curve $x = y^2 = z^3$ at the point $(1, 1, 1)$ and having no components along the line $x = y = z$.

P1.13. By considering two differential length vectors tangential to the surface $x^2 + y^2 + 2z^2 = 4$ at the point $(1, 1, 1)$, find the unit vector normal to the surface.

P1.14. Consider the differential surface lying on the plane $2x + y = 2$ and having as its projection on the xz-plane the rectangular differential surface of sides dx and dz

in the x- and z-directions, respectively. Obtain the expression for the vector $d\mathbf{S}$ associated with that surface.

Section 1.3

P1.15. Three points are given in cylindrical coordinates by $A(2, \pi/3, 1)$, $B(2\sqrt{3}, \pi/6, -2)$, and $C(2, 5\pi/6, 0)$. **(a)** Find the volume of the parallelepiped having the lines from the origin to the three points as one set of its contiguous edges. **(b)** Determine if the point $D(\sqrt{3}, \pi/2, 2.5)$ in cylindrical coordinates lies in the plane containing $A, B,$ and C.

P1.16. Four points are given in spherical coordinates by $A(1, \pi/2, 0), B(\sqrt{8}, \pi/4, \pi/3)$, $C(1, 0, 0)$, and $D(\sqrt{12}, \pi/6, \pi/2)$. Show that these four points are situated at the corners of a parallelogram and find the area of the parallelogram.

P1.17. Three unit vectors are given in cylindrical coordinates as follows: $\mathbf{A} = \mathbf{a}_r$ at $(2, \pi/6, 0)$, $\mathbf{B} = \mathbf{a}_\phi$ at $(1, \pi/3, 2)$, and $\mathbf{C} = \mathbf{a}_\phi$ at $(3, 5\pi/6, 1)$. Find: **(a)** $\mathbf{A} \cdot \mathbf{B}$; **(b)** $\mathbf{B} \cdot \mathbf{C}$; and **(c)** $\mathbf{B} \times \mathbf{C}$.

P1.18. Three unit vectors are given in spherical coordinates as follows: $\mathbf{A} = \mathbf{a}_r$ at $(2, \pi/6, \pi/2)$, $\mathbf{B} = \mathbf{a}_\theta$ at $(1, \pi/3, 0)$, and $\mathbf{C} = \mathbf{a}_\phi$ at $(3, \pi/4, 3\pi/2)$. Find: **(a)** $\mathbf{A} \cdot \mathbf{B}$; **(b)** $\mathbf{A} \cdot \mathbf{C}$; **(c)** $\mathbf{B} \cdot \mathbf{C}$; and **(d)** $\mathbf{A} \times \mathbf{B} \cdot \mathbf{C}$.

P1.19. Convert the vector $\mathbf{a}_x + \mathbf{a}_y - \sqrt{2}\,\mathbf{a}_z$ at the point $(1, 1, \sqrt{2})$ to one in spherical coordinates.

P1.20. Determine if the vector $(\mathbf{a}_{rc} - \sqrt{3}\,\mathbf{a}_\phi + 3\mathbf{a}_z)$ at the point $(3, \pi/3, 5)$ in cylindrical coordinates is equal to the vector $(3\mathbf{a}_{rs} - \sqrt{3}\,\mathbf{a}_\theta - \mathbf{a}_\phi)$ at the point $(1, \pi/3, \pi/6)$ in spherical coordinates.

P1.21. Find the expression for the unit vector tangential to the curve given in cylindrical coordinates by $r^2 \sin 2\phi = 1, z = 0$. Then obtain the unit vectors tangential to the curve at the points: **(a)** $(1, \pi/4, 0)$ and **(b)** $(\sqrt{2}, \pi/12, 0)$.

P1.22. Find the expression for the unit vector tangential to the curve given in spherical coordinates by $r = 1, \phi = 2\theta, 0 \le \theta \le \pi$. Then obtain the unit vectors tangential to the curve at the points: **(a)** $(1, \pi/4, \pi/2)$ and **(b)** $(1, \pi/2, \pi)$.

Section 1.4

P1.23. An otherwise hemispherical dome of radius 2 m has a symmetrically situated hemispherical trough of radius 1 m, as shown by the cross-sectional view in Fig. 1.30. Assuming the origin to be at the center of the base of the dome, obtain the expression for the two-dimensional scalar field describing the height h of the dome in each of the two coordinate systems: **(a)** rectangular (x, y) and **(b)** polar (r, ϕ).

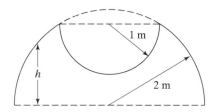

FIGURE 1.30   For Problem P1.23.

P1.24. Assuming the origin to be at the center of Earth and the z axis to be passing through the poles, write vector functions for the force experienced by a mass m

in the gravitational field of Earth (mass M) in each of the three coordinate systems: **(a)** Cartesian, **(b)** cylindrical, and **(c)** spherical. Describe the constant-magnitude surfaces and the direction lines.

P1.25. Assuming the origin to be at the center of Earth and the z-axis to be passing through the poles, write vector functions for the linear velocity of points inside Earth due to its spin motion in each of the three coordinate systems: **(a)** Cartesian; **(b)** cylindrical; and **(c)** spherical. Describe the constant-magnitude surfaces and the direction lines.

P1.26. Obtain the equations for the direction lines for the following vector fields and passing through the point $(1, 2, 3)$: **(a)** $2y\mathbf{a}_x - xa_y$ and **(b)** $xa_x + ya_y + za_z$.

P1.27. Obtain the equation for the direction line for the vector field given in cylindrical coordinates by $(\sin \phi \, \mathbf{a}_r + \cos \phi \, \mathbf{a}_\phi)$ and passing through the point $(2, \pi/3, 1)$.

P1.28. Obtain the equation for the direction line for the vector field given in spherical coordinates by $(2 \cos \theta \, \mathbf{a}_r - \sin \theta \, \mathbf{a}_\theta)$ and passing through the point $(2, \pi/4, \pi/6)$.

Section 1.5

P1.29. Given $f(z, t) = 10 \cos (3\pi \times 10^7 t + 0.1\pi z)$, draw sketches of f versus z for $t = 0, \frac{1}{6} \times 10^{-7}$ s, and $\frac{1}{3} \times 10^{-7}$ s on the same graph. Discuss the nature of the function.

P1.30. Given $f(z, t) = 10 \cos 0.1 \, \pi z \cos 3\pi \times 10^7 t$, draw sketches of f versus z for $t = 0, \frac{1}{12} \times 10^{-7}$ s, $\frac{1}{6} \times 10^{-7}$ s, $\frac{1}{4} \times 10^{-7}$ s, and $\frac{1}{3} \times 10^{-7}$ s on the same graph. Discuss the nature of the function.

P1.31. Three sinusoidally time-varying linearly polarized vector fields are given at a point by

$$\mathbf{F}_1 = \sqrt{3} \, \mathbf{a}_x \cos (2\pi \times 10^6 t + 30°)$$

$$\mathbf{F}_2 = \mathbf{a}_z \cos (2\pi \times 10^6 t + 30°)$$

$$\mathbf{F}_3 = \left(\frac{1}{2} \mathbf{a}_x + \sqrt{3} \, \mathbf{a}_y + \frac{\sqrt{3}}{2} \mathbf{a}_z \right) \cos (2\pi \times 10^6 t - 60°)$$

Determine the polarizations of the following: **(a)** $\mathbf{F}_1 + \mathbf{F}_2$; **(b)** $\mathbf{F}_1 + \mathbf{F}_2 + \mathbf{F}_3$; and **(c)** $\mathbf{F}_1 - \mathbf{F}_2 + \mathbf{F}_3$.

P1.32. Two sinusoidally time-varying, linearly polarized vector fields are given at a point by

$$\mathbf{F}_1 = (C\mathbf{a}_x + C\mathbf{a}_y + \mathbf{a}_z) \cos 2\pi \times 10^6 t$$

$$\mathbf{F}_2 = (C\mathbf{a}_x + \mathbf{a}_y - 2\mathbf{a}_z) \sin 2\pi \times 10^6 t$$

where C is a constant. **(a)** Determine the polarization of the vector $\mathbf{F}_1 + \mathbf{F}_2$ for $C = 2$. **(b)** Find the value(s) of C for which the tip of the vector $\mathbf{F}_1 + \mathbf{F}_2$ traces a circle with time.

P1.33. Consider an analog watch that keeps accurate time and assume the origin to be at the center of the dial, the x-axis passing through the 12 mark, and the y-axis passing through the 3 mark. **(a)** Write the expression for the time-varying unit vector directed along the hour hand of the watch. **(b)** Write the expression for the time-varying unit vector directed along the minute hand of the watch. **(c)** Obtain the specific expression for these unit vectors when the hour hand and the minute hand are aligned exactly and between the 5 and 6 marks.

P1.34. The components of a sinusoidally time-varying vector field are given at a point by

$$F_x = 1 \cos \omega t$$
$$F_y = 1 \cos (\omega t + 60°)$$

Show that the field is elliptically polarized in the xy-plane with the equation of the ellipse given by $x^2 - xy + y^2 = 3/4$. Further show that the axial ratio (ratio of the major axis to the minor axis) of the ellipse is $\sqrt{3}$ and the tilt angle (angle made by the major axis with the x-axis) is 45°.

Section 1.6

P1.35. Find

$$\sqrt{\frac{(24 + j7)(13e^{j0.2\pi})}{5 - j12}} + 9\underline{/150°}$$

and express in rectangular form.

P1.36. Find all values of $\sqrt[4]{-8 + j8\sqrt{3}}$ and express in rectangular form.

P1.37. Find the value of $\ln (1 + j1)$ having the smallest magnitude and express in polar form.

P1.38. Find the steady-state solution for the differential equation

$$\frac{dI}{dt} + 12I = 65 \cos (5t - 30°) + 45 \sin (9t + 30°)$$

P1.39. Find the steady-state solution for the integrodifferential equation

$$\frac{dI}{dt} + I + 5 \int I \, dt = 10 \sin 3t$$

P1.40. Find the steady-state solution for the differential equation

$$\frac{d^2 I}{dt^2} + 2\frac{dI}{dt} + I = 4 \cos^3 t$$

REVIEW PROBLEMS

R1.1. Using the equality

$$A \times (B \times C) = (A \cdot C)B - (A \cdot B)C$$

show that if $A \times F = C$ and $B \times F = D$, then

$$F = \frac{C \times D}{A \cdot D} = -\frac{C \times D}{B \cdot C}$$

Find F if $A = (a_x + a_y)$, $B = (a_x + 2a_y - 2a_z)$, $C = (a_x - a_y)$, and $D = (6a_x - 5a_y - 2a_z)$.

R1.2. The tips of three vectors $A, B,$ and C originating from a common point determine a plane. **(a)** Show that the shortest distance from the common point to the plane is $|A \cdot B \times C|/|A \times B + B \times C + C \times A|$. **(b)** Compute its value for $A = 2a_z$, $B = \frac{1}{2} (a_{rc} - \sqrt{3} a_\phi)$ in cylindrical coordinates, and $C = \frac{1}{4}(3a_{rs} + \sqrt{3} a_\theta + 2a_\phi)$ in spherical coordinates, at the point $(\sqrt{3}, 3, 2)$ in Cartesian coordinates.

R1.3.   Find the edge length of the largest equilateral tetrahedron that can be fit inside a sphere of radius unity. Then find the radius of the largest sphere that can be fit inside that tetrahedron.

R1.4.   Consider an observer always moving in the southeast direction on the surface of a spherical Earth of radius a, starting at the Greenwich meridian on the equator. **(a)** Find the equation for the curve traversed by the observer, using a spherical coordinate system with the origin at the center of Earth, the North Pole at $\theta = 0$, and $\phi = 0$, corresponding to the Greenwich meridian. **(b)** Find the first two values of the south latitude when the observer is again on the Greenwich meridian. **(c)** Does the observer ever reach the South Pole?

R1.5.   Three sinusoidally time-varying, linearly polarized vector fields are given at a point by

$$\mathbf{F}_1 = (A\mathbf{a}_x + B\mathbf{a}_y + C\mathbf{a}_z) \cos 2\pi \times 10^6 t$$
$$\mathbf{F}_2 = (D\mathbf{a}_x + E\mathbf{a}_y) \cos 2\pi \times 10^6 t$$
$$\mathbf{F}_3 = (4\mathbf{a}_x + 4\mathbf{a}_y + 2\mathbf{a}_z) \sin 2\pi \times 10^6 t$$

Find the values of A, B, C, D, and E that satisfy simultaneously the following conditions: **(i)** $\mathbf{F}_2$ is polarized perpendicular to the plane $x + 2y = 0$; **(ii)** $(\mathbf{F}_1 + \mathbf{F}_2 + \mathbf{F}_3)$ is circularly polarized; and **(iii)** $(\mathbf{F}_1 - \mathbf{F}_2 + \mathbf{F}_3)$ is linearly polarized.

R1.6.   Find the steady-state solution to the differential equation

$$\frac{dI}{dt} + 3I = 4 \sin t \sin 2t \sin 3t$$

CHAPTER 2

Fields and Materials

In Chapter 1, we provided a general introduction to vectors and fields. Basic to our study of the elements of engineering electromagnetics is an understanding of the concepts of electric and magnetic fields. Hence, in this chapter, we first study these concepts from considerations of the experimental laws of Coulomb and Ampère. We will learn how to compute the electric and magnetic fields due to charge and current distributions, respectively. Combining the electric and magnetic field concepts, we introduce the Lorentz force equation and use it to discuss charged particle motion in electric and magnetic fields. We devote the remainder of the chapter to materials.

Materials contain charged particles that respond to applied electric and magnetic fields to produce secondary fields. We will learn that there are three basic phenomena resulting from the interaction of the charged particles with the electric and magnetic fields. These are conduction, polarization, and magnetization. Although a given material may exhibit all three properties, it is classified as a conductor (including semiconductor), a dielectric, or a magnetic material, depending on whether conduction, polarization, or magnetization is the predominant phenomenon. Thus, we introduce these materials one at a time and develop a set of relations known as the constitutive relations that enable us to avoid the necessity of explicitly taking into account the interaction of the charged particles with the fields.

2.1 THE ELECTRIC FIELD

From our study of Newton's law of gravitation in elementary physics, we are familiar with the gravitational force field associated with material bodies by virtue of their physical property known as *mass*. Newton's experiments showed that the gravitational force of attraction between two bodies of masses m_1 and m_2 separated by a distance R, which is very large compared to their sizes, is equal to

$m_1 m_2 G/R^2$, where G is the universal constant of gravitation. In a similar manner, a force field known as the *electric field* is associated with bodies that are *charged*. A material body may be charged positively or negatively or may possess no net charge. In the International System of Units that we use throughout this book, the unit of charge is the coulomb, abbreviated C. The charge of an electron is -1.60219×10^{-19} C. Alternatively, approximately 6.24×10^{18} electrons represent a charge of one negative coulomb.

Coulomb's
law

Experiments conducted by Coulomb showed that the following hold for two charged bodies that are very small in size compared to their separation so that they can be considered as *point charges:*

1. The magnitude of the force is proportional to the product of the magnitudes of the charges.
2. The magnitude of the force is inversely proportional to the square of the distance between the charges.
3. The magnitude of the force depends on the medium.
4. The direction of the force is along the line joining the charges.
5. Like charges repel; unlike charges attract.

For free space, the constant of proportionality is $1/4\pi\varepsilon_0$, where ε_0 is known as the permittivity of free space, having a value 8.854×10^{-12}, or approximately equal to $10^{-9}/36\pi$. (For convenience, we shall use a value of $10^{-9}/36\pi$ for ε_0 throughout this book.) Thus, if we consider two point charges Q_1C and Q_2C separated R m in free space, as shown in Fig. 2.1, then the forces $\mathbf{F}_1$ and $\mathbf{F}_2$ experienced by Q_1 and Q_2, respectively, are given by

$$\mathbf{F}_1 = \frac{Q_1 Q_2}{4\pi\varepsilon_0 R^2}\, \mathbf{a}_{21} \tag{2.1a}$$

and

$$\mathbf{F}_2 = \frac{Q_2 Q_1}{4\pi\varepsilon_0 R^2}\, \mathbf{a}_{12} \tag{2.1b}$$

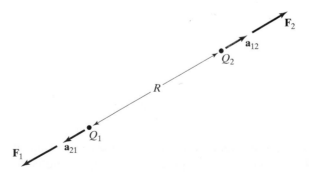

FIGURE 2.1   Forces experienced by two point charges Q_1 and Q_2.

where $\mathbf{a}_{21}$ and $\mathbf{a}_{12}$ are unit vectors along the line joining Q_1 and Q_2, as shown in Fig. 2.1. Equations (2.1a) and (2.1b) represent Coulomb's law. Since the units of force are newtons, we note that ε_0 has the units (coulomb)^2 per (newton-meter^2). These are commonly known as *farads per meter*, where a farad is (coulomb)^2 per newton-meter.

Electric field defined

In the case of the gravitational field of a material body, we define the gravitational field intensity as the force per unit mass experienced by a small test mass placed in that field. In a similar manner, the force per unit charge experienced by a small test charge placed in an electric field is known as the *electric field intensity*, denoted by the symbol $\mathbf{E}$. Alternatively, if in a region of space, a test charge q experiences a force $\mathbf{F}$, then the region is said to be characterized by an electric field of intensity $\mathbf{E}$ given by

$$\mathbf{E} = \frac{\mathbf{F}}{q} \tag{2.2}$$

The unit of electric field intensity is newton per coulomb, or more commonly volt per meter, where a volt is newton-meter per coulomb. The test charge should be so small that it does not alter the electric field in which it is placed. Ideally, $\mathbf{E}$ is defined in the limit that q tends to zero; that is,

$$\mathbf{E} = \lim_{q \to 0} \frac{\mathbf{F}}{q} \tag{2.3}$$

Equation (2.3) is the defining equation for the electric field intensity irrespective of the source of the electric field. Just as one body by virtue of its mass is the source of a gravitational field acting on other bodies by virtue of their masses, a charged body is the source of an electric field acting on other charged bodies. We will, however, learn in Chapter 3 that there exists another source for the electric field, namely, a time-varying magnetic field.

Electrostatic separation of minerals

Equation (2.2) or (2.3) tells us that the force experienced by a charged particle placed at a point in an external electric field is in the same direction as that of the electric field if the charge is positive, but opposite to that of the electric field if the charge is negative, as shown in Fig. 2.2. This phenomenon is the basis behind *electrostatic separation*, a process widely used in industry to separate minerals.[1] An example is illustrated in Fig. 2.3. Phosphate ore composed of

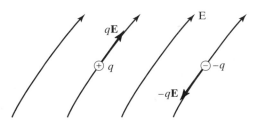

FIGURE 2.2   Forces experienced by positive and negative charges in an electric field.

---

[1]See, for example, A. D. Moore, ed., *Electrostatics and Its Applications* (New York: John Wiley & Sons, 1973), Chap. 10.

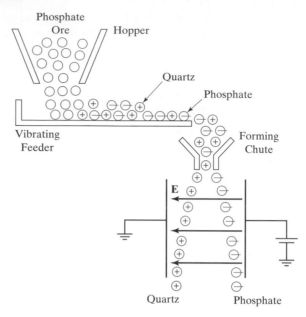

FIGURE 2.3   Example for illustrating electrostatic separation of minerals.

granules of quartz and phosphate rock is dropped through a hopper onto a vibrating feeder. The friction between the two types of granules resulting from the vibration causes the quartz particles to be positively charged and the phosphate particles to be negatively charged. The oppositely charged particles are then passed through a chute into the electric field region between two parallel plates, where they are separated and subsequently collected separately.

Cathode ray tube

There are many other devices based on the electric force on a charged particle. We shall, however, discuss only one other application, the cathode ray tube, which is widely used in oscilloscopes, TV receivers, computer display terminals, and so on. The schematic of a cathode ray tube is shown in Fig. 2.4. Electrons are emitted from the heated cathode and are accelerated toward the anode by an electric field directed from the anode toward the cathode. After passing through the anode, they enter a region between two orthogonal pairs of parallel plates, one pair being horizontal and the other vertical. A voltage applied to the horizontal set of plates produces an electric field between the plates directed vertically, thereby deflecting the electrons vertically and imparting to them a vertical component of velocity as they leave the region between the plates. Similarly, a voltage applied to the vertical set of plates deflects the electrons horizontally sideways and imparts to them a sideways component of velocity as they leave the region between the plates. Thus, by varying the voltages applied to the two sets of plates, the electron beam can be made to strike the fluorescent screen and produce a bright spot at any point on the screen.

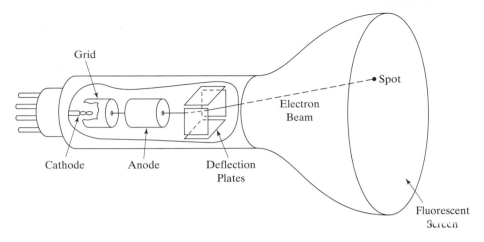

FIGURE 2.4 Schematic diagram of a cathode ray tube.

Returning now to Coulomb's law and letting one of the two charges in Fig. 2.1, say, Q_2, be a small test charge q, we have

Electric field due to a point charge

$$\mathbf{F}_2 = \frac{Q_1 q}{4\pi\varepsilon_0 R^2} \mathbf{a}_{12} \tag{2.4}$$

The electric field intensity $\mathbf{E}_2$ at the test charge due to the point charge Q_1 is then given by

$$\mathbf{E}_2 = \frac{\mathbf{F}_2}{q} - \frac{Q_1}{4\pi\varepsilon_0 R^2} \mathbf{a}_{12} \tag{2.5}$$

Generalizing this result by making R a variable, that is, by moving the test charge around in the medium, writing the expression for the force experienced by it, and dividing the force by the test charge, we obtain the electric field intensity $\mathbf{E}$ due to a point charge Q to be

$$\boxed{\mathbf{E} = \frac{Q}{4\pi\varepsilon_0 R^2} \mathbf{a}_R} \tag{2.6}$$

where R is the distance from the point charge to the point at which the field intensity is to be computed and $\mathbf{a}_R$ is the unit vector along the line joining the two points under consideration and directed away from the point charge. The electric field intensity due to a point charge is thus directed everywhere radially away from the point charge and its constant-magnitude surfaces are spherical surfaces centered at the point charge, as shown by the cross-sectional view in Fig. 2.5.

Using (2.6) in conjunction with (1.25) and (1.26), we can obtain the expression for the electric field intensity at a point $P(x, y, z)$ due to a point charge Q located at a point $P'(x', y', z')$. Thus, noting that the vector $\mathbf{R}$ from P' to P is

FIGURE 2.5   Direction lines and cross sections of constant-magnitude surfaces of electric field due to a point charge.

given by $[(x - x')\mathbf{a}_x + (y - y')\mathbf{a}_y + (z - z')\mathbf{a}_z]$ and the unit vector $\mathbf{a}_R$ is equal to $\mathbf{R}/R$, we obtain

$$
\begin{aligned}
\mathbf{E} &= \frac{Q\mathbf{R}}{4\pi\varepsilon_0 R^3} \\
&= \frac{Q}{4\pi\varepsilon_0} \frac{(x - x')\mathbf{a}_x + (y - y')\mathbf{a}_y + (z - z')\mathbf{a}_z}{[(x - x')^2 + (y - y')^2 + (z - z')^2]^{3/2}}
\end{aligned}
\tag{2.7}
$$

For a numerical example, if P and P' are $(3, 1, 1)$ and $(1, -1, 0)$, respectively, then

$$
\mathbf{E} = \frac{Q}{108\pi\varepsilon_0} (2\mathbf{a}_x + 2\mathbf{a}_y + \mathbf{a}_z)
$$

If we now have several point charges $Q_1, Q_2, \ldots$, as shown in Fig. 2.6, the force experienced by a test charge situated at a point P is the vector sum of the forces experienced by the test charge due to the individual charges. It then follows that the electric field intensity at point P is the superposition of the electric

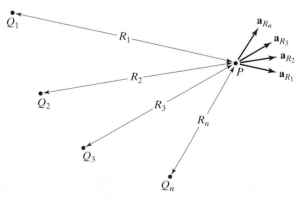

FIGURE 2.6   Collection of point charges and unit vectors along the directions of their electric fields at a point P.

field intensities due to the individual charges; that is,

$$\mathbf{E} = \frac{Q_1}{4\pi\varepsilon_0 R_1^2}\,\mathbf{a}_{R_1} + \frac{Q_2}{4\pi\varepsilon_0 R_2^2}\,\mathbf{a}_{R_2} + \cdots + \frac{Q_n}{4\pi\varepsilon_0 R_n^2}\,\mathbf{a}_{R_n} \qquad (2.8)$$

We shall illustrate the application of (2.8) by means of an example involving two point charges.

Example 2.1

Let us consider two point charges $Q_1 = 8\pi\varepsilon_0$ C and $Q_2 = -4\pi\varepsilon_0$ C situated at $(-1,0,0)$ and $(1,0,0)$, respectively. We wish to (a) find the electric field intensity at the point $(0, 0, 1)$ and (b) discuss computer generation of the direction line of $\mathbf{E}$ passing through that point.

a. Using (2.8) and (2.7) in conjunction with the geometry in Fig. 2.7(a), we obtain

$$[\mathbf{E}]_{(0,0,1)} = [\mathbf{E}_1]_{(0,0,1)} + [\mathbf{E}_2]_{(0,0,1)}$$

$$= \frac{8\pi\varepsilon_0}{4\pi\varepsilon_0}\frac{(\mathbf{a}_x + \mathbf{a}_z)}{2^{3/2}} - \frac{4\pi\varepsilon_0}{4\pi\varepsilon_0}\frac{(-\mathbf{a}_x + \mathbf{a}_z)}{2^{3/2}} \qquad (2.9)$$

$$= 1.118\left(\frac{3\mathbf{a}_x + \mathbf{a}_z}{\sqrt{10}}\right)$$

Note that the direction of $\mathbf{E}$ is given by the unit vector $(3\mathbf{a}_x + \mathbf{a}_z)/\sqrt{10}$ pointing away from the positive charge Q_1. The field vectors $\mathbf{E}_1$ and $\mathbf{E}_2$, and the resultant field vector $\mathbf{E}$, are shown in Fig. 2.7(a).

b. To discuss the computer generation of the direction line of $\mathbf{E}$, we recall that a direction line is a curve such that at any given point on the curve, the field is tangential to the curve. For the case of the electric field, it is also the path followed by an

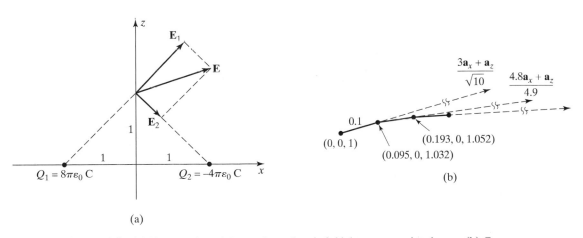

(a)

(b)

FIGURE 2.7 (a) Computation of the resultant electric field due to two point charges. (b) Generation of the direction line of the electric field of (a).

infinitesimal test charge when released at a point on the curve. To obtain the direction line through the point $(0, 0, 1)$, we go by an incremental distance from $(0, 0, 1)$ along the direction of the electric field vector at that point to reach a new point, compute the field at the new point, and continue the process. Thus, choosing for the purpose of illustration an incremental distance of 0.1 m and going along the unit vector $(3\mathbf{a}_x + \mathbf{a}_z)/\sqrt{10}$ from $(0, 0, 1)$, we obtain the new point to be $(0.095, 0, 1.032)$, as shown in Fig. 2.7(b). The electric field at this point is

$$[\mathbf{E}]_{(0.095, 0, 1.032)} = \frac{8\pi\varepsilon_0}{4\pi\varepsilon_0} \frac{(1.095\mathbf{a}_x + 1.032\mathbf{a}_z)}{(1.095^2 + 1.032^2)^{3/2}} - \frac{4\pi\varepsilon_0}{4\pi\varepsilon_0} \frac{(-0.905\mathbf{a}_x + 1.032\mathbf{a}_z)}{(0.905^2 + 1.032^2)^{3/2}}$$

$$= 1.015\left(\frac{4.8\mathbf{a}_x + \mathbf{a}_z}{4.9}\right) \tag{2.10}$$

Note that the direction of this electric field, which is along the unit vector $(4.8\mathbf{a}_x + \mathbf{a}_z)/4.9$, is slanted more toward the negative charge Q_2 than that of the electric field at the point $(0, 0, 1)$, as shown in Fig. 2.7(b), indicating the swing of the direction line toward Q_2. The procedure is continued by going the incremental distance of 0.1 m from $(0.095, 0, 1.032)$ along the unit vector $(4.8\mathbf{a}_x + \mathbf{a}_z)/4.9$ to the new point $(0.193, 0, 1.052)$ and computing the field vector at that point, and so on, until the direction line is terminated close to the point charge Q_2. The same can be done to obtain the portion of the direction line from $(0, 0, 1)$ toward the point charge Q_1, by moving opposite to $\mathbf{E}$. Values of the coordinates of the beginning point (X and Z), the magnitude of the electric field at that point (E), and the components of the unit vector along the electric field (UX and UZ), pertinent to the steps along the direction line computed in this manner are listed in Table 2.1, parts (a) and (b), corresponding to the segments of the direction line from $(0, 0, 1)$ toward Q_2 and Q_1, respectively. It can be seen that the test charge takes 17 steps toward Q_2 but only 14 steps back toward Q_1.

In the simple procedure employed in Example 2.1, there is a (cumulative) error associated with each step. This error can be reduced by employing a modified procedure as follows: Instead of moving the test charge by 0.1 m from its current location, say, point A, to a new location, say, point B, along the direction of the electric field at point A, it is moved by 0.1 m to a point C along a direction that bisects the directions of the fields at points A and B. Computer-plotted field maps in the xz-plane for pairs of point charges Q_1 and Q_2 located at $(-1, 0)$ and $(1, 0)$, respectively, by using this modified procedure are shown in Fig. 2.8. In each map, plotting of a direction line begins at one of the point charges and terminates when the line reaches to within a distance of 0.1 m from the second point charge, or if it goes beyond a specified rectangular region. In this manner, direction lines beginning at points around each point charge and at 30° intervals on a circle of radius 0.1 m are plotted, with the 0° angle corresponding to the $+x$-direction.

For Fig. 2.8(a), $Q_1 = 2Q$ at $(-1, 0)$ and $Q_2 = -Q$ at $(1, 0)$. The rectangular region is one having corners at $(-3, 2)$, $(3, 2)$, $(3, -2)$, and $(-3, -2)$. The direction lines beginning at each point charge either end on the second charge or go out of the boundary of the rectangular region. For Figs. 2.8(b)–(d), region of map is rectangle having corners at $(-3, 4)$, $(3, 4)$, $(3, 0)$, and $(-3, 0)$, taking

TABLE 2.1   Values of Parameters Pertinent to the Steps in the Computer Generation of the Direction Line of **E** in Fig. 2.7 for (a) the Segment from (0, 0, 1) Toward the Charge Q_2 and (b) the Segment from (0, 0, 1) Back Toward the Charge Q_1.

X = 0.000	Z = 1.000	E =	1.118	UX = 0.949	UZ = 0.316
X = 0.095	Z = 1.032	E =	1.015	UX = 0.979	UZ = 0.204
X = 0.193	Z = 1.052	E =	0.942	UX = 0.997	UZ = 0.076
X = 0.292	Z = 1.060	E =	0.898	UX = 0.998	UZ = -0.065
X = 0.392	Z = 1.053	E =	0.882	UX = 0.977	UZ = -0.215
X = 0.490	Z = 1.032	E =	0.898	UX = 0.930	UZ = -0.368
X = 0.583	Z = 0.995	E =	0.951	UX = 0.858	UZ = -0.513
X = 0.669	Z = 0.944	E =	1.051	UX = 0.766	UZ = -0.643
X = 0.745	Z = 0.879	E =	1.212	UX = 0.660	UZ = -0.751
X = 0.811	Z = 0.804	E =	1.459	UX = 0.548	UZ = -0.836
X = 0.866	Z = 0.721	E =	1.837	UX = 0.439	UZ = -0.899
X = 0.910	Z = 0.631	E =	2.426	UX = 0.337	UZ = -0.942
X = 0.944	Z = 0.536	E =	3.391	UX = 0.246	UZ = -0.969
X = 0.968	Z = 0.440	E =	5.100	UX = 0.167	UZ = -0.986
X = 0.985	Z = 0.341	E =	8.537	UX = 0.101	UZ = -0.995
X = 0.995	Z = 0.241	E =	17.101	UX = 0.049	UZ = -0.999
X = 1.000	Z = 0.142	E =	49.846	UX = 0.010	UZ = -1.000
X = 1.001	Z = 0.042	E =	577.540	UX = -0.023	UZ = -1.000

Number of steps = 17

(a)

X = 0.000	Z = 1.000	E =	1.118	UX = 0.949	UZ = 0.316
X = -.095	Z = 0.968	E =	1.243	UX = 0.908	UZ = 0.420
X = -.186	Z = 0.926	E =	1.411	UX = 0.862	UZ = 0.507
X = -.272	Z = 0.876	E =	1.634	UX = 0.815	UZ = 0.580
X = -.353	Z = 0.818	E =	1.931	UX = 0.768	UZ = 0.640
X = -.430	Z = 0.754	E =	2.333	UX = 0.724	UZ = 0.689
X = -.503	Z = 0.685	E =	2.888	UX = 0.684	UZ = 0.730
X = -.571	Z = 0.612	E =	3.681	UX = 0.648	UZ = 0.762
X = -.636	Z = 0.536	E =	4.871	UX = 0.616	UZ = 0.788
X = -.697	Z = 0.457	E =	6.769	UX = 0.590	UZ = 0.808
X = -.756	Z = 0.376	E =	10.074	UX = 0.568	UZ = 0.823
X = -.813	Z = 0.294	E =	16.616	UX = 0.551	UZ = 0.835
X = -.868	Z = 0.210	E =	32.588	UX = 0.538	UZ = 0.843
X = -.922	Z = 0.126	E =	91.176	UX = 0.529	UZ = 0.849
X = -.975	Z = 0.041	E =	860.610	UX = 0.522	UZ = 0.853

Number of steps = 14

(b)

advantage of the symmetry of the field map about the axis through the charges, illustrated in Fig. 2.8(a). For Fig. 2.8(b), $Q_1 = 4Q$ at $(-1, 0)$ and $Q_2 = Q$ at $(1, 0)$. A zero field point exists within the region at $(\frac{1}{3}, 0)$, between the two charges. For direction lines passing through this point, the test charge gets trapped at that point and the procedure used is to untrap it by displacing it by 0.01 perpendicular to the axis and continue plotting of the line until it terminates at a point on the boundary of the region. For Fig. 2.8(c), $Q_1 = 9Q$ at $(-1, 0)$ and $Q_2 = -Q$ at $(1, 0)$. A zero field point exists within the region at $(2, 0)$, to the right of Q_2. Also, three additional field lines are shown plotted. Two of these are from Q_1 at angles of 40° and 45° and the third is from Q_2 at 10°. For Fig. 2.8(d), $Q_1 = 81Q$ at $(-1, 0)$ and $Q_2 = Q$ at $(1, 0)$. A zero field point exists just to the left of Q_2 between the

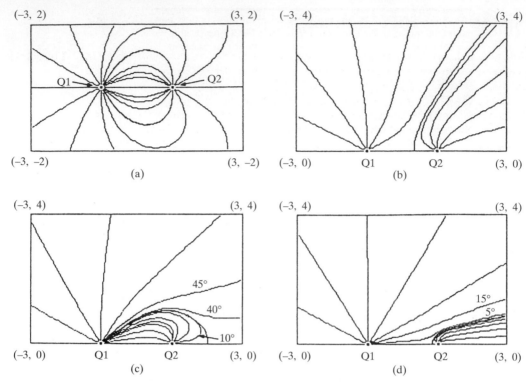

FIGURE 2.8 Computer-generated maps of direction lines of electric field for pairs of point charges Q_1 and Q_2 at $(-1, 0)$ and $(1, 0)$, respectively, in the xz-plane. (a) $Q_1 = 2Q, Q_2 = -Q$; (b) $Q_1 = 4Q$, $Q_2 = Q$; (c) $Q_1 = 9Q, Q_2 = -Q$; and (d) $Q_1 = 81Q, Q_2 = Q$.

two charges. The map also includes two additional field lines originating from Q_1 at 5° and 15° angles.

Types of charge distributions   The foregoing illustration of the computation of the electric field intensity due to two point charges can be extended to the computation of the field intensity due to continuous charge distributions. Continuous charge distributions are of three types: line charges, surface charges, and volume charges, depending on whether the charge is distributed along a line like chalk powder along a thin line drawn on the blackboard, on a surface like chalk powder on the erasing surface of a blackboard eraser, or in a volume like chalk powder in the chalk itself. The corresponding charge densities are the line charge density ρ_L, the surface charge density ρ_S, and the volume charge density ρ, having the units of charge per unit length (coulombs per meter), charge per unit area (coulombs per meter squared), and charge per unit volume (coulombs per meter cubed), respectively. The technique of finding the electric field intensity due to a given charge distribution consists of dividing the region of the charge distribution into a number of differential lengths, surfaces, or volumes, depending on the type of the distribution, considering the charge in each differential element to be a point charge, and using superposition. We shall illustrate the procedure by means of two examples.

Example 2.2

Charge Q C is distributed with uniform density along a circular ring of radius a lying in *Ring charge* the xy-plane and having its center at the origin, as shown in Fig. 2.9. We wish to find the electric field intensity at a point on the z-axis.

Let us divide the ring into a large number of segments so that the charge in each segment can be considered to be a point charge located at the center of the segment. Let the segments be of equal length and numbered $1, 2, 3, \ldots, 2n$, as shown in Fig. 2.9. Then the electric field intensity at the point $(0, 0, z)$ due to the charge in the jth segment is given by

$$\mathbf{E}_j = \frac{Q_j}{4\pi\varepsilon_0 R_j^2} \mathbf{a}_{R_j}$$

where Q_j is the charge in the jth segment and R_j and $\mathbf{a}_{R_j}$ are as shown in the figure. Since the charge is uniformly distributed, Q_j is the same for all j and is equal to the charge density times the length of the segment. Thus,

$$Q_j = \left(\frac{Q}{2\pi a}\right)\left(\frac{2\pi a}{2n}\right) = \frac{Q}{2n}$$

Furthermore, since the point $(0, 0, z)$ is along the axis of the ring, it is equidistant from all segments so that R_j is the same for all j. Hence,

$$R_j = \sqrt{z^2 + a^2}$$

Now, from symmetry considerations, we note that for every segment $1, 2, 3, \ldots, n$, there is a corresponding segment diametrically opposite to it in the other half of the ring such that the electric field intensity due to the two segments together is directed along the z-axis, as illustrated for segment j in Fig. 2.9. Hence, to find $\mathbf{E}$ due to the entire ring charge, it is sufficient if we consider the z-component of $\mathbf{E}_j$, multiply it by 2, and sum

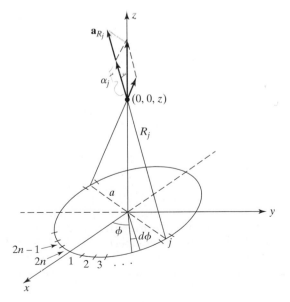

FIGURE 2.9 Determination of electric field due to a circular ring of charge of uniform density.

from $j = 1$ to $j = n$. Thus, we obtain the required electric field intensity to be

$$
\begin{aligned}
[\mathbf{E}]_{(0,0,z)} &= \sum_{j=1}^{n} \frac{2Q_j}{4\pi\varepsilon_0 R_j^2} (\mathbf{a}_{R_j} \cdot \mathbf{a}_z) \mathbf{a}_z \\
&= \sum_{j=1}^{n} \frac{Q_j}{2\pi\varepsilon_0 R_j^2} \cos \alpha_j \mathbf{a}_z \\
&= \sum_{j=1}^{n} \frac{Q_j z}{2\pi\varepsilon_0 R_j^3} \mathbf{a}_z \\
&= \sum_{j=1}^{n} \frac{Qz}{4\pi\varepsilon_0 n(z^2 + a^2)^{3/2}} \mathbf{a}_z \\
&= \frac{Qz}{4\pi\varepsilon_0 (z^2 + a^2)^{3/2}} \mathbf{a}_z
\end{aligned}
\tag{2.11}
$$

Note that $[\mathbf{E}]_{(0,0,z)}$ is directed in the $+z$-direction above the origin ($z > 0$) and in the $-z$-direction below the origin ($z < 0$), as to be expected.

Alternative to the summation procedure just employed, we can obtain E_z at $(0, 0, z)$ by setting up an integral expression and evaluating it. Thus, considering a differential length $a\, d\phi$ of the ring charge at the point $(a, \phi, 0)$, as shown in Fig. 2.9, and making use of symmetry considerations as discussed in connection with the summation procedure, we obtain

$$
\begin{aligned}
[E_z]_{(0,0,z)} &= \int_{\phi=0}^{\pi} \frac{2(Q/2\pi a)a\, d\phi}{4\pi\varepsilon_0 (a^2 + z^2)} \frac{z}{(a^2 + z^2)^{1/2}} \\
&= \frac{Qz}{4\pi^2 \varepsilon_0 (a^2 + z^2)^{3/2}} \int_{\phi=0}^{\pi} d\phi \\
&= \frac{Qz}{4\pi\varepsilon_0 (a^2 + z^2)^{3/2}}
\end{aligned}
\tag{2.12}
$$

For this example, the two results given by (2.11) and (2.12) are identical. In general, however, the summation procedure gives an approximate result for any finite value of n, and the integral gives the exact result, provided it can be evaluated in closed form. The summation procedure is, however, more illuminating as to the application of superposition and is convenient for computer solution.

Example 2.3

Infinite plane sheet of charge

Let us consider an infinite plane sheet of charge in the xy-plane with uniform surface charge density ρ_{S0} C/m^2 and find the electric field intensity due to it everywhere.

Let us first consider a point $(0, 0, z)$ on the z-axis, as shown in Fig. 2.10(a). Then the solution can be carried out by dividing the sheet into a number of infinitesimal surfaces in Cartesian coordinates and using superposition. An alternate procedure consists of using the result of Example 2.2 by dividing the sheet into concentric rings centered at the origin and each having infinitesimal width dr in the radial direction. One such ring having the arbitrary radius r and width dr is shown in Fig. 2.10(a). The charge in that ring is equal to $\rho_{S0}(2\pi r\, dr)$, the product of the uniform surface charge density and the area of the ring. According to the result obtained in Example 2.2, the electric field intensity at

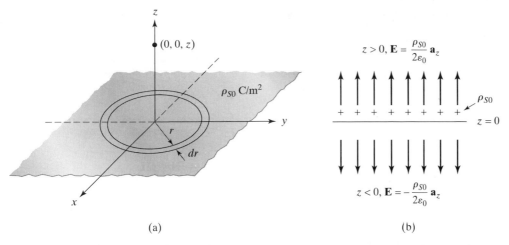

FIGURE 2.10   (a) Determination of electric field due to an infinite plane sheet of uniform surface charge density ρ_{S0} C/m^2 (b) Electric field due to the infinite plane sheet of charge of (a).

$(0, 0, z)$ due to this ring charge is given by

$$[d\mathbf{E}]_{(0,0,z)} = \frac{(\rho_{S0} 2\pi r \, dr)z}{4\pi\varepsilon_0(r^2 + z^2)^{3/2}} \mathbf{a}_z$$

The electric field intensity due to the entire sheet of charge is then given by

$$[\mathbf{E}]_{(0,0,z)} = \int_{r=0}^{\infty} [d\mathbf{E}]_{(0,0,z)}$$

$$= \int_{r=0}^{\infty} \frac{\rho_{S0} r z \, dr}{2\varepsilon_0(r^2 + z^2)^{3/2}} \mathbf{a}_z$$

$$= \frac{\rho_{S0} z}{2\varepsilon_0}\left[-\frac{1}{\sqrt{r^2 + z^2}} \right]_{r=0}^{\infty} \mathbf{a}_z$$

$$= \frac{\rho_{S0} z}{2\varepsilon_0 |z|} \mathbf{a}_z$$

Finally, since the charge density is uniform and the origin of the coordinate system can be chosen anywhere on the infinite sheet without changing the geometry, this result is valid everywhere. Thus, the required electric field intensity is

$$\mathbf{E} = \pm \frac{\rho_{S0}}{2\varepsilon_0} \mathbf{a}_z \qquad \text{for} \quad z \gtrless 0 \qquad (2.13)$$

which has the magnitude $\rho_{S0}/2\varepsilon_0$ everywhere and directed normally away from the sheet, as shown by the cross-sectional view in Fig. 2.10(b). Defining $\mathbf{a}_n$ to be the unit normal vector directed away from the sheet, that is,

$$\mathbf{a}_n = \pm \mathbf{a}_z \qquad \text{for} \quad z \gtrless 0$$

we have

$$\boxed{\mathbf{E} = \frac{\rho_{S0}}{2\varepsilon_0} \mathbf{a}_n} \qquad (2.14)$$

K2.1. Coulomb's law; Electric field intensity; $\mathbf{E}$ due to a point charge; Computation of $\mathbf{E}$ due to charge distributions; $\mathbf{E}$ due to an infinite plane sheet of charge of uniform density.

D2.1. Point charges, each of value $\sqrt{4\pi\varepsilon_0}$ C, are located at the vertices of an n-sided regular polygon circumscribed by a circle of radius a. Find the electric force on each charge for **(a)** $n = 3$; **(b)** $n = 4$; and **(c)** $n = 6$.

Ans. **(a)** $0.577/a^2$ N; **(b)** $0.957/a^2$ N; **(c)** $1.827/a^2$ N; all directed away from the center of the polygon.

D2.2. In Fig. 2.7, let the point charges be $Q_1 = 8\pi\varepsilon_0$ C at $(-1, 0, 0)$ and $Q_2 = 4\pi\varepsilon_0$ C at $(1, 0, 0)$. Find the following: **(a)** $\mathbf{E}$ at $(0, 0, 1)$; **(b)** the coordinates of the point at the end of the second step; and **(c)** the unit vector along $\mathbf{E}$ at the point computed in **(b)**.

Ans. **(a)** $(0.353\mathbf{a}_x + 1.061\mathbf{a}_z)$; **(b)** $(0.060, 0, 1.191)$; **(c)** $(0.264\mathbf{a}_x + 0.965\mathbf{a}_z)$.

D2.3. In Fig. 2.9, let there be a second ring of charge $-Q$, uniformly distributed along a circle of radius a, having its center at $(0, 0, 2a)$ and lying parallel to the xy-plane. Find $\mathbf{E}$ due to the two rings of charge together at each of the following points: **(a)** $(0, 0, 0)$; **(b)** $(0, 0, a)$; and **(c)** $(0, 0, 3a)$.

Ans. **(a)** $(0.0142Q/\varepsilon_0 a^2)\mathbf{a}_z$; **(b)** $(0.0563Q/\varepsilon_0 a^2)\mathbf{a}_z$; **(c)** $(-0.0206Q/\varepsilon_0 a^2)\mathbf{a}_z$.

D2.4. Infinite plane sheets of charge lie in the $z = 0$, $z = 2$, and $z = 4$ planes with uniform surface charge densities ρ_{S1}, ρ_{S2}, and ρ_{S3}, respectively. Given that the resulting electric field intensities at the points $(3, 5, 1), (1, -2, 3)$, and $(3, 4, 5)$ are $\mathbf{0}, 6\mathbf{a}_z$, and $4\mathbf{a}_z$ V/m, respectively, find the following: **(a)** ρ_{S1}; **(b)** ρ_{S2}; **(c)** ρ_{S3}; and **(d)** $\mathbf{E}$ at $(-2, 1, -6)$.

Ans. **(a)** $4\varepsilon_0$ C/m^2; **(b)** $6\varepsilon_0$ C/m^2; **(c)** $-2\varepsilon_0$ C/m^2; **(d)** $-4\mathbf{a}_z$ V/m.

2.2 THE MAGNETIC FIELD

In the preceding section, we presented an experimental law known as Coulomb's law having to do with the electric force associated with two charged bodies, and we introduced the electric field intensity vector as the force per unit charge experienced by a test charge placed in the electric field. In this section, we present another experimental law known as *Ampère's law of force,* analogous to Coulomb's law, and use it to introduce the magnetic field concept.

Ampère's law of force

Ampère's law of force is concerned with magnetic forces associated with two loops of wire carrying currents by virtue of motion of charges in the loops. Figure 2.11 shows two loops of wire carrying currents I_1 and I_2 and each of which is divided into a large number of elements having infinitesimal lengths. The total force experienced by a loop is the vector sum of forces experienced by the infinitesimal current elements comprising the loop. The force experienced by each of these current elements is the vector sum of the forces exerted on it by the infinitesimal current elements comprising the second loop. If the number of elements in loop 1 is m and the number of elements in loop 2 is n, then there are $m \times n$ pairs of elements. A pair of magnetic forces is associated with each pair of these elements just as a pair of electric forces is associated with a pair of point charges. Thus, if we consider an element $d\mathbf{l}_1$ in loop 1 and an element $d\mathbf{l}_2$

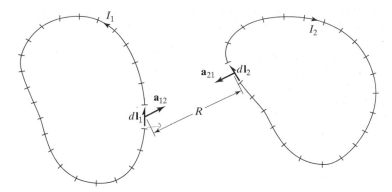

FIGURE 2.11 Two loops of wire carrying currents I_1 and I_2.

in loop 2, then the forces $d\mathbf{F}_1$ and $d\mathbf{F}_2$ experienced by the elements $d\mathbf{l}_1$ and $d\mathbf{l}_2$, respectively, are given by

$$d\mathbf{F}_1 = I_1 \, d\mathbf{l}_1 \times \left(\frac{k I_2 \, d\mathbf{l}_2 \times \mathbf{a}_{21}}{R^2} \right) \tag{2.15a}$$

$$d\mathbf{F}_2 = I_2 \, d\mathbf{l}_2 \times \left(\frac{k I_1 \, d\mathbf{l}_1 \times \mathbf{a}_{12}}{R^2} \right) \tag{2.15b}$$

where $\mathbf{a}_{21}$ and $\mathbf{a}_{12}$ are unit vectors along the line joining the two current elements, R is the distance between them, and k is a constant of proportionality that depends on the medium. For free space, k is equal to $\mu_0/4\pi$, where μ_0 is known as the permeability of free space, having a value $4\pi \times 10^{-7}$. From (2.15a) or (2.15b), we note that the units of μ_0 are newtons per ampere squared. These are commonly known as *henrys per meter*, where a henry is a newton-meter per ampere squared.

Equations (2.15a) and (2.15b) represent Ampère's force law as applied to a pair of current elements. Some of the features evident from these equations are as follows:

1. The magnitude of the force is proportional to the product of the two currents and to the product of the lengths of the two current elements.
2. The magnitude of the force is inversely proportional to the square of the distance between the current elements.
3. To determine the direction of the force acting on the current element $d\mathbf{l}_1$, we first find the cross product $d\mathbf{l}_2 \times \mathbf{a}_{21}$ and then cross $d\mathbf{l}_1$ into the resulting vector. Similarly, to determine the direction of the force acting on the current element $d\mathbf{l}_2$, we first find the cross product $d\mathbf{l}_1 \times \mathbf{a}_{12}$ and then cross $d\mathbf{l}_2$ into the resulting vector. For the general case of arbitrary orientations of $d\mathbf{l}_1$ and $d\mathbf{l}_2$, these operations yield $d\mathbf{F}_1$ and $d\mathbf{F}_2$, which are not equal and opposite. To illustrate by means of an example, let us consider

$I_1\,d\mathbf{l}_1 = I_1\,dx\,\mathbf{a}_x$ at $(1,0,0)$ and $I_2\,d\mathbf{l}_2 = I_2\,dy\,\mathbf{a}_y$ at $(0,1,0)$. Then

$$\mathbf{a}_{12} = -\mathbf{a}_{21} = \frac{1}{\sqrt{2}}(-\mathbf{a}_x + \mathbf{a}_y) \quad R = \sqrt{2}$$

$$I_2\,d\mathbf{l}_2 \times \mathbf{a}_{21} = (I_2\,dy\,\mathbf{a}_y) \times \frac{1}{\sqrt{2}}(\mathbf{a}_x - \mathbf{a}_y) = -\frac{I_2}{\sqrt{2}}\,dy\,\mathbf{a}_z$$

$$d\mathbf{F}_1 = (I_1\,dx\,\mathbf{a}_x) \times \left(\frac{-kI_2\,dy\,\mathbf{a}_z}{2\sqrt{2}}\right) = \frac{kI_1I_2}{2\sqrt{2}}\,dx\,dy\,\mathbf{a}_y$$

$$I_1\,d\mathbf{l}_1 \times \mathbf{a}_{12} = (I_1\,dx\,\mathbf{a}_x) \times \frac{1}{\sqrt{2}}(-\mathbf{a}_x + \mathbf{a}_y) = \frac{I_1}{\sqrt{2}}\,dx\,\mathbf{a}_z$$

$$d\mathbf{F}_2 = (I_2\,dy\,\mathbf{a}_y) \times \left(\frac{kI_1\,dx\,\mathbf{a}_z}{2\sqrt{2}}\right) = \frac{kI_1I_2}{2\sqrt{2}}\,dx\,dy\,\mathbf{a}_x$$

Thus, $d\mathbf{F}_2 \neq -d\mathbf{F}_1$. This is not a violation of Newton's third law since isolated current elements do not exist without sources and sinks of charges at their ends. Newton's third law, however, must and does hold for complete current loops.

Magnetic flux density   The forms of (2.15a) and (2.15b) suggest that each current element is acted on by a field which is due to the other current element. By definition, this field is the magnetic field and is characterized by a quantity known as the *magnetic flux density vector,* denoted by the symbol $\mathbf{B}$. Thus, we note from (2.15b) that the magnetic flux density at the element $d\mathbf{l}_2$ due to the element $d\mathbf{l}_1$ is given by

$$\mathbf{B}_1 = \frac{\mu_0}{4\pi}\frac{I_1\,d\mathbf{l}_1 \times \mathbf{a}_{12}}{R^2} \tag{2.16}$$

and that this flux density acting on $d\mathbf{l}_2$ results in a force on it given by

$$d\mathbf{F}_2 = I_2\,d\mathbf{l}_2 \times \mathbf{B}_1 \tag{2.17}$$

Similarly, we note from (2.15a) that the magnetic flux density at the element $d\mathbf{l}_1$ due to the element $d\mathbf{l}_2$ is given by

$$\mathbf{B}_2 = \frac{\mu_0}{4\pi}\frac{I_2\,d\mathbf{l}_2 \times \mathbf{a}_{21}}{R^2} \tag{2.18}$$

and that this flux density acting on $d\mathbf{l}_1$ results in a force on it given by

$$d\mathbf{F}_1 = I_1\,d\mathbf{l}_1 \times \mathbf{B}_2 \tag{2.19}$$

From (2.17) and (2.19), we see that the units of $\mathbf{B}$ are newtons per ampere-meter, commonly known as *webers per meter squared* (or tesla), where a weber is a newton-meter per ampere. The units of webers per unit area give the character of flux density to the quantity $\mathbf{B}$, unlike the character of field intensity as that of $\mathbf{E}$ for the electric field case.

Generalizing (2.17) and (2.19), we say that an infinitesimal current element of length $d\mathbf{l}$ and current I placed in a magnetic field of flux density $\mathbf{B}$

experiences a force $d\mathbf{F}$ given by

$$\boxed{d\mathbf{F} = I \, d\mathbf{l} \times \mathbf{B}} \qquad\qquad (2.20)$$

as shown in Fig. 2.12. Alternatively, if a current element experiences a force in a region of space, then the region is said to be characterized by a magnetic field.

There are many devices using the principle of magnetic force on a current-carrying wire. One such device in everyday life is the loudspeaker. As shown by the cross-sectional view in Fig. 2.13, the loudspeaker consists of a permanent magnet between the poles of which is a coil wound around a cylinder attached to the apex of a movable cone-shaped diaphragm. Current through the coil varies in accordance with the audio signal from the output stage of the hi-fi amplifier or radio receiver. A magnetic force is thus exerted on the coil, vibrating it back and forth in step with the changes in the current. Since the coil assembly is attached to the cone, the cone also vibrates, thereby producing sound waves in the air.

Principle of loudspeaker

Returning now to (2.16) and (2.18) and generalizing, we obtain the magnetic flux density due to an infinitesimal current element of length $d\mathbf{l}$ and carrying current I to be

Magnetic field due to a current element

$$\boxed{\mathbf{B} = \frac{\mu_0}{4\pi} \frac{I \, d\mathbf{l} \times \mathbf{a}_R}{R^2}} \qquad\qquad (2.21)$$

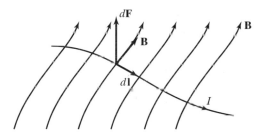

FIGURE 2.12   Force experienced by a current element in a magnetic field.

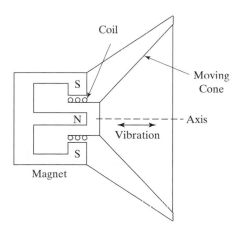

FIGURE 2.13   Cross-sectional view of a loudspeaker.

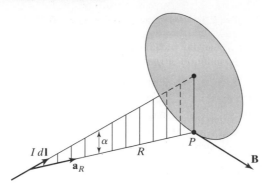

FIGURE 2.14 Magnetic flux density due to an infinitesimal current element.

where R is the distance from the current element to the point at which the flux density is to be computed and $\mathbf{a}_R$ is the unit vector along the line joining the current element and the point under consideration and directed away from the current element, as shown in Fig. 2.14. Equation (2.21) is known as the *Biot–Savart law* and is analogous to the expression for the electric field intensity due to a point charge. The Biot–Savart law tells us that the magnitude of $\mathbf{B}$ at a point P is proportional to the current I, the element length dl, and the sine of the angle α between the current element and the line joining it to the point P, and is inversely proportional to the square of the distance from the current element to the point P. Hence, the magnetic flux density is zero at points along the axis of the current element and increases in magnitude as the point P is moved away from the axis on a spherical surface centered at the current element, becoming a maximum for α equal to 90°. This is in contrast to the behavior of the electric field intensity due to a point charge, which remains the same in magnitude at points on a spherical surface centered at the point charge. The direction of $\mathbf{B}$ at point P is normal to the plane containing the current element and the line joining the current element to P as given by the cross product operation $d\mathbf{l} \times \mathbf{a}_R$, that is, right circular to the axis of the wire. Thus, the direction lines of the magnetic flux density due to a current element are circles centered at points on the axis of the current element and lying in planes normal to the axis. This is in contrast to the direction lines of the electric field intensity due to a point charge, which are radial lines emanating from the point charge.

Example 2.4

Let us consider an infinitesimal length 10^{-3} m of wire located at the point $(1, 0, 0)$ and carrying current 2 A in the direction of the unit vector $\mathbf{a}_x$. We wish to find the magnetic flux density due to the current element at the point $(0, 2, 2)$.

Noting that the current element is given by

$$I\,d\mathbf{l} = (2)(10^{-3})\mathbf{a}_x = 0.002\mathbf{a}_x$$

and the vector **R** from the location $(1, 0, 0)$ of the current element to the point $(0, 2, 2)$ is given by

$$\mathbf{R} = (0 - 1)\mathbf{a}_x + (2 - 0)\mathbf{a}_y + (2 - 0)\mathbf{a}_z = -\mathbf{a}_x + 2\mathbf{a}_y + 2\mathbf{a}_z$$

and using Biot–Savart law, we obtain

$$[\mathbf{B}]_{(0,2,2)} = \frac{\mu_0}{4\pi} \frac{I \, d\mathbf{l} \times \mathbf{a}_R}{R^2}$$

$$= \frac{\mu_0}{4\pi} \frac{I \, d\mathbf{l} \times \mathbf{R}}{R^3}$$

$$= \frac{\mu_0}{4\pi} \frac{0.002 \, \mathbf{a}_x \times (-\mathbf{a}_x + 2\mathbf{a}_y + 2\mathbf{a}_z)}{27}$$

$$= \frac{0.001 \mu_0}{27\pi} (-\mathbf{a}_y + \mathbf{a}_z) \text{ Wb/m}^2$$

The Biot–Savart law can be used to find the magnetic flux density due to a current carrying filamentary wire of any length and shape by dividing the wire into a number of infinitesimal elements and using superposition. We shall illustrate the procedure by means of an example.

Example 2.5

Let us consider an infinitely long, straight wire situated along the z-axis and carrying current I A in the $+z$-direction. We wish to find the magnetic flux density everywhere.   *Infinitely long, straight wire*

Let us consider a point on the xy-plane specified by the cylindrical coordinates $(r, \phi, 0)$, as shown in Fig. 2.15(a). Then the solution for the magnetic flux density at $(r, \phi, 0)$ can be obtained by considering a differential length dz of the wire at the point $(0, 0, z)$

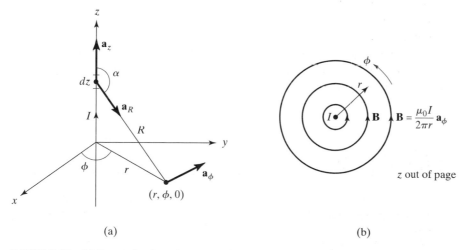

FIGURE 2.15 (a) Determination of magnetic field due to an infinitely long, straight wire of current I A. (b) Magnetic field due to the wire of (a).

and using superposition. Applying Biot–Savart law (2.21) to the geometry in Fig. 2.15(a), we obtain the magnetic flux density at $(r, \phi, 0)$ due to the current element $I\, dz\, \mathbf{a}_z$ at $(0, 0, z)$ to be

$$[d\mathbf{B}]_{(r,\phi,0)} = \frac{\mu_0}{4\pi} \frac{I\, dz\, \mathbf{a}_z \times \mathbf{a}_R}{R^2}$$

$$= \frac{\mu_0 I\, dz}{4\pi} \frac{\sin \alpha}{R^2} \mathbf{a}_\phi$$

$$= \frac{\mu_0 I\, dz}{4\pi} \frac{r}{R^3} \mathbf{a}_\phi$$

$$= \frac{\mu_0 I r\, dz}{4\pi (z^2 + r^2)^{3/2}} \mathbf{a}_\phi$$

The magnetic flux density due to the entire wire is then given by

$$[\mathbf{B}]_{(r,\phi,0)} = \int_{z=-\infty}^{\infty} d\mathbf{B}$$

$$= \int_{z=-\infty}^{\infty} \frac{\mu_0 I r}{4\pi (z^2 + r^2)^{3/2}}\, dz\, \mathbf{a}_\phi$$

$$= \frac{\mu_0 I r}{4\pi} \left[\frac{z}{r^2 \sqrt{z^2 + r^2}} \right]_{z=-\infty}^{\infty} \mathbf{a}_\phi$$

$$= \frac{\mu_0 I}{2\pi r} \mathbf{a}_\phi$$

Now, since the origin can be chosen to be anywhere on the wire without changing the geometry, this result is valid everywhere. Thus, the required magnetic flux density is

$$\mathbf{B} = \frac{\mu_0 I}{2\pi r} \mathbf{a}_\phi \tag{2.22}$$

which has the magnitude $\mu_0 I / 2\pi r$ and surrounds the wire, as shown by the cross-sectional view in Fig. 2.15(b).

Types of current distributions   The magnetic field computation illustrated in Example 2.5 can be extended to current distributions. Current distributions are of two types: surface currents and volume currents, depending on whether current flows on a surface like rain water flowing down a smooth wall or in a volume like rainwater flowing down a gutter downspout. The corresponding current densities are the surface current density $\mathbf{J}_S$ and the volume current density, or simply the current density $\mathbf{J}$, having the units of current crossing unit length (amperes per meter) and current crossing unit area (amperes per meter squared), respectively. Note that the current densities are vector quantities since flow is involved. Assuming for simplicity surface current of uniform density flowing on a plane sheet, as shown in Fig. 2.16(a), one obtains the current I on the sheet by multiplying the magnitude of $\mathbf{J}_S$ by the dimension w of the sheet normal to the direction of $\mathbf{J}_S$. Similarly, for volume current of uniform density flowing in a straight wire, as shown in Fig. 2.16(b), the current I in the wire is given by the product of the magnitude

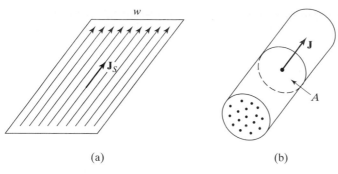

FIGURE 2.16   Determination of currents due to (a) surface current and
(b) volume current distributions of uniform densities.

of **J** and the area of cross section A of the wire normal to the direction of **J**. If the
current density is nonuniform, the current can be obtained by performing an appropriate integration along the width of the sheet or over the cross section of the
wire, depending on the case. We shall illustrate the determination of the magnetic
field due to a current distribution by means of an example.

Example 2.6

Let us consider an infinite plane sheet of current in the xz-plane with uniform surface
current density $\mathbf{J}_S = J_{S0}\mathbf{a}_z$ A/m and find the magnetic flux density everywhere.
  Let us first consider a point $(0, y, 0)$ on the positive y-axis, as shown in Fig. 2.17(a).
Then the solution can be carried out by dividing the sheet into a number of thin vertical

*Infinite plane
sheet of
current*

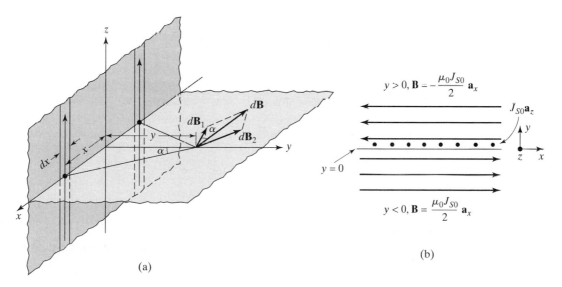

FIGURE 2.17   (a) Determination of magnetic field due to an infinite plane sheet of current density $J_{S0}\mathbf{a}_z$ A/m. (b) Magnetic field due to the current sheet of (a).

strips and using superposition. Two such strips, which are on either side of the z-axis and equidistant from it, are shown in Fig. 2.17(a). Each strip is an infinitely long filamentary wire of current $J_{S0} \, dx$. Then applying the result of Example 2.5 to each strip and noting that the resultant magnetic flux density at $(0, y, 0)$ due to the two strips together has only an x-component, we obtain

$$
\begin{aligned}
d\mathbf{B} &= d\mathbf{B}_1 + d\mathbf{B}_2 \\
&= -2 \, dB_1 \cos \alpha \, \mathbf{a}_x \\
&= -2 \frac{\mu_0 J_{S0} \, dx}{2\pi \sqrt{x^2 + y^2}} \frac{y}{\sqrt{x^2 + y^2}} \mathbf{a}_x \\
&= -\frac{\mu_0 J_{S0} y \, dx}{\pi (x^2 + y^2)} \mathbf{a}_x
\end{aligned}
$$

The magnetic flux density due to the entire sheet is then given by

$$
[\mathbf{B}]_{(0,y,0)} = \int_{x=0}^{\infty} d\mathbf{B}
$$

$$
= -\int_{x=0}^{\infty} \frac{\mu_0 J_{S0} y}{\pi (x^2 + y^2)} \, dx \, \mathbf{a}_x
$$

$$
= -\frac{\mu_0 J_{S0} y}{\pi} \left[\frac{1}{y} \tan^{-1} \frac{x}{y} \right]_{x=0}^{\infty} \mathbf{a}_x
$$

$$
= -\frac{\mu_0 J_{S0}}{2} \mathbf{a}_x \qquad \text{for} \quad y > 0
$$

Since the magnetic field due to each strip is circular to that strip, a similar result applies for a point on the negative y-axis except for $+x$-direction for the field. Thus,

$$
[\mathbf{B}]_{(0,y,0)} = \frac{\mu_0 J_{S0}}{2} \mathbf{a}_x \qquad \text{for} \quad y < 0
$$

Now, since the origin can be chosen to be anywhere on the sheet without changing the geometry, the foregoing results are valid everywhere in the respective regions. Thus, the required magnetic flux density is

$$
\mathbf{B} = \mp \frac{\mu_0 J_{S0}}{2} \mathbf{a}_x \qquad \text{for} \quad y \gtrless 0 \tag{2.23}
$$

which has the magnitude $\mu_0 J_{S0}/2$ everywhere and is directed in the $\mp \mathbf{a}_x$ direction for $y \gtrless 0$, as shown in Fig. 2.17(b). Defining $\mathbf{a}_n$ to be the unit normal vector directed away from the sheet, that is,

$$
\mathbf{a}_n = \pm \mathbf{a}_y \qquad \text{for} \quad y \gtrless 0
$$

and noting that

$$
\mathbf{B} = \frac{\mu_0}{2} (J_{S0} \, \mathbf{a}_z) \times (\pm \mathbf{a}_y) \qquad \text{for} \quad y \gtrless 0
$$

we can write

$$
\boxed{\mathbf{B} = \frac{\mu_0}{2} \mathbf{J}_S \times \mathbf{a}_n} \tag{2.24}
$$

Returning now to (2.20), we can formulate the magnetic force in terms of moving charge, since current is due to flow of charges. Thus, if the time taken by the charge dq contained in the length $d\mathbf{l}$ of the current element to flow with a velocity $\mathbf{v}$ across the infinitesimal cross-sectional area of the element is dt, then $I = dq/dt$, and $d\mathbf{l} = \mathbf{v}\, dt$, so that

Magnetic force in terms of charge

$$dF = \frac{dq}{dt}\, \mathbf{v}\, dt \times \mathbf{B} = dq\, \mathbf{v} \times \mathbf{B} \qquad (2.25)$$

It then follows that the force $\mathbf{F}$ experienced by a test charge q moving with a velocity $\mathbf{v}$ in a magnetic field of flux density $\mathbf{B}$ is given by

$$\boxed{\mathbf{F} = q\mathbf{v} \times \mathbf{B}} \qquad (2.26)$$

We may now obtain a defining equation for $\mathbf{B}$ in terms of the moving test charge. To do this, we note from (2.26) that the magnetic force is directed normally to both $\mathbf{v}$ and $\mathbf{B}$, as shown in Fig. 2.18, and that its magnitude is equal to $qvB \sin \delta$, where δ is the angle between $\mathbf{v}$ and $\mathbf{B}$. A knowledge of the force $\mathbf{F}$ acting on a test charge moving with an arbitrary velocity $\mathbf{v}$ provides only the value of $B \sin \delta$. To find $\mathbf{B}$, we must determine the maximum force qvB that occurs for δ equal to $90°$ by trying out several directions of $\mathbf{v}$, keeping its magnitude constant. Thus, if this maximum force is $\mathbf{F}_m$ and it occurs for a velocity $v\mathbf{a}_m$, then

$$\mathbf{B} = \frac{\mathbf{F}_m \times \mathbf{a}_m}{qv} \qquad (2.27)$$

As in the case of defining the electric field intensity, we assume that the test charge does not alter the magnetic field in which it is placed. Ideally, $\mathbf{B}$ is defined in the limit that qv tends to zero; that is,

$$\boxed{\mathbf{B} = \lim_{qv \to 0} \frac{\mathbf{F}_m \times \mathbf{u}_m}{qv}} \qquad (2.28)$$

Equation (2.28) is the defining equation for the magnetic flux density irrespective of the source of the magnetic field. We have learned in this section that an electric current or a charge in motion is a source of the magnetic field. We will learn in Chapter 3 that there exists another source for the magnetic field, namely, a time-varying electric field.

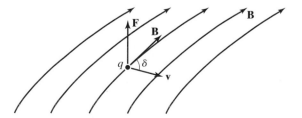

FIGURE 2.18   Force experienced by a test charge q moving with a velocity $\mathbf{v}$ in a magnetic field $\mathbf{B}$.

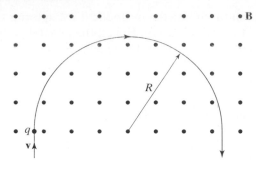

FIGURE 2.19   Circular motion of a charged particle entering a uniform magnetic field region.

Charged particle motion in uniform magnetic field

There are many devices based on the magnetic force on a moving charge. Of particular interest is the motion of a charged particle in a uniform magnetic field, as shown in Fig. 2.19. In this figure, a particle of mass m and charge q entering the magnetic field region with velocity $\mathbf{v}$ perpendicular to $\mathbf{B}$ experiences a force qvB perpendicular to $\mathbf{v}$. Hence, the particle describes a circular path of radius R, equal to mv/qB, obtained by equating the centripetal force mv^2/R to the magnetic force qvB. The fact that the radius is equal to mv/qB is used in several different applications. In the mass spectrograph, the mass-to-charge ratio of the particles is obtained by measuring the radius of the circular orbit for known values of v and B. For ions of the same charge but of different masses, the radii of the circular paths are directly proportional to their masses and to their velocities. This forms the basis for electromagnetic separation of isotopes, two or more forms of a chemical element having the same chemical properties and the same atomic number but different atomic weights. In the cyclotron, a particle accelerator, the particle undergoes a series of semicircular orbits of successively increasing velocities and hence radii before it exits the field region with high energy.

K2.2. Ampère's law of force; Magnetic flux density; Biot–Savart law; Computation of $\mathbf{B}$ due to current distributions; $\mathbf{B}$ due to an infinitely long, straight wire; $\mathbf{B}$ due to an infinite plane sheet of current of uniform density.

D2.5. For $I_1\, d\mathbf{l}_1 = I_1\, dy\, \mathbf{a}_y$ located at $(1,0,0)$ and $I_2\, d\mathbf{l}_2 = I_2\, dx\, \mathbf{a}_x$ located at $(0,1,0)$, find: **(a)** $d\mathbf{F}_1$ and **(b)** $d\mathbf{F}_2$.
Ans.   **(a)** $-(\mu_0 I_1 I_2/8\sqrt{2}\pi)\, dx\, dy\, \mathbf{a}_x$;   **(b)** $-(\mu_0 I_1 I_2/8\sqrt{2}\pi)\, dx\, dy\, \mathbf{a}_y$.

D2.6. A current I flows in a wire along the curve $x = 2y = z^2 + 2$ and in the direction of increasing z. If the wire is situated in a magnetic field $\mathbf{B} = (y\mathbf{a}_x - x\mathbf{a}_y)/(x^2 + y^2)$, find the magnetic force acting on an infinitesimal length of the wire having the projection dz on the z-axis at each of the following points: **(a)** $(2,1,0)$; **(b)** $(3,1.5,1)$; and **(c)** $(6,3,2)$.
Ans.   **(a)** $I\, dz\, (2\mathbf{a}_x + \mathbf{a}_y)/5$;   **(b)** $I\, dz\, (2\mathbf{a}_x + \mathbf{a}_y - 5\mathbf{a}_z)/7.5$;   **(c)** $I\, dz\, (2\mathbf{a}_x + \mathbf{a}_y - 10\mathbf{a}_z)/15$.

D2.7. Given $\mathbf{B} = (B_0/3)(2\mathbf{a}_x + 2\mathbf{a}_y - \mathbf{a}_z)$, find the magnitude of the magnetic force acting on a test charge q moving with velocity v_0 at the point $(2, 2, -1)$ for each of the following paths of the test charge: **(a)** $x = y = -2z$; **(b)** $4x = 4y = z + 9$; and **(c)** $x = y = 2z^2$.
Ans.   **(a)** 0;   **(b)** $qv_0 B_0$;   **(c)** $0.1641 qv_0 B_0$.

D2.8. Infinite plane sheets of current lie in the $x = 0$, $y = 0$, and $z = 0$ planes with uniform surface current densities $J_{S0}\mathbf{a}_z$, $2J_{S0}\mathbf{a}_x$, and $-J_{S0}\mathbf{a}_x$ A/m, respectively. Find the resulting magnetic flux densities at the following points: **(a)** $(1, 2, 2)$; **(b)** $(2, -2, -1)$; and **(c)** $(-2, 1, -2)$.

Ans. **(a)** $\mu_0 J_{S0}(\mathbf{a}_y + \mathbf{a}_z)$; **(b)** $-\mu_0 J_{S0}\mathbf{a}_z$; **(c)** $\mu_0 J_{S0}(-\mathbf{a}_y + \mathbf{a}_z)$.

2.3 LORENTZ FORCE EQUATION

In Section 2.1, we learned that a test charge q placed in an electric field of intensity $\mathbf{E}$ experiences a force

$$\boxed{\mathbf{F}_E = q\mathbf{E}} \tag{2.29}$$

and in Section 2.2, we learned that a test charge q moving with a velocity $\mathbf{v}$ in a magnetic field of flux density $\mathbf{B}$ experiences a force

$$\boxed{\mathbf{F}_M = q\mathbf{v} \times \mathbf{B}} \tag{2.30}$$

Combining (2.29) and (2.30), we can write the expression for the total force acting on a test charge q moving with velocity $\mathbf{v}$ in a region characterized by electric field of intensity $\mathbf{E}$ and magnetic field of flux density $\mathbf{B}$ to be

$$\boxed{\mathbf{F} = \mathbf{F}_E + \mathbf{F}_M = q(\mathbf{E} + \mathbf{v} \times \mathbf{B})} \tag{2.31}$$

Equation (2.31) is known as the *Lorentz force equation.*

    We observe from (2.31) that the electric and magnetic fields at a point can be determined from a knowledge of the forces experienced by a test charge at that point for several different velocities. For a given $\mathbf{B}$, $\mathbf{E}$ can be found from the force for one velocity, since $\mathbf{F}_E$ acts in the direction of $\mathbf{E}$. For a given $\mathbf{E}$, $\mathbf{B}$ can be found from two forces for two noncollinear velocities, since $\mathbf{F}_M$ acts perpendicular to both $\mathbf{v}$ and $\mathbf{B}$. Thus, to find both $\mathbf{E}$ and $\mathbf{B}$, the knowledge of a minimum of three forces is necessary. We shall illustrate the determination of $\mathbf{E}$ and $\mathbf{B}$ from three forces by means of an example.

Determination of electric and magnetic fields from forces on a test charge

Example 2.7

The forces experienced by a test charge q for three different velocities at a point in a region of electric and magnetic fields are given by

$$\mathbf{F}_1 = qE_0\mathbf{a}_x \qquad \text{for} \quad \mathbf{v}_1 = v_0\mathbf{a}_x$$
$$\mathbf{F}_2 = qE_0(2\mathbf{a}_x + \mathbf{a}_y) \qquad \text{for} \quad \mathbf{v}_2 = v_0\mathbf{a}_y$$
$$\mathbf{F}_3 = qE_0(\mathbf{a}_x + \mathbf{a}_y) \qquad \text{for} \quad \mathbf{v}_3 = v_0\mathbf{a}_z$$

where v_0 and E_0 are constants. We wish to find $\mathbf{E}$ and $\mathbf{B}$ at that point.

From the Lorentz force equation, we have

$$q\mathbf{E} + qv_0\mathbf{a}_x \times \mathbf{B} = qE_0\mathbf{a}_x \tag{2.32a}$$

$$q\mathbf{E} + qv_0\mathbf{a}_y \times \mathbf{B} = q(2E_0\mathbf{a}_x + E_0\mathbf{a}_y) \tag{2.32b}$$

$$q\mathbf{E} + qv_0\mathbf{a}_z \times \mathbf{B} = q(E_0\mathbf{a}_x + E_0\mathbf{a}_y) \tag{2.32c}$$

Eliminating $\mathbf{E}$ by subtracting (2.32a) from (2.32b) and (2.32c) from (2.32b), we obtain

$$v_0(\mathbf{a}_y - \mathbf{a}_x) \times \mathbf{B} = E_0(\mathbf{a}_x + \mathbf{a}_y) \tag{2.33a}$$

$$v_0(\mathbf{a}_y - \mathbf{a}_z) \times \mathbf{B} = E_0\mathbf{a}_x \tag{2.33b}$$

Since the cross product of two vectors is perpendicular to the two vectors, it follows from (2.33a) that $(\mathbf{a}_x + \mathbf{a}_y)$ is perpendicular to $\mathbf{B}$ and from (2.33b) that $\mathbf{a}_x$ is perpendicular to $\mathbf{B}$. Thus, $\mathbf{B}$ is perpendicular to both $(\mathbf{a}_x + \mathbf{a}_y)$ and $\mathbf{a}_x$. But the cross product of $(\mathbf{a}_x + \mathbf{a}_y)$ and $\mathbf{a}_x$ is perpendicular to both of them. Therefore, $\mathbf{B}$ must be directed parallel to $(\mathbf{a}_x + \mathbf{a}_y) \times \mathbf{a}_x$. Thus, we can write

$$\mathbf{B} = C(\mathbf{a}_x + \mathbf{a}_y) \times \mathbf{a}_x = -C\mathbf{a}_z \tag{2.34}$$

where C is a proportionality constant to be determined. To do this, we substitute (2.34) into (2.33b) to obtain

$$v_0(\mathbf{a}_y - \mathbf{a}_z) \times (-C\mathbf{a}_z) = E_0\mathbf{a}_x$$

$$-v_0C\mathbf{a}_x = E_0\mathbf{a}_x$$

or $C = -E_0/v_0$. Thus, we get

$$\mathbf{B} = \frac{E_0}{v_0}\mathbf{a}_z$$

Alternatively, we can obtain this result by assuming $\mathbf{B} = B_x\mathbf{a}_x + B_y\mathbf{a}_y + B_z\mathbf{a}_z$, substituting in (2.33a) and (2.33b), equating the like components, and solving the resulting algebraic equations. Thus, substituting in (2.33a), we have

$$v_0 \begin{vmatrix} \mathbf{a}_x & \mathbf{a}_y & \mathbf{a}_z \\ -1 & 1 & 0 \\ B_x & B_y & B_z \end{vmatrix} = E_0(\mathbf{a}_x + \mathbf{a}_y)$$

or

$$v_0[B_z\mathbf{a}_x + B_z\mathbf{a}_y - (B_y + B_x)\mathbf{a}_z] = E_0\mathbf{a}_x + E_0\mathbf{a}_y$$

$$B_z = \frac{E_0}{v_0} \quad \text{and} \quad (B_y + B_x) = 0$$

Substituting in (2.33b), we have

$$v_0 \begin{vmatrix} \mathbf{a}_x & \mathbf{a}_y & \mathbf{a}_z \\ 0 & 1 & -1 \\ B_x & B_y & B_z \end{vmatrix} = E_0\mathbf{a}_x$$

or

$$v_0[(B_z + B_y)\mathbf{a}_x - B_x\mathbf{a}_y - B_x\mathbf{a}_z] = E_0\mathbf{a}_x$$

$$B_z + B_y = \frac{E_0}{v_0} \quad \text{and} \quad B_x = 0$$

Thus, we obtain $B_z = E_0/v_0, B_x = 0, B_y = 0$, and, hence,

$$\mathbf{B} = \frac{E_0}{v_0}\mathbf{a}_z$$

Finally, we can find $\mathbf{E}$ by substituting the result obtained for $\mathbf{B}$ in any one of the three equations (2.32a)–(2.32c). Thus, substituting $\mathbf{B} = (E_0/v_0)\mathbf{a}_z$ in (2.32c), we obtain

$$\mathbf{E} = E_0(\mathbf{a}_x + \mathbf{a}_y)$$

The Lorentz force equation is a fundamental equation in electromagnetics. Together with the pertinent laws of mechanics, it constitutes the starting point for the study of charged particle motion in electric and/or magnetic fields. Devices based on charged particle motion in fields are abundant in practice. Examples, some of which we discussed in Sections 2.1 and 2.2, are cathode ray tubes, ink-jet printers, electron microscopes, mass spectrographs, particle accelerators, and microwave tubes such as klystrons, magnetrons, and traveling wave tubes. Interaction between charged particles and fields is the basis for the study of the electromagnetic properties of materials and for the study of radio-wave propagation in gaseous media such as Earth's ionosphere, in which the constituent gasses are partially ionized by the solar radiation. *Lorentz force applications*

Tracing the path of a charged particle in a region of electric and magnetic fields involves setting the mechanical force, as given by the product of the mass of the test charge and its acceleration, equal to the electromagnetic force, as given by the Lorentz force equation, and solving the resulting differential equation(s) subject to initial condition(s). For simplicity, we shall consider a two-dimensional situation in which the motion is confined to the xy-plane in a region of uniform, crossed electric and magnetic fields, $\mathbf{E} = E_0\mathbf{a}_y$ and $\mathbf{B} = B_0\mathbf{a}_z$, as shown in Fig. 2.20, where E_0 and B_0 are constants. We shall assume that a test charge q having mass m starts at $t = 0$ at the point $(x_0, y_0, 0)$ with initial velocity $\mathbf{v} = v_{x0}\mathbf{a}_x + v_{y0}\mathbf{a}_y$. *Tracing of charged particle motion in electric and magnetic fields*

From the Lorentz force equation (2.31), the force exerted by the crossed electric and magnetic fields on the test charge is given by

$$\begin{aligned}
\mathbf{F} &= q(\mathbf{E} + \mathbf{v} \times \mathbf{B}) \\
&= qE_0\mathbf{a}_y + q(v_x\mathbf{a}_x + v_y\mathbf{a}_y + v_z\mathbf{a}_z) \times B_0\mathbf{a}_z \quad (2.35)\\
&= qB_0v_y\mathbf{a}_x + (qE_0 - qB_0v_x)\mathbf{a}_y
\end{aligned}$$

FIGURE 2.20   Test charge q in a region of crossed electric and magnetic fields.

The equations of motion of the test charge can then be written as

$$\frac{dv_x}{dt} = \frac{qB_0}{m} v_y \tag{2.36a}$$

$$\frac{dv_y}{dt} = \frac{qE_0}{m} - \frac{qB_0}{m} v_x \tag{2.36b}$$

$$\frac{dv_z}{dt} = 0 \tag{2.36c}$$

Equation (2.36c) together with the initial conditions $v_z = 0$ and $z = 0$ at $t = 0$ simply tells us that the path of the test charge is confined to the $z = 0$ plane.

Eliminating v_y from (2.36a) and (2.36b), we obtain

$$\frac{d^2 v_x}{dt^2} + \left(\frac{qB_0}{m}\right)^2 v_x = \left(\frac{q}{m}\right)^2 B_0 E_0 \tag{2.37}$$

the solution for which is

$$v_x = \frac{E_0}{B_0} + C_1 \cos \omega_c t + C_2 \sin \omega_c t \tag{2.38a}$$

where C_1 and C_2 are constants to be determined from the initial conditions and $\omega_c = qB_0/m$. From (2.36a), the solution for v_y is then given by

$$v_y = -C_1 \sin \omega_c t + C_2 \cos \omega_c t \tag{2.38b}$$

Using initial conditions $v_x = v_{x0}$ and $v_y = v_{y0}$ at $t = 0$ to evaluate C_1 and C_2 in (2.38a) and (2.38b), we obtain

$$v_x = \frac{E_0}{B_0} + \left(v_{x0} - \frac{E_0}{B_0}\right) \cos \omega_c t + v_{y0} \sin \omega_c t \tag{2.39a}$$

$$v_y = -\left(v_{x0} - \frac{E_0}{B_0}\right) \sin \omega_c t + v_{y0} \cos \omega_c t \tag{2.39b}$$

Integrating (2.39a) and (2.39b) with respect to t and using initial conditions $x = x_0$ and $y = y_0$ at $t = 0$ to evaluate the constants of integration, we then obtain

$$x = x_0 + \frac{E_0}{B_0} t + \frac{1}{\omega_c}\left(v_{x0} - \frac{E_0}{B_0}\right) \sin \omega_c t + \frac{v_{y0}}{\omega_c}(1 - \cos \omega_c t) \tag{2.40a}$$

$$y = y_0 - \frac{1}{\omega_c}\left(v_{x0} - \frac{E_0}{B_0}\right)(1 - \cos \omega_c t) + \frac{v_{y0}}{\omega_c} \sin \omega_c t \tag{2.40b}$$

Equations (2.40a) and (2.40b) give the position of the test charge versus time, whereas (2.39a) and (2.39b) give the corresponding velocity components.

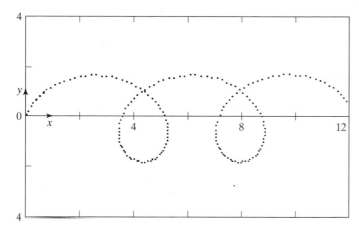

FIGURE 2.21 An example of tracing the path of an electron in crossed electric and magnetic fields.

For $B_0 = 0, \omega_c \to 0$, and the solutions reduce to

$$x = x_0 + v_{x0}t \tag{2.41a}$$

$$y = y_0 + v_{y0}t + \frac{1}{2}\frac{qE_0}{m}t^2 \tag{2.41b}$$

$$v_x = v_{x0} \tag{2.41c}$$

$$v_y = v_{y0} + \frac{qE_0}{m}t \tag{2.41d}$$

These can also be obtained directly from (2.36a) and (2.36b) with B_0 set equal to zero.

The path of a test charge in the crossed electric and magnetic fields may now be traced by using (2.40a) and (2.40b) for B_0 not equal to zero, and (2.41a) and (2.41b) for B_0 equal to zero. For example, the path of an electron ($q/m = -1.7578 \times 10^{11}$ C/kg) for $x_0 = 0$, $y_0 = 0$, $E_0 = -10^3$ V/m, $B_0 = 10^{-4}$ Wb/m², $v_{x0} = 10^7$ m/s, and $v_{y0} = 3 \times 10^7$ m/s is shown in Fig. 2.21, in which the spacing betwen the dots corresponds to a time interval of 5×10^{-9} s.

K2.3. Lorentz force equation; Determination of **E** and **B** from forces on a test charge; Charged-particle motion in electric and magnetic fields.

D2.9. A magnetic field $\mathbf{B} = (B_0/3)(\mathbf{a}_x + 2\mathbf{a}_y - 2\mathbf{a}_z)$ exists at a point. For each of the following velocities of a test charge q, find the electric field **E** at that point for which the acceleration experienced by the test charge is zero: **(a)** $v_0(\mathbf{a}_x - \mathbf{a}_y + \mathbf{a}_z)$; **(b)** $v_0(2\mathbf{a}_x + \mathbf{a}_y + 2\mathbf{a}_z)$; and **(c)** v_0 along the line $y = -z = 2x$.
Ans. **(a)** $-v_0 B_0(\mathbf{a}_y + \mathbf{a}_z)$; **(b)** $v_0 B_0(2\mathbf{a}_x - 2\mathbf{a}_y - \mathbf{a}_z)$; **(c)** **0.**

D2.10.   In a region of uniform electric and magnetic fields $\mathbf{E} = E_0\mathbf{a}_y$ and $\mathbf{B} = B_0\mathbf{a}_z$, respectively, a test charge q of mass m moves in the manner

$$x = \frac{E_0}{\omega_c B_0} (\omega_c t - \sin \omega_c t)$$

$$y = \frac{E_0}{\omega_c B_0} (1 - \cos \omega_c t)$$

$$z = 0$$

where $\omega_c = qB_0/m$. Find the forces acting on the test charge for the following values of t: **(a)** $t = 0$; **(b)** $t = \pi/2\omega_c$; and **(c)** $t = \pi/\omega_c$.
Ans.   **(a)** $qE_0\mathbf{a}_y$;   **(b)** $qE_0\mathbf{a}_x$;   **(c)** $-qE_0\mathbf{a}_y$.

2.4   CONDUCTORS AND SEMICONDUCTORS

Conduction

In our discussion of electric and magnetic fields thus far, we considered the medium to be free space. We shall now introduce materials. Materials contain charged particles that respond to applied electric and magnetic fields. Depending on their response to an applied electric field, they may be classified as conductors, semiconductors, or dielectrics. According to the classical model, an atom consists of a tightly bound, positively charged nucleus surrounded by a diffuse electron cloud having an equal and opposite charge to the nucleus. While the electrons for the most part are less tightly bound, the majority of them are associated with the nucleus and are known as *bound* electrons. These bound electrons can be displaced but not removed from the influence of the nucleus on application of an electric field. Not taking part in this bonding mechanism are the *free*, or *conduction*, electrons. These electrons are constantly under thermal agitation, being released from the parent atom at one point and recaptured at another point. In the absence of an applied electric field, their motion is completely random; that is, the average thermal velocity on a macroscopic scale is zero, so that there is no net current and the electron cloud maintains a fixed position. When an electric field is applied, an additional velocity is superimposed on the random velocities, thereby causing a *drift* of the average position of the electrons along the direction opposite to that of the electric field. This process is known as *conduction*. In certain materials, a large number of electrons may take part in this process. These materials are known as *conductors*. In certain other materials, only very few or a negligible number of electrons may participate in conduction. These materials are known as *dielectrics*, or insulators. A class of materials for which conduction occurs not only by electrons but also by another type of carriers known as *holes*—vacancies created by detachment of electrons due to breaking of covalent bonds with other atoms—is intermediate to that of conductors and dielectrics. These materials are called *semiconductors*.

The quantum theory describes the motion of the current carriers in terms of energy levels. According to this theory, the electrons in an atom can have associated with them only certain discrete values of energy. When a large number of atoms are packed together, as in a crystalline solid, each energy level in the individual atom splits into a number of levels with slightly different energies, with the degree of splitting governed by the interatomic spacing, thereby giving rise to allowed bands of energy levels that may be widely separated, may be close together, or may even overlap. Four possible energy band diagrams are shown in Fig. 2.22, in which a forbidden band consists of energy levels that no electron in any atom of the solid can occupy. For case (a), the lower allowed band is only partially filled at the temperature of absolute zero. At higher temperatures, the electron population in the band spreads out somewhat, but only very few electrons reach higher energy levels. Thus, since there are many unfilled levels in the same band, it is possible to increase the energy of the system by moving the electrons to these unoccupied levels very easily by the application of an electric field, thereby resulting in drift of the electrons. The material is then classified as a conductor. For cases (b) and (c), the lower band is completely filled, whereas the next-higher band is completely empty at the temperature of absolute zero. If the width of the forbidden band is very large as in (b), the situation at normal temperatures is essentially the same as at absolute zero, and, hence, there are no neighboring empty energy levels for the electrons to move. The only way for conduction to take place is for the electrons in the filled band to get excited and move to the next higher band. But this is very difficult to achieve with reasonable electric fields and the material is then classified as a dielectric. Only by supplying a very large amount of energy can an electron be excited to move from the lower band to the higher band, where it has available neighboring empty levels for causing conduction. The dielectric is said to break down under such conditions. If, on the other hand, the width of the forbidden band in which the Fermi level lies is not too large, as in (c), some of the electrons in the lower band move into the upper band at normal temperatures, so that conduction can take place under the influence of an electric field, not only in the upper band, but also in the lower band because of the vacancies (holes) left by the electrons that moved into the upper band. The material is then classified as a semiconductor. A semiconductor crystal in pure form is known as an intrinsic semiconductor. The properties of an intrinsic crystal can be altered by introducing impurities into it. The crystal is then said to be an extrinsic semiconductor.

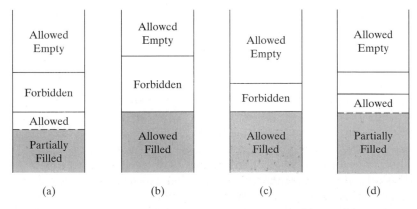

FIGURE 2.22 Energy band diagrams for different cases: (a) and (d) conductor; (b) dielectric; and (c) semiconductor.

For case (d), two allowed bands overlap; one or both of the bands is only partially filled and the situation corresponds to a conductor.

In the foregoing discussion, we classified materials on the basis of their ability to permit conduction of electrons under the application of an external electric field. For conductors, we are interested in knowing about the relationship between the *drift velocity* of the electrons and the applied electric field, since the predominant process is conduction. But for collisions with the atomic lattice, the electric field continuously accelerates the electrons in the direction opposite to it as they move about at random. Collisions with the atomic lattice, however, provide the frictional mechanism by means of which the electrons lose some of the momentum gained between collisions. The net effect is as though the electrons drift with an average drift velocity $\mathbf{v}_d$, under the influence of the force exerted by the applied electric field and an opposing force due to the frictional mechanism. This opposing force is proportional to the momentum of the electron and inversely proportional to the average time τ between collisions. Thus, the equation of motion of an electron is given by

$$m \frac{d\mathbf{v}_d}{dt} = e\mathbf{E} - \frac{m\mathbf{v}_d}{\tau} \tag{2.42}$$

where e and m are the charge and mass of an electron.

Rearranging (2.42), we have

$$m \frac{d\mathbf{v}_d}{dt} + \frac{m}{\tau} \mathbf{v}_d = e\mathbf{E} \tag{2.43}$$

For the sudden application of a constant electric field $\mathbf{E}_0$ at $t = 0$, the solution for (2.43) is given by

$$\mathbf{v}_d = \frac{e\tau}{m} \mathbf{E}_0 - \frac{e\tau}{m} \mathbf{E}_0 e^{-t/\tau} \tag{2.44}$$

where we have evaluated the arbitrary constant of integration by using the initial condition that $\mathbf{v}_d = \mathbf{0}$ at $t = 0$. The values of τ for typical conductors such as copper are of the order of 10^{-14} s, so that the exponential term on the right side of (2.44) decays to a negligible value in a time much shorter than that of practical interest. Thus, neglecting this term, we have

$$\mathbf{v}_d = \frac{e\tau}{m} \mathbf{E}_0 \tag{2.45}$$

and the drift velocity is proportional in magnitude and opposite in direction to the applied electric field since the value of e is negative.

In fact, since we can represent a time-varying field as a superposition of step functions starting at appropriate times, the exponential term in (2.44) may be neglected as long as the electric field varies slowly compared to τ. For fields varying sinusoidally with time, this means that as long as the period T of the sinusoidal variation is several times the value of τ, or the radian frequency $\omega \ll 2\pi/\tau$, the drift velocity follows the variations in the electric field. Since $1/\tau \approx 10^{14}$, this condition is satisfied even at frequencies up to several hundred gigahertz (a gigahertz is 10^9 Hz). Thus, for all practical purposes, we can assume that

$$\mathbf{v}_d = \frac{e\tau}{m}\mathbf{E} \tag{2.46}$$

Now, we define the *mobility*, μ_e, of the electron as the ratio of the magni-   *Mobility*
tudes of the drift velocity and the applied electric field. Then we have

$$\mu_e = \frac{|\mathbf{v}_d|}{|\mathbf{E}|} = \frac{|e|\tau}{m} \tag{2.47}$$

and

$$\mathbf{v}_d = -\mu_e\mathbf{E} \quad \text{for electrons} \tag{2.48a}$$

For values of τ typically of the order of 10^{-14} s, we note by substituting for $|e|$ and m on the right side of (2.47) that the electron mobilities are of the order of 10^{-3} C-s/kg. Alternative units for the mobility are square meters per volt-second. In semiconductors, conduction is due not only to the movement of electrons, but also to the movement of holes. We can define the mobility μ_h of a hole similarly to μ_e as the ratio of the drift velocity of the hole to the applied electric field. Thus, we have

$$\mathbf{v}_d = \mu_h\mathbf{E} \quad \text{for holes} \tag{2.48b}$$

Note from (2.48b) that conduction of a hole takes place along the direction of the applied electric field since a hole is a vacancy created by the removal of an electron and, hence, a hole movement is equivalent to the movement of a positive charge of value equal to the magnitude of the charge of an electron. In general, the mobility of holes is lower than the mobility of electrons for a particular semiconductor. For example, for silicon, the values of μ_e and μ_h are 0.135 m²/V-s and 0.048 m²/V-s, respectively. Semiconductors are denoted n-type or p-type, depending on whether the conduction is predominantly due to the movement of electrons or holes.

The drift of electrons in a conductor and that of electrons and holes in a   *Conduction*
semiconductor is equivalent to a current flow. This current is known as the *con-*   *current*
duction current. The conduction current density may be obtained in the following manner. If there are N_e free electrons per cubic meter of the material, then the amount of charge ΔQ passing through an infinitesimal area ΔS normal to the drift velocity at a point in the material in a time Δt is given by

$$\Delta Q = N_e e(\Delta S)(v_d \, \Delta t) \tag{2.49}$$

The current ΔI flowing across ΔS is given by

$$\Delta I = \frac{|\Delta Q|}{\Delta t} = N_e|e|v_d \, \Delta S \tag{2.50}$$

The magnitude of the current density at the point is the ratio of ΔI to ΔS in the limit ΔS tends to zero, and the direction is opposite to that of $\mathbf{v}_d$. Thus, the conduction current density $\mathbf{J}_c$ resulting from the drift of electrons in the conductor is given by

$$\mathbf{J}_c = -N_e|e|\mathbf{v}_d \tag{2.51}$$

Substituting for $\mathbf{v}_d$ from (2.48a), we have

$$\mathbf{J}_c = \mu_e N_e|e|\mathbf{E} \tag{2.52}$$

Conductivity   Defining a quantity σ as

$$\boxed{\sigma = \mu_e N_e|e|} \tag{2.53}$$

we obtain the simple and important relationship between $\mathbf{J}_c$ and $\mathbf{E}$:

$$\boxed{\mathbf{J}_c = \sigma\mathbf{E}} \tag{2.54}$$

The quantity σ is known as the electrical conductivity of the material, and (2.54) is known as Ohm's law valid at a point. We shall show later that the well-known Ohm's law in circuit theory follows from it. In a semiconductor, the current density is the sum of the contributions due to the drifts of electrons and holes. If the densities of holes and electrons are N_h and N_e, respectively, the conduction current density is given by

$$\boxed{\mathbf{J}_c = (\mu_h N_h|e| + \mu_e N_e|e|)\mathbf{E}} \tag{2.55}$$

Thus, the conductivity of a semiconducting material is given by

$$\boxed{\sigma = \mu_h N_h|e| + \mu_e N_e|e|} \tag{2.56a}$$

For an intrinsic semiconductor, $N_h = N_e$, so that (2.56a) reduces to

$$\boxed{\sigma = (\mu_h + \mu_e)N_e|e|} \tag{2.56b}$$

The units of conductivity are $(\text{meter}^2/\text{volt-second})(\text{coulomb}/\text{meter}^3)$ or ampere/volt-meter, also commonly known as siemens per meter (S/m), where a siemen is an ampere per volt. The ranges of conductivities for conductors, semiconductors, and dielectrics are shown in Fig. 2.23. Values of conductivities for a few materials are listed in Table 2.2. The constant values of conductivities do not imply that the conduction current density is proportional to the applied electric field intensity for all values of current density and field intensity. However, the range of current densities for which the material is linear, that is, for which the conductivity is a constant, is very large for conductors.

Conductor in   Let us now consider a conductor placed in a static electric field, as shown
a static   in Fig. 2.24(a). The free electrons in the conductor move opposite to the direc-
electric field   tion lines of the electric field. If there is a way by means of which the flow of

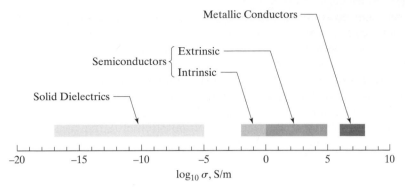

FIGURE 2.23 Ranges of conductivities for conductors, semiconductors, and dielectrics.

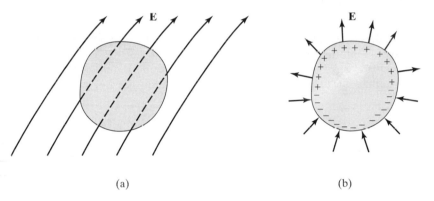

(a)   (b)

FIGURE 2.24 For illustrating the surface charge formation at the boundary of a conductor placed in a static electric field.

TABLE 2.2   Conductivities of Some Materials

Material	Conductivity (S/m)	Material	Conductivity (S/m)
Silver	6.1×10^7	Seawater	4
Copper	5.8×10^7	Intrinsic germanium	2.2
Gold	4.1×10^7	Intrinsic silicon	1.6×10^{-3}
Aluminum	3.5×10^7	Fresh water	10^{-3}
Tungsten	1.8×10^7	Distilled water	2×10^{-4}
Brass	1.5×10^7	Dry earth	10^{-5}
Solder	7.0×10^6	Bakelite	10^{-9}
Lead	4.8×10^6	Glass	$10^{-10} - 10^{-14}$
Constantin	2.0×10^6	Mica	$10^{-11} - 10^{-15}$
Mercury	1.0×10^6	Fused quartz	0.4×10^{-17}

electrons can be continued to form a closed circuit, then a continuous flow of current takes place. Since the conductor is bounded by free space, the electrons are held at the boundary from moving further. Thus, a negative surface charge forms on the boundary, accompanied by an equal amount of positive surface charge, as shown in Fig. 2.24(b), since the conductor as a whole is neutral. The surface charge distribution formed in this manner produces a secondary electric field which, together with the applied electric field, makes the field inside the conductor zero. We shall illustrate the computation of the surface charge densities by means of a simple example.

Example 2.8

Let us consider an infinite plane conducting slab of thickness d occupying the region between $z = 0$ and $z = d$ and in a uniform electric field $\mathbf{E} = E_0 \mathbf{a}_z$ produced by two infinite plane sheets of equal and opposite uniform charge densities on either side of the slab, as shown in Fig. 2.25(a). We wish to find the charge densities induced on the surfaces of the slab.

Since the applied electric field is uniform and is directed along the z-direction, a negative charge of uniform density forms on the surface $z = 0$ due to the accumulation of free electrons at that surface. A positive charge of equal and opposite uniform density forms on the surface $z = d$ due to a deficiency of electrons at that surface. Let these surface charge

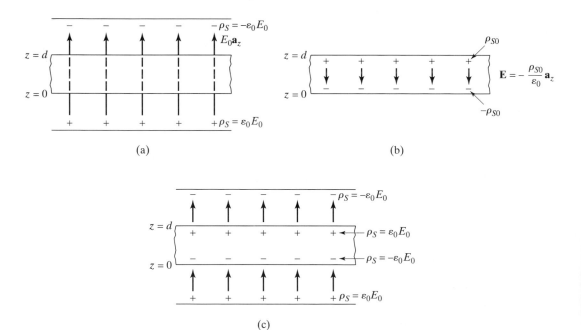

FIGURE 2.25   (a) Infinite plane slab conductor in a uniform applied field. (b) Induced surface charge at the boundaries of the conductor and the secondary field. (c) Sum of the applied and the secondary fields.

densities be $-\rho_{S0}$ and ρ_{S0}, respectively. To satisfy the property that the field in the interior of the conductor is zero, the secondary field produced by the surface charges must be equal and opposite to the applied field; that is, it must be equal to $-E_0\mathbf{a}_z$. Now, each surface charge produces a field intensity directed normally from it and having a magnitude $1/2\varepsilon_0$ times the charge density so that the field due to the two surface charges together is equal to $-(\rho_{S0}/\varepsilon_0)\mathbf{a}_z$ inside the conductor and zero outside the conductor, as shown in Fig. 2.25(b). Thus, for zero field inside the conductor,

$$-\frac{\rho_{S0}}{\varepsilon_0}\mathbf{a}_z = -E_0\mathbf{a}_z$$

or

$$\rho_{S0} = \varepsilon_0 E_0$$

The field outside the conductor remains the same as the applied field since the secondary field in that region due to the surface charges is zero. The induced surface charge distribution and the fields inside and outside the conductor are shown in Fig. 2.25(c). In the general case, the induced surface charge produces a secondary field outside the conductor also, thereby changing the applied field.

Returning now to (2.54), we shall show that the well-known Ohm's law in circuit theory follows from it. To do this, let us consider a bar of conducting material of conductivity σ, length l, and uniform cros-sectional area A and between the ends of which a voltage V is applied, as shown in Fig. 2.26. The voltage sets up an electric field directed along the length of the conductor, thereby giving rise to conduction current. We shall define voltage rigorously in Section 3.1, but for the purpose of discussion here, the voltage between the two ends of the conductor is simply the electric field intensity times the length of the conductor, that is, *Ohm's law, resistance*

$$V = El \tag{2.57}$$

Then from (2.54) and (2.57), the conduction current density magnitude is given by

$$J_c = \sigma E = \frac{\sigma V}{l} \tag{2.58}$$

Assuming uniformity of the field and hence of the conduction current density in the cross-sectional area of the conductor, we then obtain the conduction current to be

$$I = J_c A = \frac{\sigma A}{l}V \tag{2.59}$$

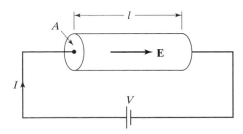

FIGURE 2.26 For the derivation of Ohm's law in circuit theory.

Upon rearrangement, we get

$$V = I \frac{l}{\sigma A} \tag{2.60}$$

which is in the form of the familiar Ohm's law,

$$\boxed{V = IR} \tag{2.61}$$

From (2.60), the resistance R of the conducting bar can now be identified as

$$\boxed{R = \frac{l}{\sigma A}} \tag{2.62}$$

the units of R being ohms.

Hall effect   We shall conclude this section with a discussion of the Hall effect, an important phenomenon employed in the determination of charge densities in conducting and semiconducting materials, as well as in other techniques such as the measurement of fluid flow using electromagnetic flow meters. Let us consider the p-type semiconducting material in the form of a rectangular bar shown in Fig. 2.27, in which holes drift in the x-direction with a velocity $\mathbf{v} = v_x \mathbf{a}_x$ due to an applied voltage between the two ends of the bar. If a magnetic field $\mathbf{B} = B_z \mathbf{a}_z$ is applied in a perpendicular direction, then the drifting holes will experience a magnetic force $\mathbf{F}_m$ that deflects them in the $\mathbf{a}_x \times \mathbf{a}_z$ or $-\mathbf{a}_y$-direction. This deflection of holes toward the $-y$-direction establishes an electric field $\mathbf{E}_H = E_y \mathbf{a}_y$ in the material, resulting in the development of a voltage between the two sides of the bar. This phenomenon is known as the *Hall effect*, and the voltage developed is known as the *Hall voltage*. If it were not for the establishment of the Hall electric field, the holes will continually deflect toward the $-y$-direction as they drift in the x-direction. The Hall electric field exerts force $\mathbf{F}_H$ on the holes in

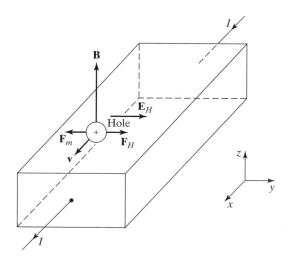

FIGURE 2.27   For illustrating the Hall effect phenomenon.

the $+y$-direction, which in the steady-state balances exactly the magnetic force $\mathbf{F}_m$ in the $-y$-direction so that the net y-directed force is zero. According to the Lorentz force equation (2.31), the Hall electric field that achieves this balance is given by

$$q(\mathbf{E}_H + \mathbf{v} \times \mathbf{B}) = q(E_y \mathbf{a}_y + v_x \mathbf{a}_x \times B_z \mathbf{a}_z)$$
$$= q(E_y - v_x B_z)\mathbf{a}_y = \mathbf{0} \tag{2.63}$$

or $E_y = v_x B_z$. By using this result, the hole density can be computed from a measurement of the Hall voltage for known values of the magnetic field B_z, the currrent I, and the cross-sectional dimensions of the bar. If the material is n-type instead of p-type, then the charge carriers are electrons, and $\mathbf{v}$ would be in the $-x$-direction. The deflection of the charge carriers will still be toward the $-y$-direction since the charge is negative. This results in an electric field in the $-y$-direction and hence in a Hall voltage of opposite polarity to that in the case of the p-type material. Thus, the polarity of the Hall voltage can be used to determine if the charge carriers are holes or electrons.

K2.4. Conduction; Conduction current density; Conductivity; Ohm's law; Conductor in a static electric field; Resistance; Hall effect.

D2.11. Find the magnitude of the electric field intensity required to establish the flow of a conduction current of 0.1 A across an area of 1 cm^2 normal to the field for each of the following cases: **(a)** in copper; **(b)** in an intrinsic semiconductor material with electron and hole mobilities of 3600 cm^2/V-s and 1700 cm^2/V-s, respectively, and electron and hole densities of 2.5×10^{13} cm^{-3}; and **(c)** in a metallic wire of circular cross section of radius 1 mm, length 1 m, and resistance 1 ohm.
Ans.   **(a)** 17.24 μV/m;   **(b)** 471.1 V/m;   **(c)** 3.14 mV/m.

D2.12. An infinite plane conducting slab lies between and parallel to two infinite plane sheets of charge of uniform surface charge densities ρ_{SA} and ρ_{SB}, as shown by the cross-sectional view in Fig. 2.28. Find the surface charge densities on the two surfaces of the slab: **(a)** ρ_{S1} and **(b)** ρ_{S2}.
Ans.   **(a)** $\frac{1}{2}(\rho_{SB} - \rho_{SA})$;   **(b)** $\frac{1}{2}(\rho_{SA} - \rho_{SB})$.

FIGURE 2.28   For Problem D2.12.

2.5 DIELECTRICS

In the preceding section, we learned that conductors are characterized by an abundance of *conduction*, or *free*, electrons that give rise to conduction current under the influence of an applied electric field. In this section, we turn our attention to dielectric materials in which the *bound* electrons are predominant. Under the application of an external electric field, the bound electrons of an

Polarization, electric dipole

atom are displaced such that the centroid of the electron cloud is separated from the centroid of the nucleus. The atom is then said to be *polarized*, thereby creating an *electric dipole,* as shown in Fig. 2.29(a). This kind of polarization is called *electronic polarization*. The schematic representation of an electric dipole is shown in Fig. 2.29(b). The strength of the dipole is defined by the electric dipole moment **p** given by

$$\boxed{\mathbf{P} = Q\mathbf{d}}$$

(2.64)

where **d** is the vector displacement between the centroids of the positive and negative charges, each of magnitude Q coulombs.

In certain dielectric materials, polarization may exist in the molecular structure of the material even under the application of no external electric field. The polarization of individual atoms and molecules, however, is randomly oriented, and hence the net polarization on a *macroscopic* scale is zero. The application of an external field results in torques acting on the *microscopic* dipoles, as shown in Fig. 2.30, to convert the initially random polarization into a partially coherent one along the field, on a macroscopic scale. This kind of polarization is known as *orientational polarization*. A third kind of polarization known as *ionic polarization* results from the separation of positive and negative ions in molecules formed by the transfer of electrons from one atom to another in the molecule. Certain materials exhibit permanent polarization, that is, polarization even in the absence of an applied electric field. Electrets, when allowed to solidify in

FIGURE 2.29   (a) Electric dipole. (b) Schematic representation of an electric dipole.

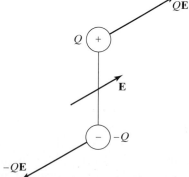

FIGURE 2.30   Torque acting on an electric dipole in an external electric field.

the applied electric field, become permanently polarized, and ferroelectric materials exhibit spontaneous, permanent polarization.

On a macroscopic scale, we define a vector $\mathbf{P}$, called the *polarization vector,* as the *electric dipole moment per unit volume.* Thus, if N denotes the number of molecules per unit volume of the material, then there are $N\,\Delta v$ molecules in a volume Δv and

$$\mathbf{P} = \frac{1}{\Delta v} \sum_{j=1}^{N\,\Delta v} \mathbf{p}_j = N\mathbf{p} \tag{2.65}$$

where $\mathbf{p}$ is the average dipole moment per molecule. The units of $\mathbf{P}$ are coulomb-meter/meter^3 or coulombs per square meter. It is found that for many dielectric materials, the polarization vector is related to the electric field $\mathbf{E}$ in the dielectric in the simple manner given by

$$\mathbf{P} = \varepsilon_0 \chi_e \mathbf{E} \tag{2.66}$$

where χ_e, a dimensionless parameter, is known as the *electric susceptibility.* The quantity χ_e is a measure of the ability of the material to become polarized and differs from one dielectric to another.

When a dielectric material is placed in an electric field, the induced dipoles produce a secondary electric field such that the resultant field, that is, the sum of the originally applied field and the secondary field, and the polarization vector satisfy (2.66). We shall illustrate this by means of a simple example.

Dielectric in an electric field

Example 2.9

Let us consider an infinite plane dielectric slab of thickness d sandwiched between two infinite plane sheets of equal and opposite uniform charge densities ρ_{S0} and $-\rho_{S0}$ in the $z = 0$ and $z = d$ planes, respectively, as shown in Fig. 2.31(a). We wish to investigate the effect of polarization in the dielectric.

In the absence of the dielectric, the electric field between the sheets of charge is given by

$$\mathbf{E}_a = \frac{\rho_{S0}}{\varepsilon_0} \mathbf{a}_z \tag{2.67}$$

In the presence of the dielectric, this field acts as the applied electric field inducing dipole moments in the dielectric with the negative charges separated from the positive charges and pulled away from the direction of the field. Since the electric field and the electric susceptibility are uniform, the density of the induced dipole moments, that is, the polarization vector $\mathbf{P}$, is uniform, as shown in Fig. 2.31(b). Such a distribution results in exact neutralization of all the charges except at the boundaries of the dielectric since, for each positive (or negative) charge not on the surface, there is the same amount of negative (or positive) charge associated with the dipole adjacent to it, thereby canceling its effect. Thus, the net result is the formation of a positive surface charge at the boundary $z = d$ and a negative surface charge at the boundary $z = 0$, as shown in Fig. 2.31(c). These surface

FIGURE 2.31   For investigating the effect of polarization induced in a dielectric material sandwiched between two infinite plane sheets of charge.

charges are known as polarization surface charges since they are due to the polarization in the dielectric. In view of the uniform density of the dipole moments, the surface charge densities are uniform. Also, in the absence of a net charge in the interior of the dielectric, the surface charge densities must be equal in magnitude to preserve the charge neutrality of the dielectric.

Let us therefore denote the surface charge densities as

$$\rho_{pS} = \begin{cases} \rho_{pS0} & \text{for} \quad z = d \\ -\rho_{pS0} & \text{for} \quad z = 0 \end{cases} \tag{2.68}$$

where the subscript p in addition to the other subscripts stands for polarization. If we now consider a vertical column of infinitesimal rectangular cross-sectional area ΔS cut out from the dielectric, as shown in Fig. 2.31(d), the equal and opposite surface charges make the column appear as a dipole of moment $(\rho_{pS0} \, \Delta S) d\mathbf{a}_z$. On the other hand, writing

$$\mathbf{P} = P_0 \mathbf{a}_z \tag{2.69}$$

where P_0 is a constant in view of the uniformity of the induced polarization, the dipole moment of the column is equal to $\mathbf{P}$ times the volume of the column, or $P_0(d \, \Delta S)\mathbf{a}_z$. Equating the dipole moments computed in the two different ways, we have

$$\rho_{pS0} = P_0 \tag{2.70}$$

Thus, we have related the surface charge density to the magnitude of the polarization vector. Now, the surface charge distribution produces a secondary field $\mathbf{E}_s$ given by

$$\mathbf{E}_s = \begin{cases} -\dfrac{\rho_{pS0}}{\varepsilon_0} \mathbf{a}_z = -\dfrac{P_0}{\varepsilon_0} \mathbf{a}_z & \text{for} \quad 0 < z < d \\ 0 & \text{otherwise} \end{cases} \tag{2.71}$$

Denoting the total field in the dielectric to be $\mathbf{E}_t$, we have

$$\mathbf{E}_t = \mathbf{E}_a + \mathbf{E}_s = \frac{\rho_{S0}}{\varepsilon_0} \mathbf{a}_z - \frac{P_0}{\varepsilon_0} \mathbf{a}_z \tag{2.72}$$

But from (2.66),

$$\mathbf{P} = \varepsilon_0 \chi_{e0} \mathbf{E}_t \tag{2.73}$$

Substituting (2.69) and (2.72) into (2.73), we obtain

$$P_0 = \chi_{e0}(\rho_{S0} - P_0)$$

or

$$P_0 = \frac{\chi_{e0} \rho_{S0}}{1 + \chi_{e0}} \tag{2.74}$$

Thus, the polarization surface charge densities are given by

$$\rho_{pS} = \begin{cases} \dfrac{\chi_{e0} \rho_{S0}}{1 + \chi_{e0}} & \text{for} \quad z - d \\ -\dfrac{\chi_{e0} \rho_{S0}}{1 + \chi_{e0}} & \text{for} \quad z = 0 \end{cases} \tag{2.75}$$

and the electric field intensity in the dielectric is

$$\mathbf{E}_t = \frac{\rho_{S0}}{\varepsilon_0(1 + \chi_{e0})} \mathbf{a}_z \tag{2.76}$$

as shown in Fig. 2.31(e).

We have just learned that the electric field in a dielectric material is the superposition of an applied field $\mathbf{E}_a$ and a secondary field $\mathbf{E}_s$, which results from the polarization $\mathbf{P}$, which in turn is induced by the total field $(\mathbf{E}_a + \mathbf{E}_s)$. Thus, we have

Displacement flux density, permittivity

$$\mathbf{P} = \varepsilon_0 \chi_e (\mathbf{E}_a + \mathbf{E}_s) \tag{2.77a}$$

$$\mathbf{E}_s = f(\mathbf{P}) \tag{2.77b}$$

where $f(\mathbf{P})$ denotes a function of $\mathbf{P}$. Determination of the secondary field $\mathbf{E}_s$ and hence the total field $(\mathbf{E}_a + \mathbf{E}_s)$ for a given applied field $\mathbf{E}_a$ requires a simultaneous solution of (2.77a) and (2.77b). To eliminate the need for the explicit determination of $\mathbf{P}$, we now define a new vector field $\mathbf{D}$, known as the displacement flux density, or the electric flux density, as

$$\mathbf{D} = \varepsilon_0 \mathbf{E} + \mathbf{P} \tag{2.78}$$

Note that the units of $\mathbf{D}$ are the same as those of $\mathbf{P}$, that is, coulombs per square meter, having the character of flux density for the electric field case, as $\mathbf{B}$ does in the magnetic field case.

Substituting for $\mathbf{P}$ in (2.78) by using (2.66), we obtain

$$\begin{aligned} \mathbf{D} &= \varepsilon_0 \mathbf{E} + \varepsilon_0 \chi_e \mathbf{E} \\ &= \varepsilon_0 (1 + \chi_e) \mathbf{E} \\ &= \varepsilon_0 \varepsilon_r \mathbf{E} \end{aligned} \tag{2.79}$$

or

$$\mathbf{D} = \varepsilon \mathbf{E} \tag{2.80}$$

where we define

$$\varepsilon_r = 1 + \chi_e \tag{2.81}$$

and

$$\varepsilon = \varepsilon_0 \varepsilon_r \tag{2.82}$$

The quantity ε_r is known as the *relative permittivity* or *dielectric constant* of the dielectric, and ε is the *permittivity* of the dielectric. The permittivity ε takes into account the effects of polarization, and there is no need to consider them when we use ε for ε_0! The relative permittivity is an experimentally measurable parameter, as we shall discuss in Section 3.3. Its values for several dielectric materials are listed in Table 2.3.

TABLE 2.3   Relative Permittivities of Some Materials

Material	Relative permittivity	Material	Relative permittivity
Air	1.0006	Dry earth	5
Paper	2.0–3.0	Mica	6
Teflon	2.1	Neoprene	6.7
Polystyrene	2.56	Wet earth	10
Plexiglass	2.6–3.5	Ethyl alcohol	24.3
Nylon	3.5	Glycerol	42.5
Fused quartz	3.8	Distilled water	81
Bakelite	4.9	Titanium dioxide	100

Returning now to Example 2.9, we observe that in the absence of the dielectric between the sheets of charge,

$$\mathbf{E} = \mathbf{E}_a = \frac{\rho_{S0}}{\varepsilon_0}\,\mathbf{a}_z \tag{2.83a}$$

$$\mathbf{D} = \varepsilon_0\mathbf{E}_a = \rho_{S0}\mathbf{a}_z \tag{2.83b}$$

since $\mathbf{P}$ is equal to zero. In the presence of the dielectric between the sheets of charge,

$$\mathbf{E} = \mathbf{E}_t = \frac{\rho_{S0}}{\varepsilon_0(1 + \chi_{e0})}\,\mathbf{a}_z = \frac{\rho_{S0}}{\varepsilon}\,\mathbf{a}_z \tag{2.84a}$$

$$\mathbf{D} = \varepsilon\mathbf{E} = \rho_{S0}\mathbf{a}_z \tag{2.84b}$$

Thus, the $\mathbf{D}$ fields are the same in both cases, independent of the permittivity of the medium, whereas the expressions for the $\mathbf{E}$ fields differ in the permittivities, that is, with ε_0 replaced by ε. The situation in general is however not so simple because the dielectric alters the original field distribution. In the case of Example 2.9, the geometry is such that the original field distribution is not altered by the dielectric. Also in the general case of a nonuniform susceptibility, the situation is equivalent to having a polarization volume charge inside the dielectric in addition to polarization surface charges on its boundaries. Furthermore, for time-varying fields, the electric dipoles oscillate with time, creating the equivalent of a polarization current in the dielectric. However, all these are implicitly taken into account by the permittivity ε of the dielectric.

The nature of (2.54), which is characteristic of conductors, and of (2.80), which is characteristic of dielectrics, implies that $\mathbf{J}_c$ in the case of conductors and $\mathbf{D}$ in the case of dielectrics are in the same direction as that of $\mathbf{E}$. Such materials are said to be *isotropic* materials. For *anisotropic* materials, this is not necessarily the case. To explain, we shall consider *anisotropic dielectric materials*. Then $\mathbf{D}$ is not in general in the same direction as that of $\mathbf{E}$. This arises because the induced polarization is such that the polarization vector $\mathbf{P}$ is not necessarily in the same direction as that of $\mathbf{E}$. In fact, the angle between the directions of the applied $\mathbf{E}$ and the resulting $\mathbf{P}$ depends on the direction of $\mathbf{E}$. The relationship between $\mathbf{D}$ and $\mathbf{E}$ is then expressed in the form of a matrix equation as

Anisotropic dielectric materials

$$\begin{bmatrix} D_x \\ D_y \\ D_z \end{bmatrix} = \begin{bmatrix} \varepsilon_{xx} & \varepsilon_{xy} & \varepsilon_{xz} \\ \varepsilon_{yx} & \varepsilon_{yy} & \varepsilon_{yz} \\ \varepsilon_{zx} & \varepsilon_{zy} & \varepsilon_{zz} \end{bmatrix} \begin{bmatrix} E_x \\ E_y \\ E_z \end{bmatrix} \tag{2.85}$$

Thus, each component of $\mathbf{D}$ is in general dependent on each component of $\mathbf{E}$. The square matrix in (2.85) is known as the *permittivity tensor* of the anisotropic dielectric.

Although $\mathbf{D}$ is not in general parallel to $\mathbf{E}$ for anisotropic dielectrics, there are certain polarizations of $\mathbf{E}$ for which $\mathbf{D}$ is parallel to $\mathbf{E}$. These are said to correspond to the characteristic polarizations, where the word *polarization* here refers to the direction of the field, not to the creation of electric dipoles. We shall consider an example to investigate the characteristic polarizations.

Example 2.10

An anisotropic dielectric material is characterized by the permittivity tensor

$$[\varepsilon] = \varepsilon_0 \begin{bmatrix} 7 & 2 & 0 \\ 2 & 4 & 0 \\ 0 & 0 & 3 \end{bmatrix}$$

Let us find $\mathbf{D}$ for several cases of $\mathbf{E}$.

Substituting the given permittivity matrix into (2.85), we obtain

$$D_x = 7\varepsilon_0 E_x + 2\varepsilon_0 E_y$$
$$D_y = 2\varepsilon_0 E_x + 4\varepsilon_0 E_y$$
$$D_z = 3\varepsilon_0 E_z$$

For $\mathbf{E} = E_0 \mathbf{a}_z$, $\mathbf{D} = 3\varepsilon_0 E_0 \mathbf{a}_z = 3\varepsilon_0 \mathbf{E}$; $\mathbf{D}$ is parallel to $\mathbf{E}$.

For $\mathbf{E} = E_0 \mathbf{a}_x$, $\mathbf{D} = 7\varepsilon_0 E_0 \mathbf{a}_x + 2\varepsilon_0 E_0 \mathbf{a}_y$; $\mathbf{D}$ is not parallel to $\mathbf{E}$.

For $\mathbf{E} = E_0 \mathbf{a}_y$, $\mathbf{D} = 2\varepsilon_0 E_0 \mathbf{a}_x + 4\varepsilon_0 E_0 \mathbf{a}_y$; $\mathbf{D}$ is not parallel to $\mathbf{E}$.

For $\mathbf{E} = E_0 (\mathbf{a}_x + 2\mathbf{a}_y)$, $\mathbf{D} = 11\varepsilon_0 E_0 \mathbf{a}_x + 10\varepsilon_0 E_0 \mathbf{a}_y$; $\mathbf{D}$ is not parallel to $\mathbf{E}$.

For $\mathbf{E} = E_0 (2\mathbf{a}_x + \mathbf{a}_y)$, $\mathbf{D} = 16\varepsilon_0 E_0 \mathbf{a}_x + 8\varepsilon_0 E_0 \mathbf{a}_y = 8\varepsilon_0 E_0 (2\mathbf{a}_x + \mathbf{a}_y) = 8\varepsilon_0 \mathbf{E}$; $\mathbf{D}$ is parallel to $\mathbf{E}$.

When $\mathbf{D}$ is parallel to $\mathbf{E}$, that is, for the characteristic polarizations of $\mathbf{E}$, one can define an *effective permittivity* as the ratio of $\mathbf{D}$ to $\mathbf{E}$. Thus, for the case of $\mathbf{E} = E_0 \mathbf{a}_z$, the effective permittivity is $3\varepsilon_0$, and for the case of $\mathbf{E} = E_0(2\mathbf{a}_x + \mathbf{a}_y)$, the effective permittivity is $8\varepsilon_0$. For the characteristic polarizations, the anisotropic material behaves effectively as an isotropic dielectric having the permittivity equal to the corresponding effective permittivity.

K2.5.   Polarization; Electric dipole; Polarization vector; Displacement flux density; Permittivity; Relative permittivity; Anisotropic dielectric; Characteristic polarizations; Effective permittivity.

D2.13.   Infinite plane sheets of uniform charge densities $1 \ \mu C/m^2$ and $-1 \ \mu C/m^2$ occupy the planes $z = 0$ and $z = d$, respectively. The region $0 < z < d$ is a dielectric of permittivity $4\varepsilon_0$. Find the values of **(a)** $\mathbf{D}$, **(b)** $\mathbf{E}$, and **(c)** $\mathbf{P}$ in the region $0 < z < d$.
Ans.   **(a)** $10^{-6} \mathbf{a}_z \ C/m^2$;   **(b)** $9000\pi \mathbf{a}_z \ V/m$;   **(c)** $0.75 \times 10^{-6} \mathbf{a}_z \ C/m^2$.

D2.14.   For an anisotropic dielectric material characterized by the $\mathbf{D}$ to $\mathbf{E}$ relationship

$$\begin{bmatrix} D_x \\ D_y \\ D_z \end{bmatrix} = \varepsilon_0 \begin{bmatrix} 8 & 2 & 0 \\ 2 & 5 & 0 \\ 0 & 0 & 9 \end{bmatrix} \begin{bmatrix} E_x \\ E_y \\ E_z \end{bmatrix}$$

find the value of the effective relative permittivity for each of the following electric field intensities corresponding to the characteristic polarizations:
(a) $\mathbf{E} = E_0 \mathbf{a}_z$; **(b)** $\mathbf{E} = E_0(\mathbf{a}_x - 2\mathbf{a}_y)$; and **(c)** $\mathbf{E} = E_0(2\mathbf{a}_x + \mathbf{a}_y)$.
Ans.   **(a)** 9;   **(b)** 4;   **(c)** 9.

2.6 MAGNETIC MATERIALS

In the preceding two sections, we have been concerned with the response of materials to electric fields. We now turn our attention to materials known as magnetic materials, which, as the name implies, are classified according to their magnetic behavior. According to a simplified atomic model, the electrons associated with a particular nucleus orbit around the nucleus in circular paths while spinning about themselves. In addition, the nucleus itself has a spin motion associated with it. Since the movement of charge constitutes a current, these orbital and spin motions are equivalent to current loops of atomic dimensions. A current loop is the magnetic analog of the electric dipole. Thus, each atom can be characterized by a superposition of magnetic dipole moments corresponding to the electron orbital motions, electron spin motions, and the nuclear spin. However, owing to the heavy mass of the nucleus, the angular velocity of the nuclear spin is much smaller than that of an electron spin and hence the equivalent current associated with the nuclear spin is much smaller than the equivalent current associated with an electron spin. The dipole moment due to the nuclear spin can therefore be neglected in comparison with the other two effects. The schematic representations of a magnetic dipole as seen from along its axis and from a point in its plane are shown in Figs. 2.32(a) and (b), respectively. The strength of the dipole is defined by the magnetic dipole moment **m** given by

Magnetization, magnetic dipole

$$\boxed{\mathbf{m} = I A \mathbf{a}_n} \tag{2.86}$$

where A is the area enclosed by the current loop, and $\mathbf{a}_n$ is the unit vector normal to the plane of the loop and directed in the right-hand sense.

In many materials, the net magnetic moment of each atom is zero; that is, on the average, the magnetic dipole moments corresponding to the various electronic orbital and spin motions add up to zero. An external magnetic field has the effect of inducing a net dipole moment by changing the angular velocities of the electronic orbits, thereby magnetizing the material. This kind of magnetization, known as *diamagnetism,* is in fact prevalent in all materials. In certain materials known as *paramagnetic materials,* the individual atoms possess net nonzero magnetic moments even in the absence of an external magnetic field. These *permanent* magnetic moments of the individual atoms are, however, randomly oriented so that the net magnetization on a macroscopic scale is zero. An applied magnetic field has the effect of exerting torques on the individual

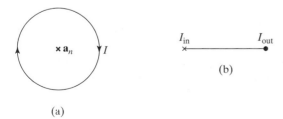

(a)

(b)

FIGURE 2.32 Schematic representation of a magnetic dipole as seen from (a) along its axis and (b) a point in its plane.

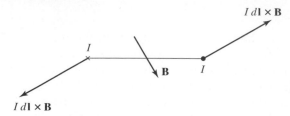

FIGURE 2.33   Torque acting on a magnetic dipole in an external magnetic field.

permanent dipoles, as shown in Fig. 2.33, to convert, on a macroscopic scale, the initially random alignment into a partially coherent one along the magnetic field, that is, with the normal to the current loop directed along the magnetic field. This kind of magnetization is known as *paramagnetism*. Certain materials known as *ferromagnetic, antiferromagnetic,* and *ferrimagnetic* materials exhibit permanent magnetization, that is, magnetization even in the absence of an applied magnetic field.

On a macroscopic scale, we define a vector **M**, called the *magnetization vector,* as the *magnetic dipole moment per unit volume.* Thus, if N denotes the number of molecules per unit volume of the material, then there are $N\,\Delta v$ molecules in a volume Δv and

$$\mathbf{M} = \frac{1}{\mathrm{D}v} \sum_{j=1}^{N\,\mathrm{D}v} \mathbf{m}_j = N\mathbf{m} \tag{2.87}$$

where **m** is the average dipole moment per molecule. The units of **M** are ampere-meter^2/meter^3 or amperes per meter. It is found that for many magnetic materials, the magnetization vector is related to the magnetic field **B** in the material in the simple manner given by

$$\mathbf{M} = \frac{\chi_m}{1 + \chi_m} \frac{\mathbf{B}}{\mu_0} \tag{2.88}$$

where χ_m, a dimensionless parameter, is known as the *magnetic susceptibility.* The quantity χ_m is a measure of the ability of the material to become magnetized and differs from one magnetic material to another.

Magnetic material in a magnetic field   When a magnetic material is placed in a magnetic field, the induced dipoles produce a secondary magnetic field such that the resultant field, that is, the sum of the originally applied field and the secondary field, and the magnetization vector satisfy (2.88). We shall illustrate this by means of an example.

Example 2.11

Let us consider an infinite plane magnetic material slab of thickness d sandwiched between two infinite plane sheets of equal and opposite uniform current densities $J_{S0}\mathbf{a}_y$ and $-J_{S0}\mathbf{a}_y$ in the $z = 0$ and $z = d$ planes, respectively, as shown in Fig. 2.34(a). We wish to investigate the effect of magnetization in the magnetic material.

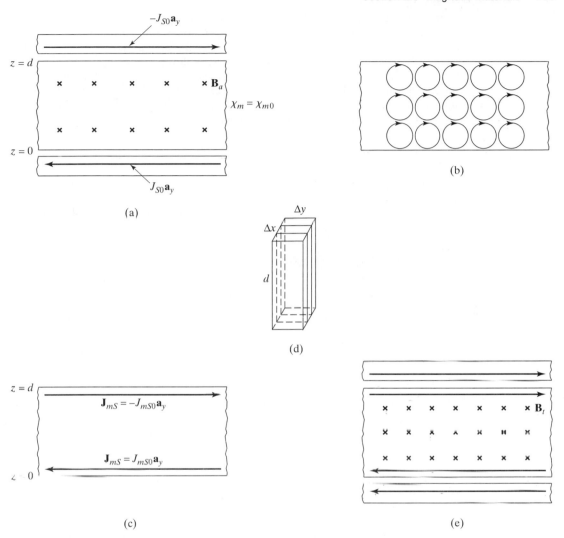

FIGURE 2.34   For investigating the effect of magnetization induced in a magnetic material sandwiched between two infinite plane sheets of current.

In the absence of the magnetic material, the magnetic field between the sheets of current is given by

$$\mathbf{B}_a = \mu_0 J_{S0} \mathbf{a}_y \times \mathbf{a}_z$$
$$= \mu_0 J_{S0} \mathbf{a}_x \qquad (2.89)$$

In the presence of the magnetic material, this field acts as the applied magnetic field resulting in magnetic dipole moments in the material that are oriented along the field. Since the magnetic field and the magnetic susceptibility are uniform, the density of the dipole moments, that is, the magnetization vector **M**, is uniform as shown in Fig. 2.34(b). Such a distribution results in exact cancelation of currents everywhere except at the boundaries of the material since, for each current segment not on the surface, there is a current segment associated with the dipole adjacent to it and carrying the same amount of current in the opposite direction, thereby canceling its effect. Thus, the net result is the formation of a negative y-directed surface current at the boundary $z = d$ and a positive y-directed surface current at the boundary $z = 0$, as shown in Fig. 2.34(c). These surface currents are known as magnetization surface currents since they are due to the magnetization in the material. In view of the uniform density of the dipole moments, the surface current densities are uniform. Also, in the absence of a net current in the interior of the magnetic material, the surface current densities must be equal in magnitude so that whatever current flows on one surface returns via the other surface.

Let us therefore denote the surface current densities as

$$\mathbf{J}_{mS} = \begin{cases} J_{mS0}\mathbf{a}_y & \text{for} \quad z = 0 \\ -J_{mS0}\mathbf{a}_y & \text{for} \quad z = d \end{cases} \tag{2.90}$$

where the subscript m in addition to the other subscripts stands for magnetization. If we now consider a vertical column of infinitesimal rectangular cross-sectional area $\Delta S = (\Delta x)(\Delta y)$ cut out from the magnetic material, as shown in Fig. 2.34(d), the rectangular current loop of width Δx makes the column appear as a dipole of moment $(J_{mS0}\,\Delta x)$ $(d\,\Delta y)\mathbf{a}_x$. On the other hand, writing

$$\mathbf{M} = M_0\mathbf{a}_x \tag{2.91}$$

where M_0 is a constant in view of the uniformity of the magnetization, the dipole moment of the column is equal to $\mathbf{M}$ times the volume of the column, or $M_0(d\,\Delta x\,\Delta y)\mathbf{a}_x$. Equating the dipole moments computed in the two different ways, we have

$$J_{mS0} = M_0 \tag{2.92}$$

Thus, we have related the surface current density to the magnitude of the magnetization vector. Now, the surface current distribution produces a secondary field $\mathbf{B}_s$ given by

$$\mathbf{B}_s = \begin{cases} \mu_0 J_{mS0}\mathbf{a}_x = \mu_0 M_0\mathbf{a}_x & \text{for} \quad 0 < z < d \\ \mathbf{0} & \text{otherwise} \end{cases} \tag{2.93}$$

Denoting the total field inside the magnetic material to be $\mathbf{B}_t$, we have

$$\begin{aligned} \mathbf{B}_t = \mathbf{B}_a + \mathbf{B}_s &= \mu_0 J_{S0}\mathbf{a}_x + \mu_0 M_0\mathbf{a}_x \\ &= \mu_0(J_{S0} + M_0)\mathbf{a}_x \end{aligned} \tag{2.94}$$

But, from (2.88),

$$\mathbf{M} = \frac{\chi_{m0}}{1 + \chi_{m0}}\frac{\mathbf{B}_t}{\mu_0} \tag{2.95}$$

Substituting (2.91) and (2.94) into (2.95), we have

$$M_0 = \frac{\chi_{m0}}{1 + \chi_{m0}}(J_{S0} + M_0)$$

or

$$M_0 = \chi_{m0}J_{S0} \tag{2.96}$$

Thus, the magnetization surface current densities are given by

$$\mathbf{J}_{mS} = \begin{cases} \chi_{m0}J_{S0}\mathbf{a}_y & \text{for} \quad z = 0 \\ -\chi_{m0}J_{S0}\mathbf{a}_y & \text{for} \quad z = d \end{cases} \tag{2.97}$$

and the magnetic flux density in the magnetic material is

$$\mathbf{B}_t = \mu_0(1 + \chi_{m0})J_{S0}\mathbf{a}_x \tag{2.98}$$

as shown in Fig. 2.34(e).

We have just learned that the magnetic field in a magnetic material is the superposition of an applied field $\mathbf{B}_a$ and a secondary field $\mathbf{B}_s$ that results from the magnetization $\mathbf{M}$, which in turn is induced by the total field $(\mathbf{B}_a + \mathbf{B}_s)$. Thus, we have

Magnetic field intensity, permeability

$$\mathbf{M} = \frac{\chi_m}{1 + \chi_m} \frac{\mathbf{B}_a + \mathbf{B}_s}{\mu_0} \tag{2.99a}$$

$$\mathbf{B}_s = f(\mathbf{M}) \tag{2.99b}$$

where $f(\mathbf{M})$ denotes a function of $\mathbf{M}$. Determination of the secondary field $\mathbf{B}_s$ and hence the total field $(\mathbf{B}_a + \mathbf{B}_s)$ for a given applied field $\mathbf{B}_a$ requires a simultaneous solution of (2.99a) and (2.99b). To eliminate the need for the explicit determination of $\mathbf{M}$, we now define a new vector field $\mathbf{H}$, known as the *magnetic field intensity*, as

$$\boxed{\mathbf{H} = \frac{\mathbf{B}}{\mu_0} - \mathbf{M}} \tag{2.100}$$

Note that the units of $\mathbf{H}$ are the same as those of $\mathbf{M}$, that is, amperes per meter, having the character of field intensity for the magnetic field case, as $\mathbf{E}$ does for the electric field case.

Substituting for $\mathbf{M}$ by using (2.88), we obtain

$$\mathbf{H} = \frac{\mathbf{B}}{\mu_0} - \frac{\chi_m}{1 + \chi_m} \frac{\mathbf{B}}{\mu_0}$$

$$= \frac{\mathbf{B}}{\mu_0(1 + \chi_m)} \tag{2.101}$$

$$= \frac{\mathbf{B}}{\mu_0 \mu_r}$$

or

$$\boxed{\mathbf{H} = \frac{\mathbf{B}}{\mu}} \tag{2.102}$$

where we define

$$\boxed{\mu_r = 1 + \chi_m} \tag{2.103}$$

and

$$\boxed{\mu = \mu_0 \mu_r} \tag{2.104}$$

The quantity μ_r is known as the *relative permeability* of the magnetic material and μ is the *permeability* of the magnetic material. The permeability μ takes into account the effects of magnetization, and there is no need to consider them when we use μ for μ_0!

Returning now to Example 2.11, we observe that in the absence of the magnetic material between the sheets of current,

$$\mathbf{B} = \mathbf{B}_a = \mu_0 J_{S0} \mathbf{a}_x \tag{2.105a}$$

$$\mathbf{H} = \frac{\mathbf{B}}{\mu_0} = J_{S0} \mathbf{a}_x \tag{2.105b}$$

since $\mathbf{M}$ is equal to zero. In the presence of the magnetic material between the sheets of current,

$$\mathbf{B} = \mathbf{B}_t = \mu_0(1 + \chi_m)J_{S0}\mathbf{a}_x = \mu J_{S0}\mathbf{a}_x \tag{2.106a}$$

$$\mathbf{H} = \frac{\mathbf{B}}{\mu} = J_{S0}\mathbf{a}_x \tag{2.106b}$$

Thus, the $\mathbf{H}$ fields are the same in both cases independent of the permeability of the medium, whereas the expressions for the $\mathbf{B}$ fields differ in the permeabilities, that is, with μ_0 replaced by μ. The situation in general is however not so simple because the magnetic material alters the original field distribution. In the case of Example 2.11, the geometry is such that the original field distribution is not altered by the magnetic material. Also in the general case of a nonuniform susceptibility, the situation is equivalent to having a magnetization volume current inside the material in addition to the surface current at the boundaries. However, all of these are implicitly taken into account by the permeability μ of the magnetic material. For anisotropic magnetic materials, $\mathbf{H}$ is not in general parallel to $\mathbf{B}$ and the relationship between the two quantities is expressed in the form of a matrix equation, as given by

$$\begin{bmatrix} B_x \\ B_y \\ B_z \end{bmatrix} = \begin{bmatrix} \mu_{xx} & \mu_{xy} & \mu_{xz} \\ \mu_{yx} & \mu_{yy} & \mu_{yz} \\ \mu_{zx} & \mu_{zy} & \mu_{zz} \end{bmatrix} \begin{bmatrix} H_x \\ H_y \\ H_z \end{bmatrix} \tag{2.107}$$

just as in the case of the relationship between $\mathbf{D}$ and $\mathbf{E}$ for anistropic dielectric materials.

Ferromagnetic materials

For many materials for which the relationship between $\mathbf{H}$ and $\mathbf{B}$ is linear, the relative permeability does not differ appreciably from unity, unlike the case of linear dielectric materials, for which the relative permittivity can be very large, as shown in Table 2.3. In fact, for diamagnetic materials, the magnetic susceptibility χ_m is a small negative number of the order -10^{-4} to -10^{-8}, whereas for paramagnetic materials, χ_m is a small positive number of the order 10^{-3} to 10^{-7}. Ferromagnetic materials, however, possess large values of relative permeability on the order of several hundreds, thousands, or more. The relationship between $\mathbf{B}$ and $\mathbf{H}$ for these materials is nonlinear, resulting in a nonunique value of μ_r, for a given material. In fact, these materials are characterized by hysteresis, that is, the relationship between $\mathbf{B}$ and $\mathbf{H}$ dependent on the past history of the material.

Ferromagnetic materials possess strong dipole moments owing to the predominance of the electron spin moments over the electron orbital moments. The theory of ferromagnetism is based on the concept of magnetic *domains*, as formulated by Weiss in 1907. A magnetic domain is a small region in the material in which the atomic dipole moments are all aligned in one direction, due to strong interaction fields arising from the neighboring dipoles. In the absence of an external magnetic field, although each domain is magnetized to saturation, the magnetizations in various domains are randomly oriented, as shown in Fig. 2.35(a) for a single crystal specimen. The random orientation results from minimization of the associated energy. The net magnetization is therefore zero on a macroscopic scale. With the application of a weak external magnetic field, the volumes of the domains in which the original magnetizations are favorably oriented relative to the applied field grow at the expense of the volumes of the other domains, as shown in Fig. 2.35(b). This feature is known as domain wall motion. Upon removal of the applied field, the domain wall motion reverses, bringing the material close to its original state of magnetization. With the application of stronger external fields, the domain wall motion continues to such an extent that it becomes irreversible; that is, the material does not return to its original unmagnetized state on a macroscopic scale upon removal of the field. With the application of still stronger fields, the domain wall motion is accompanied by domain rotation, that is, alignment of the magnetizations in the individual domains with the applied field, as shown in Fig. 2.35(c), thereby magnetizing the material to saturation. The material retains some magnetization along the direction of the applied field even after removal of the field. In fact, an external field opposite to the original direction has to be applied to bring the net magnetization back to zero.

We may now discuss the relationship between **B** and **H** for a ferromagnetic material, which is depicted graphically as shown by a typical curve in Fig. 2.36. This curve is known as the hysteresis curve, or the *B–H* curve. To trace the development of the hysteresis effect, we start with an unmagnetized sample of ferromagnetic material in which both **B** and **H** are initially zero, corresponding to point *a* on the curve. As *H* is increased, the magnetization builds up, thereby increasing *B* gradually along the curve *ab* and finally to saturation at *b*, according

Hysteresis curve

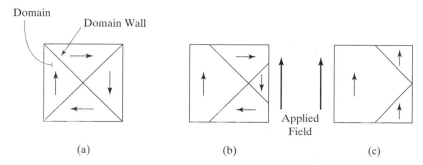

(a) (b) (c)

FIGURE 2.35 For illustrating the different steps in the magnetization of a ferromagnetic specimen: (a) unmagnetized state; (b) domain wall motion; and (c) domain rotation.

FIGURE 2.36   Hysteresis curve for a ferromagnetic material.

to the following sequence of events as discussed earlier: (1) reversible motion of domain walls, (2) irreversible motion of domain walls, and (3) domain rotation. The regions corresponding to these events along the curve *ab* as well as other portions of the hysteresis curve are shown marked 1, 2, and 3, respectively, in Fig. 2.36. If the value of *H* is now decreased to zero, the value of *B* does not retrace the curve *ab* backward, but instead follows the curve *bc*, which indicates that a certain amount of magnetization remains in the material even after the magnetizing field is completely removed. In fact, it requires a magnetic field intensity in the opposite direction to bring *B* back to zero, as shown by the portion *cd* of the curve. The value of *B* at the point *c* is known as the *remanence,* or *retentivity,* whereas the value of *H* at *d* is known as the *coercivity* of the material. Further increase in **H** in this direction results in the saturation of **B** in the direction opposite to that corresponding to *b*, as shown by the portion *de* of the curve. If **H** is now decreased to zero, reversed in direction, and increased, the resulting variation of **B** occurs in accordance with the curve *efgb*, thereby completing the hysteresis loop.

The nature of the hysteresis curve suggests that the hysteresis phenomenon can be used to distinguish between two states, for example, "1" and "0" in a binary number magnetic memory system. There are several kinds of magnetic memories. Although differing in details, all these are based on the principles of storing and retrieving information in regions on a magnetic medium. In disk, drum, and tape memories, the magnetic medium moves, whereas in bubble and core memories, the medium is stationary. We shall briefly discuss here only the floppy disk, or diskette, commonly used as secondary memory in personal computers.[2]

Floppy disk   The floppy disk consists of a coating of ferrite material applied over a thin flexible nonmagnetic substrate for physical support. Ferrites are a class of magnetic materials characterized by almost rectangular-shaped hysteresis loops so

[2]See, for example, Robert M. White, "Disk-Storage Technology," *Scientific American,* August 1980, pp. 138–148.

that the two remanent states are well-defined. The disk is divided into many circular tracks, and each track is subdivided into regions called sectors, as shown in Fig. 2.37. To access a sector, an electromagnetic read/write head moves across the spinning disk to the appropriate track and waits for the correct sector to rotate beneath it. The head consists of a ferrite core around which a coil is wound and with a gap at the bottom, as shown in Fig. 2.38. Writing data on the disk is done by passing current through the coil. The current generates a magnetic field that in the core confines essentially to the material, but in the air gap spreads out into the magnetic medium below it, thereby magnetizing the region to represent the 0 state. To store the 1 state in a region, the current in the coil is reversed to magnetize the medium in the reverse direction. Reading of data from the disk is accomplished by the changing magnetic field from the magnetized regions on the disk inducing a voltage in the coil of the head as the disk rotates under the head. The voltage is induced in accordance with Faraday's law (which we shall study in Section 3.2) whenever there is a change in magnetic flux linked by the coil. We have here only discussed the basic principles behind storing data on the

FIGURE 2.37   Arrangement of sectors on a floppy disk.

FIGURE 2.38   Writing of data on a floppy disk.

disk and retrieving data from it. There are a number of ways in which bits can be encoded on the disk. We shall however not pursue the topic here.

K2.6. Magnetization; Magnetic dipole; Magnetization vector; Magnetic field intensity; Permeability; Relative permeability; Ferromagnetic materials; Hysteresis.

D2.15. Find the magnetic dipole moment for each of the following cases: **(a)** 1 μC of charge in a circular orbit of radius $1/\sqrt{\pi}$ mm in the xy-plane around the z-axis in the sense of increasing ϕ with angular velocity of 1 revolution per millisecond; **(b)** a square current loop having the vertices at the points $A(10^{-3}, 0, 0)$, $B(0, 10^{-3}, 0)$, $C(-10^{-3}, 0, 0)$, and $D(0, -10^{-3}, 0)$ with current 0.1 A flowing in the sense $ABCDA$; and **(c)** an equilateral triangular current loop having vertices at the points $A(10^{-3}, 0, 0)$, $B(0, 10^{-3}, 0)$, and $C(0, 0, 10^{-3})$ with current 0.1 A flowing in the sense $ABCA$.
Ans. **(a)** $10^{-9}\mathbf{a}_z$ A-m^2; **(b)** $2 \times 10^{-7}\mathbf{a}_z$ A-m^2; **(c)** $5 \times 10^{-8}(\mathbf{a}_x + \mathbf{a}_y + \mathbf{a}_z)$ A-m^2.

D2.16. Infinite plane sheets of current densities $0.1\mathbf{a}_y$ A/m and $-0.1\mathbf{a}_y$ A/m occupy the planes $z = 0$ and $z = d$, respectively. The region $0 < z < d$ is a magnetic material of permeability $100\mu_0$. Find **(a)** H, **(b)** B, and **(c)** M in the region $0 < z < d$.
Ans. **(a)** $0.1\mathbf{a}_x$ A/m; **(b)** $4\pi \times 10^{-6}\mathbf{a}_x$ Wb/m^2; **(c)** $9.9\mathbf{a}_x$ A/m.

2.7 SUMMARY

In this chapter, we first introduced the electric field concept from consideration of an experimental law known as Coulomb's law, having to do with the electric forces between two charges. We learned that electric force acts on charges merely by virtue of the property of charge. The electric force acting on a test charge q at a point in the field region is given by

$$\mathbf{F} = q\mathbf{E}$$

where **E** is the electric field intensity at that point. The electric field intensity due to a point charge Q in free space is given by

$$\mathbf{E} = \frac{Q}{4\pi\varepsilon_0 R^2}\mathbf{a}_R$$

where ε_0 is the permittivity of free space, R is the distance from the point charge to the point at which the field intensity is to be computed, and $\mathbf{a}_R$ is the unit vector along the line joining the two points and directed away from the point charge. Using superposition in conjunction with the electric field due to a point charge, we discussed the computation of the electric field due to two point charges and the computer generation of the direction lines of the electric field. We then extended the determination of electric field intensity to continuous charge distributions.

Next we introduced the magnetic field concept from considerations of Ampère's law of force, having to do with the magnetic forces between two current loops. We learned that the magnetic field exerts force only on moving charges. The magnetic force acting on a test charge q moving with a velocity **v** at a point in the field region is given by

$$\mathbf{F} = q\mathbf{v} \times \mathbf{B}$$

where $\mathbf{B}$ is the magnetic flux density at that point. In terms of current flowing in a wire, the magnetic force acting on a current element of length $d\mathbf{l}$ and current I at a point in the field region is given by

$$\mathbf{F} = I\,d\mathbf{l} \times \mathbf{B}$$

The magnetic flux density due to a current element $I\,d\mathbf{l}$ in free space is given by the Biot–Savart law

$$\mathbf{B} = \frac{\mu_0}{4\pi}\frac{I\,d\mathbf{l} \times \mathbf{a}_R}{R^2}$$

where μ_0 is the permeability of free space, and R and $\mathbf{a}_R$ have the same meanings as in the expression for $\mathbf{E}$ due to a point charge. Using superposition in conjunction with the Biot–Savart law, we discussed the computation of the magnetic field due to current distributions.

Combining the electric and magnetic field concepts, we then introduced the Lorentz force equation

$$\mathbf{F} = q(\mathbf{E} + \mathbf{v} \times \mathbf{B})$$

which gives the force acting on a test charge q moving with velocity $\mathbf{v}$ at a point in a region characterized by electric field of intensity $\mathbf{E}$ and magnetic field of flux density $\mathbf{B}$. We used the Lorentz force equation to discuss (1) the determination of $\mathbf{E}$ and $\mathbf{B}$ at a point from a knowledge of forces acting on a test charge at that point for three different velocities and (2) the tracing of charged particle motion in a region of crossed electric and magnetic fields.

We devoted the rest of the chapter to introduce materials. We learned that materials can be classified as (1) conductors, (2) semiconductors, (3) dielectrics, and (4) magnetic materials, depending on the nature of the response of the charged particles in the materials to applied fields. Conductors are characterized by conduction, which is the phenomenon of steady drift of free electrons under the influence of an applied electric field, thereby resulting in a conduction current. In semiconductors, also characterized by conduction, the charge carriers are not only electrons, but also holes. We learned that the conduction current density is related to the electric field intensity in the manner

$$\mathbf{J}_c = \sigma\mathbf{E} \tag{2.108}$$

where σ is the conductivity of the material. We discussed (1) the formation of surface charge at the boundaries of a conductor placed in a static electric field, (2) the derivation of Ohm's law in circuit theory, and (3) the Hall effect.

Dielectrics are characterized by polarization, which is the phenomenon of the creation and net alignment of electric dipoles, formed by the displacement of the centroids of the electron clouds from the centroids of the nucleii of the atoms, along the direction of an applied electric field. Magnetic materials are characterized by magnetization, which is the phenomenon of net alignment of the axes of the magnetic dipoles, formed by the electron orbital and spin motion around the nucleii of the atoms, along the direction of an applied magnetic field. To eliminate the need for explicitly taking into account the effects of polarization and magnetization, we defined two new vector fields known as the displacement flux density and the magnetic field intensity, given by

$$\mathbf{D} = \varepsilon_0 \mathbf{E} + \mathbf{P}$$

$$\mathbf{H} = \frac{\mathbf{B}}{\mu_0} - \mathbf{M}$$

respectively, where $\mathbf{P}$ is the polarization vector, and $\mathbf{M}$ is the magnetization vector. We learned that for isotropic materials, these expressions simplify to

$$\mathbf{D} = \varepsilon \mathbf{E} \qquad (2.109)$$

$$\mathbf{H} = \frac{\mathbf{B}}{\mu} \qquad (2.110)$$

where

$$\varepsilon = \varepsilon_0 \varepsilon_r$$

$$\mu = \mu_0 \mu_r$$

are the permittivity and the permeability, respectively, of the material and the quantities ε_r and μ_r are the relative permittivity and the relative permeability, respectively, which take into account implicitly the effects of polarization and magnetization, respectively. Equations (2.108), (2.109), and (2.110) are known as the constitutive relations. Finally, we discussed the hysteresis phenomenon associated with ferromagnetic materials and discussed an application based on the use of the hysteresis curve.

REVIEW QUESTIONS

Q2.1.   State Coulomb's law. To what law in mechanics is Coulomb's law analogous?

Q2.2.   What is the value of the permittivity of free space? What are its units?

Q2.3.   What is the definition of electric field intensity? What are its units?

Q2.4.   Discuss two applications based on the electric force on a charged particle.

Q2.5.   Describe the electric field due to a point charge.

Q2.6.   Discuss the computer generation of the direction lines of the electric field due to two point charges.

Q2.7.   Discuss the different types of charge distributions. How do you determine the electric field intensity due to a charge distribution?

Q2.8.   Describe the electric field due to an infinite plane sheet of uniform surface charge density.

Q2.9.   State Ampère's force law as applied to current elements. Why is it not necessary for Newton's third law to hold for current elements?

Q2.10.   What are the units of magnetic flux density? How is magnetic flux density defined in terms of force on a current element?

Q2.11.   What is the value of the permeability of free space? What are its units?

Q2.12.   Describe the magnetic field due to a current element.

Q2.13. Discuss the different types of current distributions. How do you determine the magnetic flux density due to a current distribution?

Q2.14. Describe the magnetic field due to an infinite plane sheet of uniform surface current density.

Q2.15. How is magnetic flux density defined in terms of force on a moving charge?

Q2.16. Discuss two applications based on the magnetic force on a current-carrying wire or on a moving charge.

Q2.17. State the Lorentz force equation.

Q2.18. Discuss the determination of **E** and **B** at a point from the knowledge of forces experienced by a test charge at that point for several velocities. What is the minimum required number of forces?

Q2.19. Give some examples of devices based on charged particle motion in electric and magnetic fields.

Q2.20. Discuss the tracing of the path of a charged particle in a region of crossed electric and magnetic fields.

Q2.21. Distinguish between bound electrons and free electrons in an atom and briefly describe the phenomenon of conduction.

Q2.22. Discuss the classification of a material as a conductor, semiconductor, or dielectric with the aid of energy band diagrams.

Q2.23. What is mobility? Give typical values of mobilities for electrons and holes.

Q2.24. State Ohm's law valid at a point, defining the conductivities for conductors and semiconductors.

Q2.25. Discuss the formation of surface charge at the boundaries of a conductor placed in a static electric field.

Q2.26. Discuss the derivation of Ohm's law in circuit theory from the Ohm's law valid at a point.

Q2.27. Discuss the Hall effect.

Q2.28. Briefly describe the phenomenon of polarization in a dielectric material. What are the different kinds of polarization?

Q2.29. What is an electric dipole? How is its strength defined?

Q2.30. What is a polarization vector? How is it related to the electric field intensity?

Q2.31. Discuss the effect of polarization in a dielectric material using the example of polarization surface charge.

Q2.32. Discuss the definition of displacement flux density and the permittivity concept.

Q2.33. What is an anisotropic dielectric material? When can an effective permittivity be defined for an anisotropic dielectric material?

Q2.34. Briefly describe the phenomenon of magnetization in a magnetic material. What are the different kinds of magnetic materials?

Q2.35. What is a magnetic dipole? How is its strength defined?

Q2.36. What is the magnetization vector? How is it related to the magnetic flux density?

Q2.37. Discuss the effect of magnetization in a magnetic material using the example of magnetization surface current.

Q2.38. Discuss the definition of magnetic field intensity and the permeability concept.

Q2.39. Discuss the phenomenon of hysteresis associated with ferromagnetic materials.

Q2.40. Discuss the principles behind storing data on a floppy disk and retrieving the data from it.

PROBLEMS

Section 2.1

P2.1. Point charges, each of value Q, are situated at the corners of a regular tetrahedron of edge length L. Find the electric force on each point charge.

⁎ **P2.2.** Six point charges, each of value Q, are situated at $(d, 0, 0)$, $(-d, 0, 0)$, $(0, d, 0)$, $(0, -d, 0)$, $(0, 0, d)$, and $(0, 0, -d)$. A test charge q located at the origin is displaced by a distance $\Delta \ll d$ along the positive x-axis. Find an approximate expression for the electric force acting on the charge.

P2.3. **Design Problem.** For each of the following pairs of electric field intensities, find if possible the location and the value of a point charge that produces both fields: **(a)** $\mathbf{E}_1 = (2\mathbf{a}_x + 2\mathbf{a}_y + \mathbf{a}_z)$ V/m at $(2, 2, 3)$ and $\mathbf{E}_2 = (\mathbf{a}_x + 2\mathbf{a}_y + 2\mathbf{a}_z)$ V/m at $(-1, 0, 3)$; and **(b)** $\mathbf{E}_1 = (2\mathbf{a}_x + 2\mathbf{a}_y + \mathbf{a}_z)$ V/m at $(1, 1, 1)$ and $\mathbf{E}_2 = (2\mathbf{a}_x + \mathbf{a}_y + 2\mathbf{a}_z)$ V/m at $(1, 2, 0)$.

⁎ **P2.4.** Two equal and opposite point charges Q and $-Q$ are located at $(0, 0, d/2)$ and $(0, 0, -d/2)$, respectively. Such an arrangement is known as the electric dipole. Show that the electric field intensity due to the electric dipole at very large distances from the origin compared to the spacing d is given approximately by $(Qd/4\pi\varepsilon_0 r^3)(2 \cos \theta \, \mathbf{a}_r + \sin \theta \, \mathbf{a}_\theta)$.

P2.5. A point charge q of mass m is in static equilibrium at the origin, in the presence of a circular ring charge Q in the xy-plane and two point charges, each of value Q, on the z-axis. The ring charge is uniformly distributed, with radius a and center at the origin. The two point charges are located at $(0, 0, -a)$ and $(0, 0, a)$. All charges are of the same sign. The point charge q is constrained to move along the z-axis. It is given a slight displacement $z_0 \ll a$ and released at $t = 0$. Obtain the approximate differential equation for the motion of the charge and find the frequency of oscillation.

P2.6. **Design Problem.** Design an arrangement of a circular ring charge of uniform density and total charge Q equal to 1 μC that produces electric field intensities of $10^3 \mathbf{a}_z$ V/m at the two points $(0, 0, 1)$ and $(0, 0, 2)$. If Q is not equal to 1 μC, determine, if any, the restriction on its value for a solution to exist.

⁎ **P2.7.** Assuming that the circular ring of Example 2.2 is coated with charge such that the charge density is given by $\rho_L = \rho_{L0} \cos \phi$ C/m, find the electric field intensity at a point on the z-axis by setting up the integral expression and evaluating it.

⁎ **P2.8.** Consider a circular disc of radius a lying in the xy-plane with its center at the origin and carrying charge of density $4\pi\varepsilon_0/r$ C/m². Obtain the expression for the electric field intensity at the point $(0, 0, z)$ by setting up the integral and evaluating it.

P2.9. Consider the charge distributed with density $4\pi\varepsilon_0|z|$ C/m along the line between $(0, 0, -a)$ and $(0, 0, a)$. Obtain the expression for the electric field intensity at $(r, \phi, 0)$ in cylindrical coordinates, by considering a differential length element along the line charge, setting up the field as an integral and evaluating it.

P2.10. Consider the volume charge distributed uniformly with density ρ_0 C/m³ between the planes $z = -a$ and $z = a$. Using superposition in conjunction with the result of Example 2.3, show that the electric field intensity due to the slab of charge is given by

$$\mathbf{E} = \begin{cases} -(\rho_0 a/\varepsilon_0)\mathbf{a}_z & \text{for} \quad z < -a \\ (\rho_0 z/\varepsilon_0)\mathbf{a}_z & \text{for} \quad -a < z < a \\ (\rho_0 a/\varepsilon_0)\mathbf{a}_z & \text{for} \quad z > a \end{cases}$$

Section 2.2

P2.11. Three identical current elements $I\,dz\,\mathbf{a}_z$ A are located at equally spaced points on a circle of radius 1 m centered at the origin and lying on the xy-plane. The first point is $(1, 0, 0)$. Find the magnetic force on each current element.

P2.12. For the current element $I\,dx\,(\mathbf{a}_x + \mathbf{a}_y)$ A situated at the point $(1, -2, 2)$, find the magnetic flux densities at three points: **(a)** $(2, -1, 3)$, **(b)** $(2, -3, 4)$, and **(c)** $(3, 0, 2)$.

P2.13. **Design Problem.** For each of the following pairs of magnetic flux densities, find, if possible, the orientation of an infinitely long filamentary wire and the current in it required to produce both fields: **(a)** $\mathbf{B}_1 = 10^{-7}\mathbf{a}_y$ Wb/m² at $(3, 0, 0)$ and $\mathbf{B}_2 = -10^{-7}\mathbf{a}_x$ Wb/m² at $(0, 4, 0)$; and **(b)** $\mathbf{B}_1 = 10^{-7}(\mathbf{a}_y - \mathbf{a}_z)$ Wb/m² at $(\sqrt{2}, 0, 0)$ and $\mathbf{B}_2 = -10^{-7}\mathbf{a}_x$ Wb/m² at $(0, \sqrt{2}, 0)$.

P2.14. Two long identical rigid filamentary wires, each of length l and weight w, are suspended horizontally from the ceiling by long weightless threads, each of length L. The wires are arranged to be parallel and separated by a distance d, small compared to l and L. A current I is passed through both wires via flexible connections so as to cause the wires to be attracted to each other. **(a)** Should the currents be in the same sense or in opposite senses for attraction to occur? **(b)** If the current is gradually increased from zero, the wires will gradually approach each other. A condition may be reached when any further increase of current will cause the wires to swing and touch each other. Determine the critical value of I at which this happens.

P2.15. A circular loop of wire of radius a is situated in the xy-plane with its center at the origin. It carries a current I in the clockwise sense as seen along the positive z-axis, that is, in the sense of increasing values of ϕ. Obtain the expression for $\mathbf{B}$ due to the current loop at a point on the z-axis.

P2.16. A straight wire along the z-axis carries current I in the positive z-direction. Consider the portion of the wire between $(0, 0, a_1)$ and $(0, 0, a_2)$, where $a_2 > a_1$. Show that the magnetic flux density at an arbitrary point $P(r, \phi, z)$ due to this portion of the wire is given by

$$\mathbf{B} = \frac{\mu_0 I}{4\pi r}(\cos\alpha_1 - \cos\alpha_2)\mathbf{a}_\phi$$

where α_1 and α_2 are the angles subtended by the lines from P to $(0, 0, a_1)$ and $(0, 0, a_2)$, respectively, with the z-axis. Verify your result in the limit $a_1 \to -\infty$ and $a_2 \to \infty$.

P2.17. A square loop of wire lies in the xy-plane with its corners at $(1, 1, 0), (-1, 1, 0), (-1, -1, 0),$ and $(1, -1, 0)$. A current of 1 A flows in the loop in the sense defined by connecting the specified points in succession. Applying the result of Problem P2.16 to each side of the loop, find the magnetic flux densities at two points: **(a)** $(0, 0, 0)$ and **(b)** $(2, 0, 0)$.

P2.18. **Design Problem.** Design an arrangement of a pair of infinitely long, straight, filamentary wires parallel to the z-direction and in a plane parallel to the xz-plane, each carrying current I equal to 1 A but in opposite directions, which produce magnetic flux densities of $10^{-7}\mathbf{a}_y$ Wb/m² and $0.5 \times 10^{-7}\mathbf{a}_y$ Wb/m² at the points $(0, 1, 0)$ and $(0, 2, 0)$, respectively. If I is not equal to 1 A, determine, if any, the restriction on its value for a solution to exist.

P2.19. Three infinite plane current sheets, each of a uniform current density, exist in the coordinate planes of a Cartesian coordinate system. The magnetic flux densities due to these current sheets are given at three points as follows: at $(1, 2, 3)$, $\mathbf{B} = 3B_0\mathbf{a}_x$; at $(7, -5, 6)$, $\mathbf{B} = B_0(-\mathbf{a}_x + 2\mathbf{a}_z)$; at $(8, 9, -4)$, $\mathbf{B} = B_0(\mathbf{a}_x + 2\mathbf{a}_y)$.

Find the magnetic flux densities at the following points: **(a)** $(-6, -2, -3)$; **(b)** $(-4, -5, 7)$; and **(c)** $(6, -3, -5)$.

P2.20. Consider current distribution with uniform density $J_0 \mathbf{a}_z$ A/m^2 in the volume between the planes $y = -a$ and $y = 0$, and with uniform density $-J_0 \mathbf{a}_z$ A/m^2 in the volume between the planes $y = 0$ and $y = a$. Using superposition in conjunction with the result of Example 2.6, show that the magnetic flux density due to the current distribution is given by

$$\mathbf{B} = \begin{cases} \mu_0 J_0 (|y| - a)\, \mathbf{a}_x & \text{for } |y| \le a \\ 0 & \text{otherwise} \end{cases}$$

P2.21. Show that the ratio of the radii of orbits of two charged particles of the same charge but different masses entering a region of uniform magnetic field perpendicular to the field and with equal kinetic energies is equal to the ratio of the square roots of their masses.

Section 2.3

P2.22. Show that in a region of uniform, crossed electric and magnetic fields $\mathbf{E}$ and $\mathbf{B}$, respectively, a test charge released at a point in the field region with initial velocity $\mathbf{v} = (\mathbf{E} \times \mathbf{B})/B^2$ moves with constant velocity equal to the initial value. Compute $\mathbf{v}$ for $\mathbf{E}$ and $\mathbf{B}$ equal to $E_0(2\mathbf{a}_y + 2\mathbf{a}_y + \mathbf{a}_z)$ and $B_0(\mathbf{a}_x - 2\mathbf{a}_y + 2\mathbf{a}_z)$, respectively.

P2.23. Let $\mathbf{F}_1, \mathbf{F}_2$, and $\mathbf{F}_3$ be the forces experienced by a test charge q at a point in a region of electric and magnetic fields $\mathbf{E}$ and $\mathbf{B}$, respectively, for velocities $\mathbf{v}_1, \mathbf{v}_2$, and $\mathbf{v}_3$, respectively, of the charge. If $\mathbf{F}_1, \mathbf{F}_2$, and $\mathbf{F}_3$ are such that $\mathbf{A} = \mathbf{F}_1 \times \mathbf{F}_2 + \mathbf{F}_2 \times \mathbf{F}_3 + \mathbf{F}_3 \times \mathbf{F}_1 \ne 0$, that is, their tips do not lie on a straight line when drawn from a common point, show that

$$\mathbf{B} = \frac{1}{q}\left[\frac{\mathbf{F}_2 - \mathbf{F}_1}{(\mathbf{v}_2 - \mathbf{v}_1) \times \mathbf{A}} \right] \mathbf{A}$$

P2.24. The forces experienced by a test charge q at a point in a region of electric and magnetic fields $\mathbf{E}$ and $\mathbf{B}$, respectively, are given as follows for three different velocities of the test charge, where v_0 and E_0 are constants.

$$\begin{aligned}
\mathbf{F}_1 &= qE_0(\mathbf{a}_x - \mathbf{a}_y + \mathbf{a}_z) && \text{for } \mathbf{v}_1 = v_0\mathbf{a}_x \\
\mathbf{F}_2 &= qE_0(\mathbf{a}_x - \mathbf{a}_y - \mathbf{a}_z) && \text{for } \mathbf{v}_2 = v_0\mathbf{a}_y \\
\mathbf{F}_3 &= 0 && \text{for } \mathbf{v}_3 = v_0\mathbf{a}_z
\end{aligned}$$

Find $\mathbf{E}$ and $\mathbf{B}$ at that point.

P2.25. Three forces $\mathbf{F}_1, \mathbf{F}_2$, and $\mathbf{F}_3$ experienced by a test charge q at a point in a region of electric and magnetic fields for three different velocities of the test charge are given as follows:

$$\begin{aligned}
\mathbf{F}_1 &= 0 && \text{for } \mathbf{v}_1 = v_0\mathbf{a}_x \\
\mathbf{F}_2 &= 0 && \text{for } \mathbf{v}_2 = v_0\mathbf{a}_y \\
\mathbf{F}_3 &= qE_0\mathbf{a}_z && \text{for } \mathbf{v}_3 = v_0(\mathbf{a}_x + 2\mathbf{a}_y)
\end{aligned}$$

Find the forces $\mathbf{F}_4, \mathbf{F}_5$, and $\mathbf{F}_6$ experienced by the test charge at that point for three other velocities: **(a)** $\mathbf{v}_4 = 0$, **(b)** $\mathbf{v}_5 = v_0(\mathbf{a}_x + \mathbf{a}_y)$, and **(c)** $\mathbf{v}_6 = (v_0/4)(3\mathbf{a}_x + \mathbf{a}_y)$.

P2.26. Uniform electric and magnetic fields exist in a region of space. A test charge q released with an initial velocity $\mathbf{v}_1$ or $\mathbf{v}_2$ moves with constant velocity equal to the initial value. Show that the test charge moves with constant velocity equal to the initial value when released with an initial velocity $(m\mathbf{v}_1 + n\mathbf{v}_2)/(m + n)$ for any nonzero $(m + n)$.

Section 2.4

P2.27. Consider two electrons moving under thermal agitation with velocities equal in magnitude and opposite in direction. A uniform electric field is applied along the direction of motion of one of the electrons. Show that the gain in kinetic energy by the accelerating electron is greater than the loss in kinetic energy by the decelerating electron.

P2.28. **(a)** For a sinusoidally time-varying electric field $\mathbf{E} = \mathbf{E}_0 \cos \omega t$, where $\mathbf{E}_0$ is a constant, show that the steady-state solution to (2.43) is given by

$$\mathbf{v}_d = \frac{\tau e}{m\sqrt{1 + \omega^2\tau^2}} \mathbf{E}_0 \cos (\omega t - \tan^{-1} \omega\tau)$$

(b) Based on the assumption of one free electron per atom, the free electron density N_e in silver is 5.86×10^{28} m^{-3}. Using the conductivity for silver given in Table 2.2, find the frequency at which the drift velocity lags the applied field by $\pi/4$. What is the ratio of the mobility at this frequency to the mobility at zero frequency?

P2.29. **(a)** An infinite plane conducting slab carries uniformly distributed surface charges on both of its surfaces. If the net surface charge density, that is, the sum of the surface charge densities on the two surfaces, is ρ_{S0} C/m^2, find the surface charge densities on the two surfaces. **(b)** Two infinite plane parallel conducting slabs 1 and 2 carry uniformly distributed surface charges on all four of their surfaces. If the net surface charge densities are ρ_{S1} and ρ_{S2} C/m^2, respectively, for the slabs 1 and 2, find the surface charge densities on all four surfaces.

P2.30. The region $x < 0$ is occupied by a conductor. An infinitely long line charge of uniform density ρ_{L0} is situated along the line passing through $(d, 0, 0)$ and parallel to the z-axis, where $d > 0$. From the secondary field required to make the total electric field inside the conductor equal to zero and from symmetry considerations, as shown by the cross-sectional view in Fig. 2.39, show that the field

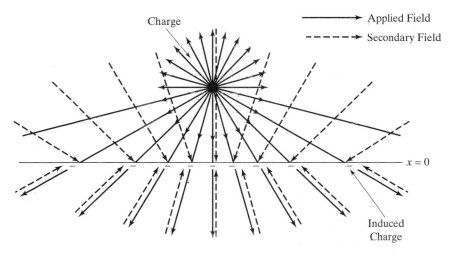

FIGURE 2.39   For Problem P2.30.

outside the conductor is the same as the field due to the line charge passing through $(d, 0, 0)$ and a parallel "image" line charge of uniform density $-\rho_{L0}$ along the line passing through $(-d, 0, 0)$. Find the expression for the electric field outside the conductor. *Hint:* Use the expression for the electric field intensity due to an infinitely long line charge of uniform density ρ_{L0} along the z-axis given by $(\rho_{L0}/2\pi\varepsilon_0 r)\mathbf{a}_r$.

Section 2.5

P2.31. Show that the torque acting on an electric dipole of moment **p** due to an applied electric field **E** is $\mathbf{p} \times \mathbf{E}$. Compute the torque for a dipole consisting of 1 μC of charge at $(0, 0, 10^{-3})$ and -1 μC of charge at $(0, 0, -10^{-3})$ in an electric field $\mathbf{E} = 10^3 (2\mathbf{a}_x - \mathbf{a}_y + 2\mathbf{a}_z)$ V/m.

P2.32. A point charge Q is situated at the origin surrounded by a spherical dielectric shell of uniform permittivity $4\varepsilon_0$ and having inner and outer radii a and b, respectively. Find the following: **(a)** the **D** and **E** fields in the three regions $0 < r < a$, $a < r < b$, and $r > b$ and **(b)** the polarization vector inside the dielectric shell.

P2.33. An anisotropic dielectric material is characterized by the **D** to **E** relationship

$$\begin{bmatrix} D_x \\ D_y \\ D_z \end{bmatrix} = \varepsilon_0 \begin{bmatrix} 3 & 1 & 1 \\ 1 & 3 & 1 \\ 1 & 1 & 3 \end{bmatrix} \begin{bmatrix} E_x \\ E_y \\ E_z \end{bmatrix}$$

(a) Find **D** for $\mathbf{E} = E_0(\mathbf{a}_x + \mathbf{a}_y)$. **(b)** Find **D** for $\mathbf{E} = E_0(\mathbf{a}_x - \mathbf{a}_y)$. **(c)** Find **E** for $\mathbf{D} = D_0(\mathbf{a}_x + \mathbf{a}_y - 2\mathbf{a}_z)$. Comment on your result for each case.

P2.34. An anisotropic dielectric material is characterized by the **D** to **E** relationship

$$\begin{bmatrix} D_x \\ D_y \\ D_z \end{bmatrix} = \begin{bmatrix} \varepsilon_{xx} & \varepsilon_{xy} & 0 \\ \varepsilon_{yx} & \varepsilon_{yy} & 0 \\ 0 & 0 & \varepsilon_{zz} \end{bmatrix} \begin{bmatrix} E_x \\ E_y \\ E_z \end{bmatrix}$$

For $\mathbf{E} = E_x \mathbf{a}_x + E_y \mathbf{a}_y$, find the value(s) of E_y/E_x for which **D** is parallel to **E**. Find the effective permittivity for each case.

Section 2.6

P2.35. Charge Q is distributed with uniform density on a circular disk of radius a lying in the xy-plane and rotating around the z-axis with angular velocity ω in the sense of increasing ϕ. Find the magnetic dipole moment of the rotating charge.

P2.36. Considering for simplicity a rectangular current loop in the xy-plane, show that the torque acting on a magnetic dipole of moment **m** due to an applied magnetic field **B** is $\mathbf{m} \times \mathbf{B}$. Then find the torque acting on a circular current loop of radius 1 mm, in the xy-plane, centered at the origin and with current 0.1 A flowing in the sense of increasing ϕ in a magnetic field $\mathbf{B} = 10^{-5}(2\mathbf{a}_x - 2\mathbf{a}_y + \mathbf{a}_z)$ Wb/m².

P2.37. A portion of the B–H curve for a ferromagnetic material can be approximated by the analytical expression

$$\mathbf{B} = \mu_0 kHH$$

where k is a constant having units of meter per ampere. Find μ, μ_r, χ_m, and **M**.

P2.38. An anisotropic magnetic material is characterized by the **B** to **H** relationship

$$\begin{bmatrix} B_x \\ B_y \\ B_z \end{bmatrix} = k\mu_0 \begin{bmatrix} 7 & 6 & 0 \\ 6 & 12 & 0 \\ 0 & 0 & 3 \end{bmatrix} \begin{bmatrix} H_x \\ H_y \\ H_z \end{bmatrix}$$

where k is a constant. Find the effective permeability for $\mathbf{H} = H_0(3\mathbf{a}_x - 2\mathbf{a}_y)$.

REVIEW PROBLEMS

R2.1. Consider the arrangement of three point charges Q_1, $kQ_1(k > 0)$, and Q_2, as shown in Fig. 2.40, where Q_1 and kQ_1 are fixed and Q_2 is constrained to move on the semicircle. **(a)** Find the value of α in terms of k for which Q_2 is in equilibrium. **(b)** Find the numerical value of α for $k = 8$.

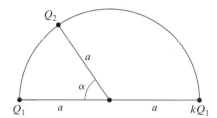

Q_1 a a kQ_1 **FIGURE 2.40**   For Problem R2.1.

R2.2. Consider a line charge distribution along the z-axis between $z = -a$ and $z = a$. **(a)** Show that if the charge density is an even function $f_1(z)$, the electric field intensity at a point $(r, \phi, 0)$ has only an r-component and set up the integral expression for it. **(b)** Show that if the charge density is an odd function $f_2(z)$, the electric field intensity at $(r, \phi, 0)$ has only a z-component and set up the integral expression for it. **(c)** Given that the charge density is

$$f(z) = \begin{cases} 4\pi\varepsilon_0(a + 2z) & \text{for } -a \le z \le 0 \\ 4\pi\varepsilon_0 a & \text{for } 0 \le z \le a \end{cases}$$

express $f(z)$ as the sum of even and odd functions $f_1(z)$ and $f_2(z)$, and evaluate the electric field components.

R2.3. Current I flows along a wire which is straight from ∞ to a on the x-axis, circular from $(a, 0, 0)$ to $(0, a, 0)$ and lying on the xy-plane in the sense of increasing ϕ, and then from a to ∞ on the y-axis. Find **B** at $(0, 0, a)$.

R2.4. Current is distributed with density $J_0(y/a)\mathbf{a}_z$ A/m^2 in the volume between the planes $y = -a$ and $y = a$. Show that the magnetic flux density due to the current distribution is given by

$$\mathbf{B} = \begin{cases} \dfrac{\mu_0 J_0}{2a}(a^2 - y^2)\mathbf{a}_x & \text{for } -a < y < a \\ 0 & \text{otherwise} \end{cases}$$

R2.5. Consider a test charge moving with constant velocity $\mathbf{v} = v_0(2\mathbf{a}_x - 2\mathbf{a}_y + \mathbf{a}_z)$ in a region of a uniform electric field of intensity $\mathbf{E} = E_0(2\mathbf{a}_x + \mathbf{a}_y - 2\mathbf{a}_z)$ and a uniform magnetic field of flux density $\mathbf{B} = B_{x0}\mathbf{a}_x + B_{y0}\mathbf{a}_y + B_{z0}\mathbf{a}_z$. Is this information sufficient to find uniquely B_{x0}, B_{y0}, and B_{z0}? If not, given that **v** is perpendicular to **B**, find B_{x0}, B_{y0}, and B_{z0} in terms of E_0 and v_0.

R2.6. Two infinite plane conducting slabs lie between and parallel to two infinite plane sheets of uniform surface charge densities ρ_{SA} and ρ_{SB}, as shown by the cross-sectional view in Fig. 2.41. Find the surface charge densities on all four surfaces of the slabs.

FIGURE 2.41   For Problem R2.6.

R2.7. An anisotropic dielectric material is characterized by the **D** to **E** relationship

$$\begin{bmatrix} D_x \\ D_y \\ D_z \end{bmatrix} = \varepsilon_0 \begin{bmatrix} 6.5 & 1.5 & 0 \\ 1.5 & 2.5 & 0 \\ 0 & 0 & 2 \end{bmatrix} \begin{bmatrix} E_x \\ E_y \\ E_z \end{bmatrix}$$

Express $\mathbf{E} = E_0(\mathbf{a}_x - \mathbf{a}_y)$ as the linear combination of $\mathbf{E}_1$ and $\mathbf{E}_2$, which correspond to two of the characteristic polarizations of the material.

R2.8. Charge Q is distributed with density proportional to r on a circular disk of radius a lying on the xy-plane with its center at the origin and rotating around the z-axis with angular velocity ω in the sense of increasing ϕ. Find the magnetic dipole moment.

B. MAXWELL'S EQUATIONS AND UNIFORM PLANE WAVES

Maxwell's Equations in Integral Form and Boundary Conditions

In Chapter 1, we learned the simple rules of vector algebra and familiarized ourselves with the basic concepts of fields in general. In Chapter 2, we introduced electric and magnetic fields in terms of forces on charged particles and extended the discussion to fields in materials. We now have the necessary background to introduce the additional tools required for the understanding of the various quantities associated with Maxwell's equations and then discuss Maxwell's equations. In particular, our goal in this chapter is to learn Maxwell's equations in integral form as a prerequisite to the derivation of their differential forms in the next chapter. Maxwell's equations in integral form govern the interdependence of certain field and source quantities associated with regions in space, that is, contours, surfaces, and volumes. The differential forms of Maxwell's equations, however, relate the characteristics of the field vectors at a given point to one another and to the source densities at that point.

Maxwell's equations in integral form are a set of four laws resulting from several experimental findings and a purely mathematical contribution. We shall, however, consider them as postulates and learn to understand their physical significance as well as their mathematical formulation. The source quantities involved in their formulation are charges and currents. The field quantities have to do with the line and surface integrals of the electric and magnetic field vectors. We shall therefore first introduce line and surface integrals and then consider successively the four Maxwell's equations in integral form.

3.1   LINE AND SURFACE INTEGRALS

Line integral   To introduce the line integral, let us consider in a region of electric field $\mathbf{E}$ the movement of a test charge q from the point A to the point B along the path C as shown in Fig. 3.1(a). At each and every point along the path the electric field exerts a force on the test charge and, hence, does a certain amount of work in moving the charge to another point an infinitesimal distance away. To find the total amount of work done from A to B, we divide the path into a number of infinitesimal segments $\Delta\mathbf{l}_1, \Delta\mathbf{l}_2, \Delta\mathbf{l}_3, \ldots \Delta\mathbf{l}_n$, as shown in Fig. 3.1(b), find the infinitesimal amount of work done for each segment, and then add up the contributions from all the segments. Since the segments are infinitesimal in length, we can consider each of them to be straight and the electric field at all points within a segment to be the same and equal to its value at the start of the segment.

If we now consider one segment, say, the jth segment, and take the component of the electric field for that segment along the length of that segment, we obtain the result $E_j \cos \alpha_j$, where α_j is the angle between the direction of the electric field vector $\mathbf{E}_j$ at the start of that segment and the direction of that segment. Since the electric field intensity has the meaning of force per unit charge, the electric force along the direction of the jth segment is then equal to $qE_j \cos \alpha_j$. To obtain the work done in carrying the test charge along the length of the jth segment, we then multiply this electric force component by the length Δl_j of that segment. Thus, for the jth segment, we obtain the result for the work done by the electric field as

$$\Delta W_j = qE_j \cos \alpha_j \, \Delta l_j \tag{3.1}$$

If we do this for all the infinitesimal segments and add up all the contributions, we get the total work done by the electric field in moving the test charge from A to B along the path C to be

$$
\begin{aligned}
W_{AB} &= \Delta W_1 + \Delta W_2 + \Delta W_3 + \cdots + \Delta W_n \\
&= qE_1 \cos \alpha_1 \, \Delta l_1 + qE_2 \cos \alpha_2 \, \Delta l_2 + qE_3 \cos \alpha_3 \, \Delta l_3 + \cdots \\
&\quad + qE_n \cos \alpha_n \, \Delta l_n \\
&= q \sum_{j=1}^{n} E_j \cos \alpha_j \, \Delta l_j \\
&= q \sum_{j=1}^{n} (E_j)(\Delta l_j) \cos \alpha_j
\end{aligned}
\tag{3.2}
$$

Using the dot product operation between two vectors, we obtain

$$\boxed{W_{AB} = q \sum_{j=1}^{n} \mathbf{E}_j \cdot \Delta\mathbf{l}_j} \tag{3.3}$$

For a numerical example, let us consider the electric field given by

$$\mathbf{E} = y\mathbf{a}_y$$

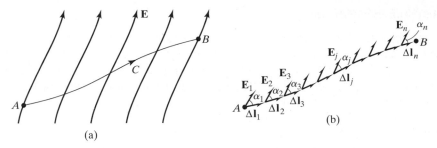

FIGURE 3.1 For evaluating the total amount of work done in moving a test charge along a path C from point A to point B in a region of electric field.

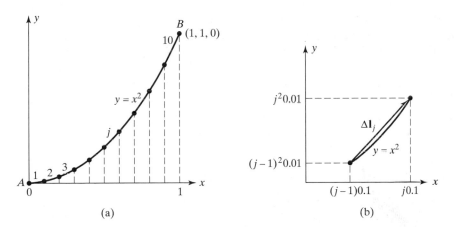

FIGURE 3.2 (a) Division of the path $y = x^2$ from $A(0, 0, 0)$ to $B(1, 1, 0)$ into 10 segments. (b) Length vector corresponding to the jth segment of part (a) approximated as a straight line.

and determine the work done by the field in the movement of 3 μC of charge from the point $A(0, 0, 0)$ to the point $B(1, 1, 0)$ along the parabolic path $y = x^2$, $z = 0$ shown in Fig. 3.2(a).

For convenience, we shall divide the path into 10 segments having equal projections along the x-axis, as shown in Fig. 3.2(a). We shall number the segments 1, 2, 3, ..., 10. The coordinates of the starting and ending points of the jth segment are as shown in Fig. 3.2(b). The electric field at the start of the jth segment is given by

$$\mathbf{E}_j = (j - 1)^2 0.01 \mathbf{a}_y$$

The length vector corresponding to the jth segment, approximated as a straight line connecting its starting and ending points, is

$$\Delta \mathbf{l}_j = 0.1\mathbf{a}_x + [j^2 - (j - 1)^2]0.01\mathbf{a}_y$$
$$= 0.1\mathbf{a}_x + (2j - 1)0.01\mathbf{a}_y$$

The required work is then given by

$$W_{AB} = 3 \times 10^{-6} \sum_{j=1}^{10} \mathbf{E}_j \cdot \Delta \mathbf{l}_j$$

$$= 3 \times 10^{-6} \sum_{j=1}^{10} [(j-1)^2 0.01 \mathbf{a}_y] \cdot [0.1 \mathbf{a}_x + (2j-1)0.01 \mathbf{a}_y]$$

$$= 3 \times 10^{-10} \sum_{j=1}^{10} (j-1)^2 (2j-1)$$

$$= 3 \times 10^{-10} [0 + 3 + 20 + 63 + 144 + 275 + 468 + 735$$
$$+ 1088 + 1539]$$

$$= 3 \times 10^{-10} \times 4335 \text{ J} = 1.3005 \ \mu\text{J}$$

The result that we have just obtained for W_{AB} is approximate since we divided the path from A to B into a finite number of segments. By dividing it into larger and larger numbers of segments, we can obtain more and more accurate results. In the limit that $n \to \infty$, the result converges to the exact value. The summation in (3.3) then becomes an integral, which represents exactly the work done by the field and is given by

$$\boxed{W_{AB} = q \int_A^B \mathbf{E} \cdot d\mathbf{l}} \tag{3.4}$$

The integral on the right side of (3.4) is known as the *line integral of* $\mathbf{E}$ *from A to B*, along the specified path.

Evaluation of line integral

We shall illustrate the evaluation of the line integral by computing the exact value of the work done by the electric field in the movement of the 3 μC charge for the path in Fig. 3.2(a). To do this, we note that at any arbitrary point on the curve $y = x^2, z = 0$,

$$dy = 2x \, dx \qquad dz = 0$$

so that the differential length vector tangential to the curve is given by

$$d\mathbf{l} = dx \, \mathbf{a}_x + dy \, \mathbf{a}_y + dz \, \mathbf{a}_z$$
$$= dx \, \mathbf{a}_x + 2x \, dx \, \mathbf{a}_y$$

The value of $\mathbf{E} \cdot d\mathbf{l}$ at the point is

$$\mathbf{E} \cdot d\mathbf{l} = y\mathbf{a}_y \cdot (dx \, \mathbf{a}_x + 2x \, dx \, \mathbf{a}_y)$$
$$= x^2 \mathbf{a}_y \cdot (dx \, \mathbf{a}_x + 2x \, dx \, \mathbf{a}_y)$$
$$= 2x^3 \, dx$$

Thus, the required work is given by

$$W_{AB} = q \int_{(0,0,0)}^{(1,1,0)} \mathbf{E} \cdot d\mathbf{l} = 3 \times 10^{-6} \int_0^1 2x^3 \, dx$$

$$= 3 \times 10^{-6} \left[\frac{2x^4}{4} \right]_0^1 = 1.5 \ \mu\text{J}$$

Note that we have evaluated the line integral by using x as the variable of integration. Alternatively, using y as the variable of integration, we obtain

$$\mathbf{E} \cdot d\mathbf{l} = y\mathbf{a}_y \cdot (dx\,\mathbf{a}_x + dy\,\mathbf{a}_y)$$

$$= y\,dy$$

$$W_{AB} = q \int_{(0,0,0)}^{(1,1,0)} \mathbf{E} \cdot d\mathbf{l} = 3 \times 10^{-6} \int_0^1 y\,dy$$

$$= 3 \times 10^{-6} \left[\frac{y^2}{2}\right]_0^1 = 1.5\ \mu\text{J}$$

Thus, the integration can be performed with respect to x or y (or z in the three-dimensional case). What is important, however, is that the integrand must be expressed as a function of the variable of integration and the limits appropriate to that variable must be employed.

Returning now to (3.4) and dividing both sides by q, we note that the line integral of $\mathbf{E}$ from A to B has the physical meaning of work per unit charge done by the field in moving the test charge from A to B. This quantity is known as the *voltage between A and B* along the specified path and is denoted by the symbol V_{AB}, having the units of volts. Thus, *Voltage defined*

$$\boxed{V_{AB} = \int_A^B \mathbf{E} \cdot d\mathbf{l}} \qquad (3.5)$$

When the path under consideration is a closed path, that is, one that has no beginning or ending, such as a rubber band, as shown in Fig. 3.3, the line integral is written with a circle associated with the integral sign in the manner $\oint_C \mathbf{E} \cdot d\mathbf{l}$. The line integral of a vector around a closed path is known as the *circulation* of that vector. In particular, the line integral of $\mathbf{E}$ around a closed path is the work per unit charge done by the field in moving a test charge around the closed path. It is the voltage around the closed path and is also known as the *electromotive force*. We shall now consider an example of evaluating the line integral of a vector around a closed path.

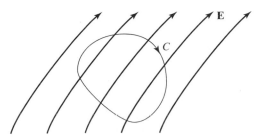

FIGURE 3.3 Closed path C in a region of electric field.

Example 3.1

Let us consider the force field

$$\mathbf{F} = x\mathbf{a}_y$$

and evaluate $\oint_C \mathbf{F} \cdot d\mathbf{l}$, where C is the closed path $ABCDA$ shown in Fig. 3.4.
Noting that

$$\oint_{ABCDA} \mathbf{F} \cdot d\mathbf{l} = \int_A^B \mathbf{F} \cdot d\mathbf{l} + \int_B^C \mathbf{F} \cdot d\mathbf{l} + \int_C^D \mathbf{F} \cdot d\mathbf{l} + \int_D^A \mathbf{F} \cdot d\mathbf{l}$$

we simply evaluate each of the line integrals on the right side and add them up to obtain the required quantity.

First, we observe that since the entire closed path lies in the $z = 0$ plane, $dz = 0$ and $d\mathbf{l} = dx\,\mathbf{a}_x + dy\,\mathbf{a}_y$ for all four straight lines. Then for the side AB,

$$y = 1, \qquad dy = 0, \qquad d\mathbf{l} = dx\,\mathbf{a}_x + (0)\mathbf{a}_y = dx\,\mathbf{a}_x$$

$$\mathbf{F} \cdot d\mathbf{l} = (x\mathbf{a}_y) \cdot (dx\,\mathbf{a}_x) = 0$$

$$\int_A^B \mathbf{F} \cdot d\mathbf{l} = 0$$

For the side BC,

$$x = 3, \qquad dx = 0, \qquad d\mathbf{l} = (0)\mathbf{a}_x + dy\,\mathbf{a}_y = dy\,\mathbf{a}_y$$

$$\mathbf{F} \cdot d\mathbf{l} = (3\mathbf{a}_y) \cdot (dy\,\mathbf{a}_y) = 3\,dy$$

$$\int_B^C \mathbf{F} \cdot d\mathbf{l} = \int_1^5 3\,dy = 12$$

For the side CD,

$$y = 2 + x, \qquad dy = dx, \qquad d\mathbf{l} = dx\,\mathbf{a}_x + dx\,\mathbf{a}_y$$

$$\mathbf{F} \cdot d\mathbf{l} = (x\mathbf{a}_y) \cdot (dx\,\mathbf{a}_x + dx\,\mathbf{a}_y) = x\,dx$$

$$\int_C^D \mathbf{F} \cdot d\mathbf{l} = \int_3^1 x\,dx = -4$$

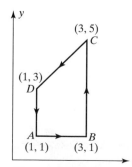

FIGURE 3.4   For evaluating the line integral of a vector field around a closed path.

For the side DA,

$$x = 1, \qquad dx = 0, \qquad d\mathbf{l} = (0)\mathbf{a}_x + dy\,\mathbf{a}_y$$

$$\mathbf{F} \cdot d\mathbf{l} = (\mathbf{a}_y) \cdot (dy\,\mathbf{a}_y) = dy$$

$$\int_D^A \mathbf{F} \cdot d\mathbf{l} = \int_3^1 dy = -2$$

Finally,

$$\oint_{ABCDA} \mathbf{F} \cdot d\mathbf{l} = 0 + 12 - 4 - 2 = 6$$

In this example, we found that the line integral of $\mathbf{F}$ around the closed path *Conservative* C is nonzero. The field is then said to be a *nonconservative field*. For a noncon- *vs. noncon-* servative field, the line integral between two points, say, A and B, is dependent *servative* on the path followed from A to B. To show this, let us consider the two paths *fields* ACB and ADB, as shown in Fig. 3.5. Then we can write

$$\oint_{ACBDA} \mathbf{F} \cdot d\mathbf{l} = \int_{ACB} \mathbf{F} \cdot d\mathbf{l} + \int_{BDA} \mathbf{F} \cdot d\mathbf{l}$$

$$= \int_{ACB} \mathbf{F} \cdot d\mathbf{l} - \int_{ADB} \mathbf{F} \cdot d\mathbf{l}$$

$\qquad(3.6)$

It can be easily seen that if $\oint_{ACBDA} \mathbf{F} \cdot d\mathbf{l}$ is not equal to zero, then $\int_{ACB} \mathbf{F} \cdot d\mathbf{l}$ is not equal to $\int_{ADB} \mathbf{F} \cdot d\mathbf{l}$. The two integrals are equal only if $\oint_{ACBDA} \mathbf{F} \cdot d\mathbf{l}$ is equal to zero, which is the case for *conservative fields*. Examples of conservative fields are Earth's gravitational field and the static electric field. An example of nonconservative fields is the time-varying electric field. Thus, in a time-varying electric field, the voltage between two points A and B is dependent on the path followed to evaluate the line integral of $\mathbf{E}$ from A to B, whereas in a static electric field, the voltage, more commonly known as the *potential difference*, between two points A and B is uniquely defined because the line integral of $\mathbf{E}$ from A to B is independent of the path followed from A to B.

To introduce the surface integral, let us consider a region of magnetic field *Surface* and an infinitesimal surface at a point in that region. Since the surface is infini- *integral* tesimal, we can assume the magnetic flux density to be uniform on the surface,

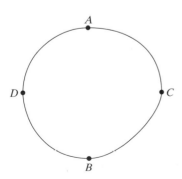

FIGURE 3.5 Two different paths from point A to point B.

FIGURE 3.6 Infinitesimal surface ΔS in a magnetic field **B** oriented (a) normal to the field. (b) parallel to the field, and (c) with its normal making an angle α to the field.

although it may be nonuniform over a wider region. If the surface is oriented normal to the magnetic field lines, as shown in Fig. 3.6(a), then the magnetic flux (webers) crossing the surface is simply given by the product of the surface area (meters squared) and the magnetic flux density (Wb/m^2) on the surface, that is, $B\,\Delta S$. If, however, the surface is oriented parallel to the magnetic field lines, as shown in Fig. 3.6(b), there is no magnetic flux crossing the surface. If the surface is oriented in such a manner that the normal to the surface makes an angle α with the magnetic field lines as shown in Fig. 3.6(c), then the amount of magnetic flux crossing the surface can be determined by considering that the component of **B** normal to the surface is $B \cos \alpha$ and the component tangential to the surface is $B \sin \alpha$. The component of **B** normal to the surface results in a flux of $(B \cos \alpha)\,\Delta S$ crossing the surface, whereas the component tangential to the surface does not contribute at all to the flux crossing the surface. Thus, the magnetic flux crossing the surface in this case is $(B \cos \alpha)\,\Delta S$. We can obtain this result alternatively by noting that the projection of the surface onto the plane normal to the magnetic field lines is $\Delta S \cos \alpha$. Hence, the magnetic flux crossing the surface ΔS is the same as that crossing normal to the area $\Delta S \cos \alpha$, that is, $B(\Delta S \cos \alpha)$ or $(B \cos \alpha)\,\Delta S$.

To aid further in the understanding of this concept, imagine raindrops falling vertically downward uniformly. If you hold a rectangular loop horizontally, the number of drops falling through the loop is simply equal to the area of the loop multiplied by the density (number of drops per unit area) of the drops. If the loop is held vertically, no rain falls through the loop. If the loop is held at some angle to the horizontal, the number of drops falling through the loop is the same as that which would fall through another (smaller) loop, which is the projection of the slanted loop on to the horizontal plane.

Let us now consider a large surface S in the magnetic field region, as shown in Fig. 3.7. The magnetic flux crossing this surface can be found by dividing the surface into a number of infinitesimal surfaces $\Delta S_1, \Delta S_2, \Delta S_3, \ldots, \Delta S_n$ and applying the result just obtained for each infinitesimal surface and adding up the contributions from all the surfaces. To obtain the contribution from the jth surface, we draw the normal vector to that surface and find the angle α_j between the

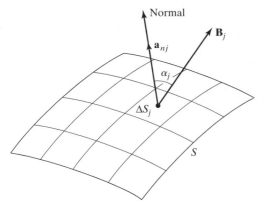

FIGURE 3.7 Division of a large surface S in a magnetic field region into a number of infinitesimal surfaces.

normal vector and the magnetic flux density vector $\mathbf{B}_j$ associated with that surface. Since the surface is infinitesimal, we can assume $\mathbf{B}_j$ to be the value of $\mathbf{B}$ at the centroid of the surface, and we can also erect the normal vector at that point. The contribution to the total magnetic flux from the jth infinitesimal surface is then given by

$$\Delta \psi_j = B_j \cos \alpha_j \, \Delta S_j \tag{3.7}$$

where the symbol ψ represents magnetic flux. The total magnetic flux crossing the surface S is then given by

$$
\begin{aligned}
[\psi]_S &= \Delta\psi_1 + \Delta\psi_2 + \Delta\psi_3 + \cdots + \Delta\psi_n \\
&= B_1 \cos \alpha_1 \, \Delta S_1 + B_2 \cos \alpha_2 \, \Delta S_2 + B_3 \cos \alpha_3 \, \Delta S_3 + \cdots \\
&\quad + B_n \cos \alpha_n \, \Delta S_n \\
&= \sum_{j=1}^{n} B_j \cos \alpha_j \, \Delta S_j \\
&= \sum_{j=1}^{n} B_j(\Delta S_j) \cos \alpha_j
\end{aligned}
\tag{3.8}
$$

Using the dot product operation between two vectors, we obtain

$$[\psi]_S = \sum_{j=1}^{n} \mathbf{B}_j \cdot \Delta S_j \, \mathbf{a}_{nj} \tag{3.9}$$

where $\mathbf{a}_{nj}$ is the unit vector normal to the surface ΔS_j. Furthermore, by using the concept of an infinitesimal surface vector as one having magnitude equal to the area of the surface and direction normal to the surface, that is,

$$\Delta \mathbf{S}_j = \Delta S_j \, \mathbf{a}_{nj} \tag{3.10}$$

we can write (3.9) as

$$\boxed{[\psi]_S = \sum_{j=1}^{n} \mathbf{B}_j \cdot \Delta \mathbf{S}_j} \tag{3.11}$$

For a numerical example, let us consider the magnetic field given by

$$\mathbf{B} = 3xy^2\mathbf{a}_z \text{ Wb/m}^2$$

and determine the magnetic flux crossing the portion of the xy-plane lying between $x = 0, x = 1, y = 0$, and $y = 1$. For convenience, we shall divide the surface into 25 equal areas, as shown in Fig. 3.8 (a). We shall designate the squares as 11, 12, . . . , 15, 21, 22, . . . , 55, where the first digit represents the number of the square in the x-direction and the second digit represents the number of the square in the y-direction. The x- and y-coordinates of the midpoint of the ijth square are $(2i - 1)0.1$ and $(2j - 1)0.1$, respectively, as shown in Fig. 3.8(b). The magnetic field at the center of the ijth square is then given by

$$\mathbf{B}_{ij} = 3(2i - 1)(2j - 1)^2 0.001\mathbf{a}_z$$

Since we have divided the surface into equal areas and since all areas are in the xy-plane,

$$\Delta\mathbf{S}_{ij} = 0.04\mathbf{a}_z \qquad \text{for all } i \text{ and } j$$

The required magnetic flux is then given by

$$[\psi]_S = \sum_{i=1}^{5} \sum_{j=1}^{5} \mathbf{B}_{ij} \cdot \Delta\mathbf{S}_{ij}$$

$$= \sum_{i=1}^{5} \sum_{j=1}^{5} 3(2i - 1)(2j - 1)^2 0.001\mathbf{a}_z \cdot 0.04\mathbf{a}_z$$

$$= 0.00012 \sum_{i=1}^{5} \sum_{j=1}^{5} (2i - 1)(2j - 1)^2$$

$$= 0.00012(1 + 3 + 5 + 7 + 9)(1 + 9 + 25 + 49 + 81)$$

$$= 0.495 \text{ Wb}$$

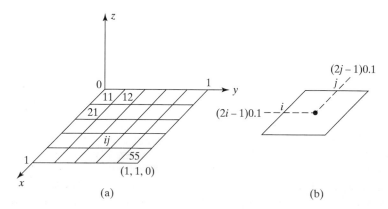

(a)   (b)

FIGURE 3.8   (a) Division of the portion of the xy-plane lying between $x = 0$, $x = 1, y = 0$, and $y = 1$ into 25 squares. (b) Area corresponding to the ijth square.

The result that we have just obtained for $[\psi]_S$ is approximate since we have divided the surface S into a finite number of areas. By dividing it into larger and larger numbers of squares, we can obtain more and more accurate results. In the limit that $n \to \infty$, the result converges to the exact value. The summation in (3.11) then becomes an integral that represents exactly the magnetic flux crossing the surface and is given by

$$[\psi]_S = \int_S \mathbf{B} \cdot d\mathbf{S} \tag{3.12}$$

where the symbol S associated with the integral sign denotes that the integration is performed over the surface S. The integral on the right side of (3.12) is known as the *surface integral of* **B** *over* S. The surface integral is a double integral since dS is equal to the product of two differential lengths.

Evaluation of surface integral

We shall illustrate the evaluation of the surface integral by computing the exact value of the magnetic flux crossing the surface in Fig. 3.8(a). To do this, we note that at any arbitrary point on the surface, the differential surface vector is given by

$$d\mathbf{S} = dx\,dy\,\mathbf{a}_z$$

The value of $\mathbf{B} \cdot d\mathbf{S}$ at the point is

$$\mathbf{B} \cdot d\mathbf{S} = 3xy^2\mathbf{a}_z \cdot dx\,dy\,\mathbf{a}_z$$
$$= 3xy^2\,dx\,dy$$

Thus, the required magnetic flux is given by

$$[\psi]_S = \int_S \mathbf{B} \cdot d\mathbf{S}$$

$$= \int_{x=0}^{1} \int_{y=0}^{1} 3xy^2\,dx\,dy = 0.5 \text{ Wb}$$

When the surface under consideration is a closed surface, the surface integral is written with a circle associated with the integral sign in the manner $\oint_S \mathbf{B} \cdot d\mathbf{S}$. A closed surface is one that encloses a volume. Hence, if you are anywhere in that volume, you can get out of it only by making a hole in the surface, and vice versa. A simple example is the surface of a balloon inflated and tied up at the mouth. The surface integral of **B** over the closed surface S is simply the magnetic flux *emanating* from the volume bounded by the surface. Thus, whenever a closed surface integral is evaluated, the unit vectors normal to the differential surfaces are chosen to be pointing out of the volume, so as to give the outward flux of the vector field, unless specified otherwise. We shall now consider an example of evaluating $\oint_S \mathbf{B} \cdot d\mathbf{S}$.

Example 3.2

Let us consider the magnetic field

$$\mathbf{B} = (x + 2)\mathbf{a}_x + (1 - 3y)\mathbf{a}_y + 2z\mathbf{a}_z$$

and evaluate $\oint_S \mathbf{B} \cdot d\mathbf{S}$, where S is the surface of the cubical box bounded by the planes

$$
\begin{array}{ll}
x = 0 & x = 1 \\
y = 0 & y = 1 \\
z = 0 & z = 1
\end{array}
$$

as shown in Fig. 3.9.

Noting that

$$\oint_S \mathbf{B} \cdot d\mathbf{S} = \int_{abcd} \mathbf{B} \cdot d\mathbf{S} + \int_{efgh} \mathbf{B} \cdot d\mathbf{S} + \int_{adhe} \mathbf{B} \cdot d\mathbf{S} + \int_{bcgf} \mathbf{B} \cdot d\mathbf{S}$$

$$+ \int_{aefb} \mathbf{B} \cdot d\mathbf{S} + \int_{dhgc} \mathbf{B} \cdot d\mathbf{S}$$

we simply evaluate each of the surface integrals on the right side and add them up to obtain the required quantity. In doing so, we recognize that since the quantity we want is the magnetic flux out of the box, we should direct the unit normal vectors toward the outside of the box. Thus, for the surface $abcd$,

$$x = 0, \qquad \mathbf{B} = 2\mathbf{a}_x + (1 - 3y)\mathbf{a}_y + 2z\mathbf{a}_z, \qquad d\mathbf{S} = -dy\,dz\,\mathbf{a}_x$$

$$\mathbf{B} \cdot d\mathbf{S} = -2\,dy\,dz$$

$$\int_{abcd} \mathbf{B} \cdot d\mathbf{S} = \int_{z=0}^{1} \int_{y=0}^{1} (-2)\,dy\,dz = -2$$

For the surface $efgh$,

$$x = 1, \qquad \mathbf{B} = 3\mathbf{a}_x + (1 - 3y)\mathbf{a}_y + 2z\mathbf{a}_z, \qquad d\mathbf{S} = dy\,dz\,\mathbf{a}_x$$

$$\mathbf{B} \cdot d\mathbf{S} = 3\,dy\,dz$$

$$\int_{efgh} \mathbf{B} \cdot d\mathbf{S} = \int_{z=0}^{1} \int_{y=0}^{1} 3\,dy\,dz = 3$$

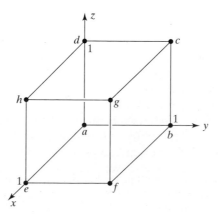

FIGURE 3.9   For evaluating the surface integral of a vector field over a closed surface.

For the surface *adhe*,

$$y = 0, \qquad \mathbf{B} = (x + 2)\mathbf{a}_x + 1\mathbf{a}_y + 2z\mathbf{a}_z, \qquad d\mathbf{S} = -dz\,dx\,\mathbf{a}_y$$

$$\mathbf{B} \cdot d\mathbf{S} = -dz\,dx$$

$$\int_{aehd} \mathbf{B} \cdot d\mathbf{S} = \int_{x=0}^{1} \int_{z=0}^{1} (-1)\,dz\,dx = -1$$

For the surface *bcgf*,

$$y = 1, \qquad \mathbf{B} = (x + 2)\mathbf{a}_x - 2\mathbf{a}_y + 2z\mathbf{a}_z, \qquad d\mathbf{S} = dz\,dx\,\mathbf{a}_y$$

$$\mathbf{B} \cdot d\mathbf{S} = -2\,dz\,dx$$

$$\int_{bfgc} \mathbf{B} \cdot d\mathbf{S} = \int_{x=0}^{1} \int_{z=0}^{1} (-2)\,dz\,dx = -2$$

For the surface *aefb*,

$$z = 0, \qquad \mathbf{B} = (x + 2)\mathbf{a}_x + (1 - 3y)\mathbf{a}_y + 0\mathbf{a}_z, \qquad d\mathbf{S} = -dx\,dy\,\mathbf{a}_z$$

$$\mathbf{B} \cdot d\mathbf{S} = 0$$

$$\int_{aefb} \mathbf{B} \cdot d\mathbf{S} = 0$$

For the surface *dhgc*,

$$z = 1, \qquad \mathbf{B} = (x + 2)\mathbf{a}_x + (1 - 3y)\mathbf{a}_y + 2\mathbf{a}_z, \qquad d\mathbf{S} = dx\,dy\,\mathbf{a}_z$$

$$\mathbf{B} \cdot d\mathbf{S} = 2\,dx\,dy$$

$$\int_{dhgc} \mathbf{B} \cdot d\mathbf{S} = \int_{y=0}^{1} \int_{x=0}^{1} 2\,dx\,dy = 2$$

Finally,

$$\oint_S \mathbf{B} \cdot d\mathbf{S} = -2 + 3 - 1 - 2 + 0 + 2 = 0$$

K3.1. Line integral; Line integral of **E**, Voltage; Surface integral; Surface integral of **B**; Magnetic flux.

D3.1. For each of the curves **(a)** $y = x^2$, $z = 0$, **(b)** $x^2 + y^2 = 2$, $z = 0$, and **(c)** $y = \sin 0.5\pi x$, $z = 0$ in a region of electric field $\mathbf{E} = y\mathbf{a}_x + x\mathbf{a}_y$, find the approximate value of the work done by the field in carrying a charge of 1 μC from the point $(1, 1, 0)$ to the neighboring point on the curve, whose x coordinate is 1.1, by evaluating $\mathbf{E} \cdot \Delta\mathbf{l}$ along a straight line path.
Ans. **(a)** 0.31 μJ; **(b)** -0.0112 μJ; **(c)** 0.0877 μJ.

D3.2. For $\mathbf{F} = y(\mathbf{a}_x + \mathbf{a}_y)$, find $\int \mathbf{F} \cdot d\mathbf{l}$ for the straight-line paths between the following pairs of points from the first point to the second point: **(a)** $(0, 0, 0)$ to $(2, 0, 0)$; **(b)** $(0, 2, 0)$ to $(2, 2, 0)$; and **(c)** $(2, 0, 0)$ to $(2, 2, 0)$.
Ans. **(a)** 0; **(b)** 4; **(c)** 2.

D3.3. Given $\mathbf{B} = (y\mathbf{a}_x - x\mathbf{a}_y)$ Wb/m^2, find by evaluating $\mathbf{B} \cdot \Delta\mathbf{S}$ the approximate absolute value of the magnetic flux crossing from one side to the other side of an infinitesimal surface of area 0.001 m^2 at the point $(1, 2, 1)$ for each of the following orientations of the surface: **(a)** in the $x = 1$ plane; **(b)** on the surface $2x^2 + y^2 = 6$; and **(c)** normal to the unit vector $\frac{1}{3}(2\mathbf{a}_x + \mathbf{a}_y + 2\mathbf{a}_z)$.

Ans.   **(a)** 2×10^{-3} Wb;   **(b)** $(1/\sqrt{2}) \times 10^{-3}$ Wb;   **(c)** 10^{-3} Wb.

D3.4. For the vector field $\mathbf{A} = x(\mathbf{a}_x + \mathbf{a}_y)$, find the absolute value of $\int \mathbf{A} \cdot d\mathbf{S}$ over the following plane surfaces: **(a)** square having the vertices at $(0,0,0), (0,2,0), (0,2,2)$, and $(0,0,2)$; **(b)** square having the vertices at $(2,0,0), (2,2,0), (2,2,2)$, and $(2, 0, 2)$; **(c)** square having the vertices at $(0, 0, 0), (2, 0, 0), (2, 0, 2)$, and $(0, 0, 2)$; and **(d)** triangle having the vertices at $(0,0,0), (2,0,0)$, and $(0,0,2)$.

Ans.   **(a)** 0;   **(b)** 8;   **(c)** 4;   **(d)** $\frac{4}{3}$.

3.2 FARADAY'S LAW

In the preceding section, we introduced the line and surface integrals. We are now ready to consider Maxwell's equations in integral form. The first equation, which we shall discuss in this section, is a consequence of an experimental finding by Michael Faraday in 1831 that time-varying magnetic fields give rise to electric fields and, hence, it is known as *Faraday's law*. Faraday discovered that when the magnetic flux enclosed by a loop of wire changes with time, a current is produced in the loop, indicating that a voltage or an *electromotive force*, abbreviated as emf, is induced around the loop. The variation of the magnetic flux can result from the time variation of the magnetic flux enclosed by a fixed loop or from a moving loop in a static magnetic field or from a combination of the two, that is, a moving loop in a time-varying magnetic field.

Statement of Faraday's law

In mathematical form, Faraday's law is given by

$$\oint_C \mathbf{E} \cdot d\mathbf{l} = -\frac{d}{dt} \int_S \mathbf{B} \cdot d\mathbf{S} \tag{3.13}$$

where S is a surface bounded by the closed path C, as shown in Fig. 3.10. In words, Faraday's law states that *the electromotive force around a closed path is equal to the negative of the time rate of change of the magnetic flux enclosed by that path*. There are certain procedures and observations of interest pertinent to the application of (3.13). We shall discuss these next.

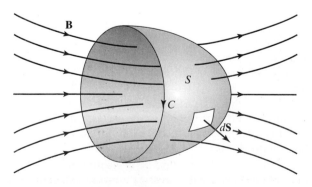

FIGURE 3.10   For illustrating Faraday's law.

1. The magnetic flux on the right side is to be evaluated in accordance with the *right-hand screw rule* (R.H.S. rule), a convention that is applied consistently for all electromagnetic field laws involving integration over surfaces bounded by closed paths. The right-hand screw rule consists of imagining a right-hand screw being turned around the closed path, as illustrated in Fig. 3.11 for two opposing senses of paths, and using the resulting direction of advance of the screw to evaluate the surface integral. The application of this rule to the geometry of Fig. 3.10 means that in evaluating the surface integral of **B** over S, the normal vector to the differential surface dS should be directed as shown in that figure.

Right-hand screw rule

2. In evaluating the surface integral of **B**, any surface S *bounded by C* can be employed. For example, if the loop C is a planar loop, it is not necessary to consider the plane surface having the loop as its perimeter. One can consider a curved surface bounded by C or any combination of plane (or plane and curved) surfaces which together are bounded by C, and which is sometimes a more desirable choice. To illustrate this point, consider the planar loop $PQRP$ in Fig. 3.12 (a). The most obvious surface bounded by this loop is the plane surface

(a)   (b)

FIGURE 3.11   Right-hand screw rule convention employed in the formulation of electomagnetic field laws.

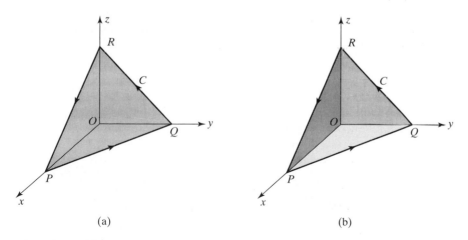

(a)   (b)

FIGURE 3.12   (a) A plane surface and (b) a combination of three plane surfaces, bounded by the closed path C.

PQR inclined to the coordinate planes. Now imagine this plane surface to be an elastic sheet glued to the perimeter and pushed in toward the origin so as to conform to the coordinate planes. Then we obtain the combination of the plane surfaces *OPQ*, *OQR*, and *ORP*, as shown in Fig. 3.12(b), which together constitute a surface also bounded by the loop. To evaluate the surface integral of **B** for the surface in Fig. 3.12(a), we need to make use of the *d***S** vector on that slant surface. On the other hand, for the geometry in Fig. 3.12(b), we can use the (simpler) *d***S** vectors associated with the coordinate planes. The fact that any surface *S* bounded by a closed path *C* can be employed to evaluate the magnetic flux enclosed by *C* implies that the magnetic flux through all such surfaces is the same in order for the emf around *C* to be unique. As we shall learn in Section 3.4, it is a fundamental property of the magnetic field that the magnetic flux is the same through all surfaces bounded by a given closed path.

3. The closed path *C* on the left side need not represent a loop of wire, but can be an imaginary contour. It means that the time-varying magnetic flux induces an electric field in the region and this results in an emf around the closed path. If a wire is placed in the position occupied by the closed path, the emf will produce a current in the loop simply because the charges in the wire are constrained to move along the wire.

Lenz's law

4. The minus sign on the right side together with the right-hand screw rule ensures that *Lenz's law* is always satisfied. Lenz's law states that *the sense of the induced emf is such that any current it produces tends to* oppose the change *in the magnetic flux producing it*. It is important to note that the induced emf acts to oppose the change in the flux and not the flux itself. To clarify this, let us consider that the flux is into the paper and increasing with time. Then the induced emf acts to produce flux out of the paper. On the other hand, if the same flux is decreasing with time, then the induced emf acts to produce flux into the paper.

Faraday's law for N-turn coil

5. If the loop *C* contains more than one turn, such as in an *N*-turn coil, then the surface *S* bounded by the periphery of the loop takes the shape of a spiral ramp, as shown in Fig. 3.13 (a) for *N* equal to 2. This surface can be visualized by taking two paper plates, cutting each of them along a radius, as shown in Figs. 3.13(b) and (c), and joining the edge *BO* of the plate in (c) to the edge *A'O*

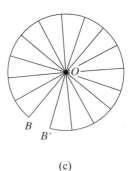

(a)	(b)	(c)

FIGURE 3.13   For illustrating the surface bounded by a loop containing two turns.

of the plate in (b). For a tightly wound coil, this is equivalent to the situation in which N separate, identical, single-turn loops are stacked so that the emf induced in the N-turn coil is N times that induced in one turn. Thus, for an N-turn coil,

$$\text{emf} = -N\frac{d\psi}{dt} \qquad (3.14)$$

where ψ is the magnetic flux computed as though the coil is a one-turn coil.

We shall now consider two examples to illustrate the determination of induced emf using Faraday's law, the first involving a stationary loop in a time-varying magnetic field and the second involving a moving conductor in a static magnetic field.

Example 3.3

A time-varying magnetic field is given by

$$\mathbf{B} = B_0 \cos \omega t \, \mathbf{a}_y$$

Stationary loop in a time-varying magnetic field

where B_0 is a constant. It is desired to find the induced emf around the rectangular loop C in the xz-plane bounded by the lines $x = 0$, $x = a$, $z = 0$, and $z = b$, as shown in Fig. 3.14.

Choosing $d\mathbf{S} = dx \, dz \, \mathbf{a}_y$ in accordance with the right-hand screw rule and using the plane surface S bounded by the loop, we obtain the magnetic flux enclosed by the loop to be

$$\psi = \int_S \mathbf{B} \cdot d\mathbf{S} = \int_{z=0}^{b} \int_{x=0}^{a} B_0 \cos \omega t \, \mathbf{a}_y \cdot dx \, dz \, \mathbf{a}_y$$

$$= B_0 \cos \omega t \int_{z=0}^{b} \int_{x=0}^{a} dx \, dz = ab B_0 \cos \omega t$$

Note that since the magnetic flux density is uniform and normal to the plane of the loop, this result could have been obtained by simply multiplying the area ab of the loop by the

FIGURE 3.14   Rectangular loop in the xz-plane situated in a time-varying magnetic field.

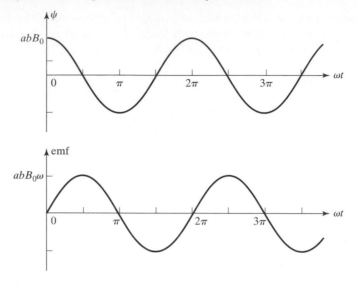

FIGURE 3.15   Time variations of magnetic flux ψ enclosed by the loop of Fig. 3.14 and the resulting induced emf around the loop.

component $B_0 \cos \omega t$ of the flux density vector. The induced emf around the loop is then given by

$$\oint_C \mathbf{E} \cdot d\mathbf{l} = -\frac{d}{dt} \int_S \mathbf{B} \cdot d\mathbf{S}$$

$$= -\frac{d}{dt} [abB_0 \cos \omega t] = abB_0 \omega \sin \omega t$$

The time variations of the magnetic flux enclosed by the loop and the induced emf around the loop are shown in Fig. 3.15. It can be seen that when the magnetic flux enclosed by the loop into the paper is *decreasing* with time, the induced emf is positive, thereby producing a clockwise current if the loop were a wire. This polarity of the current gives rise to a magnetic field directed into the paper inside the loop and, hence, acts to *oppose the decrease* of the magnetic flux enclosed by the loop. When the magnetic flux enclosed by the loop into the paper is *increasing* with time, the induced emf is negative, thereby producing a counterclockwise current around the loop. This polarity of the current gives rise to a magnetic field directed out of the paper inside the loop and hence acts to *oppose the increase* of the magnetic flux enclosed by the loop. These observations are consistent with Lenz's law.

Example 3.4

Moving conductor in a static magnetic field

A rectangular loop of wire with three sides fixed and the fourth side movable is situated in a plane perpendicular to a uniform magnetic field $\mathbf{B} = B_0 \mathbf{a}_z$, as illustrated in Fig. 3.16. The movable side consists of a conducting bar moving with a velocity v_0 in the y-direction. It is desired to find the emf induced around the closed path C of the loop.

Letting the position of the movable side at any time t be $y_0 + v_0 t$ and considering $d\mathbf{S} = dx\, dy\, \mathbf{a}_z$ in accordance with the right-hand screw rule and using the plane surface

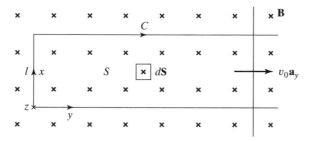

FIGURE 3.16 Rectangular loop of wire with a movable side situated in a uniform magnetic field.

S bounded by the loop, we obtain the magnetic flux enclosed by the loop to be

$$\int_S \mathbf{B} \cdot d\mathbf{S} = \int_S B_0 \mathbf{a}_z \cdot dx \, dy \, \mathbf{a}_z$$

$$= \int_{x=0}^{l} \int_{y=0}^{y_0 + v_0 t} B_0 \, dx \, dy$$

$$= B_0 l(y_0 + v_0 t)$$

Note that this result could also have been obtained as the product of the area of the loop $l(y_0 + v_0 t)$ and the flux density B_0 because of the uniformity of the flux density within the area of the loop and its perpendicularity to the plane of the loop. The emf induced around C is given by

$$\oint_C \mathbf{E} \cdot d\mathbf{l} = -\frac{d}{dt}\int_S \mathbf{B} \cdot d\mathbf{S}$$

$$= -\frac{d}{dt}[B_0 l(y_0 + v_0 t)]$$

$$= -B_0 l v_0$$

Note that if the bar is moving to the right, the induced emf is negative and produces a current in the sense opposite to that of C. This polarity of the current is such that it gives rise to a magnetic field directed out of the paper inside the loop. The flux of this magnetic field is in opposition to the flux of the original magnetic field and hence tends to *oppose the increase* in the magnetic flux enclosed by the loop. On the other hand, if the bar is moving to the left, v_0 is negative, the induced emf is positive, and produces current in the same sense as that of C. This polarity of current is such that it gives rise to a magnetic field directed into the paper inside the loop. The flux of this magnetic field is in augmentation to the flux of the original magnetic field and hence tends to *oppose the decrease* in the magnetic flux enclosed by the loop. These observations are once again consistent with Lenz's law.

It is also of interest to note that the induced emf can also be interpreted as being due to the electric field induced in the moving bar by virtue of its motion perpendicular to the magnetic field. Thus, a charge Q in the bar experiences a force $\mathbf{F} = Q\mathbf{v} \times \mathbf{B}$ or $Qv_0\mathbf{a}_y \times B_0\mathbf{a}_z = Qv_0 B_0 \mathbf{a}_x$. To an observer moving with the bar, this force appears as an electric force due to an electric field $\mathbf{F}/Q = v_0 B_0 \mathbf{a}_x$. Viewed from inside the loop, this electric field is in the counterclockwise sense. Hence, the induced emf, which is the line integral of $\mathbf{E}$ along the bar, is given by

$$\int_{x=0}^{l} v_0 B_0 \mathbf{a}_x \cdot dx\, \mathbf{a}_x = \int_{x=0}^{l} v_0 B_0 \, dx = v_0 B_0 l$$

in the counterclockwise sense (i.e., opposite to C), consistent with the result deduced from Faraday's law. This concept of induced emf is known as the *motional emf concept*, which is employed widely in the study of electromechanics.

In the two examples we just discussed, we have implicitly illustrated the principles behind two of the practical applications of Faraday's law. These are pertinent to the reception of radio and TV signals using a loop antenna and electromechanical energy conversion.

Principle of loop antenna

That the arrangement considered in Example 3.3 illustrates the principle of a loop antenna can be seen by noting that if the loop C were in the xy-plane or in the yz-plane, no emf would be induced in it since the magnetic flux density is then parallel to the plane of the loop and no flux is enclosed by the loop. In fact, for any arbitrary orientation of the loop, only that component of $\mathbf{B}$ normal to the plane of the loop contributes to the magnetic flux enclosed by the loop and hence to the emf induced in the loop. Thus, for a given magnetic field, the voltage induced in the loop varies as the orientation of the loop is changed, with the maximum occurring when the loop is in the plane perpendicular to the magnetic field. Pocket AM radios generally employ inside them a type of loop antenna consisting of many turns of wire wound around a bar of magnetic material, and TV receivers generally employ a single-turn circular loop for UHF channels. Thus, for maximum signals to be received, the AM radios and the TV loop antennas need to be oriented appropriately. Another point of interest evident from Example 3.3 is that the induced emf is proportional to ω, the radian frequency of the source of the magnetic field. Hence, for the same voltage to be induced for a given amplitude B_0 of the magnetic flux density, the area of the loop times the number of turns is inversely proportional to the frequency.

Locating a radio transmitter

What is undesirable for one purpose can sometimes be used to advantage for another purpose. The fact that no voltage is induced in the loop antenna when the magnetic field is parallel to the plane of the loop is useful for locating the transmitter of a radio wave. Since the magnetic field of an incoming radio wave is perpendicular to its direction of propagation, no voltage is induced in the loop when its axis is along the direction of the transmitter. For a transmitter on Earth's surface, it is then sufficient to use two spaced vertical loop antennas and find their orientations for which no signals are received. By then producing backward along the axes of the two loop antennas, as shown by the top view in Fig. 3.17, the location of the transmitter can be determined.

Electro-mechanical energy conversion

That the arrangement considered in Example 3.4 is a simple example of an electromechanical energy converter can be seen by recognizing that in view of the current flow in the moving bar, the bar is acted on by a magnetic force. Since for positive v_0, the current flows in the loop in the sense opposite to that of C and hence in the positive x-direction in the moving bar, and since the magnetic field is in the z-direction, the magnetic force is exerted in the $\mathbf{a}_x \times \mathbf{a}_z$ or

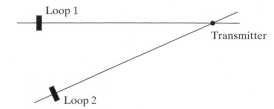

FIGURE 3.17   Top view of an arrangement consisting of two loop antennas for locating a transmitter of radio waves.

$-\mathbf{a}_y$-direction. Thus, to keep the bar moving, an external force must be exerted in the $+\mathbf{a}_y$-direction, thereby requiring mechanical work to be done by an external agent. It is this mechanical work that is converted into electrical energy in the loop.

Principle of rotating generator

What we have just discussed is the principle of generation of electric power by linear motion of a conductor in a magnetic field. Practical electric generators are of the rotating type. The principle of a rotating generator can be illustrated by considering a rectangular loop of wire situated symmetrically about the z-axis and rotating with angular velocity ω around the z-axis in a constant magnetic field $\mathbf{B} = B_0\mathbf{a}_x$, as shown in Fig. 3.18(a). Then noting from the view in Fig. 3.18(b) that the magnetic flux ψ enclosed by the loop at any arbitrary value of time is the same as that enclosed by its projection onto the yz-plane at that time, we obtain $\psi = B_0 A \cos \omega t$, where A is the area of the loop and the situation shown in Fig. 3.18(a) is assumed for $t = 0$. The emf induced in the loop is $-d\psi/dt$, or $\omega B_0 A \sin \omega t$. Thus, the rotating loop in the constant magnetic field produces an alternating voltage. The same result can be achieved by a stationary loop in a rotating magnetic field. In fact, in most generators, a stationary member, or stator, carries the coils in which the voltage is induced, and a rotating member, or rotor, provides the magnetic field. As in the case of the arrangement of Example 3.4, a certain amount of mechanical work must be done to keep the loop rotating. It is this mechanical work, which is supplied by the prime mover (such as a turbine in the case of a hydroelectric generator or the engine of an automobile in the case of its alternator) turning the rotor, that is converted into electrical energy.

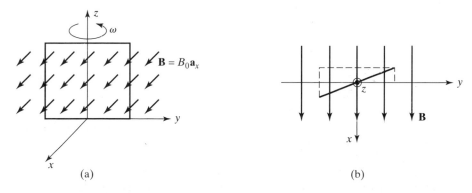

(a)   (b)

FIGURE 3.18   For illustrating the principle of a rotating generator.

Magnetic levitation

There are numerous other applications of Faraday's law, but we shall discuss only one more before we conclude this section. This is the phenomenon of *magnetic levitation,* a basis for rapid transit systems employing trains that hover over their guideways and do not touch the rail, among other applications. Magnetic levitation arises from a combination of Faraday's law and Ampère's force law. It can be explained and demonstrated through a series of simple experiments, culminating in a current-carrying coil lifting up above a metallic plate, as described in the following:

1. Consider a pair of coils (30 to 50 turns of No. 26 wire of about 4-in. diameter) attached to nails on a piece of wood, as shown in Fig. 3.19. Set to zero the output of a variable power supply obtained by connecting a variac to the 110-V ac mains. Connect one output terminal (A) of the variac to the beginnings (C_1 and C_2) of both coils and the second output terminal (B) to the ends (D_1 and D_2) of both coils so that currents flow in the two coils in the same sense. Apply some voltage to the coils by turning up the variac and note the attraction between the coils. Repeat the experiment by connecting A to C_1 and D_2 and B to C_2 and D_1, so that currents in the two coils flow in opposite senses, and note repulsion this time. What we have just described is Ampère's force law at work. If the currents flow in the same sense, the magnetic force is one of attraction, and if the currents flow in opposite senses, it is one of repulsion, as shown in Figs. 3.20(a) and (b), respectively, for straight wires, for the sake of simplicity.[1]

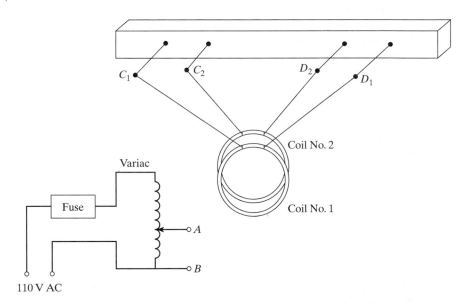

FIGURE 3.19   Experimental setup for demonstration of Ampère's force law, Faraday's law, and the principle of magnetic levitation.

---

[1]See L. Pearce Williams, "André-Marie Ampère," *Scientific American,* January 1989, pp. 90–97, for an interesting account of Ampère's experiments involving helical and spiral coils.

(a) (b)

FIGURE 3.20 For explaining (a) force of attraction for currents flowing in the same sense and (b) force of repulsion for currents flowing in opposite senses.

2. Connect coil No. 2 to the variac and coil No. 1 to an oscilloscope to observe the induced voltage in coil No. 1, thereby demonstrating Faraday's law. Note change in induced voltage as variac voltage is changed. Also note the change in induced voltage by keeping the variac voltage constant and moving coil No. 1 away from coil No. 2 and/or turning it about the vertical.

3. Connect coil No. 2 to the variac and leave coil No. 1 open-circuited. Observe that no action takes place as the variac voltage is applied to coil No. 2. This is because although a voltage is induced in coil No. 1, no current flows in it. Now short circuit coil No. 1 and repeat the experiment to note repulsion. This is due to the induced voltage in coil No. 1 causing a current flow in it in the sense opposite to that in coil No. 2, and, hence, is a result of the combination of Faraday's law and Ampère's force law. That the force is one of repulsion can be deduced by writing circuit equations and showing that the current in the short-circuited coil does indeed flow in the sense opposite to that in the excited coil. However, it can be explained with the aid of physical reasoning as follows. When both coils are excited in the same sense in part (1) of the demonstration, the magnetic flux linking each coil is the sum of two fluxes in the same sense, due to the two currents. When the two coils are excited in opposite senses, the magnetic flux linking each coil is the algebraic sum of two fluxes in opposing senses, due to the two currents. Therefore, for the same source voltage and for the same pair of coils, the currents that flow in the coils in the second case have to be greater than those in the first case, for the induced voltage in each coil to equal the applied voltage. Thus, the force of repulsion in the second case is greater than the force of attraction in the first case. Consider now the case of one of the coils excited by source voltage, say, V_g, and the other short-circuited. Then the situation can be thought of as the first coil excited by $V_g/2$ and $V_g/2$ in series, and the second coil excited by $V_g/2$ and $-V_g/2$ in series, thereby resulting in a force of attraction and a force of repulsion. Since the force of repulsion is greater than the force of attraction, the net force, according to superposition, is one of repulsion.

4. Now to demonstrate actual levitation, place a smaller coil (about 30 turns of No. 28 wire of about 2-in. diameter) on a heavy aluminum plate (5 in. × 5 in. × $\frac{1}{2}$ in.), as shown in Fig. 3.21. Applying only the minimum necessary voltage and turning the variac only momentarily to avoid overheating, pass current through the coil from the variac to see the coil levitate. This levitation is due to the repulsive action between the current in the coil and the induced

FIGURE 3.21   Setup for demonstrating magnetic levitation.

currents in the metallic plate. Since the plate is heavy and cannot move, the alternative is for the coil to lift up.

K3.2.   Faraday's law; Right-hand screw rule; Lenz's law; Faraday's law for N-turn coil.

D3.5.   Given $\mathbf{B} = B_0(\sin \omega t\, \mathbf{a}_x - \cos \omega t\, \mathbf{a}_y)$ Wb/m², find the induced emf around each of the following closed paths: **(a)** the rectangular path from $(0, 0, 0)$ to $(0, 1, 0)$ to $(0, 1, 1)$ to $(0, 0, 1)$ to $(0, 0, 0)$; **(b)** the triangular path from $(1, 0, 0)$ to $(0, 1, 0)$ to $(0, 0, 1)$ to $(1, 0, 0)$; and **(c)** the rectangular path from $(0, 0, 0)$ to $(1, 1, 0)$ to $(1, 1, 1)$ to $(0, 0, 1)$ to $(0, 0, 0)$.

Ans.   **(a)** $-\omega B_0 \cos \omega t$ V;   **(b)** $-\dfrac{\omega B_0}{\sqrt{2}} \cos (\omega t - \pi/4)$ V;

(c) $-\sqrt{2}\, \omega B_0 \cos (\omega t + \pi/4)$ V.

D3.6.   A square loop lies in the xy-plane forming the closed path C connecting the points $(0, 0, 0)$, $(1, 0, 0)$, $(1, 1, 0)$, $(0, 1, 0)$, and $(0, 0, 0)$, in that order. A magnetic field $\mathbf{B}$ exists in the region. From considerations of Lenz's law, determine whether the induced emf around the closed path C at $t = 0$ is positive, negative, or zero for each of the following magnetic fields, where B_0 is a positive constant: **(a)** $\mathbf{B} = B_0 t\, \mathbf{a}_z$; **(b)** $\mathbf{B} = B_0 \cos (2\pi t + 60°)\mathbf{a}_z$; and **(c)** $\mathbf{B} = B_0 e^{-t^2}\, \mathbf{a}_z$.

Ans.   **(a)** negative;   **(b)** positive;   **(c)** zero.

D3.7.   For $\mathbf{B} = B_0 \cos \omega t\, \mathbf{a}_z$ Wb/m², find the induced emf around the following closed paths: **(a)** the closed path comprising the straight lines successively connecting the points $(0, 0, 0)$, $(1, 0, 0)$, $(1, 1, 0)$, $(0, 1, 0)$, $(0, 0, 0.001)$, and $(0, 0, 0)$; **(b)** the closed path comprising the straight lines successively connecting the points $(0, 0, 0)$, $(1, 0, 0)$, $(1, 1, 0)$, $(0, 1, 0)$, $(0, 0, 0.001)$, $(1, 0, 0.001)$, $(1, 1, 0.001)$, $(0, 1, 0.001)$, $(0, 0, 0.002)$, and $(0, 0, 0)$ with a slight kink in the straight line at the point $(0, 0, 0.001)$ to avoid touching the point; and **(c)** the closed path comprising the helical path $r = 1/\sqrt{\pi}, \phi = 1000\pi z$ from $(1/\sqrt{\pi}, 0, 0)$ to $(1/\sqrt{\pi}, 0, 0.01)$ and the straight-line path from $(1/\sqrt{\pi}, 0, 0.01)$ to $(1/\sqrt{\pi}, 0, 0)$ with slight kinks to avoid touching the helical path.

Ans.   **(a)** $\omega B_0 \sin \omega t$ V;   **(b)** $2\omega B_0 \sin \omega t$ V;   **(c)** $5\omega B_0 \sin \omega t$ V.

3.3   AMPÈRE'S CIRCUITAL LAW

In the preceding section, we introduced Faraday's law, one of Maxwell's equations, in integral form. In this section, we introduce another of Maxwell's equations in integral form. This equation, known as *Ampère's circuital law,* is a combination of an experimental finding of Oersted that electric currents generate magnetic fields and a mathematical contribution of Maxwell that time-varying electric fields give rise to magnetic fields. It is this contribution of Maxwell that led to the prediction of electromagnetic wave propagation even before the phenomenon was discovered experimentally. In mathematical form, Ampere's circuital law is analogous to Faraday's law and is given by

$$\oint_C \mathbf{H} \cdot d\mathbf{l} = [I_c]_S + \frac{d}{dt} \int_S \mathbf{D} \cdot d\mathbf{S} \qquad (3.15)$$

where S is a surface bounded by C.

The quantity $\oint_C \mathbf{H} \cdot d\mathbf{l}$ on the left side of (3.15) is the line integral of the vector field $\mathbf{H}$ around the closed path C. We learned in Section 3.1 that the quantity $\oint_C \mathbf{E} \cdot d\mathbf{l}$ has the physical meaning of work per unit charge associated with the movement of a test charge around the closed path C. The quantity $\oint_C \mathbf{H} \cdot d\mathbf{l}$ does not have a similar physical meaning. This is because the magnetic force on a moving charge is directed perpendicular to the direction of motion of the charge as well as to the direction of the magnetic field and hence does not do work in the movement of the charge. By recalling that $\mathbf{H}$ has the units of amperes per meter, we obtain the units of current (A) for $\oint_C \mathbf{H} \cdot d\mathbf{l}$. However, by analogy with the name *electromotive force* for $\oint_C \mathbf{E} \cdot d\mathbf{l}$, the quantity $\oint_C \mathbf{H} \cdot d\mathbf{l}$ is known as the *magnetomotive force,* abbreviated as mmf.

The quantity $[I_c]_S$ on the right side of (3.15) is the current due to flow of free charges crossing the surface S. It can be a convection current such as due to motion of a charged cloud in space or a conduction current due to motion of charges in a conductor. It cannot be due to the polarization and magnetization phenomena in a material because the effects of these phenomena are taken into account implicitly in the definitions of $\mathbf{D}$ and $\mathbf{H}$, respectively. Although $[I_c]_S$ can be filamentary current, surface current, or volume current, or a combination of these, it is formulated in terms of the volume current density vector, $\mathbf{J}$, in the manner

$$[I_c]_S = \int_S \mathbf{J} \cdot d\mathbf{S} \qquad (3.16)$$

Just as the surface integral of the magnetic flux density vector $\mathbf{B}$ (Wb/m^2) over a surface S gives the magnetic flux (Wb) crossing that surface, the surface integral of $\mathbf{J}$ (A/m^2) over a surface S gives the current (A) crossing that surface.

The quantity $\int_S \mathbf{D} \cdot d\mathbf{S}$ on the right side of (3.15) is the flux of the vector field $\mathbf{D}$ crossing S. Hence, it is the displacement flux, or the electric flux, crossing the surface S. By recalling that $\mathbf{D}$ has the units of coulombs per meter squared, we obtain the units of charge (C) for the displacement flux. Hence, the quantity $d/dt (\int_S \mathbf{D} \cdot d\mathbf{S})$ has the units of d/dt (charge) or current and is known as the *displacement current*. While in materials, it includes currents resulting from the polarization phenomena, in free space, it is physically not a current in the sense that it does not represent the flow of charges, but it gives rise to the same effect as a current does, namely, producing a magnetic field.

Displacement current

We may now write (3.15) in the manner

$$\oint_C \mathbf{H} \cdot d\mathbf{l} = \int_S \mathbf{J} \cdot d\mathbf{S} + \frac{d}{dt} \int_S \mathbf{D} \cdot d\mathbf{S} \tag{3.17}$$

where we recall that

$$\mathbf{D} = \varepsilon \mathbf{E} \tag{3.18a}$$

$$\mathbf{H} = \frac{\mathbf{B}}{\mu} \tag{3.18b}$$

and it is understood that $\int_S \mathbf{J} \cdot d\mathbf{S}$, although formulated in terms of the volume current density vector $\mathbf{J}$, represents the algebraic sum of all currents due to flow of charges crossing the surface S. If the current is due to conduction, then

$$\mathbf{J} = \mathbf{J}_c = \sigma \mathbf{E} \tag{3.18c}$$

In words, (3.17) states that *the magnetomotive force around a closed path C is equal to the algebraic sum of the current due to flow of charges and the displacement current bounded by C.* The situation is illustrated in Fig. 3.22.

As in the case of Faraday's law, there are certain procedures and observations pertinent to the application of (3.17). These are as follows.

1. The surface integrals on the right side of (3.17) are to be evaluated in accordance with the R.H.S. rule, which means that for the geometry of Fig. 3.22, the normal vector to the differential surface dS should be directed as shown in the figure.

2. In evaluating the surface integrals, *any surface S bounded by C can be employed.* However, *the same surface* must be employed for the two surface integrals. It is not correct to consider two different surfaces to evaluate the two surface integrals, although both surfaces may be bounded by C.

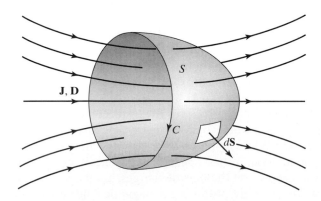

FIGURE 3.22   For illustrating Ampère's circuital law.

Observation 2 implies that for the mmf around C to be unique, the sum of the two currents (current due to flow of charges and displacement current) through all possible surfaces bounded by C is the same. Let us now consider two surfaces S_1 and S_2 bounded by the closed paths C_1 and C_2, respectively, as shown in Fig. 3.23, where C_1 and C_2 are traversed in opposite senses and touch each other so that S_1 and S_2 together form a closed surface. The situation may be imagined by considering the closed surface to be that of a potato and C_1 and C_2 to be two rubber bands around the potato.

Applying Ampère's circuital law to C_1 and S_1 and noting that dS_1 is chosen in accordance with the R.H.S. rule, we have

$$\oint_{C_1} \mathbf{H} \cdot d\mathbf{l} = \int_{S_1} \mathbf{J} \cdot d\mathbf{S}_1 + \frac{d}{dt} \int_{S_1} \mathbf{D} \cdot d\mathbf{S}_1 \qquad (3.19a)$$

Similarly, applying Ampère's circuital law to C_2 and S_2 and noting again that dS_2 is chosen in accordance with the R.H.S. rule, we have

$$\oint_{C_2} \mathbf{H} \cdot d\mathbf{l} = \int_{S_1} \mathbf{J} \cdot d\mathbf{S}_2 + \frac{d}{dt} \int_{S_2} \mathbf{D} \cdot d\mathbf{S}_2 \qquad (3.19b)$$

Now adding (3.19a) and (3.19b), we obtain

$$0 = \oint_{S_1 + S_2} \mathbf{J} \cdot d\mathbf{S} + \frac{d}{dt} \oint_{S_1 + S_2} \mathbf{D} \cdot d\mathbf{S} \qquad (3.20)$$

where the left side results from the fact that C_1 and C_2 are actually the same path but traversed in opposite senses, so that the two line integrals are the negatives of each other.

Since the closed surface $S_1 + S_2$ can be of any size and shape, we can generalize (3.20) to write

$$\oint_S \mathbf{J} \cdot d\mathbf{S} + \frac{d}{dt} \oint_S \mathbf{D} \cdot d\mathbf{S} = 0$$

or

$$\boxed{\frac{d}{dt} \oint_S \mathbf{D} \cdot d\mathbf{S} = -\oint_S \mathbf{J} \cdot d\mathbf{S}} \qquad (3.21)$$

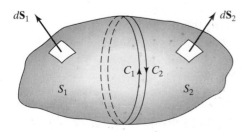

FIGURE 3.23. Two closed paths C_1 and C_2 touching each other and bounding the surfaces S_1 and S_2, respectively, which together form a closed surface.

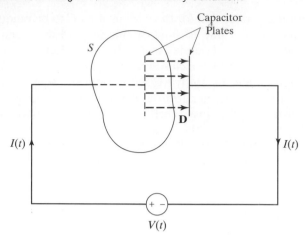

FIGURE 3.24   Capacitor circuit for illustrating that the displacement current from one plate to the other is equal to the wire current.

Thus, the displacement current emanating from a closed surface is equal to the current due to charges flowing into the volume bounded by that closed surface.

Capacitor circuit
An important example of the property given by (3.21) at work is in a capacitor circuit, as shown in Fig. 3.24 . In this circuit, the time-varying voltage source sets up a time-varying electric field between the plates of the capacitor and directed from one plate to the other. Therefore, one can talk about displacement current crossing a surface between the plates. According to (3.21) applied to a closed surface S enclosing one of the plates, as shown in the figure,

$$\frac{d}{dt} \oint_S \mathbf{D} \cdot d\mathbf{S} = I(t) \tag{3.22}$$

where $I(t)$ is the current (due to flow of charges in the wire) drawn from the voltage source. Neglecting fringing effects and assuming that the electric field is normal to the plates and uniform, we have, from (3.22),

$$\frac{d}{dt} \oint_S \mathbf{D} \cdot d\mathbf{S} = \frac{d}{dt} (DA) = I(t) \tag{3.23}$$

where A is the area of each plate. Thus, where the wire current ends on one of the plates, the displacement current takes over and completes the circuit to the second plate.

Measurement of dielectric permittivity
Since for a given electric field intensity, the displacement flux density between the plates of the capacitor varies with the permittivity of the dielectric, (3.23) can be used as a basis for the measurement of the relative permittivity of a dielectric material. We shall discuss the details of a simple experimental demonstration of this technique in the following.

A schematic of the experimental setup is shown in Fig. 3.25. It consists of a parallel-plate capacitor arrangement, made up of two 12 in. × 12 in. aluminum sheets attached to wooden blocks spaced about $\frac{1}{2}$ in. apart and driven by an ac

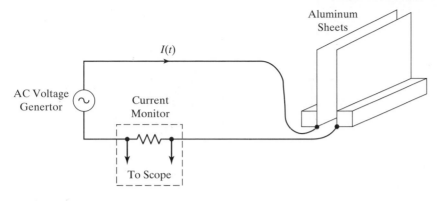

FIGURE 3.25 Experimental setup for demonstration of measurement of the relative permittivity of a dielectric material.

voltage generator in series with a resistor. By observing the voltage across the resistor, the current through the circuit can be monitored. With no dielectric between the plates, $\mathbf{D} = \varepsilon_0 \mathbf{E}$, and

$$I(t) = \frac{d}{dt} \int_A \varepsilon_0 \mathbf{E} \cdot d\mathbf{S} = \varepsilon_0 A \frac{dE}{dt}$$

With a dielectric between the plates, $\mathbf{D} = \varepsilon \mathbf{E} = \varepsilon_0 \varepsilon_r \mathbf{E}$, and

$$I(t) = \frac{d}{dt} \int_A \varepsilon_0 \varepsilon_r \mathbf{E} \cdot d\mathbf{S} = \varepsilon_0 \varepsilon_r A \frac{dE}{dt}$$

Thus, if the electric field between the plates is the same in both cases,

$$\varepsilon_r = \frac{I(t) \text{ with the dielectric}}{I(t) \text{ without the dielectric}}$$

To maintain the same electric field between the plates, the voltage $(= \int \mathbf{E} \cdot d\mathbf{l})$ across the plates is insured to be the same in both cases. The procedure therefore consists of the following steps:

1. With the dielectric sample of the same size as the plates inserted between the plates and the voltage source applied, measure the voltage across the resistor and the voltage across the capacitor plates.
2. Without changing the spacing between the plates, pull the dielectric sample out. Adjust the voltage source value such that the voltage across the capacitor plates is the same as in step 1. This can be omitted if the value of the resistor is chosen such that the voltage drop across it is only a small fraction of the source voltage. Measure the voltage across the resistor.
3. Compute the ratio of the voltage across the resistor measured in step 1 to that measured in step 2. This ratio is the relative permittivity ε_r of the dielectric sample, at the frequency of the voltage source and to within the assumptions made concerning the field between the plates.

FIGURE 3.26   Schematic of a condenser microphone.

In the experimental demonstration we just discussed, we maintained the electric field intensity **E** between the plates constant and made use of the variation of **D** and hence of $I(t)$ with ε. Alternatively, If ε remains the same and **E** changes, once again **D** and hence $I(t)$ changes. This is the basis behind the operation of a condenser microphone.

Condenser microphone

A condenser microphone consists of a tightly stretched but movable metallic-plate diaphragm electrically insulated from a second fixed plate, as shown in Fig. 3.26 . A dc voltage is applied between the plates via a resistor. Sound waves impinging on the diaphragm cause it to vibrate, thereby changing the spacing between the two plates. This, in turn, changes the electric field between the plates, resulting in a change in the displacement current and hence in the wire current through the resistor. The voltage fluctuations across the resistor are picked up and amplified.

Radiation from antenna

Let us now return to Ampère's circuital law (3.17) and examine it together with Faraday's law (3.13). To do this, we repeat the two laws

$$\oint_C \mathbf{E} \cdot d\mathbf{l} = -\frac{d}{dt}\int_S \mathbf{B} \cdot d\mathbf{S} \tag{3.24}$$

$$\oint_C \mathbf{H} \cdot d\mathbf{l} = \int_S \mathbf{J} \cdot d\mathbf{S} + \frac{d}{dt}\int_S \mathbf{D} \cdot d\mathbf{S} \tag{3.25}$$

and observe that time-varying electric and magnetic fields are interdependent, since according to Faraday's law (3.24), a time-varying magnetic field produces an electric field, whereas according to Ampère's circuital law (3.25), a time-varying electric field gives rise to a magnetic field. In addition, Ampère's circuital law tells us that an electric current generates a magnetic field. These properties from the basis for the phenomena of radiation and propagation of electromagnetic waves. To provide a simplified, qualitative explanation of radiation from an antenna, we begin with a piece of wire carrying a time-varying current, $I(t)$, as shown in Fig. 3.27. Then, the time-varying current generates a time-varying magnetic field $\mathbf{H}(t)$, which surrounds the wire. Time-varying electric and magnetic

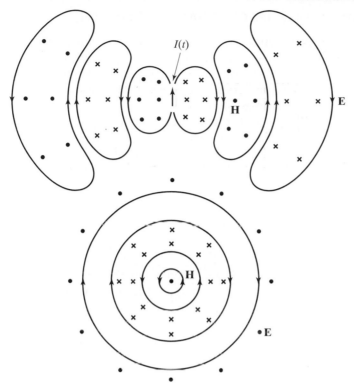

FIGURE 3.27. Two views of a simplified depiction of electromagnetic wave radiation from a piece of wire carrying a time-varying current.

fields, $\mathbf{E}(t)$ and $\mathbf{H}(t)$, are then produced in succession, as shown by two views in Fig. 3.27, thereby giving rise to electromagnetic waves. Thus, just as water waves are produced when a rock is thrown in a pool of water, electromagnetic waves are radiated when a piece of wire in space is excited by a time-varying current.

K3.3. Ampère's circuital law; Displacement current; Capacitor circuit.

D3.8. For $\mathbf{E} = E_0 t e^{-t^2}\mathbf{a}_z$ in free space, find the displacement current crossing an area of 0.1 m² in the xy-plane from the $-z$-side to the $+z$-side for each of the following values of t: **(a)** $t = 0$; **(b)** $t = 1/\sqrt{2}$ s; and **(c)** $t = 1$ s.
Ans. **(a)** $0.1\,\varepsilon_0 E_0$ A; **(b)** 0; **(c)** $-0.1 e^{-1}\varepsilon_0 E_0$ A.

D3.9. Three point charges $Q_1(t)$, $Q_2(t)$, and $Q_3(t)$ situated at the corners of an equilateral triangle of sides 1 m are connected to each other by wires along the sides of the triangle. Currents of I A and $3I$ A flow from Q_1 to Q_2 and Q_1 to Q_3, respectively. The displacement current emanating from a spherical surface of radius 0.1 m and centered at Q_2 is $-2I$ A. Find the following: **(a)** the current flowing from Q_2 to Q_3; **(b)** the displacement current emanating from the spherical surface of radius 0.1 m and centered at Q_1; and **(c)** the displacement current emanating from the spherical surface of radius 0.1 m and centered at Q_3.
Ans. **(a)** $3I$ A; **(b)** $-4I$ A; **(c)** $6I$ A.

3.4   GAUSS' LAWS AND THE LAW OF CONSERVATION OF CHARGE

In the previous two sections, we learned two of the four Maxwell's equations. These two equations have to do with the line integrals of the electric and magnetic fields around closed paths. The remaining two Maxwell's equations are pertinent to the surface integrals of the electric and magnetic fields over closed surfaces. These are known as *Gauss' laws*.

Gauss' law for the electric field   Gauss' law for the electric field states that *the displacement flux emanating from a closed surface S is equal to the charge contained within the volume V bounded by that surface.* This statement, although familiarly known as Gauss' law, has its origin in experiments conducted by Faraday. In mathematical form, it is given by

$$\oint_S \mathbf{D} \cdot d\mathbf{S} = [Q]_V \qquad (3.26)$$

The quantity $[Q]_V$ is the free charge contained within the volume V bounded by S. It can be free charge such as in a charged cloud or in a conductor. It cannot be due to the polarization phenomenon in a dielectric material since the effect of this phenomenon is taken into account implicitly in the definition of **D**. Although $[Q]_V$ can be a point charge, surface charge, or volume charge, or a combination of these, it is formulated as the volume integral of the volume charge density ρ, that is, in the manner

$$[Q]_V = \int_V \rho \, dv \qquad (3.27)$$

Evaluation of volume integral   The volume integral is a triple integral since dv is the product of three differential lengths. For an illustration of the evaluation of a volume integral, let us consider

$$\rho = (x + y + z) \text{ C/m}^3$$

and the cubical volume V bounded by the planes $x = 0$, $x = 1$, $y = 0$, $y = 1$, $z = 0$, and $z = 1$. Then the charge Q contained within the cubical volume is given by

$$
\begin{aligned}
Q &= \int_V \rho \, dv = \int_{x=0}^{1} \int_{y=0}^{1} \int_{z=0}^{1} (x + y + z) \, dx \, dy \, dz \\
&= \int_{x=0}^{1} \int_{y=0}^{1} \left[xz + yz + \frac{z^2}{2} \right]_{z=0}^{1} dx \, dy \\
&= \int_{x=0}^{1} \int_{y=0}^{1} \left(x + y + \frac{1}{2} \right) dx \, dy \\
&= \int_{x=0}^{1} \left[xy + \frac{y^2}{2} + \frac{y}{2} \right]_{y=0}^{1} dx
\end{aligned}
$$

$$= \int_{x=0}^{1} (x + 1)\, dx$$

$$= \left[\frac{x^2}{2} + x \right]_{x=0}^{1}$$

$$= \tfrac{3}{2}\ \mathrm{C}$$

We may now write Gauss' law for the electric field (3.26) in the manner

$$\oint_{S} \mathbf{D} \cdot d\mathbf{S} = \int_{V} \rho\, dv \qquad\qquad (3.28)$$

where we recall that

$$\mathbf{D} = \varepsilon \mathbf{E}$$

and it is understood that $\int_{V} \rho\, dv$, although formulated in terms of the volume charge density ρ, represents the algebraic sum of all free charges contained within V. The situation is illustrated in Fig. 3.28 .

Gauss' law for the magnetic field is analogous to Gauss' law for the electric field and is given by

Gauss' law for the magnetic field

$$\oint_{S} \mathbf{B} \cdot d\mathbf{S} = 0 \qquad\qquad (3.29)$$

In words, (3.29) states that *the magnetic flux emanating from a closed surface is equal to zero.* In physical terms, (3.29) signifies that magnetic charges do not exist and magnetic flux lines are closed. Whatever magnetic flux enters (or leaves) a certain part of a closed surface must leave (or enter) through the remainder of the closed surface, as illustrated in Fig. 3.29.

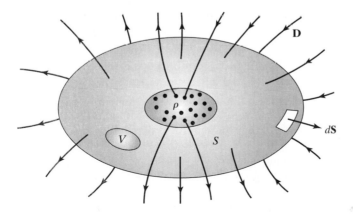

FIGURE 3.28   For illustrating Gauss' law for the electric field.

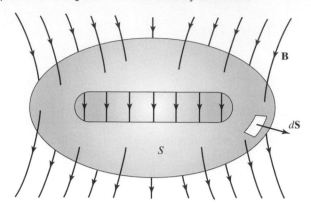

FIGURE 3.29   For illustrating Gauss' law for the magnetic field.

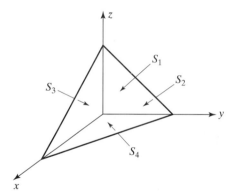

FIGURE 3.30   Slanted plane surface S_1 and surfaces $S_2, S_3,$ and S_4 in the coordinate planes.

This property of the magnetic field is sometimes useful in the computation of magnetic flux crossing a given surface (which is not closed). For example, to find the magnetic flux crossing the slanted plane surface S_1 in Fig. 3.30, it is not necessary to evaluate formally the surface integral of **B** over that surface. Since the slant surface S_1 and the three surfaces $S_2, S_3,$ and S_4 in the coordinate planes together form a closed surface, the required flux is the same as the net flux crossing the surfaces $S_2, S_3,$ and S_4. In fact, the net flux crossing the surfaces $S_2, S_3,$ and S_4 is the same as that crossing any nonplanar surface having the same periphery as that of S_1. Thus, as already pointed out in Section 3.2, it is a fundamental property of the magnetic field that the magnetic flux is the same through all surfaces bounded by a closed path, and hence *any surface S bounded by closed path C* can be used in Faraday's law.

In view of the foregoing discussion, it can be seen that Gauss' law for the magnetic field is not independent of Faraday's law. To show this mathematically, we consider the geometry shown in Fig. 3.23 and apply Faraday's law to the two closed paths to write

$$\oint_{C_1} \mathbf{E} \cdot d\mathbf{l} = -\frac{d}{dt} \int_{S_1} \mathbf{B} \cdot d\mathbf{S}_1$$

$$\oint_{C_2} \mathbf{E} \cdot d\mathbf{l} = -\frac{d}{dt} \int_{S_2} \mathbf{B} \cdot d\mathbf{S}_2$$

Adding the two equations, we obtain

$$0 = -\frac{d}{dt} \oint_{S_1 + S_2} \mathbf{B} \cdot d\mathbf{S}$$

or

$$\oint_{S_1 + S_2} \mathbf{B} \cdot d\mathbf{S} = \text{constant with time} \qquad (3.30)$$

Since there is no experimental evidence that the right side of (3.30) is nonzero, it follows that

$$\oint_S \mathbf{B} \cdot d\mathbf{S} = 0$$

where we have replaced $S_1 + S_2$ by S.

In a similar manner, Gauss' law for the electric field is not independent of Ampère's circuital law in view of the *law of conservation of charge*. The law of conservation of charge states that *the net current due to flow of charges emanating from a closed surface S is equal to the time rate of decrease of the charge within the volume V bounded by S.* It is given in mathematical form by

Law of conservation of charge

$$\boxed{\oint_S \mathbf{J} \cdot d\mathbf{S} = -\frac{d}{dt} \int_V \rho \, dv} \qquad (3.31)$$

As illustrated in Fig. 3.31, this law follows from the property that electric charge is conserved. If the charge in a given volume is decreasing with time at a certain rate, there must be a net outflow of the charge at the same rate. Since current is defined to be the rate of flow of charge, (3.31) then follows. As in the case of (3.17), it is understood that $\oint_S \mathbf{J} \cdot d\mathbf{S}$ in (3.31), although formulated in terms of $\mathbf{J}$, represents the algebraic sum of all currents due to flow of charges crossing S.

Comparing (3.21) and (3.31), we obtain

$$\frac{d}{dt} \oint_S \mathbf{D} \cdot d\mathbf{S} = \frac{d}{dt} \int_V \rho \, dv$$

$$\frac{d}{dt} \left(\oint_S \mathbf{D} \cdot d\mathbf{S} - \int_V \rho \, dv \right) = 0$$

$$\oint_S \mathbf{D} \cdot d\mathbf{S} - \int_V \rho \, dv = \text{constant with time} \qquad (3.32)$$

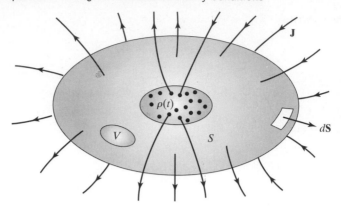

FIGURE 3.31   For illustrating the law of conservation of charge.

Since there is no experimental evidence that the right side of (3.32) is nonzero, it follows that

$$\oint_S \mathbf{D} \cdot d\mathbf{S} = \int_V \rho \, dv$$

Thus, since (3.21) follows from Ampère's circuital law, Gauss' law for the electric field follows from Ampère's circuital law with the aid of the law of conservation of charge.

   We shall now illustrate the combined application of Gauss' law for the electric field, the law of conservation of charge, and Ampère's circuital law by means of an example.

Example 3.5

Let us consider current I A flowing from a point charge $Q(t)$ at the origin to infinity along a semi-infinitely long straight wire occupying the positive z-axis, and find $\oint_C \mathbf{H} \cdot d\mathbf{l}$, where C is a circular path of radius a lying in the xy-plane and centered at the point charge, as shown in Fig. 3.32.

   Considering the hemispherical surface S bounded by C, and above the xy-plane, as shown in Fig. 3.32, and applying Ampère's circuital law, we obtain

$$\oint_C \mathbf{H} \cdot d\mathbf{l} = I + \frac{d}{dt} \int_S \mathbf{D} \cdot d\mathbf{S} \tag{3.33}$$

From Gauss' law for the electric field, the displacement flux emanating from a spherical surface centered at the point charge is equal to Q. In view of the spherical symmetry of the electric field about the point charge, half of the flux goes through the hemispherical surface. Thus,

$$\int_S \mathbf{D} \cdot d\mathbf{S} = \frac{Q}{2} \tag{3.34}$$

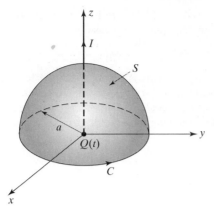

FIGURE 3.32   Semi-infinitely long wire of current I, with a point charge $Q(t)$ at the origin.

From the law of conservation of charge applied to a spherical surface centered at the point charge,

$$I = -\frac{dQ}{dt} \qquad (3.35)$$

Substituting (3.34) into (3.33) and then using (3.35), we obtain

$$\oint_C \mathbf{H} \cdot d\mathbf{l} = I + \frac{d}{dt}\left(\frac{Q}{2}\right)$$

$$= I + \frac{1}{2}\frac{dQ}{dt}$$

$$= I + \frac{1}{2}(-I)$$

$$= \frac{I}{2}$$

It should be noted that the same result holds for any contour C lying in any plane passing through the origin and surrounding the point charge $Q(t)$ and the wire in the right-hand sense as seen looking along the positive z-axis.

K3.4. Volume integral; Gauss' law for the electric field; Gauss' law for the magnetic field; Law of conservation of charge.

D3.10. Several types of charge are located, in Cartesian coordinates, as follows: a point charge of $1\ \mu C$ at $(1, 1, -1.5)$, a line charge of uniform density $2\ \mu C/m$ along the straight line from $(-1, -1, -1)$ to $(3, 3, 3)$, and a surface charge of uniform density $-1\ \mu C/m^2$ on that part of the plane $x = 0$ between $z = -1$ and $z = 1$. Find the displacement flux emanating from each of the following closed surfaces: **(a)** surface of the cubical box bounded by the planes $x = \pm 2$, $y = \pm 2$, and $z = \pm 2$; **(b)** surface of the cylindrical box of radius 2 m, having the z-axis as its axis and lying between $z = -2$ and $z = 2$; and **(c)** surface of the octahedron having its vertices at $(3, 0, 0), (-3, 0, 0), (0, 3, 0), (0, -3, 0), (0, 0, 3)$, and $(0, 0, -3)$. *Ans.* **(a)** $3.3923\ \mu C$; **(b)** $1.3631\ \mu C$; **(c)** $-3.0718\ \mu C$.

D3.11. Magnetic fluxes of absolute values ψ_1, ψ_2, and ψ_3 cross three surfaces S_1, S_2, and S_3, respectively, comprising a closed surface S. If $\psi_1 + \psi_2 + \psi_3 = \psi_0$, find the smallest of ψ_1, ψ_2, and ψ_3 for each of the following cases: **(a)** ψ_1, ψ_2, and ψ_3 are in arithmetic progression; **(b)** $1/\psi_1$, $1/\psi_2$, and $1/\psi_3$ are in arithmetic progression; and **(c)** $\ln \psi_1$, $\ln \psi_2$, and $\ln \psi_3$ are in arithmetic progression.

Ans.   **(a)** $\dfrac{1}{6}\psi_0$;   **(b)** $\dfrac{1}{2 + 2\sqrt{2}}\psi_0$;   **(c)** $\dfrac{1}{3 + \sqrt{5}}\psi_0$.

D3.12.   Three point charges $Q_1(t)$, $Q_2(t)$, and $Q_3(t)$ are situated at the vertices of a triangle and are connected by means of wires carrying currents. A current I A flows from Q_1 to Q_2 and $3I$ A flows from Q_2 to Q_3. The charge Q_3 is increasing with time at the rate of $5I$ C/s. Find the following: **(a)** dQ_1/dt; **(b)** dQ_2/dt; and **(c)** the current flowing from Q_1 to Q_3.
    Ans.   **(a)** $-3I$ C/s;   **(b)** $-2I$ C/s;   **(c)** $2I$ A.

3.5   APPLICATION TO STATIC FIELDS

Maxwell's equations in integral form for static fields

For static fields, that is, for $d/dt = 0$, Maxwell's equations in integral form become

$$\oint_C \mathbf{E} \cdot d\mathbf{l} = 0 \tag{3.36a}$$

$$\oint_C \mathbf{H} \cdot d\mathbf{l} = \int_S \mathbf{J} \cdot d\mathbf{S} \tag{3.36b}$$

$$\oint_S \mathbf{D} \cdot d\mathbf{S} = \int_V \rho \, dv \tag{3.36c}$$

$$\oint_S \mathbf{B} \cdot d\mathbf{S} = 0 \tag{3.36d}$$

whereas the law of conservation of charge becomes

$$\oint_S \mathbf{J} \cdot d\mathbf{S} = 0 \tag{3.37}$$

It can be immediately seen from (3.36a)–(3.36d) that the interdependence between the electric and magnetic fields no longer exists. Equation (3.36a) tells us simply that the static electric field is a conservative field. Similarly, (3.36d) tells us that the magnetic flux is the same through all surfaces bounded by a closed path. On the other hand, (3.36c) and (3.36b) enable us to find the static electric and magnetic fields for certain time-invariant charge and current distributions, respectively. These distributions must be such that the resulting electric and magnetic fields possess symmetry to be able to replace the integrals on the left sides of (3.36c) and (3.36b) by algebraic expressions involving the components of electric and magnetic fields, respectively.

In addition, in the case of (3.36b), the current on the right side must be uniquely given for a given closed path C, which property is ensured by (3.37). An example in which this current is uniquely given is that of the infinitely long wire in

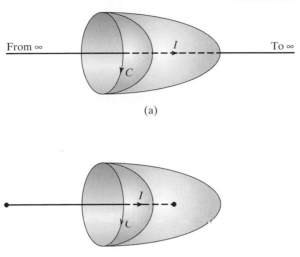

From ∞ I To ∞

C

(a)

I

C

(b)

FIGURE 3.33   For illustrating that the current enclosed by a
closed path C is uniquely given in (a) but not in (b).

Fig. 3.33(a). This is because the current crossing all possible surfaces bounded by
the closed path C is equal to I since the wire, being infinitely long, pierces through
all such surfaces. This can also be seen in a different manner by imagining the
closed path to be a rigid loop and visualizing that the loop cannot be moved to one
side of the wire without cutting the wire. On the other hand, if the wire is finitely
long, as shown in Fig. 3.33(b), it can be seen that for some surfaces bounded by C,
the wire pierces through the surface, whereas for some other surfaces, it does not.
Alternatively, a rigid loop occupying the closed path can be moved to one side of
the wire without cutting the wire. Thus, for this case, there is no unique value of the
wire current enclosed by C and hence (3.36b) cannot be used to determine **H**. The
problem here is that (3.37) is not satisfied since for current to flow in the finitely
long wire, there must be time-varying charges at the two ends, thereby giving rise
to time-varying electric field. Hence, a displacement current exists in addition to
the wire current such that the algebraic sum of the two currents crossing all surfaces bounded by C is the same and requires the use of (3.17).

    We shall now illustrate the application of (3.36c) and (3.36b) by means of
some examples.

Example 3.6

Let us consider charge distributed uniformly with density ρ_{L0} C/m along the z-axis and   **D** *due to a*
find the electric field due to the infinitely long line charge using (3.36c).   *line charge*

    Let us consider the closed surface S of a cylinder of radius r, with the line charge as
its axis and extending from $z = 0$ to $z = l$, as shown in Fig. 3.34. Then according to (3.36c),

$$\oint_S \mathbf{D} \cdot d\mathbf{S} = \rho_{L0} l \qquad (3.38)$$

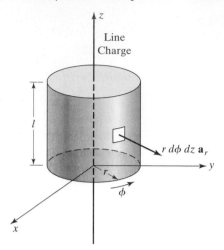

FIGURE 3.34   For the determination of electric field due to an infinitely long line charge of uniform density ρ_{L0} C/m.

Although this result is valid for any closed surface enclosing the portion of the line charge from $z = 0$ to $z = l$, we have chosen the particular surface in Fig. 3.34 to be able to reduce the surface integral of $\mathbf{D}$ in (3.36c) and hence in (3.38) to an algebraic quantity. To do this, we note the following:

a. In view of the uniform charge density, the entire line charge can be thought of as the superposition of pairs of equal point charges located at equal distances above and below any given point on the z-axis. Hence the field due to the entire line charge has only a radial component independent of ϕ and z.

b. In view of (a), the contribution to the closed surface integral from the top and bottom surfaces of the cylindrical box is zero.

Thus, we have

$$\mathbf{D} = D_r(r)\mathbf{a}_r$$

and

$$\oint_S \mathbf{D} \cdot d\mathbf{S} = \int_{\phi=0}^{2\pi} \int_{z=0}^{l} D_r(r)\mathbf{a}_r \cdot r \, d\phi \, dz \, \mathbf{a}_r$$

$$= 2\pi r l D_r(r)$$

(3.39)

Comparing (3.38) and (3.39), we obtain

$$2\pi r l D_r(r) = \rho_{L0} l$$

$$D_r(r) = \frac{\rho_{L0}}{2\pi r}$$

$$\boxed{\mathbf{D} = \frac{\rho_{L0}}{2\pi r} \mathbf{a}_r}$$

(3.40)

The field varies inversely with the radial distance away from the line charge.

Example 3.7

Let us consider charge distributed uniformly with density ρ_0 C/m^3 in the spherical region $r \leq a$, as shown by the cross-sectional view in Fig. 3.35, and find the electric field due to the spherical charge by using (3.36c).

D due to a spherical volume charge

As in Example 3.6, we shall once again choose a surface S that enables the replacement of the surface integral in (3.36c) by an algebraic quantity. To do this, we note from considerations of symmetry, and of the spherical charge as a superposition of point charges, that **D** possesses only an r-component dependent on r only. Thus,

$$\mathbf{D} = D_r(r)\mathbf{a}_r$$

Choosing then a spherical surface of radius r centered at the origin, we obtain

$$\oint_S \mathbf{D} \cdot d\mathbf{S} = \int_{\theta=0}^{\pi} \int_{\phi=0}^{2\pi} D_r(r)\mathbf{a}_r \cdot r^2 \sin\theta \, d\theta \, d\phi \, \mathbf{a}_r \tag{3.41}$$

$$= 4\pi r^2 D_r(r)$$

Noting that the charge exists only for $r < a$, and with uniform density, we obtain the charge enclosed by the spherical surface to be

$$\int_V \rho \, dv = \begin{cases} \frac{4}{3}\pi r^3 \rho_0 & \text{for} \quad r \leq a \\ \frac{4}{3}\pi a^3 \rho_0 & \text{for} \quad r \geq a \end{cases} \tag{3.42}$$

Substituting (3.41) and (3.42) into (3.36c), we get

$$4\pi r^2 D_r(r) = \begin{cases} \frac{4}{3}\pi r^3 \rho_0 & \text{for} \quad r \leq a \\ \frac{4}{3}\pi a^3 \rho_0 & \text{for} \quad r \geq a \end{cases}$$

$$D_r(r) = \begin{cases} \dfrac{\rho_0 r}{3} & \text{for} \quad r \leq a \\[2mm] \dfrac{\rho_0 a^3}{3r^2} & \text{for} \quad r \geq a \end{cases}$$

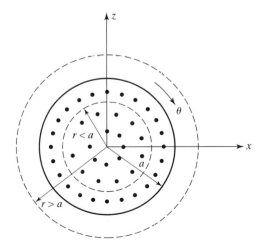

FIGURE 3.35   For the determination of electric field due to a spherical charge of uniform density ρ_0 C/m^3.

FIGURE 3.36   Variation of D_r with r for the spherical charge of Fig. 3.35.

$$\mathbf{D} = \begin{cases} \dfrac{\rho_0 r}{3}\,\mathbf{a}_r & \text{for} \quad r \leq a \\[2mm] \dfrac{\rho_0 a^3}{3r^2}\,\mathbf{a}_r & \text{for} \quad r \geq a \end{cases} \qquad (3.43)$$

The variation of D_r with r is shown plotted in Fig. 3.36.

Example 3.8

H *due to a*
cylindrical
wire of
current

Let us consider current flowing with uniform density $\mathbf{J} = J_0 \mathbf{a}_z$ A/m^2 in an infinitely long solid cylindrical wire of radius a with its axis along the z-axis, as shown by the cross-sectional view in Fig. 3.37. We wish to find the magnetic field everywhere using (3.36b).
      The current distribution can be thought of as the superposition of infinitely long filamentary wires parallel to the z-axis. Then in view of the symmetry about the z-axis and from the nature of the magnetic field due to an infinitely long wire given by (2.22), we can say that the required **H** has only a ϕ component dependent on r only. Thus,

$$\mathbf{H} = H_\phi(r)\mathbf{a}_\phi$$

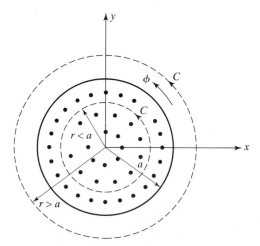

FIGURE 3.37   For the determination of magnetic field due to an infinitely long solid cylindrical wire of uniform current density $J_0 \mathbf{a}_z$ A/m^2.

Choosing then a circular closed path C of radius r lying in the xy-plane and centered at the origin, we obtain

$$\oint_C \mathbf{H} \cdot d\mathbf{I} = \int_{\phi=0}^{2\pi} H_\phi(r)\mathbf{a}_\phi \cdot r \, d\phi \, \mathbf{a}_\phi \qquad (3.44)$$

$$= 2\pi r H_\phi(r)$$

Considering the plane surface bounded by C, and noting that the current exists only for $r < a$, we obtain the current enclosed by the closed path to be

$$\int_S \mathbf{J} \cdot d\mathbf{S} = \begin{cases} \displaystyle\int_{r=0}^{r} \int_{\phi=0}^{2\pi} J_0 \mathbf{a}_z \cdot r \, dr \, d\phi \, \mathbf{a}_z & \text{for } r \leq a \\[3mm] \displaystyle\int_{r=0}^{a} \int_{\phi=0}^{2\pi} J_0 \mathbf{a}_z \cdot r \, dr \, d\phi \, \mathbf{a}_z & \text{for } r \geq a \end{cases} \qquad (3.45)$$

$$= \begin{cases} J_0 \pi r^2 & \text{for } r \leq a \\ J_0 \pi a^2 & \text{for } r \geq a \end{cases}$$

Substituting (3.44) and (3.45) into (3.36b), we get

$$2\pi r H_\phi = \begin{cases} J_0 \pi r^2 & \text{for } r \leq a \\ J_0 \pi a^2 & \text{for } r \geq a \end{cases}$$

$$H_\phi = \begin{cases} \dfrac{J_0 r}{2} & \text{for } r \leq a \\[3mm] \dfrac{J_0 a^2}{2r} & \text{for } r \geq a \end{cases}$$

$$\boxed{\mathbf{H} = \begin{cases} \dfrac{J_0 r}{2} \mathbf{a}_\phi & \text{for } r \leq a \\[3mm] \dfrac{J_0 a^2}{2r} \mathbf{a}_\phi & \text{for } r \geq a \end{cases}} \qquad (3.46)$$

The variation of H_ϕ with r is shown plotted in Fig. 3.38.

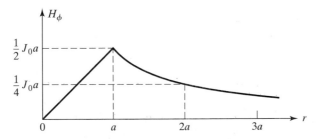

FIGURE 3.38 Variation of H_ϕ with r for the cylindrical wire of current of Fig. 3.37.

K3.5. Maxwell's equations in integral form for static fields; Uniqueness of current enclosed by a closed path; **D** due to symmetrical charge distributions; **H** due to symmetrical current distributions.

D3.13. Charge is distributed with uniform density ρ_0 C/m^3 inside a regular solid of edges a. Find the displacement flux emanating from one side of the solid for each of the following shapes of the solid: **(a)** tetrahedron; **(b)** cube; and **(c)** octahedron.
Ans.   **(a)** $0.0295\rho_0 a^3$ C;   **(b)** $0.1667\rho_0 a^3$ C;   **(c)** $0.0589\rho_0 a^3$ C.

D3.14. The cross section of an infinitely long solid wire having the z-axis as its axis is a regular polygon of sides a. Current flows in the wire with uniform density $J_0\mathbf{a}_z$ A/m^2. Find the line integral of **H** along one side of the polygon and traversed in the sense of increasing ϕ for each of the following shapes of the polygon: **(a)** equilateral triangle; **(b)** square; and **(c)** octagon.
Ans.   **(a)** $0.1443a^2 J_0$ A;   **(b)** $0.25a^2 J_0$ A;   **(c)** $0.6036a^2 J_0$ A.

3.6   BOUNDARY CONDITIONS

Boundary condition explained

In our study of electromagnetics, we will be considering many problems involving more than one medium. Examples are reflections of waves at an air–dielectric interface, determination of capacitance for a multiple-dielectric capacitor, and guiding of waves in a metallic waveguide. To solve a problem involving a boundary surface between different media, we need to know the conditions satisfied by the field components at the boundary. These are known as the *boundary conditions*. They are a set of relationships relating the field components at a point adjacent to and on one side of the boundary to the field components at a corresponding point adjacent to and on the other side of the boundary. These relationships arise from the fact that Maxwell's equations in integral form involve closed paths and surfaces and they must be satisfied for all possible closed paths and surfaces whether they lie entirely in one medium or encompass a portion of the boundary between two different media. In the latter case, Maxwell's equations in integral form must be satisfied collectively by the fields on either side of the boundary, thereby resulting in the boundary conditions.

We shall derive the boundary conditions by considering the Maxwell's equations

$$\oint_C \mathbf{E} \cdot d\mathbf{l} = -\frac{d}{dt}\int_S \mathbf{B} \cdot d\mathbf{S} \tag{3.47a}$$

$$\oint_C \mathbf{H} \cdot d\mathbf{l} = \int_S \mathbf{J} \cdot d\mathbf{S} + \frac{d}{dt}\int_S \mathbf{D} \cdot d\mathbf{S} \tag{3.47b}$$

$$\oint_S \mathbf{D} \cdot d\mathbf{S} = \int_V \rho\, dv \tag{3.47c}$$

$$\oint_S \mathbf{B} \cdot d\mathbf{S} = 0 \tag{3.47d}$$

and applying them one at a time to a closed path or a closed surface encompassing the boundary, and in the limit that the area enclosed by the closed path,

Medium 1

Medium 2

FIGURE 3.39 For deriving the boundary conditions resulting from Faraday's law and Ampère's circuital law.

or the volume bounded by the closed surface goes to zero. Thus, let us consider two semi-infinite media separated by a plane boundary, as shown in Fig. 3.39. Let us denote the quantities pertinent to medium 1 by subscript 1 and the quantities pertinent to medium 2 by subscript 2. Let $\mathbf{a}_n$ be the unit normal vector to the surface and directed into medium 1, as shown in Fig. 3.39, and let all normal components of fields at the boundary in both media denoted by an additional subscript n be directed along $\mathbf{a}_n$. Let the surface charge density (C/m^2) and the surface current density (A/m) on the boundary be ρ_S and $\mathbf{J}_S$, respectively. Note that, in general, the fields at the boundary in both media and the surface charge and current densities are functions of position on the boundary.

First, we consider a rectangular closed path *abcda* of infinitesimal area in the plane normal to the boundary and with its sides *ab* and *cd* parallel to and on either side of the boundary, as shown in Fig. 3.39. Applying Faraday's law (3.47a) to this path in the limit that *ad* and *bc* $\rightarrow 0$ by making the area *abcd* tend to zero, but with *ab* and *cd* remaining on either side of the boundary, we have

Boundary condition for $E_{tangential}$

$$\lim_{\substack{ad \to 0 \\ bc \to 0}} \oint_{abcda} \mathbf{E} \cdot d\mathbf{l} = -\lim_{\substack{ad \to 0 \\ bc \to 0}} \frac{d}{dt} \int_{\substack{area \\ abcd}} \mathbf{B} \cdot d\mathbf{S} \qquad (3.48)$$

In this limit, the contributions from *ad* and *bc* to the integral on the left side of (3.48) approach zero. Since *ab* and *cd* are infinitesimal, the sum of the contributions from *ab* and *cd* becomes $[E_{ab}(ab) + E_{cd}(cd)]$, where E_{ab} and E_{cd} are the components of $\mathbf{E}_1$ and $\mathbf{E}_2$ along *ab* and *cd*, respectively. The right side of (3.48) is equal to zero since the magnetic flux crossing the area *abcd* approaches zero as the area *abcd* tends to zero. Thus, (3.48) gives

$$E_{ab}(ab) + E_{cd}(cd) = 0$$

or, since *ab* and *cd* are equal, and $E_{dc} = -E_{cd}$,

$$E_{ab} - E_{dc} = 0 \qquad (3.49)$$

Let us now define $\mathbf{a}_s$ to be the unit vector normal to the area *abcd* and in the direction of advance of a right-hand screw as it is turned in the sense of the closed path *abcda*. Noting then that $\mathbf{a}_s \times \mathbf{a}_n$ is the unit vector along *ab*, we can write (3.49) as

$$\mathbf{a}_s \times \mathbf{a}_n \cdot (\mathbf{E}_1 - \mathbf{E}_2) = 0$$

Rearranging the order of the scalar triple product, we obtain

$$\mathbf{a}_s \cdot \mathbf{a}_n \times (\mathbf{E}_1 - \mathbf{E}_2) = 0 \tag{3.50}$$

Since we can choose the rectangle *abcd* to be in any plane normal to the boundary, (3.50) must be true for all orientations of $\mathbf{a}_s$. It then follows that

$$\boxed{\mathbf{a}_n \times (\mathbf{E}_1 - \mathbf{E}_2) = \mathbf{0}} \tag{3.51a}$$

or, in scalar form,

$$\boxed{E_{t1} - E_{t2} = 0} \tag{3.51b}$$

where E_{t1} and E_{t2} are the components of $\mathbf{E}_1$ and $\mathbf{E}_2$, respectively, tangential to the boundary. In words, (3.51a) and (3.51b) state that *at any point on the boundary, the components of* $\mathbf{E}_1$ *and* $\mathbf{E}_2$ *tangential to the boundary are equal.*

Boundary condition for H*tangential*   Similarly, applying Ampère's circuital law (3.47b) to the closed path in the limit that *ad* and *bc* → 0, we have

$$\lim_{\substack{ad \to 0 \\ bc \to 0}} \oint_{abcda} \mathbf{H} \cdot d\mathbf{l} = \lim_{\substack{ad \to 0 \\ bc \to 0}} \int_{\substack{area \\ abcd}} \mathbf{J} \cdot d\mathbf{S} + \lim_{\substack{ad \to 0 \\ bc \to 0}} \frac{d}{dt} \int_{\substack{area \\ abcd}} \mathbf{D} \cdot d\mathbf{S} \tag{3.52}$$

Using the same argument as for the left side of (3.48), we obtain the quantity on the left side of (3.52) to be equal to $[H_{ab}(ab) + H_{cd}(cd)]$, where H_{ab} and H_{cd} are the components of $\mathbf{H}_1$ and $\mathbf{H}_2$ along *ab* and *cd*, respectively. The second integral on the right side of (3.52) is zero since the displacement flux crossing the area *abcd* approaches zero as the area *abcd* tends to zero. The first integral on the right side of (3.52) would also be equal to zero but for a contribution from the surface current on the boundary because letting the area *abcd* tend to zero with *ab* and *cd* on either side of the boundary reduces only the volume current, if any, enclosed by it to zero, keeping the surface current still enclosed by it. This contribution is the surface current flowing normal to the line that *abcd* approaches as it tends to zero, that is, $[\mathbf{J}_S \cdot \mathbf{a}_s](ab)$. Thus, (3.52) gives

$$H_{ab}(ab) + H_{cd}(cd) = (\mathbf{J}_S \cdot \mathbf{a}_s)(ab)$$

or, since *ab* and *cd* are equal and $H_{dc} = -H_{cd}$,

$$H_{ab} - H_{dc} = \mathbf{J}_S \cdot \mathbf{a}_s \tag{3.53}$$

In terms of $\mathbf{H}_1$ and $\mathbf{H}_2$, we have

$$\mathbf{a}_s \times \mathbf{a}_n \cdot (\mathbf{H}_1 - \mathbf{H}_2) = \mathbf{J}_S \cdot \mathbf{a}_s$$

or

$$\mathbf{a}_s \cdot \mathbf{a}_n \times (\mathbf{H}_1 - \mathbf{H}_2) = \mathbf{a}_s \cdot \mathbf{J}_S \tag{3.54}$$

Since (3.54) must be true for all orientations of $\mathbf{a}_s$, that is, for a rectangle *abcd* in any plane normal to the boundary, it follows that

$$\mathbf{a}_n \times (\mathbf{H}_1 - \mathbf{H}_2) = \mathbf{J}_S \qquad (3.55a)$$

or, in scalar form,

$$H_{t1} - H_{t2} = J_S \qquad (3.55b)$$

where H_{t1} and H_{t2} are the components of $\mathbf{H}_1$ and $\mathbf{H}_2$, respectively, tangential to the boundary. In words, (3.55a) and (3.55b) state that *at any point on the boundary, the components of* $\mathbf{H}_1$ *and* $\mathbf{H}_2$ *tangential to the boundary are discontinuous by the amount equal to the surface current density at that point.* It should be noted that the information concerning the direction of $\mathbf{J}_S$ relative to that of $(\mathbf{H}_1 - \mathbf{H}_2)$, which is contained in (3.55a), is not present in (3.55b). Thus, in general, (3.55b) is not sufficient, and it is necessary to use (3.55a).

Now, we consider a rectangular box *abcdefgh* of infinitesimal volume enclosing an infinitesimal area of the boundary and parallel to it, as shown in Fig. 3.40. Applying Gauss' law for the electric field (3.47c) to this box in the limit that the side surfaces (abbreviated *ss*) tend to zero by making the volume of the box tend to zero but with the sides *abcd* and *efgh* remaining on either side of the boundary, we have *Boundary condition for D_{normal}*

$$\lim_{ss \to 0} \oint_{\substack{\text{surface} \\ \text{of the box}}} \mathbf{D} \cdot d\mathbf{S} = \lim_{ss \to 0} \int_{\substack{\text{volume} \\ \text{of the box}}} \rho \, dv \qquad (3.56)$$

In this limit, the contributions from the side surfaces to the integral on the left side of (3.56) approach zero. The sum of the contributions from the top and bottom surfaces becomes $[D_{n1}(abcd) - D_{n2}(efgh)]$ since *abcd* and *efgh* are infinitesimal. The quantity on the right side of (3.56) would be zero but for the surface charge on the boundary since letting the volume of the box tend to zero with the sides *abcd* and *efgh* on either side of it reduces only the volume charge, if any,

Medium 1

Medium 2

FIGURE 3.40 For deriving the boundary conditions resulting from the two Gauss' laws.

enclosed by it to zero, keeping the surface charge still enclosed by it. This surface charge is equal to $\rho_S(abcd)$. Thus, (3.56) gives

$$D_{n1}(abcd) - D_{n2}(efgh) = \rho_S(abcd)$$

or, since $abcd$ and $efgh$ are equal,

$$\boxed{D_{n1} - D_{n2} = \rho_S} \tag{3.57a}$$

In terms of $\mathbf{D}_1$ and $\mathbf{D}_2$, (3.57a) is given by

$$\boxed{\mathbf{a}_n \cdot (\mathbf{D}_1 - \mathbf{D}_2) = \rho_S} \tag{3.57b}$$

In words, (3.57a) and (3.57b) state that *at any point on the boundary, the components of $\mathbf{D}_1$ and $\mathbf{D}_2$ normal to the boundary are discontinuous by the amount of the surface charge density at that point.*

Boundary condition for B_{normal}   Similarly, applying Gauss' law for the magnetic field (3.47d) to the box $abcdefgh$ in the limit that the side surfaces tend to zero, we have

$$\lim_{ss \to 0} \oint_{\substack{\text{surface} \\ \text{of the box}}} \mathbf{B} \cdot d\mathbf{S} = 0 \tag{3.58}$$

Using the same argument as for the left side of (3.56), we obtain the quantity on the left side of (3.58) to be equal to $[B_{n1}(abcd) - B_{n2}(efgh)]$. Thus, (3.58) gives

$$B_{n1}(abcd) - B_{n2}(efgh) = 0$$

or, since $abcd$ and $efgh$ are equal

$$\boxed{B_{n1} - B_{n2} = 0} \tag{3.59a}$$

In terms of $\mathbf{B}_1$ and $\mathbf{B}_2$, (3.59a) is given by

$$\boxed{\mathbf{a}_n \cdot (\mathbf{B}_1 - \mathbf{B}_2) = 0} \tag{3.59b}$$

In words, (3.59a) and (3.59b) state that *at any point on the boundary, the components of $\mathbf{B}_1$ and $\mathbf{B}_2$ normal to the boundary are equal.*

Summarizing the boundary conditions, we have

$$
\boxed{
\begin{aligned}
\mathbf{a}_n \times (\mathbf{E}_1 - \mathbf{E}_2) &= \mathbf{0} & \text{(3.60a)} \\
\mathbf{a}_n \times (\mathbf{H}_1 - \mathbf{H}_2) &= \mathbf{J}_S & \text{(3.60b)} \\
\mathbf{a}_n \cdot (\mathbf{D}_1 - \mathbf{D}_2) &= \rho_S & \text{(3.60c)} \\
\mathbf{a}_n \cdot (\mathbf{B}_1 - \mathbf{B}_2) &= 0 & \text{(3.60d)}
\end{aligned}
}
$$

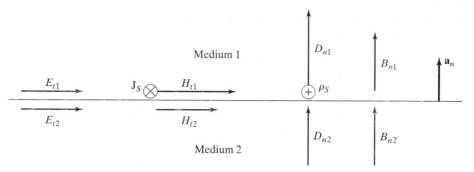

FIGURE 3.41 For illustrating the boundary conditions at an interface between two different media.

or, in scalar form,

$$E_{t1} - E_{t2} = 0 \tag{3.61a}$$

$$H_{t1} - H_{t2} = J_S \tag{3.61b}$$

$$D_{n1} - D_{n2} = \rho_S \tag{3.61c}$$

$$B_{n1} - B_{n2} = 0 \tag{3.61d}$$

as illustrated in Fig. 3.41. Although we have derived these boundary conditions by considering a plane interface between the two media, it should be obvious that we can consider any arbitrary-shaped boundary and obtain the same results by letting the sides ab and cd of the rectangle and the top and bottom surfaces of the box tend to zero, in addition to the limits that the sides ad and bc of the rectangle and the side surfaces of the box tend to zero. The boundary conditions given by (3.60a)–(3.60d) are general. When they are applied to particular cases, the special properties of the pertinent media come into play. Two such properties are that (1) no time-varying fields can exist inside a perfect conductor ($\sigma = \infty$) and (2) there can be no free charge and no current due to motion of free charges associated with the surface of a perfect dielectric ($\sigma = 0$). We shall now consider an example.

Example 3.9

In Fig. 3.42, the region $x < 0$ is a perfect conductor, the region $0 < x < d$ is a perfect dielectric of $\varepsilon = 2\varepsilon_0$ and $\mu = \mu_0$, and the region $x > d$ is free space. The electric and magnetic fields in the region $0 < x < d$ are given at a particular instant of time by

$$\mathbf{E} = E_1 \cos \pi x \sin 2\pi z \, \mathbf{a}_x + E_2 \sin \pi x \cos 2\pi z \, \mathbf{a}_z$$

$$\mathbf{H} = H_1 \cos \pi x \sin 2\pi z \, \mathbf{a}_y$$

We wish to find (a) ρ_S and $\mathbf{J}_S$ on the surface $x = 0$ and (b) $\mathbf{E}$ and $\mathbf{H}$ for $x = d+$, that is, immediately adjacent to the $x = d$-plane and on the free-space side, at that instant of time.

FIGURE 3.42   For illustrating the application of boundary conditions.

a. Denoting the perfect dielectric medium ($0 < x < d$) to be medium 1 and the perfect conductor medium ($x < 0$) to be medium 2, we have $\mathbf{a}_n = \mathbf{a}_x$, and all fields with subscript 2 are equal to zero. Then from (3.60c) and (3.60b), we obtain

$$[\rho_S]_{x=0} = \mathbf{a}_n \cdot [\mathbf{D}_1]_{x=0} = \mathbf{a}_x \cdot 2\varepsilon_0 E_1 \sin 2\pi z \,\mathbf{a}_x$$

$$= 2\varepsilon_0 E_1 \sin 2\pi z$$

$$[\mathbf{J}_S]_{x=0} = \mathbf{a}_n \times [\mathbf{H}_1]_{x=0} = \mathbf{a}_x \times H_1 \sin 2\pi z \,\mathbf{a}_y$$

$$= H_1 \sin 2\pi z \,\mathbf{a}_z$$

Note that the remaining two boundary conditions (3.60a) and (3.60d) are already satisfied by the given fields since E_y and B_x do not exist and for $x = 0$, $E_z = 0$.

b. Denoting the perfect dielectric medium ($0 < x < d$) to be medium 1 and the free-space medium ($x > d$) to be medium 2 and setting $\rho_S = 0$, we obtain from (3.61a) and (3.61c)

$$[E_y]_{x=d+} = [E_y]_{x=d-} = 0$$

$$[E_z]_{x=d+} = [E_z]_{x=d-} = E_2 \sin \pi d \cos 2\pi z$$

$$[D_x]_{x=d+} = [D_x]_{x=d-} = 2\varepsilon_0 [E_x]_{x=d-}$$

$$= 2\varepsilon_0 E_1 \cos \pi d \sin 2\pi z$$

$$[E_x]_{x=d+} = \frac{1}{\varepsilon_0}[D_x]_{x=d+}$$

$$= 2E_1 \cos \pi d \sin 2\pi z$$

Thus

$$[\mathbf{E}]_{x=d+} = 2E_1 \cos \pi d \sin 2\pi z \,\mathbf{a}_x + E_2 \sin \pi d \cos 2\pi z \,\mathbf{a}_z$$

Setting $\mathbf{J}_S = \mathbf{0}$ and using (3.61b) and (3.61d), we obtain

$$[H_y]_{x=d+} = [H_y]_{x=d-} = H_1 \cos \pi d \sin 2\pi z$$

$$[H_z]_{x=d+} = [H_z]_{x=d-} = 0$$

$$[B_x]_{x=d+} = [B_x]_{x=d-} = 0$$

Thus,

$$[\mathbf{H}]_{x=d+} = H_1 \cos \pi d \sin 2\pi z \,\mathbf{a}_y$$

K3.6. Boundary conditions; Tangential component of $\mathbf{E}$; Tangential component of $\mathbf{H}$; Normal component of $\mathbf{D}$; Normal component of $\mathbf{B}$.

D3.15. For each of the following values of the displacement flux density at a point on the surface of a perfect conductor (no electric field inside and hence $E_t = 0$ on the surface), find the surface charge density at that point: **(a)** $\mathbf{D} = D_0(\mathbf{a}_x - 2\mathbf{a}_y + 2\mathbf{a}_z)$ and pointing away from the surface; **(b)** $\mathbf{D} = D_0(\mathbf{a}_x + \sqrt{3}\mathbf{a}_z)$ and pointing toward the surface; and **(c)** $\mathbf{D} = D_0(0.8\mathbf{a}_x + 0.6\mathbf{a}_y)$ and pointing away from the surface. Assume D_0 to be positive for all cases.

Ans. **(a)** $3D_0$; **(b)** $-2D_0$; **(c)** D_0.

D3.16. The region $x > 0$ is a perfect dielectric of permittivity $2\varepsilon_0$ and the region $x < 0$ is a perfect dielectric of permittivity $3\varepsilon_0$. Consider the field components at point 1 on the $+x$-side of the boundary to be denoted by subscript 1 and the field components at the adjacent point 2 on the $-x$-side of the boundary to be denoted by subscript 2. If $\mathbf{E}_1 = E_0(2\mathbf{a}_x + \mathbf{a}_y)$, find the following: **(a)** E_{x1}/E_{x2}; **(b)** E_1/E_2; and **(c)** D_1/D_2.

Ans. **(a)** 1.5; **(b)** $3/\sqrt{5}$; **(c)** $2/\sqrt{5}$.

D3.17. The plane $z = 0$ forms the boundary between free space ($z > 0$) and another medium. Find the following: **(a)** $\mathbf{J}_S(0,0,0)$ at $t = 0$ if $z < 0$ is a perfect conductor and $\mathbf{H}(0, 0, 0+) = H_0(3\mathbf{a}_x - 4\mathbf{a}_y) \cos \omega t$; **(b)** $\mathbf{H}(0, 0, 0+)$ if $z < 0$ is a magnetic material of $\mu = 20\mu_0$ and $\mathbf{H}(0, 0, 0-) = H_0(10\mathbf{a}_x + \mathbf{a}_z)$; and **(c)** the ratio of $B(0,0,0-)$ to $B(0,0,0+)$ for the case of **(b)**.

Ans. **(a)** $H_0(4\mathbf{a}_x + 3\mathbf{a}_y)$; **(b)** $10H_0(\mathbf{a}_x + 2\mathbf{a}_z)$; **(c)** 8.989.

3.7 SUMMARY

We first learned in this chapter how to evaluate line and surface integrals of vector quantities and then we introduced Maxwell's equations in integral form. These equations, which form the basis of electromagnetic field theory, are given as follows in words and in mathematical form:

Faraday's law. The electromotive force around a closed path C is equal to the negative of the time rate of change of the magnetic flux enclosed by that path; that is,

$$\oint_C \mathbf{F} \cdot d\mathbf{l} = -\frac{d}{dt} \int_S \mathbf{B} \cdot d\mathbf{S} \tag{3.62}$$

Ampère's circuital law. The magnetomotive force around a closed path C is equal to the sum of the current enclosed by that path due to the actual flow of charges and the displacement current due to the time rate of change of the displacement flux enclosed by that path; that is,

$$\oint_C \mathbf{H} \cdot d\mathbf{l} = \int_S \mathbf{J} \cdot d\mathbf{S} + \frac{d}{dt} \int_S \mathbf{D} \cdot d\mathbf{S} \tag{3.63}$$

Gauss' law for the electric field. The displacement flux emanating from a closed surface S is equal to the charge enclosed by that surface; that is,

$$\oint_S \mathbf{D} \cdot d\mathbf{S} = \int_V \rho \, dv \tag{3.64}$$

Gauss' law for the magnetic field. The magnetic flux emanating from a closed surface S is equal to zero; that is,

$$\oint_S \mathbf{B} \cdot d\mathbf{S} = 0 \tag{3.65}$$

An auxiliary equation, the law of conservation of charge, is given by

$$\oint_S \mathbf{J} \cdot d\mathbf{S} = -\frac{d}{dt} \int_V \rho \, dv \tag{3.66}$$

In words, (3.66) states that the current due to flow of charges emanating from a closed surface is equal to the time rate of decrease of the charge enclosed by that surface.

In using (3.62)–(3.66), we recall that

$$\mathbf{D} = \varepsilon \mathbf{E} \tag{3.67}$$

$$\mathbf{H} = \frac{\mathbf{B}}{\mu} \tag{3.68}$$

where ε and μ are the permittivity and permeability, respectively, of the medium. In addition, if the current density $\mathbf{J}$ is due to conduction, then

$$\mathbf{J} = \mathbf{J}_c = \sigma \mathbf{E} \tag{3.69}$$

where σ is the conductivity of the medium. In evaluating the right sides of (3.62) and (3.63), the normal vectors to the surfaces must be chosen such that they are directed in the right-hand sense, that is, toward the side of advance of a right-hand screw as it is turned around C. In (3.64), (3.65), and (3.66), it is understood that the surface integrals are evaluated so as to find the flux outward from the volume bounded by the surface. We also learned that (3.65) is not independent of (3.62) and that (3.64) follows from (3.63) with the aid of (3.66).

We discussed several applications of Maxwell's equations, including the computation of static electric and magnetic fields due to symmetrical charge and current distributions, respectively. Finally, we derived the boundary conditions resulting from the application of Maxwell's equations to closed paths and closed surfaces encompassing the boundary between two media, and in the limits that the areas enclosed by the closed paths and the volumes bounded by the closed surfaces go to zero. These boundary conditions are given by

$$\mathbf{a}_n \times (\mathbf{E}_1 - \mathbf{E}_2) = \mathbf{0} \tag{3.70a}$$

$$\mathbf{a}_n \times (\mathbf{H}_1 - \mathbf{H}_2) = \mathbf{J}_S \tag{3.70b}$$

$$\mathbf{a}_n \cdot (\mathbf{D}_1 - \mathbf{D}_2) = \rho_S \tag{3.70c}$$

$$\mathbf{a}_n \cdot (\mathbf{B}_1 - \mathbf{B}_2) = 0 \tag{3.70d}$$

where the subscripts 1 and 2 refer to media 1 and 2, respectively, and $\mathbf{a}_n$ is unit vector normal to the boundary at the point under consideration and directed into medium 1. In words, the boundary conditions state that at a point on the boundary, the tangential components of $\mathbf{E}$ and the normal components of $\mathbf{B}$ are continuous, whereas the tangential components of $\mathbf{H}$ are discontinuous by the amount equal to J_S at that point, and the normal components of **D** are discontinuous by the amount equal to ρ_S at that point.

REVIEW QUESTIONS

Q3.1. How do you find the work done in moving a test charge by an infinitesimal distance in an electric field? What is the amount of work involved in moving the test charge normal to the electric field?

Q3.2. What is the physical interpretation of the line integral of **E** between two points A and B?

Q3.3. How do you find the approximate value of the line integral of a vector field along a given path? How do you find the exact value of the line integral?

Q3.4. Discuss conservative versus nonconservative fields, giving examples.

Q3.5. How do you find the magnetic flux crossing an infinitesimal surface?

Q3.6. What is the magnetic flux crossing an infinitesimal surface oriented parallel to the magnetic flux density vector? For what orientation of the infinitesimal surface relative to the magnetic flux density vector is the magnetic flux crossing the surface a maximum?

Q3.7. How do you find the approximate value of the surface integral of a vector field over a given surface? How do you find the exact value of the surface integral?

Q3.8. Provide physical interpretations for the closed surface integrals of any two vectors of your choice.

Q3.9. State Faraday's law.

Q3.10. What are the different ways in which an emf is induced around a loop?

Q3.11. Discuss the right-hand screw rule convention associated with the application of Faraday's law.

Q3.12. To find the induced emf around a planar loop, is it necessary to consider the magnetic flux crossing the plane surface bounded by the loop? Explain.

Q3.13. What is Lenz's law?

Q3.14. Discuss briefly the motional emf concept.

Q3.15. How would you orient a loop antenna to obtain maximum signal from an incident electromagnetic wave that has its magnetic field linearly polarized in the north–south direction?

Q3.16. State three applications of Faraday's law.

Q3.17. State Ampère's circuital law.

Q3.18. What is displacement current? Compare and contrast displacement current with current due to flow of charges.

Q3.19. Is it meaningful to consider two different surfaces bounded by a closed path to compute the two different currents on the right side of Ampère's circuital law to find $\oint \mathbf{H} \cdot d\mathbf{l}$ around the closed path?

Q3.20. Discuss the relationship between the displacement current emanating from a closed surface and the current due to flow of charges emanating from the same closed surface.

Q3.21. Give an example involving displacement current.

Q3.22. Discuss briefly the principle of radiation from a wire carrying a time-varying current.

Q3.23. State Gauss' law for the electric field.

Q3.24. How do you evaluate a volume integral?

Q3.25. State Gauss' law for the magnetic field.

Q3.26. What is the physical interpretation of Gauss' law for the magnetic field?

Q3.27. Discuss the dependence of Gauss' law for the magnetic field on Faraday's law.

Q3.28. State the law of conservation of charge.

Q3.29. How is Gauss' law for the electric field dependent on Ampère's circuital law?

Q3.30. Summarize Maxwell's equations in integral form for time-varying fields.

Q3.31. Summarize Maxwell's equations in integral form for static fields.

Q3.32. Are static electric and magnetic fields interdependent? Explain.

Q3.33. Discuss briefly the application of Gauss' law for the electric field to determine the electric field due to charge distributions.

Q3.34. When can you say that the current in a wire enclosed by a closed path is uniquely defined? Give two examples.

Q3.35. Give an example in which the current in a wire enclosed by a closed path is not uniquely defined. Is it correct to apply Ampère's circuital law for the static case in such a situation? Explain.

Q3.36. Discuss briefly the application of Ampère's circuital law to determine the magnetic field due to current distributions.

Q3.37. What is a boundary condition? How do boundary conditions arise?

Q3.38. Summarize the boundary conditions for the general case of a boundary between two arbitrary media, indicating correspondingly the Maxwell's equations in integral form from which they are derived.

Q3.39. Discuss the boundary conditions on the surface of a perfect conductor.

Q3.40. Discuss the boundary conditions at the interface between two perfect dielectric media.

PROBLEMS

Section 3.1

P3.1. For the vector field $\mathbf{F} = y\mathbf{a}_x - z\mathbf{a}_y + x\mathbf{a}_z$, find $\int_{(0,0,0)}^{(1,1,1)} \mathbf{F} \cdot d\mathbf{l}$ for each of the following paths from $(0,0,0)$ to $(1,1,1)$: **(a)** $x = y = z$ and **(b)** $x = y = z^3$.

P3.2. Given $\mathbf{F} = xy\mathbf{a}_x + yz\mathbf{a}_y + zx\mathbf{a}_z$, find $\oint_C \mathbf{F} \cdot d\mathbf{l}$, where C is the closed path comprising the straight lines from $(0,0,0)$ to $(1,1,1)$, from $(1,1,1)$ to $(1,1,0)$, and from $(1,1,0)$ to $(0,0,0)$.

P3.3. For the vector field $\mathbf{F} = \cos y \, \mathbf{a}_x - x \sin y \, \mathbf{a}_y$, find $\int_{(0,0,0)}^{(1,2\pi,1)} \mathbf{F} \cdot d\mathbf{l}$ in each of the following ways: **(a)** along the straight-line path between the two points; **(b)** along the curved path $x = z = \sin(y/4)$ between the two points; and **(c)** without choosing any particular path. Is the vector field conservative or nonconservative? Explain.

P3.4. Given $\mathbf{A} = 2r \sin \phi \, \mathbf{a}_r + r^2 \mathbf{a}_\phi + z\mathbf{a}_z$ in cylindrical coordinates, find $\oint_C \mathbf{A} \cdot d\mathbf{l}$, where C is the closed path comprising the straight line from $(0,0,0)$ to $(1,0,0)$, the circular arc from $(1,0,0)$ to $(1,\pi/2,0)$ through $(1,\pi/4,0)$, the straight line from $(1,\pi/2,0)$ to $(1,\pi/2,1)$, and the straight line from $(1,\pi/2,1)$ to $(0,0,0)$.

P3.5. Given $\mathbf{A} = e^{-r}(\cos \theta \, \mathbf{a}_r + \sin \theta \, \mathbf{a}_\theta) + r \sin \theta \, \mathbf{a}_\phi$ in spherical coordinates, find $\int \mathbf{A} \cdot d\mathbf{l}$ for each of the following paths: **(a)** straight-line path from $(0,0,0)$ to $(2,0,0)$; **(b)** circular arc from $(2,0,\pi/4)$ to $(2,\pi/2,\pi/4)$ through $(2,\pi/4,\pi/4)$; and **(c)** circular arc from $(2,\pi/6,0)$ to $(2,\pi/6,\pi/2)$ through $(2,\pi/6,\pi/4)$.

P3.6. Given $\mathbf{A} = x^2yz\mathbf{a}_x + y^2zx\mathbf{a}_y + z^2xy\mathbf{a}_z$, evaluate $\oint_S \mathbf{A} \cdot d\mathbf{S}$, where S is the surface of the cubical box bounded by the planes $x = 0, x = 1, y = 0, y = 1, z = 0$, and $z = 1$.

P3.7. Given $\mathbf{A} = (x^2y + 2)\mathbf{a}_x + 3\mathbf{a}_y - 2xyz\mathbf{a}_z$, evaluate $\oint_S \mathbf{A} \cdot d\mathbf{S}$, where S is the surface of the rectangular box bounded by the planes $x = 0, x = 1, y = 0, y = 2$, $z = 0$, and $z = 3$.

P3.8. Given $\mathbf{A} = r \cos \phi \, \mathbf{a}_r - r \sin \phi \, \mathbf{a}_\phi$ in cylindrical coordinates, evaluate $\oint_S \mathbf{A} \cdot d\mathbf{S}$, where S is the surface of the box bounded by the plane surfaces $\phi = 0, \phi = \pi/2$, $z = 0, z = 1$, and the cylindrical surface $r = 2, 0 < \phi < \pi/2$.

P3.9. Given $\mathbf{A} = r^2\mathbf{a}_r + r \sin \theta \, \mathbf{a}_\theta$ in spherical coordinates, find $\oint_S \mathbf{A} \cdot d\mathbf{S}$, where S is the surface of that part of the spherical volume of radius unity and lying in the first octant.

Section 3.2

P3.10. Find the induced emf around the rectangular closed path C connecting the points $(0,0,0), (a,0,0), (a,b,0), (0,b,0)$, and $(0,0,0)$ in that order for each of the following magnetic fields:

(a) $\mathbf{B} = \dfrac{B_0a^2}{(x + a)^2}e^{-t}\mathbf{a}_z$

(b) $\mathbf{B} = B_0 \sin \dfrac{\pi x}{a} \cos \omega t \, \mathbf{a}_z$

P3.11. A magnetic field is given in the xz-plane by $\mathbf{B} = (B_0/x)\mathbf{a}_y$ Wb/m^2, where B_0 is a constant. A rigid rectangular loop is situated in the xz-plane and with its corners at the points $(x_0, z_0), (x_0, z_0 + b), (x_0 + a, z_0 + b)$, and $(x_0 + a, z_0)$. If the loop is moving in that plane with a velocity $\mathbf{v} = v_0\mathbf{a}_x$ m/s, where v_0 is a constant, find by using Faraday's law the induced emf around the loop in the sense defined by connecting the above points in succession. Discuss your result by using the motional emf concept.

P3.12. A magnetic field is given in the xz-plane by $\mathbf{B} = B_0 \cos \pi(x - v_0t) \, \mathbf{a}_y$ Wb/m^2. Consider a rigid square loop situated in the xz-plane with its vertices at $(x, 0, 1)$, $(x, 0, 2), (x + 1, 0, 2)$, and $(x + 1, 0, 1)$. **(a)** Find the expression for the emf induced around the loop in the sense defined by connecting the above points in succession. **(b)** What would be the induced emf if the loop is moving with the velocity $\mathbf{v} = v_0\mathbf{a}_x$ m/s instead of being stationary?

P3.13. A rigid rectangular loop of metallic wire is hung by pivoting one side along the x-axis, as shown in Fig. 3.43. The loop is free to swing about the pivoted side

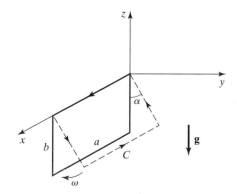

FIGURE 3.43   For Problem P3.13.

without friction under the influence of gravity and in the presence of a uniform magnetic field $\mathbf{B} = B_0 \mathbf{a}_z$ Wb/m². If the loop is given a slight angular displacement and released, show that the emf induced around the closed path C of the loop is approximately equal to $B_0 ab\omega$, where ω is the angular velocity of swing of the loop toward the vertical. Does the loop swing faster or slower than in the absence of the magnetic field? Explain.

P3.14. A rigid conducting bar of length L, mass M, and electrical resistance R rolls without friction down two parallel conducting rails that are inclined at an angle α with the horizontal, as shown in Fig. 3.44. The rails are of negligible resistance and are joined at the bottom by another conductor, also of negligible resistance, so that the total resistance of the loop formed by the rolling bar and the three other sides is R. The entire arrangement is situated in a region of uniform static magnetic field $\mathbf{B} = B_0 \mathbf{a}_z$ Wb/m², directed vertically downward. Assume the bar to be rolling down with uniform velocity $\mathbf{v}$ parallel to the rails under the influence of Earth's gravity (acting in the positive z-direction) and the magnetic force due to the current in the loop produced by the induced emf. Show that v is equal to $(MgR/B_0^2 L^2) \tan \alpha \sec \alpha$.

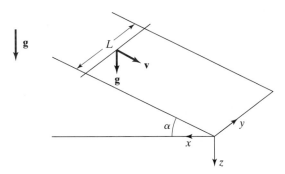

FIGURE 3.44   For Problem P3.14.

P3.15. A rigid rectangular loop of base b and height h situated normal to the xy-plane and with its sides pivoted to the z-axis revolves about the z-axis with angular velocity ω rad/s in the sense of increasing ϕ, as shown in Fig. 3.45. Find the induced emf around the closed path C of the loop for each of the following magnetic fields: **(a)** $\mathbf{B} = B_0 \mathbf{a}_y$ Wb/m² and **(b)** $\mathbf{B} = B_0(y\mathbf{a}_x - x\mathbf{a}_y)$ Wb/m². Assume the loop to be in the xz-plane at $t = 0$.

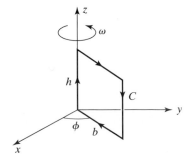

FIGURE 3.45   For Problem P3.15.

P3.16. A rigid rectangular loop of area A is situated in the xz-plane and symmetrically about the z-axis, as shown in Fig. 3.46, in a region of magnetic field $\mathbf{B} = B_0(\sin \omega t \, \mathbf{a}_x + \cos \omega t \, \mathbf{a}_y)$ Wb/m². Find the induced emf around the closed path C of the loop for each of the following cases: **(a)** the loop is stationary; **(b)** the loop revolves around the z-axis in the sense of increasing ϕ with uniform angular velocity of ω rad/s; and **(c)** the loop revolves around the z-axis in the sense of decreasing ϕ with uniform angular velocity of ω rad/s. For parts (b) and (c), assume that the loop is in the xz-plane at $t = 0$.

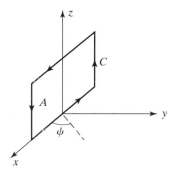

FIGURE 3.46   For Problem P3.16.

Section 3.3

P3.17. Given that $\mathbf{H} = \pm H_0(t \mp \sqrt{\mu_0 \varepsilon_0} \, z)^2 \mathbf{a}_y$ and $\mathbf{D} = \sqrt{\mu_0 \varepsilon_0} \, H_0(t \mp \sqrt{\mu_0 \varepsilon_0} \, z)^2 \mathbf{a}_x$ for $z \gtrless 0$, find the current due to flow of charges enclosed by the rectangular closed path from $(0, 0, 1)$ to $(0, 1, 1)$ to $(0, 1, -1)$ to $(0, 0, -1)$ to $(0, 0, 1)$.

P3.18. A current density due to flow of charges is given by $\mathbf{J} = -(x\mathbf{a}_x + y\mathbf{a}_y + z^2\mathbf{a}_z)$ A/m². Find the displacement current emanating from each of the following closed surfaces: **(a)** the surface of the cubical box bounded by the planes $x = \pm 2, y = \pm 2$, and $z = \pm 2$, and **(b)** the surface of the cylindrical box bounded by the surfaces $r = 1, z = 0$, and $z = 2$.

P3.19. A voltage source connected to a parallel-plate capacitor by means of wires sets up a uniform electric field of $E = 180 \sin 2\pi \times 10^6 t \sin 4\pi \times 10^6 t$ V/m between the plates of the capacitor and normal to the plates. Assume that no field exists outside the region between the plates. If the area of each plate is 0.1 m² and the medium between the plates is free space, find the root-mean-square value of the current drawn from the voltage source.

P3.20. Assume that the time variation of the electric field in Problem P3.19 is as shown in Fig. 3.47. Find and plot versus time the current drawn from the voltage source. What is the root-mean-square value of the current?

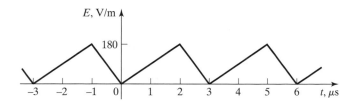

FIGURE 3.47   For Problem P3.20.

P3.21. A voltage source connected to a parallel-plate capacitor by means of wires sets up a uniform electric field of $E = 180 \cos 2\pi \times 10^6 t$ V/m between the plates of the capacitor and normal to the plates. Assume that no field exists outside the region between the plates. If the area of each plate is 0.1 m^2 and the medium between the plates is a material characterized by $\sigma = 10^{-3}$ S/m and $\varepsilon = 18\varepsilon_0$, find the expression for the current drawn from the voltage source as a single cosinusoidal function of time.

Section 3.4

P3.22. For each of the following charge distributions, find the displacement flux emanating from the surface enclosing the charge: **(a)** $\rho(x, y, z) = \rho_0(x + y + z)^2$ for $0 < x < 1, 0 < y < 1$, and $0 < z < 1$; **(b)** $\rho(r, \phi, z) = \rho_0 \cos^2 \phi$ for $r < 1, 0 < \phi < \pi/2$, and $0 < z < 1$ in cylindrical coordinates; and **(c)** $\rho(r, \theta, \phi) = (\rho_0/r) \sin \theta$ for $r < 1, 0 < \theta < \pi/2$ in spherical coordinates.

P3.23. Using the property that $\oint_S \mathbf{B} \cdot d\mathbf{S} = 0$, find the absolute value of the magnetic flux crossing that portion of the surface $y = \sin x$ bounded by $x = 0, x = \pi$, $z = 0$, and $z = 1$ for $\mathbf{B} = B_0(y\mathbf{a}_x - x\mathbf{a}_y)$ Wb/m^2.

P3.24. Given $\mathbf{J} = (x\mathbf{a}_x + y\mathbf{a}_y + z\mathbf{a}_z)$ A/m^2, find the time rate of decrease of the charge contained within each of the following volumes: **(a)** volume bounded by the planes $x = 0, x = 1, y = 0, y = 1, z = 0$, and $z = 1$; **(b)** volume bounded by the cylinders $r = 1$ and $r = 2$ and the planes $z = 0$ and $z = 1$; and **(c)** volume bounded by the spherical surfaces $r = 1$ and $r = 2$ and the conical surface $\theta = \pi/3$.

P3.25. Current I flows along a straight wire form a point charge $Q_1(t)$ located at the origin to a point charge $Q_2(t)$ located at $(0, 0, 1)$. Find the line integral of $\mathbf{H}$ along the square closed path having the vertices at $(1, 1, 0), (-1, 1, 0), (-1, -1, 0)$, and $(1, -1, 0)$ and traversed in that order.

P3.26. Current I flows along a straight wire from a point charge $Q_1(t)$ at the origin to a point charge $Q_2(t)$ at the point $(2, 2, 2)$. Find the line integral of $\mathbf{H}$ around the triangular closed path having the vertices at $(3, 0, 0), (0, 3, 0)$, and $(0, 0, 3)$ and traversed in that order.

Section 3.5

P3.27. Charge is distributed with density $\rho(x, y, z)$ in a cubical box bounded by the planes $x = \pm 1$ m, $y = \pm 1$ m, and $z = \pm 1$ m. Find the displacement flux emanating from one side of the box for each of the following cases: **(a)** $\rho(x, y, z) = (3 - x^2 - y^2 - z^2)$ C/m^3 and **(b)** $\rho(x, y, z) = \sqrt{|xyz|}$ C/m^3.

P3.28. Two concentric, spherical conducting shells of inner radii a and c, respectively, and outer radii $b(< c)$ and d, respectively, carry net surface charges of Q_1C and Q_2C, respectively. Find the surface charge densities on all four surfaces of the shells.

P3.29. Charge is distributed with density $\rho_0 e^{-r^2}$ C/m^3 in the cylindrical region $r < 1$. Find $\mathbf{D}$ everywhere.

P3.30. Charge is distributed with uniform density ρ_0 C/m^3 in the region $a < r < 2a$ in spherical coordinates. Find $\mathbf{D}$ everywhere and plot D_r versus r.

P3.31. Current flows with density $\mathbf{J}(x, y)$ in an infinitely long thick wire having the z-axis as its axis. The cross section of the wire in the xy-plane is the square bounded by $x = \pm 1$ m and $y = \pm 1$ m. Find the line integral of $\mathbf{H}$ along one side of the square

and traversed in the sense of increasing ϕ for each of the following cases:
(a) $\mathbf{J}(x, y) = (|x| + |y|)\mathbf{a}_z$ A/m² and **(b)** $\mathbf{J}(x, y) = x^2 y^2 \mathbf{a}_z$ A/m².

P3.32. Current flows with density $\mathbf{J} = J_0(r/a)\mathbf{a}_z$ A/m² along an infinitely long solid cylindrical wire of radius a having the z-axis as its axis. Find $\mathbf{H}$ everywhere and plot H_ϕ versus r.

P3.33. A coaxial cable consists of an inner conductor of radius $3a$ and an outer conductor of inner radius $4a$ and outer radius $5a$. Assume the cable to be infinitely long and its axis to be along the z-axis. Current I flows with uniform density in the $+z$-direction in the inner conductor and returns with uniform density in the $-z$-direction in the outer conductor. Find $\mathbf{H}$ everywhere and plot H_ϕ versus r.

Section 3.6

P3.34. Show that the results obtained for the electric field due to the sheet of charge in Example 2.3 and for the magnetic field due to the sheet of current in Example 2.6 are consistent with the boundary conditions.

P3.35. The electric field in free space outside a conducting sphere of radius a and centered at the origin resulting from an applied field is given by

$$\mathbf{E} = E_0\left(1 + \frac{2a^3}{r^3}\right)\cos\theta\,\mathbf{a}_r - E_0\left(1 - \frac{a^3}{r^3}\right)\sin\theta\,\mathbf{a}_\theta$$

(a) Show that the tangential component of $\mathbf{E}$ is zero on the conductor surface.
(b) Find the charge density on the conductor surface. **(c)** Find the applied and secondary fields. (*Hint:* Consider $a \to 0$.)

P3.36. In Problem P2.30, show that the applied and secondary fields together satisfy the boundary condition of zero tangential component of electric field on the conductor surface. From the boundary condition for the normal component of $\mathbf{D}$, find the charge density on the conductor surface and show that the total induced surface charge per unit width in the z-direction is $-\rho_{L0}$.

P3.37. The rectangular cavity resonator is a box comprising the region $0 < x < a$, $0 < y < b$, and $0 < z < d$, and bounded by perfectly conducting walls on all of its six sides. The time-varying electric and magnetic fields inside the resonator are given by

$$\mathbf{E} = E_0 \sin\frac{\pi x}{a}\sin\frac{\pi z}{d}\cos\omega t\,\mathbf{a}_y$$

$$\mathbf{H} = H_{01}\sin\frac{\pi x}{a}\cos\frac{\pi z}{d}\sin\omega t\,\mathbf{a}_x - H_{02}\cos\frac{\pi x}{a}\sin\frac{\pi z}{d}\sin\omega t\,\mathbf{a}_z$$

where E_0, H_{01}, and H_{02} are constants. Find ρ_S and $\mathbf{J}_S$ on all six walls, assuming the medium inside the box to be a perfect dielectric of $\varepsilon = 4\varepsilon_0$.

P3.38. Medium 1, comprising the region $r < a$ in spherical coordinates, is a perfect dielectric of permittivity ε_1, whereas medium 2, comprising the region $r > a$ in spherical coordinates, is free space. The electric field intensities in the two media are given by

$$\mathbf{E}_1 = E_{01}(\cos\theta\,\mathbf{a}_r - \sin\theta\,\mathbf{a}_\theta)$$

$$\mathbf{E}_2 = E_{02}\left[\left(1 + \frac{a^3}{2r^3}\right)\cos\theta\,\mathbf{a}_r - \left(1 - \frac{a^3}{4r^3}\right)\sin\theta\,\mathbf{a}_\theta\right]$$

respectively. Find ε_1.

P3.39. A boundary separates free space from a perfect dielectric medium. At a point on the boundary, the electric field intensity on the free space side is $\mathbf{E}_1 = E_0(4\mathbf{a}_x + 2\mathbf{a}_y + 5\mathbf{a}_z)$, whereas on the dielectric side, it is $\mathbf{E}_2 = 3E_0(\mathbf{a}_x + \mathbf{a}_z)$, where E_0 is a constant. Find the permittivity of the dielectric medium.

P3.40. Medium 1, comprising the region $r < a$ in spherical coordinates, is a magnetic material of permeability μ_1, whereas medium 2, comprising the region $r > a$ in spherical coordinates, is free space. The magnetic flux densities in the two media are given by

$$\mathbf{B}_1 = B_{01}(\cos\theta\, \mathbf{a}_r - \sin\theta\, \mathbf{a}_\theta)$$

$$\mathbf{B}_2 = B_{02}\left[\left(1 + 1.94\frac{a^3}{r^3}\right)\cos\theta\, \mathbf{a}_r - \left(1 - 0.97\frac{a^3}{r^3}\right)\sin\theta\, \mathbf{a}_\theta\right]$$

respectively. Find μ_1.

REVIEW PROBLEMS

R3.1. Show that the vector field given by

$$\mathbf{F} = \cos\theta\sin\phi\, \mathbf{a}_r - \sin\theta\sin\phi\, \mathbf{a}_\theta + \cot\theta\cos\phi\, \mathbf{a}_\phi$$

is a conservative field. Then find the value of $\int \mathbf{F} \cdot d\mathbf{l}$ from the point $(1, \pi/6, \pi/3)$ to the point $(4, \pi/3, \pi/6)$.

R3.2. In Fig. 3.48, a rectangular loop of wire with three sides fixed and the fourth side movable is situated in a plane perpendicular to a nonuniform magnetic field $\mathbf{B} = B_0 y\mathbf{a}_z$ Wb/m², where B_0 is a constant. The position of the movable side is varied with time in the manner $y = y_0 + a\cos\omega t$, where $a < y_0$. Find the induced emf around the closed path C of the loop. Verify that Lenz's law is satisfied. Show also that the induced emf consists of two frequency components, ω and 2ω.

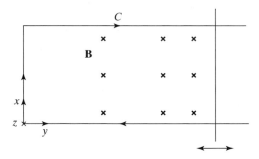

FIGURE 3.48   For Problem R3.2.

R3.3. A voltage source is connected by means of wires to a parallel-plate capacitor made up of circular plates of radii a in the $z = 0$ and $z = d$ planes and having their centers on the z-axis. The electric field between the plates is given by

$$\mathbf{E} = E_0\sin\frac{\pi r}{2a}\cos\omega t\, \mathbf{a}_z \quad \text{for}\quad r < a$$

Find the amplitude of the current drawn from the voltage source, assuming the region between the plates to be free space and that no field exists outside this region.

R3.4. Current I flows along a straight wire from a point charge $Q_1(t)$ located at one of the vertices of a cube to a point charge $Q_2(t)$ at the center of the cube. Find the absolute value of the line integral of **H** around the periphery of one of the three sides of the cube not containing the vertex at which Q_1 is located.

R3.5. Charge is distributed with density $\rho = \rho_0(r/a)^2$, where ρ_0 is a constant, in the spherical region $r < a$. Find **D** everywhere and plot D_r versus r.

R3.6. Current flows axially with uniform density $\mathbf{J}_0$ A/m^2 in the region between two infinitely long parallel, cylindrical surfaces of radii a and b ($< a$), and with their axes separated by the vector distance **c**, where $|\mathbf{c}| < (a - b)$. Find the magnetic field intensity in the current-free region inside the cylindrical surface of radius b.

R3.7. The region $3x + 4y + 12z < 12$ is occupied by a perfect conductor. If at a point on the perfect conductor surface, the surface charge and current densities at a particular instant of time are ρ_{S0} C/m^2 and $J_{S0}(4\mathbf{a}_x - 3\mathbf{a}_y)$ A/m, find **D** and **H** at that point at that instant of time.

R3.8. Medium 1, comprising the region $r < a$ in spherical coordinates, is a perfect dielectric of permittivity $\varepsilon_1 = 2\varepsilon_0$, whereas medium 2, comprising the region $r > a$, is free space. The electric field intensity in medium 1 is given by $\mathbf{E}_1 = E_0\mathbf{a}_z$. Find the electric field intensity at the points **(a)** $(0, 0, a)$, **(b)** $(0, a, 0)$, and **(c)** $(0, a/\sqrt{2}, a/\sqrt{2})$, in Cartesian coordinates, in medium 2.

Maxwell's Equations in Differential Form, Potential Functions, and Energy Storage

In Chapter 3, we introduced Maxwell's equations in integral form. We learned that the quantities involved in the formulation of these equations are the scalar quantities, electromotive force, magnetomotive force, magnetic flux, displacement flux, charge, and current, which are related to the field vectors and source densities through line, surface, and volume integrals. Thus, the integral forms of Maxwell's equations, while containing all the information pertinent to the interdependence of the field and source quantities over a given region in space, do not permit us to study directly the interaction between the field vectors and their relationships with the source densities at individual points. It is our goal in this chapter to derive the differential forms of Maxwell's equations that apply directly to the field vectors and source densities at a given point.

We shall derive Maxwell's equations in differential form by applying Maxwell's equations in integral form to infinitesimal closed paths, surfaces, and volumes, in the limit that they shrink to points. We will find that the differential equations relate the spatial variations of the field vectors at a given point to their temporal variations and to the charge and current densities at that point. Using Maxwell's equations in differential form, we introduce the electromagnetic potential functions, derive differential equations for the potential functions, and consider the potential functions for the static field case. Finally, we consider the topic of energy storage in electric and magnetic fields.

In the process of deriving the Maxwell's equations in differential form, we shall become familiar with the operations of curl and divergence and then learn the Stokes and divergence theorems. In introducing the potential functions and the associated differential equations, we make use of the operations of gradient and Laplacian.

4.1 FARADAY'S LAW AND AMPÈRE'S CIRCUITAL LAW

We recall from Chapter 3 that Faraday's law is given in integral form by

Faraday's law, special case

$$\oint_C \mathbf{E} \cdot d\mathbf{l} = -\frac{d}{dt} \int_S \mathbf{B} \cdot d\mathbf{S} \qquad (4.1)$$

where S is any surface bounded by the closed path C. In the most general case, the electric and magnetic fields have all three components (x, y, and z) and are dependent on all three coordinates (x, y, and z) in addition to time (t). For simplicity, we shall, however, first consider the case in which the electric field has an x component only, which is dependent only on the z coordinate, in addition to time. Thus,

$$\mathbf{E} = E_x(z, t)\mathbf{a}_x \qquad (4.2)$$

In other words, this simple form of time-varying electric field is everywhere directed in the x-direction and it is uniform in planes parallel to the xy-plane.

Let us now consider a rectangular path C of infinitesimal size lying in a plane parallel to the xz-plane and defined by the points (x, z), $(x, z + \Delta z)$, $(x + \Delta x, z + \Delta z)$, and $(x + \Delta x, z)$, as shown in Fig. 4.1. According to Faraday's law, the emf around the closed path C is equal to the negative of the time rate of change of the magnetic flux enclosed by C. The emf is given by the line integral of $\mathbf{E}$ around C. Thus, evaluating the line integrals of $\mathbf{E}$ along the four sides of the

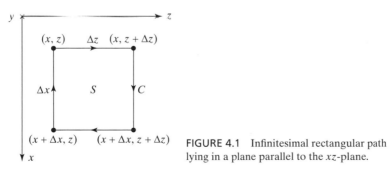

FIGURE 4.1   Infinitesimal rectangular path lying in a plane parallel to the xz-plane.

rectangular path, we obtain

$$\int_{(x,z)}^{(x,z+\Delta z)} \mathbf{E} \cdot d\mathbf{l} = 0 \qquad \text{since } E_z = 0 \tag{4.3a}$$

$$\int_{(x,z+\Delta z)}^{(x+\Delta x,z+\Delta z)} \mathbf{E} \cdot d\mathbf{l} = [E_x]_{z+\Delta z} \, \Delta x \tag{4.3b}$$

$$\int_{(x+\Delta x,z+\Delta z)}^{(x+\Delta x,z)} \mathbf{E} \cdot d\mathbf{l} = 0 \qquad \text{since } E_z = 0 \tag{4.3c}$$

$$\int_{(x+\Delta x,z)}^{(x,z)} \mathbf{E} \cdot d\mathbf{l} = -[E_x]_z \, \Delta x \tag{4.3d}$$

Adding up (4.3a)–(4.3d), we obtain

$$\oint_C \mathbf{E} \cdot d\mathbf{l} = [E_x]_{z+\Delta z} \, \Delta x - [E_x]_z \, \Delta x$$
$$= \{[E_x]_{z+\Delta z} - [E_x]_z\} \, \Delta x \tag{4.4}$$

In (4.3a)–(4.3d) and (4.4), $[E_x]_z$ and $[E_x]_{z+\Delta z}$ denote values of E_x evaluated along the sides of the path for which $z = z$ and $z = z + \Delta z$, respectively.

To find the magnetic flux enclosed by C, let us consider the plane surface S bounded by C. According to the right-hand screw rule, we must use the magnetic flux crossing S toward the positive y-direction, that is, into the page, since the path C is traversed in the clockwise sense. The only component of $\mathbf{B}$ normal to the area S is the y-component. Also since the area is infinitesimal in size, we can assume B_y to be uniform over the area and equal to its value at (x, z). The required magnetic flux is then given by

$$\int_S \mathbf{B} \cdot d\mathbf{S} = [B_y]_{(x,z)} \, \Delta x \, \Delta z \tag{4.5}$$

Substituting (4.4) and (4.5) into (4.1) to apply Faraday's law to the rectangular path C under consideration, we get

$$\{[E_x]_{z+\Delta z} - [E_x]_z\} \, \Delta x = -\frac{d}{dt} \{[B_y]_{(x,z)} \, \Delta x \, \Delta z\}$$

or

$$\frac{[E_x]_{z+\Delta z} - [E_x]_z}{\Delta z} = -\frac{\partial [B_y]_{(x, z)}}{\partial t} \tag{4.6}$$

If we now let the rectangular path shrink to the point (x, z) by letting Δx and Δz tend to zero, we obtain

$$\lim_{\substack{\Delta x \to 0 \\ \Delta z \to 0}} \frac{[E_x]_{z+\Delta z} - [E_x]_z}{\Delta z} = -\lim_{\substack{\Delta x \to 0 \\ \Delta z \to 0}} \frac{\partial [B_y]_{(x,z)}}{\partial t}$$

or

$$\boxed{\frac{\partial E_x}{\partial z} = -\frac{\partial B_y}{\partial t}} \tag{4.7}$$

Equation (4.7) is Faraday's law in differential form for the simple case of $\mathbf{E}$ given by (4.2). It relates the variation of E_x with z (space) at a point to the variation of B_y with t (time) at that point. Since this derivation can be carried out for any arbitrary point (x, y, z), it is valid for all points. It tells us in particular that an E_x associated with a time-varying B_y has a differential in the z-direction. This is to be expected since if this is not the case, $\oint \mathbf{E} \cdot d\mathbf{l}$ around the infinitesimal rectangular path would be zero.

Example 4.1

Given $\mathbf{E} = 10 \cos (6\pi \times 10^8 t - 2\pi z) \, \mathbf{a}_x$ V/m, let us find $\mathbf{B}$ that satisfies (4.7).
From (4.7), we have

$$\frac{\partial B_y}{\partial t} = -\frac{\partial E_x}{\partial z}$$

$$= -\frac{\partial}{\partial z} [10 \cos (6\pi \times 10^8 t - 2\pi z)]$$

$$= -20\pi \sin (6\pi \times 10^8 t - 2\pi z)$$

$$B_y = \frac{10^{-7}}{3} \cos (6\pi \times 10^8 t - 2\pi z)$$

$$\mathbf{B} = \frac{10^{-7}}{3} \cos (6\pi \times 10^8 t - 2\pi z) \, \mathbf{a}_y$$

We shall now proceed to derive the differential form of (4.1) for the general case of the electric field having all three components (x, y, z), each of them depending on all three coordinates $(x, y, \text{and } z)$, in addition to time (t); that is, *Faraday's law, general case*

$$\boxed{\mathbf{E} = E_x(x, y, z, t)\mathbf{a}_x + E_y(x, y, z, t)\mathbf{a}_y + E_z(x, y, z, t)\mathbf{a}_z} \tag{4.8}$$

To do this, let us consider the three infinitesimal rectangular paths in planes parallel to the three mutually orthogonal planes of the Cartesian coordinate system, as shown in Fig. 4.2. Evaluating $\oint \mathbf{E} \cdot d\mathbf{l}$ around the closed paths *abcda*, *adefa*, and *afgba*, we get

$$\oint_{abcda} \mathbf{E} \cdot d\mathbf{l} = [E_y]_{(x,z)} \Delta y + [E_z]_{(x,y+\Delta y)} \Delta z \tag{4.9a}$$
$$- [E_y]_{(x,z+\Delta z)} \Delta y - [E_z]_{(x,y)} \Delta z$$

$$\oint_{adefa} \mathbf{E} \cdot d\mathbf{l} = [E_z]_{(x,y)} \Delta z + [E_x]_{(y,z+\Delta z)} \Delta x \tag{4.9b}$$
$$- [E_z]_{(x+\Delta x,y)} \Delta z - [E_x]_{(y,z)} \Delta x$$

$$\oint_{afgba} \mathbf{E} \cdot d\mathbf{l} = [E_x]_{(y,z)} \Delta x + [E_y]_{(x+\Delta x,z)} \Delta y \tag{4.9c}$$
$$- [E_x]_{(y+\Delta y,z)} \Delta x - [E_y]_{(x,z)} \Delta y$$

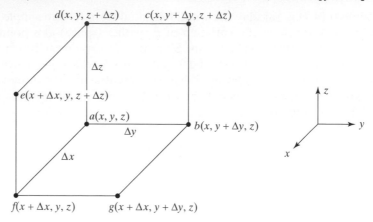

FIGURE 4.2 Infinitesimal rectangular paths in three mutually orthogonal planes.

In (4.9a)–(4.9c), the subscripts associated with the field components in the various terms on the right sides of the equations denote the values of the coordinates that remain constant along the sides of the closed paths corresponding to the terms. Now, evaluating $\int \mathbf{B} \cdot d\mathbf{S}$ over the surfaces *abcd*, *adef*, and *afgb*, keeping in mind the right-hand screw rule, we have

$$\int_{abcd} \mathbf{B} \cdot d\mathbf{S} = [B_x]_{(x,y,z)} \, \Delta y \, \Delta z \tag{4.10a}$$

$$\int_{adef} \mathbf{B} \cdot d\mathbf{S} = [B_y]_{(x,y,z)} \, \Delta z \, \Delta x \tag{4.10b}$$

$$\int_{afgb} \mathbf{B} \cdot d\mathbf{S} = [B_z]_{(x,y,z)} \, \Delta x \, \Delta y \tag{4.10c}$$

Applying Faraday's law to each of the three paths by making use of (4.9a)–(4.9c) and (4.10a)–(4.10c) and simplifying, we obtain

$$\frac{[E_z]_{(x,y+\Delta y)} - [E_z]_{(x,y)}}{\Delta y} - \frac{[E_y]_{(x,z+\Delta z)} - [E_y]_{(x,z)}}{\Delta z} = -\frac{\partial [B_x]_{(x,y,z)}}{\partial t} \tag{4.11a}$$

$$\frac{[E_x]_{(y,z+\Delta z)} - [E_x]_{(y,z)}}{\Delta z} - \frac{[E_z]_{(x+\Delta x,y)} - [E_z]_{(x,y)}}{\Delta x} = -\frac{\partial [B_y]_{(x,y,z)}}{\partial t} \tag{4.11b}$$

$$\frac{[E_y]_{(x+\Delta x,z)} - [E_y]_{(x,z)}}{\Delta x} - \frac{[E_x]_{(y+\Delta y,z)} - [E_x]_{(y,z)}}{\Delta y} = -\frac{\partial [B_z]_{(x,y,z)}}{\partial t} \tag{4.11c}$$

If we now let all three paths shrink to the point *a* by letting Δx, Δy, and Δz tend to zero, (4.11a)–(4.11c) reduce to

$$\frac{\partial E_z}{\partial y} - \frac{\partial E_y}{\partial z} = -\frac{\partial B_x}{\partial t} \qquad (4.12a)$$

$$\frac{\partial E_x}{\partial z} - \frac{\partial E_z}{\partial x} = -\frac{\partial B_y}{\partial t} \qquad (4.12b)$$

$$\frac{\partial E_y}{\partial x} - \frac{\partial E_x}{\partial y} = -\frac{\partial B_z}{\partial t} \qquad (4.12c)$$

Equations (4.12a)–(4.12c) are the differential equations governing the relationships between the space variations of the electric field components and the time variations of the magnetic field components at a point. In particular, we note that the space derivatives are all lateral derivatives, that is, derivatives evaluated along directions lateral to the directions of the field components and not along the directions of the field components. An examination of one of the three equations is sufficient to reveal the physical meaning of these relationships. For example, (4.12a) tells us that a time-varying B_x at a point results in an electric field at that point having y- and z-components such that their net right-lateral differential normal to the x-direction is nonzero. The right-lateral differential of E_y normal to the x-direction is its derivative in the $\mathbf{a}_y \times \mathbf{a}_x$, or $-\mathbf{a}_z$-direction, that is, $\partial E_y/\partial(-z)$ or $-\partial E_y/\partial z$. The right-lateral differential of E_z normal to the x-direction is its derivative in the $\mathbf{a}_z \times \mathbf{a}_x$, or $\mathbf{a}_y$-direction, that is, $\partial E_z/\partial y$. Thus, the net right-lateral differential of the y- and z-components of the electric field normal to the x-direction is $(-\partial E_y/\partial z) + (\partial E_z/\partial y)$, or $(\partial E_z/\partial y - \partial E_y/\partial z)$. Figure 4.3(a) shows an example in which the net right-lateral differential is zero although the individual derivatives are nonzero. This is because $\partial E_z/\partial y$ and $\partial E_y/\partial z$ are both positive and equal so that their difference is zero. On the other hand, for the example in Fig. 4.3(b), $\partial E_z/\partial y$ is positive and $\partial E_y/\partial z$ is negative so that their difference, that is, the net right-lateral differential, is nonzero.

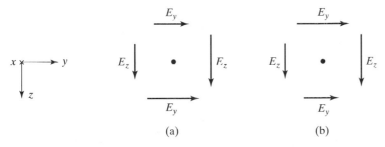

FIGURE 4.3   For illustrating (a) zero and (b) nonzero net right-lateral differential of E_y and E_z normal to the x-direction.

Curl (del cross)   Equations (4.12a)–(4.12c) can be combined into a single vector equation as given by

$$\left(\frac{\partial E_z}{\partial y} - \frac{\partial E_y}{\partial z}\right)\mathbf{a}_x + \left(\frac{\partial E_x}{\partial z} - \frac{\partial E_z}{\partial x}\right)\mathbf{a}_y + \left(\frac{\partial E_y}{\partial x} - \frac{\partial E_x}{\partial y}\right)\mathbf{a}_z$$
$$= -\frac{\partial B_x}{\partial t}\mathbf{a}_x - \frac{\partial B_y}{\partial t}\mathbf{a}_y - \frac{\partial B_z}{\partial t}\mathbf{a}_z \tag{4.13}$$

This can be expressed in determinant form as

$$\begin{vmatrix} \mathbf{a}_x & \mathbf{a}_y & \mathbf{a}_z \\ \dfrac{\partial}{\partial x} & \dfrac{\partial}{\partial y} & \dfrac{\partial}{\partial z} \\ E_x & E_y & E_z \end{vmatrix} = -\frac{\partial \mathbf{B}}{\partial t} \tag{4.14}$$

or as

$$\left(\mathbf{a}_x \frac{\partial}{\partial x} + \mathbf{a}_y \frac{\partial}{\partial y} + \mathbf{a}_z \frac{\partial}{\partial z}\right) \times (E_x\mathbf{a}_x + E_y\mathbf{a}_y + E_z\mathbf{a}_z) = -\frac{\partial \mathbf{B}}{\partial t} \tag{4.15}$$

The left side of (4.14) or (4.15) is known as the *curl of* **E**, denoted as $\nabla \times \mathbf{E}$ (del cross **E**), where ∇ (del) is the vector operator given by

$$\nabla = \mathbf{a}_x \frac{\partial}{\partial x} + \mathbf{a}_y \frac{\partial}{\partial y} + \mathbf{a}_z \frac{\partial}{\partial z} \tag{4.16}$$

Thus, we have

$$\nabla \times \mathbf{E} = -\frac{\partial \mathbf{B}}{\partial t} \tag{4.17}$$

Equation (4.17) is Maxwell's equation in differential form corresponding to Faraday's law. It tells us that at a point in an electromagnetic field, the curl of the electric field intensity is equal to the time rate of decrease of the magnetic flux density. We shall discuss curl further in Section 4.3, but note that for static fields, $\nabla \times \mathbf{E}$ is equal to the null vector. Thus, for a static vector field to be realized as an electric field, the components of its curl must all be zero.

Although we have deduced (4.17) from (4.1) by considering the Cartesian coordinate system, it is independent of the coordinate system since (4.1) is independent of the coordinate system. The expressions for the curl of a vector in cylindrical and spherical coordinate systems are derived in Appendix A. They are reproduced here together with that in (4.14) for the Cartesian coordinate system.

CARTESIAN

$$\nabla \times \mathbf{A} = \begin{vmatrix} \mathbf{a}_x & \mathbf{a}_y & \mathbf{a}_z \\ \dfrac{\partial}{\partial x} & \dfrac{\partial}{\partial y} & \dfrac{\partial}{\partial z} \\ A_x & A_y & A_z \end{vmatrix} \tag{4.18a}$$

CYLINDRICAL

$$\nabla \times \mathbf{A} = \begin{vmatrix} \dfrac{\mathbf{a}_r}{r} & \mathbf{a}_\phi & \dfrac{\mathbf{a}_z}{r} \\ \dfrac{\partial}{\partial r} & \dfrac{\partial}{\partial \phi} & \dfrac{\partial}{\partial z} \\ A_r & rA_\phi & A_z \end{vmatrix} \tag{4.18b}$$

SPHERICAL

$$\nabla \times \mathbf{A} = \begin{vmatrix} \dfrac{\mathbf{a}_r}{r^2 \sin \theta} & \dfrac{\mathbf{a}_\theta}{r \sin \theta} & \dfrac{\mathbf{a}_\phi}{r} \\ \dfrac{\partial}{\partial r} & \dfrac{\partial}{\partial \theta} & \dfrac{\partial}{\partial \phi} \\ A_r & rA_\theta & r \sin \theta \, A_\phi \end{vmatrix} \tag{4.18c}$$

Example 4.2

Find the curls of the following vector fields: **(a)** $y\mathbf{a}_x - x\mathbf{a}_y$ and **(b)** $\mathbf{a}_\phi$ in cylindrical coordinates.

a. Using (4.18a), we have

$$\nabla \times (y\mathbf{a}_x - x\mathbf{a}_y) = \begin{vmatrix} \mathbf{a}_x & \mathbf{a}_y & \mathbf{a}_z \\ \dfrac{\partial}{\partial x} & \dfrac{\partial}{\partial y} & \dfrac{\partial}{\partial z} \\ y & -x & 0 \end{vmatrix}$$

$$= \mathbf{a}_x \left[-\frac{\partial}{\partial z}(-x) \right] + \mathbf{a}_y \left[\frac{\partial}{\partial z}(y) \right] + \mathbf{a}_z \left[\frac{\partial}{\partial x}(-x) - \frac{\partial}{\partial y}(y) \right]$$

$$= -2\mathbf{a}_z$$

b. Using (4.18b), we obtain

$$\nabla \times \mathbf{a}_\phi = \begin{vmatrix} \dfrac{\mathbf{a}_r}{r} & \mathbf{a}_\phi & \dfrac{\mathbf{a}_z}{r} \\ \dfrac{\partial}{\partial r} & \dfrac{\partial}{\partial \phi} & \dfrac{\partial}{\partial z} \\ 0 & r & 0 \end{vmatrix} = \frac{\mathbf{a}_r}{r} \left[-\frac{\partial}{\partial z}(r) \right] + \frac{\mathbf{a}_z}{r} \left[\frac{\partial}{\partial r}(r) \right] = \frac{1}{r} \mathbf{a}_z$$

We shall now consider the derivation of the differential form of Ampère's circuital law given in integral form by

$$\oint_C \mathbf{H} \cdot d\mathbf{l} = \int_S \mathbf{J} \cdot d\mathbf{S} + \frac{d}{dt}\int_S \mathbf{D} \cdot d\mathbf{S} \tag{4.19}$$

where S is any surface bounded by the closed path C. To do this, we need not repeat the procedure employed in the case of Faraday's law. Instead, we note from (4.1) and (4.17) that in converting to the differential form from integral form, the line integral of $\mathbf{E}$ around the closed path C is replaced by the curl of $\mathbf{E}$, the surface integral of $\mathbf{B}$ over the surface S bounded by C is replaced by $\mathbf{B}$ itself, and the total time derivative is replaced by partial derivative, as shown:

$$\oint_C \mathbf{E} \cdot d\mathbf{l} = -\frac{d}{dt}\int_S \mathbf{B} \cdot d\mathbf{S}$$

$$\nabla \times \mathbf{E} = -\frac{\partial}{\partial t}(\mathbf{B})$$

Then using the analogy between Ampère's circuital law and Faraday's law, we can write the following:

$$\oint_C \mathbf{H} \cdot d\mathbf{l} = \int_S \mathbf{J} \cdot d\mathbf{S} + \frac{d}{dt}\int_S \mathbf{D} \cdot d\mathbf{S}$$

$$\nabla \times \mathbf{H} = \mathbf{J} + \frac{\partial}{\partial t}(\mathbf{D})$$

Thus, for the general case of the magnetic field having all three components (x, y, and z), each of them depending on all three coordinates (x, y, and z), in addition to time (t), that is, for

$$\mathbf{H} = H_x(x, y, z, t)\mathbf{a}_x + H_y(x, y, z, t)\mathbf{a}_y + H_z(x, y, z, t)\mathbf{a}_z \tag{4.20}$$

the differential form of Ampère's circuital law is given by

$$\nabla \times \mathbf{H} = \mathbf{J} + \frac{\partial \mathbf{D}}{\partial t} \tag{4.21}$$

The quantity $\partial \mathbf{D}/\partial t$ is known as the *displacement current density*. Equation (4.21) tells us that at a point in an electromagnetic field, the curl of the magnetic field intensity is equal to the sum of the current density due to flow of charges and the displacement current density. In Cartesian coordinates, (4.21) becomes

$$
\begin{vmatrix}
\mathbf{a}_x & \mathbf{a}_y & \mathbf{a}_z \\
\dfrac{\partial}{\partial x} & \dfrac{\partial}{\partial y} & \dfrac{\partial}{\partial z} \\
H_x & H_y & H_z
\end{vmatrix} = \mathbf{J} + \frac{\partial \mathbf{D}}{\partial t} \tag{4.22}
$$

This is equivalent to three scalar equations relating the lateral space derivatives of the components of $\mathbf{H}$ to the components of the current density and the time derivatives of the electric field components. These scalar equations can be interpreted in a manner similar to the interpretation of (1.12a)–(1.12c) in the case of Faraday's law. Also, expressions similar to (4.22) can be written in the cylindrical and spherical coordinate systems by using the determinant expansions for the curl in those coordinate systems, given by (1.18b) and (1.18c), respectively.

Having obtained the differential form of Ampère's circuital law for the general case, we can now simplify it for any particular case. Let us consider the particular case of

Ampère's circuital law, special case

$$
\boxed{\mathbf{H} - H_y(z, t)\mathbf{a}_y} \tag{4.23}
$$

that is, a magnetic field directed everywhere in the y-direction and uniform in planes parallel to the xy-plane. Then since $\mathbf{H}$ does not depend on x and y, we can replace $\partial/\partial x$ and $\partial/\partial y$ in the determinant expansion for $\nabla \times \mathbf{H}$ by zeros. In addition, setting $H_x = H_z = 0$, we have

$$
\begin{vmatrix}
\mathbf{a}_x & \mathbf{a}_y & \mathbf{a}_z \\
0 & 0 & \dfrac{\partial}{\partial z} \\
0 & H_y & 0
\end{vmatrix} = \mathbf{J} + \frac{\partial \mathbf{D}}{\partial t} \tag{4.24}
$$

Equating like components on the two sides and noting that the y- and z-components on the left side are zero, we obtain

$$
-\frac{\partial H_y}{\partial z} = J_x + \frac{\partial D_x}{\partial t}
$$

or

$$
\boxed{\frac{\partial H_y}{\partial z} = -J_x - \frac{\partial D_x}{\partial t}} \tag{4.25}
$$

Equation (4.25) is Ampère's circuital law in differential form for the simple case of $\mathbf{H}$ given by (4.23). It relates the variation of H_y with z (space) at a point to the current density J_x and to the variation of D_x with t (time) at that point. It tells us in particular that an H_y associated with a current density J_x or a time-varying D_x or a nonzero combination of the two quantities has a differential in the z-direction.

Example 4.3

Given $\mathbf{E} = E_0 z^2 e^{-t} \mathbf{a}_x$ in free space ($\mathbf{J} = \mathbf{0}$). We wish to determine if there exists a magnetic field such that both Faraday's law and Ampère's circuital law are satisfied simultaneously.

Using Faraday's law and Ampère's circuital law in succession, we have

$$\frac{\partial B_y}{\partial t} = -\frac{\partial E_x}{\partial z} = -2E_0 z e^{-t}$$

$$B_y = 2E_0 z e^{-t}$$

$$H_y = \frac{2E_0}{\mu_0} z e^{-t}$$

$$\frac{\partial D_x}{\partial t} = -\frac{\partial H_y}{\partial z} = -\frac{2E_0}{\mu_0} e^{-t}$$

$$D_x = \frac{2E_0}{\mu_0} e^{-t}$$

$$E_x = \frac{2E_0}{\mu_0 \varepsilon_0} e^{-t}$$

$$\mathbf{E} = \frac{2E_0}{\mu_0 \varepsilon_0} e^{-t} \mathbf{a}_x$$

which is not the same as the original $\mathbf{E}$. Hence, a magnetic field does not exist which together with the given $\mathbf{E}$ satisfies both laws simultaneously. The pair of fields $\mathbf{E} = E_0 z^2 e^{-t} \mathbf{a}_x$ and $\mathbf{B} = 2E_0 z e^{-t} \mathbf{a}_y$ satisfies only Faraday's law, whereas the pair of fields $\mathbf{B} = 2E_0 z e^{-t} \mathbf{a}_y$ and $\mathbf{E} = (2E_0/\mu_0 \varepsilon_0) e^{-t} \mathbf{a}_x$ satisfies only Ampère's circuital law.

Lumped-circuit theory approximations

To generalize the observation made in the example just discussed, there are certain pairs of time-varying electric and magnetic fields that satisfy only Faraday's law as given by (4.17) and certain other pairs that satisfy only Ampère's circuital law as given by (4.21). In the strictest sense, every physically realizable pair of time-varying electric and magnetic fields must satisfy simultaneously both laws as given by (4.17) and (4.21). However, under the low-frequency approximation, it is valid for the fields to satisfy the laws with certain terms neglected in one or both laws. Lumped-circuit theory is based on such approximations. Thus, the terminal voltage-to-current relationship $V(t) = d[LI(t)]/dt$ for an inductor is obtained by ignoring the effect of the time-varying electric field, that is, $\partial \mathbf{D}/\partial t$ term in Ampère's circuital law. The terminal current-to-voltage relationship $I(t) = d[CV(t)]/dt$ for a capacitor is obtained by ignoring the effect of the time-varying magnetic field, that is, $\partial \mathbf{B}/\partial t$ term in Faraday's law. The terminal voltage-to-current relationship $V(t) = RI(t)$ for a resistor is obtained by ignoring the effects of both time-varying electric field and time-varying magnetic field, that is, both $\partial \mathbf{D}/\partial t$ term in Ampère's circuital law and $\partial \mathbf{B}/\partial t$ term in Faraday's law. In contrast to these approximations, electromagnetic wave propagation phenomena and transmission-line (distributed circuit) theory are based on the simultaneous application of the two laws with all terms included, that is, as given by (4.17) and (4.21).

K4.1. Faraday's law in differential form; Ampère's circuital law in differential form; Curl of a vector.

D4.1. Given $\mathbf{E} = E_0 \cos (6\pi \times 10^8 t - 2\pi z) \mathbf{a}_x$ V/m, find the time rate of increase of B_y at $t = 10^{-8}$ s for each of the following values of z: **(a)** 0; **(b)** $\frac{1}{4}$ m; and **(c)** $\frac{2}{3}$ m.
Ans. **(a)** 0; **(b)** $2\pi E_0$; **(c)** $-\sqrt{3}\pi E_0$.

D4.2. For the vector field $\mathbf{A} = xy^2 \mathbf{a}_x + xz\mathbf{a}_y + x^2 yz\mathbf{a}_z$, find the following: **(a)** the net right-lateral differential of A_x and A_y normal to the z-direction at the point $(1,1,1)$; **(b)** the net right lateral differential of A_y and A_z normal to the x-direction at the point $(1,2,1)$; and **(c)** the net right-lateral differential of A_z and A_x normal to the y-direction at the point $(1, 1, -1)$.
Ans. **(a)** -1; **(b)** 0; **(c)** 2.

D4.3. Given $\mathbf{J} = \mathbf{0}$ and $\mathbf{H} = H_0 e^{-(3 \times 10^8 t - z)^2} \mathbf{a}_y$ A/m, find the time rate of increase of D_x for each of the following cases: **(a)** $z = 2$ m, $t = 10^{-8}$ s; **(b)** $z = 3$ m, $t = \frac{1}{3} \times 10^{-8}$ s; and **(c)** $z = 3$ m, $t = 10^{-8}$ s.
Ans. **(a)** $-0.7358 H_0$; **(b)** $0.0733 H_0$; **(c)** 0.

4.2 GAUSS' LAWS AND THE CONTINUITY EQUATION

Thus far, we have derived Maxwell's equations in differential form corresponding to the two Maxwell's equations in integral form involving the line integrals of $\mathbf{E}$ and $\mathbf{H}$ around the closed path, that is, Faraday's law and Ampère's circuital law, respectively. The remaining two Maxwell's equations in integral form, namely, Gauss' law for the electric field and Gauss' law for the magnetic field, are concerned with the closed surface integrals of $\mathbf{D}$ and $\mathbf{B}$, respectively. In this section, we shall derive the differential forms of these two equations.

We recall from Section 3.4 that Gauss' law for the electric field is given by   *Gauss' law for the electric field*

$$\oint_S \mathbf{D} \cdot d\mathbf{S} = \int_V \rho \, dv \tag{4.26}$$

where V is the volume enclosed by the closed surface S. To derive the differential form of this equation, let us consider a rectangular box of edges of infinitesimal lengths Δx, Δy, and Δz and defined by the six surfaces $x = x$, $x = x + \Delta x$, $y = y$, $y = y + \Delta y$, $z = z$, and $z = z + \Delta z$, as shown in Fig. 4.4, in a region of electric field

$$\mathbf{D} = \mathbf{D}_x(x, y, z, t)\mathbf{a}_x + D_y(x, y, z, t)\mathbf{a}_y + D_z(x, y, z, t)\mathbf{a}_z \tag{4.27}$$

and charge of density $\rho(x, y, z, t)$. According to Gauss' law for the electric field, the displacement flux emanating from the box is equal to the charge enclosed by

FIGURE 4.4   Infinitesimal rectangular box.

the box. The displacement flux is given by the surface integral of $\mathbf{D}$ over the surface of the box, which is comprised of six plane surfaces. Thus, evaluating the displacement flux emanating from the box through each of the six plane surfaces of the box, we have

$$\int \mathbf{D} \cdot d\mathbf{S} = -[D_x]_x \, \Delta y \, \Delta z \qquad \text{for the surface } x = x \qquad (4.28\text{a})$$

$$\int \mathbf{D} \cdot d\mathbf{S} = [D_x]_{x+\Delta x} \, \Delta y \, \Delta z \qquad \text{for the surface } x = x + \Delta x \qquad (4.28\text{b})$$

$$\int \mathbf{D} \cdot d\mathbf{S} = -[D_y]_y \, \Delta z \, \Delta x \qquad \text{for the surface } y = y \qquad (4.28\text{c})$$

$$\int \mathbf{D} \cdot d\mathbf{S} = [D_y]_{y+\Delta y} \, \Delta z \, \Delta x \qquad \text{for the surface } y = y + \Delta y \qquad (4.28\text{d})$$

$$\int \mathbf{D} \cdot d\mathbf{S} = -[D_z]_z \, \Delta x \, \Delta y \qquad \text{for the surface } z = z \qquad (4.28\text{e})$$

$$\int \mathbf{D} \cdot d\mathbf{S} = [D_z]_{z+\Delta z} \, \Delta x \, \Delta y \qquad \text{for the surface } z = z + \Delta z \qquad (4.28\text{f})$$

Adding up (4.28a)–(4.28f), we obtain the total displacement flux emanating from the box to be

$$\oint_S \mathbf{D} \cdot d\mathbf{S} = \{[D_x]_{x+\Delta x} - [D_x]_x\} \, \Delta y \, \Delta z$$
$$+ \{[D_y]_{y+\Delta y} - [D_y]_y\} \, \Delta z \, \Delta x \qquad (4.29)$$
$$+ \{[D_z]_{z+\Delta z} - [D_z]_z\} \, \Delta x \, \Delta y$$

Now the charge enclosed by the rectangular box is given by

$$\int_V \rho \, dv = \rho(x, y, z, t) \cdot \Delta x \, \Delta y \, \Delta z = \rho \, \Delta x \, \Delta y \, \Delta z \qquad (4.30)$$

where we have assumed ρ to be uniform throughout the volume of the box and equal to its value at (x, y, z) since the box is infinitesimal in volume.

Substituting (4.29) and (4.30) into (4.26), we get

$$\{[D_x]_{x+\Delta x} - [D_x]_x\} \, \Delta y \, \Delta z + \{[D_y]_{y+\Delta y} - [D_y]_y\} \, \Delta z \, \Delta x$$

$$+ \{[D_z]_{z+\Delta z} - [D_z]_z\} \, \Delta x \, \Delta y = \rho \, \Delta x \, \Delta y \, \Delta z$$

or, dividing throughout by the volume Δv_1

$$\frac{[D_x]_{x+\Delta x} - [D_x]_x}{\Delta x} + \frac{[D_y]_{y+\Delta y} - [D_y]_y}{\Delta y} + \frac{[D_z]_{z+\Delta z} - [D_z]_z}{\Delta z} = \rho \quad (4.31)$$

If we now let the box shrink to the point (x, y, z) by letting Δx, Δy, and Δz tend to zero, we obtain

$$\lim_{\Delta x \to 0} \frac{[D_x]_{x+\Delta x} - [D_x]_x}{\Delta x} + \lim_{\Delta y \to 0} \frac{[D_y]_{y+\Delta y} - [D_y]_y}{\Delta y}$$

$$+ \lim_{\Delta z \to 0} \frac{[D_z]_{z+\Delta z} - [D_z]_z}{\Delta z} = \lim_{\substack{\Delta x \to 0 \\ \Delta y \to 0 \\ \Delta z \to 0}} \rho$$

or

$$\boxed{\frac{\partial D_x}{\partial x} + \frac{\partial D_y}{\partial y} + \frac{\partial D_z}{\partial z} = \rho} \qquad (4.32)$$

Equation (4.32) is the differential equation governing the relationship between the space variations of the components of **D** to the charge density. In particular, we note that the derivatives are all longitudinal derivatives, that is, derivatives evaluated along the directions of the field components, in contrast to the lateral derivatives encountered in Section 4.1. Thus, (4.32) tells us that the net longitudinal differential, that is, the algebraic sum of the longitudinal derivatives, of the components of **D** at a point in space is equal to the charge density at that point. Conversely, a charge density at a point results in an electric field having components of **D** such that their net longitudinal differential is nonzero. Figure 4.5(a) shows an example in which the net longitudinal differential is zero. This is because $\partial D_x/\partial x$ and $\partial D_y/\partial y$ are equal in magnitude but opposite in sign, whereas $\partial D_z/\partial z$ is zero. On the other hand, for the example in Fig. 4.5(b), both

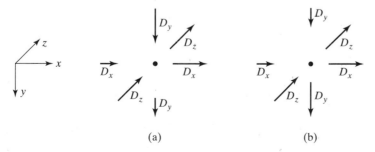

(a)   (b)

FIGURE 4.5   For illustrating (a) zero and (b) nonzero net longitudinal differential of the components of **D**.

$\partial D_x/\partial x$ and $\partial D_y/\partial y$ are positive and $\partial D_z/\partial z$ is zero, so that the net longitudinal differential is nonzero.

Divergence
(del dot)

Equation (4.32) can be written in vector notation as

$$\left(\mathbf{a}_x\frac{\partial}{\partial x} + \mathbf{a}_y\frac{\partial}{\partial y} + \mathbf{a}_z\frac{\partial}{\partial z}\right) \cdot (D_x\mathbf{a}_x + D_y\mathbf{a}_y + D_z\mathbf{a}_z) = \rho \qquad (4.33)$$

The left side of (4.33) is known as the *divergence of* $\mathbf{D}$, denoted as $\nabla \cdot \mathbf{D}$ (del dot $\mathbf{D}$). Thus, we have

$$\boxed{\nabla \cdot \mathbf{D} = \rho} \qquad (4.34)$$

Equation (4.34) is Maxwell's equation in differential form corresponding to Gauss' law for the electric field. It tells us that the divergence of the displacement flux density at a point is equal to the charge density at that point. We shall discuss divergence further in Section 4.3.

Although we have deduced (4.34) from (4.26) by considering the Cartesian coordinate system, it is independent of the coordinate system since (4.26) is independent of the coordinate system. The expressions for the divergence of a vector in cylindrical and spherical coordinate systems are derived in Appendix A. They are reproduced here together with that in (4.32) for the Cartesian coordinate system.

CARTESIAN

$$\nabla \cdot \mathbf{A} = \frac{\partial A_x}{\partial x} + \frac{\partial A_y}{\partial y} + \frac{\partial A_z}{\partial z} \qquad (4.35a)$$

CYLINDRICAL

$$\nabla \cdot \mathbf{A} = \frac{1}{r}\frac{\partial}{\partial r}(rA_r) + \frac{1}{r}\frac{\partial A_\phi}{\partial \phi} + \frac{\partial A_z}{\partial z} \qquad (4.35b)$$

SPHERICAL

$$\nabla \cdot \mathbf{A} = \frac{1}{r^2}\frac{\partial}{\partial r}(r^2A_r) + \frac{1}{r \sin \theta}\frac{\partial}{\partial \theta}(A_\theta \sin \theta) + \frac{1}{r \sin \theta}\frac{\partial A_\phi}{\partial \phi} \qquad (4.35c)$$

Example 4.4

Find the divergences of the following vector fields: **(a)** $3x\mathbf{a}_x + (y - 3)\mathbf{a}_y + (2 - z)\mathbf{a}_z$ and **(b)** $r^2 \sin \theta\, \mathbf{a}_\theta$ in spherical coordinates.

a. Using (4.35a), we have

$$\nabla \cdot [3x\mathbf{a}_x + (y - 3)\mathbf{a}_y + (2 - z)\mathbf{a}_z] = \frac{\partial}{\partial x}(3x) + \frac{\partial}{\partial y}(y - 3) + \frac{\partial}{\partial z}(2 - z)$$

$$= 3 + 1 - 1 = 3$$

b. Using (4.35c), we obtain

$$\nabla \cdot r^2 \sin \theta \, \mathbf{a}_\theta = \frac{1}{r \sin \theta} \frac{\partial}{\partial \theta} (r^2 \sin^2 \theta)$$

$$= \frac{1}{r \sin \theta} (2r^2 \sin \theta \cos \theta)$$

$$= 2r \cos \theta$$

We shall now consider the derivation of the differential form of Gauss' law for the magnetic field given in integral form by

Gauss' law for the magnetic field

$$\oint_S \mathbf{B} \cdot d\mathbf{S} = 0 \tag{4.36}$$

where S is any closed surface. To do this, we need not repeat the procedure employed in the case of Gauss' law for the electric field. Instead, we note from (4.26) and (4.34) that in converting to the differential form from integral form, the surface integral of $\mathbf{D}$ over the closed surface S is replaced by the divergence of $\mathbf{D}$ and the volume integral of ρ is replaced by ρ itself, as shown:

$$\oint_S \mathbf{D} \cdot d\mathbf{S} = \int_V \rho \, dv$$

$$\nabla \cdot \mathbf{D} = \rho$$

Then using the analogy between the two Gauss' laws, we can write the following:

$$\oint_S \mathbf{B} \cdot d\mathbf{S} = 0 = \int_V 0 \, dv$$

$$\nabla \cdot \mathbf{B} = 0$$

Thus, Gauss' law in differential form for the magnetic field

$$\mathbf{B} = B_x(x, y, z, t)\mathbf{a}_x + B_y(x, y, z, t)\mathbf{a}_y + B_z(x, y, z, t)\mathbf{a}_z \tag{4.37}$$

is given by

$$\nabla \cdot \mathbf{B} = 0 \tag{4.38}$$

which tells us that the divergence of the magnetic flux density at a point is equal to zero. Conversely, for a vector field to be realized as a magnetic field, its divergence must be zero. In Cartesian coordinates, (4.38) becomes

$$\frac{\partial B_x}{\partial x} + \frac{\partial B_y}{\partial y} + \frac{\partial B_z}{\partial z} = 0 \qquad (4.39)$$

pointing out that the net longitudinal differential of the components of **B** is zero. Also, expressions similar to (4.39) can be written in cylindrical and spherical coordinate systems by using the expressions for the divergence in those coordinate systems, given by (4.35b) and (4.35c), respectively.

Example 4.5

Determine if the vector $\mathbf{A} = (1/r^2)(\cos \phi \, \mathbf{a}_r + \sin \phi \, \mathbf{a}_\phi)$ in cylindrical coordinates can represent a magnetic field **B**.

Noting that

$$\nabla \cdot \mathbf{A} = \frac{1}{r} \frac{\partial}{\partial r}\left(\frac{\cos \phi}{r}\right) + \frac{1}{r} \frac{\partial}{\partial \phi}\left(\frac{\sin \phi}{r^2}\right)$$

$$= -\frac{\cos \phi}{r^3} + \frac{\cos \phi}{r^3} = 0$$

we conclude that the given vector can represent a **B**.

Continuity equation   We shall conclude this section by deriving the differential form of the law of conservation of charge given in integral form by

$$\oint_S \mathbf{J} \cdot d\mathbf{S} = -\frac{d}{dt} \int_V \rho \, dv \qquad (4.40)$$

Using analogy with Gauss' law for the electric field, we can write the following:

$$\oint_S \mathbf{J} \cdot d\mathbf{S} = -\frac{d}{dt} \int_V \rho \, dv$$

$$\nabla \cdot \mathbf{J} = -\frac{\partial}{\partial t}(\rho)$$

Thus, the differential form of the law of conservation of charge is given by

$$\nabla \cdot \mathbf{J} = -\frac{\partial \rho}{\partial t} \qquad (4.41)$$

Equation (4.41) is familiarly known as the *continuity equation*. It tells us that the divergence of the current density due to flow of charges at a point is equal to the time rate of decrease of the charge density at that point. It can be expanded in a given coordinate system by using the expression for the divergence in that coordinate system.

K4.2. Gauss' law for the electric field in differential form; Gauss' law for the magnetic field in differential form; Divergence of a vector; Continuity equation.

D4.4. For the vector field $\mathbf{A} = yz\mathbf{a}_x + xy\mathbf{a}_y + xyz^2\mathbf{a}_z$, find the net longitudinal differential of the components of $\mathbf{A}$ at the following points: **(a)** $(1, 1, -1)$; **(b)** $(1, 1, -\frac{1}{2})$; and **(c)** $(1, 1, 1)$.
    Ans. **(a)** -1; **(b)** 0; **(c)** 3.

D4.5. The following hold at a point in a charge-free region: (i) the sum of the longitudinal differentials of D_x and D_y is D_0 and (ii) the longitudinal differential of D_y is three times the longitudinal differential of D_z. Find: **(a)** $\partial D_x/\partial x$; **(b)** $\partial D_y/\partial y$; and **(c)** $\partial D_z/\partial z$.
    Ans. **(a)** $4D_0$; **(b)** $-3D_0$; **(c)** $-D_0$

D4.6. In a small region around the origin, the current density due to flow of charges is given by $\mathbf{J} = J_0(x^2\mathbf{a}_x + y^2\mathbf{a}_y + z^2\mathbf{a}_z)$ A/m^2, where J_0 is a constant. Find the time rate of increase of the charge density at each of the following points: **(a)** $(0.02, 0.01, 0.01)$; **(b)** $(0.02, -0.01, -0.01)$; and **(c)** $(-0.02, \quad 0.01, 0.01)$.
    Ans. **(a)** $-0.08J_0$ (C/m^3)/s; **(b)** 0; **(c)** $0.04J_0$ (C/m^3)/s.

4.3 CURL AND DIVERGENCE

In Sections 4.1 and 4.2, we derived the differential forms of Maxwell's equations and the law of conservation of charge from their integral forms. Maxwell's equations are given by

$$\nabla \times \mathbf{E} = -\frac{\partial \mathbf{B}}{\partial t} \tag{4.42a}$$

$$\nabla \times \mathbf{H} = \mathbf{J} + \frac{\partial \mathbf{D}}{\partial t} \tag{4.42b}$$

$$\nabla \cdot \mathbf{D} = \rho \tag{4.42c}$$

$$\nabla \cdot \mathbf{B} = 0 \tag{4.42d}$$

whereas the continuity equation is given by

$$\nabla \cdot \mathbf{J} = -\frac{\partial \rho}{\partial t} \tag{4.43}$$

These equations contain two new vector (differential) operations, namely, the curl and the divergence. The curl of a vector is a vector quantity, whereas the divergence of a vector is a scalar quantity. In this section, we shall introduce the basic definitions of curl and divergence and then discuss physical interpretations of these quantities. We shall also derive two associated theorems.

A.   Curl

Curl, basic definition

To discuss curl first, let us consider Ampère's circuital law without the displacement current density term; that is,

$$\nabla \times \mathbf{H} = \mathbf{J} \tag{4.44}$$

We wish to express $\nabla \times \mathbf{H}$ at a point in the current region in terms of $\mathbf{H}$ at that point. If we consider an infinitesimal surface $\Delta\mathbf{S}$ at the point and take the dot product of both sides of (4.44) with $\Delta\mathbf{S}$, we get

$$(\nabla \times \mathbf{H}) \cdot \Delta\mathbf{S} = \mathbf{J} \cdot \Delta\mathbf{S} \tag{4.45}$$

But $\mathbf{J} \cdot \Delta\mathbf{S}$ is simply the current crossing the surface $\Delta\mathbf{S}$, and according to Ampère's circuital law in integral form without the displacement current term,

$$\oint_C \mathbf{H} \cdot d\mathbf{l} = \mathbf{J} \cdot \Delta\mathbf{S} \tag{4.46}$$

where C is the closed path bounding $\Delta\mathbf{S}$. Comparing (4.45) and (4.46), we have

$$(\nabla \times \mathbf{H}) \cdot \Delta\mathbf{S} = \oint_C \mathbf{H} \cdot d\mathbf{l}$$

or

$$(\nabla \times \mathbf{H}) \cdot \Delta S \, \mathbf{a}_n = \oint_C \mathbf{H} \cdot d\mathbf{l} \tag{4.47}$$

where $\mathbf{a}_n$ is the unit vector normal to ΔS and directed toward the side of advance of a right-hand screw as it is turned around C. Dividing both sides of (4.47) by ΔS, we obtain

$$(\nabla \times \mathbf{H}) \cdot \mathbf{a}_n = \frac{\oint_C \mathbf{H} \cdot d\mathbf{l}}{\Delta S} \tag{4.48}$$

The maximum value of $(\nabla \times \mathbf{H}) \cdot \mathbf{a}_n$, and hence that of the right side of (4.48), occurs when $\mathbf{a}_n$ is oriented parallel to $\nabla \times \mathbf{H}$, that is, when the surface ΔS is oriented normal to the current density vector $\mathbf{J}$. This maximum value is simply $|\nabla \times \mathbf{H}|$. Thus,

$$|\nabla \times \mathbf{H}| = \left[\frac{\oint_C \mathbf{H} \cdot d\mathbf{l}}{\Delta S} \right]_{\max}$$

Since the direction of $\nabla \times \mathbf{H}$ is the direction of $\mathbf{J}$, or that of the unit vector normal to ΔS, we can then write

$$\nabla \times \mathbf{H} = \left[\frac{\oint_C \mathbf{H} \cdot d\mathbf{l}}{\Delta S} \right]_{\max} \mathbf{a}_n$$

This result is however approximate since (4.47) is exact only in the limit that ΔS tends to zero. Thus,

$$\nabla \times \mathbf{H} = \lim_{\Delta S \to 0} \left[\frac{\oint_C \mathbf{H} \cdot d\mathbf{l}}{\Delta S} \right]_{\max} \mathbf{a}_n \tag{4.49}$$

which is the expression for $\nabla \times \mathbf{H}$ at a point in terms of $\mathbf{H}$ at that point. Although we have derived this for the $\mathbf{H}$ vector, it is a general result and, in fact, is often the starting point for the introduction of curl. Thus, for any vector field $\mathbf{A}$,

$$
\nabla \times \mathbf{A} = \lim_{\Delta S \to 0} \left[\frac{\oint_C \mathbf{A} \cdot d\mathbf{l}}{\Delta S} \right]_{\max} \mathbf{a}_n
\tag{4.50}
$$

Equation (4.50) tells us that to find the curl of a vector at a point in that vector field, we first consider an infinitesimal surface at that point and compute the closed line integral or circulation of the vector around the periphery of this surface by orienting the surface such that the circulation is maximum. We then divide the circulation by the area of the surface to obtain the maximum value of the circulation per unit area. Since we need this maximum value of the circulation per unit area in the limit that the area tends to zero, we do this by gradually shrinking the area and making sure that each time we compute the circulation per unit area an orientation for the area that maximizes this quantity is maintained. The limiting value to which the maximum circulation per unit area approaches is the magnitude of the curl. The limiting direction to which the normal vector to the surface approaches is the direction of the curl. The task of computing the curl is simplified if we consider one component at a time and compute that component since then it is sufficient if we always maintain the orientation of the surface normal to that component axis. In fact, this is what we did in Section 4.1, which led us to the determinant expression for the curl in Cartesian coordinates, by choosing for convenience rectangular surfaces whose sides are all parallel to the coordinate planes.

We are now ready to discuss the physical interpretation of the curl. We do this with the aid of a simple device known as the *curl meter*, which responds to the circulation of the vector field. Although the curl meter may take several forms, we shall consider one consisting of a circular disk that floats in water with a paddle wheel attached to the bottom of the disk, as shown in Fig. 4.6. A dot at the periphery on top of the disk serves to indicate any rotational motion of the curl meter about its axis (i.e., the axis of the paddle wheel). Let us now consider a stream of rectangular cross section carrying water in the z-direction, as shown in Fig. 4.6(a). Let us assume the velocity $\mathbf{v}$ of the water to be independent of height but increasing sinusoidally from a value of zero at the banks to a maximum value v_0 at the center, as shown in Fig. 4.6(b), and investigate the behavior of the curl meter when it is placed vertically at different points in the stream. We assume that the size of the curl meter is vanishingly small so that it does not disturb the flow of water as we probe its behavior at different points.

Since exactly in midstream the blades of the paddle wheel lying on either side of the centerline are hit by the same velocities, the paddle wheel does not rotate. The curl meter simply slides down the stream without any rotational motion, that is, with the dot on top of the disk maintaining the same position relative to the center of the disk, as shown in Fig. 4.6(c). At a point to the left of the

Physical interpretation of curl

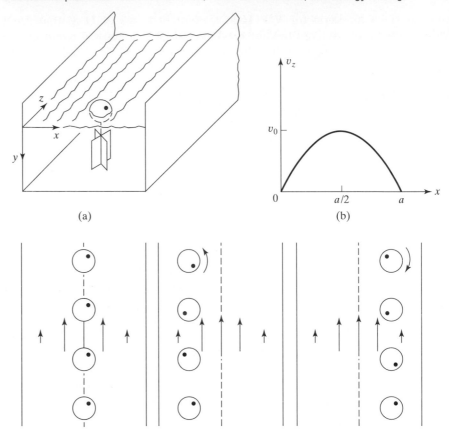

FIGURE 4.6   For explaining the physical interpretation of curl using the curl meter.

midstream, the blades of the paddle wheel are hit by a greater velocity on the right side than on the left side so that the paddle wheel rotates in the counterclockwise sense, as seen looking along the positive y-axis. The curl meter rotates in the counterclockwise direction about its axis as it slides down the stream, as indicated by the changing position of the dot on top of the disk relative to the center of the disk, as shown in Fig. 4.6(d). At a point to the right of midstream, the blades of the paddle wheel are hit by a greater velocity on the left side than on the right side so that the paddle wheel rotates in the clockwise sense, as seen looking along the positive y-axis. The curl meter rotates in the clockwise direction about its axis as it slides down the stream, as indicated by the changing position of the dot on top of the disk relative to the center of the disk, as shown in Fig 4.6(e).

If we now pick up the curl meter and insert it in the water with its axis parallel to the x-axis, the curl meter does not rotate because its blades are hit with the same force above and below its axis. If the curl meter is inserted in the water with its axis parallel to the z-axis, it does not rotate since the water flow is then parallel to the blades.

To relate the behavior of the curl meter with the curl of the velocity vector field of the water flow, we note that since the velocity vector is given by

$$\mathbf{v} = v_z(x)\mathbf{a}_z = v_0 \sin \frac{\pi x}{a} \mathbf{a}_z$$

its curl is given by

$$\nabla \times \mathbf{v} = \begin{vmatrix} \mathbf{a}_x & \mathbf{a}_y & \mathbf{a}_z \\ \dfrac{\partial}{\partial x} & \dfrac{\partial}{\partial y} & \dfrac{\partial}{\partial z} \\ 0 & 0 & v_z \end{vmatrix}$$

$$= -\frac{\partial v_z}{\partial x}\mathbf{a}_y$$

$$= -\frac{\pi v_0}{a} \cos \frac{\pi x}{a} \mathbf{a}_y$$

Therefore, the x- and z-components of the curl are zero, whereas the y-component is nonzero varying with x in a cosinusoidal manner, from negative values left of midstream, to zero at midstream, to positive values right of midstream. Thus, no rotation of the curl meter corresponds to zero value for the component of the curl along its axis. Rotation of the curl meter in the counterclockwise or left-hand sense as seen looking along its axis corresponds to a nonzero negative value, and rotation in the clockwise or right-hand sense corresponds to a nonzero positive value for the component of the curl. It can further be visualized that the rate of rotation of the curl meter is a measure of the magnitude of the pertinent nonzero component of the curl.

The foregoing illustration of the physical interpretation of the curl of a vector field can be used to visualize the behavior of electric and magnetic fields. Thus, from

$$\nabla \times \mathbf{E} = -\frac{\partial \mathbf{B}}{\partial t}$$

we know that at a point in an electromagnetic field, the circulation of the electric field per unit area in a given plane is equal to the component of $-\partial\mathbf{B}/\partial t$ along the unit vector normal to that plane and directed in the right-hand sense. Similarly, from

$$\nabla \times \mathbf{H} = \mathbf{J} + \frac{\partial \mathbf{D}}{\partial t}$$

we know that at a point in an electromagnetic field, the circulation of the magnetic field per unit area in a given plane is equal to the component of $\mathbf{J} + \partial\mathbf{D}/\partial t$ along the unit vector normal to that plane and directed in the right-hand sense.

B. Divergence

Turning now to the discussion of divergence, let us consider Gauss' law for the electric field in differential form; that is,

$$\nabla \cdot \mathbf{D} = \rho \qquad (4.51)$$

Divergence, basic definition

We wish to express $\nabla \cdot \mathbf{D}$ at a point in the charge region in terms of $\mathbf{D}$ at that point. If we consider an Infinitesimal volume Δv at that point and multiply both sides of (4.51) by Δv, we get

$$(\nabla \cdot \mathbf{D}) \, \Delta v = \rho \, \Delta v \tag{4.52}$$

But $\rho \, \Delta v$ is simply the charge contained in the volume Δv, and according to Gauss' law for the electric field in integral form,

$$\oint_S \mathbf{D} \cdot d\mathbf{S} = \rho \, \Delta v \tag{4.53}$$

where S is the closed surface bounding Δv. Comparing (4.52) and (4.53), we have

$$(\nabla \cdot \mathbf{D}) \, \Delta v = \oint_S \mathbf{D} \cdot d\mathbf{S} \tag{4.54}$$

Dividing both sides of (4.54) by Δv, we obtain

$$\nabla \cdot \mathbf{D} = \frac{\oint_S \mathbf{D} \cdot d\mathbf{S}}{\Delta v} \tag{4.55}$$

This result is however approximate since (4.54) is exact only in the limit that Δv tends to zero. Thus,

$$\nabla \cdot \mathbf{D} = \lim_{\Delta v \to 0} \frac{\oint_S \mathbf{D} \cdot d\mathbf{S}}{\Delta v} \tag{4.56}$$

which is the expression for $\nabla \cdot \mathbf{D}$ at a point in terms of $\mathbf{D}$ at that point. Although we have derived this for the $\mathbf{D}$ vector, it is a general result and, in fact, is often the starting point for the introduction of divergence. Thus, for any vector field, $\mathbf{A}$,

$$\boxed{\nabla \cdot \mathbf{A} = \lim_{\Delta v \to 0} \frac{\oint_S \mathbf{A} \cdot d\mathbf{S}}{\Delta v}} \tag{4.57}$$

Equation (4.57) tells us that to find the divergence of a vector at a point in that vector field, we first consider an infinitesimal volume at that point and compute the surface integral of the vector over the surface bounding that volume, that is, the outward flux of the vector field from that volume. We then divide the flux by the volume to obtain the flux per unit volume. Since we need this flux per unit volume in the limit that the volume tends to zero, we do this by gradually shrinking the volume. The limiting value to which the flux per unit volume approaches is the value of the divergence of the vector field at the point to which the volume is shrunk. In fact, this is what we did in Section 4.2, which led to the expression for the divergence in Cartesian coordinates, by choosing for convenience the volume of a rectangular box whose surfaces are parallel to the coordinate planes.

Physical interpretation of divergence   We are now ready to discuss the physical interpretation of the divergence. To simplify this task, we shall consider the continuity equation given by

$$\nabla \cdot \mathbf{J} = -\frac{\partial \rho}{\partial t} \tag{4.58}$$

Let us investigate three different cases: (1) positive value, (2) negative value, and (3) zero value of the time rate of decrease of the charge density at a point, that is, the divergence of the current density vector at that point. We shall do this with the aid of a simple device, which we shall call the *divergence meter*. The divergence meter can be imagined to be a tiny elastic balloon enclosing the point and that expands when hit by charges streaming outward from the point and contracts when acted on by charges streaming inward toward the point. For case 1, that is, when the time rate of decrease of the charge density at the point is positive, there is a net amount of charge streaming out of the point in a given time, resulting in a net current flow outward from the point that will make the imaginary balloon expand. For case 2, that is, when the time rate of decrease of the charge density at the point is negative or the time rate of increase of the charge density is positive, there is a net amount of charge streaming toward the point in a given time, resulting in a net current flow toward the point and the imaginary balloon will contract. For case 3, that is, when the time rate of decrease of the charge density at the point is zero, the balloon will remain unaffected since the charge is streaming out of the point at exactly the same rate as it is streaming into the point. The situation corresponding to case 1 is illustrated in Figs. 4.7(a) and (b), whereas that corresponding to case 2 is illustrated in Figs. 4.7(c) and (d), and that corresponding to case 3 is illustrated in Fig. 4.7(e). Note that in Figs. 4.7(a), (c), and (e), the imaginary balloon slides along the lines of current flow while responding to the divergence by expanding, contracting, or remaining unaffected.

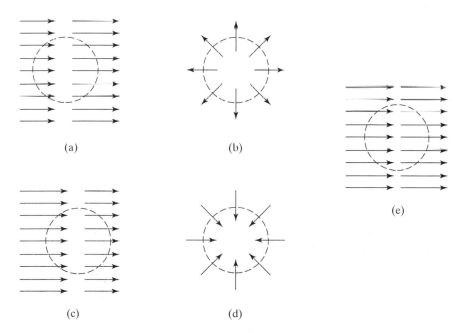

(a)

(b)

(c)

(d)

(e)

FIGURE 4.7   For explaining the physical interpretation of divergence using the divergence meter.

Generalizing the foregoing discussion to the physical interpretation of the divergence of any vector field at a point, we can imagine the vector field to be a velocity field of streaming charges acting on the divergence meter and obtain in most cases a qualitative picture of the divergence of the vector field. If the divergence meter expands, the divergence is positive and a source of the flux of the vector field exists at that point. If the divergence meter contracts, the divergence is negative and a sink of the flux of the vector field exists at that point. It can be further visualized that the rate of expansion or contraction of the divergence meter is a measure of the magnitude of the divergence. If the divergence meter remains unaffected, the divergence is zero, and neither a source nor a sink of the flux of the vector field exists at that point; alternatively, there can exist at the point pairs of sources and sinks of equal strengths.

C.　Stokes' and Divergence Theorems

We shall now derive two useful theorems in vector calculus, *Stokes' theorem* and the *divergence theorem*. Stokes' theorem relates the closed line integral of a vector field to the surface integral of the curl of that vector field, whereas the divergence theorem relates the closed surface integral of a vector field to the volume integral of the divergence of that vector field.

Stokes'
theorem
　　　　To derive Stokes' theorem, let us consider an arbitrary surface S in a magnetic field region and divide this surface into a number of infinitesimal surfaces $\Delta S_1, \Delta S_2, \Delta S_3, \ldots$, bounded by the contours $C_1, C_2, C_3, \ldots$, respectively. Then, applying (4.45) to each one of these infinitesimal surfaces and adding up, we get

$$\sum_j (\nabla \times \mathbf{H})_j \cdot \Delta S_j \, \mathbf{a}_{nj} = \oint_{C_1} \mathbf{H} \cdot d\mathbf{l} + \oint_{C_2} \mathbf{H} \cdot d\mathbf{l} + \cdots \qquad (4.59)$$

where $\mathbf{a}_{nj}$ are unit vectors normal to the surfaces ΔS_j chosen in accordance with the right-hand screw rule. In the limit that the number of infinitesimal surfaces tends to infinity, the left side of (4.59) approaches to the surface integral of $\nabla \times \mathbf{H}$ over the surface S. The right side of (4.59) is simply the closed line integral of $\mathbf{H}$ around the contour C since the contributions to the line integrals from the portions of the contours interior to C cancel, as shown in Fig. 4.8. Thus,

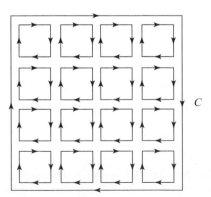

C

FIGURE 4.8　For deriving Stokes' theorem.

we get

$$\int_S (\nabla \times \mathbf{H}) \cdot d\mathbf{S} = \oint_C \mathbf{H} \cdot d\mathbf{l} \qquad (4.60)$$

Equation (4.60) is Stokes' theorem. Although we have derived it by considering the **H** field, it is general and can be derived from the application of (4.50) to a geometry such as that in Fig. 4.8. Thus, for any vector field **A**,

$$\boxed{\oint_C \mathbf{A} \cdot d\mathbf{l} = \int_S (\nabla \times \mathbf{A}) \cdot d\mathbf{S}} \qquad (4.61)$$

where S is any surface bounded by C.

Example 4.6

Let us evaluate the line integral of Example 3.1 by using Stokes' theorem.
For $\mathbf{F} = x\mathbf{a}_y$,

$$\nabla \times \mathbf{F} = \begin{vmatrix} \mathbf{a}_x & \mathbf{a}_y & \mathbf{a}_z \\ \dfrac{\partial}{\partial x} & \dfrac{\partial}{\partial y} & \dfrac{\partial}{\partial z} \\ 0 & x & 0 \end{vmatrix} = \mathbf{a}_z$$

With reference to Fig. 3.4, we then have

$$\oint_{ABCDA} \mathbf{F} \cdot d\mathbf{l} = \int_{\substack{\text{area} \\ ABCDA}} (\nabla \times \mathbf{F}) \cdot d\mathbf{S}$$

$$= \int_{\substack{\text{area} \\ ABCDA}} \mathbf{a}_z \cdot dx\, dy\, \mathbf{a}_z$$

$$= \int_{\substack{\text{area} \\ ABCDA}} dx\, dy$$

$$= \text{area } ABCDA$$

$$= 6$$

which agrees with the result obtained in Example 3.1.

To derive the divergence theorem, let us consider an arbitrary volume V in an electric field region and divide this volume into a number of infinitesimal volumes $\Delta v_1, \Delta v_2, \Delta v_3, \ldots$, bounded by the surfaces $S_1, S_2, S_3, \ldots$, respectively. Then, applying (4.54) to each one of these infinitesimal volumes and adding up, we get

Divergence theorem

$$\sum_j (\nabla \cdot \mathbf{D})_j \, \Delta v_j = \oint_{S_1} \mathbf{D} \cdot d\mathbf{S} + \oint_{S_2} \mathbf{D} \cdot d\mathbf{S} + \cdots \qquad (4.62)$$

In the limit that the number of the infinitesimal volumes tends to infinity, the left side of (4.62) approaches to the volume integral of $\nabla \cdot \mathbf{D}$ over the volume V.

FIGURE 4.9   For deriving the divergence theorem.

The right side of (4.62) is simply the closed surface integral of $\mathbf{D}$ over S since the contribution to the surface integrals from the portions of the surfaces interior to S cancel, as shown in Fig. 4.9. Thus, we get

$$\int_V (\boldsymbol{\nabla} \cdot \mathbf{D}) \, dv = \oint_S \mathbf{D} \cdot d\mathbf{S} \tag{4.63}$$

Equation (4.63) is the divergence theorem. Although we have derived it by considering the $\mathbf{D}$ field, it is general and can be derived from the application of (4.57) to a geometry such as that in Fig. 4.9. Thus, for any vector field $\mathbf{A}$,

$$\boxed{\oint_S \mathbf{A} \cdot d\mathbf{S} = \int_V (\boldsymbol{\nabla} \cdot \mathbf{A}) \, dv} \tag{4.64}$$

where V is the volume bounded by S.

Example 4.7

Divergence of the curl of a vector

By using the Stokes and divergence theorems, show that for any vector field $\mathbf{A}$, $\boldsymbol{\nabla} \cdot \boldsymbol{\nabla} \times \mathbf{A} = 0$.

Let us consider volume V bounded by the closed surface $S_1 + S_2$, where S_1 and S_2 are bounded by the closed paths C_1 and C_2, respectively, as shown in Fig.4.10. Note that C_1 and C_2 touch each other and are traversed in opposite senses and that $d\mathbf{S}_1$ and $d\mathbf{S}_2$ are

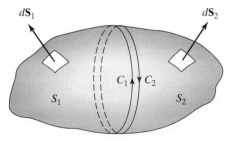

FIGURE 4.10   For proving the identity $\boldsymbol{\nabla} \cdot \boldsymbol{\nabla} \times \mathbf{A} = 0$.

directed in the right-hand sense relative to C_1 and C_2, respectively. Then, using divergence and Stokes' theorems in succession, we obtain

$$\int_V (\nabla \cdot \nabla \times \mathbf{A})\, dv = \oint_{S_1+S_2} (\nabla \times \mathbf{A}) \cdot d\mathbf{S}$$

$$= \int_{S_1} (\nabla \times \mathbf{A}) \cdot d\mathbf{S}_1 + \int_{S_2} (\nabla \times \mathbf{A}) \cdot d\mathbf{S}_2$$

$$= \oint_{C_1} \mathbf{A} \cdot d\mathbf{l} + \oint_{C_2} \mathbf{A} \cdot d\mathbf{l}$$

$$= 0$$

Since this result holds for any arbitrary volume V, it follows that

$$\boxed{\nabla \cdot \nabla \times \mathbf{A} = 0} \qquad (4.65)$$

We shall make use of (4.65) in the following section.

K4.3. Basic definition of curl; Physical interpretation of curl; Basic definition of divergence; Physical interpretation of divergence; Stokes' theorem; Divergence theorem; Divergence of the curl of a vector.

D4.7. With the aid of the curl meter, determine if the z-component of the curl of the vector field $\mathbf{A} = (x^2 - 4)\mathbf{a}_y$ is positive, zero, or negative at each of the following points: **(a)** $(2, -3, 1)$; **(b)** $(0, 2, 4)$; and **(c)** $(-1, 2, -1)$.
Ans.   **(a)** positive;   **(b)** zero;   **(c)** negative.

D4.8. With the aid of the divergence meter, determine if the divergence of the vector field $\mathbf{A} = (x - 2)^2 \mathbf{a}_x$ is positive, zero, or negative at each of the following points: **(a)** $(2, 4, 3)$; **(b)** $(1, 1, -1)$; and **(c)** $(3, -1, 4)$.
Ans.   **(a)** zero;   **(b)** negative;   **(c)** positive.

D4.9. Using Stokes' theorem, find the absolute value of the line integral of the vector field $(x\mathbf{a}_y + \sqrt{3}y\mathbf{a}_z)$ around each of the following closed paths: **(a)** the perimeter of a square of sides 2 m lying in the xy-plane; **(b)** a circular path of radius $1/\sqrt{\pi}$ m lying in the xy-plane; and **(c)** the perimeter of an equilateral triangle of sides 2 m lying in the yz-plane.
Ans.   **(a)** 4;   **(b)** 1;   **(c)** 3.

D4.10. Using the divergence theorem, find the surface integral of the vector field $(x\mathbf{a}_x + y\mathbf{a}_y + z\mathbf{a}_z)$ over each of the following closed surfaces: **(a)** the surface of a cube of sides 1 m; **(b)** the surface of a cylinder of radius $1/\sqrt{\pi}$ m and length 2 m; and **(c)** the surface of a sphere of radius $1/(\pi)^{1/3}$ m.
Ans.   **(a)** 3;   **(b)** 6;   **(c)** 4.

4.4.   GRADIENT, LAPLACIAN, AND THE POTENTIAL FUNCTIONS

In Example 4.7, we showed that for any vector $\mathbf{A}$, $\nabla \cdot \nabla \times \mathbf{A} = 0$ (for alternate proof, see Problem P4.18). It then follows from Gauss' law for the magnetic field in differential form, $\nabla \cdot \mathbf{B} = 0$, that the magnetic flux density vector $\mathbf{B}$ can be

Magnetic vector potential

expressed as the curl of another vector $\mathbf{A}$; that is,

$$\boxed{\mathbf{B} = \nabla \times \mathbf{A}} \tag{4.66}$$

The vector $\mathbf{A}$ in (4.66) is known as the *magnetic vector potential.*

Gradient   Substituting (4.66) into Faraday's law in differential form, $\nabla \times \mathbf{E} = -\partial \mathbf{B}/\partial t$, and rearranging, we then obtain

$$\nabla \times \mathbf{E} + \frac{\partial}{\partial t}(\nabla \times \mathbf{A}) = \mathbf{0}$$

or

$$\nabla \times \left(\mathbf{E} + \frac{\partial \mathbf{A}}{\partial t}\right) = \mathbf{0} \tag{4.67}$$

If the curl of a vector is equal to the null vector, that vector can be expressed as the *gradient* of a scalar, since the curl of the gradient of a scalar function is identically equal to the null vector. The gradient of a scalar, say, Φ, denoted $\nabla\Phi$ (del Φ) is defined in such a manner that the increment $d\Phi$ in Φ from a point P to a neighboring point Q is given by

$$d\Phi = \nabla\Phi \cdot d\mathbf{l} \tag{4.68}$$

where $d\mathbf{l}$ is the differential length vector from P to Q. Applying Stokes' theorem to the vector $\nabla \times \nabla\Phi$ and a surface S bounded by closed path C, we then have

$$\int_S (\nabla \times \nabla\Phi) \cdot d\mathbf{S} = \oint_C \nabla\Phi \cdot d\mathbf{l}$$

$$= \oint_C d\Phi \tag{4.69}$$

$$= 0$$

for any single-valued function Φ. Since (4.69) holds for an arbitrary S, it follows that

$$\boxed{\nabla \times \nabla\Phi = \mathbf{0}} \tag{4.70}$$

To obtain the expression for the gradient in the Cartesian coordinate system, we write

$$d\Phi = \frac{\partial \Phi}{\partial x} dx + \frac{\partial \Phi}{\partial y} dy + \frac{\partial \Phi}{\partial z} dz$$

$$= \left(\frac{\partial \Phi}{\partial x} \mathbf{a}_x + \frac{\partial \Phi}{\partial y} \mathbf{a}_y + \frac{\partial \Phi}{\partial z} \mathbf{a}_z\right) \cdot (dx\, \mathbf{a}_x + dy\, \mathbf{a}_y + dz\, \mathbf{a}_z) \tag{4.71}$$

Then comparing with (4.68), we observe that

$$\boxed{\nabla\Phi = \frac{\partial \Phi}{\partial x} \mathbf{a}_x + \frac{\partial \Phi}{\partial y} \mathbf{a}_y + \frac{\partial \Phi}{\partial z} \mathbf{a}_z} \tag{4.72}$$

Note that the right side of (4.72) is simply the vector obtained by applying the del operator to the scalar function Φ. It is for this reason that the gradient of Φ is written as $\nabla\Phi$. Expressions for the gradient in cylindrical and spherical coordinate systems are derived in Appendix A. These are as follows:

CYLINDRICAL

$$\nabla\Phi = \frac{\partial\Phi}{\partial r}\,\mathbf{a}_r + \frac{1}{r}\frac{\partial\Phi}{\partial\phi}\,\mathbf{a}_\phi + \frac{\partial\Phi}{\partial z}\,\mathbf{a}_z \qquad (4.73a)$$

SPHERICAL

$$\nabla\Phi = \frac{\partial\Phi}{\partial r}\,\mathbf{a}_r + \frac{1}{r}\frac{\partial\Phi}{\partial\theta}\,\mathbf{a}_\theta + \frac{1}{r\sin\theta}\frac{\partial\Phi}{\partial\phi}\,\mathbf{a}_\phi \qquad (4.73b)$$

To discuss the physical interpretation of the gradient, let us consider a surface on which Φ is equal to a constant, say, Φ_0, and a point P on that surface, as shown in Fig. 4.11(a). If we now consider another point Q_1 on the same surface and an infinitesimal distance away from P, $d\Phi$ between these two points is zero since Φ is constant on the surface. Thus, for the vector $d\mathbf{l}_1$ drawn from P to Q_1, $[\nabla\Phi]_P \cdot d\mathbf{l}_1 = 0$ and hence $[\nabla\Phi]_P$ is perpendicular to $d\mathbf{l}_1$. Since this is true for all points $Q_1, Q_2, Q_3, \ldots$ on the constant Φ surface, it follows that $[\nabla\Phi]_P$ must be normal to all possible infinitesimal length vectors $d\mathbf{l}_1, d\mathbf{l}_2, d\mathbf{l}_3, \ldots$ drawn at P and hence is normal to the surface. Denoting $\mathbf{a}_n$ to be the unit normal vector to the surface at P, we then have

Physical interpretation of gradient

$$[\nabla\Phi]_P = |\nabla\Phi|_P\,\mathbf{a}_n \qquad (4.74)$$

Let us now consider two surfaces on which Φ is constant, having values Φ_0 and $\Phi_0 + d\Phi$, as shown in Fig. 4.11(b). Let P and Q be points on the $\Phi = \Phi_0$ and

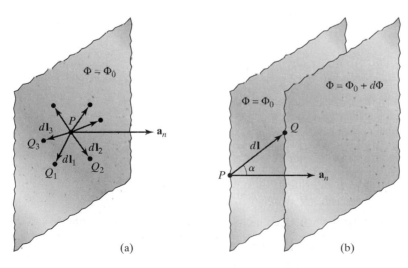

FIGURE 4.11 For discussing the physical interpretation of the gradient of a scalar function.

$\Phi = \Phi_0 + d\Phi$ surfaces, respectively, and $d\mathbf{l}$ be the vector drawn from P to Q. Then from (4.68) and (4.74),

$$
\begin{aligned}
d\Phi &= [\nabla\Phi]_P \cdot d\mathbf{l} \\
&= |\nabla\Phi|_P \, \mathbf{a}_n \cdot d\mathbf{l} \\
&= |\nabla\Phi|_P \, dl \cos\alpha
\end{aligned}
$$

where α is the angle between $\mathbf{a}_n$ at P and $d\mathbf{l}$. Thus,

$$
|\nabla\Phi|_P = \frac{d\Phi}{dl \cos\alpha} \tag{4.75}
$$

Since $dl \cos\alpha$ is the distance between the two surfaces along $\mathbf{a}_n$ and hence is the shortest distance between them, it follows that $|\nabla\Phi|_P$ is the maximum rate of increase of Φ at the point P. Thus, the gradient of a scalar function Φ at a point is a vector having magnitude equal to the maximum rate of increase of Φ at that point and is directed along the direction of the maximum rate of increase, which is normal to the constant Φ surface passing through that point; that is,

$$
\boxed{\nabla\Phi = \frac{d\Phi}{dn} \, \mathbf{a}_n} \tag{4.76}
$$

where dn is a differential length along $\mathbf{a}_n$. The concept of the gradient of a scalar function we just discussed is often utilized to find a unit vector normal to a given surface. We shall illustrate this by means of an example.

Example 4.8

Let us find the unit vector normal to the surface $y = x^2$ at the point $(2, 4, 1)$ by using the concept of the gradient of a scalar.

Writing the equation for the surface as

$$
x^2 - y = 0
$$

we note that the scalar function that is constant on the surface is given by

$$
\Phi(x, y, z) = x^2 - y
$$

The gradient of the scalar function is then given by

$$
\begin{aligned}
\nabla\Phi &= \nabla(x^2 - y) \\
&= \frac{\partial(x^2 - y)}{\partial x} \, \mathbf{a}_x + \frac{\partial(x^2 - y)}{\partial y} \, \mathbf{a}_y + \frac{\partial(x^2 - y)}{\partial z} \, \mathbf{a}_z \\
&= 2x\mathbf{a}_x - \mathbf{a}_y
\end{aligned}
$$

The value of the gradient at the point $(2, 4, 1)$ is $[2(2)\mathbf{a}_x - \mathbf{a}_y] = (4\mathbf{a}_x - \mathbf{a}_y)$. Thus, the required unit vector is

$$
\mathbf{a}_n = \pm\frac{4\mathbf{a}_x - \mathbf{a}_y}{|4\mathbf{a}_x - \mathbf{a}_y|} = \pm\left(\frac{4}{\sqrt{17}} \, \mathbf{a}_x - \frac{1}{\sqrt{17}} \, \mathbf{a}_y\right)
$$

Returning now to (4.67), we write

$$\mathbf{E} + \frac{\partial \mathbf{A}}{\partial t} = -\nabla \Phi \qquad (4.77)$$

where we have chosen the scalar to be $-\Phi$, the reason for the minus sign to be explained in Section 4.5. Rearranging (4.77), we obtain

$$\boxed{\mathbf{E} = -\nabla \Phi - \frac{\partial \mathbf{A}}{\partial t}} \qquad (4.78)$$

The quantity Φ in (4.78) is known as the electric scalar potential.

The electric scalar potential Φ and the magnetic vector potential $\mathbf{A}$ are known as the electromagnetic potentials. As we shall show later in this section, the electric scalar potential is related to the source charge density ρ, whereas the magnetic vector potential is related to the source current density $\mathbf{J}$. For the time-varying case, the two are not independent since the charge and current densities are related through the continuity equation. For a given $\mathbf{J}$, it is sufficient to determine $\mathbf{A}$, since $\mathbf{B}$ can be found from (4.66) and then $\mathbf{E}$ can be found by using Ampère's circuital law $\nabla \times \mathbf{H} = \mathbf{J} + \partial \mathbf{D}/\partial t$. For static fields, that is, for $\partial/\partial t = 0$, the two potentials are independent. Equation (4.66) remains unaltered, whereas (4.78) reduces to $\mathbf{E} = -\nabla \Phi$. We shall consider the static field case in Section 4.5.

To proceed further, we recall that the Maxwell's equations in differential form are given by

$$\nabla \times \mathbf{E} = -\frac{\partial \mathbf{B}}{\partial t} \qquad (4.79a)$$

$$\nabla \times \mathbf{H} = \mathbf{J} + \frac{\partial \mathbf{D}}{\partial t} \qquad (4.79b)$$

$$\nabla \cdot \mathbf{D} = \rho \qquad (4.79c)$$

$$\nabla \cdot \mathbf{B} = 0 \qquad (4.79d)$$

From (4.79d), we expressed $\mathbf{B}$ in the manner

$$\mathbf{B} = \nabla \times \mathbf{A} \qquad (4.80)$$

and then from (4.79a), we obtained

$$\mathbf{E} = -\nabla \Phi - \frac{\partial \mathbf{A}}{\partial t} \qquad (4.81)$$

We now substitute (4.81) and (4.80) into (4.79c) and (4.79b), respectively, to obtain

$$\nabla \cdot \left(-\nabla \Phi - \frac{\partial \mathbf{A}}{\partial t} \right) = \frac{\rho}{\varepsilon} \qquad (4.82a)$$

$$\nabla \times \nabla \times \mathbf{A} - \mu\varepsilon \frac{\partial}{\partial t} \left(-\nabla \Phi - \frac{\partial \mathbf{A}}{\partial t} \right) = \mu \mathbf{J} \qquad (4.82b)$$

Laplacian of
a scalar

We now define the Laplacian of a scalar quantity Φ, denoted $\nabla^2\Phi$ (del squared Φ) as

$$\boxed{\nabla^2\Phi = \nabla \cdot \nabla\Phi}$$ (4.83)

In Cartesian coordinates,

$$\nabla\Phi = \frac{\partial\Phi}{\partial x}\mathbf{a}_x + \frac{\partial\Phi}{\partial y}\mathbf{a}_y + \frac{\partial\Phi}{\partial z}\mathbf{a}_z$$

$$\nabla \cdot \mathbf{A} = \frac{\partial A_x}{\partial x} + \frac{\partial A_y}{\partial y} + \frac{\partial A_z}{\partial z}$$

so that

$$\nabla^2\Phi = \frac{\partial}{\partial x}\left(\frac{\partial\Phi}{\partial x}\right) + \frac{\partial}{\partial y}\left(\frac{\partial\Phi}{\partial y}\right) + \frac{\partial}{\partial z}\left(\frac{\partial\Phi}{\partial z}\right)$$

or

$$\boxed{\nabla^2\Phi = \frac{\partial^2\Phi}{\partial x^2} + \frac{\partial^2\Phi}{\partial y^2} + \frac{\partial^2\Phi}{\partial z^2}}$$ (4.84)

Note that the Laplacian of a scalar is a scalar quantity. Expressions for the Laplacian of a scalar in cylindrical and spherical coordinates are derived in Appendix A. These are as follows:

CYLINDRICAL

$$\nabla^2\Phi = \frac{1}{r}\frac{\partial}{\partial r}\left(r\frac{\partial\Phi}{\partial r}\right) + \frac{1}{r^2}\frac{\partial^2\Phi}{\partial\phi^2} + \frac{\partial^2\Phi}{\partial z^2}$$ (4.85a)

SPHERICAL

$$\nabla^2\Phi = \frac{1}{r^2}\frac{\partial}{\partial r}\left(r^2\frac{\partial\Phi}{\partial r}\right) + \frac{1}{r^2\sin\theta}\frac{\partial}{\partial\theta}\left(\sin\theta\frac{\partial\Phi}{\partial\theta}\right) + \frac{1}{r^2\sin^2\theta}\frac{\partial^2\Phi}{\partial\phi^2}$$ (4.85b)

Before proceeding further, it is interesting to note that the four vector differential operations that we have learned thus far in this chapter are such that

The *curl* of a *vector* is a *vector*.
The *divergence* of a *vector* is a *scalar*.
The *gradient* of a *scalar* is a *vector*.
The *Laplacian* of a *scalar* is a *scalar*.

Thus, all four combinations of vector and scalar are involved in the four operations.

Laplacian of
a vector

Next, we define the Laplacian of a vector, denoted $\nabla^2\mathbf{A}$ as

$$\boxed{\nabla^2\mathbf{A} = \nabla(\nabla \cdot \mathbf{A}) - \nabla \times \nabla \times \mathbf{A}}$$ (4.86)

Expanding the right side of (4.86) in Cartesian coordinates and simplifying, we obtain in the Cartesian coordinate system,

$$\nabla^2 \mathbf{A} = (\nabla^2 A_x)\mathbf{a}_x + (\nabla^2 A_y)\mathbf{a}_y + (\nabla^2 A_z)\mathbf{a}_z \qquad (4.87)$$

Thus, in the Cartesian coordinate system, the Laplacian of a vector is a vector whose components are the Laplacians of the corresponding components of $\mathbf{A}$. It should however be cautioned that this simple observation does not hold in the cylindrical and spherical coordinate systems (see, e.g., Problem P4.23).

Using (4.83) and (4.86), we now write (4.82a) and (4.82b) as

$$\nabla^2 \Phi + \frac{\partial}{\partial t}(\nabla \cdot \mathbf{A}) = -\frac{\rho}{\varepsilon} \qquad (4.88a)$$

$$\nabla^2 \mathbf{A} - \nabla\left(\nabla \cdot \mathbf{A} + \mu\varepsilon\frac{\partial \Phi}{\partial t}\right) - \mu\varepsilon\frac{\partial^2 \mathbf{A}}{\partial t^2} = -\mu\mathbf{J} \qquad (4.88b)$$

Equations (4.88a) and (4.88b) are a pair of coupled differential equations for Φ and $\mathbf{A}$. To uncouple the equations, we make use of a theorem known as Helmholtz's theorem, which states that a vector field is completely specified by its curl and divergence. Therefore, since the curl of $\mathbf{A}$ is given by (4.80), we are at liberty to specify the divergence of $\mathbf{A}$. We do this by setting

Potential function equations

$$\nabla \cdot \mathbf{A} = -\mu\varepsilon\frac{\partial \Phi}{\partial t} \qquad (4.89)$$

which is known as the Lorenz condition.{(FR 6 4)}[2] This uncouples (4.88a) and (4.88b) to give us

$$\nabla^2 \Phi - \mu\varepsilon\frac{\partial^2 \Phi}{\partial t^2} = -\frac{\rho}{\varepsilon} \qquad (4.90)$$

$$\nabla^2 \mathbf{A} - \mu\varepsilon\frac{\partial^2 \mathbf{A}}{\partial t^2} = -\mu\mathbf{J} \qquad (4.91)$$

These are the differential equations relating the electromagnetic potentials Φ and $\mathbf{A}$ to the source charge and current densities ρ and $\mathbf{J}$, respectively.

Before proceeding further, we shall show that the continuity equation is implied by the Lorenz condition. To do this, we take the Laplacian of both sides of (4.89). We then have

$$\nabla^2(\nabla \cdot \mathbf{A}) = -\mu\varepsilon\nabla^2\frac{\partial \Phi}{\partial t}$$

[1] Note "Lorenz condition" and not "Lorentz condition." In previous editions of this book, as well as extensively in books by other authors, this condition has been mistakenly attributed to Lorentz instead of Lorenz. See the note "Lorentz or Lorenz," by J. Van Bladel in *IEEE Antennas and Propagation Magazine*, Vol. 33, No. 2, April 1991, p. 69.

or

$$\mathbf{\nabla} \cdot \nabla^2 \mathbf{A} = -\mu\varepsilon \frac{\partial}{\partial t} \nabla^2 \Phi \tag{4.92}$$

Substituting for $\nabla^2 \mathbf{A}$ and $\nabla^2 \Phi$ in (4.92) from (4.91) and (4.90), respectively, we get

$$\mathbf{\nabla} \cdot \left(\mu\varepsilon \frac{\partial^2 \mathbf{A}}{\partial t^2} - \mu \mathbf{J} \right) = -\mu\varepsilon \frac{\partial}{\partial t} \left(\mu\varepsilon \frac{\partial^2 \Phi}{\partial t^2} - \frac{\rho}{\varepsilon} \right)$$

or

$$\mu\varepsilon \frac{\partial^2}{\partial t^2} \left(\mathbf{\nabla} \cdot \mathbf{A} + \mu\varepsilon \frac{\partial \Phi}{\partial t} \right) = \mu \left(\mathbf{\nabla} \cdot \mathbf{J} + \frac{\partial \rho}{\partial t} \right) \tag{4.93}$$

Thus, by assuming the Lorenz condition (4.89), we imply $\mathbf{\nabla} \cdot \mathbf{J} + \partial\rho/\partial t = 0$, which is the continuity equation.

As pointed out in the previous section, it is sufficient to determine $\mathbf{A}$ for the time-varying case for a given $\mathbf{J}$. Hence, we shall be concerned only with (4.91), which we shall refer to in Section 11.1 in connection with obtaining the electromagnetic field due to an elemental antenna.

K4.4.   Magnetic vector potential; Gradient of a scalar; Physical interpretation of gradient; Electric scalar potential; Laplacian of a scalar; Potential function equations.

D4.11.   Find the outward pointing unit vectors normal to the closed surface $2x^2 + 2y^2 + z^2 = 8$ at the following points: **(a)** $(\sqrt{2}, \sqrt{2}, 0)$; **(b)** $(1, 1, 2)$; and **(c)** $(1, \sqrt{2}, \sqrt{2})$.

Ans.   **(a)** $\dfrac{\mathbf{a}_x + \mathbf{a}_y}{\sqrt{2}}$;   **(b)** $\dfrac{\mathbf{a}_x + \mathbf{a}_y + \mathbf{a}_z}{\sqrt{3}}$ **(c)** $\dfrac{\sqrt{2}\mathbf{a}_x + 2\mathbf{a}_y + \mathbf{a}_z}{\sqrt{7}}$

D4.12.   Two scalar functions are given by

$$\Phi_1(x, y, z) = x^2 + y^2 + z^2$$

$$\Phi_2(x, y, z) = x + 2y + 2z$$

Find the following at the point $(3, 4, 12)$: **(a)** the maximum rate of increase of Φ_1; **(b)** the maximum rate of increase of Φ_2; and **(c)** the rate of increase of Φ_1 along the direction of the maximum rate of increase of Φ_2.

Ans.   **(a)** 26;   **(b)** 3;   **(c)** $23\frac{1}{3}$.

D4.13.   Find the Laplacians of the following functions: **(a)** x^2yz^3; **(b)** $(1/r) \sin \phi$ in cylindrical coordinates; and **(c)** $r^2 \cos \theta$ in spherical coordinates.

Ans.   **(a)** $2yz^3 + 6x^2yz$;   **(b)** 0;   **(c)** $4 \cos \theta$.

4.5   POTENTIAL FUNCTIONS FOR STATIC FIELDS

Potential difference

As already pointed out in the preceding section, Eq. (4.78) reduces to

$$\boxed{\mathbf{E} = -\mathbf{\nabla}\Phi} \tag{4.94}$$

for the static field case. We observe from (4.94) that the potential function Φ then is such that the electric field lines are orthogonal to the equipotential

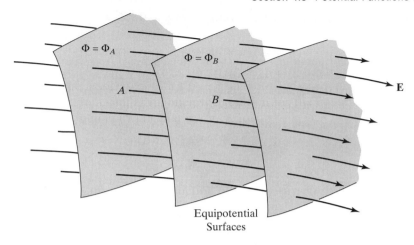

FIGURE 4.12 Set of equipotential surfaces in a region of static electric field.

surfaces, that is, to the surfaces on which the potential remains constant, as shown in Fig. 4.12. If we consider two such equipotential surfaces corresponding to $\Phi = \Phi_A$ and $\Phi = \Phi_B$, as shown in the figure, the potential difference $\Phi_A - \Phi_B$ is given, according to the definition of the gradient, by

$$\Phi_A - \Phi_B = \int_B^A d\Phi = \int_B^A \boldsymbol{\nabla}\Phi \cdot d\mathbf{l}$$

$$= -\int_A^B \boldsymbol{\nabla}\Phi \cdot d\mathbf{l} \tag{4.95}$$

Using (4.94), we obtain

$$\boxed{\Phi_A - \Phi_B = \int_A^B \mathbf{E} \cdot d\mathbf{l}} \tag{4.96}$$

We now recall from Section 3.1 that $\int_A^B \mathbf{E} \cdot d\mathbf{l}$ is the voltage between points A and B. Thus, the potential difference in the static field case has the same meaning as the voltage. The reason for the minus sign in (4.78) and hence in (4.94) is now evident, since without it the voltage between A and B would be the negative of the potential difference between A and B.

Before proceeding further, we recall that the voltage between two points A and B in a time-varying electric field is in general dependent on the path followed from A to B to evaluate $\int_A^B \mathbf{E} \cdot d\mathbf{l}$ since according to Faraday's law,

Potential difference versus voltage

$$\oint_C \mathbf{E} \cdot d\mathbf{l} = -\frac{d}{dt} \int_S \mathbf{B} \cdot d\mathbf{S}$$

is not in general equal to zero. On the other hand, the potential difference (or voltage) between two points A and B in a static electric field is independent of the path followed from A to B to evaluate $\int_A^B \mathbf{E} \cdot d\mathbf{l}$ since for static fields,

$$\oint_C \mathbf{E} \cdot d\mathbf{l} = 0$$

Thus, the potential difference (or voltage) between two points in a static electric field has a unique value. Since the potential difference and voltage have the same meaning for static fields, we shall hereafter replace Φ in (4.94) by V, thereby writing

$$\boxed{\mathbf{E} = -\nabla V} \tag{4.97}$$

Electric
potential due
to a point
charge

      Let us now consider the electric field of a point charge and investigate the electric potential due to the point charge. To do this, we recall that the electric field intensity due to a point charge Q is directed radially away from the point charge and its magnitude is $Q/4\pi\varepsilon R^2$, where R is the radial distance from the point charge. Since the equipotential surfaces are everywhere orthogonal to the field lines, it then follows that they are spherical surfaces centered at the point charge, as shown by the cross-sectional view in Fig. 4.13. If we now consider two equipotential surfaces of radii R and $R + dR$, the potential drop from the surface of radius R to the surface of radius $R + dR$ is $(Q/4\pi\varepsilon R^2)dR$ or the incremental potential rise dV is given by

$$dV = -\frac{Q}{4\pi\varepsilon R^2}\,dR$$
$$= d\left(\frac{Q}{4\pi\varepsilon R} + C\right) \tag{4.98}$$

where C is a constant. Thus,

$$V(R) = \frac{Q}{4\pi\varepsilon R} + C \tag{4.99}$$

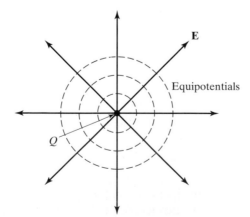

FIGURE 4.13   Cross-sectional view of equipotential surfaces and electric field lines for a point charge.

Since the potential difference between two points does not depend on the value of C, we can choose C such that V is zero at some arbitrary reference point. Here we can conveniently set C equal to zero by noting that it is equal to $V(\infty)$ and by choosing $R = \infty$ for the reference point. Thus, we obtain the electric potential due to a point charge Q to be

$$V = \frac{Q}{4\pi\varepsilon R} \qquad\qquad (4.100)$$

We note that the potential drops off inversely with the radial distance away from the point charge.

Equation (4.100) is often the starting point for the computation of the potential field due to static charge distributions and the subsequent determination of the electric field by using (4.97). We shall illustrate this by considering the case of the electric dipole in the following example.

Example 4.9

As we have learned in Section 2.5, the electric dipole consists of two equal and opposite point charges. Let us consider a static electric dipole consisting of point charges Q and $-Q$ situated on the z-axis at $z = d/2$ and $z = -d/2$, respectively, as shown in Fig. 4.14(a) and find the potential and hence the electric field at a point P far from the dipole. *Electric dipole*

First, we note that in view of the symmetry associated with the dipole around the z-axis, it is convenient to use the spherical coordinate system. Denoting the distance from

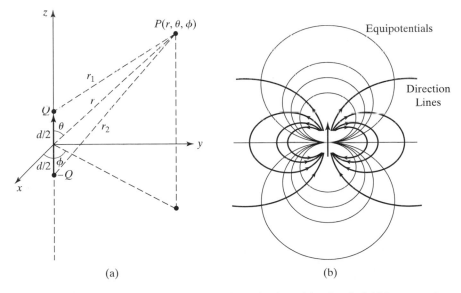

(a)　　　　　　　　　　　　(b)

FIGURE 4.14 (a) Geometry pertinent to the determination of the electric field due to an electric dipole. (b) Cross sections of equipotential surfaces and direction lines of the electric field for the electric dipole.

the point charge Q to P to be r_1 and the distance from the point charge $-Q$ to P to be r_2, we write the expression for the electric potential at P due to the electric dipole as

$$V = \frac{Q}{4\pi\varepsilon r_1} + \frac{-Q}{4\pi\varepsilon r_2} = \frac{Q}{4\pi\varepsilon}\left(\frac{1}{r_1} - \frac{1}{r_2}\right)$$

For a point P far from the dipole, that is, for $r \gg d$, the lines drawn from the two charges to the point are almost parallel. Hence,

$$r_1 \approx r - \frac{d}{2}\cos\theta$$

$$r_2 \approx r + \frac{d}{2}\cos\theta$$

and

$$\frac{1}{r_1} - \frac{1}{r_2} = \frac{r_2 - r_1}{r_1 r_2} \approx \frac{d\cos\theta}{r^2}$$

so that

$$V \approx \frac{Qd\cos\theta}{4\pi\varepsilon r^2} = \frac{\mathbf{p} \cdot \mathbf{a}_r}{4\pi\varepsilon r^2} \tag{4.101}$$

where $\mathbf{p} = Qd\mathbf{a}_z$ is the dipole moment of the electric dipole. Thus, the potential field of the electric dipole drops off inversely with the square of the distance from the dipole. Proceeding further, we obtain the electric field intensity due to the dipole to be

$$\mathbf{E} = -\nabla V = -\frac{\partial}{\partial r}\left(\frac{Qd\cos\theta}{4\pi\varepsilon r^2}\right)\mathbf{a}_r - \frac{1}{r}\frac{\partial}{\partial\theta}\left(\frac{Qd\cos\theta}{4\pi\varepsilon r^2}\right)\mathbf{a}_\theta$$

$$= \frac{Qd}{4\pi\varepsilon r^3}(2\cos\theta\,\mathbf{a}_r + \sin\theta\,\mathbf{a}_\theta) \tag{4.102}$$

Equation (4.101) shows that the equipotential surfaces are given by $r^2 \sec\theta = $ constant, whereas from (4.102), it can be shown that the direction lines of the electric field are given by $r\, \mathrm{cosec}^2\,\theta = $ constant and $\phi = $ constant. These are shown sketched in Fig. 4.14(b). Alternative to using the equation for the direction lines, they can be sketched by recognizing that (1) they must originate from the positive charge and end on the negative charge and (2) they must be everywhere perpendicular to the equipotential surfaces.

Electro-cardiography

A technique in everyday life in which the potential field of an electric dipole is relevant is electrocardiography. This technique is based on the characterization of the electrical activity of the heart by using a dipole model.[2] The dipole moment, $\mathbf{p}$, referred to in medical literature as the *electric force vector* or the *activity* of the heart, sets up an electric potential within the chest cavity and a characteristic pattern of equipotentials on the body surface. The potential differences between various points on the body are measured as a function of time and are used to deduce the temporal evolution of the dipole moment during the cardiac cycle, thereby monitoring changes in the electrical activity of the heart.

[2]See, for example, R. K. Hobbie, "The Electrocardiogram as an Example in Electrostatics," *American Journal of Physics,* June 1973, pp. 824–831.

We shall now consider an example for illustrating a method of computer plotting of equipotentials when a closed form expression such as that for the electric dipole of Example 4.9 is not available.

Example 4.10

Let us consider two point charges $Q_1 = 8\pi\varepsilon_0\,C$ and $Q_2 = -4\pi\varepsilon_0\,C$ situated at $(-1,0,0)$ and $(1,0,0)$, respectively, as shown in Fig. 4.15. We wish to discuss the computer plotting of the equipotentials due to the two point charges.   *Computer plotting of equipotentials*

First, we recognize that since the equipotential surfaces are surfaces of revolution about the axis of the two charges, it is sufficient to consider the equipotential lines in any plane containing the two charges. Here we shall consider the xz-plane. The equipotential lines are also symmetrical about the x-axis, and, hence, we shall plot them only on one side of the x-axis and inside the rectangular region having corners at $(-4,0), (4,0)\ (4,5)$, and $(-4,5)$.

As we go from Q_1 to Q_2 along the x-axis, the potential varies from $+\infty$ to $-\infty$ and is given by

$$V = \frac{8\pi\varepsilon_0}{4\pi\varepsilon_0(1+x)} - \frac{4\pi\varepsilon_0}{4\pi\varepsilon_0(1-x)}$$

$$= \frac{1-3x}{1-x^2}$$

The value of x lying between -1 and 1 for a given potential V_0 is then given by

$$V_0 = \frac{1-3x}{1-x^2}$$

or

$$x = \begin{cases} \dfrac{3 - \sqrt{9 - 4V_0(1 - V_0)}}{2V_0} & \text{for} \quad V_0 \neq 0 \\[3mm] \dfrac{1}{3} & \text{for} \quad V_0 = 0 \end{cases}$$

We shall begin the equipotential line at this value of x on the x-axis for a given value of V_0. To plot the line, we make use of the property that the equipotential lines are

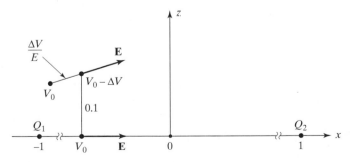

FIGURE 4.15   For illustrating the procedure for the computer plotting of equipotentials due to two point charges.

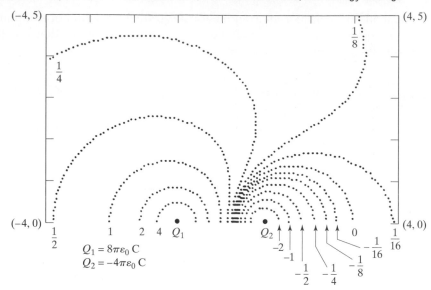

FIGURE 4.16　Personal computer-generated plot of equipotentials for the arrangement of two point charges of Fig. 4.15. The values of potentials are in volts.

orthogonal to the direction lines of $\mathbf{E}$ so that they are tangential to the unit vector $(E_z \mathbf{a}_x - E_x \mathbf{a}_z)/E$. We shall step along this unit vector by a small distance (chosen here to be 0.1), and if necessary, correct the position by repeatedly moving along the electric field until the potential is within a specified value (chosen here to be 0.001 V) of that for which the line is being plotted. To correct the position, we make use of the fact that $\nabla V = -\mathbf{E}$. Thus, the incremental distance required to be moved opposite to the electric field to increase the potential by ΔV is $\Delta V/E$, and, hence, the distances required to be moved opposite to the x- and z-directions are $(\Delta V/E)(E_x/E)$ and $(\Delta V/E)(E_z/E)$, respectively. The plotting of the line is terminated when the point goes out of the rectangular region.

　　The computer plot obtained from a run of a PC program that carries out this procedure for values of potentials ranging from -2 V to 4 V is shown in Fig. 4.16. It should, however, be pointed out that for a complete plot, those equipotential lines that surround both point charges should also be considered.

　　The computation of potential can be extended to continuous charge distributions by using superposition in conjunction with the expression for the potential due to a point charge, as in the case of electric field computation in Section 2.1. We shall illustrate by means of an example.

Example 4.11

Potential due to a line charge

An infinitely long line charge of uniform density ρ_{L0} C/m is situated along the z-axis. It is desired to obtain the potential field due to this charge.

　　First, we divide the line into a number of infinitesimal segments each of length dz, as shown in Fig. 4.17, such that the charge $\rho_{L0}\,dz$ in each segment can be considered as a

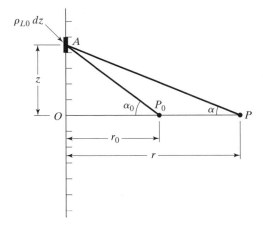

FIGURE 4.17 Geometry for the computation of the potential field of an infinitely long line charge of uniform density ρ_{L0} C/m.

point charge. Let us consider a point P at a distance r from the z-axis, with the projection of P onto the z-axis being O. For the sake of generality, we consider the point P_0 at a distance r_0 from O along OP as the reference point for zero potential and write the potential dV at P due to the infinitesimal charge $\rho_{L0}\,dz$ at A as

$$dV = \frac{\rho_{L0}\,dz}{4\pi\varepsilon_0(AP)} - \frac{\rho_{L0}\,dz}{4\pi\varepsilon_0(AP_0)}$$

$$= \frac{\rho_{L0}\,dz}{4\pi\varepsilon_0\sqrt{r^2 + z^2}} - \frac{\rho_{L0}\,dz}{4\pi\varepsilon_0\sqrt{r_0^2 + z^2}} \qquad (4.103)$$

We will, however, find later that we have to choose the reference point for zero potential at a finite value of r, in contrast to the case of the point charge for which the reference point can be chosen to be infinity. The potential V at P due to the entire line charge is now given by the integral of (4.103), where the integration is to be performed between the limits $z = -\infty$ and $z = \infty$ Thus,

$$V = \int_{z=-\infty}^{\infty} dV = \int_{z=-\infty}^{\infty} \left(\frac{\rho_{L0}\,dz}{4\pi\varepsilon_0\sqrt{r^2 + z^2}} - \frac{\rho_{L0}\,dz}{4\pi\varepsilon_0\sqrt{r_0^2 + z^2}} \right)$$

$$= \frac{\rho_{L0}}{2\pi\varepsilon_0} \int_{z=0}^{\infty} \left(\frac{dz}{\sqrt{r^2 + z^2}} - \frac{dz}{\sqrt{r_0^2 + z^2}} \right) \qquad (4.104)$$

Introducing $z = r\tan\alpha$ and $z = r_0\tan\alpha_0$ in the first and second terms, respectively, in the integrand on the right side of (4.104), we have

$$V = \frac{\rho_{L0}}{2\pi\varepsilon_0} \left(\int_{\alpha=0}^{\pi/2} \sec\alpha\,d\alpha - \int_{\alpha_0=0}^{\pi/2} \sec\alpha_0\,d\alpha_0 \right)$$

$$= \frac{\rho_{L0}}{2\pi\varepsilon_0} \{ [\ln(\sec\alpha + \tan\alpha)]_{\alpha=0}^{\pi/2} - [\ln(\sec\alpha_0 + \tan\alpha_0)]_{\alpha_0=0}^{\pi/2} \} \qquad (4.105)$$

$$= \frac{\rho_{L0}}{2\pi\varepsilon_0} \left[\ln \frac{(\sqrt{r^2 + z^2} + z)r_0}{(\sqrt{r_0^2 + z^2} + z)r} \right]_{z=0}^{\infty}$$

$$= -\frac{\rho_{L0}}{2\pi\varepsilon_0} \ln \frac{r}{r_0}$$

In view of the cylindrical symmetry about the line charge, (4.105) is the general expression in cylindrical coordinates for the potential field of the infinitely long line charge of uniform density. It can be seen from (4.105) that a choice of $r_0 = \infty$ is not a good choice, since then the potential would be infinity at all points. The difficulty lies in the fact that infinity plus a finite number is still infinity. We also note from (4.105) that the equipotential surfaces are $\ln r/r_0 = $ constant or $r = $ constant, that is, surfaces of cylinders with the line charge as their axis. The result of Example 3.6 shows that the electric field due to the line charge is directed radially away from the line charge. Thus, the direction lines of **E** and the equipotential surfaces are indeed orthogonal to each other.

Magnetic vector potential due to a current element

We shall now turn out attention to the magnetic vector potential for the static field case. Thus, let us consider a current element of length $d\mathbf{l}$ situated at the origin, as shown in Fig. 4.18, and carrying current I A. We shall obtain the magnetic vector potential due to this current element. To do this, we recall from Section 2.2 that the magnetic field due to it at a point $P(r, \theta, \phi)$ is given by

$$\mathbf{B} = \frac{\mu}{4\pi} \frac{I\, d\mathbf{l} \times \mathbf{a}_r}{r^2} \qquad (4.106)$$

Expressing **B** as

$$\mathbf{B} = \frac{\mu}{4\pi} I\, d\mathbf{l} \times \left(-\mathbf{\nabla} \frac{1}{r}\right) \qquad (4.107)$$

and using the vector identity

$$A \times \mathbf{\nabla}\Phi = \Phi\mathbf{\nabla} \times A - \mathbf{\nabla} \times \Phi A \qquad (4.108)$$

we obtain

$$\mathbf{B} = -\frac{\mu I}{4\pi r}\mathbf{\nabla} \times d\mathbf{l} + \mathbf{\nabla} \times \frac{\mu I\, d\mathbf{l}}{4\pi r} \qquad (4.109)$$

Since $d\mathbf{l}$ is a constant, $\mathbf{\nabla} \times d\mathbf{l} = \mathbf{0}$, and (4.109) reduces to

$$\mathbf{B} = \mathbf{\nabla} \times \frac{\mu I\, d\mathbf{l}}{4\pi r} \qquad (4.110)$$

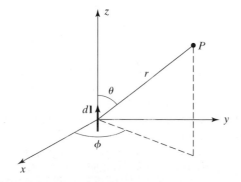

FIGURE 4.18   For finding the magnetic vector potential due to a current element.

Comparing (4.110) with (4.66), we see that the magnetic vector potential due to the current element situated at the origin is given by

$$\mathbf{A} = \frac{\mu I \, d\mathbf{l}}{4\pi r} \tag{4.111}$$

It follows from (4.111) that for a current element $I \, d\mathbf{l}$ situated at an arbitrary point, the magnetic vector potential is given by

$$\boxed{\mathbf{A} = \frac{\mu I \, d\mathbf{l}}{4\pi R}} \tag{4.112}$$

where R is the distance from the current element. Thus, it has a magnitude inversely proportional to the radial distance from the element (similar to the inverse distance dependence of the electric scalar potential due to a point charge) and direction parallel to the element. We shall make use of this result in Section 12.1.

K4.5. Potential difference; Potential due to a point charge; Computation of potential due to charge distributions; Electric dipole; Plotting of equipotential lines; Magnetic vector potential due to a current element.

D4.14. In a region of static electric field $\mathbf{E} = yz\mathbf{a}_x + zx\mathbf{a}_y + xy\mathbf{a}_z$, find the potential difference $V_A - V_B$ for each of the following pairs of points: **(a)** $A(2, 1, 1)$ and $B(1, 4, 0.5)$; **(b)** $A(2, 2, 2)$ and $B(1, 1, 1)$; and **(c)** $A(5, 1, 0.2)$ and $B(1, 2, 3)$.
Ans. **(a)** 0 V; **(b)** -7 V; **(c)** 5 V.

D4.15. Three point charges are located as follows: $30\pi\varepsilon_0$ C at $(3, 4, 0)$, $10\pi\varepsilon_0$ C at $(3, -4, 0)$, and $-40\pi\varepsilon_0$ C at $(-5, 0, 0)$. Find the following: **(a)** the potential at the point $(0, 0, 3.2)$; **(b)** the coordinate x to three decimal places of the point on the x-axis at which the potential is a maximum; and **(c)** the potential at the point found in **(b)**.
Ans. **(a)** 0 V; **(b)** 3.872 m; **(c)** 1.3155 V.

D4.16. For each of the following arrangements of point charges, find the first significant term in the expression for the electric potential at distances far from the origin $(r \gg d)$: **(a)** Q at $(0, 0, d)$, $2Q$ at $(0, 0, 0)$, and Q at $(0, 0, -d)$ and **(b)** Q at $(0, 0, d)$, $-2Q$ at $(0, 0, 0)$, and Q at $(0, 0, -d)$.
Ans **(a)** $Q/\pi\varepsilon_0 r$; **(b)** $(Qd^2/4\pi\varepsilon_0 r^3)(3\cos^2\theta - 1)$.

4.6 ENERGY STORAGE IN ELECTRIC AND MAGNETIC FIELDS

In the preceding section, we learned that the potential difference between two points A and B in a static electric field has the meaning of work done per unit charge in moving a test charge from point A to point B. If we transfer a test charge from a point of higher potential to a point of lower potential, the field does the work and hence there is loss in potential energy of the system, which is supplied to the test charge. Where in the system does this energy come from? Alternatively, if we transfer the test charge from a point of lower potential to a point of higher potential, an external agent moving the charge has to do work, thus increasing the potential energy of the system. Where in the system does this energy expended by the external agent reside? Wherever in the system the energy may reside, a convenient way is to think of the energy as being stored in the electric field. In the first case, part of the stored energy in the field is expended in moving the test charge, whereas in the second case, the energy expended by the external agent increases the stored energy.

Work
required to
assemble a
system of n
point charges

Let us then consider a system of two point charges Q_1 and Q_2 situated an infinite distance apart so that no forces are exerted on either charge and, hence, the charges are in equilibrium. Then recalling that the potential at a point is the potential difference between that point and a reference point, which for a point charge is conveniently chosen to be at infinity, we note that to bring Q_2 close to Q_1, as shown in Fig. 4 .19(a), an amount of work equal to Q_2 times the potential of Q_1 at Q_2 must be expended by an external agent. Thus, the potential energy of the system is increased by the amount

$$W_2 = Q_2 V_2^1 \tag{4.113}$$

where V_2^1 is the potential of Q_1 at the location of Q_2. If we start with a system of three charges Q_1, Q_2, and Q_3 situated an infinite distance apart from each other, then the amount of work required to bring Q_2 and Q_3 close to Q_1 can be determined in two steps. First, we bring Q_2 close to Q_1, for which the work required is given by (4.113). Then we bring Q_3 close to Q_1, as shown in Fig. 4.19(b). But, this time, we have to overcome not only the force exerted on Q_3 by Q_1, but also the force exerted by Q_2. Hence, the required work is given by

$$W_3 = Q_3 V_3^1 + Q_3 V_3^2 \tag{4.114}$$

Thus, the total work required to bring Q_2 and Q_3 close to Q_1 is

$$W_e = W_2 + W_3 = Q_2 V_2^1 + (Q_3 V_3^1 + Q_3 V_3^2) \tag{4.115}$$

The potential energy of the system is increased by the amount given by (4.115).

We can proceed in this manner and consider a system of n point charges $Q_1, Q_2, Q_3, \ldots, Q_n$ initially located infinitely far apart from each other. The total work required in bringing the charges close to each other is given by

$$
\begin{aligned}
W_e &= W_2 + W_3 + \cdots + W_n \\
&= Q_2 V_2^1 + (Q_3 V_3^1 + Q_3 V_3^2) + (Q_4 V_4^1 + Q_4 V_4^2 + Q_4 V_4^3) + \cdots \quad (4.116) \\
&= \sum_{i=2}^{n} \sum_{j=1}^{i-1} Q_i V_i^j
\end{aligned}
$$

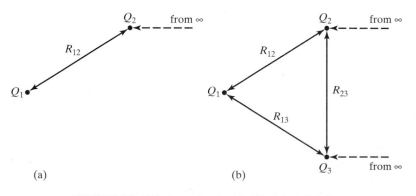

(a)   (b)

FIGURE 4.19   Bringing point charges closer from infinity.

where V_i^j is the potential of Q_j at the location of Q_i. However, we note that

$$Q_i V_i^j = Q_i \frac{Q_j}{4\pi\varepsilon_0 R_{ji}} = Q_j \frac{Q_i}{4\pi\varepsilon_0 R_{ij}} = Q_j V_j^i \tag{4.117}$$

Hence, (4.116) may be written as

$$W_e = Q_1 V_1^2 + (Q_1 V_1^3 + Q_2 V_2^3) + (Q_1 V_1^4 + Q_2 V_2^4 + Q_3 V_3^4) + \cdots$$

$$= \sum_{i=2}^{n} \sum_{j=1}^{i-1} Q_j V_j^i \tag{4.118}$$

Adding (4.116) and (4.118), we have

$$2W_e = Q_1(V_1^2 + V_1^3 + V_1^4 + \cdots)$$
$$+ Q_2(V_2^1 + V_2^3 + V_2^4 + \cdots)$$
$$+ Q_3(V_3^1 + V_3^2 + V_3^4 + \cdots)$$
$$+ \cdots$$
$$= Q_1(\text{potential at } Q_1 \text{ due to all other charges})$$
$$+ Q_2(\text{potential at } Q_2 \text{ due to all other charges}) \tag{4.119}$$
$$+ Q_3(\text{potential at } Q_3 \text{ due to all other charges})$$
$$+ \cdots$$
$$= Q_1 V_1 + Q_2 V_2 + Q_3 V_3 + \cdots$$
$$= \sum_{i=1}^{n} Q_i V_i$$

where V_i is the potential at Q_i due to all other charges. Dividing both sides of (4.119) by 2, we have

$$\boxed{W_e = \frac{1}{2} \sum_{i=1}^{n} Q_i V_i} \tag{4.120}$$

Thus, the potential energy stored in the system of n point charges is given by (4.120).

Example 4.12

Three point charges of values 1, 2, and 3 C are situated at the corners of an equilateral triangle of sides 1 m. It is desired to find the work required to move these charges to the corners of an equilateral triangle of shorter sides $\frac{1}{2}$ m, as shown in Fig. 4.20.

The potential energy stored in the system of three charges at the corners of the smaller equilateral triangle is given by

$$\frac{1}{2} \sum_{i=1}^{3} Q_i V_i = \frac{1}{2} \left[1\left(\frac{2}{4\pi\varepsilon_0} + \frac{3}{4\pi\varepsilon_0}\right) + 2\left(\frac{1}{4\pi\varepsilon_0} + \frac{3}{4\pi\varepsilon_0}\right) + 3\left(\frac{1}{4\pi\varepsilon_0} + \frac{2}{4\pi\varepsilon_0}\right) \right]$$

$$= \frac{1}{2} \left[\frac{5 + 8 + 9}{4\pi\varepsilon_0} \right] = \frac{11}{4\pi\varepsilon_0} \text{ N-m}$$

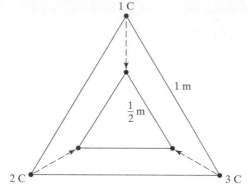

FIGURE 4.20   Bringing three point charges from the corners of a larger equilaterial triangle to the corners of a smaller equilateral triangle.

The potential energy stored in the system of three charges at the corners of the smaller equilateral triangle is equal to twice this value since all distances are halved. The increase in potential energy of the system in going from the larger to the smaller equilateral triangle is equal to $11/4\pi\varepsilon_0$ N-m. Obviously, this increase in energy must be supplied by an external agent, and, hence, the work required to move the charges to the corners of the equilateral triangle of sides $\frac{1}{2}$ m from the corners of the equilateral triangle of sides 1 m is equal to $11/4\pi\varepsilon_0$ N-m.

Potential energy in a continuous charge distribution

If we have a continuous distribution of charge with density $\rho(x, y, z)$ instead of an assembly of discrete charges, we can treat it as a continuous collection of infinitesimal charges of value $\rho(x, y, z)\,\Delta v$, each of which can be considered as a point charge, and obtain the potential energy of the system as

$$W_e = \frac{1}{2}\lim_{\Delta v \to 0}\sum[\rho(x, y, z)\,\Delta v]V(x, y, z)$$

$$= \frac{1}{2}\int_{\substack{\text{volume} \\ \text{containing }\rho}}\rho V\,dv \tag{4.121}$$

Thus far, we have found the potential energy of the charge distribution by considering the work done in assembling the system. We stated at the beginning of this section that the potential energy can be thought of as being stored in the electric field set up by the system of charges. If so, we should be able to express the energy in terms of the electric field. To do this, we substitute for ρ in (4.121) from (4.34) and obtain

$$W_e = \frac{1}{2}\int_{\substack{\text{volume} \\ \text{containing }\rho}}(\nabla \cdot \mathbf{D})V\,dv \tag{4.122}$$

Since $\nabla \cdot \mathbf{D} = 0$ in the region not containing ρ, the value of the integral on the right side of (4.122) is not altered if we change the volume of integration from the volume containing ρ to the entire space. Thus,

$$W_e = \frac{1}{2}\int_{\text{all space}}(\varepsilon\nabla \cdot \mathbf{E})V\,dv \tag{4.123}$$

where we have also replaced $\mathbf{D}$ by $\varepsilon\mathbf{E}$. We now use the vector identity

$$\nabla \cdot \Phi\mathbf{A} = \Phi\nabla \cdot \mathbf{A} + \mathbf{A} \cdot \nabla\Phi \qquad (4.124)$$

to replace $V\nabla \cdot \mathbf{E}$ on the right side of (4.123) by $\nabla \cdot V\mathbf{E} - \mathbf{E} \cdot \nabla V$ and obtain

$$W_e = \frac{1}{2}\varepsilon \int_{\text{all space}} (\nabla \cdot V\mathbf{E} - \mathbf{E} \cdot \nabla V)\, dv$$

$$= \frac{1}{2}\varepsilon \int_{\text{all space}} \nabla \cdot V\mathbf{E}\, dv + \frac{1}{2}\varepsilon \int_{\text{all space}} \mathbf{E} \cdot \mathbf{E}\, dv \qquad (4.125)$$

where we have replaced ∇V by $-\mathbf{E}$ in accordance with (4.97). Using the divergence theorem, we equate the first integral on the right side of (4.125) to a surface integral; thus,

$$\int_{\text{all space}} \nabla \cdot V\mathbf{E}\, dv = \int_{\substack{\text{surface} \\ \text{bounding} \\ \text{all space}}} V\mathbf{E} \cdot \mathbf{a}_n\, dS \qquad (4.126)$$

However, as viewed from a surface bounding all space, a charge distribution of finite volume appears as a point charge, say, Q. Hence, as $r \to \infty$, we can write

$$\mathbf{E} \longrightarrow \frac{Q}{4\pi\varepsilon r^2}\mathbf{a}_r$$

$$V \longrightarrow \frac{Q}{4\pi\varepsilon r}$$

$$\int_{\substack{\text{surface} \\ \text{bounding} \\ \text{all space}}} V\mathbf{E} \cdot \mathbf{a}_n\, dS = \lim_{r\to\infty} \int_{\theta=0}^{\pi} \int_{\phi=0}^{2\pi} \frac{Q}{4\pi\varepsilon r}\, \frac{Q}{4\pi\varepsilon r^2}\mathbf{a}_r \cdot r^2 \sin\theta\, d\theta\, d\phi\, \mathbf{a}_r$$

$$= \int_{\theta=0}^{\pi} \int_{\phi=0}^{2\pi} \lim_{r\to\infty} \frac{Q^2}{(4\pi\varepsilon)^2 r} \sin\theta\, d\theta\, d\phi = 0 \qquad (4.127)$$

Equation (4.127) also holds for a charge distribution of infinite extent, provided the electric field due to the charge distribution falls off at least as $(1/r^2)\mathbf{a}_r$ and hence the potential falls off at least as $1/r$. Thus, (4.125) reduces to

$$W_e = \frac{1}{2}\varepsilon \int_{\text{all space}} \mathbf{E} \cdot \mathbf{E}\, dv = \int_{\text{all space}} \left(\frac{1}{2}\varepsilon E^2\right) dv \qquad (4.128)$$

Equation (4.128) indicates clearly that the idea of energy residing in the electric field is a valid one provided we integrate $\frac{1}{2}\varepsilon E^2$ throughout the entire space. The quantity $\frac{1}{2}\varepsilon E^2$ is evidently the energy density associated with the electric field; that is,

Electric stored energy density

$$\boxed{w_e = \frac{1}{2}\varepsilon E^2} \qquad (4.129)$$

Example 4.13

A volume charge is distributed throughout a sphere of radius a m, and centered at the origin, with uniform density ρ_0 C/m^3. We wish to find the energy stored in the electric field of this charge distribution.

From Example 3.7, the electric field of the uniformly distributed spherical charge, having its center at the origin, is given by

$$
\mathbf{E} = \frac{\mathbf{D}}{\varepsilon} = \begin{cases} \dfrac{\rho_0 r}{3\varepsilon_0} \mathbf{a}_r & \text{for} \quad r < a \\[2mm] \dfrac{\rho_0 a^3}{3\varepsilon_0 r^2} \mathbf{a}_r & \text{for} \quad r > a \end{cases}
$$

Hence, the energy density in the electric field is given by

$$
\frac{1}{2} \varepsilon E_r^2 = \begin{cases} \dfrac{\rho_0^2 r^2}{18\varepsilon_0} & \text{for} \quad r < a \\[2mm] \dfrac{\rho_0^2 a^6}{18\varepsilon_0 r^4} & \text{for} \quad r > a \end{cases}
$$

The energy stored in the electric field is

$$
W_e = \int_{r=0}^{a} \int_{\theta=0}^{\pi} \int_{\phi=0}^{2\pi} \frac{\rho_0^2 r^2}{18\varepsilon_0} r^2 \sin\theta \, dr \, d\theta \, d\phi
$$

$$
+ \int_{r=a}^{\infty} \int_{\theta=0}^{\pi} \int_{\phi=0}^{2\pi} \frac{\rho_0^2 a^6}{18\varepsilon_0 r^4} r^2 \sin\theta \, dr \, d\theta \, d\phi
$$

$$
= \frac{4\pi \rho_0^2 a^5}{15\varepsilon_0}
$$

Magnetic stored energy density

Just as energy is stored in an electric field, there is energy storage associated with magnetic field. The expression for the magnetic stored energy density, w_m, can be derived by considering the building up of a current distribution.[3] We shall, however, simply present here the result, which will also be evident when we consider power flow in an electromagnetic field in Section 5.5. It is given by

$$
\boxed{w_m = \tfrac{1}{2} \mu H^2} \tag{4.130}
$$

Let us now consider an example.

---

[3] See, for example, N. N. Rao, *Basic Electromagnetics with Applications* (Englewood Cliffs, N.J.: Prentice Hall, 1972), pp. 226–230.

Example 4.14

Current I flows in the $+z$-direction with uniform density on the cylindrical surface $r = a$ and returns in the $-z$-direction with uniform density on a second cylindrical surface $r = b$ so that the surface current distribution is given by

$$\mathbf{J}_S = \begin{cases} \dfrac{I}{2\pi a}\,\mathbf{a}_z & \text{for} \quad r = a \\[2mm] -\dfrac{I}{2\pi b}\,\mathbf{a}_z & \text{for} \quad r = b \end{cases}$$

We wish to find the energy stored in the magnetic field per unit length of the current distribution.

From application of Ampère's circuital law in integral form, we obtain the magnetic field due to the given current distribution as

$$\mathbf{H} = \begin{cases} 0 & \text{for} \quad r < a \\[2mm] \dfrac{I}{2\pi r}\,\mathbf{a}_\phi & \text{for} \quad a < r < b \\[2mm] 0 & \text{for} \quad r > b \end{cases}$$

Hence, the energy density in the magnetic field is given by

$$\frac{1}{2}\mu H_\phi^2 = \begin{cases} 0 & \text{for} \quad r < a \\[2mm] \dfrac{\mu_0 I^2}{8\pi^2 r^2} & \text{for} \quad a < r < b \\[2mm] 0 & \text{for} \quad r > b \end{cases}$$

The energy stored in the magnetic field per unit length of the current distribution is

$$\begin{aligned} W_m &= \int_{r=a}^{b} \int_{\phi=0}^{2\pi} \int_{z=0}^{1} \frac{\mu_0 I^2}{8\pi^2 r^2}\, r\, dr\, d\phi\, dz \\[2mm] &= \frac{\mu_0 I^2}{4\pi} \ln \frac{b}{a} \end{aligned}$$

K4.6. Potential energy associated with a charge distribution; Energy density in electric field; Energy density in magnetic field.

D4.17. Four point charges, each of value $4\pi\varepsilon_0$ C are situated at the vertices A, B, C, and D (located in that sequence) of a square of sides 1 m. Find the following: **(a)** the work required to assemble the charge distribution; **(b)** the additional work required to move the charge at D to the center of the square; and **(c)** the additional work required to then move the charge at C to the center of the side AB.
Ans. **(a)** $68.037\varepsilon_0$ J;   **(b)** $19.2961\varepsilon_0$ J;   **(c)** $36.1746\varepsilon_0$ J.

D4.18. Two spherical charges, each of the same radius a and the same uniform charge density ρ_0 C/m^3, are situated infinitely apart. The two spherical charges are brought together and made into a single spherical charge of uniform density. Find the following: **(a)** the work required if the radius of the new spherical charge is a; **(b)** the work required if the charge density of the new spherical charge is ρ_0; and **(c)** the radius and the charge density of the new spherical charge for which the work required is zero.
Ans. **(a)** $1.6755\rho_0^2 a^5/\varepsilon_0$ J;   **(b)** $0.9842\rho_0^2 a^5/\varepsilon_0$ J;   **(c)** $2a, 0.25\rho_0$.

4.7 SUMMARY

We have in this chapter derived the differential forms of Maxwell's equations from their integral forms, which we introduced in Chapter 3. For the general case of electric and magnetic fields having all three components, each of them dependent on all coordinates and time, Maxwell's equations in differential form are given as follows in words and in mathematical form.

Faraday's law.   The curl of the electric field intensity is equal to the negative of the time derivative of the magnetic flux density; that is,

$$\nabla \times \mathbf{E} = -\frac{\partial \mathbf{B}}{\partial t} \qquad (4.131)$$

Ampère's circuital law.   The curl of the magnetic field intensity is equal to the sum of the current density due to flow of charges and the displacement current density, which is the time derivative of the displacement flux density; that is,

$$\nabla \times \mathbf{H} = \mathbf{J} + \frac{\partial \mathbf{D}}{\partial t} \qquad (4.132)$$

Gauss' law for the electric field.   The divergence of the displacement flux density is equal to the charge density; that is,

$$\nabla \cdot \mathbf{D} = \rho \qquad (4.133)$$

Gauss' law for the magnetic field.   The divergence of the magnetic flux density is equal to zero; that is,

$$\nabla \cdot \mathbf{B} = 0 \qquad (4.134)$$

Auxiliary to (4.131)–(4.134), the continuity equation is given by

$$\nabla \cdot \mathbf{J} = -\frac{\partial \rho}{\partial t} \qquad (4.135)$$

This equation, which is the differential form of the law of conservation of charge, states that the sum of the divergence of the current density due to flow of charges and the time derivative of the charge density is equal to zero.

In using (4.131)–(4.135), we recall that

$$\mathbf{D} = \varepsilon \mathbf{E}$$

$$\mathbf{H} = \frac{\mathbf{B}}{\mu}$$

where ε and μ are the permittivity and the permeability, respectively, of the medium. In addition, if the current density $\mathbf{J}$ is due to conduction, then

$$\mathbf{J} = \mathbf{J}_c = \sigma \mathbf{E}$$

where σ is the conductivity of the medium.

We have learned that the basic definitions of curl and divergence, which have enabled us to discuss their physical interpretations with the aid of the curl and divergence meters, are

$$\nabla \times \mathbf{A} = \lim_{\Delta S \to 0} \left[\frac{\oint_C \mathbf{A} \cdot d\mathbf{l}}{\Delta S} \right]_{max} \mathbf{a}_n$$

$$\nabla \cdot \mathbf{A} = \lim_{\Delta v \to 0} \frac{\oint_S \mathbf{A} \cdot d\mathbf{S}}{\Delta v}$$

Thus, the curl of a vector field at a point is a vector whose magnitude is the circulation of that vector field per unit area with the area oriented so as to maximize this quantity and in the limit that the area shrinks to the point. The direction of the vector is normal to the area in the aforementioned limit and in the right-hand sense. The divergence of a vector field at a point is a scalar quantity equal to the net outward flux of that vector field per unit volume in the limit that the volume shrinks to the point. In Cartesian coordinates, the expansions for curl and divergence are

$$\nabla \times \mathbf{A} = \begin{vmatrix} \mathbf{a}_x & \mathbf{a}_y & \mathbf{a}_z \\ \dfrac{\partial}{\partial x} & \dfrac{\partial}{\partial y} & \dfrac{\partial}{\partial z} \\ A_x & A_y & A_z \end{vmatrix}$$

$$= \left(\frac{\partial A_z}{\partial y} - \frac{\partial A_y}{\partial z} \right) \mathbf{a}_x + \left(\frac{\partial A_x}{\partial z} - \frac{\partial A_z}{\partial x} \right) \mathbf{a}_y + \left(\frac{\partial A_y}{\partial x} - \frac{\partial A_x}{\partial y} \right) \mathbf{a}_z$$

$$\nabla \cdot \mathbf{A} = \frac{\partial A_x}{\partial x} + \frac{\partial A_y}{\partial y} + \frac{\partial A_z}{\partial z}$$

Thus, Maxwell's equations in differential form relate the spatial variations of the field vectors at a point to their temporal variations and to the charge and current densities at that point.

We have also learned two theorems associated with curl and divergence. These are the Stokes' theorem and the divergence theorem given, respectively, by

$$\oint_C \mathbf{A} \cdot d\mathbf{l} = \int_S (\nabla \times \mathbf{A}) \cdot d\mathbf{S}$$

$$\oint_S \mathbf{A} \cdot d\mathbf{S} = \int_V (\nabla \cdot \mathbf{A}) \, dv$$

Stokes' theorem enables us to replace the line integral of a vector around a closed path by the surface integral of the curl of that vector over any surface bounded by that closed path, and vice versa. The divergence theorem enables us to replace the surface integral of a vector over a closed surface by the volume integral of the divergence of that vector over the volume bounded by the closed surface and vice versa.

We then introduced the electric scalar and magnetic vector potential functions, Φ and $\mathbf{A}$, respectively. In view of (4.134), we have

$$\mathbf{B} = \nabla \times \mathbf{A} \qquad\qquad (4.136)$$

and then in view of (4.131), we obtain

$$\mathbf{E} = -\nabla\Phi - \frac{\partial \mathbf{A}}{\partial t} \tag{4.137}$$

In (4.137), $\nabla\Phi$ is the gradient of the scalar function Φ. We learned that the gradient of a scalar Φ is a vector having magnitude equal to the maximum rate of increase of Φ at that point, and its direction is the direction in which the maximum rate of increase occurs, that is, normal to the constant Φ surface passing through that point; that is,

$$\nabla\Phi = \frac{\partial \Phi}{\partial n}\mathbf{a}_n$$

In Cartesian coordinates, the expansion for the gradient is

$$\nabla\Phi = \frac{\partial \Phi}{\partial x}\mathbf{a}_x + \frac{\partial \Phi}{\partial y}\mathbf{a}_y + \frac{\partial \Phi}{\partial z}\mathbf{a}_z$$

Next, we derived two differential equations for the potential functions. These are given by

$$\nabla^2\Phi - \mu\varepsilon\frac{\partial^2 \Phi}{\partial t^2} = -\frac{\rho}{\varepsilon} \tag{4.138a}$$

$$\nabla^2\mathbf{A} - \mu\varepsilon\frac{\partial^2 \mathbf{A}}{\partial t^2} = -\mu\mathbf{J} \tag{4.138b}$$

where $\nabla^2\Phi$ is the Laplacian of the scalar Φ and $\nabla^2\mathbf{A}$ is the Laplacian of the vector $\mathbf{A}$. In Cartesian coordinates,

$$\nabla^2\Phi = \frac{\partial^2 \Phi}{\partial x^2} + \frac{\partial^2 \Phi}{\partial y^2} + \frac{\partial^2 \Phi}{\partial z^2}$$

and

$$\nabla^2\mathbf{A} = (\nabla^2 A_x)\mathbf{a}_x + (\nabla^2 A_y)\mathbf{a}_y + (\nabla^2 A_z)\mathbf{a}_z$$

In deriving (4.138a) and (4.138b), we made use of the Lorenz condition

$$\nabla \cdot \mathbf{A} = -\mu\varepsilon\frac{\partial \Phi}{\partial t}$$

which is consistent with the continuity equation.

We then considered the potential functions for the static field case, for which (4.137) reduces to

$$\mathbf{E} = -\nabla\Phi = -\nabla V \tag{4.139}$$

whereas (4.136) remains unaltered. In (4.139), the symbol Φ is replaced by the symbol V since the electric potential difference between two points in a static electric field has the same meaning as the voltage between the two points. We considered the potential field of a point charge and found that for the point charge

$$V = \frac{Q}{4\pi\varepsilon R} \tag{4.140}$$

where R is the radial distance away from the point charge. The equipotential surfaces for the point charge are thus spherical surfaces centered at the point charge. We illustrated the application of the potential concept in the determination of electric field due to charge distributions by considering the examples of an electric dipole and a line charge. We also discussed a procedure for computer plotting of equipotentials. We then derived the expression for the magnetic vector potential due to a current element. For a current element $I\,d\mathbf{l}$, the magnetic vector potential is given by

$$\mathbf{A} = \frac{\mu I\,d\mathbf{l}}{4\pi R} \tag{4.141}$$

where R is the distance from the current element.

Finally, we discussed energy storage in electric and magnetic fields. We found that the work required to assemble a system of n point charges is given by

$$W_e = \frac{1}{2}\sum_{i=1}^{n} Q_i V_i \tag{4.142}$$

where V_i is the electric potential at the point charge Q_i due to all the other charges. For a continuous charge distribution of density ρ,

$$W_e = \frac{1}{2}\int_{\substack{\text{volume} \\ \text{containing } \rho}} \rho V\,dv$$

which reduces to

$$W_e = \int_{\substack{\text{all} \\ \text{space}}} \left(\frac{1}{2}\varepsilon E^2\right) dv$$

so that the energy density associated with the electric field can be identified to be

$$w_e = \tfrac{1}{2}\varepsilon E^2 \tag{4.143}$$

Similarly, the energy density associated with the magnetic field is given by

$$w_m = \tfrac{1}{2}\mu H^2 \tag{4.144}$$

We illustrated by means of examples the determination of energy stored in the electric field of a charge distribution and the magnetic field of a current distribution.

REVIEW QUESTIONS

Q4.1. State Faraday's law in differential form for the simple case of $\mathbf{E} = E_x(z, t)\mathbf{a}_x$. How is it derived from Faraday's law in integral form?

Q4.2. State Faraday's law in differential form for the general case of an arbitrary electric field. How is it derived from its integral form?

Q4.3. What is meant by the net right-lateral differential of the x- and y-components of a vector normal to the z-direction? Give an example in which the net right-lateral differential of E_x and E_y normal to the z-direction is zero although the individual derivatives are nonzero.

Q4.4. What is the determinant expansion for the curl of a vector in Cartesian coordinates?

Q4.5. State Ampère's circuital law in differential form for the general case of an arbitrary magnetic field. How is it derived from its integral form?

Q4.6. State Ampère's circuital law in differential form for the simple case of $\mathbf{H} = H_y(z, t)\mathbf{a}_y$. How is it derived from Ampère's circuital law in differential form for the general case?

Q4.7. If a pair of $\mathbf{E}$ and $\mathbf{B}$ at a point satisfies Faraday's law in differential form, does it necessarily follow that it also satisfies Ampère's circuital law in differential form, and vice versa?

Q4.8. State Gauss' law for the electric field in differential form. How is it derived from its integral form?

Q4.9. What is meant by the net longitudinal differential of the components of a vector field? Give an example in which the net longitudinal differential of the components of a vector field is zero, although the individual derivatives are nonzero.

Q4.10. What is the expansion for the divergence of a vector in Cartesian coordinates?

Q4.11. State Gauss' law for the magnetic field in differential form. How is it derived from its integral form?

Q4.12. How can you determine if a given vector can represent a magnetic field?

Q4.13. State the continuity equation and discuss its physical interpretation.

Q4.14. Summarize Maxwell's equations in differential form.

Q4.15. State and briefly discuss the basic definition of the curl of a vector.

Q4.16. What is a curl meter? How does it help visualize the behavior of the curl of a vector field?

Q4.17. Provide two examples of physical phenomena in which the curl of a vector field is nonzero.

Q4.18. State and briefly discuss the basic definition of the divergence of a vector.

Q4.19. What is a divergence meter? How does it help visualize the behavior of the divergence of a vector field?

Q4.20. Provide two examples of physical phenomena in which the divergence of a vector field is nonzero.

Q4.21. State Stokes' theorem and discuss its application.

Q4.22. State the divergence theorem and discuss its application.

Q4.23. What is the divergence of the curl of a vector?

Q4.24. What are electromagnetic potentials? How do they arise?

Q4.25. What is the expansion for the gradient of a scalar in Cartesian coordinates? When can a vector be expressed as the gradient of a scalar?

Q4.26. Discuss the physical interpretation for the gradient of a scalar function and the application of the gradient concept for the determination of unit vector normal to a surface.

Q4.27. How is the Laplacian of a scalar defined? What is its expansion in Cartesian coordinates?

Q4.28. Compare and contrast the operations of curl of a vector, divergence of a vector, gradient of a scalar, and Laplacian of a scalar.

Q4.29. How is the Laplacian of a vector defined? What is its expansion in Cartesian coordinates?

Q4.30. Outline the derivation of the differential equations for the electromagnetic potentials.

Q4.31. What is the relationship between the static electric field intensity and the electric scalar potential?

Q4.32. Distinguish between voltage, as applied to time-varying fields, and the potential difference in a static electric field.

Q4.33. Describe the electric potential field of a point charge.

Q4.34. Discuss the determination of the electric field intensity due to a charge distribution by using the potential concept.

Q4.35. Discuss the procedure for the computer plotting of equipotentials due to two (or more) point charges.

Q4.36. Compare the magnetic vector potential field due to a current element to the electric scalar potential due to a point charge.

Q4.37. Discuss the concept of energy storage in electric and magnetic fields.

Q4.38. Outline the derivation of the expression for the work required to assemble a system of n point charges.

Q4.39. State and discuss the expression for the energy density associated with the electric field of a charge distribution.

Q4.40. State and discuss the expression for the energy density associated with the magnetic field of a current distribution.

PROBLEMS

Section 4.1

P4.1. Find the curls of the following vector fields:
  (a) $zx\mathbf{a}_x + xy\mathbf{a}_y + yz\mathbf{a}_z$
  (b) $\cos y\, \mathbf{a}_x - x \sin y\, \mathbf{a}_y$
  (c) $(e^{-r^2}/r)\mathbf{a}_\phi$ in cylindrical coordinates
  (d) $2r \cos \theta\, \mathbf{a}_r + r\mathbf{a}_\theta$ in spherical coordinates

P4.2. For each of the following electric fields, find $\mathbf{B}$ that satisfies Faraday's law in differential form:
  (a) $\mathbf{E} = E_0 \cos 3\pi z \cos 9\pi \times 10^8 t\, \mathbf{a}_x$
  (b) $\mathbf{E} = E_0 \mathbf{a}_y \cos [3\pi \times 10^8 t + 0.2\pi(4x + 3z)]$

P4.3. Obtain the simplified differential equations for the following cases: (a) Ampère's circuital law for $\mathbf{H} = H_x(z, t)\mathbf{a}_x$ and (b) Faraday's law for $\mathbf{E} = E_\phi(r, t)\mathbf{a}_\phi$ in cylindrical coordinates.

P4.4. For the electric field $\mathbf{E} = E_0 e^{-\alpha z} \cos \omega t\, \mathbf{a}_x$ in free space ($\mathbf{J} = 0$), find $\mathbf{B}$ that satisfies Faraday's law in differential form and then determine if the pair of $\mathbf{E}$ and $\mathbf{B}$ satisfy Ampère's circuital law in differential form.

P4.5. For the electric field $\mathbf{E} = E_0 \cos (\omega t - \alpha y - \beta z)\, \mathbf{a}_x$ in free space ($\mathbf{J} = 0$), find the necessary condition relating $\alpha, \beta, \omega, \mu_0$, and ε_0 for the field to satisfy both of Maxwell's curl equations.

P4.6. For the electric field $\mathbf{E} = E_0 e^{-kx} \cos (2 \times 10^8 t - y)\, \mathbf{a}_z$ in free space ($\mathbf{J} = 0$), find the value(s) of k for which the field satisfies both of Maxwell's curl equations.

P4.7. For each of the following current distributions, find the corresponding magnetic field intensity using Ampère's circuital law in differential form without the displacement current density term and plot both the current density and the magnetic field intensity components versus the appropriate coordinate:

(a) $\mathbf{J} = \begin{cases} J_0 \dfrac{z}{a} \, \mathbf{a}_x & \text{for} \quad -a < z < a \\ 0 & \text{otherwise} \end{cases}$

(b) $\mathbf{J} = \begin{cases} J_0 \mathbf{a}_\phi & \text{for} \quad a < r < 2a \\ 0 & \text{otherwise} \end{cases}$

in cylindrical coordinates

Section 4.2

P4.8. Find the divergences of the following vector fields:
(a) $zx\mathbf{a}_x + xy\mathbf{a}_y + yz\mathbf{a}_z$
(b) $3\mathbf{a}_x + (y - 3)\mathbf{a}_y + (2 + z)\mathbf{a}_z$
(c) $r \sin \phi \, \mathbf{a}_\phi$ in cylindrical coordinates
(d) $r \cos \theta \, (\cos \theta \, \mathbf{a}_r - \sin \theta \, \mathbf{a}_\theta)$ in spherical coordinates

P4.9. For each of the following charge distributions, find the corresponding displacement flux density using Gauss' law for the electric field in differential form and plot both the charge density and the displacement flux density component versus the appropriate coordinate:

(a) $\rho = \begin{cases} \rho_0 \dfrac{x}{a} & \text{for} \quad -a < x < a \\ 0 & \text{otherwise} \end{cases}$

(b) $\rho = \begin{cases} \rho_0 \dfrac{r}{a} & \text{for} \quad 0 < r < a \\ 0 & \text{otherwise} \end{cases}$

in cylindrical coordinates

P4.10. For each of the following vector fields, find the value of the constant k for which the vector field can be realized as a magnetic field or as a current density in the absence of charge accumulation (or depletion):
(a) $(1/y^k) \, (2x\mathbf{a}_x + y\mathbf{a}_y)$
(b) $r(\sin k\phi \, \mathbf{a}_r + \cos k\phi \, \mathbf{a}_\phi)$ in cylindrical coordinates
(c) $[1 + (2/r^3)] \cos \theta \, \mathbf{a}_r + k[1 - (1/r^3)] \sin \theta \, \mathbf{a}_\theta$ in spherical coordinates

P4.11. Determine which of the following static fields can be realized both as an electric field in a charge-free region and a magnetic field in a current-free region:
(a) $y\mathbf{a}_x + x\mathbf{a}_y$
(b) $[1 + (1/r^2)] \cos \phi \, \mathbf{a}_r - [1 - (1/r^2)] \sin \phi \, \mathbf{a}_\phi$ in cylindrical coordinates
(c) $r \sin \theta \, \mathbf{a}_\theta$ in spherical coordinates

P4.12. Use the continuity equation, Ohm's law, and Gauss' law for the electric field in differential form to show that the time variation of the charge density at a point inside a conductor of uniform conductivity σ and permittivity ε is governed by the differential equation

$$\frac{\partial \rho}{\partial t} + \frac{\sigma}{\varepsilon} \rho = 0$$

Then show that the charge density inside the conductor decays exponentially with a time constant ε/σ. Compute the value of the time constant for copper.

Section 4.3

P4.13. With the aid of the curl meter and also by expansion in the Cartesian coordinate system, discuss the curl of the velocity vector field associated with the flow of water in the stream of Fig. 4.6(a), except that the velocity v_z varies in a nonlinear manner from zero at the banks to a maximum of v_0 at the center given by

$$\mathbf{v} = \frac{4v_0}{a^2} (ax - x^2)\mathbf{a}_z$$

P4.14. With the aid of the curl meter and also by expansion in the cylindrical coordinate system, discuss the curl of the linear velocity vector field associated with points inside Earth due to its spin motion.

P4.15. Discuss the divergences of the following vector fields with the aid of the divergence meter and also by expansion in the appropriate coordinate system: **(a)** the position vector field associated with points in three-dimensional space and **(b)** the linear velocity vector field associated with points inside Earth due to its spin motion.

P4.16. Verify Stokes' theorem for the following cases: **(a)** the vector field $\mathbf{A} = zx\mathbf{a}_x + xy\mathbf{a}_y + yz\mathbf{a}_z$ and the closed path comprising the straight lines from $(0, 0, 0)$ to $(0, 1, 0)$, from $(0, 1, 0)$ to $(0, 1, 1)$, and from $(0, 1, 1)$ to $(0, 0, 0)$ and **(b)** the vector field $\mathbf{A} = \cos y\,\mathbf{a}_x - x \sin y\,\mathbf{a}_y$ independent of a closed path.

P4.17. Verify the divergence theorem for the following cases: **(a)** the vector field $xy\mathbf{a}_x + yz\mathbf{a}_y + zx\mathbf{a}_z$ and the cubical box bounded by the planes $x = 0, x = 1$, $y = 0, y = 1, z = 0$, and $z = 1$ and **(b)** the vector field $y^2\mathbf{a}_y - 2yz\mathbf{a}_z$ and the closed surface bounded by the planes $x = 0, y = 0, z = 0, z = 1$ and $x + y = 1$.

Section 4.4

P4.18. Show by expansion in the Cartesian coordinate system that: **(a)** $\nabla \cdot \nabla \times \mathbf{A} = 0$ for any $\mathbf{A}$ and **(b)** $\nabla \times \nabla \Phi = \mathbf{0}$ for any Φ.

P4.19. Determine which of the following vectors can be expressed as the curl of another vector and which of them can be expressed as the gradient of a scalar:

(a) $xy\mathbf{a}_x + yz\mathbf{a}_y + zx\mathbf{a}_z$

(b) $(1/r^2)(\cos \phi\, \mathbf{a}_r + \sin \phi\, \mathbf{a}_\phi)$ in cylindrical coordinates

(c) $(1/r) \sin \theta\, \mathbf{a}_\phi$ in spherical coordinates.

P4.20. Find the scalar functions whose gradients are given by the following vector functions:

(a) $e^{-y}(\cos x\, \mathbf{a}_x - \sin x\, \mathbf{a}_y)$

(b) $(\cos \phi\, \mathbf{a}_r - \sin \phi\, \mathbf{a}_\phi)$ in cylindrical coordinates

(c) $(1/r^3)(2 \cos \theta\, \mathbf{a}_r + \sin \theta\, \mathbf{a}_\theta)$ in spherical coordinates

P4.21. By using the gradient concept, show that the unit vector along the line of intersection of two planes

$$a_1 x + a_2 y + a_3 z = c_1$$
$$b_1 x + b_2 y + b_3 z = c_2$$

which are not parallel is given by

$$\pm \frac{(a_2 b_3 - a_3 b_2)\mathbf{a}_x + (a_3 b_1 - a_1 b_3)\mathbf{a}_y + (a_1 b_2 - a_2 b_1)\mathbf{a}_z}{\sqrt{(a_2 b_3 - a_3 b_2)^2 + (a_3 b_1 - a_1 b_3)^2 + (a_1 b_2 - a_2 b_1)^2}}$$

Then find the unit vector along the intersection of the planes $x + y + z = 3$ and $y = x$.

P4.22. By using the gradient concept, show that the equation of the plane passing through the point (x_0, y_0, z_0) and normal to the vector $(a\mathbf{a}_x + b\mathbf{a}_y + c\mathbf{a}_z)$ is given by

$$a(x - x_0) + b(y - y_0) + c(z - z_0) = 0$$

Then find the equation of the plane tangential to the surface $xyz = 1$ at the point $(\frac{1}{2}, \frac{1}{4}, 8)$.

P4.23. Show that the Laplacian of a vector in cylindrical coordinates is given by

$$\nabla^2 \mathbf{A} = \left(\nabla^2 A_r - \frac{A_r}{r^2} - \frac{2}{r^2} \frac{\partial A_\phi}{\partial \phi} \right) \mathbf{a}_r + \left(\nabla^2 A_\phi - \frac{A_\phi}{r^2} + \frac{2}{r^2} \frac{\partial A_r}{\partial \phi} \right) \mathbf{a}_\phi + (\nabla^2 A_z)\mathbf{a}_z$$

Section 4.5

P4.24. For the static electric potential function $V(x, y) = xy$, discuss the equipotential surfaces and the direction lines of the electric field with the aid of sketches.

P4.25. An arrangement of point charges known as the rectangular quadrupole consists of the point charges Q, $-Q$, Q, and $-Q$, at the points $(0, 0, 0)$, $(\Delta x, 0, 0)$, $(\Delta x, 0, \Delta z)$, and $(0, 0, \Delta z)$, respectively. Obtain the approximate expression for the electric potential and hence for the electric field intensity due to the rectangular quadrupole at distances r from the origin large compared to Δx and Δz.

P4.26. For a finitely long line charge of uniform density ρ_{L0} C/m situated along the line between $(0, 0, -a)$ and $(0, 0, a)$, obtain the expression for the electric potential at an arbitrary point (r, ϕ, z) in cylindrical coordinates. Further show that the equipotential surfaces are ellipsoids with the ends of the line as their foci.

P4.27. Show that for two infinitely long line charges parallel to the z-axis, having uniform densities $\rho_{L1} = 2k\pi\varepsilon_0$ C/m and $\rho_{L2} = -2\pi\varepsilon_0$ C/m and passing through $(-1, 0, 0)$ and $(1, 0, 0)$, respectively, the potential is given by $V = \ln(r_2 / r_1^k)$, where r_1 and r_2 are distances to the point from the line charges 1 and 2, respectively.

P4.28. Consider the surface charge distributed uniformly with density ρ_{S0} C/m² on a rectangular-shaped surface of sides a and b. Show that the electric potential at the center of the rectangle is

$$\frac{\rho_{S0}}{2\pi\varepsilon_0} \left(a \ln \frac{\sqrt{a^2 + b^2} + b}{a} + b \ln \frac{\sqrt{a^2 + b^2} + a}{b} \right)$$

Further show that for a square-shaped surface of sides a, the potential at the center is $(\rho_{S0} a / \pi\varepsilon_0) \ln(1 + \sqrt{2})$.

P4.29. Consider surface charge of uniform density ρ_{S0} C/m² distributed on an infinitely long strip lying between the straight lines $x = -a$, $y = 0$ and $x = a$, $y = 0$. Noting that the electric potential is independent of z, show that the potential difference between two points in the first quadrant of the xy-plane $(x_1, y_1, 0)$ and $(x_2, y_2, 0)$ is given by

$$\frac{\rho_{S0}}{4\pi\varepsilon_0}\left\{(a - x_2)\ln[(a - x_2)^2 + y_2^2] + (a + x_2)\ln[(a + x_2)^2 + y_2^2]\right.$$

$$- (a - x_1)\ln[(a - x_1)^2 + y_1^2] - (a + x_1)\ln[(a + x_1)^2 + y_1^2]$$

$$+ 2y_2\left(\tan^{-1}\frac{a - x_2}{y_2} + \tan^{-1}\frac{a + x_2}{y_2}\right)$$

$$\left.- 2y_1\left(\tan^{-1}\frac{a - x_1}{y_1} + \tan^{-1}\frac{a + x_1}{y_1}\right)\right\}$$

P4.30. Consider a circular current loop of radius a lying in the xy-plane with its center at the origin and with current I flowing in the sense of increasing ϕ, so that the magnetic dipole moment $\mathbf{m}$ is $I\pi a^2\, \mathbf{a}_z$. Show that far from the dipole such that $r \gg a$, the magnetic vector potential is given by

$$\mathbf{A} \approx \frac{\mu\mathbf{m} \times \mathbf{a}_r}{4\pi r^2}$$

and hence the magnetic flux density is given by

$$\mathbf{B} = \frac{\mu m}{4\pi r^3}(2\cos\theta\,\mathbf{a}_r + \sin\theta\,\mathbf{a}_\theta)$$

P4.31. By expansion in Cartesian coordinates, show that

$$\mathbf{A} \times \nabla\Phi = \Phi\nabla \times \mathbf{A} - \nabla \times (\Phi\mathbf{A})$$

Section 4.6

P4.32. Three point charges Q, Q, and kQ ($k > 0$) are constrained to move on a circle of radius a. Let α be the angle subtended at the center of the circle by the two equal charges. Show that for the work required to assemble the charge distribution to be a minimum, the relationship between α and k is given by

$$4k\sin^3\frac{\alpha}{4} = \cos\frac{\alpha}{2}$$

Find the values of α for values of k equal to $1, 2$, and $1/2$.

P4.33. Charges Q and $-Q$ are distributed uniformly on the surfaces of two concentric spheres of radii a and b ($> a$), respectively. The outer sphere is made up of two separable hemispheres. Find the work required to carry the two hemispheres to an infinite distance from the inner sphere and join them to form a sphere of uniformly distributed charge $-Q$.

P4.34. A volume charge distribution is given in spherical coordinates by

$$\rho = \begin{cases} \rho_0(r/a)^2 & \text{for } r < a \\ 0 & \text{for } r > a \end{cases}$$

(a) Find the work required to rearrange the charge with uniform density within the region $r < a$. **(b)** Instead of as in part (a), find the radius of the spherical region and the charge density for which neither additional work is required nor work is made available in redistributing the charge uniformly within the region.

P4.35. A current distribution is given in cylindrical coordinates by

$$\mathbf{J} = \begin{cases} J_0 \mathbf{a}_z & \text{for} \quad r < 3a \\ -J_0 \mathbf{a}_z & \text{for} \quad 4a < r < 5a \end{cases}$$

Find the energy stored in the magnetic field of the current distribution per unit length in the z-direction.

P4.36. Current I_0 flows with density $(2I_0 r^2 / \pi a^4) \mathbf{a}_z$ A/m² in an infinitely long solid cylindrical wire of radius a with its axis along the z-axis, and returns with uniform density $-(I_0 / 4\pi a)\mathbf{a}_z$ A/m on a cylindrical surface of radius $2a$ and coaxial with the solid wire. Find the energy stored in the magnetic field of the current distribution per unit length in the z-direction.

REVIEW PROBLEMS

R4.1. Find the numerical value(s) of k, if any, such that the electric field in free space ($\mathbf{J} = 0$) given by

$$\mathbf{E} = E_0 \sin 6x \sin (3 \times 10^9 t - kz) \, \mathbf{a}_y$$

satisfies both of Maxwell's curl equations.

R4.2. The x-variation of charge density independent of y and z is shown in Fig.4.21. Find and sketch the resulting D_x versus x.

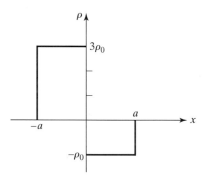

FIGURE 4.21   For Problem R4.2.

R4.3. With respect to the properties of physical realizibility as electric and magnetic fields, static vector fields can be classified into four groups: **(i)** electric field only; **(ii)** magnetic field only; **(iii)** electric field in a charge-free region or a magnetic field in a current-free region; and **(iv)** none of the preceding three. For each of the following fields, determine the group to which it belongs:

(a) $x\mathbf{a}_x + y\mathbf{a}_y$

(b) $(x^2 - y^2)\mathbf{a}_x - 2xy\mathbf{a}_y + 4\mathbf{a}_z$

(c) $\dfrac{e^{-r}}{r} \mathbf{a}_\phi$ in cylindrical coordinates

(d) $\dfrac{1}{r} (\cos \phi \, \mathbf{a}_r + \sin \phi \, \mathbf{a}_\phi)$ in cylindrical coordinates

R4.4. By using the gradient concept, show that the angle α between two planes

$$a_1x + a_2y + a_3z = c_1$$
$$b_1x + b_2y + b_3z = c_2$$

is given by

$$\alpha = \cos^{-1}\frac{a_1b_1 + a_2b_2 + a_3b_3}{\sqrt{(a_1^2 + a_2^2 + a_3^2)(b_1^2 + b_2^2 + b_3^2)}}$$

Then find the angle between the planes $x + y + z = 1$ and $z = 0$.

R4.5. Consider a circular disk of radius a lying in the xy-plane with its center at the origin and carrying charge of uniform density ρ_{S0} C/m². Obtain the expression for the potential V due to the charged disk at a point $(0, 0, z)$ on the z-axis. Verify your answer by considering the limiting cases of $V(z)$ for $|z| \gg a$ and $E_z(z)$ for $|z| \ll a$.

R4.6. Obtain the magnetic vector potential at an arbitrary point due to an infinitely long straight filamentary wire lying along the z-axis and carrying a current I in the $+z$-direction. Then evaluate **B** by performing the curl operation on the magnetic vector potential.

R4.7. Find the amount of work required for rearranging a uniformly distributed surface charge Q of radius a into a uniformly distributed volume charge of radius a.

R4.8. Current I_0 flows in the $+z$-direction with density proportional to r in an infinitely long solid cylindrical wire of radius a with its axis along the z-axis and returns in the $-z$-direction with uniform density on a cylindrical surface of radius $b\ (> a)$ and coaxial with the solid wire. Find the value of b/a for which the magnetic energy stored per unit length in the z-direction external to the solid wire is equal to six times that inside the wire.

CHAPTER 5

Uniform Plane Waves and Power Flow in an Electromagnetic Field

In Chapters 3 and 4, we studied Maxwell's equations in integral and differential forms, as well as several associated topics. Our goal in this chapter is to gain an understanding of the basic principles of electromagnetic wave propagation, the remaining fundamental prerequisite to the study of the elements pertinent to the engineering applications of electromagnetics. We first consider wave propagation in free space and then extend the discussion to material media, thereby learning how the characteristics of wave propagation are modified from those associated with free space.

We shall employ an approach in this chapter that will enable us not only to learn how the coupling between space variations and time variations of the electric and magnetic fields, as indicated by Maxwell's equations, results in wave motion but also to illustrate the basic principle of radiation of waves from an antenna, which is treated in detail in Chapter 11. We augment our discussion of radiation and propagation of waves by considering such examples as the principle of an antenna array and the Doppler effect. We discuss power flow and energy storage associated with the wave motion and introduce the Poynting vector. Finally, we consider the topic of reflection of waves at a boundary between two different media.

5.1 UNIFORM PLANE WAVES IN TIME DOMAIN IN FREE SPACE

Uniform plane wave defined

In Chapter 4, we learned that the space variations of the electric- and magnetic-field components are related to the time variations of the magnetic- and electric-field components, respectively, through Maxwell's equations. This interdependence gives rise to the phenomenon of electromagnetic wave propagation. In the general case, electromagnetic wave propagation involves electric and magnetic fields having more than one component, each dependent on all three coordinates, in addition to time. However, a simple and very useful type of wave that serves as a building block in the study of electromagnetic waves consists of electric and magnetic fields that are perpendicular to each other and to the direction of propagation and are *uniform* in *planes* perpendicular to the direction of propagation. These waves are known as *uniform plane waves*. By orienting the coordinate axes such that the electric field is in the x-direction, the magnetic field is in the y-direction, and the direction of propagation is in the z-direction, as shown in Fig. 5.1, we have

$$\mathbf{E} = E_x(z, t)\mathbf{a}_x \qquad (5.1a)$$

$$\mathbf{H} = H_y(z, t)\mathbf{a}_y \qquad (5.1b)$$

Uniform plane waves do not exist in practice because they cannot be produced by finite-sized antennas. At large distances from physical antennas and ground, however, the waves can be approximated as uniform plane waves. Furthermore, the principles of guiding of electromagnetic waves along transmission lines and waveguides and the principles of many other wave phenomena can be studied basically in terms of uniform plane waves. Hence, it is very important that we understand the principles of uniform plane wave propagation.

To illustrate the phenomenon of interaction of electric and magnetic fields giving rise to uniform plane electromagnetic wave propagation and the principle of radiation of electromagnetic waves from an antenna, we shall consider a simple, idealized, hypothetical source. This source consists of an infinite sheet lying in the xy-plane, as shown in Fig. 5.2. On this infinite plane sheet, a uniformly distributed current flows in the negative x-direction, as given by

Infinite plane current sheet source

$$\boxed{\mathbf{J}_S = -J_S(t)\mathbf{a}_x \qquad \text{for} \quad z = 0} \qquad (5.2)$$

where $J_S(t)$ is a given function of time. Because of the uniformity of the surface current density on the infinite sheet, if we consider any line of width w parallel

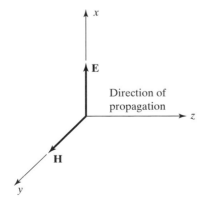

FIGURE 5.1 Directions of electric and magnetic fields and direction of propagation for a simple case of uniform plane wave.

FIGURE 5.2   Infinite plane sheet in the
xy-plane carrying surface current of uniform
density.

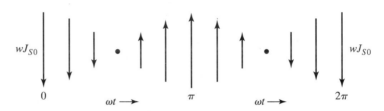

FIGURE 5.3   Time history of current flow across a line of width w parallel to the
y-axis for the current sheet of Fig. 5.2, for $\mathbf{J}_S = -J_{S0} \cos \omega t\, \mathbf{a}_x$.

to the y-axis, as shown in Fig. 5.2, the current crossing that line is simply given by
w times the current density, that is, $wJ_S(t)$. If $J_S(t) = J_{S0} \cos \omega t$, then the current
$wJ_{S0} \cos \omega t$, crossing the width w, actually alternates between negative x- and
positive x-directions, that is, downward and upward. The time history of this current flow for one period of the sinusoidal variation is illustrated in Fig. 5.3, with
the lengths of the lines indicating the magnitudes of the current. We shall consider the medium on either side of the current sheet to be free space.

   To find the electromagnetic field due to the time-varying current sheet, we
shall begin with Faraday's law and Ampère's circuital law given, respectively, by

$$\nabla \times \mathbf{E} = -\frac{\partial \mathbf{B}}{\partial t} \tag{5.3a}$$

$$\nabla \times \mathbf{H} = \mathbf{J} + \frac{\partial \mathbf{D}}{\partial t} \tag{5.3b}$$

and use a procedure that consists of the following steps:

1. Obtain the particular differential equations for the case under consideration.
2. Derive the general solution to the differential equations of step 1 without
   regard to the current on the sheet.

3. Show that the solution obtained in step 2 is a superposition of traveling waves propagating in the $+z$- and $-z$-directions.

4. Extend the general solution of step 2 to take into account the current on the sheet, thereby obtaining the required solution.

Although the procedure may be somewhat lengthy, we shall in the process learn several useful concepts and techniques.

1. To obtain the particular differential equations for the case under consideration, we first note that since (5.2) can be thought of as a current *distribution* having only an x-component of the current density that varies only with z, we can set J_y, J_z, and all derivatives with respect to x and y in (5.3a) and (5.3b) equal to zero. Hence, (5.3a) and (5.3b) reduce to

$$-\frac{\partial E_y}{\partial z} = -\frac{\partial D_x}{\partial t} \quad (5.4a) \qquad\qquad -\frac{\partial H_y}{\partial z} = J_x + \frac{\partial D_x}{\partial t} \quad (5.5a)$$

$$\frac{\partial E_x}{\partial z} = -\frac{\partial B_y}{\partial t} \quad (5.4b) \qquad\qquad \frac{\partial H_x}{\partial z} = \frac{\partial D_y}{\partial t} \quad (5.5b)$$

$$0 = -\frac{\partial B_z}{\partial t} \quad (5.4c) \qquad\qquad 0 = \frac{\partial D_z}{\partial t} \quad (5.5c)$$

In these six equations, there are only two equations involving J_x and the pertinent electric- and magnetic-field components, namely, the simultaneous pair (5.4b) and (5.5a). Thus, the equations of interest are

$$\boxed{\begin{aligned} \frac{\partial E_x}{\partial z} &= -\frac{\partial B_y}{\partial t} \end{aligned}} \qquad\qquad (5.6a)$$

$$\boxed{\begin{aligned} \frac{\partial H_y}{\partial z} &= -J_x - \frac{\partial D_x}{\partial t} \end{aligned}} \qquad\qquad (5.6b)$$

which are the same as (4.7) and (4.25), the simplified forms of Faraday's law and Ampère's circuital law, respectively, for the special case of electric and magnetic fields characterized by (5.1a) and (5.1b), respectively.

2. In applying (5.6a) and (5.6b) to (5.2), we note that J_x in (5.6b) is a volume current density, whereas (5.2) represents a surface current density. Hence, we shall solve (5.6a) and (5.6b) by setting $J_x = 0$ and then extend the solution to take into account the current on the sheet. For $J_x = 0$, (5.6a) and (5.6b) become

Derivation of wave equation

$$\boxed{\begin{aligned} \frac{\partial E_x}{\partial z} &= -\frac{\partial B_y}{\partial t} = -\mu_0 \frac{\partial H_y}{\partial t} \end{aligned}} \qquad\qquad (5.7a)$$

$$\boxed{\begin{aligned} \frac{\partial H_y}{\partial z} &= -\frac{\partial D_x}{\partial t} = -\varepsilon_0 \frac{\partial E_x}{\partial t} \end{aligned}} \qquad\qquad (5.7b)$$

Differentiating (5.7a) with respect to z and then substituting for $\partial H_y / \partial z$ from (5.7b), we obtain

$$\frac{\partial^2 E_x}{\partial z^2} = -\mu_0 \frac{\partial}{\partial z}\left(\frac{\partial H_y}{\partial t}\right) = -\mu_0 \frac{\partial}{\partial t}\left(\frac{\partial H_y}{\partial z}\right) = -\mu_0 \frac{\partial}{\partial t}\left(-\varepsilon_0 \frac{\partial E_x}{\partial t}\right)$$

or

$$\boxed{\frac{\partial^2 E_x}{\partial z^2} = \mu_0 \varepsilon_0 \frac{\partial^2 E_x}{\partial t^2}} \tag{5.8}$$

We have thus eliminated H_y from (5.7a) and (5.7b) and obtained a single second-order partial differential equation involving E_x only. Equation (5.8) is known as the *wave equation*. In particular, it is a one-dimensional wave equation in time-domain form, that is, for arbitrary time dependence of E_x.

Solution of
wave
equation

To obtain the solution for (5.8), we introduce a change of variable by defining $\tau = z\sqrt{\mu_0 \varepsilon_0}$. Substituting for z in (5.8) in terms of τ, we then have

$$\frac{\partial^2 E_x}{\partial \tau^2} = \frac{\partial^2 E_x}{\partial t^2} \tag{5.9}$$

or

$$\frac{\partial^2 E_x}{\partial \tau^2} - \frac{\partial^2 E_x}{\partial t^2} = 0$$

$$\left(\frac{\partial}{\partial \tau} + \frac{\partial}{\partial t}\right)\left(\frac{\partial}{\partial \tau} - \frac{\partial}{\partial t}\right) E_x = 0 \tag{5.10}$$

where the quantities in parentheses are operators operating on one another and on E_x. Equation (5.10) is satisfied if

$$\left(\frac{\partial}{\partial \tau} \pm \frac{\partial}{\partial t}\right) E_x = 0$$

or

$$\frac{\partial E_x}{\partial \tau} = \mp \frac{\partial E_x}{\partial t} \tag{5.11}$$

Let us first consider the equation corresponding to the upper sign in (5.11); that is,

$$\frac{\partial E_x}{\partial \tau} = -\frac{\partial E_x}{\partial t}$$

This equation says that the partial derivative of $E_x(\tau, t)$ with respect to τ is equal to the negative of the partial derivative of $E_x(\tau, t)$ with respect to t. The simplest function that satisfies this requirement is the function $(t - \tau)$. It then follows that any arbitrary function of $(t - \tau)$, say, $f(t - \tau)$ satisfies the requirement since

$$\frac{\partial}{\partial t}[f(t - \tau)] = f'(t - \tau)\frac{\partial}{\partial t}(t - \tau) = f'(t - \tau)$$

and

$$\frac{\partial}{\partial \tau}[f(t - \tau)] = f'(t - \tau)\frac{\partial}{\partial \tau}(t - \tau) = -f'(t - \tau) = -\frac{\partial}{\partial t}[f(t - \tau)]$$

where the prime associated with $f'(t - \tau)$ denotes differentiation of f with respect to $(t - \tau)$. In a similar manner, the solution for the equation corresponding to the lower sign in (5.11), that is, for

$$\frac{\partial E_x}{\partial \tau} = \frac{\partial E_x}{\partial t}$$

can be seen to be any arbitrary function of $(t + \tau)$, say, $g(t + \tau)$. Combining the two solutions, we write the solution for (5.11) to be

$$E_x(\tau, t) = Af(t - \tau) + Bg(t + \tau) \tag{5.12}$$

where A and B are arbitrary constants.

Substituting now for τ in (5.12) in terms of z, we obtain the solution for (5.8) to be

$$\boxed{E_x(z, t) = Af(t - z\sqrt{\mu_0\varepsilon_0}) + Bg(t + z\sqrt{\mu_0\varepsilon_0})} \tag{5.13}$$

The corresponding solution for $H_y(z, t)$ can be obtained by substituting (5.13) into (5.7a) or (5.7b). Thus, using (5.7a),

$$\frac{\partial H_y}{\partial t} = \sqrt{\frac{\varepsilon_0}{\mu_0}}[Af'(t - z\sqrt{\mu_0\varepsilon_0}) - Bg'(t + z\sqrt{\mu_0\varepsilon_0})]$$

$$\boxed{H_y(z, t) = \frac{1}{\sqrt{\mu_0/\varepsilon_0}}[Af(t - z\sqrt{\mu_0\varepsilon_0}) - Bg(t + z\sqrt{\mu_0\varepsilon_0})]} \tag{5.14}$$

The fields given by (5.13) and (5.14) are the general solutions to the differential equations (5.7a) and (5.7b).

3. To proceed further, we need to know the meanings of the functions f and g in (5.13) and (5.14). To discuss the meaning of f, let us consider a specific example

Traveling wave functions explained

$$f(t - z\sqrt{\mu_0\varepsilon_0}) = (t - z\sqrt{\mu_0\varepsilon_0})^2$$

Plots of this function versus z for two values of $t, t = 0$ and $t = \sqrt{\mu_0\varepsilon_0}$, are shown in Fig. 5.4(a). An examination of these plots reveals that as time increases from 0 to $\sqrt{\mu_0\varepsilon_0}$, every point on the plot for $t = 0$ moves by one unit in the $+z$-direction, thereby making the plot for $t = \sqrt{\mu_0\varepsilon_0}$ an exact replica of the plot for $t = 0$, except displaced by one unit in the $+z$-direction. The function f is therefore said to represent a *traveling wave propagating in the +z-direction*, or simply a $(+)$ *wave*. In particular, it is a *uniform plane wave* since its value does not vary with position in a given constant z-plane. By dividing the distance traveled by the

(a)

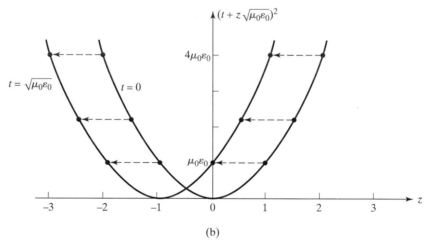

(b)

FIGURE 5.4 (a) Plots of the function $(t - z\sqrt{\mu_0\varepsilon_0})^2$ versus z for $t = 0$ and $t = \sqrt{\mu_0\varepsilon_0}$. (b) Plots of the function $(t + z\sqrt{\mu_0\varepsilon_0})^2$ versus z for $t = 0$ and $t = \sqrt{\mu_0\varepsilon_0}$.

time taken, the velocity of propagation of the wave can be obtained to be

$$v_p = \frac{1}{\sqrt{\mu_0\varepsilon_0}} = 3 \times 10^8 \text{ m/s} \qquad (5.15)$$

which is equal to c, the velocity of light in free space. Similarly, to discuss the meaning of g, we shall consider

$$g(t + z\sqrt{\mu_0\varepsilon_0}) = (t + z\sqrt{\mu_0\varepsilon_0})^2$$

Then plotting the function versus z for $t = 0$ and $t = \sqrt{\mu_0\varepsilon_0}$, as shown in Fig. 5.4(b), we can see that the plot for $t = \sqrt{\mu_0\varepsilon_0}$ is an exact replica of the plot for $t = 0$, except displaced by one unit in the $-z$-direction. The function g is therefore said to represent a *traveling wave propagating in the $-z$-direction*, or simply a $(-)$ *wave*. Once again, it is a uniform plane wave with the velocity of propagation equal to $1/\sqrt{\mu_0 \varepsilon_0}$.

To generalize the foregoing discussion of the functions f and g, let us consider two pairs of t and z, say, t_1 and z_1, and $t_1 + \Delta t$ and $z_1 + \Delta z$. Then for the function f to maintain the same value for these two pairs of z and t, we must have

$$t_1 - z_1\sqrt{\mu_0 \varepsilon_0} = (t_1 + \Delta t) - (z_1 + \Delta z)\sqrt{\mu_0 \varepsilon_0}$$

or

$$\Delta z = \frac{1}{\sqrt{\mu_0 \varepsilon_0}} \Delta t$$

Since $\sqrt{\mu_0 \varepsilon_0}$, is a positive quantity, this indicates that as time progresses, a given value of the function moves forward in z with the velocity $1/\sqrt{\mu_0 \varepsilon_0}$, thereby giving the characteristic of a $(+)$ wave for f. Similarly, for the function g to maintain the same value for the two pairs of t and z, we must have

$$t_1 + z_1\sqrt{\mu_0 \varepsilon_0} = (t_1 + \Delta t) + (z_1 + \Delta z)\sqrt{\mu_0 \varepsilon_0}$$

or

$$\Delta z = -\frac{1}{\sqrt{\mu_0 \varepsilon_0}} \Delta t$$

The minus sign associated with $1/\sqrt{\mu_0 \varepsilon_0}$ indicates that as time progresses, a given value of the function moves backward in z with the velocity $1/\sqrt{\mu_0 \varepsilon_0}$, giving the characteristic of a $(-)$ wave for g.

We shall now define the intrinsic impedance of free space, η_0, to be

$$\eta_0 = \sqrt{\frac{\mu_0}{\varepsilon_0}} \approx 120\pi \ \Omega = 377 \ \Omega \qquad (5.16)$$

From (5.13) and (5.14), we see that η_0 is the ratio of E_x to H_y for the $(+)$ wave or the negative of the same ratio for the $(-)$ wave. Since the units of E_x are volts per meter and the units of H_y are amperes per meter, the units of E_x/H_y are volts per ampere or ohms, thereby giving the character of impedance for η_0. Replacing $\sqrt{\mu_0/\varepsilon_0}$ in (5.14) by η_0 and substituting v_p for $1/\sqrt{\mu_0 \varepsilon_0}$ in the arguments of the functions f and g in both (5.13) and (5.14), we can now write (5.13) and (5.14) as

$$E_x(z, t) = Af\left(t - \frac{z}{v_p}\right) + Bg\left(t + \frac{z}{v_p}\right) \qquad (5.17a)$$

$$H_y(z, t) = \frac{1}{\eta_0}\left[Af\left(t - \frac{z}{v_p}\right) - Bg\left(t + \frac{z}{v_p}\right)\right] \qquad (5.17b)$$

4. Having learned that the solution to (5.7a) and (5.7b) consists of super-position of traveling waves propagating in the $+z$- and $-z$-directions, we now make use of this solution together with other considerations to find the electromagnetic field due to the infinite plane current sheet of Fig. 5.2, and with the current density given by (5.2). To do this, we observe the following:

a. Since the current sheet, which is the source of waves, is in the $z = 0$ plane, there can be only a $(+)$ wave in the region $z > 0$ and only a $(-)$ wave in the region $z < 0$. Thus,

$$\mathbf{E}(z, t) = \begin{cases} A f\left(t - \dfrac{z}{v_p}\right)\mathbf{a}_x & \text{for} \quad z > 0 \\[3mm] B g\left(t + \dfrac{z}{v_p}\right)\mathbf{a}_x & \text{for} \quad z < 0 \end{cases} \tag{5.18a}$$

$$\mathbf{H}(z, t) = \begin{cases} \dfrac{A}{\eta_0} f\left(t - \dfrac{z}{v_p}\right)\mathbf{a}_y & \text{for} \quad z > 0 \\[3mm] -\dfrac{B}{\eta_0} g\left(t + \dfrac{z}{v_p}\right)\mathbf{a}_y & \text{for} \quad z < 0 \end{cases} \tag{5.18b}$$

b. From the boundary condition (3.51b) applied to the surface $z = 0$, we have

$$[E_x]_{z=0+} - [E_x]_{z=0-} = 0 \tag{5.19}$$

or $A f(t) = B g(t)$. Thus, (5.18a) and (5.18b) reduce to

$$\mathbf{E}(z, t) = F\left(t \mp \dfrac{z}{v_p}\right)\mathbf{a}_x \qquad \text{for} \quad z \gtrless 0 \tag{5.20a}$$

$$\mathbf{H}(z, t) = \pm\dfrac{1}{\eta_0} F\left(t \mp \dfrac{z}{v_p}\right)\mathbf{a}_y \qquad \text{for} \quad z \gtrless 0 \tag{5.20b}$$

where we have used $A f(t) = B g(t) = F(t)$.

c. From the boundary condition (3.55a) applied to the surface $z = 0$, we have

$$\mathbf{a}_z \times \{[\mathbf{H}]_{z=0+} - [\mathbf{H}]_{z=0-}\} = -J_S(t)\mathbf{a}_x \tag{5.21}$$

or $(2/\eta_0)F(t) = J_S(t)$. Thus, $F(t) = (\eta_0/2)J_S(t)$, and (5.20a) and (5.20b) become

$$\mathbf{E}(z, t) = \dfrac{\eta_0}{2} J_S\left(t \mp \dfrac{z}{v_p}\right)\mathbf{a}_x \qquad \text{for} \quad z \gtrless 0 \tag{5.22a}$$

$$\mathbf{H}(z, t) = \pm\dfrac{1}{2} J_S\left(t \mp \dfrac{z}{v_p}\right)\mathbf{a}_y \qquad \text{for} \quad z \gtrless 0 \tag{5.22b}$$

Equations (5.22a) and (5.22b) represent the complete solution for the electromagnetic field due to the infinite plane current sheet of surface current density given by

$$J_S(t) = -J_S(t)\mathbf{a}_x \qquad \text{for} \quad z = 0 \tag{5.23}$$

The solution corresponds to uniform plane waves having their field components uniform in planes parallel to the current sheet and propagating to either side of the current sheet with the velocity $v_p(= c)$. The time variation of the electric field component E_x in a given $z =$ constant plane is the same as the current density variation delayed by the time $|z|/v_p$ and multiplied by $\eta_0/2$. The time variation of the magnetic field component in a given $z =$ constant plane is the same as the current density variation delayed by $|z|/v_p$ and multiplied by $\pm\frac{1}{2}$ depending on $z \gtrless 0$. Using these properties, one can construct plots of the field components versus time for fixed values of z and versus z for fixed values of t. We shall illustrate by means of an example.

Example 5.1

Let us consider the function $J_S(t)$ in (5.23) to be that given in Fig. 5.5. We wish to find and sketch (a) E_x versus t for $z = 300$ m, (b) H_y versus t for $z = -450$ m, (c) E_x versus z for $t = 1$ μs, and (d) H_y versus z for $t = 2.5$ μs.

  a. Since $v_p = c = 3 \times 10^8$ m/s, the time delay corresponding to 300 m is 1 μs. Thus, the plot of E_x versus t for $z = 300$ m is the same as that of $J_S(t)$ multiplied by $\eta_0/2$, or 188.5, and delayed by 1 μs, as shown in Fig. 5.6(a).

  b. The time delay corresponding to 450 m is 1.5 μs. Thus, the plot of H_y versus t for $z = -450$ m is the same as that of $J_S(t)$ multiplied by $-1/2$ and delayed by 1.5 μs, as shown in Fig. 5.6(b).

  c. To sketch E_x versus z for a fixed value of t, say, t_1, we use the argument that a given value of E_x existing at the source at an earlier value of time, say, t_2, travels away from the source by the distance equal to $(t_1 - t_2)$ times v_p. Thus, at $t = 1$ μs, the values of E_x corresponding to points A and B in Fig. 5.5 move to the locations $z = \pm 300$ m and $z = \pm 150$ m, respectively, and the value of E_x corresponding to point C exists right at the source. Hence, the plot of E_x versus z for $t = 1$ μs is as shown in Fig. 5.6(c). Note that points beyond C in Fig. 5.5 correspond to $t > 1$ μs, and therefore they do not appear in the plot of Fig. 5.6(c).

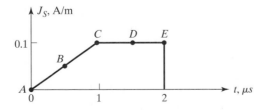

FIGURE 5.5 Plot of J_S versus t for Example 5.1.

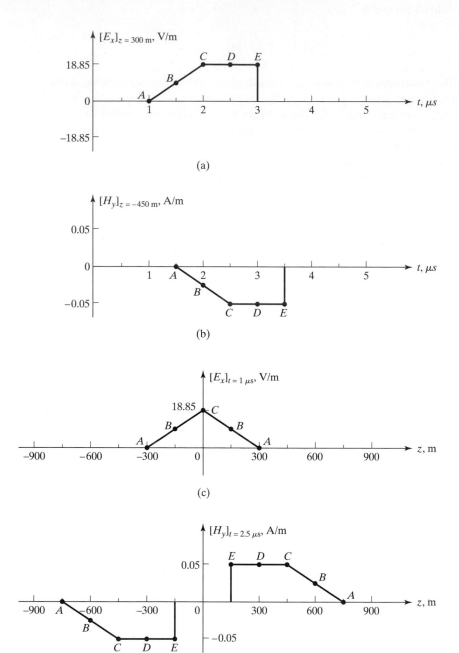

FIGURE 5.6 Plots of field components versus t for fixed values of z and versus z for fixed values of t for Example 5.1.

d. Using arguments as in part (c), we see that at $t = 2.5$ μs, the values of H_y corresponding to points A, B, C, D, and E in Fig. 5.5 move to the locations $z = \pm750$ m, ±600 m, ±450 m, ±300 m, and ±150 m, respectively, as shown in Fig. 5.6(d). Note that the plot is an odd function of z since the factor by which J_{S0} is multiplied to obtain H_y is $\pm\frac{1}{2}$ depending on $z \gtrless 0$.

K5.1. Infinite plane current sheet; Uniform plane wave; Wave equation; Time domain; Traveling-wave functions; Velocity of propagation; Intrinsic impedance of free space; Time delay.

D5.1. For each of the following traveling-wave functions, find the velocity of propagation both in magnitude and direction: **(a)** $(0.05y - t)^2$; **(b)** $u(t + 0.02x)$; and **(c)** $\cos(2\pi \times 10^8 t - 2\pi z)$.
Ans. **(a)** $20\mathbf{a}_y$ m/s; **(b)** $-50\mathbf{a}_x$ m/s; **(c)** $10^8\mathbf{a}_z$ m/s.

D5.2 The time variation for $z = 0$ of a function $f(z, t)$ representing a traveling wave propagating in the $+z$-direction with velocity 200 m/s is shown in Fig. 5.7. Find the value of the function for each of the following cases: **(a)** $z = 300$ m, $t = 2.0$ s; **(b)** $z = -200$ m, $t = 0.4$ s; and **(c)** $z = 100$ m, $t = 0.5$ s.
Ans. **(a)** $0.25A$; **(b)** $0.6A$; **(c)** 0.

FIGURE 5.7   For Problem D5.2.

D5.3. The time variation for $z = 0$ of a function $g(z, t)$ representing a traveling wave propagating in the $-z$-direction with velocity 100 m/s is shown in Fig. 5.8. Find the value of the function for each of the following cases: **(a)** $z = 200$ m, $t - 0.2$ s; **(b)** $z = -300$ m, $t = 3.4$ s; and **(c)** $z = 100$ m, $t - 0.6$ s.
Ans. **(a)** $0.9A$; **(b)** $0.4A$; **(c)** A.

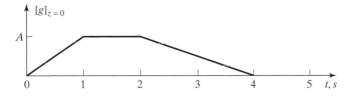

FIGURE 5.8   For Problem D5.3.

5.2 SINUSOIDALLY TIME-VARYING UNIFORM PLANE WAVES IN FREE SPACE

In the previous section, we considered the current density on the infinite plane current sheet to be an arbitrary function of time and obtained the solution for the electromagnetic field. As already pointed out in Section 1.5, of particular interest are fields varying sinusoidally with time. These are produced by a source

Solution for the field for the sinusoidal case

whose current density varies sinusoidally with time. Thus, assuming the current density on the infinite plane sheet of Fig. 5.2 to be

$$\mathbf{J}_S = -J_{S0} \cos \omega t \, \mathbf{a}_x \qquad \text{for} \quad z = 0 \tag{5.24}$$

where J_{S0} is the amplitude and ω is the radian frequency, we obtain the corresponding solution for the electromagnetic field by substituting $J_S(t) = J_{S0} \cos \omega t$ in (5.22a) and (5.22b):

$$\mathbf{E} = \frac{\eta_0 J_{S0}}{2} \cos (\omega t \mp \beta z) \, \mathbf{a}_x \qquad \text{for} \quad z \gtrless 0 \tag{5.25a}$$

$$\mathbf{H} = \pm \frac{J_{S0}}{2} \cos (\omega t \mp \beta z) \, \mathbf{a}_y \qquad \text{for} \quad z \gtrless 0 \tag{5.25b}$$

where

$$\beta = \frac{\omega}{v_p} \tag{5.26}$$

Properties and parameters of sinusoidal waves

Equations (5.25a) and (5.25b) represent sinusoidally time-varying uniform plane waves propagating away from the current sheet. The phenomenon is illustrated in Fig. 5.9, which shows sketches of the current density on the sheet and the distance variation of the electric and magnetic fields on either side of the current sheet for three values of t. It should be understood that in these sketches the field variations depicted along the z-axis hold also for any other line parallel to the z-axis. We shall now discuss in detail several important parameters and properties associated with the sinusoidal waves:

1. The argument $(\omega t \mp \beta z)$ of the cosine functions is the phase of the fields. We shall denote the phase by the symbol ϕ. Thus,

$$\phi = \omega t \mp \beta z \tag{5.27}$$

Note that ϕ is a function of t and z.

Frequency

2. Since

$$\frac{\partial \phi}{\partial t} = \omega \tag{5.28}$$

the rate of change of phase with time for a fixed value of z is equal to ω, the radian frequency of the wave. The linear frequency given by

$$f = \frac{\omega}{2\pi} \tag{5.29}$$

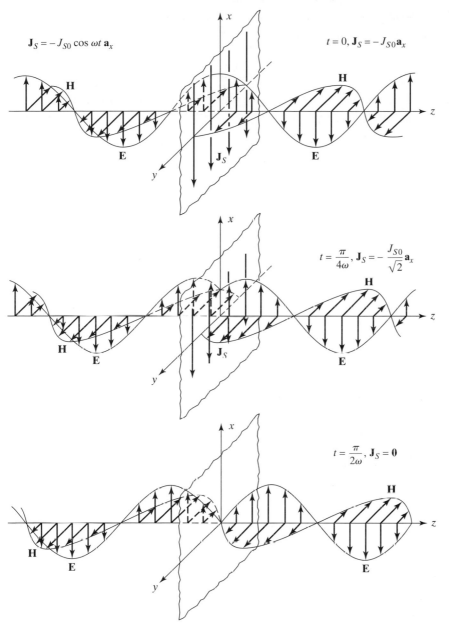

FIGURE 5.9 Time history of uniform plane electromagnetic wave radiating away from an infinite plane current sheet in free space.

is the number of times the phase changes by 2π radians in one second for a fixed value of z. The situation is pertinent to an observer at a point in the field region watching a *movie* of the field variations with time and counting the number of times in one second the field goes through a certain phase point, say, the positive maximum.

Phase
constant

3. Since

$$\frac{\partial \phi}{\partial z} = \mp \beta \tag{5.30}$$

the magnitude of the rate of change of phase with distance z for a fixed value of time is equal to β, known as the *phase constant*. The situation is pertinent to taking a *still photograph* of the phenomenon at any given time along the z-axis, counting the number of radians of phase change in one meter.

Wavelength

4. It follows from property 3 that the distance, along the z-direction, in which the phase changes by 2π radians for a fixed value of time is equal to $2\pi/\beta$. This distance is known as the *wavelength*, denoted by the symbol λ. Thus,

$$\lambda = \frac{2\pi}{\beta} \tag{5.31}$$

It is the distance between two consecutive positive maximum points on the sinusoid, or between any other two points that are displaced from these two positive maximum points by the same distance and to the same side, as shown in Fig. 5.10.

Phase
velocity

5. From (5.26), we note that the velocity of propagation of the wave is given by

$$v_p = \frac{\omega}{\beta} \tag{5.32}$$

Here, it is known as the phase velocity, since a constant value of phase progresses with that velocity along the z-direction. It is the velocity with which an observer has to move along the direction of propagation of the wave to be associated with a particular phase point on the moving sinusoid. Thus, it follows from (5.27) that

$$d(\omega t \mp \beta z) = 0$$

which gives

$$\omega \, dt \mp \beta \, dz = 0$$

$$\frac{dz}{dt} = \pm \frac{\omega}{\beta}$$

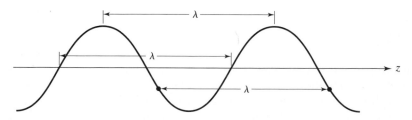

FIGURE 5.10   For explaining wavelength.

where the $+$ and $-$ signs correspond to $(+)$ and $(-)$ waves, respectively. We recall that for free space, $v_p = 1/\sqrt{\mu_0 \varepsilon_0} = c = 3 \times 10^8$ m/s.

6. From (5.31), (5.29), and (5.32), we note that

$$\lambda f = \left(\frac{2\pi}{\beta}\right)\left(\frac{\omega}{2\pi}\right) = \frac{\omega}{\beta}$$

*Classification
of waves*

or

$$\boxed{\lambda f = v_p} \tag{5.33}$$

Thus, the wavelength and frequency of a wave are not independent of each other but are related through the phase velocity. This is not surprising because λ is a parameter governing the variation of the field with distance for a fixed time, f is a parameter governing the variation of the field with time for a fixed value of z, and we know from Maxwell's equations that the space and time variations of the fields are interdependent. For free space (5.33) gives

$$\boxed{\lambda \text{ in meters } \times f \text{ in hertz} = 3 \times 10^8} \tag{5.34a}$$

or

$$\boxed{\lambda \text{ in meters } \times f \text{ in megahertz} = 300} \tag{5.34b}$$

It can be seen from these relationships that the higher the frequency, the shorter the wavelength. Waves are classified according to frequency or wavelength. Table 5.1 lists the commonly used designations for the various bands up to 300 GHz, where 1 GHz is 10^9 Hz. The corresponding frequency ranges and wavelength ranges are also given. The frequencies above about 300 GHz fall into regions far infrared and beyond. The AM radio (550–1650 kHz) falls in the medium wave band, whereas the FM radio makes use of 88–108 MHz in the VHF band. The VHF TV channels 2–6 use 54–88 MHz, and 7–13 employ 174–216 MHz. The UHF TV channels are in the 470–890-MHz range. Microwave ovens operate

TABLE 5.1 Commonly Used Designations for the Various Frequency Ranges

Designation	Frequency Range	Wavelength Range
ELF (extremely low frequency)	30–3000 Hz	10,000–100 km
VLF (very low frequency)	3–30 kHz	100–10 km
LF (low frequency) or long waves	30–300 kHz	10–1 km
MF (medium frequency) or medium waves	300–3000 kHz	1000–100 m
HF (high frequency) or short waves	3–30 MHz	100–10 m
VHF (very high frequency)	30–300 MHz	10–1 m
UHF (ultrahigh frequency)	300–3000 MHz	100–10 cm
Microwaves	1–30 GHz	30–1 cm
Millimeter waves	30–300 GHz	10–1 mm

at 2450 MHz. Police traffic radars operate at about 10.5 and 24.1 GHz. Various other ranges in Table 5.1 are used for various other applications too numerous to mention here.

*Intrinsic
impedance*

7. The electric and magnetic fields are such that

$$\frac{\text{amplitude of } \mathbf{E}}{\text{amplitude of } \mathbf{H}} = \eta_0 \qquad (5.35)$$

We recall that η_0, the intrinsic impedance of free space, has a value approximately equal to 120π or $377 \ \Omega$.

8. The electric and magnetic fields have components lying in the planes of constant phase ($z = $ constant planes) and perpendicular to each other and to the direction of propagation. In fact, the cross product of $\mathbf{E}$ and $\mathbf{H}$ results in a vector that is directed along the direction of propagation, as can be seen by noting that

$$\mathbf{E} \times \mathbf{H} = \pm \frac{\eta_0 J_{S0}^2}{4} \cos^2 \left(\omega t \mp \beta z \right) \mathbf{a}_z \qquad \text{for} \quad z \gtrless 0 \qquad (5.36)$$

Polarization

9. From the discussion of polarization of sinusoidally time-varying fields in Section 1.5, we note that the fields given by (5.25a) and (5.25b) are linearly polarized. Hence, the wave is said to be linearly polarized. Two linearly polarized waves of equal amplitudes propagating in the same direction, with their electric fields (and, hence, their magnetic fields) perpendicular to each other and out of phase by 90°, add up to give a circularly polarized wave. Since a circle can be traversed in one of two opposite senses, we talk of right-handed or clockwise circular polarization and left-handed or counterclockwise circular polarization. The convention is that if in a given constant phase plane, the tip of the field vector rotates with time in the clockwise sense as seen looking along the direction of propagation of the wave, the wave is said to be right circularly polarized. If the tip of the field vector rotates in the counterclockwise sense, the wave is said to be left circularly polarized. Similar considerations hold for elliptically polarized waves, which arise due to the superposition of two linearly polarized waves in the general case.

We shall now consider two examples of the application of the properties we have learned thus far in this section.

Example 5.2

The electric field of a uniform plane wave is given by

$$\mathbf{E} = 10 \sin (3\pi \times 10^8 t - \pi z) \, \mathbf{a}_x + 10 \cos (3\pi \times 10^8 t - \pi z) \, \mathbf{a}_y \text{ V/m}$$

Let us find (a) the various parameters associated with the wave, (b) the corresponding magnetic field $\mathbf{H}$, and (c) the polarization of the wave.

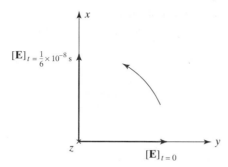

FIGURE 5.11   For Example 5.2.

a. From the argument of the sine and cosine functions, we can identify the following:

$$\omega = 3\pi \times 10^8 \text{ rad/s}$$

$$\beta = \pi \text{ rad/m}$$

Then

$$f = \frac{\omega}{2\pi} = 1.5 \times 10^8 \text{ Hz}$$

$$\lambda = \frac{2\pi}{\beta} = 2 \text{ m}$$

$$v_p = \frac{\omega}{\beta} = 3 \times 10^8 \text{ m/s}$$

Note also that $\lambda f = 3 \times 10^8 = v_p$. In view of the minus sign associated with πz, the direction of propagation of the wave is the $+z$-direction.

b. The unit vectors $\mathbf{a}_x$ and $\mathbf{a}_y$ associated with the first and second terms, respectively, tell us that the electric field contains components directed along the x- and y-directions. Using the properties 7 and 8 discussed earlier, we obtain the magnetic field of the wave to be

$$\mathbf{H} = \frac{10}{377} \sin (3\pi \times 10^8 t - \pi z) \mathbf{a}_y + \frac{10}{377} \cos (3\pi \times 10^8 t - \pi z)(-\mathbf{a}_x)$$

$$= -\frac{10}{377} \cos (3\pi \times 10^8 t - \pi z) \mathbf{a}_x + \frac{10}{377} \sin (3\pi \times 10^8 t - \pi z) \mathbf{a}_y \text{ A/m}$$

c. The two components of $\mathbf{E}$ are equal in amplitude, perpendicular, and out of phase by $90°$. Therefore, the wave is circularly polarized. To determine if the polarization is right-handed or left-handed, we look at the electric field vectors in the $z = 0$ plane for two values of time, $t = 0$ and $t = \frac{1}{6} \times 10^{-8} \text{ s}\, (3\pi \times 10^8 t = \pi/2)$. These are shown in Fig. 5.11. As time progresses, the tip of the vector rotates in the counterclockwise sense, as seen looking in the $+z$-direction. Hence, the wave is left circularly polarized.

Example 5.3

An antenna array consists of two or more antenna elements spaced appropriately and excited with currents having the appropriate amplitudes and phases in order to obtain a desired radiation characteristic. To illustrate the principle of an antenna array, let us consider two infinite plane parallel current sheets, spaced $\lambda/4$ apart and carrying currents of

Principle of antenna array

equal amplitudes but out of phase by $\pi/2$ as given by the densities

$$\mathbf{J}_{S1} = -J_{S0} \cos \omega t \, \mathbf{a}_x \qquad \text{for} \quad z = 0$$
$$\mathbf{J}_{S2} = -J_{S0} \sin \omega t \, \mathbf{a}_x \qquad \text{for} \quad z = \lambda/4$$

and find the electric field due to the array of the two current sheets.

We apply the result given by (5.25a) to each current sheet separately and then use superposition to find the required total electric field due to the array of the two current

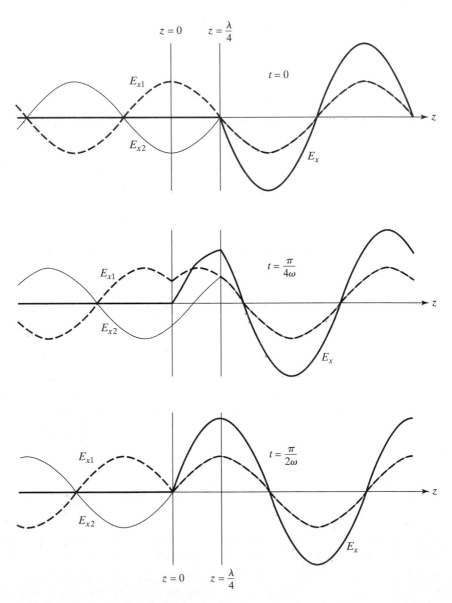

FIGURE 5.12   Time history of individual fields and the total field due to an array of two infinite plane parallel current sheets.

sheets. Thus, for the current sheet in the $z = 0$ plane, we have

$$
\mathbf{E}_1 = \begin{cases} \dfrac{\eta_0 J_{S0}}{2} \cos(\omega t - \beta z)\, \mathbf{a}_x & \text{for} \quad z > 0 \\[2ex] \dfrac{\eta_0 J_{S0}}{2} \cos(\omega t + \beta z)\, \mathbf{a}_x & \text{for} \quad z < 0 \end{cases}
$$

For the current sheet in the $z = \lambda/4$ plane, we have

$$
\mathbf{E}_2 = \begin{cases} \dfrac{\eta_0 J_{S0}}{2} \sin\left[\omega t - \beta\left(z - \dfrac{\lambda}{4}\right)\right] \mathbf{a}_x & \text{for} \quad z > \dfrac{\lambda}{4} \\[2ex] \dfrac{\eta_0 J_{S0}}{2} \sin\left[\omega t + \beta\left(z - \dfrac{\lambda}{4}\right)\right] \mathbf{a}_x & \text{for} \quad z < \dfrac{\lambda}{4} \end{cases}
$$

$$
= \begin{cases} \dfrac{\eta_0 J_{S0}}{2} \sin\left(\omega t - \beta z + \dfrac{\pi}{2}\right) \mathbf{a}_x & \text{for} \quad z > \dfrac{\lambda}{4} \\[2ex] \dfrac{\eta_0 J_{S0}}{2} \sin\left(\omega t + \beta z - \dfrac{\pi}{2}\right) \mathbf{a}_x & \text{for} \quad z < \dfrac{\lambda}{4} \end{cases}
$$

$$
= \begin{cases} \dfrac{\eta_0 J_{S0}}{2} \cos(\omega t - \beta z)\, \mathbf{a}_x & \text{for} \quad z > \dfrac{\lambda}{4} \\[2ex] -\dfrac{\eta_0 J_{S0}}{2} \cos(\omega t + \beta z)\, \mathbf{a}_x & \text{for} \quad z < \dfrac{\lambda}{4} \end{cases}
$$

Now, using superposition, we find the total electric field due to the two current sheets to be

$$
\mathbf{E} = \mathbf{E}_1 + \mathbf{E}_2
$$

$$
= \begin{cases} \eta_0 J_{S0} \cos(\omega t - \beta z)\, \mathbf{a}_x & \text{for} \quad z > \lambda/4 \\ \eta_0 J_{S0} \sin \omega t \sin \beta z\, \mathbf{a}_x & \text{for} \quad 0 < z < \lambda/4 \\ 0 & \text{for} \quad z < 0 \end{cases}
$$

Thus, the total field is zero in the region $z < 0$ due to the phase opposition of the individual fields, and, hence, there is no radiation toward that side of the array. In the region $z > \lambda/4$, the total field is twice that of the field of a single sheet due to the individual fields being in phase. The phenomenon is illustrated in Fig. 5.12, which shows the individual fields E_{x1} and E_{x2} and the total field $E_x = E_{x1} + E_{x2}$ for a few values of t. The result that we have obtained here for the total field due to the array of two current sheets, spaced $\lambda/4$ apart and fed with currents of equal amplitudes but out of phase by $\pi/2$, is said to correspond to an "endfire" radiation pattern.

Returning now to the solution for the electromagnetic field given by *Doppler* (5.25a) and (5.25b), let us ask ourselves the question, "How does the phase as- *effect* sociated with the wave change with time as viewed by a moving observer?" To answer this question, let us consider the $(+)$ wave and an observer moving along the positive z-direction with a velocity v_0 m/s, starting at $z = z_0$ at $t = 0$. Then the position of the observer as a function of time is given by $z = z_0 + v_0 t$ and the phase of the wave at that position is given by

$$\phi_{obs} = \omega t - \beta(z_0 + v_0 t)$$
$$= (\omega - \beta v_0)t - \beta z_0$$

Ignoring relativistic effects, the rate of change of phase with time or the radian frequency of the wave viewed by the moving observer is

$$\omega_{obs} = \frac{d}{dt}[(\omega - \beta v_0)t - \beta z_0]$$

$$= \omega - \beta v_0 = \omega - \frac{\omega}{v_p}v_0$$

$$= \omega\left(1 - \frac{v_0}{v_p}\right)$$

or

$$\boxed{f_{obs} = f\left(1 - \frac{v_0}{v_p}\right)} \tag{5.37}$$

Thus, the moving observer views a frequency that is different from that of the source of the wave. This phenomenon of a shift in the frequency of the wave is known as the *Doppler effect*. For an observer moving along the direction of propagation, the Doppler-shifted frequency is less than the actual frequency by the amount $f v_0/v_p$ or v_0/λ. For an observer moving opposite to the direction of propagation of the wave, the Doppler-shifted frequency is higher than the actual frequency by the same amount. The situation is illustrated in Fig. 5.13, which depicts the wave motion as viewed by a stationary observer (A) and two moving observers (O and R), one moving along and the other moving opposite to the direction of propagation of the wave with a velocity which for simplicity of illustration, is assumed to be one-half the phase velocity of the wave. From the series of sketches for one period of the wave, it can be seen that observer A views a complete cycle of the wave, whereas observer O views only one-half cycle of the wave and observer R views one and one-half cycles of the wave during that period. Thus, the stationary observer A views the same frequency as that of the wave, but moving observer O views a frequency that is one-half that of the wave and moving observer R views a frequency that is one and one-half that of the wave.

Doppler shift also occurs when the current sheet is in motion and the observer is at rest. Assuming that the Doppler-shifted radian frequency as viewed by the stationary observer in the field of a moving current sheet is ω_{obs}, we can write the phase of the ($+$) wave to be

$$\phi_{obs} = \omega_{obs}\left(t - \frac{z}{v_p}\right)$$

If the current sheet is moving with velocity v_s in the $+z$-direction, starting at $z = 0$ at $t = 0$, then the phase on the current sheet is $\omega_{obs}(t - v_s t/v_p)$. Since the

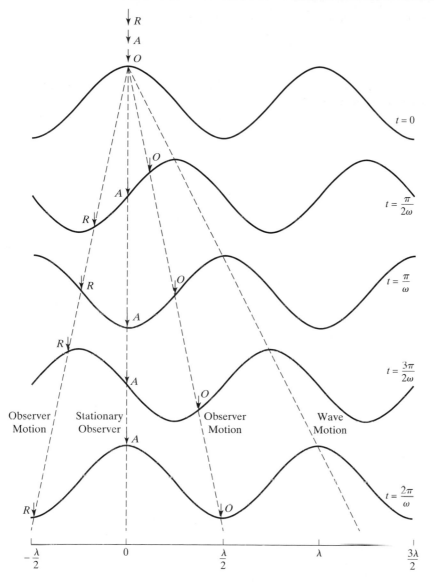

FIGURE 5.13 Wave motion as viewed by a stationary observer (A) and two moving observers (O and R).

radian frequency on the current sheet is ω, we have

$$\frac{d}{dt}\left[\omega_{\text{obs}}\left(t - \frac{v_s t}{v_p}\right)\right] = \omega$$

$$\omega_{\text{obs}}\left(1 - \frac{v_s}{v_p}\right) = \omega$$

$$\omega_{\text{obs}} = \frac{\omega}{1 - v_s/v_p}$$

or

$$f_{obs} = \frac{f}{1 - v_s/v_p} \tag{5.38}$$

For $v_s \ll v_p$, (5.38) reduces to

$$f_{obs} \approx f\left(1 + \frac{v_s}{v_p}\right) \tag{5.39}$$

the right side of which is the same as that in (5.37), with v_0 replaced by $-v_s$. For $v_s > 0$, that is, source moving toward the observer, $f_{obs} > f$. For $v_s < 0$, that is, source receding from the observer, $f_{obs} < f$.

Once again, we have ignored relativistic effects in deriving (5.38). When relativistic effects are taken into account, (5.37) and (5.38) are modified to give an expression such that the two values of f_{obs} are exactly equal for the same relative velocity between the source and the observer. This expression is given by

$$f_{obs} = f\sqrt{\frac{1 - v_r/v_p}{1 + v_r/v_p}}$$

where v_r is the difference between v_0 and v_s. For $v_r \ll v_p$, this expression approximates to the form of (5.37).

Returning to the derivation leading to (5.37), we note that for an observer moving in an arbitrary direction with velocity $\mathbf{v}_o$, only the z-component of $\mathbf{v}_o$ contributes to the Doppler shift since the value of z corresponding to the position of the observer is governed only by the z-component of $\mathbf{v}_o$. The component of $\mathbf{v}_o$ perpendicular to the z-direction does not contribute to the Doppler shift. To generalize this for a wave propagating in an arbitrary direction given by the unit vector $\mathbf{a}_p$ and an observer moving with velocity $\mathbf{v}_o$ in the field of that wave, we can say that only the component of $\mathbf{v}_o$ along $\mathbf{a}_p$ contributes to the Doppler shift. Thus, for this general case

$$\text{Doppler shift in } f = -f\frac{\mathbf{v}_o \cdot \mathbf{a}_p}{v_p} \tag{5.40}$$

In the application of (5.40), it should be kept in mind that if the source of waves is a point source instead of an infinite plane sheet of current, the direction of propagation of the wave is along the line from the point source to the observer.

Police radar   The phenomenon of Doppler effect has many applications. An example in everyday life with which some of us might have had an experience is the police traffic radar. In this application, a microwave signal is generated in the radar unit and is transmitted toward the moving car, the speed of which is to be monitored. Part of the incident signal is reflected from the moving car back to the radar unit, Doppler shifted in frequency by an amount proportional to the speed of the car. In the radar unit, the Doppler-shifted signal is received and mixed with the original signal to generate a signal of the difference frequency, which is the Doppler shift. A display in the radar unit is calibrated to indicate the speed of the car, which is the difference frequency multiplied by a proportionality constant. Some other applications of Doppler effect, just to mention a few, are in aircraft navigation, sea-state monitoring, and radio astronomy.

Example 5.4

Let us consider a police radar operating at a frequency $f = 24.1$ GHz and determine the Doppler shift due to an automobile directly approaching the radar at a speed of $v_o = 100$ km/h.

Since the radar operates on the signal reflected from the moving automobile, the Doppler shift is due to the combination of (5.37) and (5.39). Hence, it is approximately twice that for one-way transmission. Thus, the required Doppler shift in the radar operating frequency is given by

$$\Delta f \approx 2f \frac{v_o}{v_p}$$

$$= 2 \times 24.1 \times 10^9 \times \frac{10^5}{3600 \times 3 \times 10^8}$$

$$= 4463 \text{ Hz} = 4.463 \text{ kHz}$$

which is in the audio-frequency range.

An experimental setup for demonstrating the Doppler effect is shown by the schematic diagram in Fig. 5.14. It consists of a horn radiating a UHF signal generated by a klystron oscillator and connected to it through microwave components, as shown in the figure. The signal from the oscillator goes into port 1 of

Experimental setup for Doppler effect

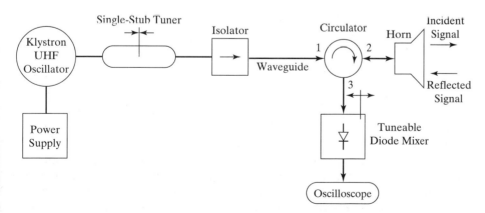

FIGURE 5.14   Schematic diagram of experimental setup for demonstrating Doppler effect.

the circulator and comes out of port 2 and is radiated from the horn. A reflective object placed in front of the horn provides a reflected signal, which is picked up by the horn and fed into port 2 of the circulator and comes out of port 3 into the mixer. Part of the original klystron signal going into port 1 also comes out of port 3. Thus, signals proportional to the incident and reflected signals are mixed in the mixer, thereby generating the beat frequency signal, which is displayed on the oscilloscope. Since the reflected signal is Doppler shifted by an amount depending upon the motion of the reflective object, the beat frequency is equal to the Doppler shift. The isolator, a one-way device, prevents any signal coming out of port 1 of the circulator from going back to the klystron oscillator and affecting its performance. By placing a metallic sheet in front of the horn and walking toward and away from the horn, the Doppler effect can be demonstrated. Walking sideways, that is, perpendicular to the direction of propagation of the wave, will of course produce very little or no effect.

K5.2. Sinusoidal waves; Phase; Frequency; Wavelength; Phase velocity; Frequency times wavelength; Intrinsic impedance; Polarization; Antenna array; Moving observer; Doppler shift.

D5.4. For a sinusoidally time-varying uniform plane wave propagating in free space, find the following: **(a)** the frequency f, if the phase of the field at a point is observed to change by 3π rad in $0.1\ \mu s$; **(b)** the wavelength λ, if the phase of the field at a particular value of time is observed to change by 0.04π in a distance of 1 m along the direction of propagation of the wave; **(c)** the frequency f, if the wavelength is 25 m; and **(d)** the wavelength λ, if the frequency is 5 MHz.
Ans. **(a)** 15 MHz; **(b)** 50 m; **(c)** 12 MHz; **(d)** 60 m.

D5.5. The magnetic field of a uniform plane wave in free space is given by

$$\mathbf{H} = H_0 \cos\left(6\pi \times 10^8 t + 2\pi y\right) \mathbf{a}_x \text{ A/m}$$

Find unit vectors along the following: **(a)** the direction of propagation of the wave; **(b)** the direction of the magnetic field at $t = 0, y = 0$; and **(c)** the direction of the electric field at $t = 0, y = 0$.
Ans. **(a)** $-\mathbf{a}_y$; **(b)** $\mathbf{a}_x$; **(c)** $-\mathbf{a}_z$.

D5.6. For the array of two infinite plane current sheets of Example 5.3, assume that

$$\mathbf{J}_{S2} = -kJ_{S0} \sin \omega t\ \mathbf{a}_x \qquad \text{for} \quad z = \lambda/4$$

where $|k| \leq 1$. Find the value of k for each of the following values of the ratio of the amplitude of the electric-field intensity for $z > \lambda/4$ to the amplitude of the electric-field intensity for $z < 0$: **(a)** 1/3; **(b)** 3; and **(c)** 7.
Ans. **(a)** $-1/2$; **(b)** 1/2; **(c)** 3/4.

D5.7. The electric-field intensity of a uniform plane propagating in a direction not along one of the coordinate axes is given by

$$\mathbf{E} = E_0(\mathbf{a}_x + 2\mathbf{a}_y + 2\mathbf{a}_z) \cos\left[9\pi \times 10^8 t - \pi(2x - 2y + z)\right]$$

Ignoring relativistic effects, find the Doppler shift in the frequency f of the wave viewed by a moving observer for each of the following velocities of the observer: **(a)** $10^3 \mathbf{a}_x$ m/s; **(b)** $10^3 \mathbf{a}_y$ m/s; **(c)** $10^3 \mathbf{a}_z$ m/s; and **(d)** $10^3(\frac{2}{3}\mathbf{a}_x + \frac{1}{3}\mathbf{a}_y - \frac{2}{3}\mathbf{a}_z)$ m/s.
Ans. **(a)** -1000 Hz; **(b)** 1000 Hz; **(c)** -500 Hz; **(d)** 0 Hz.

5.3 WAVE EQUATION AND SOLUTION FOR MATERIAL MEDIUM

In Sections 5.1 and 5.2, we discussed uniform plane wave propagation in free space. In this section, we extend the treatment to a material medium characterized by conductivity σ, permittivity ε, and permeability μ. We recall that the constitutive relations are

$$\mathbf{J}_c = \sigma\mathbf{E}$$

$$\mathbf{D} = \varepsilon\mathbf{E}$$

$$\mathbf{H} = \frac{\mathbf{B}}{\mu}$$

so that the Maxwell's equations for the material medium are

$$\nabla \times \mathbf{E} = -\frac{\partial\mathbf{B}}{\partial t} = -\mu\frac{\partial\mathbf{H}}{\partial t} \tag{5.41a}$$

$$\nabla \times \mathbf{H} = \mathbf{J} + \frac{\partial\mathbf{D}}{\partial t} = \mathbf{J}_c + \frac{\partial\mathbf{D}}{\partial t} = \sigma\mathbf{E} + \varepsilon\frac{\partial\mathbf{E}}{\partial t} \tag{5.41b}$$

To discuss electromagnetic wave propagation in the material medium, let us consider the infinite plane current sheet of Fig. 5.2, except that now the medium on either side of the sheet is a material instead of free space, as shown in Fig. 5.15.

The electric and magnetic fields for the simple case of the infinite plane current sheet in the $z = 0$ plane and carrying uniformly distributed current in the negative x-direction as given by

$$\mathbf{J}_S = -J_{S0} \cos \omega t \, \mathbf{a}_x \tag{5.42}$$

are of the form

$$\mathbf{E} = E_x(z, t)\mathbf{a}_x \tag{5.43a}$$

$$\mathbf{H} = H_y(z, t)\mathbf{a}_y \tag{5.43b}$$

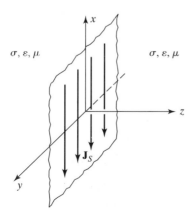

FIGURE 5.15   Infinite plane current sheet embedded in a material medium.

The corresponding simplified forms of the Maxwell's curl equations are

$$\frac{\partial E_x}{\partial z} = -\mu \frac{\partial H_y}{\partial t} \tag{5.44a}$$

$$\frac{\partial H_y}{\partial z} = -\sigma E_x - \varepsilon \frac{\partial E_x}{\partial t} \tag{5.44b}$$

Without the σE_x term on the right side of (5.44b), these two equations would be the same as (5.7a) and (5.7b) with μ_0 replaced by μ and ε_0 replaced by ε. The addition of the σE_x term complicates the solution in time domain. Hence, it is convenient to consider the solution for the sinusoidally time-varying case by using the phasor technique.

Wave
equation

Thus, letting

$$E_x(z, t) = \text{Re}[\overline{E}_x(z)e^{j\omega t}] \tag{5.45a}$$

$$H_y(z, t) = \text{Re}[\overline{H}_y(z)e^{j\omega t}] \tag{5.45b}$$

and replacing E_x and H_y in (5.44a) and (5.44b) by their phasors $\overline{E}_x$ and $\overline{H}_y$, respectively, and $\partial/\partial t$ by $j\omega$, we obtain the corresponding differential equations for the phasors $\overline{E}_x$ and $\overline{H}_y$ as

$$\frac{\partial \overline{E}_x}{\partial z} = -j\omega\mu\overline{H}_y \tag{5.46a}$$

$$\frac{\partial \overline{H}_y}{\partial z} = -\sigma\overline{E}_x - j\omega\varepsilon\overline{E}_x = -(\sigma + j\omega\varepsilon)\overline{E}_x \tag{5.46b}$$

Differentiating (5.46a) with respect to z and using (5.46b), we obtain

$$\frac{\partial^2 \overline{E}_x}{\partial z^2} = -j\omega\mu \frac{\partial \overline{H}_y}{\partial z} = j\omega\mu(\sigma + j\omega\varepsilon)\overline{E}_x \tag{5.47}$$

Defining

$$\overline{\gamma} = \sqrt{j\omega\mu(\sigma + j\omega\varepsilon)} \tag{5.48}$$

and substituting in (5.47), we have

$$\frac{\partial^2 \overline{E}_x}{\partial z^2} = \overline{\gamma}^2\, \overline{E}_x \tag{5.49}$$

which is the wave equation for $\overline{E}_x$ in the material medium.

The solution to the wave equation (5.49) is given by

$$\overline{E}_x(z) = \overline{A}e^{-\overline{\gamma}z} + \overline{B}e^{\overline{\gamma}z} \tag{5.50}$$

where $\overline{A}$ and $\overline{B}$ are arbitrary constants. Noting that $\overline{\gamma}$ is a complex number and, hence, can be written as

$$\boxed{\overline{\gamma} = \alpha + j\beta} \tag{5.51}$$

and also writing $\overline{A}$ and $\overline{B}$ in exponential form as $Ae^{j\theta}$ and $Be^{j\phi}$, respectively, we have

$$\overline{E}_x(z) = Ae^{j\theta}e^{-\alpha z}e^{-j\beta z} + Be^{j\phi}e^{\alpha z}e^{j\beta z}$$

or

$$\begin{aligned}
E_x(z, t) &= \text{Re}[\overline{E}_x(z)e^{j\omega t}] \\
&= \text{Re}[Ae^{j\theta}e^{-\alpha z}e^{-j\beta z}e^{j\omega t} + Be^{j\phi}e^{\alpha z}e^{j\beta z}e^{j\omega t}] \\
&= Ae^{-\alpha z}\cos(\omega t - \beta z + \theta) + Be^{\alpha z}\cos(\omega t + \beta z + \phi)
\end{aligned} \tag{5.52}$$

We now recognize the two terms on the right side of (5.52) as representing uniform plane waves propagating in the positive z- and negative z-directions, respectively, with phase constant β, in view of the factors $\cos(\omega t - \beta z + \theta)$ and $\cos(\omega t + \beta z + \phi)$, respectively. They are, however, multiplied by the factors $e^{-\alpha z}$ and $e^{\alpha z}$, respectively. Hence, the amplitude of the field differs from one constant phase surface to another. Since there cannot be a $(+)$ wave in the region $z < 0$, that is, to the left of the current sheet, and since there cannot be a $(-)$ wave in the region $z > 0$, that is, to the right of the current sheet, the solution for the electric field is given by

$$\mathbf{E}(z, t) = \begin{cases} Ae^{-\alpha z}\cos(\omega t - \beta z + \theta)\,\mathbf{a}_x & \text{for } z > 0 \\ Be^{\alpha z}\cos(\omega t + \beta z + \phi)\,\mathbf{a}_x & \text{for } z < 0 \end{cases} \tag{5.53}$$

To discuss how the amplitude of E_x varies with z on either side of the current sheet, we note that since σ, ε, and μ are all positive, the phase angle of $j\omega\mu(\sigma + j\omega\varepsilon)$ lies between 90° and 180°, and hence the phase angle of $\overline{\gamma}$ lies between 45° and 90°, making α and β positive quantities. This means that $e^{-\alpha z}$ decreases with increasing value of z, that is, in the positive z-direction, and $e^{\alpha z}$ decreases with decreasing value of z, that is, in the negative z-direction. Thus, the exponential factors $e^{-\alpha z}$ and $e^{\alpha z}$ associated with the solutions for E_x in (5.53) have the effect of decreasing the amplitude of the field, that is, attenuating it as it propagates away from the sheet to either side of it. For this reason, the quantity α is known as the *attenuation constant*. The attenuation per unit length is equal to e^{α}. In terms of decibels, this is equal to $20\log_{10} e^{\alpha}$, or 8.686α dB. The units of α are nepers per meter. The quantity $\overline{\gamma}$ is known as the *propagation constant* since its real and imaginary parts, α and β, together determine the propagation characteristics, that is, attenuation and phase shift of the wave.

Attenuation constant

Having found the solution for the electric field of the wave and discussed its general properties, we now turn to the solution for the corresponding magnetic field by substituting for $\bar{E}_x$ in (5.46a). Thus,

$$
\begin{aligned}
\bar{H}_y &= -\frac{1}{j\omega\mu}\frac{\partial\bar{E}_x}{\partial z} = \frac{\bar{\gamma}}{j\omega\mu}(\bar{A}e^{-\bar{\gamma}z} - \bar{B}e^{\bar{\gamma}z}) \\
&= \sqrt{\frac{\sigma + j\omega\varepsilon}{j\omega\mu}}(\bar{A}e^{-\bar{\gamma}z} - \bar{B}e^{-\bar{\gamma}z}) \\
&= \frac{1}{\bar{\eta}}(\bar{A}e^{-\bar{\gamma}z} - \bar{B}e^{\bar{\gamma}z})
\end{aligned}
\tag{5.54}
$$

where

$$
\boxed{\bar{\eta} = \sqrt{\frac{j\omega\mu}{\sigma + j\omega\varepsilon}}}
\tag{5.55}
$$

is the intrinsic impedance of the medium, which is now complex. Writing

$$
\boxed{\bar{\eta} = |\bar{\eta}|\,e^{j\tau}}
\tag{5.56}
$$

we obtain the solution for $H_y(z, t)$ as

$$
\begin{aligned}
H_y(z, t) &= \mathrm{Re}[\bar{H}_y(z)e^{j\omega t}] \\
&= \mathrm{Re}\left[\frac{1}{|\bar{\eta}|e^{j\tau}}Ae^{j\theta}e^{-\alpha z}e^{-j\beta z}e^{j\omega t} - \frac{1}{|\bar{\eta}|e^{j\tau}}Be^{j\phi}e^{\alpha z}e^{j\beta z}e^{j\omega t}\right] \\
&= \frac{A}{|\bar{\eta}|}e^{-\alpha z}\cos(\omega t - \beta z + \theta - \tau) - \frac{B}{|\bar{\eta}|}e^{\alpha z}\cos(\omega t + \beta z + \phi - \tau)
\end{aligned}
\tag{5.57}
$$

Remembering that the first and second terms on the right side of (5.57) correspond to $(+)$ and $(-)$ waves, respectively, and, hence, represent the solutions for the magnetic field in the regions $z > 0$ and $z < 0$, respectively, we write

$$
\mathbf{H}(z, t) = \begin{cases}
\dfrac{A}{|\bar{\eta}|}e^{-\alpha z}\cos(\omega t - \beta z + \theta - \tau)\,\mathbf{a}_y & \text{for}\quad z > 0 \quad (5.58a) \\[2ex]
-\dfrac{B}{|\bar{\eta}|}e^{\alpha z}\cos(\omega t + \beta z + \phi - \tau)\,\mathbf{a}_y & \text{for}\quad z < 0 \quad (5.58b)
\end{cases}
$$

Electro-
magnetic field
due to the
current sheet

To complete the solution for the electromagnetic field due to the current sheet embedded in the material medium, we need to find the values of the constants $A, B, \theta,$ and ϕ. This is done by using the boundary conditions, as in Section 5.1. Thus, from the boundary condition (3.51b) applied to the surface $z = 0$, we have

$$
[E_x]_{z=0+} - [E_x]_{z=0-} = 0
\tag{5.59}
$$

or $A \cos (\omega t + \theta) - B \cos (\omega t + \phi) = 0$, giving us $A = B$ and $\theta = \phi$. The solutions for $\mathbf{E}$ and $\mathbf{H}$ reduce to

$$\mathbf{E}(z, t) = Ae^{\mp \alpha z} \cos (\omega t \mp \beta z + \theta) \, \mathbf{a}_x \qquad \text{for} \quad z \gtrless 0 \qquad (5.60a)$$

$$\mathbf{H}(z, t) = \pm \frac{A}{|\eta|} e^{\mp \alpha z} \cos (\omega t \mp \beta z + \theta - \tau) \, \mathbf{a}_y \qquad \text{for} \quad z \gtrless 0 \qquad (5.60b)$$

Then from the boundary condition (3.55a) applied to the surface $z = 0$, we have

$$\mathbf{a}_z \times \{[\mathbf{H}]_{z=0+} - [\mathbf{H}]_{z=0-}\} = -J_{S0} \cos \omega t \, \mathbf{a}_x \qquad (5.61)$$

or

$$\frac{2A}{|\eta|} \cos (\omega t + \theta - \tau) = J_{S0} \cos \omega t$$

$$A = \frac{|\eta| J_{S0}}{2} \qquad \text{and} \qquad \theta = \tau$$

Thus, the electromagnetic field due to the infinite plane current sheet of surface current density

$$\boxed{\mathbf{J}_S = -J_{S0} \cos \omega t \, \mathbf{a}_x \qquad \text{for} \quad z = 0}$$

and with a material medium characterized by σ, ε, and μ on either side of it is given by

$$\boxed{\begin{aligned} \mathbf{E}(z, t) &= \frac{|\eta| J_{S0}}{2} e^{\mp \alpha z} \cos (\omega t \mp \beta z + \tau) \, \mathbf{a}_x \qquad \text{for} \quad z \gtrless 0 \qquad (5.62a) \\[2mm] \mathbf{H}(z, t) &= \pm \frac{J_{S0}}{2} e^{\mp \alpha z} \cos (\omega t \mp \beta z) \, \mathbf{a}_y \qquad \text{for} \quad z \gtrless 0 \qquad (5.62b) \end{aligned}}$$

As we have already discussed, (5.62a) and (5.62b) represent sinusoidally time-varying uniform plane waves, getting attenuated as they propagate away from the current sheet. The phenomenon is illustrated in Fig. 5.16, which shows sketches of current density on the sheet and the distance variation of the electric and magnetic fields on either side of the current sheet for three values of t. As in Fig. 5.9, it should be understood that in these sketches, the field variations depicted along the z-axis hold also for any other line parallel to the z-axis. We shall now discuss further the propagation characteristics associated with these waves:

Propagation characteristics

1. From (5.48) and (5.51), we have

$$\bar{\gamma}^2 = (\alpha + j\beta)^2 = j\omega\mu(\sigma + j\omega\varepsilon)$$

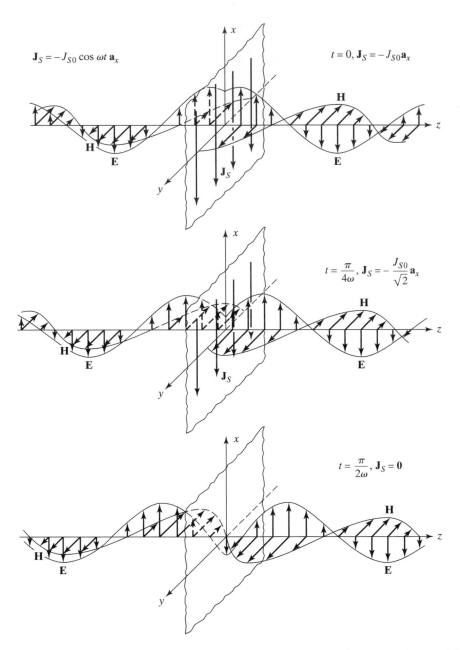

FIGURE 5.16   Time history of uniform plane electromagnetic wave radiating away from an infinite plane current sheet embedded in a material medium.

or

$$\alpha^2 - \beta^2 = -\omega^2\mu\varepsilon \tag{5.63a}$$

$$2\alpha\beta = \omega\mu\sigma \tag{5.63b}$$

Squaring (5.63a) and (5.63b) and adding and then taking the square root, we obtain

$$\alpha^2 + \beta^2 = \omega^2\mu\varepsilon\sqrt{1 + \left(\frac{\sigma}{\omega\varepsilon}\right)^2} \tag{5.64}$$

From (5.63a) and (5.64), we then have

$$\alpha^2 = \frac{1}{2}\left[-\omega^2\mu\varepsilon + \omega^2\mu\varepsilon\sqrt{1 + \left(\frac{\sigma}{\omega\varepsilon}\right)^2}\right]$$

$$\beta^2 = \frac{1}{2}\left[\omega^2\mu\varepsilon + \omega^2\mu\varepsilon\sqrt{1 + \left(\frac{\sigma}{\omega\varepsilon}\right)^2}\right]$$

Since α and β are both positive, we finally get

$$\alpha = \frac{\omega\sqrt{\mu\varepsilon}}{\sqrt{2}}\left[\sqrt{1 + \left(\frac{\sigma}{\omega\varepsilon}\right)^2} - 1\right]^{1/2} \tag{5.65}$$

$$\beta = \frac{\omega\sqrt{\mu\varepsilon}}{\sqrt{2}}\left[\sqrt{1 + \left(\frac{\sigma}{\omega\varepsilon}\right)^2} + 1\right]^{1/2} \tag{5.66}$$

We note from (5.65) and (5.66) that α and β are both dependent on σ through the factor $\sigma/\omega\varepsilon$. This factor, known as the *loss tangent*, is the ratio of the magnitude of the conduction current density $\sigma\overline{E}_x$ to the magnitude of the displacement current density $j\omega\varepsilon\overline{E}_x$ in the material medium. In practice, the loss tangent is, however, not simply inversely proportional to ω since both σ and ε are generally functions of frequency. In fact, for many materials, the dependence of $\sigma/\omega\varepsilon$ on ω is more toward constant over wide frequency ranges.

2. The phase velocity of the wave along the direction of propagation is given by

$$v_p = \frac{\omega}{\beta} = \frac{\sqrt{2}}{\sqrt{\mu\varepsilon}}\left[\sqrt{1 + \left(\frac{\sigma}{\omega\varepsilon}\right)^2} + 1\right]^{-1/2} \tag{5.67}$$

We note that the phase velocity is dependent on the frequency of the wave. Thus, waves of different frequencies travel with different phase velocities. Consequently, for a signal comprising a band of frequencies, the different frequency components do not maintain the same phase relationships as they propagate in the medium. This phenomenon is known as *dispersion*. We discuss dispersion in detail in Chapter 9.

3. The wavelength in the medium is given by

$$\lambda = \frac{2\pi}{\beta} = \frac{\sqrt{2}}{f\sqrt{\mu\varepsilon}}\left[\sqrt{1 + \left(\frac{\sigma}{\omega\varepsilon}\right)^2} + 1\right]^{-1/2} \tag{5.68}$$

In view of the attenuation of the wave with distance, the field variation with distance is not sinusoidal. Hence, the wavelength is not exactly equal to the distance between two consecutive positive maxima as in Fig. 5.10. It is, however, still exactly equal to the distance between two alternate zero crossings.

4. The ratio of the amplitude of the electric field to the amplitude of the magnetic field is equal to $|\bar{\eta}|$, the magnitude of the complex intrinsic impedance of the medium. The electric and magnetic fields are out of phase by τ, the phase angle of the intrinsic impedance. In terms of the phasor or complex field components, we have

$$\frac{\bar{E}_x}{\bar{H}_y} = \begin{cases} \bar{\eta} & \text{for the } (+) \text{ wave} \\ -\bar{\eta} & \text{for the } (-) \text{ wave} \end{cases} \tag{5.69}$$

5. From (5.48) and (5.55), we note that

$$\bar{\gamma}\,\bar{\eta} = j\omega\mu \tag{5.70a}$$

$$\frac{\bar{\gamma}}{\bar{\eta}} = \sigma + j\omega\varepsilon \tag{5.70b}$$

so that

$$\sigma = \text{Re}\left(\frac{\bar{\gamma}}{\bar{\eta}}\right) \tag{5.71a}$$

$$\varepsilon = \frac{1}{\omega} \text{Im}\left(\frac{\bar{\gamma}}{\bar{\eta}}\right) \tag{5.71b}$$

$$\mu = \frac{1}{j\omega}\,\bar{\gamma}\,\bar{\eta} \tag{5.71c}$$

Using (5.71a)–(5.71c), we can compute the material parameters $\sigma, \varepsilon,$ and μ from a knowledge of the propagation parameters $\bar{\gamma}$ and $\bar{\eta}$ at the frequency of interest.

6. To obtain the electromagnetic field due to a nonsinusoidal source, it is necessary to consider its frequency components and apply superposition, since waves of different frequencies are attenuated by different amounts and travel with different phase velocities. The nonsinusoidal signal changes shape as it propagates in the material medium, unlike in the case of free space.

We shall now consider an example of the computation of $\bar{\gamma}$ and $\bar{\eta}$ given σ, $\varepsilon, \mu,$ and f.

Example 5.5

The material parameters of a certain food item are given by $\sigma = 2.17$ S/m, $\varepsilon = 47\varepsilon_0$, and $\mu = \mu_0$ at the operating frequency $f = 2.45$ GHz of a microwave oven. We wish to find the propagation parameters $\alpha, \beta, \lambda, v_p,$ and $\bar{\eta}$.

Although explicit expressions for α and β in terms of ω, σ, ε, and μ are given by (5.65) and (5.66), it is instructive to compute their values by using complex algrebra in conjunction with the expression for $\bar{\gamma}$ given by (5.48). Thus, we have

$$\bar{\gamma} = \sqrt{j\omega\mu(\sigma + j\omega\varepsilon)}$$

$$= \sqrt{j\omega\mu \cdot j\omega\varepsilon\left(1 - j\frac{\sigma}{\omega\varepsilon}\right)}$$

$$= j\frac{\omega\sqrt{\varepsilon_r}}{c}\sqrt{1 - j\frac{\sigma}{\omega\varepsilon_r\varepsilon_0}}$$

$$= j\frac{2\pi \times 2.45 \times 10^9 \times \sqrt{47}}{3 \times 10^8}\sqrt{1 - j\frac{2.17 \times 36\pi}{2\pi \times 2.45 \times 10^9 \times 47 \times 10^{-9}}}$$

$$= j351.782\sqrt{1 - j0.3392}$$

$$= j351.782\sqrt{1.0560\underline{/-18.7369°}}$$

$$= 351.782\underline{/90°} \times 1.0276\underline{/-9.3685°}$$

$$= 361.4912\underline{/80.6315°}$$

$$= 58.85 + j356.67$$

so that

$$\alpha = 58.85 \text{ Np/m}$$

$$\beta = 356.67 \text{ rad/m}$$

$$\lambda = \frac{2\pi}{\beta} = 0.0176 \text{ m}$$

$$v_p = \frac{\omega}{\beta} = 0.4316 \times 10^8 \text{ m/s}$$

Proceeding in a similar manner with (5.55), we obtain

$$\bar{\eta} = \sqrt{\frac{j\omega\mu}{\sigma + j\omega\varepsilon}}$$

$$= \sqrt{\frac{j\omega\mu}{j\omega\varepsilon[1 - j(\sigma/\omega\varepsilon)]}}$$

$$= \frac{\eta_0}{\sqrt{\varepsilon_r}}\frac{1}{\sqrt{1 - j(\sigma/\omega\varepsilon)}}$$

$$= \frac{120\pi}{\sqrt{47}}\frac{1}{\sqrt{1 - j0.3392}}$$

$$= \frac{54.9898}{1.0276\underline{/-9.3685°}}$$

$$= 53.51\underline{/9.37°} \; \Omega$$

K5.3. Material medium; Sinusoidal waves; Material parameters; Propagation parameters; Attenuation and phase constants; Complex propagation constant; Complex intrinsic impedance; Nonsinusoidal source; Superposition.

D5.8. Compute the propagation constant and intrinsic impedance for the following cases: **(a)** $\sigma = 10^{-5}$ S/m, $\varepsilon = 5\varepsilon_0$, $\mu = \mu_0$, and $f = 10^5$ Hz; and **(b)** $\sigma = 4$ S/m, $\varepsilon = 80\varepsilon_0$, $\mu = \mu_0$, and $f = 10^9$ Hz.
Ans. **(a)** $(0.00083 + j0.00476)$ m^{-1}, $163.54\underline{/9.9°}$ Ω; **(b)** $(77.84 + j202.86)$ m^{-1}, $36.34\underline{/20.99°}$ Ω.

D5.9. For a uniform plane wave of frequency 10^6 Hz propagating in a nonmagnetic $(\mu = \mu_0)$ material medium, the propagation constant is known to be $(0.05 + j0.1)$ m^{-1}. Find the following: **(a)** the distance in which the fields are attenuated by e^{-1}; **(b)** the distance in which the fields undergo a change of phase by 1 rad; **(c)** the distance in which a constant phase of the wave travels in 1 μs; **(d)** the ratio of the amplitudes of the electric and magnetic fields; and **(e)** the phase difference between the electric and magnetic fields.
Ans. **(a)** 20 m; **(b)** 10 m; **(c)** 62.83 m; **(d)** 70.62 Ω; **(e)** 0.1476π.

5.4 UNIFORM PLANE WAVES IN DIELECTRICS AND CONDUCTORS

In the preceding section, we discussed uniform plane electromagnetic wave propagation in a material medium for the general case. In this section, we consider special cases as follows:

Case 1: Perfect dielectrics. Perfect dielectrics are characterized by $\sigma = 0$. Then

$$\bar{\gamma} = \sqrt{j\omega\mu \cdot j\omega\varepsilon} = j\omega\sqrt{\mu\varepsilon} \tag{5.72}$$

is purely imaginary, so that

$$\alpha = 0 \tag{5.73a}$$

$$\beta = \omega\sqrt{\mu\varepsilon} \tag{5.73b}$$

$$v_p = \frac{\omega}{\beta} = \frac{1}{\sqrt{\mu\varepsilon}} \tag{5.73c}$$

$$\lambda = \frac{2\pi}{\beta} = \frac{1}{f\sqrt{\mu\varepsilon}} \tag{5.73d}$$

Further,

$$\bar{\eta} = \sqrt{\frac{j\omega\mu}{j\omega\varepsilon}} = \sqrt{\frac{\mu}{\varepsilon}} \tag{5.74}$$

is purely real. Thus, the waves propagate without attenuation and with the electric and magnetic fields in phase, as in free space but with ε_0 replaced by ε and μ_0 replaced by μ. In terms of the relative permittivity ε_r and the relative permeability μ_r

of the perfect dielectric medium, the propagation parameters are

$$\beta = \beta_0 \sqrt{\mu_r \varepsilon_r} \tag{5.75a}$$

$$v_p = \frac{c}{\sqrt{\mu_r \varepsilon_r}} \tag{5.75b}$$

$$\lambda = \frac{\lambda_0}{\sqrt{\mu_r \varepsilon_r}} \tag{5.75c}$$

$$\eta = \eta_0 \sqrt{\frac{\mu_r}{\varepsilon_r}} \tag{5.75d}$$

where the quantities with subscripts "0" refer to free space.

Since $\sigma = 0, \mathbf{J}_c = \sigma \mathbf{E} = \mathbf{0}$. Thus, there cannot be any conduction current in a perfect dielectric, which in turn rules out any accumulation of free charge on the surface of a perfect dielectric. Hence, in applying the boundary conditions (3.60a)–(3.60d) to an interface between two perfect dielectric media, we set ρ_S and $\mathbf{J}_S$ equal to zero, thereby obtaining

Boundary conditions at interface between perfect dielectrics

$$\mathbf{a}_n \times (\mathbf{E}_1 - \mathbf{E}_2) = \mathbf{0} \tag{5.76a}$$

$$\mathbf{a}_n \times (\mathbf{H}_1 - \mathbf{H}_2) = \mathbf{0} \tag{5.76b}$$

$$\mathbf{a}_n \cdot (\mathbf{D}_1 - \mathbf{D}_2) = 0 \tag{5.76c}$$

$$\mathbf{a}_n \cdot (\mathbf{B}_1 - \mathbf{B}_2) = 0 \tag{5.76d}$$

These boundary conditions tell us that the tangential components of $\mathbf{E}$ and $\mathbf{H}$, and the normal components of $\mathbf{D}$ and $\mathbf{B}$ are continuous at the boundary.

Case 2: Imperfect dielectrics. Imperfect dielectrics are characterized by $\sigma \neq 0$, but $\sigma/\omega\varepsilon \ll 1$. Recalling that $\sigma \overline{E}_x$ is the conduction current density and $\omega\varepsilon \overline{E}_x$ is the displacement current density, we note that this condition is equivalent to stating that the magnitude of the conduction current density is small compared to the magnitude of the displacement current density. Using the binomial expansion

$$(1 + x)^n = 1 + nx + \frac{n(n-1)}{2!} x^2 + \cdots$$

we can then write

$$\overline{\gamma} = \sqrt{j\omega\mu(\sigma + j\omega\varepsilon)}$$

$$= \sqrt{j\omega\mu \cdot j\omega\varepsilon \left(1 - j\frac{\sigma}{\omega\varepsilon}\right)} = j\omega\sqrt{\mu\varepsilon} \left(1 - j\frac{\sigma}{\omega\varepsilon}\right)^{1/2} \tag{5.77}$$

$$\approx \frac{\sigma}{2} \sqrt{\frac{\mu}{\varepsilon}} \left(1 - \frac{\sigma^2}{8\omega^2\varepsilon^2}\right) + j\omega\sqrt{\mu\varepsilon} \left(1 + \frac{\sigma^2}{8\omega^2\varepsilon^2}\right)$$

so that

$$\alpha \approx \frac{\sigma}{2}\sqrt{\frac{\mu}{\varepsilon}}\left(1 - \frac{\sigma^2}{8\omega^2\varepsilon^2}\right) \tag{5.78a}$$

$$\beta \approx \omega\sqrt{\mu\varepsilon}\left(1 + \frac{\sigma^2}{8\omega^2\varepsilon^2}\right) \tag{5.78b}$$

$$v_p = \frac{\omega}{\beta} \approx \frac{1}{\sqrt{\mu\varepsilon}}\left(1 - \frac{\sigma^2}{8\omega^2\varepsilon^2}\right) \tag{5.78c}$$

$$\lambda = \frac{2\pi}{\beta} \approx \frac{1}{f\sqrt{\mu\varepsilon}}\left(1 - \frac{\sigma^2}{8\omega^2\varepsilon^2}\right) \tag{5.78d}$$

Further,

$$\overline{\eta} = \sqrt{\frac{j\omega\mu}{\sigma + j\omega\varepsilon}}$$

$$= \sqrt{\frac{j\omega\mu}{j\omega\varepsilon}}\left(1 - j\frac{\sigma}{\omega\varepsilon}\right)^{-1/2}$$

so that

$$\overline{\eta} \approx \sqrt{\frac{\mu}{\varepsilon}}\left[\left(1 - \frac{3}{8}\frac{\sigma^2}{\omega^2\varepsilon^2}\right) + j\frac{\sigma}{2\omega\varepsilon}\right] \tag{5.79}$$

In (5.77)–(5.79), we have retained all terms up to and including the second power in $\sigma/\omega\varepsilon$ and have neglected all higher-order terms, since $\sigma/\omega\varepsilon \ll 1$. For a value of $\sigma/\omega\varepsilon$ equal to 0.1, the quantities β, v_p, and λ are different from those for the corresponding perfect dielectric case by a factor of only 1/800, whereas the intrinsic impedance has a real part differing from the intrinsic impedance of the perfect dielectric medium by a factor of 3/800 and an imaginary part, which is 1/20 of the intrinsic impedance of the perfect dielectric medium. Thus, for all practical purposes, the only significant feature different from the perfect dielectric case is the attenuation.

Case 3: Good conductors. Good conductors are characterized by $\sigma/\omega\varepsilon \gg 1$, just the opposite of imperfect dielectrics. This condition is equivalent to stating that the magnitude of the conduction current density is large compared to the magnitude of the displacement current density. Then

$$\overline{\gamma} = \sqrt{j\omega\mu(\sigma + j\omega\varepsilon)}$$

$$\approx \sqrt{j\omega\mu\sigma}$$

$$= \sqrt{\omega\mu\sigma}\, e^{j\pi/4} \tag{5.80}$$

$$= \sqrt{\pi f\mu\sigma}\,(1 + j)$$

so that

$$\alpha \approx \sqrt{\pi f \mu \sigma} \tag{5.81a}$$

$$\beta \approx \sqrt{\pi f \mu \sigma} \tag{5.81b}$$

$$v_p = \frac{\omega}{\beta} \approx \sqrt{\frac{4\pi f}{\mu \sigma}} \tag{5.81c}$$

$$\lambda = \frac{2\pi}{\beta} \approx \sqrt{\frac{4\pi}{f \mu \sigma}} \tag{5.81d}$$

Further,

$$\bar{\eta} = \sqrt{\frac{j\omega\mu}{\sigma + j\omega\varepsilon}}$$

$$\approx \sqrt{\frac{j\omega\mu}{\sigma}}$$

or

$$\bar{\eta} \approx \sqrt{\frac{\omega\mu}{\sigma}} e^{j\pi/4}$$

$$= \sqrt{\frac{\pi f \mu}{\sigma}} (1 + j) \tag{5.82}$$

We note that α, β, v_p, and $\bar{\eta}$ are proportional to $\sqrt{f}$, provided that σ and μ are constants. This behavior is much different from the imperfect dielectric case.

Skin effect

To discuss the propagation characteristics of a wave inside a good conductor, let us consider the case of copper. The constants for copper are $\sigma = 5.80 \times 10^7$ S/m, $\varepsilon = \varepsilon_0$, and $\mu = \mu_0$. Hence, the frequency at which σ is equal to $\omega\varepsilon$ for copper is equal to $5.8 \times 10^7/2\pi\varepsilon_0$, or 1.04×10^{18} Hz. Thus, at frequencies of even several gigahertz, copper behaves like an excellent conductor. To obtain an idea of the attenuation of the wave inside the conductor, we note that the attenuation undergone in a distance of one wavelength is equal to $e^{-\alpha\lambda}$ or $e^{-2\pi}$. In terms of decibels, this is equal to $20 \log_{10} e^{2\pi} = 54.58$ dB. In fact, the field is attenuated by a factor e^{-1}, or 0.368, in a distance equal to $1/\alpha$. This distance is known as the *skin depth* and is denoted by the symbol δ. From (5.81a), we obtain

$$\delta = \frac{1}{\sqrt{\pi f \mu \sigma}} \tag{5.83}$$

The skin depth for copper is equal to

$$\frac{1}{\sqrt{\pi f \times 4\pi \times 10^{-7} \times 5.8 \times 10^7}} = \frac{0.066}{\sqrt{f}} \text{ m}$$

Thus, in copper, the fields are attenuated by a factor e^{-1} in a distance of 0.066 mm even at the low frequency of 1 MHz, thereby resulting in the concentration of the fields near to the skin of the conductor. This phenomenon is known as the *skin effect*. It also explains *shielding* by conductors.

Underwater communica- tion
To discuss further the characteristics of wave propagation in a good conductor, we note that the ratio of the wavelength in the conducting medium to the wavelength in a dielectric medium having the same ε and μ as those of the conductor is given by

$$\frac{\lambda_{\text{conductor}}}{\lambda_{\text{dielectric}}} \approx \frac{\sqrt{4\pi/f\mu\sigma}}{1/f\sqrt{\mu\varepsilon}} = \sqrt{\frac{4\pi f\varepsilon}{\sigma}} = \sqrt{\frac{2\omega\varepsilon}{\sigma}} \tag{5.84}$$

Since $\sigma/\omega\varepsilon \gg 1$, $\lambda_{\text{conductor}} \ll \lambda_{\text{dielectric}}$. For example, for seawater, $\sigma = 4$ S/m, $\varepsilon = 80\varepsilon_0$, and $\mu = \mu_0$, so that the ratio of the two wavelengths for $f = 25$ kHz ($\sigma/\omega\varepsilon = 36,000$) is equal to 0.00745. Thus, for $f = 25$ kHz, the wavelength in seawater is 1/134 of the wavelength in a dielectric having the same ε and μ as those of seawater and a still smaller fraction of the wavelength in free space. Furthermore, the lower the frequency, the smaller is this fraction. Since it is the electrical length, that is, the length in terms of the wavelength, instead of the physical length that determines the radiation characteristics of an antenna, this means that antennas of much shorter length can be used in seawater than in free space. Together with the property that $\alpha \propto \sqrt{f}$, this illustrates that the lower the frequency, the more suitable it is for underwater communication.

For a given frequency, the higher the value of σ, the greater is the value of the attenuation constant, the smaller is the value of the skin depth, and hence the less deep the waves can penetrate. For example, in the heating of malignant tissues (hyperthermia) by RF (radio-frequency) radiation, the waves penetrate much deeper into fat (low water content) than into muscle (high water content).[1]

Equation (5.82) tells us that the intrinsic impedance of a good conductor has a phase angle of 45°. Hence, the electric and magnetic fields in the medium are out of phase by 45°. The magnitude of the intrinsic impedance is given by

$$|\bar{\eta}| = \left|(1 + j)\sqrt{\frac{\pi f\mu}{\sigma}}\right| = \sqrt{\frac{2\pi f\mu}{\sigma}} \tag{5.85}$$

As a numerical example, for copper, this quantity is equal to

$$\sqrt{\frac{2\pi f \times 4\pi \times 10^{-7}}{5.8 \times 10^7}} = 3.69 \times 10^{-7}\sqrt{f}\ \Omega$$

Thus, the intrinsic impedance of copper has as low a magnitude as 0.369 Ω even at a frequency of 10^{12} Hz. In fact, by recognizing that

$$|\bar{\eta}| = \sqrt{\frac{2\pi f\mu}{\sigma}} = \sqrt{\frac{\omega\varepsilon}{\sigma}}\sqrt{\frac{\mu}{\varepsilon}} \tag{5.86}$$

---

[1]F. Sterzer et al., "RF Therapy for Malignancy," *IEEE Spectrum,* December 1980, pp. 32–37.

we note that the magnitude of the intrinsic impedance of a good conductor medium is a small fraction of the intrinsic impedance of a dielectric medium having the same ε and μ. It follows that for the same electric field, the magnetic field inside a good conductor is much larger than the magnetic field inside a dielectric having the same ε and μ as those of the conductor.

Case 4: Perfect conductors. Perfect conductors are idealizations of good conductors in the limit that $\sigma \to \infty$. From (5.83), we note that the skin depth is equal to zero, and, hence, there is no penetration of fields into the material. Thus, no time-varying fields can exist inside a perfect conductor. In view of this, the boundary conditions on a perfect conductor surface are obtained by setting the fields with subscript 2 in (3.60a)–(3.60d) equal to zero. Thus, we obtain

Boundary conditions on a perfect conductor surface

$$\mathbf{a}_n \times \mathbf{E} = \mathbf{0} \tag{5.87a}$$

$$\mathbf{a}_n \times \mathbf{H} = \mathbf{J}_S \tag{5.87b}$$

$$\mathbf{a}_n \cdot \mathbf{D} = \rho_S \tag{5.87c}$$

$$\mathbf{a}_n \cdot \mathbf{B} = 0 \tag{5.87d}$$

where we have also omitted subscripts 1, so that $\mathbf{E}, \mathbf{H}, \mathbf{D}$, and $\mathbf{B}$ are the fields on the perfect conductor surface. The boundary conditions (5.87a) and (5.87d) tell us that on a perfect conductor surface, the tangential component of the electric field intensity and the normal component of the magnetic field intensity are zero. Hence, the electric field must be completely normal, and the magnetic field must be completely tangential to the surface. The remaining two boundary conditions (5.87c) and (5.87b) tell us that the (normal) displacement flux density is equal to the surface charge density and the (tangential) magnetic field intensity is equal in magnitude to the surface current density.

Summarizing the discussion of the special cases, we observe that as σ varies from 0 to ∞, a material is classified as a perfect dielectric for $\sigma = 0$, an imperfect dielectric for $\sigma \neq 0$ but $\ll \omega\varepsilon$, a good conductor for $\sigma \gg \omega\varepsilon$, and finally a perfect conductor in the limit that $\sigma \to \infty$. This implies that a material of nonzero σ behaves as an imperfect dielectric for $f \gg f_q$ but as a good conductor for $f \ll f_q$, where f_q, the transition frequency, is equal to $\sigma/2\pi\varepsilon$. In practice, however, the situation is not so simple because, as was already mentioned in Section 5.3, σ and ε are in general functions of frequency.

K5.4. Perfect dielectric; Imperfect dielectric; Good conductor; Conduction current versus displacement current; Skin effect; Perfect conductor; Boundary conditions.

D5.10. For a nonmagnetic ($\mu = \mu_0$) perfect dielectric material, find the relative permittivity for each of the following cases: **(a)** the phase velocity in the dielectric is one-third of its value in free space; **(b)** the rate of change of phase with distance at a fixed time in the dielectric for a wave of frequency f_0 is the same as the rate of change of phase with distance at a fixed time in free space for a wave of frequency $2f_0$; **(c)** for the same frequency, the wavelength in the dielectric is

two-thirds of its value in free space; and **(d)** for the same electric-field amplitude, the magnetic-field amplitude in the dielectric is four times its value in free space.
Ans.   **(a)** 9;   **(b)** 4;   **(c)** 2.25;   **(d)** 16.

D5.11.   For a uniform plane wave of frequency $f = 10^5$ Hz propagating in a good conductor medium, the fields undergo attenuation by the factor $e^{-\pi}$ in a distance of 2.5 m. Find the following: **(a)** the distance in which the fields undergo a change of phase by 2π rad for $f = 10^5$ Hz; **(b)** the distance by which a constant phase travels in 1 μs for $f = 10^5$ Hz; and **(c)** the distance by which a constant phase travels in 1 μs for $f = 10^4$ Hz, assuming the material parameters to be the same as at $f = 10^5$ Hz.
Ans.   **(a)** 5 m;   **(b)** 0.5 m;   **(c)** 0.1581 m.

D5.12.   The electric fields of uniform plane waves of the same frequency propagating in three different materials 1, 2, and 3 are given, respectively, by

(a) $\mathbf{E}_1 = E_0 e^{-0.4\pi z} \cos (2\pi \times 10^5 t - 0.4\pi z) \, \mathbf{a}_x$

(b) $\mathbf{E}_2 = E_0 e^{-2\pi \times 10^{-5} z} \cos (2\pi \times 10^5 t - 2\pi \times 10^{-3} z) \, \mathbf{a}_x$

(c) $\mathbf{E}_3 = E_0 e^{-0.004 z} \cos (2\pi \times 10^5 t - 0.01 z) \, \mathbf{a}_x$

For each material, determine if at the frequency of operation, it can be classified as an imperfect dielectric or a good conductor or neither of the two.
Ans.   **(a)** Good conductor;   **(b)** Imperfect dielectric;   **(c)** Neither.

5.5   POWER FLOW IN AN ELECTROMAGNETIC FIELD

In the preceding section, we found the solution for the electromagnetic field due to an infinite plane current sheet situated in the $z = 0$ plane. For a surface current flowing in the negative x-direction, we found the electric field on the sheet to be directed in the positive x-direction. Since the current is flowing against the force due to the electric field, a certain amount of work must be done by the source of the current to maintain the current flow on the sheet. Let us consider a rectangular area of length Δx and width Δy on the current sheet as shown in Fig. 5. 17. Since the current density is $J_{S0} \cos \omega t$, the charge crossing the width Δy in time dt is $dq = J_{S0} \Delta y \cos \omega t \, dt$. The force exerted on this charge by the electric field is given by

$$\mathbf{F} = dq \, \mathbf{E} = J_{S0} \Delta y \cos \omega t \, dt \, E_x \mathbf{a}_x \tag{5.88}$$

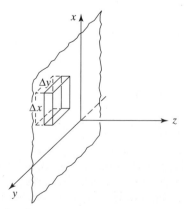

FIGURE 5.17   For the determination of power flow density associated with the electromagnetic field.

The amount of work required to be done against the electric field in displacing this charge by the distance Δx is

$$dw = F_x \, \Delta x = J_{S0} E_x \cos \omega t \, dt \, \Delta x \, \Delta y \qquad (5.89)$$

Thus, the power supplied by the source of the current in maintaining the surface current over the area $\Delta x \, \Delta y$ is

$$\frac{dw}{dt} = J_{S0} \, E_x \cos \omega t \, \Delta x \, \Delta y \qquad (5.90)$$

Recalling that E_x on the sheet is $|\bar{\eta}| (J_{S0}/2) \cos (\omega t + \tau)$, we obtain

$$\frac{dw}{dt} = |\bar{\eta}| \frac{J_{S0}^2}{2} \cos \omega t \cos (\omega t + \tau) \, \Delta x \, \Delta y \qquad (5.91)$$

We would expect the power given by (5.91) to be carried by the electromagnetic wave, half of it to either side of the current sheet. To investigate this, we note that the quantity $\mathbf{E} \times \mathbf{H}$ has the units of

$$\frac{\text{newtons}}{\text{coulomb}} \times \frac{\text{amperes}}{\text{meter}} = \frac{\text{newtons}}{\text{coulomb}} \times \frac{\text{coulomb}}{\text{second-meter}} \times \frac{\text{meter}}{\text{meter}}$$

$$= \frac{\text{newton-meters}}{\text{second}} \times \frac{1}{(\text{meter})^2} = \frac{\text{watts}}{(\text{meter})^2}$$

which represents power density. Let us then consider the rectangular box enclosing the area $\Delta x \, \Delta y$ on the current sheet and with its sides almost touching the current sheet on either side of it, as shown in Fig. 5.17. Evaluating the surface integral of $\mathbf{E} \times \mathbf{H}$ over the surface of the rectangular box, we obtain the power flow out of the box to be

$$\oint \mathbf{E} \times \mathbf{H} \cdot d\mathbf{S} = |\bar{\eta}| \frac{J_{S0}^2}{4} \cos (\omega t + \tau) \cos \omega t \, \mathbf{a}_z \cdot \Delta x \, \Delta y \, \mathbf{a}_z$$

$$+ \left[-|\bar{\eta}| \frac{J_{S0}^2}{4} \cos (\omega t + \tau) \cos \omega t \, \mathbf{a}_z \right] \cdot (-\Delta x \, \Delta y \, \mathbf{a}_z) \qquad (5.92)$$

$$= |\bar{\eta}| \frac{J_{S0}^2}{2} \cos (\omega t + \tau) \cos \omega t \, \Delta x \, \Delta y$$

This result is exactly equal to the power supplied by the current source, as given by (5.91).

We now interpret the quantity $\mathbf{E} \times \mathbf{H}$ as the power flow density vector associated with the electromagnetic field. It is known as the *Poynting vector* after J. H. Poynting and is denoted by the symbol $\mathbf{P}$. Thus,

Instantaneous Poynting vector

$$\boxed{\mathbf{P} = \mathbf{E} \times \mathbf{H}} \qquad (5.93)$$

In particular, it is the instantaneous Poynting vector, since $\mathbf{E}$ and $\mathbf{H}$ are instantaneous field vectors. Although we have here introduced the Poynting vector by considering the specific case of the electromagnetic field due to the infinite plane current sheet, the interpretation that $\oint_S \mathbf{E} \times \mathbf{H} \cdot d\mathbf{S}$ is equal to the power flow out of the closed surface S is applicable in the general case.

Let us now consider the region $z > 0$. The magnitude of the Poynting vector in this region is given by

$$
\begin{aligned}
P_z &= E_x H_y \\
&= |\bar{\eta}| \frac{J_{S0}^2}{4} e^{-2\alpha z} \cos(\omega t - \beta z + \tau) \cos(\omega t - \beta z)
\end{aligned}
\tag{5.94}
$$

The form of variation of P_z with z for $t = 0$ is shown in Fig. 5.18. If we now consider a rectangular box lying between $z = z$ and $z = z + \Delta z$ planes and having dimensions Δx and Δy in the x- and y-directions, respectively, we would in general obtain a nonzero result for the power flowing out of the box, since $\partial P_z / \partial z$ is not everywhere zero. This power flow is given by

$$
\begin{aligned}
\oint_S \mathbf{P} \cdot d\mathbf{S} &= [P_z]_{z+\Delta z} \, \Delta x \, \Delta y - [P_z]_z \, \Delta x \, \Delta y \\
&= \frac{[P_z]_{z+\Delta z} - [P_z]_z}{\Delta z} \, \Delta x \, \Delta y \, \Delta z \\
&= \frac{\partial P_z}{\partial z} \, \Delta v
\end{aligned}
\tag{5.95}
$$

where Δv is the volume of the box. Letting P_z equal $E_x H_y$ and using (5.44a) and (5.44b), we obtain

$$
\begin{aligned}
\oint_S \mathbf{P} \cdot d\mathbf{S} &= \frac{\partial}{\partial z} [E_x H_y] \, \Delta v \\
&= \left(E_x \frac{\partial H_y}{\partial z} + H_y \frac{\partial E_x}{\partial z} \right) \Delta v \\
&= \left[E_x \left(-\sigma E_x - \varepsilon \frac{\partial E_x}{\partial t} \right) + H_y \left(-\mu \frac{\partial H_y}{\partial t} \right) \right] \Delta v \\
&= -\sigma E_x^2 \, \Delta v - \frac{\partial}{\partial t} \left(\frac{1}{2} \varepsilon E_x^2 \, \Delta v \right) - \frac{\partial}{\partial t} \left(\frac{1}{2} \mu H_y^2 \, \Delta v \right)
\end{aligned}
\tag{5.96}
$$

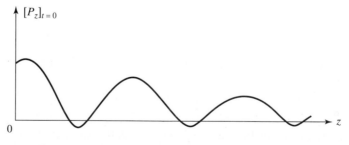

FIGURE 5.18   For the discussion of power flow associated with the electromagnetic field.

If we ignore the negative signs associated with the terms on the right side of (5.96), then each of these terms represents power flow into the box. Since the property of attenuation is associated with the parameter σ, the quantity $\sigma E_x^2 \, \Delta v$ is the power dissipated in the box due to flow of conduction current. This can also be seen by considering volume charge of density ρ moving with velocity $\mathbf{v}$ in the volume Δv under the influence of electric field $\mathbf{E}$. The volume charge experiences a force $(\rho \, \Delta v)\mathbf{E}$ and, hence, the work done in displacing it by a distance $d\mathbf{l}$ in time dt is $(\rho \, \Delta v)\mathbf{E} \cdot d\mathbf{l}$, or $(\rho \, \Delta v)\mathbf{E} \cdot \mathbf{v} \, dt$. The time rate at which the work is done or the power supplied by the field is $(\rho \, \Delta v)\mathbf{E} \cdot \mathbf{v}$, or $\mathbf{E} \cdot \mathbf{J} \, \Delta v$. In the case of conduction current, $\mathbf{J} = \sigma \mathbf{E}$ and the power supplied by the field, which is dissipated in the volume Δv, is $\sigma E^2 \, \Delta v$. Thus, the quantity σE_x^2 represents the power dissipation density (W/m^3) associated with the electromagnetic field due to the conductive property of the medium. The quantities $\frac{1}{2}\varepsilon E_x^2$ and $\frac{1}{2}\mu H_y^2$ represent the electric stored energy density (J/m^3), and the magnetic stored energy density (J/m^3), respectively, associated with the electromagnetic field due to the dielectric and magnetic properties, respectively, of the medium.

Equation (5.96) is a special case of a theorem known as *Poynting's theorem*. To derive Poynting's theorem for the general case, we make use of the vector identity

Poynting's theorem

$$\nabla \cdot (\mathbf{E} \times \mathbf{H}) = \mathbf{H} \cdot \nabla \times \mathbf{E} - \mathbf{E} \cdot \nabla \times \mathbf{H} \tag{5.97}$$

and Maxwell's curl equations

$$\nabla \times \mathbf{E} = -\frac{\partial \mathbf{B}}{\partial t} = -\mu \frac{\partial \mathbf{H}}{\partial t}$$

$$\nabla \times \mathbf{H} = \mathbf{J}_c + \frac{\partial \mathbf{D}}{\partial t} = \sigma \mathbf{E} + \varepsilon \frac{\partial \mathbf{E}}{\partial t}$$

to obtain

$$
\begin{aligned}
\nabla \cdot (\mathbf{E} \times \mathbf{H}) &= \mathbf{H} \cdot \left(-\mu \frac{\partial \mathbf{H}}{\partial t} \right) - \mathbf{E} \cdot \left(\sigma \mathbf{E} + \varepsilon \frac{\partial \mathbf{E}}{\partial t} \right) \\
&= -\mu \mathbf{H} \cdot \frac{\partial \mathbf{H}}{\partial t} - \sigma \mathbf{E} \cdot \mathbf{E} - \varepsilon \mathbf{E} \cdot \frac{\partial \mathbf{E}}{\partial t} \\
&= -\mu \frac{\partial}{\partial t} \left(\frac{1}{2} \mathbf{H} \cdot \mathbf{H} \right) - \sigma \mathbf{E} \cdot \mathbf{E} - \varepsilon \frac{\partial}{\partial t} \left(\frac{1}{2} \mathbf{E} \cdot \mathbf{E} \right) \\
&= -\sigma E^2 - \frac{\partial}{\partial t} \left(\frac{1}{2} \varepsilon E^2 \right) - \frac{\partial}{\partial t} \left(\frac{1}{2} \mu H^2 \right)
\end{aligned} \tag{5.98}
$$

Substituting $\mathbf{P}$ for $\mathbf{E} \times \mathbf{H}$ and taking the volume integral of both sides of (5.98) over the volume V, we obtain

$$
\begin{aligned}
\int_V (\nabla \cdot \mathbf{P}) \, dv = &-\int_V \sigma E^2 \, dv - \int_V \frac{\partial}{\partial t} \left(\frac{1}{2} \varepsilon E^2 \right) dv \\
&- \int_V \frac{\partial}{\partial t} \left(\frac{1}{2} \mu H^2 \right) dv
\end{aligned} \tag{5.99}
$$

Interchanging the differentiation operation with time and integration operation over volume in the second and third terms on the right side and replacing the volume integral on the left side by a closed surface integral in accordance with the divergence theorem, we get

$$\oint_S \mathbf{P} \cdot d\mathbf{S} = -\int_V \sigma E^2 \, dv - \frac{\partial}{\partial t} \int_V \frac{1}{2} \varepsilon E^2 \, dv$$
$$- \frac{\partial}{\partial t} \int_V \frac{1}{2} \mu H^2 \, dv \tag{5.100}$$

where S is the surface bounding the volume V. Equation (5.100) is the Poynting theorem for the general case. Since it should hold for any size V, it follows that the power dissipation density, the electric stored energy density, and the magnetic stored energy density are given by

$$p_d = \sigma E^2 \tag{5.101a}$$

$$w_e = \frac{1}{2} \varepsilon E^2 \tag{5.101b}$$

$$w_m = \frac{1}{2} \mu H^2 \tag{5.101c}$$

respectively.

Time-average power flow
Returning now to Fig. 5.18, we note that there are certain intervals in z for which P_z is negative, although the wave propagation is in the $+z$-direction. This is because of the phase difference between the electric and magnetic fields. There is no inconsistency here since the plot corresponds to only one value of time, namely, $t = 0$. On the other hand, the time-average value of P_z is positive everywhere, as we shall show now. The time-average value of P_z, denoted $\langle P_z \rangle$, is P_z averaged over one period of the sinusoidal time variation of the source; that is,

$$\langle P_z \rangle = \frac{1}{T} \int_0^T P_z(t) \, dt \tag{5.102}$$

where $T \, (= 1/f)$ is the period. From (5.94), we have

$$\langle P_z \rangle = \left\langle |\overline{\eta}| \frac{J_{S0}^2}{4} e^{-2\alpha z} \cos(\omega t - \beta z + \tau) \cos(\omega t - \beta z) \right\rangle$$

$$= |\overline{\eta}| \frac{J_{S0}^2}{8} e^{-2\alpha z} \langle \cos \tau + \cos(2\omega t - 2\beta z + \tau) \rangle \tag{5.103}$$

$$= |\overline{\eta}| \frac{J_{S0}^2}{8} e^{-2\alpha z} [\langle \cos \tau \rangle + \langle \cos(2\omega t - 2\beta z + \tau) \rangle]$$

Since $\cos \tau$ is independent of time, $\langle \cos \tau \rangle$ is equal to $\cos \tau$. The quantity $\langle \cos(2\omega t - 2\beta z + \tau) \rangle$ is equal to zero since the integral of a cosine or sine function over each period is zero. Thus, (5.103) reduces to

$$\langle P_z \rangle = |\overline{\eta}| \frac{J_{S0}^2}{8} e^{-2\alpha z} \cos \tau \tag{5.104}$$

which is everywhere positive.

Example 5.6

Let us consider the electric field of a uniform plane wave propagating in seawater ($\sigma = 4$ S/m, $\varepsilon = 80\varepsilon_0$, and $\mu = \mu_0$) in the positive z-direction and having the electric field

$$\mathbf{E} = 1 \cos 5 \times 10^4 \pi t \, \mathbf{a}_x \text{ V/m}$$

at $z = 0$. We wish to find the instantaneous power flow per unit area normal to the z-direction as a function of z and the time-average power flow per unit area normal to the z-direction as a function of z.

From the expression for $\mathbf{E}$, we note that the frequency of the wave is 25 kHz. At this frequency in seawater, the propagation parameters can be computed to be $\alpha = \beta \approx 0.628$ and $\overline{\eta} = 0.222\underline{/45°}$. The expressions for the instantaneous electric and magnetic fields are therefore given by

$$\mathbf{E} = 1e^{-0.628z} \cos (5 \times 10^4 \pi t - 0.628z) \, \mathbf{a}_x \text{ V/m}$$

$$\mathbf{H} = 4.502e^{-0.628z} \cos (5 \times 10^4 \pi t - 0.628z - \pi/4) \, \mathbf{a}_y \text{ A/m}$$

The instantaneous Poynting vector is then given by

$$\mathbf{P} = \mathbf{E} \times \mathbf{H}$$
$$= 4.502e^{-1.256z} \cos (5 \times 10^4 \pi t - 0.628z)$$
$$\cdot \cos (5 \times 10^4 \pi t - 0.628z - \pi/4) \, \mathbf{a}_z \text{ W/m}^2$$

Thus, the instantaneous power flow per unit area normal to the z-direction, which is simply the z-component of the instantaneous Poynting vector, is

$$P_z = 2.251e^{-1.256z} [\cos \pi/4 + \cos (10^5 \pi t - 1.256z - \pi/4)] \text{ W/m}^2$$

Finally, the time-average power flow per unit area normal to the z-direction is

$$\langle P_z \rangle = 2.251e^{-1.256z} \cos \pi/4$$
$$= 1.592e^{-1.256z} \text{ W/m}^2$$

Returning now to (5.104), we note that $\langle P_z \rangle$ can be expressed in the manner   *Time-average Poynting vector*

$$\langle P_z \rangle = |\overline{\eta}| \frac{J_{S0}^2}{8} e^{-2\alpha z} \cos \tau$$

$$= \frac{1}{2} \text{Re} \left[\left(\frac{|\overline{\eta}| J_{S0}}{2} e^{-\alpha z} e^{j(-\beta z + \tau)} \right) \left(\frac{J_{S0}}{2} e^{-\alpha z} e^{j\beta z} \right) \right] \tag{5.105}$$

$$= \text{Re}[\tfrac{1}{2} \overline{E}_x \overline{H}_y^*]$$

where $\overline{E}_x$ and $\overline{H}_y$ are the phasor electric and magnetic field components, respectively. In terms of vector quantities, we have

$$\langle \mathbf{P} \rangle = \text{Re}[\tfrac{1}{2}\overline{E}_x \mathbf{a}_x \times (\overline{H}_y\, \mathbf{a}_y)^*]$$
$$= \text{Re}[\tfrac{1}{2}\overline{\mathbf{E}} \times \overline{\mathbf{H}}^*] \tag{5.106}$$
$$= \text{Re}[\overline{\mathbf{P}}]$$

which is the general form of the time-average Poynting vector, where

$$\overline{\mathbf{P}} = \tfrac{1}{2}\overline{\mathbf{E}} \times \overline{\mathbf{H}}^* \tag{5.107}$$

is the complex Poynting vector.

K5.5.   Power flow; Poynting vector; Poynting's theorem; Power dissipation density; Electric stored energy density; Magnetic stored energy density; Time-average power flow; Time-average Poynting vector.

D5.13.   The magnetic field associated with a uniform plane wave propagating in the $+z$-direction in a nonmagnetic ($\mu = \mu_0$) material medium is given by

$$\mathbf{H} = H_0 e^{-z} \cos\left(6\pi \times 10^7 t - \sqrt{3}z\right) \mathbf{a}_y \text{ A/m}$$

Find the following: **(a)** the instantaneous power flow across a surface of area 1 m^2 in the $z = 0$ plane at $t = 0$; **(b)** the time-average power flow across a surface of area 1 m^2 in the $z = 0$ plane; and **(c)** the time-average power flow across a surface of area 1 m^2 in the $z = 1$ m plane.
Ans.   **(a)** $102.57 H_0^2 \text{ W}$;   **(b)** $51.28 H_0^2 \text{ W}$;   **(c)** $6.94 H_0^2 \text{ W}$.

D5.14.   Find the time-average values of the following: **(a)** $A \sin \omega t \sin 3\omega t$; **(b)** $A(\cos^2 \omega t - 0.5 \sin^2 2\omega t)$; and **(c)** $A \sin^6 \omega t$.
Ans.   **(a)** 0;   **(b)** $0.25A$;   **(c)** $0.3125A$.

5.6   REFLECTION OF UNIFORM PLANE WAVES

Thus far, we have considered uniform plane wave propagation in unbounded media. Practical situations are characterized by propagation involving several different media. When a wave is incident on a boundary between two different media, a reflected wave is produced. In addition, if the second medium is not a perfect conductor, a transmitted wave is set up. Together, these waves satisfy the boundary conditions at the interface between the two media. In this section, we shall consider these phenomena for waves incident normally on plane boundaries.

Normal incidence on a plane interface

To do this, let us consider the situation shown in Fig. 5.19 in which steady-state conditions are established by uniform plane waves of radian frequency ω propagating normal to the plane interface $z = 0$ between two media characterized by two different sets of values of $\sigma, \varepsilon,$ and μ, where $\sigma \neq \infty$. We shall assume that a $(+)$ wave is incident from medium 1 ($z < 0$) onto the interface, thereby setting up a reflected $(-)$ wave in that medium, and a transmitted $(+)$ wave in

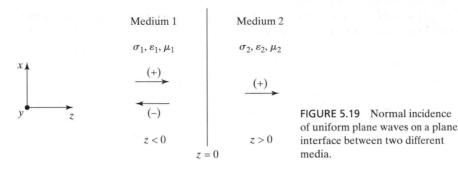

FIGURE 5.19   Normal incidence of uniform plane waves on a plane interface between two different media.

medium 2 ($z > 0$). For convenience, we shall work with the phasor or complex field components. Thus, considering the electric fields to be in the x-direction and the magnetic fields to be in the y-direction, we can write the solution for the complex field components in medium 1 to be

$$\overline{E}_{1x}(z) = \overline{E}_1^+ e^{-\overline{\gamma}_1 z} + \overline{E}_1^- e^{\overline{\gamma}_1 z} \tag{5.108a}$$

$$\overline{H}_{1y}(z) = \overline{H}_1^+ e^{-\overline{\gamma}_1 z} + \overline{H}_1^- e^{\overline{\gamma}_1 z}$$

$$= \frac{1}{\overline{\eta}_1}(\overline{F}_1^+ e^{-\overline{\gamma}_1 z} - \overline{E}_1^- e^{\overline{\gamma}_1 z}) \tag{5.108b}$$

where $\overline{E}_1^+, \overline{E}_1^-, \overline{H}_1^+$, and $\overline{H}_1^-$ are the incident and reflected wave electric and magnetic field components, respectively, at $z = 0-$ in medium 1 and

$$\overline{\gamma}_1 = \sqrt{j\omega\mu_1(\sigma_1 + j\omega\varepsilon_1)} \tag{5.109a}$$

$$\overline{\eta}_1 = \sqrt{\frac{j\omega\mu_1}{\sigma_1 + j\omega\varepsilon_1}} \tag{5.109b}$$

Recall that the real field corresponding to a complex field component is obtained by multiplying the complex field component by $e^{j\omega t}$ and taking the real part of the product. The complex field components in medium 2 are given by

$$E_{2x}(z) = \overline{E}_2^+ e^{-\overline{\gamma}_2 z} \tag{5.110a}$$

$$\overline{H}_{2y}(z) = \overline{H}_2^+ e^{-\overline{\gamma}_2 z}$$

$$= \frac{\overline{E}_2^+}{\overline{\eta}_2} e^{-\overline{\gamma}_2 z} \tag{5.110b}$$

where $\overline{E}_2^+$ and $\overline{H}_2^+$ are the transmitted wave electric- and magnetic-field components at $z = 0+$ in medium 2 and

$$\overline{\gamma}_2 = \sqrt{j\omega\mu_2(\sigma_2 + j\omega\varepsilon_2)} \tag{5.111a}$$

$$\overline{\eta}_2 = \sqrt{\frac{j\omega\mu_2}{\sigma_2 + j\omega\varepsilon_2}} \tag{5.111b}$$

To satisfy the boundary conditions at $z = 0$, we note that (1) the components of both electric and magnetic fields are entirely tangential to the interface and (2) in view of the finite conductivities of the media, no surface current exists on the interface (currents flow in the volumes of the media). Hence, from the phasor forms of the boundary conditions (3.61a) and (3.61b), we have

$$[\bar{E}_{1x}]_{z=0} = [\bar{E}_{2x}]_{z=0} \tag{5.112a}$$

$$[\bar{H}_{1y}]_{z=0} = [\bar{H}_{2y}]_{z=0} \tag{5.112b}$$

Applying these to the solution pairs given by (5.108a,b) and (5.110a,b), we have

$$\bar{E}_1^+ + \bar{E}_1^- = \bar{E}_2^+ \tag{5.113a}$$

$$\frac{1}{\eta_1}(\bar{E}_1^+ - \bar{E}_1^-) = \frac{1}{\eta_2}\bar{E}_2^+ \tag{5.113b}$$

We now define the *reflection coefficient* at the boundary, denoted by the symbol $\bar{\Gamma}$, to be the ratio of the reflected wave electric field at the boundary to the incident wave electric field at the boundary. From (5.113a) and (5.113b), we obtain

$$\boxed{\bar{\Gamma} = \frac{\bar{E}_1^-}{\bar{E}_1^+} = \frac{\eta_2 - \eta_1}{\eta_2 + \eta_1}} \tag{5.114}$$

Note that the ratio of the reflected wave magnetic field at the boundary to the incident wave magnetic field at the boundary is given by

$$\frac{\bar{H}_1^-}{\bar{H}_1^+} = \frac{-\bar{E}_1^-/\eta_1}{\bar{E}_1^+/\eta_1} = -\frac{\bar{E}_1^-}{\bar{E}_1^+} = -\bar{\Gamma} \tag{5.115}$$

The ratio of the transmitted wave electric field at the boundary to the incident wave electric field at the boundary, known as the *transmission coefficient* and denoted by the symbol $\bar{\tau}$, is given by

$$\boxed{\bar{\tau} = \frac{\bar{E}_2^+}{\bar{E}_1^+} = \frac{\bar{E}_1^+ + \bar{E}_1^-}{\bar{E}_1^+} = 1 + \bar{\Gamma}} \tag{5.116}$$

where we have used (5.113a). The ratio of the transmitted wave magnetic field at the boundary to the incident wave magnetic field at the boundary is given by

$$\frac{\bar{H}_2^+}{\bar{H}_1^+} = \frac{\bar{H}_1^+ + \bar{H}_1^-}{\bar{H}_1^+} = 1 - \bar{\Gamma} \tag{5.117}$$

The reflection and transmission coefficients given by (5.114) and (5.116), respectively, enable us to find the reflected and transmitted wave fields for a given incident wave field. We observe the following properties of $\bar{\Gamma}$ and $\bar{\tau}$:

1. For $\overline{\eta}_2 = \overline{\eta}_1$, $\overline{\Gamma} = 0$ and $\overline{\tau} = 1$. The incident wave is entirely transmitted. The situation then corresponds to a "matched" condition. A trivial case occurs when the two media have identical values of the material parameters.

2. For $\sigma_1 = \sigma_2 = 0$, that is, when both media are perfect dielectrics, $\overline{\eta}_1$ and $\overline{\eta}_2$ are real. Hence, $\overline{\Gamma}$ and $\overline{\tau}$ are real. In particular, if the two media have the same permeability μ but different permittivities ε_1 and ε_2, then

$$\overline{\Gamma} = \frac{\sqrt{\mu/\varepsilon_2} - \sqrt{\mu/\varepsilon_1}}{\sqrt{\mu/\varepsilon_2} + \sqrt{\mu/\varepsilon_1}} \tag{5.118}$$

$$= \frac{1 - \sqrt{\varepsilon_2/\varepsilon_1}}{1 + \sqrt{\varepsilon_2/\varepsilon_1}}$$

$$\overline{\tau} = \frac{2}{1 + \sqrt{\varepsilon_2/\varepsilon_1}} \tag{5.119}$$

3. For $\sigma_2 \to \infty$, $\overline{\eta}_2 \to 0$, $\overline{\Gamma} \to -1$, and $\overline{\tau} \to 0$. Thus, if medium 2 is a perfect conductor, the incident wave is entirely reflected, as it should be since there cannot be any time-varying fields inside a perfect conductor. The superposition of the reflected and incident waves would then give rise to the so-called complete standing waves in medium 1. We shall discuss complete standing waves as well as partial standing waves when we study the topic of sinusoidal steady-state analysis of waves on transmission lines in Chapter 7.

Example 5.7

Region 1 ($z < 0$) is free space, whereas region 2 ($z > 0$) is a material medium characterized by $\sigma = 10^{-4}$ S/m, $\varepsilon = 5\varepsilon_0$, and $\mu = \mu_0$. For a uniform plane wave having the electric field

$$\mathbf{E}_i = E_0 \cos (3\pi \times 10^5 t - 10^3 \pi z)\, \mathbf{a}_x \text{ V/m}$$

incident on the interface $z = 0$ from region 1, we wish to obtain the expressions for the reflected and transmitted wave electric and magnetic fields.

From computation as in Example 5.5 for $\sigma = 10^{-4}$ S/m, $\varepsilon = 5\varepsilon_0$, $\mu = \mu_0$, and $f = (3\pi \times 10^5)/2\pi = 1.5 \times 10^5$ Hz,

$$\overline{\gamma} = (6.283 + j9.425) \times 10^{-3}$$

$$\overline{\eta} = 104.559\underline{/33.69°} = 104.559\underline{/0.1872\pi}$$

Then

$$\overline{\Gamma} = \frac{\overline{\eta} - \eta_0}{\overline{\eta} + \eta_0} = \frac{104.559\underline{/33.69°} - 377}{104.559\underline{/33.69°} + 377}$$

$$= 0.6325\underline{/161.565°} = 0.6325\underline{/0.8976\pi}$$

$$\overline{\tau} = 1 + \overline{\Gamma} = 1 + 0.6325\underline{/161.565°}$$

$$= 0.4472\underline{/26.565°} = 0.4472\underline{/0.1476\pi}$$

Thus, the reflected and transmitted wave electric and magnetic fields are given by

$$\mathbf{E}_r = 0.6325E_0 \cos(3\pi \times 10^5 t + 10^{-3}\pi z + 0.8976\pi)\,\mathbf{a}_x\,V/m$$

$$\mathbf{H}_r = -\frac{0.6325E_0}{377}\cos(3\pi \times 10^5 t + 10^{-3}\pi z + 0.8976\pi)\,\mathbf{a}_y\,A/m$$

$$= -1.678 \times 10^{-3}E_0 \cos(3\pi \times 10^5 t + 10^{-3}\pi z + 0.8976\pi)\,\mathbf{a}_y\,A/m$$

$$\mathbf{E}_t = 0.4472E_0 e^{-6.283\times 10^{-3}z}$$
$$\cdot \cos(3\pi \times 10^5 t - 9.425 \times 10^{-3}z + 0.1476\pi)\,\mathbf{a}_x\,V/m$$

$$\mathbf{H}_t = \frac{0.4472E_0}{104.559}e^{-6.283\times 10^{-3}z}$$
$$\cdot \cos(3\pi \times 10^5 t - 9.425 \times 10^{-3}z + 0.1476\pi - 0.1872\pi)\,\mathbf{a}_y\,A/m$$

$$= 4.277 \times 10^{-3}E_0 e^{-6.283\times 10^{-3}z}$$
$$\cdot \cos(3\pi \times 10^5 t - 9.425 \times 10^{-3}z - 0.0396\pi)\,\mathbf{a}_y\,A/m$$

Note that at $z = 0$, the boundary conditions of $\mathbf{E}_i + \mathbf{E}_r = \mathbf{E}_t$ and $\mathbf{H}_i + \mathbf{H}_r = \mathbf{H}_t$ are satisfied, since

$$E_0 + 0.6325E_0 \cos 0.8976\pi = 0.4472E_0 \cos 0.1476\pi$$

and

$$\frac{E_0}{377} - 1.678 \times 10^3 E_0 \cos 0.8976\pi = 4.277 \times 10^3 E_0 \cos(-0.0396\pi)$$

K5.6. Plane interface between two material media; Normal incidence of uniform plane waves; Reflection; Transmission; Reflection and transmission coefficients.

D5.15. For each of the following cases of uniform plane waves of frequency $f = 1$ MHz incident normally from medium 1 $(z < 0)$ onto the interface $(z = 0)$ with medium 2 $(z > 0)$, find the values of $\overline{\Gamma}$ and $\overline{\tau}$: **(a)** Medium 1 is free space and the parameters of medium 2 are $\sigma = 10^{-3}$ S/m, $\varepsilon = 6\varepsilon_0$, and $\mu = \mu_0$; and **(b)** the parameters of medium 1 are $\sigma = 4$ S/m, $\varepsilon = 80\varepsilon_0$, and $\mu = \mu_0$, and the parameters of medium 2 are $\sigma = 10^{-3}$ S/m, $\varepsilon = 80\varepsilon_0$, and $\mu = \mu_0$.
Ans.   **(a)** $0.6909\underline{/64.177°}$, $0.3846\underline{/29.331°}$;   **(b)** $0.9486\underline{/-2.4155°}$, $1.948\underline{/-1.177°}$.

D5.16. The regions $z < 0$ and $z > 0$ are nonmagnetic $(\mu = \mu_0)$ perfect dielectrics of permittivities ε_1 and ε_2, respectively. For a uniform plane wave incident from the region $z < 0$ normally onto the boundary $z = 0$, find $\varepsilon_2/\varepsilon_1$ for each of the following to hold at $z = 0$: **(a)** the electric field of the reflected wave is $-1/3$ times the electric field of the incident wave; **(b)** the electric field of the transmitted wave is 0.4 times the electric field of the incident wave; and **(c)** the electric field of the transmitted wave is six times the electric field of the reflected wave.
Ans.   **(a)** 4;   **(b)** 16;   **(c)** 4/9.

5.7   SUMMARY

In this chapter, we studied the principles of uniform plane waves. Uniform plane waves are a building block in the study of electromagnetic wave propagation. They are the simplest type of solutions resulting from the coupling of the electric and magnetic fields in Maxwell's curl equations. Their electric and magnetic fields are perpendicular to each other and to the direction of propagation. The fields are *uniform* in the *planes* perpendicular to the direction of propagation.

We first obtained the uniform plane wave solution to Maxwell's equations in time domain in free space by considering an infinite plane current sheet in the xy-plane with uniform surface current density given by

$$\mathbf{J}_S = -J_S(t)\mathbf{a}_x \text{ A/m}$$

and deriving the electromagnetic field due to the current sheet to be

$$\mathbf{E} = \frac{\eta_0}{2} J_S\left(t \mp \frac{z}{v_p}\right)\mathbf{a}_x \qquad \text{for} \quad z \gtrless 0 \qquad (5.120a)$$

$$\mathbf{H} = \pm\frac{1}{2} J_S\left(t \mp \frac{z}{v_p}\right)\mathbf{a}_y \qquad \text{for} \quad z \gtrless 0 \qquad (5.120b)$$

where

$$v_p = \frac{1}{\sqrt{\mu_0\varepsilon_0}}$$

and

$$\eta_0 = \sqrt{\frac{\mu_0}{\varepsilon_0}}$$

are the velocity of propagation and intrinsic impedance, respectively. In (5.120a) and (5.120b), the arguments $(t - z/v_p)$ and $(t + z/v_p)$ represent wave motion in the positive z-direction and the negative z-direction, respectively, with the velocity v_p. Thus, (5.120a) and (5.120b) correspond to waves propagating away from the current sheet to either side of it. Since the fields are uniform in constant-z planes, they represent uniform plane waves. We discussed how to plot the variations of the field components versus t for fixed values of z and versus z for fixed values of t, for a given function $J_S(t)$.

We then extended the solution to sinusoidally time varying uniform plane waves by considering the current density on the infinite plane sheet to be

$$\mathbf{J}_S = -J_{S0}\cos\omega t\,\mathbf{a}_x \text{ A/m}$$

and obtaining the corresponding field to be

$$\mathbf{E} = \frac{\eta_0 J_{S0}}{2}\cos(\omega t \mp \beta z)\,\mathbf{a}_x \qquad \text{for} \quad z \gtrless 0 \qquad (5.121a)$$

$$\mathbf{H} = \pm\frac{J_{S0}}{2}\cos(\omega t \mp \beta z)\,\mathbf{a}_y \qquad \text{for} \quad z \gtrless 0 \qquad (5.121b)$$

where

$$\beta = \frac{\omega}{v_p} = \omega\sqrt{\mu_0\varepsilon_0} \qquad (5.122)$$

We discussed several important parameters and properties associated with these waves, including polarization. The quantity β is the phase constant, that is, the magnitude of the rate of change of phase with distance along the direction of propagation, for a fixed time. The velocity v_p, which from (5.122) is given by

$$v_p = \frac{\omega}{\beta} \qquad (5.123)$$

is known as the phase velocity, because it is the velocity with which a particular constant phase progresses along the direction of propagation. The wavelength λ, that is, the distance along the direction of propagation in which the phase changes by 2π radians, for a fixed time, is given by

$$\lambda = \frac{2\pi}{\beta} \qquad (5.124)$$

The wavelength is related to the frequency f in a simple manner as given by

$$v_p = \lambda f \qquad (5.125)$$

which follows from (5.123) and (5.124) and is a result of the fact that the time and space variations of the fields are interdependent. We discussed the principle of antenna array and the Doppler effect.

Next we extended the treatment of uniform plane waves to a material medium. Starting with the Maxwell's equations for a material medium given by

$$\nabla \times \mathbf{E} = -\frac{\partial \mathbf{B}}{\partial t} = -\mu \frac{\partial \mathbf{H}}{\partial t}$$

$$\nabla \times \mathbf{H} = \mathbf{J}_c + \frac{\partial \mathbf{D}}{\partial t} = \sigma \mathbf{E} + \varepsilon \frac{\partial \mathbf{E}}{\partial t}$$

and using the phasor technique, we considered the infinite plane current sheet of uniform surface current density

$$\mathbf{J}_S = -J_{S0} \cos \omega t \, \mathbf{a}_x \text{ A/m}$$

in the xy-plane and embedded in the material medium and obtained the electromagnetic field due to it to be

$$\mathbf{E} = \frac{|\overline{\eta}| J_{S0}}{2} e^{\mp \alpha z} \cos (\omega t \mp \beta z + \tau) \, \mathbf{a}_x \qquad \text{for} \quad z \gtrless 0 \qquad (5.126a)$$

$$\mathbf{H} = \pm \frac{J_{S0}}{2} e^{\mp \alpha z} \cos (\omega t \mp \beta z) \, \mathbf{a}_y \qquad \text{for} \quad z \gtrless 0 \qquad (5.126b)$$

In (5.126a, b), α and β are the attenuation and phase constants given, respectively, by the real and imaginary parts of the propagation constant, $\overline{\gamma}$. Thus,

$$\overline{\gamma} = \alpha + j\beta = \sqrt{j\omega\mu(\sigma + j\omega\varepsilon)}$$

The quantities $|\overline{\eta}|$ and τ are the magnitude and phase angle, respectively, of the intrinsic impedance, $\overline{\eta}$, of the medium. Thus,

$$\overline{\eta} = |\overline{\eta}| \, e^{j\tau} = \sqrt{\frac{j\omega\mu}{\sigma + j\omega\varepsilon}}$$

The solution given by (5.126a) and (5.126b) tells us that the wave propagation in the material medium is characterized by attenuation, as indicated by $e^{\mp\alpha z}$ and phase difference between **E** and **H** by the amount τ. We also learned that these properties as well as the phase velocity are frequency-dependent.

Having discussed uniform plane wave propagation for the general case of a medium characterized by σ, ε, and μ, we then considered several special cases. These are summarized in the following:

Perfect dielectrics.   For these materials, $\sigma = 0$. Wave propagation occurs without attenuation as in free space but with the propagation parameters governed by ε and μ instead of ε_0 and μ_0, respectively. The tangential components of **E** and **H** and the normal components of **D** and **B** are continuous at a boundary between two perfect dielectrics.

Imperfect dielectrics.   A material is classified as an imperfect dielectric for $\sigma \ll \omega\varepsilon$, that is, conduction current density small in magnitude compared to the displacement current density. The only significant feature of wave propagation in an imperfect dielectric as compared to that in a perfect dielectric is the attenuation undergone by the wave.

Good conductors.   A material is classified as a good conductor for $\sigma \gg \omega\varepsilon$, that is, conduction current density large in magnitude compared to the displacement current density. Wave propagation in a good conductor medium is characterized by attenuation and phase constants both equal to $\sqrt{\pi f \mu \sigma}$. Thus for large values of f and/or σ, the fields do not penetrate very deep into the conductor. This phenomenon is known as the skin effect. From considerations of the frequency dependence of the attenuation and wavelength for a fixed σ, we learned that low frequencies are more suitable for communication with underwater objects. We also learned that the intrinsic impedance of a good conductor medium is very low in magnitude compared to that of a dielectric medium having the same ε and μ.

Perfect conductors.   These are idealizations of good conductors in the limit $\sigma \to \infty$. For $\sigma = \infty$, the skin depth, that is, the distance in which the fields inside a conductor are attenuated by a factor e^{-1}, is zero. Hence, there can be no penetration of fields into a perfect conductor, so that on a perfect conductor surface, the tangential component of **E** and the normal component of **B** are zero, the normal component of **D** is equal to the surface charge density, and the tangential component of **H** is equal in magnitude to the surface current density.

We then learned that there is power flow, power dissipation, and energy storage associated with the wave propagation. The power flow density is given by the Poynting vector

$$\mathbf{P} = \mathbf{E} \times \mathbf{H}$$

The power dissipation density and the electric and magnetic stored energy densities are given, respectively, by

$$p_d = \sigma E^2$$
$$w_e = \tfrac{1}{2} \varepsilon E^2$$
$$w_m = \tfrac{1}{2} \mu H^2$$

The power flow out of a closed surface S plus the power dissipated in the volume V bounded by S is always equal to the sum of the time rates of decrease of electric and magnetic stored energies in the volume V, as given by Poynting's theorem

$$\oint_S \mathbf{P} \cdot d\mathbf{S} = -\int_V \sigma E^2 \, dv - \frac{\partial}{\partial t} \int_V \frac{1}{2} \varepsilon E^2 \, dv - \frac{\partial}{\partial t} \int_V \frac{1}{2} \mu H^2 \, dv$$

Finally, we considered uniform plane waves incident normally onto a plane boundary between two material media and learned how to compute the reflected and transmitted wave fields for a given incident wave field.

REVIEW QUESTIONS

Q5.1. What is a uniform plane wave? Why is the study of uniform plane waves important?

Q5.2. Outline the procedure for obtaining from the two Maxwell's curl equations the particular differential equation for the special case of $\mathbf{J} = J_x(z)\mathbf{a}_x$.

Q5.3. State the wave equation for the case of $\mathbf{E} = E_x(z, t)\mathbf{a}_x$. Describe the procedure for its solution.

Q5.4. Discuss by means of an example how a function $f(t - z\sqrt{\mu_0 \varepsilon_0})$ represents a traveling wave propagating in the positive z-direction.

Q5.5. Discuss by means of an example how a function $g(t + z\sqrt{\mu_0 \varepsilon_0})$ represents a traveling wave propagating in the negative z-direction.

Q5.6. What is the significance of the intrinsic impedance of free space? What is its value?

Q5.7. Summarize the procedure for obtaining the solution for the electromagnetic field due to the infinite plane sheet of uniform time-varying current density.

Q5.8. State and discuss the solution for the electromagnetic field due to the infinite plane sheet of current density $\mathbf{J}_S(t) = -J_S(t)\mathbf{a}_x$ for $z = 0$.

Q5.9. Discuss the parameters ω, β, and v_p associated with sinusoidally time-varying uniform plane waves.

Q5.10. Define wavelength. What is the relationship among wavelength, frequency, and phase velocity?

Q5.11. Discuss the classification of waves according to frequency, giving examples of their application in the different frequency ranges.

Q5.12. How is the direction of propagation of a uniform plane wave related to the directions of its fields?

Q5.13. Discuss right-handed and left-handed circular polarizations associated with sinusoidally time-varying uniform plane waves.

Q5.14. Discuss the principle of an antenna array with the aid of an example.

Q5.15. What is Doppler effect? How do you compute the Doppler shift in the frequency of a wave as viewed by an observer moving in an arbitrary direction?

Q5.16. Give some examples of the application of Doppler effect.

Q5.17. Discuss how the determination of the electromagnetic field due to an infinite plane current sheet of sinusoidally time-varying current density imbedded in a material medium is made convenient by using the phasor technique.

Q5.18. What is the propagation constant for a material medium? Discuss the significance of its real and imaginary parts.

Q5.19. What is the intrinsic impedance for a material medium? What is the consequence of its complex nature?

Q5.20. What is loss tangent? Discuss its significance.

Q5.21. Discuss the consequence of the frequency dependence of the phase velocity of a wave in a material medium.

Q5.22. How would you obtain the electromagnetic field due to a current sheet of non-sinusoidally time-varying current density embedded in a material medium?

Q5.23. What is the condition for a medium to be a perfect dielectric? How do the characteristics of wave propagation in a perfect dielectric medium differ from those of wave propagation in free space?

Q5.24. State the boundary conditions at the interface between two perfect dielectrics.

Q5.25. What is the criterion for a material to be an imperfect dielectric? What is the significant feature of wave propagation in an imperfect dielectric as compared to that in a perfect dielectric?

Q5.26. What is the criterion for a material to be a good conductor? Give two examples of materials that behave as good conductors for frequencies of up to several gigahertz.

Q5.27. What is skin effect? Discuss skin depth, giving some numerical values.

Q5.28. Why are low-frequency waves more suitable than high-frequency waves for communication with underwater objects?

Q5.29. Discuss the consequence of the low intrinsic impedance of a good conductor as compared to that of a dielectric medium having the same ε and μ.

Q5.30. Why can there be no fields inside a perfect conductor? What are the boundary conditions at the surface of a perfect conductor?

Q5.31. What is the Poynting vector? What is the physical interpretation of the Poynting vector over a closed surface?

Q5.32. State Poynting's theorem. How is it derived from Maxwell's curl equations?

Q5.33. What are the power dissipation density, the electric stored energy density, and the magnetic stored energy density associated with an electromagnetic field in a material medium?

Q5.34. What is time-average power flow? Discuss the time-average power flow versus instantaneous power flow associated with a uniform plane wave in a material medium.

Q5.35. Discuss the determination of the reflected and transitted wave fields from the fields of a wave incident normally onto a plane boundary between two material media.

Q5.36. What is the consequence of a wave incident on a perfect conductor?

PROBLEMS

Section 5.1

P5.1. From Maxwell's curl equations, obtain the particular differential equations for the case of $\mathbf{J} = J_z(y,t)\mathbf{a}_z$.

P5.2. For each of the following functions, plot the value of the function versus z for the two specified values of time and discuss the traveling-wave nature of the function.

(a) $e^{-|t-z|}; t = 0, t = 1\text{ s}$

(b) $(2 \times 10^8 t + z)[u(2 \times 10^8 t + z) - u(2 \times 10^8 t + z - 3)]; t = 0, t = 10^{-8}\text{ s}$

P5.3. Write expressions for traveling-wave functions corresponding to the following cases: **(a)** time variation at $x = 0$ in the manner $10u(t)$ and propagating in the $-x$-direction with velocity 0.5 m/s; **(b)** time variation at $y = 0$ in the manner $t \sin 20t$ and propagating in the $+y$-direction with velocity 4 m/s; and **(c)** distance variation at $t = 0$ in the manner $5z^3 e^{-z^2}$ and propagating in the $-z$-direction with velocity 2 m/s.

P5.4. The variation with z for $t = 0$ of a function $f(z,t)$ representing a traveling wave propagating in the $+z$-direction with velocity 100 m/s is shown in Fig. 5.20. Find and sketch: **(a)** f versus z for $t = 1$ s; **(b)** f versus t for $z = 0$; and **(c)** f versus t for $z = 200$ m.

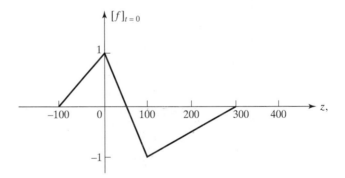

FIGURE 5.20   For Problems P5.4 and P5.5.

P5.5. Repeat Problem P5.4 if the function f represents a traveling wave propagating in the $-z$-direction with velocity 100 m/s.

P5.6. An infinite plane sheet lying in the $z = 0$ plane in free space carries a surface current of density $\mathbf{J}_S = -J_S(t)\mathbf{a}_x$, where $J_S(t)$ is as shown in Fig. 5.21. Find and sketch **(a)** E_x versus t in the $z = 300$ m plane; **(b)** E_x versus z for $t = 2\ \mu\text{s}$; and **(c)** H_y versus z for $t = 4\ \mu\text{s}$.

FIGURE 5.21   For Problem P5.6.

P5.7. An infinite plane sheet of current density $\mathbf{J}_S = -J_S(t)\mathbf{a}_x$ A/m where $J_S(t)$ is as shown in Fig. 5.22, lies in the $z = 0$ plane in free space. Find and sketch: **(a)** E_x versus t in the $z = 300$ m plane; **(b)** H_y versus t in the $z = -600$ m plane; **(c)** E_x versus z for $t = 3$ μs; and **(d)** H_y versus z for $t = 4$ μs.

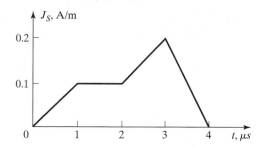

FIGURE 5.22   For Problem P5.7.

P5.8. The time variation of the electric-field intensity E_x in the $z = 300$ m plane of a uniform plane wave propagating away from an infinite plane current sheet of current density $\mathbf{J}_S(t) = -J_S(t)\mathbf{a}_x$ lying in the $z = 0$ plane in free space is given by the periodic function shown in Fig. 5.23. Find and sketch **(a)** J_S versus t; **(b)** E_x versus t in $z = -600$ m plane; **(c)** E_x versus z for $t = 2$ μs; and **(d)** H_y versus z for $t = 3$ μs.

FIGURE 5.23   For Problem P5.8.

Section 5.2

P5.9. The electric-field intensity of a uniform plane wave propagating in free space is given by

$$\mathbf{E} = 37.7 \cos (9\pi \times 10^7 t + 0.3\pi y)\, \mathbf{a}_x \text{ V/m}$$

Find: **(a)** the frequency; **(b)** the wavelength; **(c)** the direction of propagation of the wave; and **(d)** the associated magnetic-field intensity vector $\mathbf{H}$.

P5.10. Write the expressions for the electric- and magnetic-field intensities of a sinusoidally time-varying uniform plane wave propagating in free spece and having the following characteristics: **(a)** $f = 100$ MHz; **(b)** direction of propagation is the $+z$-direction; and **(c)** polarization is right circular with the electric field in the $z = 0$ plane at $t = 0$ having an x-component equal to E_0 and a y-component equal to $0.75E_0$.

P5.11. Given $\mathbf{J}_S = 0.2(\sqrt{3}\mathbf{a}_x + \mathbf{a}_y) \cos 6\pi \times 10^9 t$ A/m in the $z = 0$ plane in free space, find $\mathbf{E}$ and $\mathbf{H}$ for $z \geq 0$. Use the following three steps, which are generalizations of the solution to the electromagnetic field due to the infinite plane current sheet in the $z = 0$ plane:

1. Write the expression for **H** on the sheet and on either side of it, by noting that $[\mathbf{H}]_{z=0\pm} = \frac{1}{2}\mathbf{J}_S \times (\pm\mathbf{a}_z) = \frac{1}{2}\mathbf{J}_S \times \mathbf{a}_n$, where $\mathbf{a}_n$ is the unit vector normal to the sheet and directed toward the side of interest.

2. Extend the result of step 1 to write the expression for **H** everywhere, that is, for $z \gtrless 0$, considering the traveling wave character of the fields.

3. Write the solution for **E** everywhere by noting that **(a)** the amplitude of $\mathbf{E} = \eta_0 \times$ the amplitude of **H**, and **(b)** the direction of **E**, the direction of **H**, and the direction of propagation constitute a right-handed orthogonal set, so that $\mathbf{E} = \eta_0\mathbf{H} \times \mathbf{a}_n$.

P5.12.   Given $\mathbf{J}_S = 0.2 \sin 15\pi \times 10^7 t\,\mathbf{a}_y$ A/m in the $x = 0$ plane in free space, find **E** and **H** for $x \gtrless 0$. Use the three steps outlined in Problem P5.11, except that the current sheet is in the $x = 0$ plane.

P5.13.   For each of the following fields, determine if the polarization is right- or left-circular or elliptical.

(a) $E_0 \cos(\omega t - \beta y)\,\mathbf{a}_z + E_0 \sin(\omega t - \beta y)\,\mathbf{a}_x$

(b) $E_0 \cos(\omega t + \beta x)\,\mathbf{a}_y + E_0 \sin(\omega t + \beta x)\,\mathbf{a}_z$

(c) $E_0 \cos(\omega t + \beta y)\,\mathbf{a}_x - 2E_0 \sin(\omega t + \beta y)\,\mathbf{a}_z$

(d) $E_0 \cos(\omega t - \beta x)\,\mathbf{a}_z - E_0 \sin(\omega t - \beta x + \pi/4)\,\mathbf{a}_y$

P5.14.   Express each of the following uniform plane wave electric fields as superposition of right- and left-circularly polarized fields:

(a) $E_0\mathbf{a}_x \cos(\omega t + \beta z)$

(b) $E_0\mathbf{a}_x \cos(\omega t - \beta z + \pi/3) - E_0\mathbf{a}_y \cos(\omega t - \beta z + \pi/6)$

P5.15.   The current densities of two infinite, plane, parallel current sheets are given by

$$\mathbf{J}_{S1} = -J_{S0} \cos \omega t\,\mathbf{a}_x \qquad \text{in the } z = 0 \text{ plane}$$

$$\mathbf{J}_{S2} = -kJ_{S0} \cos \omega t\,\mathbf{a}_x \qquad \text{in the } z = \lambda/2 \text{ plane}$$

Find the electric-field intensities in the three regions: **(a)** $z < 0$; **(b)** $0 < z < \lambda/2$; and **(c)** $z > \lambda/2$.

P5.16.   The current densities of three infinite plane, parallel, current sheets are given by

$$\mathbf{J}_{S1} = -J_{S0} \cos \omega t\,\mathbf{a}_x \qquad \text{in the } z = 0 \text{ plane}$$

$$\mathbf{J}_{S2} = -kJ_{S0} \sin \omega t\,\mathbf{a}_x \qquad \text{in the } z = \lambda/4 \text{ plane}$$

$$\mathbf{J}_{S3} = -2kJ_{S0} \cos \omega t\,\mathbf{a}_x \qquad \text{in the } z = \lambda/2 \text{ plane}$$

Obtain the expression for the ratio of the amplitude of the electric field in the region $z > \lambda/2$ to the amplitude of the electric field in the region $z < 0$. Then find the ratio for each of the following values of k: **(a)** $k = -1$, **(b)** $k = 1/2$, and **(c)** $k = 1$. Find the value(s) of k for each of the following values of the ratio: **(a)** $1/3$ and **(b)** 3.

P5.17.   The electric-field intensity of a uniform plane wave is given by

$$\mathbf{E} = E_0 \cos(6\pi \times 10^9 t - 20\pi y)\,\mathbf{a}_x$$

Ignoring relativistic effects, find the Doppler shift in f for each of the following cases: **(a)** observer moving in the positive y-direction with velocity 10^3 m/s; **(b)** observer moving along the line $x = y = z$ in the sense of increasing x with velocity 10^3 m/s; and **(c)** observer moving along the line $x = -2y = 2z$ in the sense of increasing x with velocity 10^3 m/s.

P5.18. Consider an observer moving in free space along the curve $x = y = z^2$ in the direction of increasing z with velocity v_0. For a point source of frequency f located at the point $(0, 0, 1)$, obtain the expression for the Doppler shift in f as a function of z. Ignore relativistic effects. For what location(s) of the observer is the Doppler shift zero?

Section 5.3

P5.19. An infinite plane sheet in the $z = 0$ plane carries a surface current of density

$$\mathbf{J}_S = -0.2 \cos 2\pi \times 10^6 t \, \mathbf{a}_x \text{ A/m}$$

The medium on either side of the sheet is characterized by $\sigma = 10^{-3}$ S/m, $\varepsilon = 6\varepsilon_0$, and $\mu = \mu_0$. Find **E** and **H** on either side of the current sheet.

P5.20. Consider an array of two infinite plane, parallel, current sheets of uniform densities given by

$$\mathbf{J}_{S1} = -J_{S0} \cos 2\pi \times 10^6 t \, \mathbf{a}_x \qquad \text{in the } z = 0 \text{ plane}$$
$$\mathbf{J}_{S2} = -kJ_{S0} \sin 2\pi \times 10^6 t \, \mathbf{a}_x \qquad \text{in the } z = d \text{ plane}$$

situated in a medium characterized by $\sigma = 10^{-3}$ S/m, $\varepsilon = 6\varepsilon_0$, and $\mu = \mu_0$. **(a)** Find the minimum value of d (> 0) and the corresponding value of k for which the fields in the region $z < 0$ are zero. **(b)** For the values of d and k found in (a), obtain the electric-field intensity in the region $z > d$.

P5.21. A uniform plane wave of frequency 5×10^5 Hz propagating in a material medium has the following characteristics. **(i)** The fields are attenuated by the factor e^{-1} in a distance of 28.65 m. **(ii)** The fields undergo a change in phase by 2π in a distance of 111.2 m. **(iii)** The ratio of the amplitudes of the electric- and magnetic-field intensities at a point in the medium is 59.4. **(a)** What is the value of $\bar{\gamma}$? **(b)** What is the value of $\bar{\eta}$? **(c)** Find σ, ε, and μ of the medium.

P5.22. Repeat Problem P5.19 for the surface current of density

$$\mathbf{J}_S = -0.2 \cos 2\pi \times 10^6 t \cos 4\pi \times 10^6 t \, \mathbf{a}_x \text{ A/m}$$

P5.23. In Problem P5.19, assume that the region $z > 0$ is free space, whereas the region $z < 0$ is a material medium characterized by $\sigma = 10^{-3}$ S/m, $\varepsilon = 6\varepsilon_0$, and $\mu = \mu_0$. Find **E** and **H** on either side of the current sheet. (*Hint:* Make use of the complex electric and magnetic fields to satisfy the boundary conditions at $z = 0$.)

Section 5.4

P5.24. The electric field of a uniform plane wave propagating in a perfect dielectric medium having $\mu = \mu_0$ is given by

$$\mathbf{E} = 10 \cos (3\pi \times 10^7 t - 0.2\pi x) \, \mathbf{a}_z$$

Find: **(a)** the frequency; **(b)** the wavelength; **(c)** the phase velocity; **(d)** the relative permittivity of the medium; and **(e)** the associated magnetic-field vector **H**.

P5.25. An infinite plane sheet lying in the $z = 0$ plane carries a surface current of density $\mathbf{J}_S = -J_S(t)\mathbf{a}_x$ A/m, where $J_S(t)$ is as shown in Fig. 5.24. The medium on either side of the current sheet is a perfect dielectric of $\varepsilon = 2.25\varepsilon_0$ and $\mu = \mu_0$.

FIGURE 5.24   For Problem P5.25.

Find and sketch **(a)** E_x versus t for $z = 200$ m; **(b)** H_y versus t for $z = -300$ m; **(c)** E_x versus z for $t = 2$ μs; and **(d)** H_y versus z for $t = 3$ μs.

P5.26. For a uniform plane wave having $\mathbf{E} = E_x(z, t)\mathbf{a}_x$ and $\mathbf{H} = H_y(z, t)\mathbf{a}_y$ and propagating in the $+z$-direction in a perfect dielectric medium, the time variation of E_x in a constant z-plane and the distance variation of H_y for a fixed time are observed to be periodic, as shown in Figs. 5.25(a) and (b), respectively, for two complete cycles. Find the relative permittivity and the relative permeability of the medium.

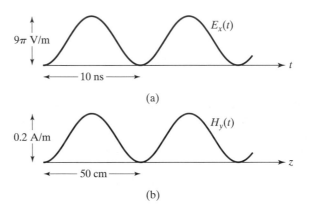

(a)

(b)

FIGURE 5.25   For Problem P5.26.

P5.27. An infinite plane sheet lying in the $z = 0$ plane carries a surface current of density

$$\mathbf{J}_S = -0.2 \cos 6\pi \times 10^8 t \, \mathbf{a}_x \text{ A/m}$$

The region $z > 0$ is a perfect dielectric of $\varepsilon = 2.25\varepsilon_0$ and $\mu = \mu_0$, whereas the region $z < 0$ is a perfect dielectric of $\varepsilon = 4\varepsilon_0$ and $\mu = \mu_0$. Find $\mathbf{E}$ and $\mathbf{H}$ on both sides of the sheet.

P5.28. For uniform plane wave propagation in ice ($\sigma \approx 10^{-6}$ S/m, $\varepsilon = 3\varepsilon_0$, and $\mu = \mu_0$), compute α, β, v_p, λ, and $\bar{\eta}$ for $f = 1$ MHz. What is the distance in which the fields are attenuated by the factor e^{-1}?

P5.29. For uniform plane wave propagation in seawater ($\sigma = 4$ S/m, $\varepsilon = 80\varepsilon_0$, and $\mu = \mu_0$), compute α, δ, β, λ, v_p, and $\bar{\eta}$ for two frequencies: **(a)** $f = 10$ GHz and **(b)** $f = 100$ kHz.

P5.30. For a uniform plane wave propagating in the $+z$-direction in a material medium, the magnetic field intensity in the $z = 0$ plane is given by

$$[\mathbf{H}]_{z=0} = 0.1 \cos^3 2\pi \times 10^8 t \, \mathbf{a}_y \text{ A/m}$$

Find $\mathbf{E}(z, t)$ for each of the following cases: **(a)** the medium is characterized by $\sigma = 0$, $\varepsilon = 9\varepsilon_0$, and $\mu = \mu_0$; **(b)** the medium is characterized by $\sigma = 10^{-3}$ S/m, $\varepsilon = 9\varepsilon_0$, and $\mu = \mu_0$; and **(c)** the medium is characterized by $\sigma = 10$ S/m, $\varepsilon = 9\varepsilon_0$, and $\mu = \mu_0$.

P5.31. Find the magnitude of the Doppler shift in f observed by an observer, moving with velocity 10 m/s along the direction of propagation of a wave propagating in seawater, for each of the following frequencies of the wave: **(a)** $f = 25$ kHz and **(b)** $f = 100$ kHz. Ignore relativistic effects.

Section 5.5

P5.32. For each of the following electric-field intensities in free space, find the instantaneous and time-average Poynting vectors:

(a) $\mathbf{E} = E_0 \cos (\omega t - \beta z) \mathbf{a}_x + 2E_0 \cos (\omega t - \beta z) \mathbf{a}_y$

(b) $\mathbf{E} = E_0 \cos (\omega t - \beta z) \mathbf{a}_x - E_0 \sin (\omega t - \beta z) \mathbf{a}_y$

(c) $\mathbf{E} = E_0 \cos (\omega t - \beta z) \mathbf{a}_x + 2E_0 \sin (\omega t - \beta z) \mathbf{a}_y$

P5.33. The electric and magnetic fields in a coaxial cable, an arrangement of two coaxial perfectly conducting cylinders of radii a and b $(> a)$ are given by

$$\mathbf{E} = \frac{V_0}{r \ln (b/a)} \cos \omega(t - z\sqrt{\mu_0\varepsilon_0}) \, \mathbf{a}_r \qquad \text{for} \quad a < r < b$$

$$\mathbf{H} = \frac{I_0}{2\pi r} \cos \omega(t - z\sqrt{\mu_0\varepsilon_0}) \, \mathbf{a}_\phi \qquad \text{for} \quad a < r < b$$

where V_0 and I_0 are constants and the axis of the cylinders is the z-axis. **(a)** Find the instantaneous and time-average Poynting vectors associated with the fields. **(b)** Find the time-average power flow along the coaxial cable.

P5.34. The electric- and magnetic-field intensities in the radiation field of an antenna located at the origin are given in spherical coordinates by

$$\mathbf{E} = E_0 \frac{\sin \theta}{r} \cos \omega(t - r\sqrt{\mu_0\varepsilon_0}) \, \mathbf{a}_\theta \text{ V/m}$$

$$\mathbf{H} = \frac{E_0}{\sqrt{\mu_0/\varepsilon_0}} \frac{\sin \theta}{r} \cos \omega(t - r\sqrt{\mu_0\varepsilon_0}) \, \mathbf{a}_\phi \text{ A/m}$$

Find: **(a)** the instantaneous Poynting vector; **(b)** the instantaneous power radiated by the antenna by evaluating the surface integral of the instantaneous Poynting vector over a spherical surface of radius r centered at the antenna and enclosing the antenna; and **(c)** the time-average power radiated by the antenna.

P5.35. The magnetic field of a uniform plane wave propagating in a nonmagnetic $(\mu - \mu_0)$ material medium is given by

$$\mathbf{H} = H_0 e^{-z} \cos (2\pi \times 10^6 t - 2z) \, \mathbf{a}_x \text{ A/m}$$

Find: **(a)** the time-average power flow per unit area normal to the z-direction and **(b)** the time-average power dissipated in the volume bounded by the planes $x = 0, x = 1, y = 0, y = 1, z = 0,$ and $z = 1$.

Section 5.6

P5.36. Region 1 $(z < 0)$ is free space, whereas region 2 $(z > 0)$ is a material medium characterized by $\sigma = 10^{-4}$ S/m, $\varepsilon = 5\varepsilon_0$, and $\mu = \mu_0$. For a uniform plane wave having the electric field

$$\mathbf{E}_i = E_0 \cos (3\pi \times 10^5 t - 10^{-3}\pi z) \, \mathbf{a}_x \text{ V/m}$$

incident on the interface $z = 0$ from region 1, obtain the expression for the reflected and transmitted wave electric fields.

P5.37. Repeat Problem P5.36 for the incident wave electric field given by

$$\mathbf{E}_i = E_0 \cos^3 (3\pi \times 10^5 t - 10^{-3}\pi z)\, \mathbf{a}_x \text{ V/m}$$

P5.38. In Fig. 5.26, medium 3 extends to infinity so that no reflected ($-$) wave exists in that medium. For a uniform plane wave having the electric field

$$\mathbf{E}_i = E_0 \cos (3 \times 10^8 \pi t - \pi z)\, \mathbf{a}_x \text{ V/m}$$

incident from medium 1 onto the interface $z = 0$, obtain the expressions for the phasor electric- and magnetic-field components in all three media.

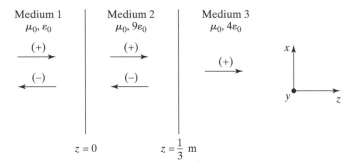

Medium 1
$\mu_0,\, \varepsilon_0$

Medium 2
$\mu_0,\, 9\varepsilon_0$

Medium 3
$\mu_0,\, 4\varepsilon_0$

$z = 0$

$z = \dfrac{1}{3}$ m

FIGURE 5.26   For Problem P5.38.

P5.39. A uniform plane wave propagating in the $+z$-direction and having the electric field $\mathbf{E}_i = E_{xi}(t)\mathbf{a}_x$, where $E_{xi}(t)$ in the $z = 0$ plane is as shown in Fig. 5.27, is incident normally from free space ($z < 0$) onto a nonmagnetic ($\mu = \mu_0$), perfect dielectric ($z > 0$) of permittivity $4\varepsilon_0$. Find and sketch the following: **(a)** E_x versus z for $t = 1\ \mu s$ and **(b)** H_y versus z for $t = 1\ \mu s$.

$[E_{xi}]_{z=0}$, V/m

E_0

0   1   2 $t, \mu s$ FIGURE 5.27   For Problem P5.39.

P5.40. The region $z < 0$ is a perfect dielectric, whereas the region $z > 0$ is a perfect conductor, as shown in Fig. 5.28. For a uniform plane wave having the electric

$z < 0$

Perfect
Dielectric

$\mu,\, \varepsilon$

$z > 0$

Perfect
Conductor

$z = 0$

FIGURE 5.28   For Problem P5.40.

and magnetic fields

$$\mathbf{E}_i = E_0 \cos(\omega t - \beta z) \mathbf{a}_x$$

$$\mathbf{H}_i = \frac{E_0}{\eta} \cos(\omega t - \beta z) \mathbf{a}_y$$

where $\beta = \omega \sqrt{\mu \varepsilon}$ and $\eta = \sqrt{\mu/\varepsilon}$, obtain the expressions for the reflected wave electric and magnetic fields and hence the expressions for the total (incident + reflected) electric and magnetic fields in the dielectric, and the current density on the surface of the perfect conductor.

REVIEW PROBLEMS

R5.1. Figures 5.29(a) and (b) show the distance variations at $t = 0$ and $t = 1$ s, respectively, of the sum of two functions $f(z, t)$ and $g(z, t)$, each of duration not exceeding 3 s, and representing traveling waves propagating in the $+z$- and $-z$-directions, respectively, with velocity 100 m/s. Find and sketch f and g versus t for $z = 0$.

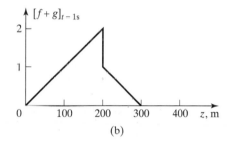

FIGURE 5.29   For Problem R5.1.

R5.2. Given $\mathbf{J}_{S1} = 0.2 \cos 6\pi \times 10^8 t \, \mathbf{a}_x$ A/m in the $y = 0$ plane and $\mathbf{J}_{S2} = 0.2 \cos 6\pi \times 10^8 t \, \mathbf{a}_z$ A/m in the $y = 0.25$ m plane, find $\mathbf{E}$ and $\mathbf{H}$ in the two regions $y < 0$ and $y > 0.25$ m, and discuss the polarizations of the fields in both regions. Note that the two current densities are directed perpendicular to each other.

R5.3. A rocket is fired from a ground station with an initial velocity v_0 m/s and making an angle α with the horizontal. Communication is maintained between the rocket and a transmitter of frequency f at the ground station. Show that the received frequency when the rocket is at its apogee is Doppler shifted down from f by the amount $(2fv_0/c)(\cos\alpha)/\sqrt{4 + \tan^2\alpha}$. Assume plane Earth and ignore relativitstic effects. For what value of α is this Doppler shift in f equal to $\frac{1}{2}(fv_0/c)$?

R5.4. The electric field of a uniform plane wave propagating in the $+z$-direction in a nonmagnetic ($\mu = \mu_0$) material medium is given by

$$\mathbf{E} = 8.4e^{-0.0432z} \cos(4\pi \times 10^6 t - 0.1829z) \, \mathbf{a}_x \text{ V/m}$$

Find the magnetic field of the wave. Further, find the values of σ and ε of the medium.

R5.5. An infinite plane current sheet of uniform density $\mathbf{J}_S = -J_S(t)\mathbf{a}_x$ is sandwiched between two perfect dielectric media, as shown in Fig. 5.30(a). If $J_S(t)$ is a

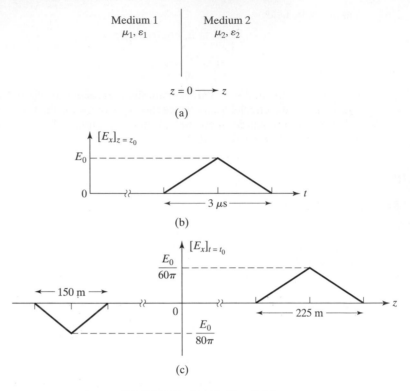

FIGURE 5.30   For Problem R5.5.

triangular pulse of duration 3 μs, the plots of $E_x(t)$ at some value of z equal to z_0 (> 0) and $H_y(z)$ for some value of t equal to t_0 (> 0) are given by Figs. 5.30(b) and (c), respectively. If $J_S(t) = J_{S0} \cos 6\pi \times 10^8 t$ A/m, instead of being a pulse, find **E** and **H** on both sides of the sheet, and the time-average power radiated by the sheet for unit area of the sheet.

R5.6.   For a sinusoidally time-varying uniform plane wave incident normally from medium 1 on to the interface $z = 0$ in Fig. 5.31, show that there is a minimum value of the frequency for which a wave at that frequency or any integer multiple of that frequency undergoes no reflection at the interface. Further, find the maximum value of the period of a nonsinusoidal periodic wave for which no reflection occurs at the interface. Note that medium 1 and medium 3 are both free space.

FIGURE 5.31   For Problem R5.6.

PART TWO

APPLIED ELEMENTS

Field and Line Essentials for Digital Electronics

After having learned the fundamental elements of engineering electromagnetics in the first five chapters, we now turn our attention to applied elements of engineering electromagnetics. In particular in this chapter, we consider topics ranging from *p-n* junction semiconductor to circuit parameters to transmission-line time-domain analysis. An understanding of these topics is considered to be particularly essential for digital electronics, while being of general importance.

We begin the chapter with two important differential equations involving the electric potential and discuss several applications based on the solution of these equations, including the analysis of a *p-n* junction semiconductor and arrangements involving two parallel conductors. We then introduce an important relationship between circuit parameters capacitance, conductance, and inductance for infinitely long, parallel conductor arrangements and consider their determination.

Next we introduce the transmission line by considering a uniform plane wave and placing two parallel plane, perfect conductors such that the fields remain unaltered by satisfying the boundary conditions on the perfect conductor surfaces. The wave is then guided between and parallel to the conductors, thus leading to the parallel-plate line. We shall learn to represent a line by the *distributed* parameter equivalent circuit and discuss wave propagation on the line in terms of voltage and current. We introduce the circuit parameters for the parallel-plate line and then discuss several techniques for the computation of line parameters. We devote the remainder of the chapter to time-domain analysis of transmission-line systems using a progressive treatment, beginning with a line terminated by a resistive load and leading to interconnections between logic gates, and culminating in crosstalk on transmission lines.

6.1   POISSON'S AND LAPLACE'S EQUATIONS

Poisson's
equation

In Section 4.5, we introduced the static electric potential as related to the static electric field in the manner

$$\mathbf{E} = -\nabla V \tag{6.1}$$

Substituting (6.1) into Maxwell's divergence equation for $\mathbf{D}$ given by

$$\nabla \cdot \mathbf{D} = \rho \tag{6.2}$$

we obtain

$$-\nabla \cdot \varepsilon \nabla V = \rho \tag{6.3}$$

where ε is the permittivity of the medium. Using the vector identity

$$\nabla \cdot \Phi \mathbf{A} = \Phi \nabla \cdot \mathbf{A} + \mathbf{A} \cdot \nabla \Phi \tag{6.4}$$

we can write (6.3) as

$$\varepsilon \nabla \cdot \nabla V + \nabla \varepsilon \cdot \nabla V = -\rho$$

or

$$\varepsilon \nabla^2 V + \nabla \varepsilon \cdot \nabla V = -\rho \tag{6.5}$$

If we assume ε to be uniform in the region of interest, then $\nabla \varepsilon = \mathbf{0}$ and (6.5) becomes

$$\nabla^2 V = \frac{-\rho}{\varepsilon} \tag{6.6}$$

This equation is known as Poisson's equation. It governs the relationship between the volume charge density ρ in a region of uniform permittivity ε to the electric scalar potential V in that region. Note that (6.6) also follows from (4.90) for $\partial/\partial t = 0$ and $\Phi = V$. In Cartesian coordinates, (6.6) becomes

$$\frac{\partial^2 V}{\partial x^2} + \frac{\partial^2 V}{\partial y^2} + \frac{\partial^2 V}{\partial z^2} = -\frac{\rho}{\varepsilon} \tag{6.7}$$

which is a three-dimensional, second-order partial differential equation. For the one-dimensional case in which V varies with x only, $\partial^2 V/\partial y^2$ and $\partial^2 V/\partial z^2$ are both equal to zero, and (6.7) reduces to

$$\frac{\partial^2 V}{\partial x^2} = \frac{d^2 V}{dx^2} = -\frac{\rho}{\varepsilon} \tag{6.8}$$

We shall illustrate the application of (6.8) by means of an example.

Example 6.1

Let us consider the space charge layer in a *p-n* junction semiconductor with zero bias, as shown in Fig. 6.1(a), in which the region $x < 0$ is doped *p*-type and the region $x > 0$ is doped *n*-type. To review briefly the formation of the space charge layer, we note that since the density of the holes on the *p* side is larger than that on the *n* side, there is a tendency

p-n junction semi-conductor

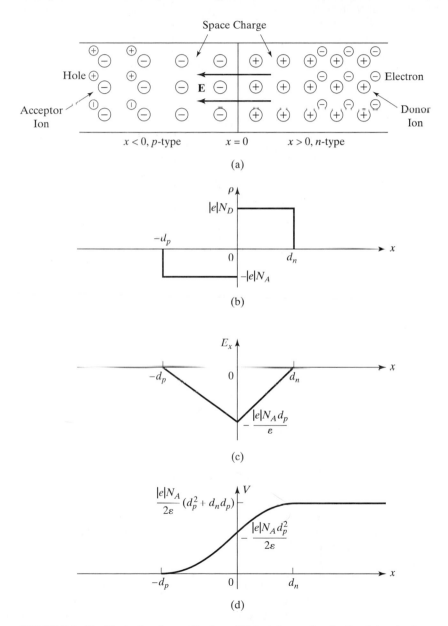

(a)

(b)

(c)

(d)

FIGURE 6.1 For illustrating the application of Poisson's equation for the determination of the potential distribution for a *p-n* junction semiconductor.

for the holes to diffuse to the n side and recombine with the electrons. Similarly, there is a tendency for the electrons on the n side to diffuse to the p side and recombine with the holes. The diffusion of holes leaves behind negatively charged acceptor atoms, and the diffusion of electrons leaves behind positively charged donor atoms. Since these acceptor and donor atoms are immobile, a space charge layer, also known as the *depletion layer*, is formed in the region of the junction with negative charges on the p side and positive charges on the n side. This space charge gives rise to an electric field directed from the n side of the junction to the p side so that it opposes diffusion of the mobile carriers across the junction thereby resulting in an equilibrium. For simplicity, let us consider an abrupt junction, that is, a junction in which the impurity concentration is constant on either side of the junction. Let N_A and N_D be the acceptor and donor ion concentrations, respectively, and d_p and d_n be the widths in the p and n regions, respectively, of the depletion layer. The space charge density ρ is then given by

$$\rho = \begin{cases} -|e|N_A & \text{for} \quad -d_p < x < 0 \\ |e|N_D & \text{for} \quad 0 < x < d_n \end{cases} \tag{6.9}$$

as shown in Fig. 6.1(b), where $|e|$ is the magnitude of the electronic charge. Since the semiconductor is electrically neutral, the total acceptor charge must be equal to the total donor charge; that is,

$$|e|N_A d_p = |e|N_D d_n \tag{6.10}$$

We wish to find the potential distribution in the depletion layer and the depletion layer width in terms of the potential difference across the depletion layer and the acceptor and donor ion concentrations.

Substituting (6.9) into (6.8), we obtain the equation governing the potential distribution to be

$$\frac{d^2V}{dx^2} = \begin{cases} \dfrac{|e|N_A}{\varepsilon} & \text{for} \quad -d_p < x < 0 \\ -\dfrac{|e|N_D}{\varepsilon} & \text{for} \quad 0 < x < d_n \end{cases} \tag{6.11}$$

To solve (6.11) for V, we integrate it once and obtain

$$\frac{dV}{dx} = \begin{cases} \dfrac{|e|N_A}{\varepsilon} x + C_1 & \text{for} \quad -d_p < x < 0 \\ -\dfrac{|e|N_D}{\varepsilon} x + C_2 & \text{for} \quad 0 < x < d_n \end{cases}$$

where C_1 and C_2 are constants of integration. To evaluate C_1 and C_2, we note that since $\mathbf{E} = -\nabla V = -(\partial V/\partial x)\mathbf{a}_x$, $\partial V/\partial x$ is simply equal to $-E_x$. Since the electric field lines begin on the positive charges and end on the negative charges, and in view of (6.10) the field and, hence, $\partial V/\partial x$ must vanish at $x = -d_p$ and $x = d_n$, giving us

$$\frac{dV}{dx} = \begin{cases} \dfrac{|e|N_A}{\varepsilon} (x + d_p) & \text{for} \quad -d_p < x < 0 \\ -\dfrac{|e|N_D}{\varepsilon} (x - d_n) & \text{for} \quad 0 < x < d_n \end{cases} \tag{6.12}$$

The field intensity, that is, $-dV/dx$, may now be sketched as a function of x as shown in Fig. 6.1(c).

Proceeding further, we integrate (6.12) and obtain

$$V = \begin{cases} \dfrac{|e|N_A}{2\varepsilon}(x + d_p)^2 + C_3 & \text{for} \quad -d_p < x < 0 \\[3mm] -\dfrac{|e|N_D}{2\varepsilon}(x - d_n)^2 + C_4 & \text{for} \quad 0 < x < d_n \end{cases}$$

where C_3 and C_4 are constants of integration. To evaluate C_3 and C_4, we first set the potential at $x = -d_p$ arbitrarily equal to zero to obtain C_3 equal to zero. Then we make use of the condition that the potential be continuous at $x = 0$, since the discontinuity in dV/dx at $x = 0$ is finite, to obtain

$$\frac{|e|N_A}{2\varepsilon}d_p^2 = -\frac{|e|N_D}{2\varepsilon}d_n^2 + C_4$$

or

$$C_4 = \frac{|e|}{2\varepsilon}(N_A d_p^2 + N_D d_n^2)$$

Substituting this value for C_4 and setting C_3 equal to zero in the expression for V, we get the required solution

$$V = \begin{cases} \dfrac{|e|N_A}{2\varepsilon}(x + d_p)^2 & \text{for} \quad -d_p < x < 0 \\[3mm] -\dfrac{|e|N_D}{2\varepsilon}(x^2 - 2xd_n) + \dfrac{|e|N_A}{2\varepsilon}d_p^2 & \text{for} \quad 0 < x < d_n \end{cases} \qquad (6.13)$$

The variation of potential with x as given by (6.13) is shown in Fig. 6.1(d).

We can proceed further and find the width $d = d_p + d_n$ of the depletion layer by setting $V(d_n)$ equal to the contact potential, V_0, that is, the potential difference across the depletion layer resulting from the electric field in the layer. Thus,

$$\begin{aligned} V_0 = V(d_n) &= \frac{|e|N_D}{2\varepsilon}d_n^2 + \frac{|e|N_A}{2\varepsilon}d_p^2 \\[2mm] &= \frac{|e|}{2\varepsilon}\frac{N_D(N_A + N_D)}{N_A + N_D}d_n^2 + \frac{|e|}{2\varepsilon}\frac{N_A(N_A + N_D)}{N_A + N_D}d_p^2 \\[2mm] &= \frac{|e|}{2\varepsilon}\frac{N_A N_D}{N_A + N_D}(d_n^2 + d_p^2 + 2d_n d_p) \\[2mm] &= \frac{|e|}{2\varepsilon}\frac{N_A N_D}{N_A + N_D}d^2 \end{aligned}$$

where we have made use of (6.10). Finally, we obtain the result that

$$d = \sqrt{\frac{2\varepsilon V_0}{|e|}\left(\frac{1}{N_A} + \frac{1}{N_D}\right)}$$

which tells us that the depletion layer width is smaller the heavier the doping is. This property is used in tunnel diodes to achieve layer widths on the order of 10^{-6} cm by heavy doping as compared to widths on the order of 10^{-4} cm in ordinary p-n junctions.

We have just illustrated an example of the application of Poisson's equation involving the solution for the potential distribution for a given charge distribution. Poisson's equation is even more useful for the solution of problems in which the charge distribution is the quantity to be determined given the functional dependence of the charge density on the potential. We shall, however, proceed to the discussion of Laplace's equation.

Laplace's equation

If the charge density in a region is zero, then Poisson's equation (6.6) reduces to

$$\boxed{\nabla^2 V = 0} \tag{6.14}$$

This equation is known as *Laplace's equation*. It governs the behavior of the potential in a charge-free region characterized by uniform permittivity. In Cartesian coordinates, it is given by

$$\frac{\partial^2 V}{\partial x^2} + \frac{\partial^2 V}{\partial y^2} + \frac{\partial^2 V}{\partial z^2} = 0 \tag{6.15}$$

Laplace's equation is also satisfied by the potential in conductors under the steady-current condition. For the steady-current condition, $\partial \rho / \partial t = 0$, and the continuity equation given for the time-varying case by

$$\nabla \cdot \mathbf{J}_c + \frac{\partial \rho}{\partial t} = 0$$

reduces to

$$\nabla \cdot \mathbf{J}_c = 0 \tag{6.16}$$

Replacing $\mathbf{J}_c$ by $\sigma \mathbf{E} = -\sigma \nabla V$, where σ is the conductivity of the conductor and assuming σ to be constant, we obtain

$$\nabla \cdot \sigma \mathbf{E} = \sigma \nabla \cdot \mathbf{E} = -\sigma \nabla \cdot \nabla V = -\sigma \nabla^2 V = 0$$

or

$$\nabla^2 V = 0$$

The problems for which Laplace's equation is applicable consist of finding the potential distribution in the region between two conductors given the charge distribution on the surfaces of the conductors or the potentials of the conductors or a combination of the two. The procedure involves the solving of Laplace's equation subject to the boundary conditions on the surfaces of the conductors. We shall illustrate this by means of an example involving variation of V in one dimension.

Example 6.2

Parallel-plate arrangement, capacitance

Let us consider two infinite, plane, parallel, perfectly conducting plates occupying the planes $x = 0$ and $x = d$ and kept at potentials $V = 0$ and $V = V_0$, respectively, as shown by the cross-sectional view in Fig. 6.2, and find the solution for Laplace's equation in the region between the plates. The arrangement may be considered an idealization of a

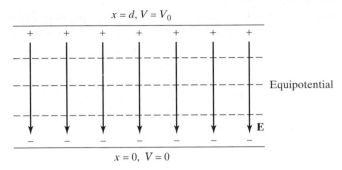

$x = d, V = V_0$

Equipotential

E

$x = 0, \ V = 0$

FIGURE 6.2   Cross-sectional view of parallel-plate capacitor for illustrating the solution of Laplace's equation in one dimension.

parallel-plate capacitor with its plates having dimensions very large compared to the spacing between them.

The potential is obviously a function of x only and hence (6.15) reduces to

$$\frac{\partial^2 V}{\partial x^2} = \frac{d^2 V}{dx^2} = 0$$

Integrating this equation twice, we obtain

$$\boxed{V(x) = Ax + B} \qquad (6.17)$$

where A and B are constants of integration. To determine the values of A and B, we make use of the boundary conditions for V; that is,

$$V = 0 \qquad \text{for} \quad x = 0$$
$$V = V_0 \qquad \text{for} \quad x = d$$

giving us

$$0 = A(0) + B \qquad \text{or} \qquad B = 0$$
$$V_0 = A(d) + B = Ad \qquad \text{or} \qquad A - \frac{V_0}{d}$$

Thus, the particular solution for the potential here is given by

$$\boxed{V = \frac{V_0}{d} x \qquad \text{for} \ \ 0 < x < d} \qquad (6.18)$$

which tells us that the equipotentials are planes parallel to the conductors, as shown in Fig. 6.2.

Proceeding further, we obtain

$$\boxed{\mathbf{E} = -\nabla V - -\frac{\partial V}{\partial x} \mathbf{a}_x = -\frac{V_0}{d} \mathbf{a}_x \qquad \text{for} \ \ 0 < x < d} \qquad (6.19)$$

This field is uniform and directed from the higher potential plate to the lower potential plate, as shown in Fig. 6.2. The surface charge densities on the two plates are given by

$$[\rho_S]_{x=0} = [\mathbf{D}]_{x=0} \cdot \mathbf{a}_x = -\frac{\varepsilon V_0}{d} \mathbf{a}_x \cdot \mathbf{a}_x = -\frac{\varepsilon V_0}{d}$$

$$[\rho_S]_{x=d} = [\mathbf{D}]_{x=d} \cdot (-\mathbf{a}_x) = -\frac{\varepsilon V_0}{d} \mathbf{a}_x \cdot (-\mathbf{a}_x) = \frac{\varepsilon V_0}{d}$$

The magnitude of the surface charge per unit area on either plate is

$$Q = |\rho_S|(1) = \frac{\varepsilon V_0}{d}$$

Finally, we can find the capacitance C per unit area of the plates, defined to be the ratio of Q to V_0. Thus,

$$C = \frac{Q}{V_0} = \frac{\varepsilon}{d} \text{ per unit area of the plates} \qquad (6.20)$$

The units of capacitance are farads (F).

Parallel-plate arrangement, conductance   If the medium between the plates in Fig. 6.2 is a conductor, then the conduction current density is given by

$$\mathbf{J}_c = \sigma \mathbf{E} = -\frac{\sigma V_0}{d} \mathbf{a}_x$$

The conduction current from the higher potential plate to the lower potential plate per unit area of the plates is

$$I_c = |\mathbf{J}_c|(1) = \frac{\sigma V_0}{d}$$

The ratio of this current to the potential difference is the conductance G (reciprocal of resistance) per unit area of the plates. Thus,

$$G = \frac{I_c}{V_0} = \frac{\sigma}{d} \text{ per unit area of the plates} \qquad (6.21)$$

The units of conductance are siemens (S).

Cylindrical and spherical capacitors   We have just illustrated the solution of Laplace's equation in one dimension by considering an example involving the variation of V with one Cartesian coordinate. In a similar manner, solutions for one-dimensional Laplace's equations involving variations of V with single coordinates in the other two coordinate systems can be obtained. Of particular interest are the case in which V is a function of the cylindrical coordinate r only, pertinent to the geometry of a capacitor made up of coaxial cylindrical conductors and the case in which V is a function of the spherical coordinate r only, pertinent to the geometry of a capacitor made up of concentric spherical conductors. These two geometries are shown in Figs. 6.3(a) and (b), respectively. The various steps in the solution of Laplace's equation and subsequent determination of capacitance for these two cases are summarized in Table 6.1, which also includes the parallel plane case of Fig. 6.2.

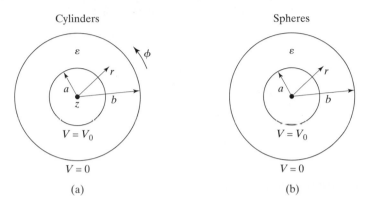

FIGURE 6.3 Cross-sectional views of capacitors made up of (a) coaxial cylindrical conductors and (b) concentric spherical conductors.

TABLE 6.1 Summary of Various Steps in the Solution of Laplace's Equation and Determination of Capacitance for Three One-Dimensional Cases

Geometry	Parallel planes	Coaxial cylinders	Concentric spheres
Figure	6.2	6.3(a)	6.3(b)
Boundary conditions	$V = 0, \quad x = 0$ $V - V_0, \quad x = d$	$V - V_0, \quad r = a$ $V - 0, \quad r = b$	$V = V_0, \quad r = a$ $V = 0, \quad r = b$
Laplace's equation	$\dfrac{\partial^2 V}{\partial x^2} = 0$	$\dfrac{1}{r}\dfrac{\partial}{\partial r}\left(r\dfrac{\partial V}{\partial r}\right) = 0$	$\dfrac{1}{r^2}\dfrac{\partial}{\partial r}\left(r^2\dfrac{\partial V}{\partial r}\right) = 0$
General solution	$V = Ax + B$	$V = A \ln r + B$	$V = \dfrac{A}{r} + B$
Particular solution	$V = V_0 \dfrac{x}{d}$	$V = V_0 \dfrac{\ln (r/b)}{\ln (a/b)}$	$V = V_0 \dfrac{1/r - 1/b}{1/a - 1/b}$
Electric field	$-\dfrac{V_0}{d}\mathbf{a}_x$	$\dfrac{V_0}{r \ln (b/a)}\mathbf{a}_r$	$\dfrac{V_0}{r^2 (1/a - 1/b)}\mathbf{a}_r$
Surface charge densities	$-\dfrac{\varepsilon V_0}{d}, \quad x = 0$ $\dfrac{\varepsilon V_0}{d}, \quad x = d$	$\dfrac{\varepsilon V_0}{a \ln (b/a)}, \quad r = a$ $\dfrac{-\varepsilon V_0}{b \ln (b/a)}, \quad r = b$	$\dfrac{\varepsilon V_0}{a^2 (1/a - 1/b)}, \quad r = a$ $\dfrac{-\varepsilon V_0}{b^2 (1/a - 1/b)}, \quad r = b$
Capacitance	$\dfrac{\varepsilon}{d}$ per unit area	$\dfrac{2\pi\varepsilon}{\ln (b/a)}$ per unit length	$\dfrac{4\pi\varepsilon}{1/a - 1/b}$

K6.1. Poisson's equation; *p-n* junction; Laplace's equation in one dimension; Parallel-plate arrangement; Capacitance; Conductance; Cylindrical and spherical capacitors.

D6.1. The potential distribution in a simplified model of a vacuum diode consisting of cathode in the plane $x = 0$ and anode in the plane $x = d$ and held at a potential V_0 relative to the cathode is given by $V = V_0(x/d)^{4/3}$ for $0 < x < d$. Find the following: **(a)** V at $x = d/8$; **(b)** $\mathbf{E}$ at $x = d/8$; **(c)** ρ at $x = d/8$; and **(d)** ρ_S on the anode.

Ans. **(a)** $V_0/16$; **(b)** $-(2V_0/3d)\mathbf{a}_x$; **(c)** $-16\varepsilon_0 V_0/9d^2$; **(d)** $4\varepsilon_0 V_0/3d$.

D6.2. Find the following: **(a)** the spacing between the plates of a parallel-plate capacitor with a dielectric of $\varepsilon = 2.25\varepsilon_0$ and having capacitance per unit area equal to 1000 pF; **(b)** the ratio of the outer radius to the inner radius for a coaxial cylindrical capacitor with a dielectric of $\varepsilon = 2.25\varepsilon_0$, and having capacitance per unit length equal to 100 pF; and **(c)** the radius of an isolated spherical conductor in free space for which the capacitance is 10 pF.

Ans. **(a)** 1.99 cm; **(b)** 3.4903; **(c)** 9 cm.

6.2 CAPACITANCE, CONDUCTANCE, AND INDUCTANCE

In the previous section, we introduced the capacitance and conductance by considering the solution of Laplace's equation in one dimension. Specifically, we derived the expressions for the capacitance per unit area and the conductance per unit area of a parallel-plate arrangement, the capacitance per unit length of a coaxial cylindrical arrangement, and the capacitance of a concentric spherical arrangement.

Coaxial cylindrical arrangement   Let us now consider the three arrangements shown in Fig. 6.4, each of which is a cross-sectional view of a pair of infinitely long coaxial perfectly conducting cylinders with a material medium between them. In Fig. 6.4 (a), the material medium is a dielectric of uniform permittivity ε; in Fig. 6.4(b), it is a conductor of uniform conductivity σ; and in Fig. 6.4 (c), it is a magnetic material

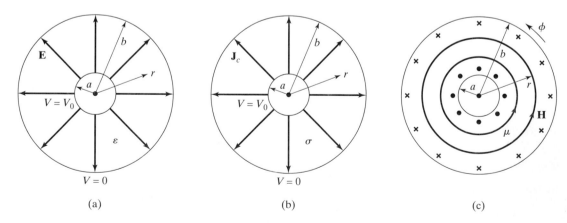

FIGURE 6.4 Cross sections of three arrangements, each consisting of two infinitely long, coaxial perfectly conducting cylinders. The medium between the cylinders is a perfect dielectric for (a), a conductor for (b), and a magnetic material for (c).

of uniform permeability μ. In (a) and (b), a potential difference of V_0 is applied between the conductors, whereas in (c), a current I flows with uniform density in the $+z$-direction on the inner cylinder and returns with uniform density in the $-z$-direction on the outer cylinder.

We know from the discussion in Section 6.1 that the arrangement of Fig. 6.4(a) is that of a coaxial cylindrical capacitor and from Table 6.1 that its capacitance (C) per unit length, defined as the magnitude of the charge per unit length on either conductor to the potential difference between the conductors, is given by

Capacitance per unit length, $\mathscr{C}$

$$\boxed{\mathscr{C} = \frac{C}{l} = \frac{2\pi\varepsilon}{\ln(b/a)}} \tag{6.22}$$

the units of C being farads (F) and, hence, those of $\mathscr{C}$ being F/m.

Just as in the case of the parallel-plate arrangement of Example 6.2, replacing the dielectric in Fig. 6.4(a) by a conductor as in Fig. 6.4(b) would result in a conduction current of density

Conductance per unit length, $\mathscr{G}$

$$\mathbf{J}_c = \sigma\mathbf{E} = \frac{\sigma V_0}{r \ln(b/a)} \mathbf{a}_r$$

in the medium and, hence, a current per unit length

$$I_c = \int_{\phi=0}^{2\pi} \mathbf{J}_c \cdot r\, d\phi\, \mathbf{a}_r = \int_{\phi=0}^{2\pi} \frac{\sigma V_0}{r \ln(b/a)} r\, d\phi$$

$$= \frac{2\pi\sigma V_0}{\ln(b/a)}$$

from the inner cylinder to the outer cylinder. Thus, the ratio of the current per unit length from the inner to the outer cylinder to the potential difference between the cylinders, that is, the conductance (G) per unit length of the arrangement, is given by

$$\boxed{\mathscr{G} = \frac{G}{l} = \frac{2\pi\sigma}{\ln(b/a)}} \tag{6.23}$$

the units of G being siemens (S) and, hence, those of $\mathscr{G}$ being S/m.

Turning now to Fig. 6.4(c), we note from Example 4.14 that the current flow on the cylinders results in a magnetic field between them as given by

Inductance per unit length, $\mathscr{L}$

$$\mathbf{H} = \frac{I}{2\pi r}\mathbf{a}_\phi \qquad \text{for} \quad a < r < b$$

The magnetic flux density is then given by

$$\mathbf{B} = \mu\mathbf{H} = \frac{\mu I}{2\pi r}\mathbf{a}_\phi \qquad \text{for} \quad a < r < b$$

The magnetic flux linking the current per unit length of the conductors is

$$\psi = \int_{r=a}^{b} \mathbf{B} \cdot dr\, \mathbf{a}_\phi = \int_{r=a}^{b} \frac{\mu I}{2\pi r}\, dr$$

$$= \frac{\mu I}{2\pi} \ln \frac{b}{a}$$

We now define the inductance (L) per unit length of the arrangement to be the ratio of the magnetic flux linking the current per unit length of the arrangement to the current. Thus,

$$\mathcal{L} = \frac{L}{l} = \frac{\mu}{2\pi} \ln \frac{b}{a} \tag{6.24}$$

The units of L are henrys (H) and, hence, those of $\mathcal{L}$ are H/m.
    An examination of (6.22), (6.23), and (6.24) reveals that

$$\frac{\mathcal{G}}{\mathcal{C}} = \frac{\sigma}{\varepsilon} \tag{6.25}$$

and

$$\mathcal{L}\mathcal{C} = \mu\varepsilon \tag{6.26}$$

TABLE 6.2   Conductance, Capacitance, and Inductance per Unit Length for Some Structures Consisting of Infinitely Long Conductors Having the Cross Sections Shown in Fig. 6.5

Description	Capacitance per unit length, $\mathcal{C}$	Conductance per unit length, $\mathcal{G}$	Inductance per unit length, $\mathcal{L}$
Parallel-plane conductors, Fig. 6.5 (a)	$\varepsilon \dfrac{w}{d}$	$\sigma \dfrac{w}{d}$	$\mu \dfrac{d}{w}$
Coaxial cylindrical conductors, Fig. 6.5 (b)	$\dfrac{2\pi\varepsilon}{\ln(b/a)}$	$\dfrac{2\pi\sigma}{\ln(b/a)}$	$\dfrac{\mu}{2\pi} \ln \dfrac{b}{a}$
Parallel cylindrical wires, Fig. 6.5 (c)	$\dfrac{\pi\varepsilon}{\cosh^{-1}(d/a)}$	$\dfrac{\pi\sigma}{\cosh^{-1}(d/a)}$	$\dfrac{\mu}{\pi} \cosh^{-1} \dfrac{d}{a}$
Eccentric inner conductor, Fig. 6.5 (d)	$\dfrac{2\pi\varepsilon}{\cosh^{-1}\left(\dfrac{a^2+b^2-d^2}{2ab}\right)}$	$\dfrac{2\pi\sigma}{\cosh^{-1}\left(\dfrac{a^2+b^2-d^2}{2ab}\right)}$	$\dfrac{\mu}{2\pi} \cosh^{-1}\left(\dfrac{a^2+b^2-d^2}{2ab}\right)$
Shielded parallel cylindrical wires, Fig. 6.5 (e)	$\dfrac{\pi\varepsilon}{\ln \dfrac{d(b^2-d^2/4)}{a(b^2+d^2/4)}}$	$\dfrac{\pi\sigma}{\ln \dfrac{d(b^2-d^2/4)}{a(b^2+d^2/4)}}$	$\dfrac{\mu}{\pi} \ln \dfrac{d(b^2-d^2/4)}{a(b^2+d^2/4)}$

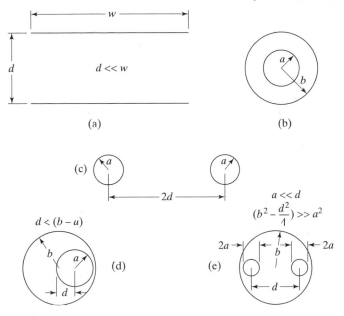

FIGURE 6.5   Cross sections of some common configurations of parallel, infinitely long conductors.

Thus, only one of the three parameters $\mathcal{C}$, $\mathcal{G}$, and $\mathcal{L}$ is independent, with the other two obtainable from it and the material parameters. Although this result is deduced here for the coaxial cylindrical arrangement, it is a general result valid for all arrangements involving two infinitely long, parallel perfect conductors embedded in a homogeneous medium (a medium of uniform material parameters). Expressions for the three quantities $\mathcal{C}$, $\mathcal{G}$, and $\mathcal{L}$ are listed in Table 6.2 for some common configurations of conductors having cross-sectional views shown in Fig. 6.5. The coaxial cylindrical arrangement is repeated for the sake of completion.

Example 6.3

It is desired to obtain the capacitance, conductance, and inductance per unit length of the parallel-cylindrical wire arrangement of Fig. 6.5(c).   *$\mathcal{C}$, $\mathcal{G}$, and $\mathcal{L}$ of parallel-cylindrical wire arrangement*

In view of (6.25) and (6.26), it is sufficient to find one of the three quantities. Hence, we choose to find the capacitance per unit length. Here we shall do this by considering the electric potential field of two parallel, infinitely long, straight-line charges of equal and opposite uniform charge densities and showing that the equipotential surfaces are cylinders having their axes parallel to the line charges. By placing conductors in two equipotential surfaces, thereby forming a parallel-wire line, we shall obtain the expression for the capacitance per unit length of the line.

Let us first consider an infinitely long, straight-line charge of uniform density ρ_{L0} C/m situated along the z-axis, as shown in Fig. 6.6(a), and obtain the electric potential

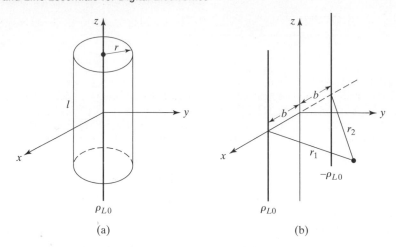

FIGURE 6.6   (a) Infinitely long line charge of uniform density along the z-axis. (b) Pair of parallel, infinitely long line charges of equal and opposite uniform densities.

due to the line charge. The symmetry associated with the problem indicates that the potential is dependent on the cylindrical coordinate r. Thus, we have

$$\nabla^2 V = \frac{1}{r}\frac{\partial}{\partial r}\left(r\frac{\partial V}{\partial r}\right) = 0 \qquad \text{for} \quad r \neq 0$$

$$V = A \ln r + B \tag{6.27}$$

where A and B are constants to be determined. We can arbitrarily set the potential to be zero at a reference value $r = r_0$, giving us $B = -A \ln r_0$ and

$$V = A \ln r - A \ln r_0 = A \ln \frac{r}{r_0} \tag{6.28}$$

To evaluate the arbitrary constant A in (6.28), we find the electric-field intensity due to the line charge is given by

$$\mathbf{E} = -\nabla V = -\frac{\partial V}{\partial r}\mathbf{a}_r = -\frac{A}{r}\mathbf{a}_r$$

The electric field is thus directed radial to the line charge. Let us now consider a cylindrical box of radius r and length l coaxial with the line charge, as shown in Fig. 6.6(a), and apply Gauss' law for the electric field in integral form to the surface of the box. For the cylindrical surface,

$$\int \mathbf{D} \cdot d\mathbf{S} = -\frac{\varepsilon A}{r}(2\pi r l)$$

For the top and bottom surfaces, $\int \mathbf{D} \cdot d\mathbf{S} = 0$ since the field is parallel to the surfaces. The charge enclosed by the box is $\rho_{L0} l$. Thus, we have

$$-\frac{\varepsilon A}{r}(2\pi r l) = \rho_{L0} l \qquad \text{or} \qquad A = -\frac{\rho_{L0}}{2\pi\varepsilon}$$

Substituting this result in (6.28), we obtain the potential field due to the line charge to be

$$V = -\frac{\rho_{L0}}{2\pi\varepsilon} \ln \frac{r}{r_0} = \frac{\rho_{L0}}{2\pi\varepsilon} \ln \frac{r_0}{r} \tag{6.29}$$

which is consistent with (4.105).

Let us now consider two infinitely long, straight-line charges of equal and opposite uniform charge densities ρ_{L0} C/m and $-\rho_{L0}$ C/m, parallel to the z-axis and passing through $x = b$ and $x = -b$, respectively, as shown in Fig. 6.6(b). Applying superposition and using (6.29), we write the potential due to the two line charges as

$$V = \frac{\rho_{L0}}{2\pi\varepsilon} \ln \frac{r_{01}}{r_1} - \frac{\rho_{L0}}{2\pi\varepsilon} \ln \frac{r_{02}}{r_2}$$

where r_1 and r_2 are the distances of the point of interest from the line charges and r_{01} and r_{02} are the distances to the reference point at which the potential is zero. By choosing the reference point to be equidistant from the two line charges, that is, $r_{01} = r_{02}$, we get

$$V = \frac{\rho_{L0}}{2\pi\varepsilon} \ln \frac{r_2}{r_1} \tag{6.30}$$

From (6.30), we note that the equipotential surfaces for the potential field of the line-charge pair are given by

$$\frac{r_2}{r_1} = \text{constant, say, } k \tag{6.31}$$

where k lies between 0 and ∞. In terms of Cartesian coordinates, (6.31) can be written as

$$\frac{(x + b)^2 + y^2}{(x - b)^2 + y^2} = k^2$$

Rearranging, we obtain

$$x^2 - 2b \frac{k^2 + 1}{k^2 - 1} x + y^2 + b^2 = 0$$

or

$$\left(x - b \frac{k^2 + 1}{k^2 - 1}\right)^2 + y^2 = \left(b \frac{2k}{k^2 - 1}\right)^2$$

This equation represents cylinders having their axes along

$$x = b \frac{k^2 + 1}{k^2 - 1}, \qquad y = 0$$

and radii equal to $b[2k/(k^2 - 1)]$. The corresponding potentials are $(\rho_{L0}/2\pi\varepsilon) \ln k$. The cross sections of the equipotential surfaces are shown in Fig. 6.7.

We can now place perfectly conducting cylinders in any two equipotential surfaces without disturbing the field configuration, as shown, for example, by the thick circles in Fig. 6.7, thereby obtaining a parallel-wire line. Letting the distance between their centers be $2d$ and their radii be a, we have

$$\pm d = b \frac{k^2 + 1}{k^2 - 1}$$

$$a = b \frac{2k}{k^2 - 1}$$

Solving these two equations for k and accepting only those solutions lying between 0

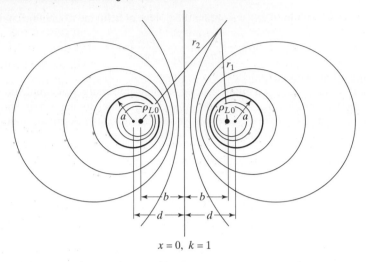

$$x = 0, \ k = 1$$

FIGURE 6.7   Cross sections of equipotential surfaces for the line charge pair of Fig 6.6(b). Thick circles represent a cross section of parallel-wire line.

and ∞, we obtain

$$k = \frac{d \pm \sqrt{d^2 - a^2}}{a}$$

The potentials of the right ($k > 1$) and left ($k < 1$) conductors are then given, respectively, by

$$V_+ = \frac{\rho_{L0}}{2\pi\varepsilon} \ln \frac{d + \sqrt{d^2 - a^2}}{a}$$

$$V_- = \frac{\rho_{L0}}{2\pi\varepsilon} \ln \frac{d - \sqrt{d^2 - a^2}}{a}$$

$$= -\frac{\rho_{L0}}{2\pi\varepsilon} \ln \frac{d + \sqrt{d^2 - a^2}}{a}$$

The potential difference between the two conductors is

$$V_0 = V_+ - V_- = \frac{\rho_{L0}}{\pi\varepsilon} \ln \frac{d + \sqrt{d^2 - a^2}}{a}$$

Finally, to find the capacitance, we note that since the electric field lines begin on the positive charge and end on the negative charge orthogonal to the equipotentials, the magnitude of the charge on either conductor, which produces the same field as the line-charge pair, must be the same as the line charge itself. Thus, considering unit length of the line, we obtain the capacitance per unit length of the parallel-wire line to be

$$\mathcal{C} = \frac{\rho_{L0}}{V_0} = \frac{\pi\varepsilon}{\ln\left[(d + \sqrt{d^2 - a^2})/a\right]}$$

$$= \frac{\pi\varepsilon}{\cosh^{-1}(d/a)}$$

(6.32)

and, hence, the expressions for $\mathcal{G}$ and $\mathcal{L}$, as given in Table 6.2.

If the conductors in a given configuration are not perfect, then the currents *Internal* flow in the volumes of the conductors instead of being confined to the surfaces. *inductance* We then have to consider the magnetic field internal to the current distribution in addition to the magnetic field external to it. The inductance associated with the internal field is known as the *internal inductance* as compared to the *external inductance* associated with the external field. The expressions for the inductance per unit length given in Table 6.2 are for the external inductance. To obtain the internal inductance, we have to take into account the fact that different flux lines in the volume occupied by the current distribution link different partial amounts of the total current. We shall illustrate this by means of an example.

Example 6.4

A current I A flows with uniform volume density $\mathbf{J} = J_0\mathbf{a}_z$ A/m^2 along an infinitely long, solid cylindrical conductor of radius a and returns with uniform surface density in the opposite direction along the surface of an infinitely long, perfectly conducting cylinder of radius $b\ (> a)$ and coaxial with the inner conductor. It is desired to find the internal inductance per unit length of the inner conductor.

The cross-sectional view of the conductor arrangement is shown in Fig. 6.8(a). From symmetry considerations, the magnetic field is entirely in the ϕ direction and independent of ϕ. Applying Ampère's circuital law to a circular contour of radius $r\ (< a)$, as shown in Fig. 6.8(a), we have

$$2\pi r H_\phi = \pi r^2 J_0$$

or

$$\mathbf{H} = H_\phi \mathbf{a}_\phi = \frac{J_0 r}{2}\mathbf{a}_\phi \qquad r < a$$

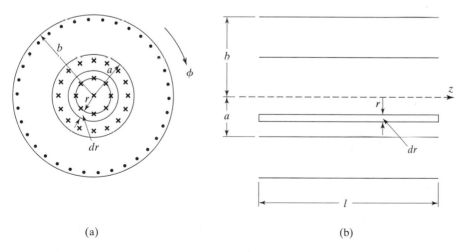

(a) (b)

FIGURE 6.8 For evaluating the internal inductance per unit length associated with a volume current of uniform density along an infinitely long cylindrical conductor.

The corresponding magnetic flux density is given by

$$\mathbf{B} = \mu\mathbf{H} = \frac{\mu J_0 r}{2}\,\mathbf{a}_\phi \qquad r < a$$

where μ is the permeability of the conductor. Let us now consider a rectangle of infinitesimal width dr in the r-direction and length l in the z-direction at a distance r from the axis, as shown in Fig. 6.8(b). The magnetic flux $d\psi_i$ crossing this rectangular surface is given by

$$d\psi_i = B_\phi \text{ (area of the rectangle)}$$

$$= \frac{\mu J_0 r l\, dr}{2}$$

where the subscript i denotes flux internal to the conductor. This flux surrounds only the current flowing within the radius r, as can be seen from Fig. 6.8(a). Let N be the fraction of the total current I linked by this flux. Then

$$N = \frac{\text{current flowing within radius } r\,(<a)}{\text{total current } I}$$

$$= \frac{J_0 \pi r^2}{J_0 \pi a^2} = \left(\frac{r}{a}\right)^2$$

The contribution from the flux $d\psi_i$ to the internal flux linkage associated with the current I is the product of N and the flux itself, that is, $N\, d\psi_i$. To obtain the internal flux linkage associated with I, we integrate $N\, d\psi_i$ between the limits $r = 0$ and $r = a$, taking into account the dependence of N on $d\psi_i$. Thus,

$$\psi_i = \int_{r=0}^{a} N\, d\psi_i = \int_{r=0}^{a} \left(\frac{r}{a}\right)^2 \frac{\mu J_0 l r}{2}\, dr = \frac{\mu J_0 l a^2}{8}$$

Finally, the required internal inductance per unit length is

$$\boxed{\mathcal{L}_i = \frac{\psi_i}{lI} = \frac{\mu J_0 a^2/8}{J_0 \pi a^2} = \frac{\mu}{8\pi}} \qquad (6.33)$$

From the steps involved in the solution of Example 6.4, we observe that the general expression for the internal inductance is

$$\boxed{L_{\text{int}} = \frac{1}{I} \int_S N\, d\psi} \qquad (6.34a)$$

where S is any surface through which the internal magnetic flux associated with I passes. We note that (6.34a) is also good for computing the external inductance since for external inductance, N is independent of $d\psi$. Hence,

$$L_{\text{ext}} = \frac{N}{I} \int_S d\psi = N\frac{\psi}{I} \qquad (6.34b)$$

In (6.34b), the value of N is unity if I is a surface current as for the arrangement of Fig. 6.4(c). On the other hand, for a filamentary wire wound on a core, N is equal to the number of turns of the winding in which case ψ represents the flux through the core, that is, the flux crossing the surface formed by one turn, according to the consideration same as that in conjunction with the discussion of Faraday's law for an N-turn coil (see Fig. 3.13).

The discussion pertaining to inductance thus far has been concerned with *self-inductance*, that is, inductance associated with a current distribution by virtue of its own flux linking it. On the other hand, if we have two independent currents I_1 and I_2, we can talk of the flux due to one current linking the second current. This leads to the concept of *mutual inductance*. The mutual inductance denoted as L_{12} is defined as

Mutual inductance

$$L_{12} = N_1 \frac{\psi_{12}}{I_2}$$
(6.35a)

where ψ_{12} is the magnetic flux produced by I_2 but linking one turn of the N_1-turn winding carrying current I_1. Similarly,

$$L_{21} = N_2 \frac{\psi_{21}}{I_1}$$
(6.35b)

where ψ_{21} is the magnetic flux produced by I_1 but linking one turn of the N_2-turn winding carrying current I_2. It can be shown that $L_{21} = L_{12}$. We shall now consider a simple example illustrating the computation of mutual inductance.

Example 6.5

A single straight wire, infinitely long and carrying current I_1, lies below to the left and parallel to a two-wire telephone line carrying current I_2, as shown by the cross-sectional and plan views in Figs. 6.9(a) and (b), respectively. It is desired to obtain the mutual inductance between the single wire and the telephone line per unit length of the wires. The thickness of the telephone wire is assumed to be negligible.

Choosing a coordinate system with the axis of the single wire as the z-axis and applying Ampère's circuital law to a circular path around the single wire, we obtain the magnetic flux density due to the single wire as

$$\mathbf{B} = \frac{\mu_0 I_1}{2\pi r} \mathbf{a}_\phi$$

The flux $d\psi_{21}$ crossing a rectangular surface of length unity and width dy lying between the telephone wires, as shown in Fig. 6.9(b), is then given by

$$d\psi_{21} = B \, dy \cos \alpha = \frac{\mu_0 I_1 y}{2\pi(h^2 + y^2)} \, dy$$

where α is the angle between the flux lines and the normal to the rectangular surface, as shown in Fig. 6.9(a). The total flux ψ_{21} crossing the rectangular surface of length unity and

FIGURE 6.9   For the computation of mutual inductance per unit length between a two-wire telephone line and a single wire parallel to it.

extending from one telephone wire to the other is

$$\psi_{21} = \int_{y=a}^{a+b} d\psi_{21} = \int_{y=a}^{a+b} \frac{\mu_0 I_1 y}{2\pi(h^2 + y^2)} \, dy$$

$$= \frac{\mu_0 I_1}{4\pi} \ln \frac{h^2 + (a + b)^2}{h^2 + a^2}$$

This is the flux due to I_1 linking I_2 per unit length along the wires. Thus, the required mutual inductance per unit length of the wires is given by

$$\mathcal{L}_{21} = \frac{\psi_{21}}{I_1} = \frac{\mu_0}{4\pi} \ln \frac{h^2 + (a + b)^2}{h^2 + a^2} \text{ H/m}$$

K6.2.   Infinitely long, coaxial cylindrical arrangement; Capacitance per unit length ($\mathcal{C}$); Conductance per unit length ($\mathcal{G}$); Inductance per unit length ($\mathcal{L}$); Relationship between $\mathcal{C}$, $\mathcal{G}$, and $\mathcal{L}$; Parallel cylindrical wire arrangement; Internal inductance; Mutual inductance.

D6.3.   A coaxial cylindrical conductor arrangement [see Fig. 6.5(b)] has the dimensions $a = 1$ cm and $b = 3$ cm. **(a)** By what value of the distance d should the inner conductor be displaced parallel to the outer conductor [see Fig. 6.5(d)] to increase the capacitance per unit length of the arrangement by 25%. **(b)** By what percentage is the inductance per unit length of the arrangement then changed from the original value?
Ans.   **(a)** 1.2368 cm;   **(b)** −20.

D6.4.   Figure 6.10 is the cross-sectional view of the coaxial cylindrical conductor arrangement in which a solid conductor of radius a is enclosed by a hollow conductor of inner radius $4a$ and outer radius $5a$. Current I_0 flows in the inner conductor in the $+z$-direction and returns on the outer conductor in the $-z$-direction with

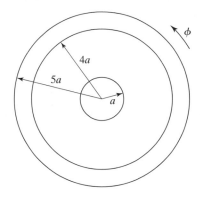

FIGURE 6.10   For Problem D6.4.

densities given by

$$
\mathbf{J} = \begin{cases}
\dfrac{I_0 e}{\pi a^2}(1 - e^{-r^2/a^2})\mathbf{a}_z & \text{for}\quad 0 < r < a \\[4mm]
-\dfrac{I_0}{9\pi a^2}\mathbf{a}_z & \text{for}\quad 4a < r < 5a
\end{cases}
$$

Find the value of N, the fraction of the current I_0 linked by the magnetic flux $d\psi_i$ at a given radius r, for each of the following values of r: **(a)** $0.8a$; **(b)** $3a$; and **(c)** $4.5a$.

Ans. **(a)** 0.4547; **(b)** 1; **(c)** 0.5278.

6.3 TRANSMISSION LINE

In the previous section, we considered a physical arrangement of two parallel, perfect conductors and discussed the circuit parameters, capacitance, conductance, and inductance per unit length of the structure. In the general case of time-varying fields, the situation corresponds to the structure characterized by the properties of all three of the circuit parameters, continuously and overlappingly along it. The arrangement is then called a *transmission line*. To introduce the transmission-line concept, we recall that in Section 5.4, we learned that the tangential component of the electric field intensity and the normal component of the magnetic field intensity are zero on a perfect conductor surface. Let us now consider the uniform plane electromagnetic wave propagating in the z-direction and having an x-component only of the electric field and a y-component only of the magnetic field, that is,

Parallel-plate line

$$
\mathbf{E} = E_x(z, t)\mathbf{a}_x
$$
$$
\mathbf{H} = H_y(z, t)\mathbf{a}_y
$$

and place perfectly conducting sheets in two planes $x = 0$ and $x = d$, as shown in Fig. 6.11. Since the electric field is completely normal and the magnetic field is completely tangential to the sheets, the two boundary conditions just referred to are satisfied, and, hence, the wave will simply propagate, as though the sheets were not present, being guided by the sheets. We then have a simple case of

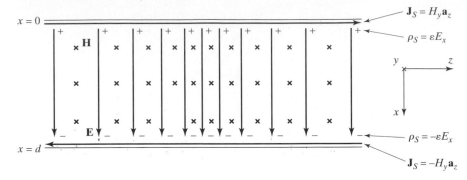

FIGURE 6.11   Uniform plane electromagnetic wave propagating between two perfectly conducting sheets, supported by charges and current on the sheets.

transmission line, namely, the parallel-plate transmission line. We shall assume the medium between the plates to be a perfect dielectric so that the waves propagate without attenuation and hence the line is lossless.

According to the remaining two boundary conditions, there must be charges and currents on the conductors. The charge densities on the two plates are

$$[\rho_S]_{x=0} = [\mathbf{a}_n \cdot \mathbf{D}]_{x=0} = \mathbf{a}_x \cdot \varepsilon E_x \mathbf{a}_x = \varepsilon E_x \qquad (6.36a)$$

$$[\rho_S]_{x=d} = [\mathbf{a}_n \cdot \mathbf{D}]_{x=d} = -\mathbf{a}_x \cdot \varepsilon E_x \mathbf{a}_x = -\varepsilon E_x \qquad (6.36b)$$

where ε is the permittivity of the medium between the two plates. The current densities on the two plates are

$$[\mathbf{J}_S]_{x=0} = [\mathbf{a}_n \times \mathbf{H}]_{x=0} = \mathbf{a}_x \times H_y \mathbf{a}_y = H_y \mathbf{a}_z \qquad (6.37a)$$

$$[\mathbf{J}_S]_{x=d} = [\mathbf{a}_n \times \mathbf{H}]_{x=d} = -\mathbf{a}_x \times H_y \mathbf{a}_y = -H_y \mathbf{a}_z \qquad (6.37b)$$

In (6.36a)–(6.37b), it is understood that the charge and current densities are functions of z and t as E_x and H_y are. Thus, the wave propagation along the transmission line is supported by charges and currents on the plates, varying with time and distance along the line, as shown in Fig. 6.11.

Let us now consider finitely sized plates having width w in the y-direction, as shown in Fig. 6.12(a), and neglect fringing of the fields at the edges or assume that the structure is part of a much larger-sized configuration. By considering a constant-z plane, that is, a plane *transverse* to the direction of propagation of the wave, as shown in Fig. 6.12(b), we can find the voltage between the two conductors in terms of the line integral of the electric field intensity evaluated along any path in that plane between the two conductors. Since the electric field is directed in the x-direction and since it is uniform in that plane, this voltage is given by

$$V(z, t) = \int_{x=0}^{d} E_x(z, t)\, dx = E_x(z, t) \int_{x=0}^{d} dx = d E_x(z, t) \qquad (6.38)$$

Thus, each transverse plane is characterized by a voltage between the two conductors, which is related simply to the electric field, as given by (6.38). Each transverse plane is also characterized by a current I flowing in the positive

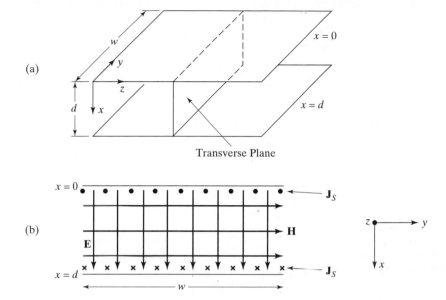

FIGURE 6.12 (a) Parallel-plate transmission line. (b) Transverse plane of the parallel-plate transmission line.

z-direction on the upper conductor and in the negative z-direction on the lower conductor. From Fig. 6.12(b), we can see that this current is given by

$$I(z, t) = \int_{y=0}^{w} J_S(z, t) \, dy = \int_{y=0}^{w} H_y(z, t) \, dy = H_y(z, t) \int_{y=0}^{w} dy \quad (6.39)$$
$$= w H_y(z, t)$$

since H_y is uniform in the cross-sectional plane. Thus, the current crossing a given transverse plane is related simply to the magnetic field in that plane, as given by (6.39).

Proceeding further, we can find the power flow down the line by evaluating the surface integral of the Poynting vector over a given transverse plane. Thus,

$$P(z, t) = \int_{\substack{\text{transverse} \\ \text{plane}}} (\mathbf{E} \times \mathbf{H}) \cdot d\mathbf{S}$$

$$= \int_{x=0}^{d} \int_{y=0}^{w} E_x(z, t) H_y(z, t) \mathbf{a}_z \cdot dx \, dy \, \mathbf{a}_z \quad (6.40)$$

$$= \int_{x=0}^{d} \int_{y=0}^{w} \frac{V(z, t)}{d} \frac{I(z, t)}{w} \, dx \, dy$$

$$= V(z, t) I(z, t)$$

which is the familiar relationship employed in circuit theory.

Transmission-
line equations

We now recall from Section 5.3 that E_x and H_y satisfy the two differential equations

$$\frac{\partial E_x}{\partial z} = -\frac{\partial B_y}{\partial t} = -\mu \frac{\partial H_y}{\partial t} \tag{6.41a}$$

$$\frac{\partial H_y}{\partial z} = -\sigma E_x - \varepsilon \frac{\partial E_x}{\partial t} = -\varepsilon \frac{\partial E_x}{\partial t} \tag{6.41b}$$

where we have set $\sigma = 0$ in view of the perfect dielectric medium. From (6.38) and (6.39), however, we have

$$E_x = \frac{V}{d} \tag{6.42a}$$

$$H_y = \frac{I}{w} \tag{6.42b}$$

Substituting for E_x and H_y in (6.41a) and (6.41b) from (6.42a) and (6.42b), respectively, we now obtain two differential equations for voltage and current along the line as

$$\frac{\partial}{\partial z}\left(\frac{V}{d}\right) = -\mu \frac{\partial}{\partial t}\left(\frac{I}{w}\right)$$

$$\frac{\partial}{\partial z}\left(\frac{I}{w}\right) = -\varepsilon \frac{\partial}{\partial t}\left(\frac{V}{d}\right)$$

or

$$\boxed{\frac{\partial V}{\partial z} = -\left(\frac{\mu d}{w}\right)\frac{\partial I}{\partial t}} \tag{6.43a}$$

$$\boxed{\frac{\partial I}{\partial z} = -\left(\frac{\varepsilon w}{d}\right)\frac{\partial V}{\partial t}} \tag{6.43b}$$

These equations are known as the *transmission-line equations.* They characterize the wave propagation along the line in terms of line voltage and line current instead of in terms of the fields.

We now denote two quantities familiarly known as the *circuit parameters,* the inductance and the capacitance per unit length of the transmission line in the z-direction by the symbols $\mathcal{L}$ and $\mathcal{C}$, respectively. We observe from Section 6.2 that the inductance per unit length, having the units henrys per meter (H/m), is the ratio of the magnetic flux per unit length at any value of z to the line current at that value of z. Noting from Fig. 6.12 that the cross-sectional area normal to the magnetic field lines and per unit length in the z-direction is $(d)(1)$, or d, we find the magnetic flux per unit length to be $B_y d$ or $\mu H_y d$. Since the line current is $H_y w$, we then have

$$\boxed{\mathcal{L} = \frac{\mu H_y d}{H_y w} = \frac{\mu d}{w}} \tag{6.44}$$

We also observe that the capacitance per unit length, having the units farads per meter (F/m), is the ratio of the magnitude of the charge per unit length on either plate at any value of z to the line voltage at that value of z. Noting from Fig 6.12 that the cross-sectional area normal to the electric field lines and per unit length in the z-direction is $(w)(1)$, or w, we find the charge per unit length to be $\rho_S w$, or $\varepsilon E_x w$. Since the line voltage is $E_x d$, we then have

$$\mathcal{C} = \frac{\varepsilon E_x w}{E_x d} = \frac{\varepsilon w}{d} \tag{6.45}$$

We note that $\mathcal{L}$ and $\mathcal{C}$ are purely dependent on the dimensions of the line and are independent of E_x and H_y. We further note that

$$\mathcal{L}\mathcal{C} = \mu\varepsilon \tag{6.46}$$

so that only one of the two parameters $\mathcal{L}$ and $\mathcal{C}$ is independent and the other can be obtained from the knowledge of ε and μ. The results given by (6.44) and (6.45) are the same as those listed in Table 6.2 for the parallel-plane conductor arrangement, whereas (6.46) is the same as given by (6.26).

Replacing now the quantities in parentheses in (6.43a) and (6.43b) by $\mathcal{L}$ and $\mathcal{C}$, respectively, we obtain the transmission-line equations in terms of these parameters as

$$\frac{\partial V}{\partial z} = -\mathcal{L}\frac{\partial I}{\partial t} \tag{6.47a}$$

$$\frac{\partial I}{\partial z} = -\mathcal{C}\frac{\partial V}{\partial t} \tag{6.47b}$$

These equations permit us to discuss wave propagation along the line in terms of circuit quantities instead of in terms of field quantities. It should, however, not be forgotten that the actual phenomenon is one of electromagnetic waves guided by the conductors of the line.

It is customary to represent a transmission line by means of its circuit equivalent, derived from the transmission-line equations (6.47a) and (6.47b). To do this, let us consider a section of infinitesimal length Δz along the line between z and $z + \Delta z$. From (6.47a), we then have *Distributed equivalent circuit*

$$\lim_{Dz \to 0} \frac{V(z + Dz, t) - V(z, t)}{Dz} = -\mathcal{L}\frac{\partial I(z, t)}{\partial t}$$

or, for $\Delta z \to 0$,

$$V(z + \Delta z, t) - V(z, t) = -\mathcal{L}\,\Delta z \frac{\partial I(z, t)}{\partial t} \tag{6.48a}$$

This equation can be represented by the circuit equivalent shown in Fig. 6.13(a) since it satisfies Kirchhoff's voltage law written around the loop $abcda$. Similarly,

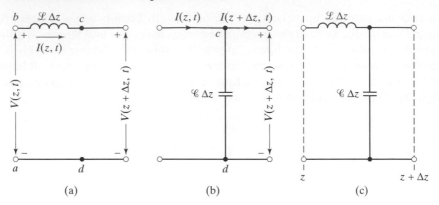

FIGURE 6.13   Development of circuit equivalent for an infinitesimal length Δz of a transmission line.

from (6.47b), we have

$$\lim_{\Delta z \to 0} \frac{I(z + \Delta z, t) - I(z, t)}{\Delta z} = \lim_{\Delta z \to 0} \left[-\mathscr{C} \frac{\partial V(z + \Delta z, t)}{\partial t} \right]$$

or, for $\Delta z \to 0$,

$$I(z + \Delta z, t) - I(z, t) = -\mathscr{C} \,\Delta z \, \frac{\partial V(z + \Delta z, t)}{\partial t} \tag{6.48b}$$

This equation can be represented by the circuit equivalent shown in Fig. 6.13(b) since it satisfies Kirchhoff's current law written for node c. Combining the two equations, we then obtain the equivalent circuit shown in Fig. 6.13(c) for a section Δz of the line. It then follows that the circuit representation for a portion of length l of the line consists of an infinite number of such sections in cascade, as shown in Fig. 6.14. Such a circuit is known as a *distributed circuit* as opposed to the *lumped circuits* that are familiar in circuit theory. The distributed circuit notion arises from the fact that the inductance and capacitance are distributed uniformly and overlappingly along the line.

   A more physical interpretation of the distributed-circuit concept follows from energy considerations. We know that the uniform plane wave propagation

FIGURE 6.14   Distributed circuit representation of a transmission line.

between the conductors of the line is characterized by energy storage in the electric and magnetic fields. If we consider a section Δz of the line, the energy stored in the electric field in this section is given by

$$W_e = \frac{1}{2}\varepsilon E_x^2(\text{volume}) = \frac{1}{2}\varepsilon E_x^2(dw\,\Delta z)$$

$$= \frac{1}{2}\frac{\varepsilon w}{d}(E_x d)^2\,\Delta z = \frac{1}{2}\mathscr{C}\,\Delta z\,V^2 \tag{6.49a}$$

The energy stored in the magnetic field in that section is given by

$$W_m = \frac{1}{2}\mu H_y^2(\text{volume}) = \frac{1}{2}\mu H_y^2(dw\,\Delta z)$$

$$= \frac{1}{2}\frac{\mu d}{w}(H_y w)^2\,\Delta z = \frac{1}{2}\mathscr{L}\,\Delta z\,I^2 \tag{6.49b}$$

Thus, we note that $\mathscr{L}$ and $\mathscr{C}$ are elements associated with energy storage in the magnetic field and energy storage in the electric field, respectively, for a given infinitesimal section of the line. Since these phenomena occur continuously and since they overlap, the inductance and capacitance must be distributed uniformly and overlappingly along the line.

Thus far, we have introduced the transmission-line equations and the distributed-circuit concept by considering the parallel-plate line in which the waves are uniform plane waves. In the general case of a line having conductors with arbitrary cross sections, the fields consist of both x- and y-components and are dependent on x- and y-coordinates in addition to the z-coordinate. Thus, the fields between the conductors are given by

TEM waves

$$\mathbf{E} = E_x(x, y, z, t)\mathbf{a}_x + \mathbf{E}_y(x, y, z, t)\mathbf{a}_y$$
$$\mathbf{H} = H_x(x, y, z, t)\mathbf{a}_x + \mathbf{H}_y(x, y, z, t)\mathbf{a}_y$$

These fields are no longer uniform in x and y but are directed entirely transverse to the direction of propagation, that is, the z-axis, which is the axis of the transmission line. Hence, they are known as *transverse electromagnetic waves,* or *TEM waves.* The uniform plane waves are simply a special case of the transverse electromagnetic waves. The transmission-line equations (6.47a) and (6.47b) and the distributed equivalent circuit of Fig. 6.14 hold for all transmission lines made of perfect conductors and perfect dielectric, that is, for all lossless transmission lines. The quantities that differ from one line to another are the line parameters $\mathscr{L}$ and $\mathscr{C}$, which depend on the geometry of the line. Since there is no z-component of $\mathbf{H}$, the electric-field distribution in any given transverse plane at any given instant of time is the same as the static electric-field distribution resulting from the application of a potential difference between the conductors equal to the line voltage in that plane at that instant of time. Similarly, since there is no z-component of $\mathbf{E}$, the magnetic-field distribution in any given transverse plane at any given instant of time is the same as the static magnetic-field distribution resulting from current flow on the conductors equal to the line current in that plane at that instant of time. Thus, the values of $\mathscr{L}$ and $\mathscr{C}$ are the same as those obtainable from static field considerations.

General solution

Before we consider several common types of lines, we shall show that the relation (6.46) is valid in general by obtaining the general solution for the transmission-line equations (6.47a) and (6.47b). To do this, we note their analogy with the field equations (5.7a) and (5.7b) in Section 5.1, as follows:

$$\frac{\partial E_x}{\partial z} = -\mu_0 \frac{\partial H_y}{\partial t} \longleftrightarrow \frac{\partial V}{\partial z} = -\mathscr{L} \frac{\partial I}{\partial t}$$

$$\frac{\partial H_y}{\partial z} = -\varepsilon_0 \frac{\partial E_x}{\partial t} \longleftrightarrow \frac{\partial I}{\partial z} = -\mathscr{C} \frac{\partial V}{\partial t}$$

The solutions to (6.47a) and (6.47b) can therefore be written by letting

$$E_x \longrightarrow V$$
$$H_y \longrightarrow I$$
$$\mu_0 \longrightarrow \mathscr{L}$$
$$\varepsilon_0 \longrightarrow \mathscr{C}$$

in the solutions (5.13) and (5.14) to the field equations. Thus, we obtain

$$V(z, t) = Af(t - z\sqrt{\mathscr{L}\mathscr{C}}) + Bg(t + z\sqrt{\mathscr{L}\mathscr{C}}) \qquad (6.50a)$$

$$I(z, t) = \frac{1}{\sqrt{\mathscr{L}/\mathscr{C}}}[Af(t - z\sqrt{\mathscr{L}\mathscr{C}}) - Bg(t + z\sqrt{\mathscr{L}\mathscr{C}})] \qquad (6.50b)$$

These solutions represent voltage and current traveling waves propagating along the $+z$- and $-z$-directions with velocity

$$v_p = \frac{1}{\sqrt{\mathscr{L}\mathscr{C}}} \qquad (6.51)$$

in view of the arguments $(t \mp z\sqrt{\mathscr{L}\mathscr{C}})$ for the functions f and g. We, however, know that the velocity of propagation in terms of the dielectric parameters is given by

$$v_p = \frac{1}{\sqrt{\mu\varepsilon}} \qquad (6.52)$$

Therefore, it follows that

$$\mathscr{L}\mathscr{C} = \mu\varepsilon \qquad (6.53)$$

We now define the *characteristic impedance* of the line to be

$$Z_0 = \sqrt{\frac{\mathscr{L}}{\mathscr{C}}} \qquad (6.54)$$

so that (6.50a) and (6.50b) become

$$V(z, t) = Af\left(t - \frac{z}{v_p}\right) + Bg\left(t + \frac{z}{v_p}\right) \qquad (6.55a)$$

$$I(z, t) = \frac{1}{Z_0}\left[Af\left(t - \frac{z}{v_p}\right) - Bg\left(t + \frac{z}{v_p}\right)\right] \qquad (6.55b)$$

where we have also substituted v_p for $1/\sqrt{\mathscr{L}\mathscr{C}}$. From (6.55a) and (6.55b), it can be seen that *the characteristic impedance is the ratio of the voltage to current in the (+) wave or the negative of the same ratio for the (−) wave.* It is analogous to the intrinsic impedance of the dielectric medium but not necessarily equal to it. For example, for the parallel-plate line,

$$Z_0 = \sqrt{\frac{\mu d/w}{\varepsilon w/d}}$$

$$= \eta\,\frac{d}{w} \qquad (6.56)$$

is not equal to η unless d/w is equal to 1. In fact for d/w equal to 1, (6.56) is strictly not valid because fringing of the fields cannot be neglected. Note also that the characteristic impedance of a lossless line is a purely real quantity. We shall find in Section 7.6 that for a lossy line, the characteristic impedance is complex just as the intrinsic impedance of a lossy medium is complex.

Equations (6.55a) and (6.55b) are the general solutions for the voltage and current along a lossless line in terms of v_p and Z_0, the parameters that characterize the propagation along the line. Whereas v_p is dependent on the dielectric as given by (6.52), Z_0 is dependent on the dielectric as well as the geometry associated with the line, in view of (6.54). Combining (6.53) and (6.54), we note that

$$Z_0 = \frac{\sqrt{\mu\varepsilon}}{\mathscr{C}} \qquad (6.57)$$

Thus, the determination of Z_0 for a given line involving a given homogeneous dielectric medium requires simply the determination of $\mathscr{C}$ of the line and then using (6.57). Since the dielectrics of common transmission lines are generally nonmagnetic, we can further express the propagation parameters in the manner

$$Z_0 = \frac{\sqrt{\varepsilon_r}}{c\mathscr{C}} \qquad (6.58a)$$

$$v_p = \frac{c}{\sqrt{\varepsilon_r}} \qquad (6.58b)$$

where ε_r is the relative permittivity of the dielectric, and $c = 1/\sqrt{\mu_0\varepsilon_0}$ is the velocity of light in free space.

Microstrip line

    If the cross section of a transmission line involves more than one dielectric, the situation corresponds to inhomogeneity. An example of this type of line is the microstrip line, used extensively in microwave integrated circuitry and digital systems. The basic microstrip line consists of a high-permittivity substrate material with a conductor strip applied to one side and a conducting ground plane applied to the other side, as shown by the cross-sectional view in Fig. 6.15(a). The approximate electric field distribution is shown in Fig. 6.15(b). Since it is not possible to satisfy the boundary condition of equal phase velocities parallel to the air-dielectric interface with pure TEM waves, the situation for the microstrip line does not correspond exactly to TEM wave propagation, as is the case with any other line involving multiple dielectrics.

    The determination of Z_0 and v_p for the case of a line with multiple dielectrics involves a modified procedure, assuming that the inhomogeneity has no effect on $\mathcal{L}$ and the propagation is TEM. Thus, if $\mathcal{C}_0$ is the capacitance per unit length of the line with all the dielectrics replaced by free space and $\mathcal{C}$ is the capacitance per unit length of the line with the dielectrics in place and computed from static field considerations, we can write

$$Z_0 = \sqrt{\frac{\mathcal{L}}{\mathcal{C}}} = \sqrt{\frac{\mathcal{L}\mathcal{C}_0}{\mathcal{C}\mathcal{C}_0}} = \frac{1}{c\sqrt{\mathcal{C}\mathcal{C}_0}} \tag{6.59a}$$

$$v_p = \frac{1}{\sqrt{\mathcal{L}\mathcal{C}}} = \frac{1}{\sqrt{\mathcal{L}\mathcal{C}_0}}\sqrt{\frac{\mathcal{C}_0}{\mathcal{C}}} = c\sqrt{\frac{\mathcal{C}_0}{\mathcal{C}}} \tag{6.59b}$$

where we have assumed nonmagnetic dielectrics. To express (6.59a) and (6.59b) in the form of (6.58a) and (6.58b), respectively, we define an effective relative

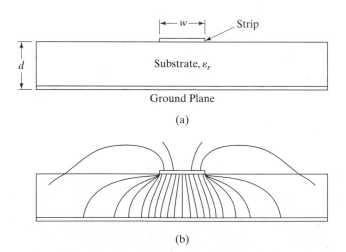

(a)

(b)

FIG. 6.15   (a) Transverse cross-sectional view of a microstrip line. (b) Approximate electric-field distribution in the transverse plane.

permittivity $\varepsilon_{r_{\text{eff}}} = \mathscr{E}/\mathscr{E}_0$ so that

$$Z_0 = \frac{\sqrt{\varepsilon_{r_{\text{eff}}}}}{c\mathscr{E}} \tag{6.60a}$$

$$v_p = \frac{c}{\sqrt{\varepsilon_{r_{\text{eff}}}}} \tag{6.60b}$$

Thus, the determination of Z_0 and v_p requires the knowledge of both $\mathscr{E}$ and $\mathscr{E}_0$.

The techniques for finding $\mathscr{E}$ (and $\mathscr{E}_0$) and, hence, the propagation parameters can be broadly divided into three categories: (1) analytical, (2) numerical, and (3) graphical.

A. Analytical Techniques

The analytical techniques are based on the closed-form solution of Laplace's equation subject to the boundary conditions or the equivalent of such a solution. We have already discussed these techniques in Sections 6.1 and 6.2 for several configurations. Hence, without further discussion, we shall simply list in Table 6.3 the expressions for Z_0 for some common types of lines, shown by cross-sectional views in Fig. 6.16. Note that in Table 6.3, $\eta = \sqrt{\mu/\varepsilon}$ is the intrinsic impedance of the dielectric medium associated with the line. *Several common types of lines*

B. Numerical Techniques

When a closed-form solution is not possible or when the approximation permitting a closed-form solution breaks down, numerical techniques can be employed. These are discussed in Chapter 12.

C. Graphical Technique

For a line with arbitrary cross section and involving a homogeneous dielectric, an approximate value of $\mathscr{E}$ and, hence, of Z_0 can be determined by constructing a *field map*, that is, a graphical sketch of the direction lines of the electric field and associated equipotential lines between the conductors. To illustrate this, let us consider the cross section shown in Fig. 6.17. Assuming that the inner conductor is positive with respect to the outer conductor, we can draw the field map from the following considerations. (1) The electric field lines originate on the *Field mapping*

TABLE 6.3 Expressions for Characteristic Impedance for the Lines of Fig. 6.16

Description	Figure	Z_0
Coaxial cable	6.16(a)	$\dfrac{\eta}{2\pi} \ln \dfrac{b}{a}$
Parallel-wire line	6.16(b)	$\dfrac{\eta}{\pi} \cosh^{-1} \dfrac{d}{a}$
Single wire above ground plane	6.16(c)	$\dfrac{\eta}{2\pi} \cosh^{-1} \dfrac{h}{a}$
Shielded parallel-wire line	6.16(d)	$\dfrac{\eta}{\pi} \ln \dfrac{d(b^2 - d^2/4)}{a(b^2 + d^2/4)}$

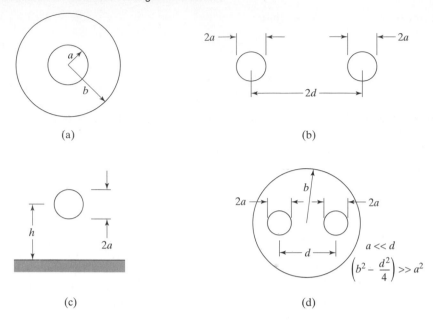

FIG. 6.16   Cross sections of some common types of transmission lines.

inner conductor and normal to it and terminate on the outer conductor and normal to it, since the tangential component of the electric field on the conductor surface must be zero. (2) The equipotential lines must be everywhere perpendicular to the electric field lines. Thus, suppose that we start with the inner conductor and draw several lines normal to it at several points on the surface, as shown in Fig. 6.17(b). We can then draw a curved line displaced from the conductor surface and such that it is perpendicular everywhere to the electric-field lines of Fig. 6.17(b), as shown in Fig. 6.17(c). This contour represents an equipotential line and forms the basis for further extension of the electric-field lines, as shown in Fig. 6.17(d). A second equipotential line can then be drawn so that it is everywhere perpendicular to the extended electric-field lines and the procedure continued until the entire cross section between the conductors is filled with two sets of orthogonal contours, as shown in Fig. 6.17(e), thereby resulting in a field map made up of curvilinear rectangles. For the actual, time-varying case, the magnetic-field lines are the same as the equipotential lines and the field map represents a sketch of the direction lines of electric and magnetic fields between the conductors.

By drawing the field lines with very small spacings, we can make the rectangles so small that each of them can be considered to be the cross section of a parallel-plate line. If we now replace the equipotential lines by perfect conductors, since it does not violate any boundary condition, it can be seen that the arrangement can be viewed as the parallel combination, in the angular direction, of m number of series combinations of n number of parallel-plate lines in the radial direction, where m is the number of rectangles in the angular direction, that is, along

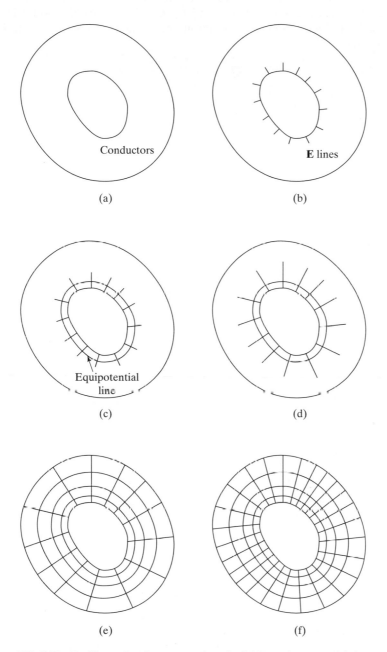

FIG. 6.17   For illustrating the construction of a field map for a transmission line of arbitrary cross section.

a magnetic-field line, and n is the number of rectangles in the radial direction, that is, along an electric field line. If $Q_1, Q_2, \ldots, Q_m$ are the charges per unit length associated with the angular direction and $V_1, V_2, \ldots, V_n$ are the potential differences associated with the radial direction, the capacitance per unit length of the line is given by

$$
\begin{aligned}
\mathscr{C} &= \frac{Q}{V} = \frac{Q_1 + Q_2 + \cdots + Q_m}{V_1 + V_2 + \cdots + V_n} \\
&= \frac{1}{\dfrac{V_1}{Q_1} + \dfrac{V_2}{Q_1} + \cdots + \dfrac{V_n}{Q_1}} \\
&\quad + \frac{1}{\dfrac{V_1}{Q_2} + \dfrac{V_2}{Q_2} + \cdots + \dfrac{V_n}{Q_2}} \\
&\quad + \cdots + \frac{1}{\dfrac{V_1}{Q_m} + \dfrac{V_2}{Q_m} + \cdots + \dfrac{V_n}{Q_m}} \\
&= \sum_{i=1}^{m} \frac{1}{\displaystyle\sum_{j=1}^{n} \dfrac{V_j}{Q_i}} \\
&= \sum_{i=1}^{m} \frac{1}{\displaystyle\sum_{j=1}^{n} \dfrac{1}{\mathscr{C}_{ij}}}
\end{aligned}
$$

where $\mathscr{C}_{ij} = Q_i/V_j$ is the capacitance per unit length corresponding to the rectangle ij. The simplicity of the field mapping technique lies in the fact that if the map consists entirely of curvilinear squares (a curvilinear rectangle becomes a curvilinear square if a circle can be inscribed in it), all $\mathscr{C}_{ij}$ are approximately equal to ε, and we obtain the simple formula

$$
\mathscr{C} \approx \varepsilon \frac{m}{n} \tag{6.61}
$$

and hence

$$
\begin{aligned}
Z_0 &= \frac{\sqrt{\mu\varepsilon}}{\mathscr{C}} \\
&\approx \frac{n}{m}
\end{aligned}
\tag{6.62}
$$

Thus, the determination of Z_0 consists of sketching a field map consisting of curvilinear squares, as shown in Fig. 6.17(f), counting the number of squares in each direction and substituting these values in (6.61). For the rough sketch of Fig. 6.17(f), $m = 26$ and $n = 4$, so that $Z_0 \approx 0.154\eta$.

K6.3. Parallel-plate line; Transmission-line equations; Circuit parameters; Distributed equivalent circuit; TEM waves; Characteristic impedance; Velocity of propagation; Z_0 and v_p for line with homogeneous dielectric; Z_0 and v_p for line with multiple dielectrics; Microstrip line; Field mapping; Curvilinear squares.

D6.5. A parallel-plate transmission line is made up of perfect conductors of width $w = 0.2$ m and separation $d = 0.01$ m. The medium between the plates is a dielectric of $\varepsilon = 2.25\varepsilon_0$ and $\mu = \mu_0$. For a uniform plane wave propagating down the line, find the power crossing a given transverse plane for each of the following cases at a given time in that plane: **(a)** the electric field between the plates is 300π V/m; **(b)** the magnetic field between the plates is 7.5 A/m; **(c)** the voltage across the plates is 4π V; and **(d)** the current along the plates is 0.5 A.
    Ans.   **(a)** 2.25π W;   **(b)** 9π W;   **(c)** 4π W;   **(d)** π W.

D6.6. Find the following: **(a)** the ratio b/a of a coaxial cable of $Z_0 = 50\ \Omega$ if $\varepsilon = 2.56\varepsilon_0$; **(b)** the ratio b/a of a coaxial cable of $Z_0 = 75\ \Omega$ if $\varepsilon = 2.25\varepsilon_0$; and **(c)** the ratio d/a of a parallel-wire line of $Z_0 = 300\ \Omega$ if $\varepsilon = \varepsilon_0$.
    Ans.   **(a)** 3.794;   **(b)** 6.521;   **(c)** 6.132.

D6.7. Two lossless transmission lines 1 and 2 have nonmagnetic ($\mu = \mu_0$), homogeneous perfect dielectrics of $\varepsilon_1 = 2.25\varepsilon_0$ and $\varepsilon_2 = 4\varepsilon_0$, respectively. The values of the ratio m/n corresponding to their curvilinear square field maps are 4 and 5 for lines 1 and 2, respectively. Find **(a)** v_{p1}/v_{p2}, **(b)** $\mathscr{C}_1/\mathscr{C}_2$, and **(c)** Z_{01}/Z_{02}, where the subscripts 1 and 2 denote lines 1 and 2, respectively.
    Ans.   **(a)** 4/3;   **(b)** 0.45;   **(c)** 5/3.

6.4 LINE TERMINATED BY RESISTIVE LOAD

In Section 6.3, we obtained the general solutions to the transmission-line equations for the lossless line, as given by (6.55a) and (6.55b). Since these solutions represent superpositions of $(+)$ and $(-)$ wave voltages and $(+)$ and $(-)$ wave currents, we now rewrite them as   *Notation*

$$V(z, t) = V^+\left(t - \frac{z}{v_p}\right) + V^-\left(t + \frac{z}{v_p}\right) \tag{6.63a}$$

$$I(z, t) = \frac{1}{Z_0}\left[V^+\left(t - \frac{z}{v_p}\right) - V^-\left(t + \frac{z}{v_p}\right)\right] \tag{6.63b}$$

or, more concisely,

$$V = V^+ + V^- \tag{6.64a}$$

$$I = \frac{1}{Z_0}(V^+ - V^-) \tag{6.64b}$$

with the understanding that V^+ is a function of $(t - z/v_p)$ and V^- is a function of $(t + z/v_p)$. In terms of $(+)$ and $(-)$ wave currents, the solution for the current may also be written as

$$I = I^+ + I^- \tag{6.65}$$

Comparing (6.64a) and (6.65), we see that

$$I^+ = \frac{V^+}{Z_0} \tag{6.66a}$$

$$I^- = -\frac{V^-}{Z_0} \tag{6.66b}$$

The minus sign in (6.62b) can be understood if we recognize that in writing (6.64a) and (6.65), we follow the notation that both V^+ and V^- have the same polarities with one conductor (say, a) positive with respect to the other conductor (say, b) and that both I^+ and I^- flow in the positive z-direction along conductor a and return in the negative z-direction along conductor b, as shown in Fig. 6.18. The power flow associated with either wave, as given by the product of the corresponding voltage and current, is then directed in the positive z-direction, as shown in Fig. 6.18. Thus,

$$P^+ = V^+ I^+ = V^+ \left(\frac{V^+}{Z_0}\right) = \frac{(V^+)^2}{Z_0} \tag{6.67a}$$

Since $(V^+)^2$ is always positive, regardless of whether V^+ is numerically positive or negative, (6.67a) indicates that the $(+)$ wave power does actually flow in the positive z-direction, as it should. On the other hand,

$$P^- = V^- I^- = V^- \left(-\frac{V^-}{Z_0}\right) = -\frac{(V^-)^2}{Z_0} \tag{6.67b}$$

Since $(V^-)^2$ is always positive, regardless of whether V^- is numerically positive or negative, the minus sign in (6.67b) indicates that P^- is negative, and, hence, the $(-)$ wave power actually flows in the negative z-direction, as it should.

Excitation by constant voltage source   Let us now consider a line of length l terminated by a load resistance R_L and driven by a constant voltage source V_0 in series with internal resistance R_g, as shown in Fig. 6.19. Note that the conductors of the transmission line are represented by double ruled lines, whereas the connections to the conductors are single ruled lines, to be treated as lumped circuits. We assume that no voltage and current exist on the line for $t < 0$ and the switch S is closed at $t = 0$. We wish to discuss the transient wave phenomena on the line for $t > 0$. The

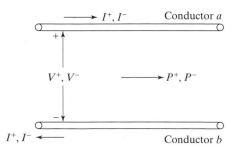

FIGURE 6.18   Polarities for voltages and currents associated with $(+)$ and $(-)$ waves.

FIGURE 6.19 Transmission line terminated by a load resistance R_L and driven by a constant voltage source in series with an internal resistance R_g.

characteristic impedance of the line and the velocity of propagation are Z_0 and v_p, respectively.

When the switch S is closed at $t = 0$, a $(+)$ wave originates at $z = 0$ and travels toward the load. Let the voltage and current of this $(+)$ wave be V^+ and I^+, respectively. Then we have the situation at $z = 0$, as shown in Fig. 6.20(a). Note that the load resistor does not come into play here since the phenomenon is one of wave propagation; hence, until the $(+)$ wave goes to the load, sets up a reflection, and the reflected wave arrives back at the source, the source does not even know of the existence of R_L. This is a fundamental distinction between ordinary (lumped-) circuit theory and transmission-line (distributed-circuit) theory. In ordinary circuit theory, no time delay is involved; the effect of a transient in one part of the circuit is felt in all branches of the circuit instantaneously. In a transmission-line system, the effect of a transient at one location on the line is felt at a different location on the line only after an interval of time that the wave takes to travel from the first location to the second. Returning now to the circuit in Fig. 6.20(a), the various quantities must satisfy the boundary condition, that is, Kirchhoff's voltage law around the loop. Thus, we have

$$V_0 - I^+ R_g - V^+ = 0 \tag{6.68a}$$

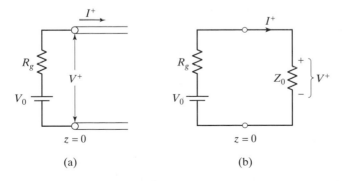

(a)   (b)

FIGURE 6.20 (a) For obtaining the $(+)$ wave voltage and current at $z = 0$ for the line of Fig. 6.19. (b) Equivalent circuit for (a).

We, however, know from (6.66a) that $I^+ = V^+/Z_0$. Hence, we get

$$V_0 - \frac{V^+}{Z_0} R_g - V^+ = 0 \tag{6.68b}$$

or

$$\boxed{\begin{aligned} V^+ &= V_0 \frac{Z_0}{R_g + Z_0} \\[2mm] I^+ &= \frac{V^+}{Z_0} = \frac{V_0}{R_g + Z_0} \end{aligned}}$$

$$\text{(6.69a)}$$
$$\text{(6.69b)}$$

Thus, we note that the situation in Fig. 6.20(a) is equivalent to the circuit shown in Fig. 6.20(b); that is, the voltage source views a resistance equal to the characteristic impedance of the line, across $z = 0$. This is to be expected since only a (+) wave exists at $z = 0$ and the ratio of the voltage to current in the (+) wave is equal to Z_0.

Reflection coefficient   The (+) wave travels toward the load and reaches the termination at $t = l/v_p$. It does not, however, satisfy the boundary condition there since this condition requires the voltage across the load resistance to be equal to the current through it times its value, R_L. But the voltage-to-current ratio in the (+) wave is equal to Z_0. To resolve this inconsistency, there is only one possibility, which is the setting up of a (−) wave, or a reflected wave. Let the voltage and current in this reflected wave be V^- and I^-, respectively. Then the total voltage across R_L is $V^+ + V^-$, and the total current through it is $I^+ + I^-$, as shown in Fig. 6.21(a). To satisfy the boundary condition, we have

$$V^+ + V^- = R_L(I^+ + I^-) \tag{6.70a}$$

But from (6.66a) and (6.66b), we know that $I^+ = V^+/Z_0$ and $I^- = -V^-/Z_0$, respectively. Hence,

$$V^+ + V^- = R_L\left(\frac{V^+}{Z_0} - \frac{V^-}{Z_0}\right) \tag{6.70b}$$

(a)   (b)

FIGURE 6.21   For obtaining the voltages and currents associated with (a) the (−) wave and (b) the (−+) wave, for the line of Fig. 6.19.

or

$$V^- = V^+ \frac{R_L - Z_0}{R_L + Z_0} \tag{6.71}$$

We now define the *voltage reflection coefficient*, denoted by the symbol Γ, as the ratio of the reflected voltage to the incident voltage. Thus,

$$\Gamma = \frac{V^-}{V^+} = \frac{R_L - Z_0}{R_L + Z_0} \tag{6.72}$$

We then note that the *current reflection coefficient* is

$$\frac{I^-}{I^+} = \frac{-V^-/Z_0}{V^+/Z_0} = -\frac{V^-}{V^+} = -\Gamma \tag{6.73}$$

Now, returning to the reflected wave, we observe that this wave travels back toward the source and it reaches there at $t = 2l/v_p$. Since the boundary condition at $z = 0$, which was satisfied by the original $(+)$ wave alone, is then violated, a reflection of the reflection, or a re-reflection, will be set up and it travels toward the load. Let us assume the voltage and current in this re-reflected wave, which is a $(+)$ wave, to be V^{-+} and I^{-+}, respectively, with the superscripts denoting that the $(+)$ wave is a consequence of the $(-)$ wave. Then the total line voltage and the line current at $z = 0$ are $V^+ + V^- + V^{-+}$ and $I^+ + I^- + I^{-+}$, respectively, as shown in Fig. 6.21(b). To satisfy the boundary condition, we have

$$V^+ + V^- + V^{-+} = V_0 - R_g(I^+ + I^- + I^{-+}) \tag{6.74a}$$

But we know that $I^+ = V^+/Z_0, I^- = -V^-/Z_0$, and $I^{-+} = V^{-+}/Z_0$. Hence,

$$V^+ + V^- + V^{-+} = V_0 - \frac{R_g}{Z_0}(V^+ - V^- + V^{-+}) \tag{6.74b}$$

Furthermore, substituting for V^+ from (6.69a), simplifying, and rearranging, we get

$$V^{-+}\left(1 + \frac{R_g}{Z_0}\right) = V^-\left(\frac{R_g}{Z_0} - 1\right)$$

or

$$V^{-+} = V^- \frac{R_g - Z_0}{R_g + Z_0} \tag{6.75}$$

Comparing (6.75) with (6.71), we note that the reflected wave views the source with internal resistance as the internal resistance alone; that is, the voltage source is equivalent to a short circuit insofar as the $(-)$ wave is concerned. A moment's thought will reveal that superposition is at work here. The effect of

Reflection coefficients for some special cases

the voltage source is taken into account by the constant outflow of the original
(+) wave from the source. Hence, for the reflection of the reflection, that is, for
the (−+) wave, we need only consider the internal resistance R_g. Thus, the voltage reflection coefficient formula (6.72) is a general formula and will be used repeatedly. In view of its importance, a brief discussion of the values of Γ for some
special cases is in order as follows:

1. $R_L = 0$, or short-circuited line.

$$\Gamma = \frac{0 - Z_0}{0 + Z_0} = -1$$

The reflected voltage is exactly the negative of the incident voltage,
thereby keeping the voltage across R_L (short circuit) always zero.

2. $R_L = \infty$, or open-circuited line.

$$\Gamma = \frac{\infty - Z_0}{\infty + Z_0} = 1$$

and the current reflection coefficient $= -\Gamma = -1$. Thus, the reflected current is exactly the negative of the incident current, thereby keeping the
current through R_L (open circuit) always zero.

3. $R_L = Z_0$, or line terminated by its characteristic impedance.

$$\Gamma = \frac{Z_0 - Z_0}{Z_0 + Z_0} = 0$$

This corresponds to no reflection, which is to be expected since $R_L(= Z_0)$
is consistent with the voltage-to-current ratio in the (+) wave alone, and,
hence, there is no violation of boundary condition and no need for the
setting up of a reflected wave. Thus, a line terminated by its characteristic
impedance is equivalent to an infinitely long line insofar as the source is
concerned.

Bounce
diagram

Returning to the discussion of the re-reflected wave, we note that this
wave reaches the load at time $t = 3l/v_p$ and sets up another reflected wave. This
process of bouncing back and forth of waves goes on indefinitely until the steady
state is reached. To keep track of this transient phenomenon, we make use of the
bounce-diagram technique. Some other names given for this diagram are *reflection diagram* and *space-time diagram*. We shall introduce the bounce diagram
through a numerical example.

Example 6.6

Let us consider the system shown in Fig. 6.22. Note that we have introduced a new quantity T, which is the one-way travel time along the line from $z = 0$ to $z = l$; that is, instead
of specifying two quantities l and v_p, we specify $T (= l/v_p)$. Using the bounce diagram
technique, we wish to obtain and plot line voltage and current versus t for fixed values z
and line voltage and current versus z for fixed values t.

FIGURE 6.22   Transmission-line system for illustrating the bounce-diagram technique of keeping track of the transient phenomenon.

Before we construct the bounce diagram, we need to compute the following quantities:

$$\text{Voltage carried by the initial } (+) \text{ wave } = 100\frac{60}{40 + 60} = 60 \text{ V}$$

$$\text{Current carried by the initial } (+) \text{ wave } = \frac{60}{60} = 1 \text{ A}$$

$$\text{Voltage reflection coefficient at load, } \Gamma_R = \frac{120 - 60}{120 + 60} = \frac{1}{3}$$

$$\text{Voltage reflection coefficient at source, } \Gamma_S = \frac{40 - 60}{40 + 60} = -\frac{1}{5}$$

Construction of bounce diagrams

The bounce diagram is essentially a two-dimensional representation of the transient waves bouncing back and forth on the line. Separate bounce diagrams are drawn for voltage and current, as shown in Fig. 6.23(a) and (b), respectively. Position (z) on the line is represented horizontally and the time (t) vertically. Reflection coefficient values for the two ends are shown at the top of the diagrams for quick reference. Note that current reflection coefficients are $-\Gamma_R = -\frac{1}{3}$ and $-\Gamma_S = \frac{1}{5}$, respectively, at the load and at the source. Crisscross lines are drawn as shown in the figures to indicate the progress of the wave as a function of both z and t, with the numerical value for each leg of travel shown beside the line corresponding to that leg and approximately at the center of the line. The arrows indicate the directions of travel. Thus, for example, the first line on the voltage bounce diagram indicates that the intial $(+)$ wave of 60 V takes a time of 1 μs to reach the load end of the line. It sets up a reflected wave of 20 V, which travels back to the source, reaching there at a time of 2 μs, which then gives rise to a $(+)$ wave of voltage -4 V, and so on, with the process continuing indefinitely.

Plots of line voltage and current versus t

Now, to sketch the line voltage and/or current versus time at any value of z, we note that since the voltage source is a constant voltage source, each individual wave voltage and current, once the wave is set up at that value of z, continues to exist there forever. Thus, at any particular time, the voltage (or current) at that value of z is a superposition of all the voltages (or currents) corresponding to the crisscross lines preceding that value of time. These values are marked on the bounce diagrams for $z = 0$ and $z = l$. Noting that each value corresponds to the 2-μs time interval between adjacent crisscross lines, we now sketch the time variations of line voltage and current at $z = 0$ and $z = l$, as shown in Figs. 6.24(a) and (b), respectively. Similarly, by observing that the numbers written along the time axis for $z = 0$ are actually valid for any pair of z and t within the triangle ($\triangleright$) inside which they lie and that the numbers written along the time axis for $z = l$

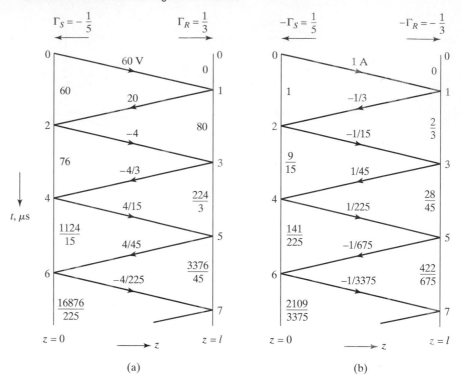

FIGURE 6.23   (a) Voltage and (b) current bounce diagrams, depicting the bouncing back and forth of the transient waves for the system of Fig. 6.22.

are actually valid for any pair of z and t within the triangle ($\lhd$) inside which they lie, we can draw the sketches of line voltage and current versus time for any other value of z. This is done for $z = l/2$ in Fig. 6.24(c).

Steady-state situation

It can be seen from the sketches of Fig. 6.24 that as time progresses, the line voltage and current tend to converge to certain values, which we can expect to be the steady-state values. In the steady state, the situation consists of a single ($+$) wave, which is actually a superposition of the infinite number of transient ($+$) waves, and a single ($-$) wave, which is actually a superposition of the infinite number of transient ($-$) waves. Denoting the steady-state ($+$) wave voltage and current to be V_{SS}^+ and I_{SS}^+, respectively, and the steady-state ($-$) wave voltage and current to be V_{SS}^- and I_{SS}^-, respectively, we obtain from the bounce diagrams

$$V_{SS}^+ = 60 - 4 + \frac{4}{15} - \cdots = 60\left(1 - \frac{1}{15} + \frac{1}{15^2} - \cdots\right) = 56.25 \text{ V}$$

$$I_{SS}^+ = 1 - \frac{1}{15} + \frac{1}{225} - \cdots = 1 - \frac{1}{15} + \frac{1}{15^2} - \cdots = 0.9375 \text{ A}$$

$$V_{SS}^- = 20 - \frac{4}{3} + \frac{4}{45} - \cdots = 20\left(1 - \frac{1}{15} + \frac{1}{15^2} - \cdots\right) = 18.75 \text{ V}$$

$$I_{SS}^- = -\frac{1}{3} + \frac{1}{45} - \frac{1}{675} + \cdots = -\frac{1}{3}\left(1 - \frac{1}{15} + \frac{1}{15^2} - \cdots\right) = -0.3125 \text{ A}$$

(a)

(b)

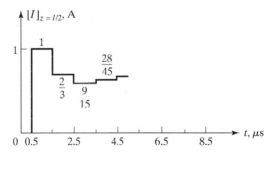

(c)

FIGURE 6.24 Time variations of line voltage and line current at (a) $z = 0$, (b) $z - l$, and (c) $z = l/2$ for the system of Fig. 6.22.

Note that $I_{SS}^+ = V_{SS}^+/Z_0$ and $I_{SS}^- = -V_{SS}^-/Z_0$, as they should be. The steady-state line voltage and current can now be obtained to be

$$V_{SS} = V_{SS}^+ + V_{SS}^- = 75 \text{ V}$$
$$I_{SS} = I_{SS}^+ + I_{SS}^- = 0.625 \text{ A}$$

These are the same as the voltage across R_L and current through R_L if the source and its internal resistance were connected directly to R_L, as shown in Fig. 6.25. This is to be expected since the series inductors and shunt capacitors of the distributed equivalent

FIGURE 6.25   Steady-state equivalent for the system of Fig. 6.22.

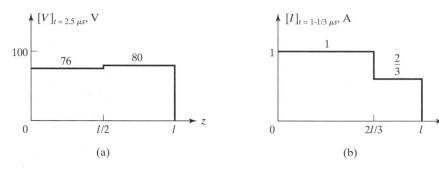

FIGURE 6.26   Variations with z of (a) line voltage for $t = 2.5$ μs and (b) line current for $t = 1\frac{1}{3}\mu$s, for the system of Fig. 6.22.

circuit behave like short circuits and open circuits, respectively, for the constant voltage source in the steady state.

Plots of line voltage and current versus z

Sketches of line voltage and current as functions of distance (z) along the line for any particular time can also be drawn from considerations similar to those employed for the sketches of Fig. 6.24. For example, suppose we wish to draw the sketch of line voltage versus z for $t = 2.5$ μs. Then we note from the voltage bounce diagram that for $t = 2.5$ μs, the line voltage is 76 V from $z = 0$ to $z = l/2$ and 80 V from $z = l/2$ to $z = l$. This is shown in Fig. 6.26(a). Similarly, Fig. 6.26(b) shows the variation of line current versus z for $t = 1\frac{1}{3}\mu$s.

Excitation by pulse voltage source

In Example 6.6, we introduced the bounce-diagram technique for a constant-voltage source. The technique can also be applied if the voltage source is a pulse. In the case of a rectangular pulse, this can be done by representing the pulse as the superposition of two step functions, as shown in Fig. 6.27, and superimposing the bounce diagrams for the two sources one on another. In doing so, we should note that the bounce diagram for one source begins at a value of time greater than zero. Alternatively, the time variation for each wave can be drawn alongside the time axes beginning at the time of start of the wave. These can then be used to plot the required sketches. An example is in order to illustrate this technique, which can also be used for a pulse of arbitrary shape.

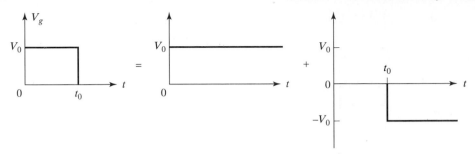

FIGURE 6.27 Representation of a rectangular pulse as the superposition of two step functions.

Example 6.7

Let us assume that the voltage source in the system of Fig. 6.22 is a 100-V rectangular pulse extending from $t = 0$ to $t = 1$ μs and extend the bounce-diagram technique.

Considering, for example, the voltage bounce diagram, we reproduce in Fig. 6.28 part of the voltage bounce diagram of Fig. 6.23(a) and draw the time variations of the individual pulses alongside the time axes, as shown in the figure. Note that voltage axes are chosen such that positive values are to the left at the left end ($z = 0$) of the diagram, but to the right at the right end ($z = l$) of the diagram.

From the voltage bounce diagram, sketches of line voltage versus time at $z = 0$ and $z = l$ can be drawn, as shown in Figs. 6.29(a) and (b), respectively. To draw the sketch of line voltage versus time for any other value of z, we note that as time progresses, the (+) wave pulses slide down the crisscross lines from left to right, whereas the (−) wave pulses

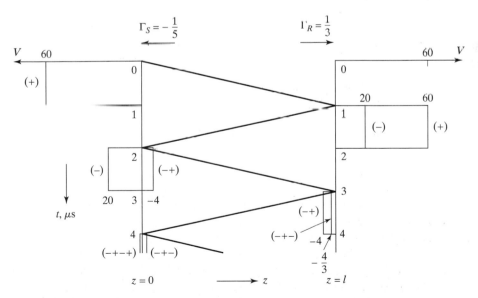

FIGURE 6.28 Voltage bounce diagram for the system of Fig. 6.22 except that the voltage source is a rectangular pulse of 1-μs duration from $t = 0$ to $t = 1$ μs.

(a)

(b)

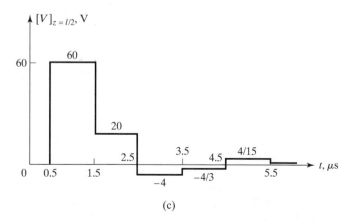

(c)

FIGURE 6.29   Time variations of line voltage at (a) $z = 0$, (b) $z = l$, and (c) $z = l/2$ for the system of Fig. 6.22, except that the voltage source is a rectangular pulse of 1-μs duration from $t = 0$ to $t = 1$ μs.

FIGURE 6.30   Variations with z of (a) the
$(-+)$ wave voltage, (b) the $(-)$ wave
voltage, and (c) the total line voltage, at
$t - 2.25\ \mu s$ for the system of Fig. 6.22,
except that the voltage source is a
rectangular pulse of 1-μs duration from
$t = 0$ to $t = 1\ \mu s$.

slide down from right to left. Thus, to draw the sketch for $z = l/2$, we displace the time plots of the $(+)$ waves at $z = 0$ and of the $(-)$ waves at $z = l$ forward in time by 0.5 μs, that is, delay them by 0.5 μs, and add them to obtain the plot shown in Fig. 6.29(c).

Sketches of line voltage versus distance (z) along the line for fixed values of time can also be drawn from the bounce diagram based on the phenomenon of the individual pulses sliding down the crisscross lines. Thus, if we wish to sketch $V(z)$ for $t = 2.25\ \mu s$, then we take the portion from $t = 2.25\ \mu s$ back to $t = 2.25 - 1 = 1.25\ \mu s$ (since the one-way travel time on the line is 1 μs) of all the $(+)$ wave pulses at $z = 0$ and lay them on the line from $z = 0$ to $z = l$, and we take the portion from $t = 2.25\ \mu s$ back to $t = 2.25 - 1 = 1.25\ \mu s$ of all the $(-)$ wave pulses at $z = l$ and lay them on the line from $z = l$ back to $z = 0$. In this case, we have only one $(+)$ wave pulse—that of the $(-+)$ wave—and only one $(-)$ wave pulse—that of the $(-)$ wave—as shown in Figs. 6.30(a) and (b). The line voltage is then the superposition of these two waveforms, as shown in Fig. 6.30(c).

Similar considerations apply for the current bounce diagram and plots of line current versus t for fixed values of z and line current versus z for fixed values of t.

K6.4. $(+)$ wave; $(-)$ wave; Voltage reflection coefficient; Current reflection coefficient; Transient bouncing of waves; Voltage bounce diagram; Current bounce diagram; Steady-state situation; Rectangular pulse voltage source; Superposition.

D6.8. In the system shown in Fig. 6.31, the switch S is closed at $t = 0$. Find the value of R_L for each of the following cases: **(a)** $V(0.5l,\ 1.7\ \mu s) = 48$ V; **(b)** $V(0.6l,\ 2.8\ \mu s) = 76$ V; **(c)** $I(0.3l, 4.4\ \mu s) = 1$ A; and **(d)** $I(0.4l, \infty) = 2.5$ A.
  Ans.   **(a)** 40 Ω;   **(b)** 120 Ω;   **(c)** 60 Ω;   **(d)** 0 Ω.

FIGURE 6.31   For Problem D.6.8.

D6.9.   For a line of characteristic impedance 75 Ω terminated by a resistance and driven by a constant-voltage source in series with an internal resistance, the line voltage and current in the steady state are known to be 30 V and 1.2 A, respectively. Find **(a)** the $(+)$ wave voltage; **(b)** the $(-)$ wave voltage; **(c)** the $(+)$ wave current; and **(d)** the $(-)$ wave current in the steady state.
Ans.   **(a)** 60 V;   **(b)** -30 V;   **(c)** 0.8 A;   **(d)** 0.4 A.

D6.10.   In Fig 6.32, a line of characteristic impedance 50 Ω is terminated by a *passive* nonlinear element. A $(+)$ wave of constant voltage V_0 is incident on the termination. If the volt-ampere characteristic of the nonlinear element in the region of interest is $V_L = 50\, I_L^2$, find the $(-)$ wave voltage for each of the following values of V_0: **(a)** 36 V; **(b)** 50 V; and **(c)** 66 V.
Ans.   **(a)** -4 V;   **(b)** 0 V;   **(c)** 6 V.

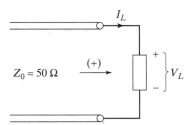

FIGURE 6.32   For Problem D.6.10.

6.5   TRANSMISSION-LINE DISCONTINUITY

Junction between two lines   We now consider the case of a junction between two lines having different values for the parameters Z_0 and v_p, as shown in Fig. 6.33. Assuming that a $(+)$ wave of voltage V^+ and a current I^+ is incident on the junction from line 1, we find that the $(+)$ wave alone cannot satisfy the boundary condition at the junction since the voltage-to-current ratio for that wave is Z_{01}, whereas the characteristic impedance of line 2 is $Z_{02} \neq Z_{01}$. Hence, a reflected wave and a transmitted wave are set up such that the boundary conditions are satisfied. Let the voltages and currents in these waves be V^-, I^-, and V^{++}, I^{++}, respectively, where the superscript $++$ denotes that the transmitted wave is a $(+)$ wave resulting from the incident $(+)$ wave. We then have the situation shown in Fig. 6.34(a).

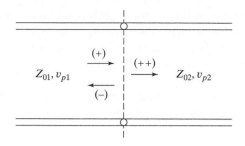

FIGURE 6.33 Transmission-line junction for illustrating reflection $(-)$ and transmission $(++)$ resulting from an incident $(+)$ wave.

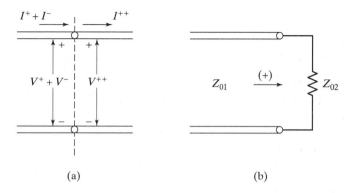

(a)　　　　　　　　　(b)

FIGURE 6.34 (a) For obtaining the reflected $(-)$ wave and transmitted $(++)$ wave voltages and currents for the system of Fig. 6.33. (b) Equivalent to (a) for using the reflection coefficient concept.

Using the boundary conditions at the junction, we then write

$$V^+ + V^- = V^{++} \tag{6.76a}$$

$$I^+ + I^- = I^{++} \tag{6.76b}$$

But we know that $I^+ = V^+/Z_{01}$, $I^- = -V^-/Z_{01}$, and $I^{++} = V^{++}/Z_{02}$. Hence, (6.76b) becomes

$$\frac{V^+}{Z_{01}} - \frac{V^-}{Z_{01}} = \frac{V^{++}}{Z_{02}} \tag{6.77}$$

Combining (6.76a) and (6.77), we have

$$V^+ + V^- = \frac{Z_{02}}{Z_{01}}(V^+ - V^-)$$

$$V^-\left(1 + \frac{Z_{02}}{Z_{01}}\right) = V^+\left(\frac{Z_{02}}{Z_{01}} - 1\right)$$

or

$$\boxed{\Gamma = \frac{V^-}{V^+} = \frac{Z_{02} - Z_{01}}{Z_{02} + Z_{01}}} \tag{6.78}$$

Thus, to the incident wave, the transmission line to the right looks like its characteristic impedance Z_{02}, as shown in Fig. 6.34(b). The difference between a resistive load of Z_{02} and a line of characteristic impedance Z_{02} is that, in the first case, power is dissipated in the load, whereas in the second case, the power is transmitted into the line.

Transmission coefficient   We now define the *voltage transmission coefficient,* denoted by the symbol τ_V, as the ratio of the transmitted wave voltage to the incident wave voltage. Thus,

$$\boxed{\tau_V = \frac{V^{++}}{V^+} = \frac{V^+ + V^-}{V^+} = 1 + \frac{V^-}{V^+} = 1 + \Gamma} \qquad (6.79)$$

The *current transmission coefficient,* τ_C, which is the ratio of the transmitted wave current to the incident wave current, is given by

$$\boxed{\tau_C = \frac{I^{++}}{I^+} = \frac{I^- + I^-}{I^+} = 1 + \frac{I^-}{I^+} = 1 - \Gamma} \qquad (6.80)$$

At this point, one may be puzzled to note that the transmitted voltage can be greater than the incident voltage if Γ is positive. However, this is not of concern, since then the transmitted current will be less than the incident current. Similarly, the transmitted current is greater than the incident current when Γ is negative, but then the transmitted voltage is less than the incident voltage. In fact, what is important is that the transmitted power P^{++} is always less than (or equal to) the incident power P^+, since

$$\begin{aligned} P^{++} &= V^{++}I^{++} = V^+(1 + \Gamma)I^+(1 - \Gamma) \\ &= V^+I^+(1 - \Gamma^2) = (1 - \Gamma^2)P^+ \end{aligned} \qquad (6.81)$$

and $(1 - \Gamma^2) \leq 1$, irrespective of whether Γ is positive or negative.

We shall illustrate the application of reflection and transmission at a junction between lines by means of an example.

Example 6.8

System of three lines   Let us consider the system of three lines in cascade, driven by a unit impulse voltage source $\delta(t)$, as shown in Fig. 6.35(a). We wish to find the output voltage V_o, thereby obtaining the unit impulse response.

To find the output voltage, we draw the voltage bounce diagram, as shown in Fig. 6.35(b). In drawing the bounce diagram, we note that since the internal resistance of the voltage source is 50 Ω, which is equal to Z_{01}, the strength of the impulse that the generator supplies to line 1 is $\frac{1}{2}$. The strengths of the various impulses propagating in the lines are then governed by the reflection and transmission coefficients indicated on the diagram. Also note that the numbers indicated beside the crisscross lines are simply the strengths of the impulses and do not represent constant voltages.

(a)

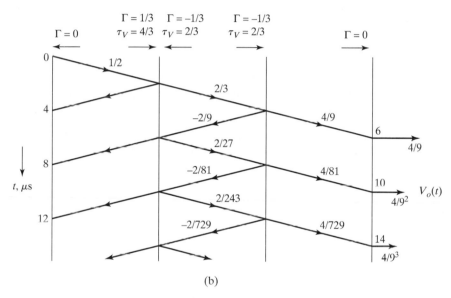

(b)

FIGURE 6.35 (a) System of three lines in cascade driven by a unit impulse voltage source. (b) Voltage bounce diagram for finding the output voltage $V_o(t)$ for the system of (a).

From the bounce diagram, we note that the output voltage is a series of impulses. *Unit impulse response*
In fact, the phenomenon can be visualized without even drawing the bounce diagram, and the strengths of the impulses can be computed. Thus, the strength of the first impulse, which occurs at $t = T_1 + T_2 + T_3 = 6\ \mu s$, is

$$1 \times \frac{50}{50 + 50} \times \left(1 + \frac{100 - 50}{100 + 50}\right) \times \left(1 + \frac{50 - 100}{50 + 100}\right) = 1 \times \frac{1}{2} \times \frac{4}{3} \times \frac{2}{3} = \frac{4}{9}$$

Each succeeding impulse is due to the additional reflection and re-reflection of the previous impulse at the right and left end, respectively, of line 2. Hence, each succeeding impulse occurs $2T_2$, or $4\ \mu s$, later than the preceding one, and its strength is

$$\left(\frac{50 - 100}{50 + 100}\right)\left(\frac{50 - 100}{50 + 100}\right) = \left(-\frac{1}{3}\right)\left(-\frac{1}{3}\right) = \frac{1}{9}$$

times the strength of the previous impulse. We can now write the output voltage as

$$V_o(t) = \frac{4}{9}\delta(t - 6 \times 10^{-6}) + \frac{4}{9^2}\delta(t - 10 \times 10^{-6})$$

$$+ \frac{4}{9^3}\delta(t - 14 \times 10^{-6}) + \cdots \qquad (6.82)$$

$$= \frac{4}{9}\sum_{n=0}^{\infty}\left(\frac{1}{9}\right)^n\delta(t - 4n \times 10^{-6} - 6 \times 10^{-6})$$

Note that $\frac{4}{9}$ is the strength of the first impulse and $\frac{1}{9}$ is the multiplication factor for each succeeding impulse. In terms of T_1, T_2, and T_3, we have

$$V_o(t) = \frac{4}{9}\sum_{n=0}^{\infty}\left(\frac{1}{9}\right)^n\delta[t - 2nT_2 - (T_1 + T_2 + T_3)]$$

$$= \frac{4}{9}\sum_{n=0}^{\infty}\left(\frac{1}{9}\right)^n\delta(t - 2nT_2 - T_0) \qquad (6.83)$$

where we have replaced $T_1 + T_2 + T_3$ by T_0.

Frequency response

Proceeding further, since the unit impulse response of the system is a series of impulses delayed in time, the response to any other excitation can be found by the superposition of time functions obtained by delaying the exciting function and multiplying by appropriate constants. In particular, by considering $V_g(t) = \cos \omega t$, we can find the frequency response of the system. Thus, assuming $V_g(t) = \cos \omega t$ and substituting the cosine function for the impulse function in (6.83), we obtain the corresponding output voltage to be

$$V_o(t) = \frac{4}{9}\sum_{n=0}^{\infty}\left(\frac{1}{9}\right)^n \cos \omega(t - 2nT_2 - T_0) \qquad (6.84)$$

The complex voltage $\overline{V}_o(\omega)$ is then given by

$$\overline{V}_o(\omega) = \frac{4}{9}\sum_{n=0}^{\infty}\left(\frac{1}{9}\right)^n e^{-j\omega(2nT_2 + T_0)}$$

$$= \frac{4}{9}e^{-j\omega T_0}\sum_{n=0}^{\infty}\left(\frac{1}{9}e^{-j2\omega T_2}\right)^n \qquad (6.85)$$

$$= \frac{(4/9)e^{-j\omega T_0}}{1 - (1/9)e^{-j2\omega T_2}}$$

Without going into a detailed discussion of the result given by (6.85), we can conclude the following: maximum amplitude occurs for $2\omega T_2 = 2m\pi, m = 0, 1, 2, \ldots$; that is, for $\omega = m\pi/T_2, m = 0, 1, 2, \ldots$, and its value is $\frac{4}{9}/(1 - \frac{1}{9}) = 0.5$. Minimum amplitude occurs for $2\omega T_2 = (2m + 1)\pi, m = 0, 1, 2, \ldots$; that is, for $\omega = (2m + 1)\pi/2T_2, m = 0, 1, 2, \ldots$, and its value is $\frac{4}{9}/(1 + \frac{1}{9}) = 0.4$. The amplitude response therefore can be roughly sketched, as shown in Fig. 6.36.

Radome

A practical situation in which the discussion of this example is applicable is in the design of a radome, which is an enclosure for protecting an antenna from the weather while allowing transparency for electromagnetic waves. A simple, idealized, planar version of the radome is a dielectric slab with free space on either side of it, as shown in Fig. 6.37. For reflection and transmission of uniform plane waves incident normally onto the dielectric slab, the arrangement is equivalent to three lines in cascade, with the characteristic impedances equal to the intrinsic impedances of the media and the velocities of

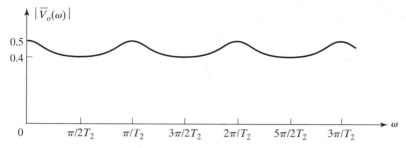

FIGURE 6.36 Rough sketch of amplitude response versus frequency for the system of Fig. 6.35(a) for sinusoidal excitation.

FIGURE 6.37 Perfect dielectric slab with free space on either side.

propagation equal to those in the media. Thus, the amplitude versus frequency response is of the same form as that in Fig. 6.36, where T_2 is the one-way travel time in the dielectric slab and the maximum is 1 instead of 0.5 (the factor of 0.5 in Fig. 6.36 is due to voltage drop across the internal resistance of the source in the transmission-line system). Hence, the lowest frequency for which the dielectric slab is completely transparent is $\omega = \pi/T_2 = \pi c/l\sqrt{\varepsilon_r\mu_r}$ or $f = c/2l\sqrt{\varepsilon_r\mu_r}$. Conversely, for a given frequency f, the minimum thickness for which the slab is transparent is $l = c/2f\sqrt{\varepsilon_r\mu_r} = \lambda/2$, where λ is the wavelength in the dielectric, corresponding to f.

We shall now discuss *time-domain reflectometry*, abbreviated TDR, a technique by means of which discontinuities in transmission-line systems can be located by making measurements with pulses. The block diagram of a typical TDR system is shown in Fig. 6.38. It consists of a pulse generator whose output is con-

Time-domain reflectometry

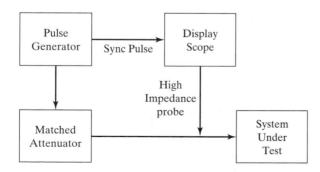

FIGURE 6.38 Block diagram of a typical time-domain reflectometer.

nected to the system under test through a matched attenuator. Voltage pulses are propagated down the transmission-line system, and the incident and re-flected pulses are monitored by the display scope using a high-impedance probe. The matched attenuator serves the purpose of absorbing the pulses arriving back from the system so that reflections of those pulses are not produced. We shall illustrate the application of a TDR system by means of an example.

Example 6.9

Let us consider a transmission line under test, as shown in Fig. 6.39, in which a disconti-nuity exists at $z = 4$ m and the line is short-circuited at the far end. We shall first analyze the system to obtain the waveform measured by a TDR system connected at the input end $z = 0$, assuming the TDR pulses to be of amplitude 1 V, duration 10 ns, and repeti-tion rate 10^5 Hz. We shall then discuss how one can deduce the information about the dis-continuity from the TDR measurement.

Assuming that a pulse from the TDR system is incident on the input of the system under test at $t = 0$, we draw the voltage-bounce diagram, as shown in Fig. 6.40. Note that

FIGURE 6.39   Transmission line with discontinuity under test by a TDR system.

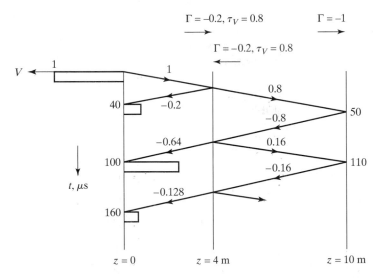

FIGURE 6.40   Voltage-bounce diagram for the system of Fig. 6.39, for TDR pulses of amplitude 1 V.

for a pulse incident on the discontinuity from either side, the resistance viewed is the parallel combination of 120 Ω and Z_0 ($= 60\ \Omega$) of the line, or 40 Ω. Hence, the reflection and transmission coefficients for the voltage are given, respectively, by

$$\Gamma = \frac{40 - 60}{40 + 60} = -0.2$$

$$\tau_V = 1 + \Gamma = 0.8$$

From the bounce diagram, the voltage pulses that would be viewed on the display scope of the TDR system up to $t = 200$ ns are shown in Fig. 6.41. Subsequent pulses become smaller and smaller in amplitude as time progresses and diminish to insignificant values well before $t = 10\ \mu$s, which is the period of the TDR pulses.

Now, to discuss how one can deduce information about the discontinuity from the TDR display of Fig. 6.41, without a priori knowledge of the discontinuity but knowing the values of Z_0 and v_p of the line and that the line is short-circuited at the far end of unknown distance from the input, we proceed in the following manner:

The first pulse is the outgoing pulse from the TDR system. The second pulse arriving at the input of the system under test at $t = 40$ ns is due to reflection from a discontinuity, since if there is no discontinuity, the voltage of the second pulse should be -1 V and there should be no subsequent pulses. From the voltage of the second pulse, we know that the reflection coefficient at the discontinuity is -0.2. The effective resistance R_L seen by the incident pulse is therefore given by the solution of

$$\frac{R_L - 60}{R_L + 60} = -0.2$$

which is 40 Ω. Since this value is less than the Z_0 of the line, the discontinuity must be due entirely to a resistance in parallel with the line or due to a combination of series and parallel resistors; it cannot be due entirely to a resistance in series with the line. Let us proceed with the assumption of a parallel resistor alone. Then the value of this resistance R must be such that

$$\frac{60R}{60 + R} = 40$$

solving which we obtain $R = 120\ \Omega$. The location of the discontinuity can be deduced by multiplying v_p by 20 ns, which is one-half of the time interval between the first and second pulses. Thus, the location is $2 \times 10^8 \times 20 \times 10^{-9} = 4$ m.

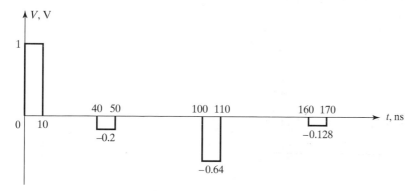

FIGURE 6.41 Voltage versus time at the input of the transmission line of Fig. 6.39, as displayed by the TDR system.

Continuing, let us postulate that the third pulse of -0.64 V at $t = 100$ ns is due to reflection occurring at a second discontinuity located at $z = 4 + 2 \times 10^8 \times (60/2) \times 10^{-9} = 10$ m. In terms of the reflection coefficient at the second discontinuity, denoted Γ_2, the voltage of the third pulse would be $\tau_{VR}\Gamma_2\tau_{VL}$, where τ_{VR} and τ_{VL} are the voltage transmission coefficients at $z = 4$ m, for pulses incident from the right and from the left, respectively. Since τ_{VR} and τ_{VL} are both equal to 0.8, we then have $0.64\,\Gamma_2 = -0.64$ or $\Gamma_2 = -1$, which corresponds to a short circuit, which would then give a fourth pulse of -0.128 V at $t = 160$ ns, and so on. From these reasonings, we confirm the assumption of a parallel resistor of 120 Ω for the discontinuity at $z = 4$ m and also conclude that the short circuit is at $z = 10$ m and that no discontinuities exist between $z = 4$ m and the short circuit. If the value of Γ_2 comes out to be different from -1, then further reasonings are necessary to deduce the information. It should also be noted that the line of reasoning depends on which of the line parameters are known.

K6.5. Voltage transmission coefficient; Current transmission coefficient; Unit impulse response; Frequency response; Time-domain reflectometry.

D6.11. Consider a $(+)$ wave incident from line 1 onto the junction between lines 1 and 2 having characteristic impedances Z_{01} and Z_{02}, respectively. Find the value of Z_{02}/Z_{01} for each of the following cases: **(a)** the reflected wave voltage is $\frac{1}{5}$ times the incident wave voltage: **(b)** the transmitted wave voltage is $\frac{1}{5}$ times the incident wave voltage; **(c)** the reflected wave voltage is $\frac{1}{5}$ times the transmitted wave voltage; and **(d)** the reflected wave current is $\frac{1}{5}$ times the transmitted wave current.

Ans. **(a)** 1.5; **(b)** $\frac{1}{9}$; **(c)** $\frac{5}{3}$; **(d)** $\frac{3}{5}$.

D6.12. The output voltage $V_o(t)$ for a system of three lines in cascade is given by

$$V_o(t) = \frac{1}{4}\sum_{n=0}^{\infty}\left(\frac{1}{3}\right)^n \delta(t - 2 \times 10^{-6}n - 3 \times 10^{-6})$$

when the input voltage $V_i(t) = \delta(t)$. If $V_i(t) = \cos \omega t$, find the amplitude of $V_o(t)$ for each of the following values of ω: **(a)** $10^6\pi$; **(b)** $1.25 \times 10^6\pi$; and **(c)** $1.5 \times 10^6\pi$.

Ans. **(a)** 0.375; **(b)** 0.2372; **(c)** 0.1875.

D6.13. Consider $(n + 1)$ lines, each of characteristic impedance Z_0, emanating from a common junction, as shown in Fig. 6.42 for $n = 2$. For a wave carrying power P

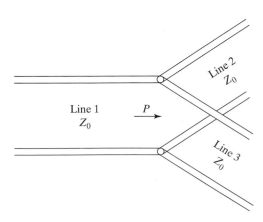

FIGURE 6.42   For Problem D.6.13.

incident on the junction from one of the lines, find the power reflected into that line and the power transmitted into each of the remaining n lines for the following cases: **(a)** $n = 2$; **(b)** $n = 3$; and **(c)** $n = 9$.

Ans. **(a)** $\frac{1}{9}P, \frac{4}{9}P$; **(b)** $\frac{1}{4}P, \frac{1}{4}P$; **(c)** $0.64P, 0.04P$.

6.6 LINES WITH REACTIVE AND NONLINEAR-RESISTIVE ELEMENTS

In this section, we consider situations in which the concept of reflection coeffi- *Inductive* cient as applicable to linear resistive terminations and discontinuities is not use- *termination* ful. These situations involve reactive elements and nonlinear resistive elements. Let us first consider the system shown in Fig. 6.43 in which a line of length l is terminated by an inductor L with zero initial current and a constant-voltage source with internal resistance equal to the characteristic impedance of the line is connected to the line at $t = 0$. The internal resistance is chosen to be equal to Z_0 so that no reflection takes place at the source end. The moment the switch S is closed at $t = 0$, a $(+)$ wave originates at $z = 0$ with voltage $V^+(= V_0/2)$ and current $I^+(= V_0/2Z_0)$ and travels down the line to reach the load end at time T. Since the inductor current cannot change instantaneously from zero to $V_0/2Z_0$, the boundary condition at $z = l$ is violated, and hence a $(-)$ wave is set up. Let the voltage and current in this $(-)$ wave be $V^-(t)$ and $I^-(t)$, respectively. Then the total voltage across L and the total current through L are $(V_0/2) + V^-$ and $(V_0/2Z_0) - (V^-/Z_0)$, respectively, as shown in Fig. 6.44. To satisfy the boundary

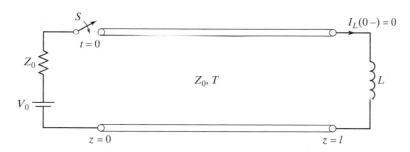

FIGURE 6.43 Line terminated by an inductor with zero initial current and driven by a constant-voltage source in series with internal resistance equal to Z_0 of the line.

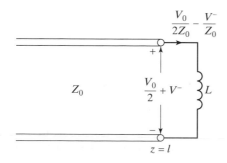

FIGURE 6.44 For obtaining the $(-)$ wave voltage and current for the system of Fig. 6.43.

condition at $z = l$, we then have

$$\frac{V_0}{2} + V^- = L\frac{d}{dt}\left(\frac{V_0}{2Z_0} - \frac{V^-}{Z_0}\right) \tag{6.86}$$

Noting that V_0 is a constant and hence that dV_0/dt is zero, and rearranging we obtain

$$\frac{L}{Z_0}\frac{dV^-}{dt} + V^- = -\frac{V_0}{2} \tag{6.87}$$

This differential equation for $[V^-]_{z=l}$ has to be solved, subject to the initial condition. This initial condition is that the current through the inductor is zero at $t = T$; that is, the inductor behaves initially like an open circuit. Thus, at $z = l$,

$$\left[\frac{V_0}{2Z_0} - \frac{V^-}{Z_0}\right]_{t=T} = 0 \tag{}$$

or

$$[V^-]_{t=T} = \frac{V_0}{2} \tag{6.88}$$

The general solution for the differential equation can be written as

$$V^- = -\frac{V_0}{2} + Ae^{-(Z_0/L)t} \tag{6.89}$$

where A is an arbitrary constant to be evaluated using (6.88). Thus, we have

$$\frac{V_0}{2} = -\frac{V_0}{2} + Ae^{-(Z_0/L)T} \tag{}$$

or

$$A = V_0 e^{(Z_0/L)T} \tag{6.90}$$

Substituting this result in (6.89), we obtain the solution for $[V^-]_{z=l}$ as

$$V^-(l, t) = -\frac{V_0}{2} + V_0 e^{-(Z_0/L)(t-T)} \qquad \text{for} \quad t > T \tag{6.91}$$

The corresponding solution for the $(-)$ wave current is given by

$$I^-(l, t) = -\frac{V^-(l, t)}{Z_0} = \frac{V_0}{2Z_0} - \frac{V_0}{Z_0}e^{-(Z_0/L)(t-T)} \qquad \text{for} \quad t > T \tag{6.92}$$

The $(-)$ wave, characterized by V^- and I^- as given by (6.91) and (6.92), respectively, travels back toward the source, and it does not set up a reflected wave, since the reflection coefficient at that end is zero. At this point, it can be seen that unlike in the case of linear resistive terminations and discontinuities, the concept of the reflection coefficient is not useful for studying transient behavior when reactive elements are involved. In fact, we note from (6.91) and (6.92) that the ratios of reflected voltage and current to the incident voltage and current, respectively, are no longer constants as in the resistive case.

We may now write the expressions for the total voltage across the inductor and the total current through the inductor as follows:

$$V(l, t) = \frac{V_0}{2} + V^-(l, t)$$

$$= \begin{cases} 0 & \text{for } t < T \\ V_0 e^{-(Z_0/l)(t-T)} & \text{for } t > T \end{cases} \tag{6.93}$$

$$I(l, t) = \frac{V_0}{2Z_0} + I^-(l, t)$$

$$= \begin{cases} 0 & \text{for } t < T \\ (V_0/Z_0)[1 - e^{-(Z_0/L)(t-T)}] & \text{for } t > T \end{cases} \tag{6.94}$$

These quantities are shown sketched in Figs. 6.45 (a) and (b), respectively. It may be seen from these sketches that in the steady state, the voltage goes to zero and

(a)

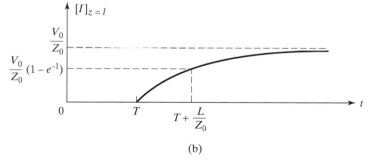

(b)

FIGURE 6.45   Time variations of (a) voltage across the inductor and (b) current through the inductor, for the system of Fig. 6.43.

the current goes to V_0/Z_0. This is consistent with the fact that the inductor behaves like a short circuit for the dc voltage in the steady state, and hence the situation in the steady state is the same as that for a short-circuited line. Note also that the variations of the voltage and current from $t = T$ to $t = \infty$ are governed by the time constant L/Z_0, which is that of the inductor L in series with Z_0 of the line. In fact, we can obtain the voltage and current sketches from considerations of initial and final behaviors of the reactive element and the time constant without out formally going through the process of setting up the differential equation and solving it. We shall illustrate this procedure by means of an example.

Example 6.10

Capacitive discontinuity

Let us consider the system shown in Fig. 6.46 consisting of a series capacitor of value 10 pF at the junction between the two lines. Note that line 2 is terminated by its own characteristic impedance, whereas the internal resistance of the voltage source is equal to the characteristic impedance of line 1 so that no reflections occur at the two ends of the system. We shall assume that the capacitor is initially uncharged and obtain the plots of line voltage and line current at the input $z = 0$ from considerations of initial and final behaviors of the capacitor.

Plots of line voltage and line current at $z = 0$ versus time are shown in Figs. 6.47(a) and (b), respectively. We shall explain the several features in these plots as follows: When the switch S is closed at $t = 0$, a $(+)$ wave of voltage 10 V and current 0.2 A goes down the line. Since the voltage across a capacitor cannot change instantaneously, the initially uncharged capacitor behaves like a short circuit when the $(+)$ wave impinges on the junction aa' at $t = 1$ ns. Therefore, the $(+)$ wave then sees a resistance of $Z_{02}(= 150\ \Omega)$ across aa' and produces a $(-)$ wave of initial voltage 5 V and initial current -0.1 A. The $(-)$ wave arrives initially at $z = 0$ at $t = 2$ ns, thereby changing the line voltage and line current there to 15 V and 0.1 A, as shown in Figs. 6.47(a) and (b), respectively. In the steady state, the capacitor behaves like an open circuit, which explains the steady-state values of 20 V and 0 A in these plots. Between $t = 2$ ns and $t = \infty$, the voltage and current vary exponentially with a time constant of $10^{-11} \times 200 = 2 \times 10^{-9}$ s $= 2$ ns, which is that of $C\ (= 10\ \text{pF})$ in series with $(Z_{01} + Z_{02})$, or 200 Ω. Hence, the voltage and current values at $t = 4$ ns are $15 + 5(1 - e^{-1}) = 18.16$ V and $0.1 - 0.1\ (1 - e^{-1}) = 0.037$ A, respectively.

FIGURE 6.46   Transmission-line system with a capacitive discontinuity.

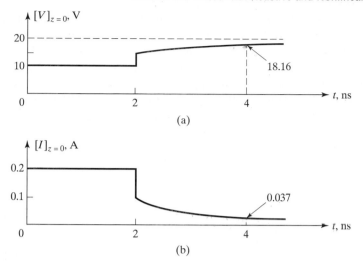

FIGURE 6.47   Plots of (a) line voltage and (b) line current at $z = 0$ for the system of Fig. 6.46.

Finally, the arguments that we have employed to explain the features in Fig. 6.47 can be used to deduce information about the nature of the discontinuity if the plots represent measurements by a time-domain reflectometer.

Another type of situation in which the concept of reflection coefficient is not useful is that involving nonlinear resistive elements. The analysis is then made convenient by a graphical technique known as the *load-line technique.* We shall introduce this technique by means of an example.

Nonlinear resistive termination

Example 6.11

Let us consider the transmission-line system shown in Fig. 6.48 in which the line is terminated by a passive nonlinear element having the indicated V-I relationship. We wish to obtain the time variations of the voltages V_S and V_L at the source and load ends, respectively, following the closure of the switch S at $t = 0$, using the load-line technique.

With reference to the notation shown in Fig. 6.48, we can write the following equations pertinent to $t = 0+$ at $z = 0$:

Load-line technique

$$50 = 200 I_S + V_S$$
$$V_S = V^+ \tag{6.95a}$$

$$I_S = I^+ = \frac{V^+}{Z_0} = \frac{V_S}{50} \tag{6.95b}$$

where V^+ and I^+ are the voltage and current, respectively, of the (+) wave set up immediately after closure of the switch. The two equations (6.95a) and (6.95b) can be solved graphically by constructing the straight lines representing them, as shown in Fig. 6.49, and

FIGURE 6.48 Line terminated by a passive nonlinear element and driven by a constant-voltage source in series with internal resistance.

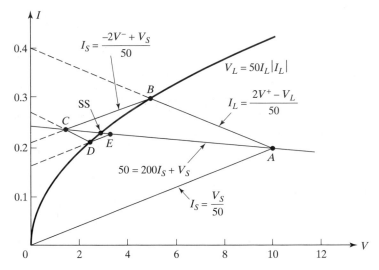

FIGURE 6.49 Graphical solution for obtaining time variations of V_S and V_L for $t > 0$ in the transmission-line system of Fig. 6.48.

obtaining the point of intersection A, which gives the values of V_S and I_S. Note in particular that (6.95b) is a straight line of slope $1/50$ and passing through the origin.

When the $(+)$ wave reaches the load end $z = l$ at $t = T$, a $(-)$ wave is set up. We can then write the following equations pertinent to $t = T+$ at $z = l$:

$$V_L = 50I_L|I_L| \tag{6.96a}$$

$$V_L = V^+ + V^-$$

$$I_L = I^+ + I^- = \frac{V^+ - V^-}{Z_0}$$

$$= \frac{V^+ - (V_L - V^+)}{50} = \frac{2V^+ - V_L}{50} \tag{6.96b}$$

where V^- and I^- are the $(-)$ wave voltage and current, respectively. The solution for V_L and I_L is then given by the intersection of the nonlinear curve representing (6.96a) and the straight line of slope $-1/50$ corresponding to (6.96b). Noting from (6.96b) that for

$V_L = V^+$, $I_L = V^+/50$, we see that the straight line passes through point A. Thus, the solution of (6.96a) and (6.96b) is given by point B in Fig. 6.49.

When the $(-)$ wave reaches the source end $z = 0$ at $t = 2T$, it sets up a reflection. Denoting this to be the $(-+)$ wave, we can then write the following equations pertinent to $t = 2T+$ at $z = 0$:

$$50 = 200I_S + V_S \qquad (6.97a)$$

$$V_S = V^+ + V^- + V^{-+}$$

$$I_S = I^+ + I^- + I^{-+} = \frac{V^+ - V^- + V^{-+}}{Z_0}$$

$$\qquad (6.97b)$$

$$= \frac{V^+ - V^- + (V_S - V^+ - V^-)}{50} = \frac{-2V^- + V_S}{50}$$

where V^{-+} and I^{-+} are the $(-+)$ wave voltage and current, respectively. Noting from (6.97a) that for $V_S = V^+ + V^-$, $I_S = (V^+ - V^-)/50$, we see that (6.97b) represents a straight line of slope $1/50$ passing through B. Thus, the solution of (6.97a) and (6.97b) is given by point C in Fig. 6.49.

Continuing in this manner, we observe that the solution consists of obtaining the points of intersection on the source and load V-I characteristics by drawing successively straight lines of slope $1/Z_0$ and $-1/Z_0$, beginning at the origin (the initial state) and with each straight line originating at the previous point of intersection, as shown in Fig. 6.49. The points $A, C, E, \ldots$, give the voltage and current at the source end for $0 < t < 2\ T$, $2T < t < 4T, 4T < t < 6T, \ldots$, whereas the points $B, D, \ldots$, give the voltage and current at the load end for $T < t < 3T, 3T < t < 5T, \ldots$. Thus, for example, the time variations of V_S and V_L are shown in Figs. 6.50(a) and (b), respectively. Finally, it can be seen from Fig. 6.49 that the steady-state values of line voltage and current are reached at the point of intersection (denoted SS) of the source and load V-I characteristics.

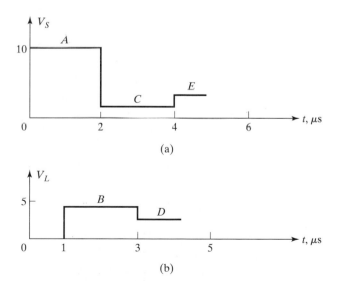

FIGURE 6.50 Time variations of (a) V_S and (b) V_L, for the transmission-line system of Fig. 6.48. The voltage levels $A, B, C, \ldots$ correspond to those in Fig. 6.49.

K6.6. Inductive termination; Capacitive discontinuity; Nonlinear resistive termination; Load-line technique.

D6.14. In the system of Fig. 6.43, assume that $V_0 = 20$ V, $Z_0 = 50\ \Omega$, and $T = 1\ \mu s$. Find the value of the voltage across the inductor at $t = 2\ \mu s$ for each of the following cases: **(a)** $L = 0.1$ mH, $I_L(0-) = 0$ A; **(b)** $L = 0.1$ mH, $I_L(0-) = 0.05$ A; **(c)** $L = 0.05$ mH, $I_L(0-) = 0.1$ A.
Ans. **(a)** 12.13 V; **(b)** 10.61 V; **(c)** 5.52 V.

D6.15. In the system shown in Fig. 6.51, the capacitor is initially uncharged. Find the values of the line voltage at $z = 0$ at the following times: **(a)** $t = 2$ ns+; **(b)** $t = \infty$; and **(c)** $t = 3$ ns.
Ans. **(a)** 0 V; **(b)** 15 V; **(c)** 7.2987 V.

D6.16. Assume that in the system of Fig. 6.48 the values of the voltage source and its internal resistance are 12 V and 10 Ω, respectively, and that Z_0 of the line is 100 Ω. By using the load-line technique, find the approximate values of: **(a)** V_L at $t = 2\ \mu s$; **(b)** V_S at $t = 3\ \mu s$; **(c)** V_L at $t = 4\ \mu s$; and **(d)** V_L at $t = \infty$.
Ans. **(a)** 2 V; **(b)** 9.3 V; **(c)** 5 V; **(d)** 8 V.

FIGURE 6.51   For Problem D6.16.

6.7 LINES WITH INITIAL CONDITIONS AND LOGIC GATES

Arbitrary initial distribution

Thus far, we have considered lines with quiescent initial conditions, that is, with no initial voltages and currents on them. As a prelude to the discussion of analysis of interconnections between logic gates, we shall now consider lines with nonzero initial conditions. We discuss first the general case of arbitrary initial voltage and current distributions by decomposing them into (+) and (−) wave voltages and currents. To do this, we consider the example shown in Fig. 6.52, in which a line open-circuited at both ends is charged initially, say, at $t = 0$, to the voltage and current distributions shown in the figure.

Writing the line voltage and current distributions as sums of (+) and (−) wave voltages and currents, we have

$$V^+(z, 0) + V^-(z, 0) = V(z, 0) \tag{6.98a}$$

$$I^+(z, 0) + I^-(z, 0) = I(z, 0) \tag{6.98b}$$

But we know that $I^+ = V^+/Z_0$ and $I^- = -V^-/Z_0$. Substituting these into (6.98b) and multiplying by Z_0, we get

$$V^+(z, 0) - V^-(z, 0) = Z_0 I(z, 0) \tag{6.99}$$

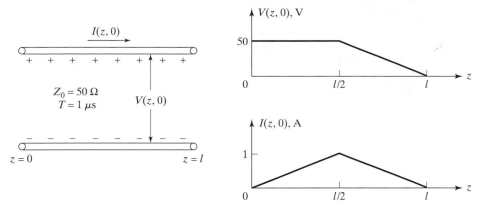

FIGURE 6.52 Line open-circuited at both ends and initially charged to the voltage and current distributions $V(z,0)$ and $I(z,0)$, respectively.

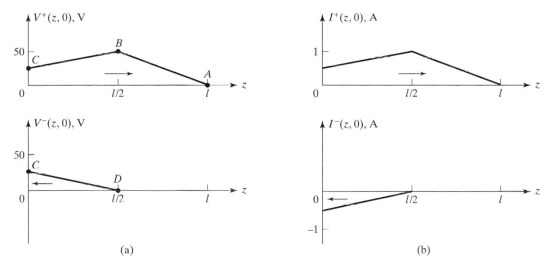

FIGURE 6.53 Distributions of (a) voltage and (b) current in the $(+)$ and $(-)$ waves obtained by decomposing the voltage and current distributions of Fig. 6.52.

Solving (6.98a) and (6.99), we obtain

$$V^+(z,0) = \tfrac{1}{2}[V(z,0) + Z_0 I(z,0)] \qquad (6.100a)$$
$$V^-(z,0) = \tfrac{1}{2}[V(z,0) - Z_0 I(z,0)] \qquad (6.100b)$$

Thus, for the distributions $V(z,0)$ and $I(z,0)$ given in Fig. 6.52, we obtain the distributions of $V^+(z,0)$ and $V^-(z,0)$ as shown by Fig. 6.53(a), and hence of $I^+(z,0)$ and $I^-(z,0)$, as shown by Fig. 6.53(b).

Suppose that we wish to find the voltage and current distributions at some later value of time, say, $t = 0.5\ \mu s$. Then, we note that as the $(+)$ and $(-)$ waves propagate and impinge on the open circuits at $z = l$ and $z = 0$, respectively, they produce the $(-)$ and $(+)$ waves, respectively, consistent with a voltage reflection coefficient of 1 and current reflection coefficient of -1 at both ends. Hence, at $t = 0.5\ \mu s$, the $(+)$ and $(-)$ wave voltage and current distributions and their sum distributions are as shown in Fig. 6.54, in which the points A, B, C, and D correspond to the points A, B, C, and D, respectively, in Fig. 6.53. Proceeding in this manner, one can obtain the voltage and current distributions for any value of time.

Suppose that we connect a resistor of value Z_0 at the end $z = l$ at $t = 0$ instead of keeping it open-circuited. Then the reflection coefficient at that end becomes zero thereafter and the $(+)$ wave, as it impinges on the resistor, gets absorbed in it instead of producing the $(-)$ wave. The line therefore completely discharges into the resistor by the time $t = 1.5\ \mu s$, with the resulting time variation of voltage across R_L, as shown in Fig. 6.55, where the points A, B, C, and D correspond to the points A, B, C, and D, respectively, in Fig. 6.53.

Uniform initial distribution   For a line with uniform initial voltage and current distributions, the analysis can be performed in the same manner as for arbitrary initial voltage and

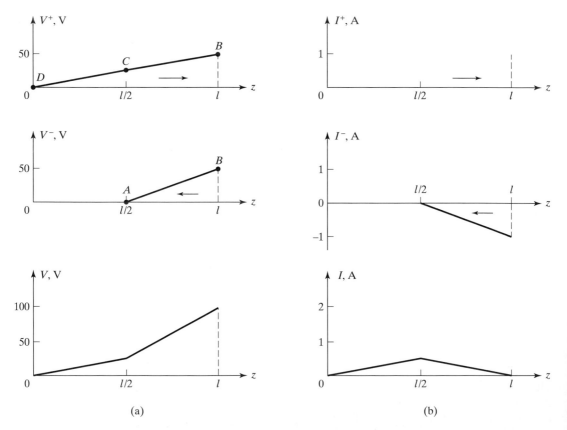

FIGURE 6.54   Distributions of (a) voltage and (b) current in the $(+)$ and $(-)$ waves and their sum for $t = 0.5\ \mu s$ for the initially charged line of Fig. 6.52.

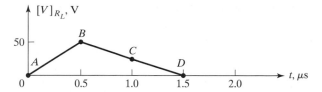

FIGURE 6.55   Voltage across $R_L(= Z_0 = 50\ \Omega)$ resulting from connecting it at $t = 0$ to the end $z = l$ of the line of Fig. 6.52.

current distributions. Alternatively and more conveniently, the analysis can be carried out with the aid of superposition and bounce diagrams, or by the load-line technique. The basis behind this method lies in the fact that the uniform distribution corresponds to a situation in which the line voltage and current remain constant with time at all points on the line until a change is made at some point on the line. The boundary condition is then violated at that point, and a transient wave of constant voltage and current is set up, to be superimposed on the initial distribution. We shall illustrate this technique of analysis by means of an example.

Example 6.12

Let us consider a line of $Z_0 = 50\ \Omega$ and $T = 1\ \mu s$ initially charged to uniform voltage $V_0 = 100$ V and zero current. A resistor $R_L = 150\ \Omega$ is connected at $t = 0$ to the end $z = 0$ of the line, as shown in Fig. 6.56(a). We wish to obtain the time variation of the voltage across R_L for $t > 0$.

Since the change is made at $z = 0$ by connecting R_L to the line, a $(+)$ wave originates at $z = 0$, so that the total line voltage at that point is $V_0 + V^+$ and the total line current is $0 + I^+$, or I^+, as shown in Fig. 6.56(b). To satisfy the boundary condition at $z = 0$, we then write

$$V_0 + V^+ = -R_L I^+ \tag{6.101}$$

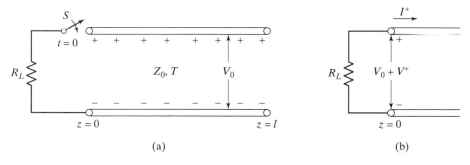

(a)   (b)

FIGURE 6.56   (a) Transmission line charged initially to uniform voltage V_0. (b) For obtaining the voltage and current associated with the transient $(+)$ wave resulting from the closure of the switch in (a).

But we know that $I^+ = V^+/Z_0$. Hence, we have

$$V_0 + V^+ = -\frac{R_L}{Z_0} V^+ \tag{6.102}$$

or

$$V^+ = -V_0 \frac{Z_0}{R_L + Z_0} \tag{6.103a}$$

$$I^+ = -V_0 \frac{1}{R_L + Z_0} \tag{6.103b}$$

For $V_0 = 100$ V, $Z_0 = 50\ \Omega$, and $R_L = 150\ \Omega$, we obtain $V^+ = -25$ V and $I^+ = -0.5$ A.

We may now draw the voltage and current bounce diagrams, as shown in Fig. 6.57. We note that in these bounce diagrams, the initial conditions are accounted for by the horizontal lines drawn at the top, with the numerical values of voltage and current indicated on them. Sketches of line voltage and current versus z for fixed values of t can be drawn from these bounce diagrams in the usual manner. Sketches of line voltage and current versus t for any fixed value of z also can be drawn from the bounce diagrams in the usual manner. Of particular interest is the voltage across R_L, which illustrates how the line discharges into the resistor. The time variation of this voltage is shown in Fig. 6.58.

Alternative to the bounce-diagram technique, the behavior of the system can be analyzed by using the load-line technique. Thus, noting that the terminal voltage-current characteristics at the ends $z = 0$ and $z = l$ are given by $V = -IR_L = -150I$ and $I = 0$, respectively, and that the characteristic impedance of the line is $50\ \Omega$, we can carry out the load-line construction, as shown in Fig. 6.59, beginning at the point A (100 V, 0 A) and

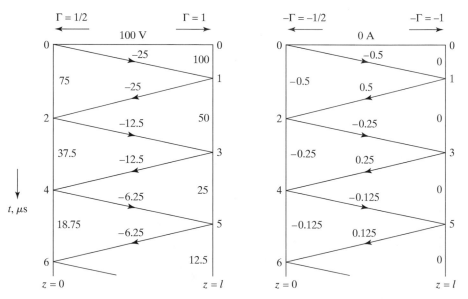

FIGURE 6.57   Voltage- and current-bounce diagrams depicting the transient phenomenon for $t > 0$ for the line of Fig. 6.56(a), for $V_0 = 100$ V, $Z_0 = 50\ \Omega$, $R_L = 150\ \Omega$, and $T = 1\ \mu s$.

FIGURE 6.58 Time variation of voltage across R_L for $t > 0$ in Fig. 6.56(a) for $V_0 = 100$ V, $Z_0 = 50$ Ω, $R_L = 150$ Ω, and $T = 1$ μs.

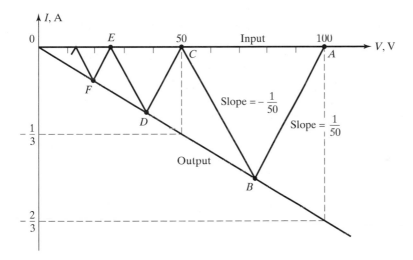

FIGURE 6.59 Load-line construction for the analysis of the system of Fig 6.56(a).

drawing alternately straight lines of slopes $1/50$ and $-1/50$ to obtain the points of inter-section $B, C, D, \ldots$. The points $B, D, F, \ldots$ give the line voltage and current values at the end $z = 0$ for intervals of 2 μs beginning at $t = 0$ μs, 2 μs, 6 μs, ..., whereas the points $C, E, \ldots$ give the line voltage and current values at the end $z = l$ for intervals of 2 μs beginning at $t = 1$ μs, 3 μs, ... For example, the time variation of the line voltage at $z = 0$ provided by the load-line construction is the same as in Fig. 6.58.

Before proceeding further, it is instructive to check the energy balance, that is, to verify that the energy dissipated in the 150-Ω resistor for $t > 0$ is indeed equal to the energy stored in the line at $t = 0-$, since the line is lossless. To do this, we note that, in general, energy is stored in both electric and magnetic fields in the line, with energy densities $\frac{1}{2}\mathcal{C}V^2$ and $\frac{1}{2}\mathcal{L}I^2$, respectively. Thus, for a line charged uniformly to voltage V_0 and current I_0, the total electric and magnetic stored energies are given, respectively, by

Energy balance

$$W_e = \tfrac{1}{2}\mathcal{C}V_0^2 l = \tfrac{1}{2}\mathcal{C}V_0^2 v_p T$$

$$= \frac{1}{2}\mathcal{C}V_0^2 \frac{1}{\sqrt{\mathcal{L}\mathcal{C}}} T = \frac{1}{2}\frac{V_0^2}{Z_0} T$$

(6.104a)

and

$$W_m = \frac{1}{2} \mathcal{L} I_0^2 l = \frac{1}{2} \mathcal{L} I_0^2 v_p T$$

$$= \frac{1}{2} \mathcal{L} I_0^2 \frac{1}{\sqrt{\mathcal{L}\mathcal{C}}} T = \frac{1}{2} I_0^2 Z_0 T \tag{6.104b}$$

Since for the example under consideration, $V_0 = 100$ V, $I_0 = 0$, and $T = 1 \, \mu s$, $W_e = 10^{-4}$ J and $W_m = 0$. Thus, the total initial stored energy in the line is 10^{-4} J. Now, denoting the power dissipated in the resistor to be P_d, we obtain the energy dissipated in the resistor to be

$$W_d = \int_{t=0}^{\infty} P_d \, dt$$

$$= \int_0^{2 \times 10^{-6}} \frac{75^2}{150} \, dt + \int_{2 \times 10^{-6}}^{4 \times 10^{-6}} \frac{37.5^2}{150} \, dt + \int_{4 \times 10^{-6}}^{6 \times 10^{-6}} \frac{18.75^2}{150} \, dt + \cdots$$

$$= \frac{2 \times 10^{-6}}{150} \times 75^2 \left(1 + \frac{1}{4} + \frac{1}{16} + \cdots \right) = 10^{-4} \, J$$

which is exactly the same as the initial stored energy in the line, thereby satisfying the energy balance.

Interconnection between logic gates

We shall now apply the procedure for the use of the load-line technique for a line with uniform initial distribution, just illustrated, to the analysis of the system in Fig. 6.60(a) in which two transistor-transistor logic (TTL) inverters are interconnected by using a transmission line of characteristic impedance Z_0 and one-way travel time T. As the name inverter implies, the gate has an output that is the inverse of the input. Thus, if the input is in the HIGH (logic 1) range, the output will be in the LOW (logic 0) range, and vice versa. Typical V-I characteristics for a TTL inverter are shown in Fig. 6.60(b). As shown in this figure, when the system is in the steady state with the output of the first inverter in the 0 state, the voltage and current along the line are given by the intersection of the output 0 characteristic and the input characteristic; when the system is in the steady state with the output of the first inverter in the 1 state, the voltage and current along the line are given by the intersection of the output 1 characteristic and the input characteristic. Thus, the line is charged to 0.2 V for the steady-state 0 condition and to 4 V for the steady-state 1 condition. We wish to study the transient phenomena corresponding to the transition when the output of the first gate switches from the 0 to the 1 state, and vice versa, assuming Z_0 of the line to be 30 Ω.

Analysis of 0-to-1 transition

Considering first the transition from the 0 state to the 1 state, and following the line of argument in Examples 6.11 and 6.12, we carry out the construction shown in Fig. 6.61(a). This construction consists of beginning at the point corresponding to the steady-state 0 (the initial state) and drawing a straight line of slope 1/30 to intersect with the output 1 characteristic at point A, then

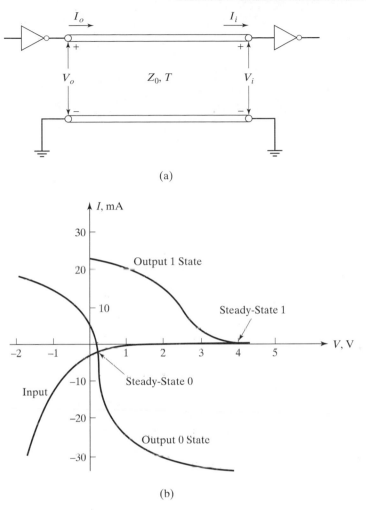

FIGURE 6.60 (a) Transmission-line interconnection between two logic gates. (b) Typical V-I characteristics for the logic gates.

drawing from point A a straight line of slope $-1/30$ to intersect the input characteristic at point B, and so on. From this construction, the variation of the voltage V_i at the input of the second gate can be sketched as shown in Fig. 6.61(b), in which the voltage levels correspond to the points $0, B, D, \ldots$, in Fig. 6.61(a). The effect of the transients on the performance of the system may now be seen by noting from Fig. 6.61(b) that depending on the value of the minimum gate voltage that will reliably be recognized as logic 1, a time delay in excess of T may be involved in the transition from 0 to 1. Thus, if this minimum voltage is 2 V, the interconnecting line will result in an extra time delay of $2T$ for the input of the second gate to switch from 0 to 1, since V_i does not exceed 2 V until $t = 3T+$.

(a)

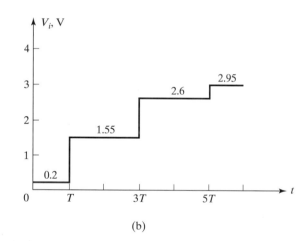

(b)

FIGURE 6.61   (a) Construction based on the load-line technique for analysis of the 0-to-1 transition for the system of Fig 6.60(a). (b) Plot of V_i versus t obtained from the construction in (a).

Analysis of 1-to-0 transition

Considering next the transition from the 1 state to the 0 state, we carry out the construction shown in Fig. 6.62(a), with the crisscross lines beginning at the point corresponding to the steady-state 1. From this construction, we obtain the plot of V_i versus t, as shown in Fig. 6.62(b), in which the voltage levels correspond to the points 1, B, D, . . . , in Fig. 6.62(a). If we assume a maximum gate input voltage that can be readily recognized as logic 0 to be 1 V, it can once again be seen that an extra time delay of $2T$ is involved in the switching of the input of the second gate from 1 to 0, since V_i does not drop below 1 V until $t = 3T+$.

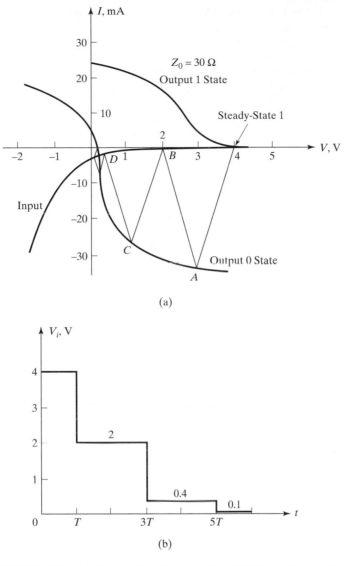

FIGURE 6.62 (a) Construction based on the load-line technique for analysis of the 1-to-0 transition for the system of Fig. 6.61(a). (b) Plot of V_i versus t obtained from the construction in (a).

K6.7. Initial conditions; Arbitrary distribution; Uniform distribution; Bounce-diagram technique; Load-line technique; Interconnection between logic gates.

D6.17. For the line of Fig. 6.52 with the initial voltage and current distributions as given in the figure, find: **(a)** $V(l/2, 0.25 \ \mu s)$; **(b)** $I(l/2, 0.25 \ \mu s)$; **(c)** $V(l/4, 1 \ \mu s)$; and **(d)** $I(l/4, 1 \ \mu s)$.
Ans. **(a)** 37.5 V; **(b)** 0.75 A; **(c)** 25 V; **(d)** −0.5 A.

D6.18. In the system shown in Fig. 6.63, a line of characteristic impedance 75 Ω and charged to 10 V is connected at $t = 0$ to another line of characteristic impedance

50 Ω and charged to 5 V. The one-way travel time T is equal to 1 μs for both lines. Find **(a)** the value of the voltage at the instant of time when both lines are charged to the same voltage throughout their lengths; **(b)** the value of the current to which the lines are charged at that instant of time; and **(c)** the energy stored in the system at any instant of time.

Ans.   **(a)** 7 V;   **(b)** 0.4 A;   **(c)** 11/12 μJ.

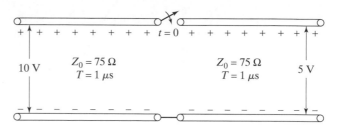

FIGURE 6.63   For Problem D6.18.

6.8   CROSSTALK ON TRANSMISSION LINES

Crosstalk explained

When two or more transmission lines are in the vicinity of one another, a wave propagating along one line, which we shall call the primary line, can induce a wave on another line, the secondary line, due to capacitive (electric field) and inductive (magnetic field) coupling between the two lines, resulting in the undesirable phenomenon of crosstalk between the lines. An example is illustrated by the arrangement of Fig. 6.64(a), which is a printed-circuit board (PCB) representa-

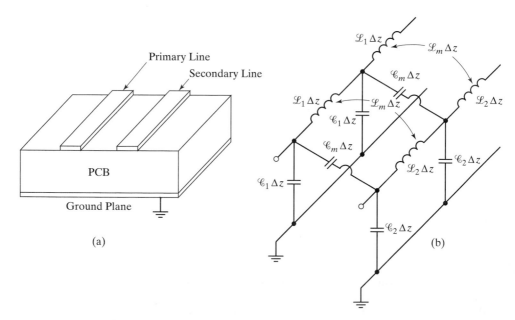

FIGURE 6.64   (a) PCB representation of two closely spaced transmission lines. (b) Distributed equivalent circuit for (a).

tion of two closely spaced transmission lines. Figure 6.64(b) represents the distributed circuit equivalent, where $\mathscr{C}_m$ and $\mathscr{L}_m$ are the coupling capacitance and coupling inductance, respectively, per unit length of the arrangement.

In this section, we shall analyze a pair of coupled transmission lines for the determination of induced waves on the secondary line for a given wave on the primary line. To keep the analysis simple, we shall consider both lines to be of the same characteristic impedance, velocity of propagation, and length, and terminated by their characteristic impedances, so that no reflections occur from the ends of either line. It is also convenient to assume the coupling to be weak so that the effects on the primary line of waves induced in the secondary line can be neglected. Thus, we shall be concerned only with the crosstalk from the primary line to the secondary line and not vice versa. Briefly, as the $(+)$ wave propagates on the primary line from source toward load, each infinitesimal length of that line induces voltage and current in the adjacent infinitesimal length of the secondary line, which set up $(+)$ and $(-)$ waves on that line. The contributions due to the infinitesimal lengths add up to give the net induced voltage and current at a given location on the secondary line. *Weak coupling analysis*

We shall represent the coupled-line pair, as shown in Fig. 6.65, with the primary line as line 1 and the secondary line as line 2. Then, when the switch S is closed at $t = 0$, a $(+)$ wave originates at $z = 0$ on line 1 and propagates toward the load. Let us consider a differential length $d\xi$ at the location $z = \xi$ of line 1 charged to the $(+)$ wave voltage and current and obtain its contributions to the induced voltages and currents in line 2.

The capacitive coupling induces a differential crosstalk current ΔI_{c2}, flowing into the nongrounded conductor of line 2, given by *Modeling for capacitive coupling*

$$\boxed{\Delta I_{c2}(\xi, t) = \mathscr{C}_m \, \Delta\xi \, \frac{\partial V_1(\xi, t)}{\partial t}} \qquad (6.105a)$$

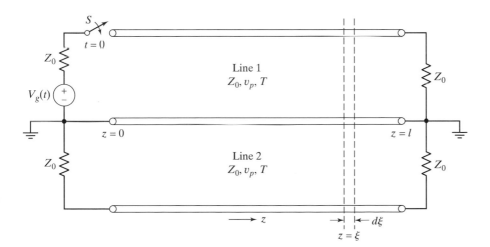

FIGURE 6.65 Coupled transmission-line pair for analysis of crosstalk.

FIGURE 6.66   (a) Modeling for capacitive coupling in crosstalk analysis. (b) Equivalent circuit for (a).

where $V_1(\xi, t)$ is the line-1 voltage. This induced current is modeled by an ideal current source, connected in parallel with line 2 at $z = \xi$ on that line, as shown in Fig. 6.66(a). The current source views the characteristic impedance of the line to either side of $z = \xi$, so that the equivalent circuit is as shown in Fig. 6.66(b). Thus, voltages of $\frac{1}{2} Z_0 \Delta I_{c2}$ are produced to the right and left of $z = \xi$ and propagate as forward-crosstalk and backward-crosstalk voltages, respectively, on line 2.

Modeling for inductive coupling   The inductive coupling induces a differential crosstalk voltage, ΔV_{c2}, which is given by

$$\Delta V_{c2}(\xi, t) = \mathscr{L}_m \, \Delta\xi \, \frac{\partial I_1(\xi, t)}{\partial t} \tag{6.105b}$$

This induced voltage is modeled by an ideal voltage source in series with line 2 at $z = \xi$ on that line, as shown in Fig. 6.67(a). The polarity of the voltage source is such that the current due to it in line 2 produces a magnetic flux, which opposes the change in the flux due to the current in line 1, in accordance with Lenz's law. The voltage source views the characteristic impedance of the line to either side of it, so that the equivalent circuit is as shown in Fig. 6.67(b). Thus,

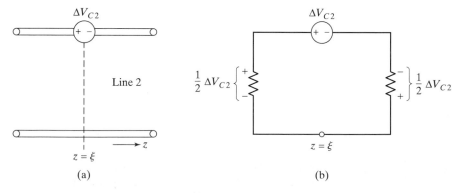

FIGURE 6.67   (a) Modeling for inductive coupling in crosstalk analysis. (b) Equivalent circuit for (a).

voltages of $\frac{1}{2}\Delta V_{c2}$ and $-\frac{1}{2}\Delta V_{c2}$ are produced to the left and right of $z = \xi$, respectively, and propagate as backward-crosstalk and forward-crosstalk voltages, respectively, on line 2.

Combining the contributions due to capacitive coupling and inductive coupling, we obtain the total differential voltages produced to the right and left of $z = \xi$ to be

$$\Delta V_2^+ = \tfrac{1}{2}Z_0\,\Delta I_{c2} - \tfrac{1}{2}\Delta V_{c2} \tag{6.106a}$$

$$\Delta V_2^- = \tfrac{1}{2}Z_0\,\Delta I_{c2} + \tfrac{1}{2}\Delta V_{c2} \tag{6.106b}$$

respectively. Substituting (6.105a) and (6.105b) into (6.106a) and (6.106b), we obtain

$$\Delta V_2^+(\xi, t) = \left[\frac{1}{2}\,\mathscr{C}_m Z_0\,\frac{\partial V_1(\xi, t)}{\partial t} - \frac{1}{2}\,\mathscr{L}_m\,\frac{\partial I_1(\xi, t)}{\partial t}\right]\Delta\xi$$
$$= \frac{1}{2}\left(\mathscr{C}_m Z_0 - \frac{\mathscr{L}_m}{Z_0}\right)\frac{\partial V_1(\xi, t)}{\partial t}\,\Delta\xi \tag{6.107a}$$

$$\Delta V_2^-(\xi, t) = \frac{1}{2}\left(\mathscr{C}_m Z_0 + \frac{\mathscr{L}_m}{Z_0}\right)\frac{\partial V_1(\xi, t)}{\partial t}\,\Delta\xi \tag{6.107b}$$

where we have substituted $I_1 = V_1/Z_0$, in accordance with the relationship between the voltage and current of a $(+)$ wave.

We are now ready to apply (6.107a) and (6.107b) in conjunction with superposition to obtain the $(+)$ and $(-)$ wave voltages at any location on line 2, due to a $(+)$ wave of voltage $V_1(t - z/v_p)$ on line 1. Thus, noting that the effect of V_1 at $z = \xi$ at a given time t is felt at a location $z > \xi$ on line 2 at time $t + (z - \xi)/v_p$, we can write

Forward-crosstalk voltage and coefficient

$$V_2^+(z, t) = \int_0^z \frac{1}{2}\left(\mathscr{C}_m Z_0 - \frac{\mathscr{L}_m}{Z_0}\right)\frac{\partial}{\partial t}\left[V_1\left(t - \frac{\xi}{v_p} - \frac{z - \xi}{v_p}\right)\right]d\xi$$
$$= \frac{1}{2}\left(\mathscr{C}_m Z_0 - \frac{\mathscr{L}_m}{Z_0}\right)\int_0^z \frac{\partial V_1(t - z/v_p)}{\partial t}\,d\xi \tag{6.108}$$

or

$$\boxed{V_2^+(z, t) = zK_f V_1'(t - z/v_p)} \tag{6.109}$$

where we have defined

$$\boxed{K_f = \frac{1}{2}\left(\mathscr{C}_m Z_0 - \frac{\mathscr{L}_m}{Z_0}\right)} \tag{6.110}$$

and the prime associated with V_1 denotes differentiation with time. The quantity K_f is called the *forward-crosstalk coefficient*. Note that the upper limit in the integral in (6.108) is z, because the line-1 voltage to the right of a given location z on that line does not contribute to the forward-crosstalk voltage on line 2 at that same location. The result given by (6.109) tells us that the forward-crosstalk voltage is proportional to z and the time derivative of the primary line voltage.

Backward-crosstalk voltage and coefficient

To obtain $V_2^-(z, t)$, we note that the effect of V_1 at $z = \xi$ at a given time t is felt at a location $z < \xi$ on line 2 at time $t + (\xi - z)/v_p$. Hence,

$$
\begin{aligned}
V_2^-(z, t) &= \int_z^l \frac{1}{2}\left(\mathscr{C}_m Z_0 + \frac{\mathscr{L}_m}{Z_0}\right)\frac{\partial}{\partial t}\left[V_1\left(t - \frac{\xi}{v_p} - \frac{\xi - z}{v_p}\right)\right] d\xi \\
&= \frac{1}{2}\left(\mathscr{C}_m Z_0 + \frac{\mathscr{L}_m}{Z_0}\right)\int_z^l \frac{\partial}{\partial t}\left[V_1\left(t + \frac{z}{v_p} - \frac{2\xi}{v_p}\right)\right] d\xi \\
&= -\frac{1}{4}v_p\left(\mathscr{C}_m Z_0 + \frac{\mathscr{L}_m}{Z_0}\right)\int_z^l \frac{\partial}{\partial \xi}\left[V_1\left(t + \frac{z}{v_p} - \frac{2\xi}{v_p}\right)\right] d\xi \\
&= -\frac{1}{4}v_p\left(\mathscr{C}_m Z_0 + \frac{\mathscr{L}_m}{Z_0}\right)\left[V_1\left(t + \frac{z}{v_p} - \frac{2\xi}{v_p}\right)\right]_{\xi=z}^l
\end{aligned}
\tag{6.111}
$$

or

$$
\boxed{V_2^-(z, t) = K_b\left[V_1\left(t - \frac{z}{v_p}\right) - V_1\left(t - \frac{2l}{v_p} + \frac{z}{v_p}\right)\right]}
\tag{6.112}
$$

where we have defined the *backward-crosstalk coefficient*

$$
\boxed{K_b = \frac{1}{4}v_p\left(\mathscr{C}_m Z_0 + \frac{\mathscr{L}_m}{Z_0}\right)}
\tag{6.113}
$$

Note that the lower limit in the integral in (6.111) is z, because the line-1 voltage to the left of a given location z on that line does not contribute to the backward-crosstalk voltage on line 2 at that same location.

We shall now consider an example to illustrate the application of (6.109) and (6.112) for a specified voltage $V_g(t)$ in Fig. 6.65.

Example 6.13

Let $V_g(t)$ in Fig. 6.65 be the function shown in Fig. 6.68, where $T_0 < T \, (= l/v_p)$. We wish to determine the $(+)$ and $(-)$ wave voltages on line 2.

Noting that

$$
V_1(t) = \frac{1}{2}V_g(t) = \begin{cases} (V_0/T_0)t & \text{for} \quad 0 < t < T_0 \\ V_0 & \text{for} \quad t > T_0 \end{cases}
$$

and hence

$$
V_1'(t) = \begin{cases} V_0/T_0 & \text{for} \quad 0 < t < T_0 \\ 0 & \text{for} \quad t > T_0 \end{cases}
$$

FIGURE 6.68   Source voltage for the system of Fig. 6.65 for
Example 6.13.

and using (6.120), we can write the (+) wave voltage on line 2 as

$$V_2^+(z, t) = zK_fV_1'(t - z/v_p)$$

$$= \begin{cases} zK_fV_0/T_0 & \text{for } 0 < (t - z/v_p) < T_0 \\ 0 & \text{otherwise} \end{cases}$$

$$= \begin{cases} zK_fV_0/T_0 & \text{for } (z/v_p) < t < (z/v_p + T_0) \\ 0 & \text{otherwise} \end{cases}$$

$$= \begin{cases} zK_fV_0/T_0 & \text{for } (z/l)T < t < [(z/l)t + T_0] \\ 0 & \text{otherwise} \end{cases}$$

This is shown in the three-dimensional plot of Fig. 6.69, in which the cross section in any constant-z plane is a pulse of voltage zK_fV_0/T_0 for $(z/l)T < t < (z/l)T + T_0$. Note that the pulse voltage is shown to be negative. This is because normally the effect of inductive coupling dominates that of the capacitive coupling, so that K_f is negative.

Using (6.110), the (−) wave voltage can be written as

$$V_2^-(z, t) = K_b[V_1(t - z/v_p) - V_1(t - 2l/v_p + z/v_p)]$$

where

$$V_1\left(t - \frac{z}{v_p}\right) = \begin{cases} \dfrac{V_0}{T_0}\left(t - \dfrac{z}{v_p}\right) & \text{for } 0 < \left(t - \dfrac{z}{v_p}\right) < T_0 \\ 0 & \text{for } \left(t - \dfrac{z}{v_p}\right) > T_0 \end{cases}$$

$$= \begin{cases} \dfrac{V_0}{T_0}\left(t - \dfrac{z}{l}T\right) & \text{for } \dfrac{z}{l}T < t < \left(\dfrac{z}{l}T + T_0\right) \\ 0 & \text{for } t > \left(\dfrac{z}{l}T + T_0\right) \end{cases}$$

$$V_1\left(t - 2l/v_p + \frac{z}{v_p}\right) = \begin{cases} \dfrac{V_0}{T_0}\left(t - \dfrac{2l}{v_p} + \dfrac{z}{v_p}\right) & \text{for } 0 < \left(t - \dfrac{2l}{v_p} + \dfrac{z}{v_p}\right) < T_0 \\ 0 & \text{for } \left(t - \dfrac{2l}{v_p} + \dfrac{z}{v_p}\right) > T_0 \end{cases}$$

$$= \begin{cases} \dfrac{V_0}{T_0}\left(t - 2T + \dfrac{z}{l}T\right) & \text{for } \left(2T - \dfrac{z}{l}T\right) < t < \left(2T - \dfrac{z}{l}T + T_0\right) \\ 0 & \text{for } t > \left(2T - \dfrac{z}{l}T + T_0\right) \end{cases}$$

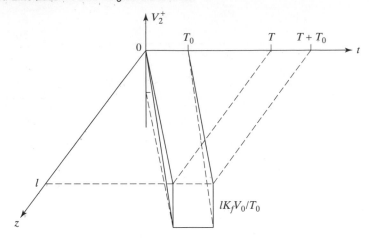

FIGURE 6.69   Three-dimensional depiction of forward-crosstalk voltage for the system of Fig. 6.65, with $V_g(t)$ as in Fig. 6.68.

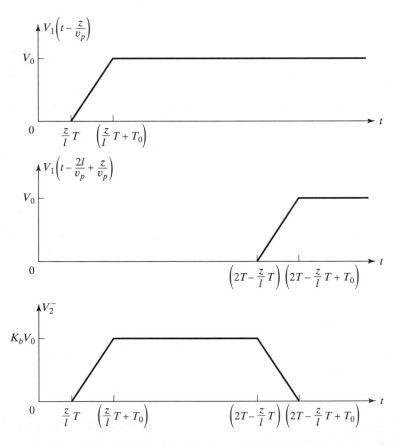

FIGURE 6.70   Determination of backward-crosstalk voltage for the system of Fig. 6.65, with $V_g(t)$ as in Fig. 6.68.

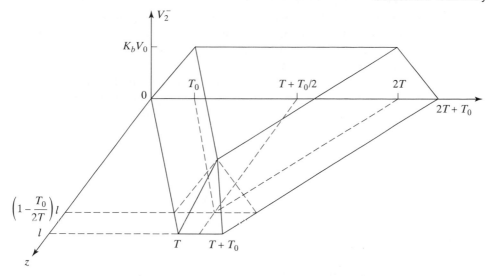

FIGURE 6.71 Three-dimensional depiction of backward-crosstalk voltage for the system of Fig. 6.65, with $V_g(t)$ as in Fig 6.68.

These two voltages and the $(-)$ wave voltage for a value of z for which $(z/l)T + T_0 < 2T - (z/l)T$ are shown in Fig. 6.70. Figure 6.71 shows the three-dimensional plot of $V^-(z, t)$ in which the cross section in any given constant-z plane gives the time variation of V_2^- for that value of z. Note that as z varies from zero to l, the shape of V_2^- changes from a trapezoidal pulse with a height of $K_b V_0$ at $z = 0$ to a triangular pulse of height $K_b V_0$ and width $2T_0$ at $z = (1 - T_0/2T)l$ and then changes to a trapezoidal pulse again but with a height continuously decreasing from $K_b V_0$ to zero at $z = l$.

K6.8. Weak coupling analysis; Capacitive coupling; Inductive coupling; Forward crosstalk; Backward crosstalk.

D6.19. In Example 6.13, assume that $\mathcal{L} = 0.9 \ \mu\text{H/m}$, $\mathcal{C} = 40 \ \text{pF/m}$, $\mathcal{L}_m = 0.093 \ \mu\text{H/m}$, and $\mathcal{C}_m = 4 \ \text{pF/m}$ for the line parameters in Fig. 6.65, and $T_0 = 0.2T$ for $V_g(t)$ in Fig. 6.68. Find the following: **(a)** the forward-crosstalk coefficient; **(b)** the backward-crosstalk coefficient; **(c)** $V_2(0, 0.01T)$; **(d)** $V_2(l, 1.1T)$; and **(e)** $V_2(0.5l, 0.6T)$. *Ans.* **(a)** -0.01 ns/m; **(b)** 0.0508; **(c)** $0.025V_0$; **(d)** $-0.0083V_0$; **(e)** $0.0212V_0$.

6.9 SUMMARY

In this chapter, we first introduced Poisson's and Laplace's equations. Poisson's equation given by

$$\nabla^2 V = -\frac{\rho}{\varepsilon}$$

is a differential equation governing the behavior of the electric scalar potential in a region of charge, whereas Laplace's equation

$$\nabla^2 V = 0$$

holds in a charge-free region. We discussed the application of Poisson's and Laplace's equations for the solution of problems involving the variation of V with one dimension only. In particular, we illustrated the solution of Poisson's equation by considering the example of a p-n junction diode and the solution of Laplace's equation by considering the determination of capacitance for several cases. We then considered the determination of circuit parameters for infinitely long, parallel conductor arrangements. Specifically, (1) we derived the expressions for the capacitance per unit length ($\mathcal{C}$), the conductance per unit length ($\mathcal{G}$), and the inductance per unit length ($\mathcal{L}$) for a coaxial cylindrical arrangement; (2) we showed that the three circuit parameters are related through the material parameters σ, ε, and μ, as given by

$$\frac{\mathcal{G}}{\mathcal{C}} = \frac{\sigma}{\varepsilon}$$

$$\mathcal{L}\mathcal{C} = \mu\varepsilon$$

and (3) we used these relationships for other geometries of the conductors. We then extended our discussions to internal inductance and mutual inductance.

Next we introduced the parallel-plate transmission line by considering a uniform plane wave propagating between two parallel perfectly conducting plates and showed that wave propagation on a transmission line can be discussed in terms of voltage and current, which are related to the electric and magnetic fields, respectively, by deriving the transmission-line equations

$$\frac{\partial V}{\partial z} = -\mathcal{L}\frac{\partial I}{\partial t} \tag{6.114a}$$

$$\frac{\partial I}{\partial z} = -\mathcal{C}\frac{\partial V}{\partial t} \tag{6.114b}$$

which then led us to the concept of the distributed circuit. We learned that propagation along a transmission line in the general case is characterized by transverse electromagnetic waves, with the parameters $\mathcal{L}$ and $\mathcal{C}$ differing from one line to another and derivable from static-field considerations. The solutions to the transmission-line equations are

$$V(z, t) = Af\left(t - \frac{z}{v_p}\right) + Bg\left(t + \frac{z}{v_p}\right) \tag{6.115a}$$

$$I(z, t) = \frac{1}{Z_0}\left[Af\left(t - \frac{z}{v_p}\right) - Bg\left(t + \frac{z}{v_p}\right)\right] \tag{6.115b}$$

where $Z_0 = \sqrt{\mathcal{L}/\mathcal{C}}$ is the characteristic impedance of the line, and $v_p = 1/\sqrt{\mathcal{L}\mathcal{C}}$ is the velocity of propagation on the line.

We discussed several techniques of determination of Z_0 and v_p for the case of a line with homogeneous dielectric, as well as for the case of a line involving more than one dielectric, an example being the microstrip line. For the former case,

$$Z_0 = \frac{\sqrt{\varepsilon_r}}{c\mathscr{C}} \qquad (6.116a)$$

$$v_p = \frac{c}{\sqrt{\varepsilon_r}} \qquad (6.116b)$$

where ε_r, c, and $\mathscr{C}$ are the relative permittivity of the dielectric, the velocity of light in free space, and the capacitance per unit length of the line computed from static field considerations, respectively. For the latter case, assuming nonmagnetic dielectrics,

$$Z_0 = \frac{1}{c\sqrt{\mathscr{C}\mathscr{C}_0}} \qquad (6.117a)$$

$$v_p = c\sqrt{\frac{\mathscr{C}_0}{\mathscr{C}}} \qquad (6.117b)$$

where $\mathscr{C}$ is the capacitance per unit length of the line with the dielectrics in place, and $\mathscr{C}_0$ is the capacitance per unit length with all dielectrics replaced by free space, both computed from static field considerations. Note that (6.117a) and (6.117b) reduce to (6.116a) and (6.116b), respectively, if all dielectrics are the same, since then $\mathscr{C} = \mathscr{C}_0\varepsilon_r$. Based on these considerations, we presented analytically obtainable expressions for Z_0 for several common types of lines, and introduced a graphical technique involving field mapping for the case of a line with arbitrary cross section.

We then discussed time-domain analysis of a transmission line terminated by a load resistance R_L and excited by a constant voltage source V_0 in series with internal resistance R_g. Writing the general solutions (6.115a) and (6.115b) concisely in the manner

$$V = V^+ + V^- \qquad (6.118a)$$
$$I = I^+ + I^- \qquad (6.118b)$$

where

$$I^+ = \frac{V^+}{Z_0} \qquad (6.119a)$$

$$I^- = -\frac{V^-}{Z_0} \qquad (6.119b)$$

we found that the situation consists of bouncing back and forth of transient $(+)$ and $(-)$ waves between the two ends of the line. The initial $(+)$ wave voltage is $V^+Z_0/(R_g + Z_0)$. All other waves are governed by the reflection coefficients at the two ends of the line, given for the voltage by

$$\Gamma_R = \frac{R_L - Z_0}{R_L + Z_0} \qquad (6.120a)$$

and

$$\Gamma_S = \frac{R_g - Z_0}{R_g + Z_0} \tag{6.120b}$$

for the load and source ends, respectively. In the steady state, the situation is the superposition of all the transient waves, equivalent to the sum of a single $(+)$ wave and a single $(-)$ wave. We discussed the bounce-diagram technique of keeping track of the transient phenomenon and extended it to a pulse voltage source.

We learned that when a wave is incident from, say, line 1 onto a junction with line 2, reflection occurs just as though line 1 is terminated by a load resistor equal to the characteristic impedance of line 2. A transmitted wave goes into line 2 in accordance with the voltage and current transmission coefficients

$$\tau_V = 1 + \Gamma \tag{6.121a}$$

and

$$\tau_C = 1 - \Gamma \tag{6.121b}$$

respectively, where Γ is the voltage reflection coefficient. Applying this to a system of three lines in cascade, we showed how to obtain the unit impulse response of the system and from it obtain the frequency response. We then extended the analysis to lines with discontinuities to discuss and illustrate by means of an example the application of time-domain reflectometry, an important experimental technique.

We then considered lines with reactive terminations and discontinuities, and nonlinear resistive terminations, where we learned that the reflection coefficient concept is not useful to study the transient behavior. For the cases of reactive elements, it is necessary to write the differential equations pertinent to the boundary conditions at the terminations and/or discontinuities, and solve them subject to the appropriate initial conditions; alternatively, the required voltages and currents can be obtained from considerations of initial and final behaviors of the reactive element(s), and associated time constant(s). For the nonlinear resistive termination case, we introduced the load-line technique of time-domain analysis.

As a prelude to the consideration of interconnections between logic gates, we discussed time-domain analysis of lines with nonzero initial conditions. For the general case, the initial voltage and current distributions $V(z, 0)$ and $I(z, 0)$ are decomposed into $(+)$ and $(-)$ wave voltages and currents as given by

$$V^+(z, 0) = \tfrac{1}{2}[V(z, 0) + Z_0 I(z, 0)]$$

$$V^-(z, 0) = \tfrac{1}{2}[V(z, 0) - Z_0 I(z, 0)]$$

$$I^+(z, 0) = \frac{1}{Z_0} V^+(z, 0)$$

$$I^-(z, 0) = -\frac{1}{Z_0} V^-(z, 0)$$

The voltage and current distributions for $t > 0$ are then obtained by keeping track of the bouncing of these waves at the two ends of the line. For the special case of uniform distribution, the analysis can be performed more conveniently by considering the situation as one in which a transient wave is superimposed on the initial distribution and using the bounce-diagram technique or the load-line technique. We then applied the load-line technique to the analysis of transmission-line interconnection between logic gates.

Finally, we studied the topic of crosstalk on transmission lines, by considering the case of weak coupling between two lines. We learned that for a given wave on the primary line, the crosstalk consists of two waves, forward and backward, induced on the secondary line and governed by the forward-crosstalk coefficient and the backward-crosstalk coefficient, respectively. We illustrated by means of an example the determination of crosstalk voltages for a specified excitation for the primary line.

REVIEW QUESTIONS

Q6.1. State Poisson's equation. How is it derived?

Q6.2. Discuss the application of Poisson's equation for the determination of potential due to the space charge layer in a *p-n* junction semiconductor.

Q6.3. State Laplace's equation. In what regions is it valid?

Q6.4. Discuss the application of Laplace's equation for a conducting medium.

Q6.5. Outline the solution of Laplace's equation in one dimension by considering the variation of potential with x only.

Q6.6. Outline the steps in the derivation of the expression for the capacitance of an arrangement of two conductors.

Q6.7. Discuss the relationship between the capacitance, conductance, and inductance per unit length for an infinitely long, parallel conductor arrangement.

Q6.8. Outline the steps in the derivation of the expressions for the capacitance, conductance, and inductance per unit length of an infinitely long parallel cylindrical-wire arrangement.

Q6.9. Distinguish between internal inductance and external inductance. Discuss the concept of flux linkage pertinent to the determination of the internal inductance.

Q6.10. Explain the concept of mutual inductance and discuss an example of its computation.

Q6.11. Describe the phenomenon of guiding of a uniform plane wave by a pair of parallel, plane, perfectly conducting sheets.

Q6.12. Discuss the derivation of the transmission-line equations from the field equations by considering the parallel-plate line.

Q6.13. Discuss the concept of the distributed circuit as compared to a lumped circuit.

Q6.14. What is a transverse electromagnetic wave? Discuss the electric and magnetic field distributions associated with a transverse electromagnetic wave.

Q6.15. Discuss the analogy between uniform plane wave parameters and transmission-line parameters.

Q6.16. Explain why the product of $\mathscr{L}$ and $\mathscr{C}$ of a line is equal to the product of μ and ε of the dielectric of the line.

Q6.17. What is the significance of the characteristic impedance of a line? Why is it not in general equal to the intrinsic impedance of the medium between the conductors of the line?

Q6.18. Discuss the geometry associated with the microstrip line and the determination of its characteristic impedance and velocity of propagation.

Q6.19. Describe the procedure for computing the transmission-line parameters by using the field mapping technique.

Q6.20. Discuss the general solutions for the line voltage and current and the notation associated with their representation in concise form.

Q6.21. What is the fundamental distinction between the occurrence of the response in one branch of a lumped circuit to the application of an excitation in a different branch of the circuit and the occurrence of the response at one location on a transmission line to the application of an excitation at a different location on the line?

Q6.22. Describe the phenomenon of the bouncing back and forth of transient waves on a transmission line excited by a constant voltage source in series with internal resistance and terminated by a resistance.

Q6.23. What is the nature of the formula for the voltage reflection coefficient? Discuss its values for some special cases.

Q6.24. What is the steady-state equivalent of a line excited by a constant voltage source? What is the actual situation in the steady state?

Q6.25. Discuss the bounce-diagram technique of keeping track of the bouncing back and forth of the transient waves on a transmission line for a constant voltage source.

Q6.26. Discuss the bounce-diagram technique of keeping track of the bouncing back and forth of the transient waves on a transmission line for a pulse voltage source.

Q6.27. How are the voltage and current transmission coefficients at the junction between two lines related to the voltage reflection coefficient?

Q6.28. Explain how it is possible for the transmitted voltage or current at a junction between two lines to exceed the incident voltage or current.

Q6.29. Discuss the determination of the unit impulse response of a system of three lines in cascade.

Q6.30. Outline the procedure for the determination of the frequency response of a system of three lines in cascade from its unit impulse response.

Q6.31. What is a radome? How is it analyzed by using transmission-line equivalent?

Q6.32. Describe the technique of locating discontinuities in a transmission-line system by using a time-domain reflectometer.

Q6.33. Discuss the transient analysis of a line driven by a constant voltage source in series with a resistance equal to the Z_0 of the line and terminated by an inductor.

Q6.34. Why is the concept of reflection coefficient not useful for studying the transient behavior of lines with reactive terminations and discontinuities?

Q6.35. Discuss the determination of the transient behavior of lines with reactive terminations and discontinuities without formally setting up the differential equations and solving them.

Q6.36. Discuss the load-line technique of obtaining the time variations of the voltages and currents at the source and load ends of a line from a knowledge of the terminal V-I characteristics.

Q6.37. Discuss the determination of the voltage and current distributions on an initially charged line for any given time from the knowledge of the initial voltage and current distributions.

Q6.38. Discuss with the aid of an example the discharging of an initially charged line into a resistor.

Q6.39. Discuss the bounce-diagram technique of transient analysis of a line with uniform initial voltage and current distributions.

Q6.40. How do you check the energy balance for the case of a line with initial voltage and/or current distribution(s) and discharged into a resistor?

Q6.41. Discuss the analysis of transmission-line interconnection between two logic gates.

Q6.42. Discuss briefly the weak-coupling analysis for crosstalk between two transmission lines.

Q6.43. Discuss the modeling of capacitive and inductive couplings for crosstalk on transmission lines.

Q6.44. Discuss and distinguish between the dependence of the forward- and backward-crosstalk coefficients on the line parameters.

Q6.45. Outline the determination of the forward- and backward-crosstalk voltages induced on a secondary line for a given excitation for the primary line.

PROBLEMS

Section 6.1

P6.1. A space-charge density distribution is given by

$$
\rho = \begin{cases}
-\rho_0\left(1 + \dfrac{x}{d}\right) & \text{for } -d < x < 0 \\[2mm]
\rho_0\left(1 - \dfrac{x}{d}\right) & \text{for } 0 < x < d \\[2mm]
0 & \text{otherwise}
\end{cases}
$$

where ρ_0 is a constant. Obtain the solution for the potential V versus x for all x. Assume $V = 0$ for $x = 0$.

P6.2. A space-charge density distribution is given by

$$
\rho = \begin{cases}
\rho_0 \sin x & \text{for } -\pi < x < \pi \\
0 & \text{otherwise}
\end{cases}
$$

where ρ_0 is a constant. Find and sketch the potential V versus x for all x. Assume $V = 0$ for $x = 0$.

P6.3. A space-charge density distribution is given in spherical coordinates by

$$
\rho = \begin{cases}
\rho_0 & \text{for } a < r < 2a \\
0 & \text{otherwise}
\end{cases}
$$

where ρ_0 is a constant. Find and sketch the potential V versus r for all r.

P6.4. The region between the two plates in Fig. 6.2 is filled with two perfect dielectric media having permittivities ε_1 for $0 < x < t$ (region 1) and ε_2 for $t < x < d$

(region 2). **(a)** Find the solutions for the potentials in the two regions $0 < x < t$ and $t < x < d$. **(b)** Find the capacitance per unit area of the plates.

P6.5. Assume that the two media in Problem P6.4 are imperfect dielectrics having conductivities σ_1 and σ_2 for $0 < x < t$ and $t < x < d$, respectively. **(a)** What are the boundary conditions to be satisfied at $x = t$? **(b)** Find the solutions for the potentials in the two regions. **(c)** Find the potential at $x = t$.

P6.6. Assume that the region between the two plates of Fig. 6.2 is filled with a perfect dielectric of nonuniform permittivity

$$\varepsilon = \frac{\varepsilon_0}{1 - (x/2d)}$$

Find the solution for the potential between the plates and obtain the expression for the capacitance per unit area of the plates.

P6.7. Assume that the region between the coaxial cylindrical conductors of Fig. 6.3(a) is filled with a dielectric of nonuniform permittivity $\varepsilon = \varepsilon_0 b/r$. Obtain the solution for the potential between the conductors and the expression for the capacitance per unit length of the cylinders.

Section 6.2

P6.8. For the parallel-wire arrangement of Fig. 6.5(c), show that for $d \gg a$, the capacitance per unit length of the line is $\pi \varepsilon / \ln (2d/a)$. Find the value of d/a for which the exact value of the capacitance per unit length is 1.05 times the value given by the approximate expression for $d \gg a$.

P6.9. For the line-charge pair of Fig. 6.7, show that the direction lines of the electric field are arcs of circles emanating from the positively charged line and terminating on the negatively charged line.

P6.10. A filamentary wire carrying current I is closely wound around a toroidal magnetic core of rectangular cross section, as shown in Fig. 6.72. The mean radius of the toroidal core is a and the number of turns per unit length along the mean circumference of the toroid is N. Find the inductance of the toroid.

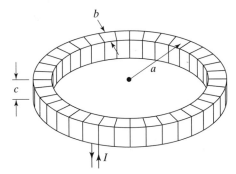

FIGURE 6.72   For Problem P6.10.

P6.11. An infinitely long, uniformly wound solenoid of radius a and having N turns per unit length carries a current I. Find the inductance per unit length of the solenoid. Assume air core ($\mu = \mu_0$).

P6.12. A current I flows with nonuniform volume density given by

$$\mathbf{J} = J_0 \left(\frac{r}{a}\right)^2 \mathbf{a}_z$$

along an infinitely long cylindrical conductor of radius a having the z-axis as its axis. The current returns with uniform surface density in the opposite direction along the surface of an infinitely long, perfectly conducting cylinder of radius $b\ (> a)$ and coaxial with the inner conductor. Find the internal inductance per unit length of the inner conductor.

P6.13. Consider the infinitely long solid cylindrical conductor of Fig. 6.8. Obtain the expression for the energy stored per unit length in the magnetic field internal to the current distribution and show that it is equal to $\frac{1}{2}\mathcal{L}_i I^2$, where I is the total current.

P6.14. An infinitely long, uniformly wound solenoid of radius a and having N_1 turns per unit length is coaxial with another infinitely long, uniformly wound solenoid of radius $b\ (> a)$ and having N_2 turns per unit length. Find the mutual inductance per unit length of the solenoids. Assume air core ($\mu = \mu_0$).

Section 6.3

P6.15. A parallel-plate transmission line is made up of perfect conductors of width $w = 0.1$ m and lying in the planes $x = 0$ and $x = 0.01$ m. The medium between the conductors is a nonmagnetic ($\mu - \mu_0$), perfect dielectric. For a uniform plane wave propagating along the line, the voltage along the line is given by

$$V(z, t) = 10 \cos (3\pi \times 10^8 t - 2\pi z) \text{ V}$$

Neglecting fringing of fields, find: **(a)** the electric field intensity $E_x(z, t)$ of the wave; **(b)** the magnetic field intensity $H_y(z, t)$ of the wave; **(c)** the current $I(z, t)$ along the line; and **(d)** the power flow $P(z, t)$ down the line.

P6.16. A parallel-plate transmission line consists of an arrangement of two perfect dielectrics, as shown by the transverse cross section in Fig. 6.73. Note that $\mu_1\varepsilon_1 = \mu_2\varepsilon_2$, so that the TEM waves propagating in the two dielectrics are in phase at all points along the interface between the dielectrics. Neglect fringing of fields and compute the values of $\mathcal{L}$, $\mathcal{C}$, and Z_0 of the line.

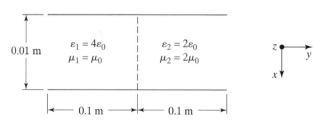

FIGURE 6.73   For Problem P6.16.

P6.17. Derive the transmission-line equations by considering the special case of two infinitely long coaxial cylindrical conductors. Also show that the power flow along the line is equal to the product of the voltage between the conductors and current along the conductors.

P6.18. By applying the curvilinear squares technique to a coaxial cable of inner radius a and outer radius b, show that the characteristic impedance of the cable is $(\eta/2\pi) \ln b/a$, where η is the intrinsic impedance of the dielectric of the cable.

P6.19. The cross section of an eccentric coaxial cable [see Fig. 6.5(d)] consists of an outer circle of radius $a = 5$ cm and an inner circle of radius $b = 2$ cm, with their centers separated by $d = 2$ cm. By constructing a field map consisting of curvilinear squares, obtain the approximate value of Z_0 in terms of η of the dielectric.

P6.20. When one microstrip line is inverted and placed on top of another microstrip line, as shown by the cross-sectional view in Fig. 6.74, a shielded strip line is obtained. Although the sandwich arrangement of this line is more difficult to fabricate than is the microstrip line, it has the advantage that the fields are confined mostly to the substrate region. Assuming for simplicity that the fields are confined to the substrate region, construct a field map consisting of curvilinear squares and compute the approximate value of Z_0 of the line, for the dimensions shown in Fig. 6.74, and considering the substrate to be a perfect dielectric having $\varepsilon = 9\varepsilon_0$ and $\mu = \mu_0$.

FIGURE 6.74   For Problem P6.20.

Section 6.4

P6.21. In the system shown in Fig. 6.75, assume that V_g is a constant voltage source of 100 V and the switch S is closed at $t = 0$. Find and sketch: (**a**) the line voltage versus z for $t = 0.2$ μs; (**b**) the line current versus z for $t = 0.4$ μs; (**c**) the line voltage versus t for $z = 30$ m; and (**d**) the line current versus t for $z = -40$ m.

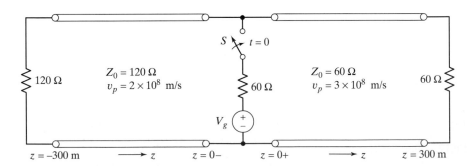

FIGURE 6.75   For Problem P6.21.

P6.22. In the system shown in Fig. 6.76(a), the switch S is closed at $t = 0$. The line voltage variations with time at $z = 0$ and $z = l$ for the first 5 μs are observed to be as shown in Fig. 6.76(b) and (c), respectively. Find the values of V_0, R_g, R_L, and T.

(a)

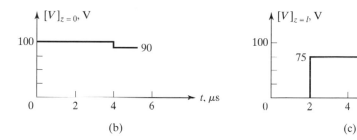

(b)   (c)

FIGURE 6.76   For Problem P6.22.

P6.23. The system shown in Fig. 6.77 is in steady state. Find **(a)** the line voltage and current, **(b)** the $(+)$ wave voltage and current, and **(c)** the $(-)$ wave voltage and current.

FIGURE 6.77   For Problem P6.23.

P6.24. In the system shown in Fig. 6.78, the switch S is closed at $t = 0$. Assume $V_g(t)$ to be a direct voltage of 90 V and draw the voltage and current bounce diagrams. From these bounce diagrams, sketch: **(a)** the line voltage and line current versus t (up to $t = 7.25$ μs) at $z = 0$, $z = l$, and $z = l/2$; and **(b)** the line voltage and line current versus z for $t - 1.2$ μs and $t = 3.5$ μs.

FIGURE 6.78   For Problem P6.24.

P6.25.   Repeat Problem P6.21 assuming V_g to be a triangular pulse, as shown in Fig. 6.79.

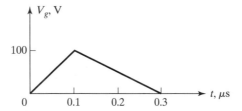

FIGURE. 6.79   For Problem P6.25.

P6.26.   For the system of Problem P6.24, assume that the voltage source is of 0.3 μs duration instead of being of infinite duration. Find and sketch the line voltage and line current versus z for $t = 1.2$ μs and $t = 3.5$ μs.

P6.27.   In the system shown in Fig. 6.80, the switch S is closed at $t = 0$. Find and sketch: **(a)** the line voltage versus z for $t = 2\frac{1}{2}$ μs; **(b)** the line current versus z for $t = 2\frac{1}{2}$ μs; and **(c)** the line voltage at $z = l$ versus t up to $t = 4$ μs.

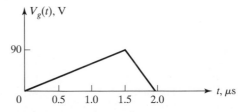

FIGURE 6.80   For Problem P6.27

P6.28. In the system shown in Fig. 6.81, the switch S is closed at $t = 0$. Draw the voltage and current-bounce diagrams and sketch **(a)** the line voltage and line current versus t for $z = 0$ and $z = l$ and **(b)** the line voltage and line current versus z for $t = 2, 9/4, 5/2, 11/4$, and 3 μs. Note that the period of the source voltage is 2 μs, which is equal to the two-way travel time on the line.

FIGURE 6.81   For Problem P6.28.

Section 6.5

P6.29. In the system shown in Fig. 6.82, an incident wave of voltage V^+ strikes the discontinuity from the left, that is, from line 1. Find the reflected wave voltage and current into line 1 and the transmitted wave voltage and current into line 2.

FIGURE 6.82   For Problem P6.29.

P6.30. In the system shown in Fig. 6.83: **(a)** find the output voltage V_o across the 300-Ω resistor for $V_g(t) = \delta(t)$; and **(b)** find and sketch the amplitude of $V_o(t)$ versus ω for $V_g(t) = \cos \omega t$.

FIGURE 6.83   For Problem P6.30.

P6.31. In Fig. 6.84(a), the plane I is the input plane from which a uniform plane wave is incident normally on the interface between medium 1 and medium 2, and the plane O is the output plane in which the response of the system is observed. For an incident wave of $E_{xi}(t) = \delta(t)$, find the minimum value of the thickness l and the corresponding value of the permittivity ε_2 of medium 2 required to obtain the electric field $E_{xo}(t)$ in the output plane, as shown in Fig. 6.84(b), in which the interval between successive impulses is 2 μs. Then find the value of A, and sketch the reflected wave electric field in the plane I.

(a)

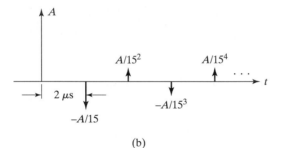

(b)

FIGURE 6.84   For Problem P6.31.

P6.32. In Fig. 6.85, a $(+)$ wave carrying power P is incident on the junction a–a' from line 1. Find **(a)** the power reflected into line 1; **(b)** the power transmitted into line 2; and **(c)** the power transmitted into line 3.

FIGURE 6.85   For Problem P6.32.

P6.33. In the system of Fig. 6.39, assume that the discontinuity at $z = 4$ m is a resistor of value 40 Ω in series with the line, instead of the 120-Ω parallel resistor. Find and sketch the waveform that the TDR system would measure up to $t = 200$ ns.

Section 6.6

P6.34. In the system shown in Fig. 6.86, the switch S is closed at $t = 0$ with no current in the relay coil and with the relay in position 1. When the relay coil current I_L reaches 1.73 A, the relay switches to position 2; when the current drops to 0.636 A, the relay switches back to position 1. **(a)** Find the time t_1 at which the relay switches to position 2. **(b)** Find the time t_2 at which the relay switches back to position 1.

FIGURE 6.86 For Problem P6.34.

P6.35. In the system shown in Fig. 6.87, the switch S is closed at $t = 0$, with the voltage across the capacitor equal to zero. **(a)** Obtain the differential equation for V^- at $z = l$. **(b)** Find the solution for $V^-(l, t)$.

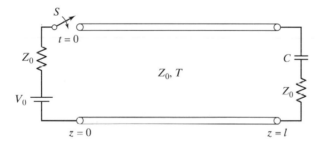

FIGURE 6.87 For Problem P6.35.

P6.36. In the system shown in Fig. 6.88, the switch S is closed at $t = 0$, with the lines uncharged and with zero current in the inductor. Obtain the solution for the line voltage versus time at $z = l+$.

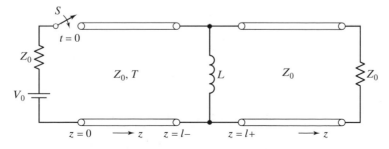

FIGURE 6.88 For Problem P6.36.

P6.37. In the system shown in Fig. 6.89(a), the network N consists of a single circuit element ($R, L,$ or C). The system is initially uncharged. The switch S is closed at $t = 0$, and the line voltage at $z = 0$ is observed to be as shown in Fig. 6.89(b). **(a)** Determine whether the circuit element is $R, L,$ or C. **(b)** Find the value of Z_{02}/Z_{01}.

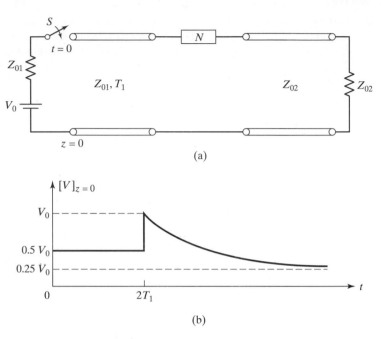

(a)

(b)

FIGURE 6.89   For Problem P6.37.

P6.38. For the system of Problem P6.24, use the load-line technique to obtain and plot line voltage and line current versus t (up to $t = 5.25$ μs) at $z = 0$ and $z = l$. Also obtain the steady-state values of line voltage and current from the load-line construction.

Section 6.7

P6.39. In the system shown in Fig. 6.90, a passive nonlinear element having the indicated volt-ampere characteristic is connected to an initially charged line at $t = 0$. Find the voltage across the nonlinear element immediately after closure of the switch.

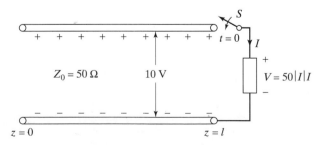

FIGURE 6.90   For Problem P6.39.

P6.40. In the system shown in Fig. 6.91, steady-state conditions are established with the switch S closed. At $t = 0$, the switch is opened. **(a)** Find and sketch the voltage across the 150-Ω resistor for $t \geq 0$, with the aid of a bounce diagram. **(b)** Show that the total energy dissipated in the 150-Ω resistor after opening the switch is exactly the same as the energy stored in the line before opening the switch.

FIGURE 6.91   For Problem P6.40.

P6.41. In the system shown in Fig. 6.92, steady-state conditions are established with the switch S closed. At $t = 0$, the switch is opened. **(a)** Sketch the voltage and current along the system for $t = 0-$. **(b)** Find the total energy stored in the lines for $t = 0-$. **(c)** Find and sketch the voltages across the two resistors for $t > 0$. **(d)** From your sketches of part (c), find the total energy dissipated in the resistors for $t > 0$.

FIGURE 6.92   For Problem P6.41.

P6.42. In the system shown in Fig. 6.93, steady-state conditions are established with the switch S open and no current in the inductor. At $t = 0$, the switch is closed. **(a)** Obtain the expressions for the line voltage and current versus t at $z = l$. **(b)** Sketch the line voltage and current versus z for $t = T/2$.

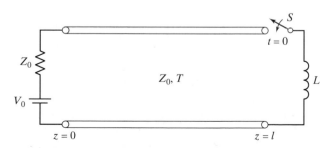

FIGURE 6.93   For Problem P6.42.

P6.43. For the system of Problem P6.39, use the load-line technique to obtain and plot line voltage versus t from $t = 0$ up to $t = 7l/v_p$ at $z = 0$ and $z = l$.

P6.44. For the example of interconnection between logic gates of Fig. 6.60(a), repeat the load-line constructions for $Z_0 = 50 \, \Omega$ and draw graphs of V_i versus t for both 0-to-1 and 1-to-0 transitions.

P6.45. For the example of interconnection between logic gates of Fig. 6.60(a), find **(a)** the minimum value of Z_0 such that for the transition from 0 to 1, the voltage V_i reaches 2 V at $t = T+$ and **(b)** the minimum value of Z_0 such that for the transition from 1 to 0, the voltage V_i reaches 1 V at $t = T+$.

Section 6.8

P6.46. In Example 6.13, assume that $V_g(t)$ is the function shown in Fig. 6.94, instead of as in Fig. 6.68. Find and sketch the following: **(a)** $V_2^+(l, t)$; **(b)** $V_2^-(0, t)$; and **(c)** $V_2^-(0.8l, t)$.

FIGURE 6.94   For Problem P6.46.

P6.47. In Example 6.13, assume that

$$V_g(t) = \begin{cases} 2V_0 \sin^2 \pi t/T & \text{for } 0 < t < T \\ 0 & \text{otherwise} \end{cases}$$

Find and sketch the following: **(a)** $V_2^+(l, t)$; **(b)** $V_2^-(0, t)$; **(c)** $V_2^-(0.75l, t)$.

P6.48. In Example 6.13, assume that $K_b/K_f = -25v_p$ and $T_0 = 0.2T$. Find and sketch the following: **(a)** $V_2^+(z, 1.1T)$; **(b)** $V_2^-(z, 1.1T)$; and **(c)** $V_2(z, 1.1T)$.

REVIEW PROBLEMS

R6.1. Assume that the region between the concentric spherical conductors of Fig. 6.3(b) is filled with a dielectric of nonuniform permittivity $\varepsilon = \varepsilon_0 b^2/r^2$. Obtain the solution for the potential between the conductors and the expression for the capacitance.

R6.2. Current I flows with uniform density along an infinitely long, hollow cylindrical conductor of inner radius a and outer radius b and returns with uniform surface density in the opposite direction along the surface of an infinitely long, perfectly conducting cylinder of radius c ($> b$) and coaxial with the hollow conductor. Find the internal inductance per unit length of the arrangement.

R6.3. Consider a transmission line having the cross section shown in Fig. 6.95. The inner conductor is a circle of radius a and the outer conductor is a square of sides $2a$. Find the approximate value of the characteristic impedance of the line, by using the method of curvilinear squares.

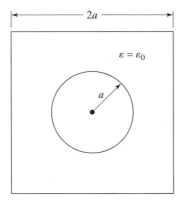

FIGURE 6.95   For Problem R6.3.

R6.4. In the system shown in Fig. 6.96, the voltage source V_g is a pulse of amplitude 10 V and duration 1 μs from $t = 0$ to $t = 1$ μs, and the switch S is closed at $t = 0$. **(a)** Find and sketch the voltage across the load resistor R_L as a function of time for $t > 0$. **(b)** Find and sketch the voltage across the internal resistance R_g of the voltage source as a function of time for $t > 0$. **(c)** Show that the total energy supplied by the voltage source is equal to the sum of the total energy dissipated in R_L and the total energy dissipated in R_g.

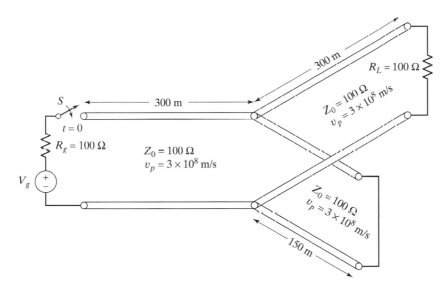

FIGURE 6.96   For Problem R6.4.

R6.5. In Fig. 6.97 the plane I is the input plane from which a uniform plane wave is incident normally on the interface between medium 1 and medium 2, and the plane O is the output plane in which the response of the system is observed. **(a)** For an incident wave of $E_{xi}(t) = \delta(t)$, a unit impulse, find and sketch the sequence of impulses for the electric field $E_{xo}(t)$ in the output plane. **(b)** If $E_{xi}(t)$ consists of a periodic sequence of unit impulses of frequency f, show that there

Plane *I*   Plane *O*   **FIGURE 6.97**   For Problem R6.5.

exists a minimum value of f for which $E_{xo}(t)$ consists of a periodic sequence of unit impulses of the same frequency f, and find that value of f.

R6.6. In the system shown in Fig. 6.98, a ($+$) wave carrying power P is incident on the junction a–a' from line 1. (**a**) Find the value of R for which there is no reflected wave into line 1. (**b**) For the value of R found in (a), find the power transmitted into each of lines 2 and 3.

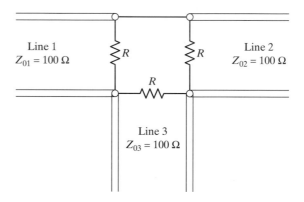

FIGURE 6.98   For Problem R6.6.

R6.7. In the system shown in Fig. 6.99, steady-state conditions are established with the switch S closed. At $t = 0$, the switch S is opened. (**a**) Find the energy stored in the

FIGURE 6.99   For Problem R6.7.

system at $t = 0-$. **(b)** Obtain the solutions for the voltages across R_{L1} and R_{L2} for $t > 0$. **(c)** Show that the total energy dissipated in R_{L1} and R_{L2} for $t > 0$ is equal to the energy stored in the system at $t = 0-$.

R6.8. For the example of interconnection between logic gates in Section 6.7, find the value of Z_0 for which the voltage reached at $t = T+$ for the transition from 0 to 1 is the same as that reached for the transition from 1 to 0. What is the value of this voltage?

R6.9. In example 6.13, assume that $V_g(t) = V_0 \cos 2\pi ft$. Show that for $f = 1/4T$, the ratio of the amplitude of $V_2(0, t)$ to V_0, the amplitude of $V_g(t)$, is equal to K_b, and the ratio of the amplitude of $V_2(l, t)$ to V_0 is equal to $\pi fl |K_f|$.

CHAPTER 7

Transmission Lines for Communications

In Chapter 6, we introduced the transmission line and studied propagation and bouncing of waves along a line, applicable to digital electronics. In this chapter, we are concerned with the steady-state analysis of transmission-line systems excited by sinusoidally time-varying sources, generally applicable to communication systems. We recall from Chapter 6 that the phenomenon on a transmission line excited by a source connected to the line at a certain instant of time, say, $t = 0$, consists of the transient bouncing of $(+)$ and $(-)$ waves along the line for $t > 0$. In the steady state, the situation is equivalent to the superposition of one $(+)$ wave, which is the sum of all the transient $(+)$ waves, and one $(-)$ wave, which is the sum of all the transient $(-)$ waves. Thus, the general solutions for the line voltage and line current in the sinusoidal steady state are superpositions of voltages and currents, respectively, of sinusoidal $(+)$ and $(-)$ waves. We shall first write these general solutions and then discuss several topics pertinent to sinusoidal steady-state analysis of transmission-line systems.

We introduce the standing-wave concept by first considering the particular case of a short-circuited line and then the general case of a line terminated by an arbitrary load. We discuss several techniques of transmission-line matching. In this connection, we introduce the Smith chart, a useful graphical aid in the solution of transmission-line problems. Finally, we extend our treatment of sinusoidal steady-state analysis to lossy lines.

Although the concepts and techniques discussed in this chapter are based on the analysis of transmission-line systems, many of these are also applicable to the analysis of other, analogous systems. Examples are uniform plane wave propagation involving multiple media, as in Section 5.6, and discontinuities in waveguides, considered in Chapters 9 and 10.

7.1 SHORT-CIRCUITED LINE

From (6.55a) and (6.55b), we write the general solutions for the line voltage and line current in the sinusoidal steady state to be

General solution in the sinusoidal steady state

$$V(z, t) = A \cos\left[\omega\left(t - \frac{z}{v_p}\right) + \theta\right] + B \cos\left[\omega\left(t + \frac{z}{v_p}\right) + \phi\right] \tag{7.1a}$$

$$I(z, t) = \frac{1}{Z_0}\left\{A \cos\left[\omega\left(t - \frac{z}{v_p}\right) + \theta\right] - B \cos\left[\omega\left(t + \frac{z}{v_p}\right) + \phi\right]\right\} \tag{7.1b}$$

The corresponding expressions for the phasor line voltage and phasor line current are

$$\overline{V}(z) = \overline{V}^+ e^{-j\beta z} + \overline{V}^- e^{j\beta z} \tag{7.2a}$$

$$\overline{I}(z) = \frac{1}{Z_0}(\overline{V}^+ e^{-j\beta z} - \overline{V}^- e^{j\beta z}) \tag{7.2b}$$

where $\overline{V}^+ = Ae^{j\theta}$ and $\overline{V}^- = Be^{j\phi}$ and we have substituted β for ω/v_p. For sinusoidal steady-state problems, it is convenient to use a distance variable d that increases as we go from the load toward the generator as opposed to z, which increases from the generator toward the load, as shown in Fig. 7.1. The wave that progresses away from the generator is still denoted as the (+) wave, and the wave that progresses toward the generator is still denoted as the (−) wave. In terms of d, the solutions for $\overline{V}$ and $\overline{I}$ are then given by

$$\overline{V}(d) = \overline{V}^+ e^{j\beta d} + \overline{V}^- e^{-j\beta d} \tag{7.3a}$$

$$\overline{I}(d) = \frac{1}{Z_0}(\overline{V}^+ e^{j\beta d} - \overline{V}^- e^{-j\beta d}) \tag{7.3b}$$

Let us now consider a lossless line short circuited at the far end $d = 0$, as shown in Fig. 7.2. We shall assume that sinusoidally time-varying traveling waves exist on the line due to a source that is not shown in the figure and that conditions have reached steady state. We wish to determine the characteristics of the waves

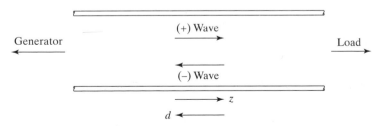

FIGURE 7.1 For illustrating the distance variable d used for sinusoidal steady-state analysis of traveling waves.

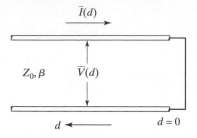

FIGURE 7.2   Transmission line short circuited at the far end.

satisfying the boundary condition at the short circuit. Since the voltage across a short circuit has to be always equal to zero, this boundary condition is given by

$$\overline{V}(0) = 0 \tag{7.4}$$

Applying it to the general solution for $\overline{V}(d)$ given by (7.3a), we obtain

$$\overline{V}(0) = \overline{V}^+ e^{j\beta(0)} + \overline{V}^- e^{-j\beta(0)} = 0$$

or

$$\overline{V}^- = -\overline{V}^+ \tag{7.5}$$

Thus, we find that the short circuit gives rise to a $(-)$ or reflected wave whose voltage is exactly the negative of the $(+)$ or incident wave voltage, at the short circuit.

Substituting (7.5) into (7.3a) and (7.3b), we get the particular solutions for the complex voltage and current on the short-circuited line to be

$$\overline{V}(d) = \overline{V}^+ e^{j\beta d} - \overline{V}^+ e^{-j\beta d} = 2j\overline{V}^+ \sin \beta d \tag{7.6a}$$

$$\overline{I}(d) = \frac{1}{Z_0}(\overline{V}^+ e^{j\beta d} + \overline{V}^+ e^{-j\beta d}) = 2\frac{\overline{V}^+}{Z_0} \cos \beta d \tag{7.6b}$$

The real voltage and current are then given by

$$
\begin{aligned}
V(d, t) &= \text{Re}[\overline{V}(d)e^{j\omega t}] \\
&= \text{Re}(2e^{j\pi/2}\,|\overline{V}^+|\,e^{j\theta} \sin \beta d\, e^{j\omega t}) \\
&= -2|\overline{V}^+|\sin \beta d \sin(\omega t + \theta)
\end{aligned}
\tag{7.7a}
$$

$$
\begin{aligned}
I(d, t) &= \text{Re}[\overline{I}(d)e^{j\omega t}] \\
&= \text{Re}\left(2\frac{|\overline{V}^+|}{Z_0}\,e^{j\theta} \cos \beta d\, e^{j\omega t}\right) \\
&= 2\frac{|\overline{V}^+|}{Z_0}\cos \beta d \cos(\omega t + \theta)
\end{aligned}
\tag{7.7b}
$$

where we have replaced $\overline{V}^+$ by $|\overline{V}^+|e^{j\theta}$ and j by $e^{j\pi/2}$. The instantaneous power flow down the line is given by

$$
\begin{aligned}
P(d, t) &= V(d, t)I(d, t) \\
&= -\frac{4|\overline{V}^+|^2}{Z_0} \sin \beta d \cos \beta d \sin(\omega t + \theta) \cos(\omega t + \theta) \\
&= -\frac{|\overline{V}^+|^2}{Z_0} \sin 2\beta d \sin 2(\omega t + \theta)
\end{aligned}
\tag{7.7c}
$$

These results for the voltage, current, and power flow on the short-circuited line are illustrated in Fig. 7.3, which shows the variation of each of these quantities with distance from the short circuit for several values of time. The numbers $1, 2, 3, \ldots, 9$ beside the curves in Fig. 7.3 represent the order of the curves corresponding to values of $(\omega t + \theta)$ equal to $0, \pi/4, \pi/2, \ldots, 2\pi$. From (7.7a), (7.7b), and (7.7c) and from the sketches of Fig. 7.3, we can infer the following:

1. The line voltage is zero for $\sin \beta d = 0$, or $\beta d = 0, \pi, 2\pi, \ldots$, or $d = 0$, $\lambda/2, \lambda, \ldots$, for all values of time. If we short circuit the line at these

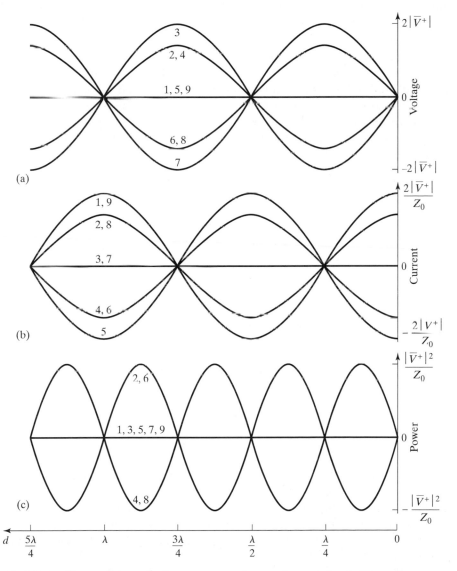

FIGURE 7.3 Time variations of voltage, current, and power flow associated with standing waves on a short-circuited transmission line.

values of d, there will be no effect on the voltage and current at any other value of d.

2. The line current is zero for $\cos \beta d = 0$, or $\beta d = \pi/2, 3\pi/2, 5\pi/2, \ldots$, or $d = \lambda/4, 3\lambda/4, 5\lambda/4, \ldots$, for all values of time. If we open circuit the line at these values of d, there will be no effect on the voltage and current at any other value of d.

3. The power flow is zero for $\sin 2\beta d = 0$, or $2\beta d = 0, \pi, 2\pi, \ldots$, or $d = 0, \lambda/4, \lambda/2, \ldots$, for all values of time.

Thus, the phenomenon on the short-circuited line is one in which the voltage, current, and power flow oscillate sinusoidally with time with different amplitudes at different locations on the line, unlike in the case of traveling waves in which a given point on the waveform progresses in distance with time. Since there is no feeling of wave motion down the line, these waves are known as *standing waves*. In particular, they represent *complete standing waves* in view of the zero amplitudes of the voltage, current, and power flow at certain locations on the line, as just discussed and as shown in Fig. 7.3. Complete standing waves are the result of $(+)$ and $(-)$ traveling waves of equal amplitudes. Whatever power is incident on the short circuit by the $(+)$ wave is reflected entirely in the form of the $(-)$ wave since the short circuit cannot absorb any power. Although there is instantaneous power flow at values of d between the voltage and current nodes, there is no time-average power flow for any value of d, as can be seen from

$$
\begin{aligned}
\langle P \rangle &= \frac{1}{T} \int_{t=0}^{T} P(d, t)\,dt = \frac{\omega}{2\pi} \int_{t=0}^{2\pi/\omega} P(d, t)\,dt \\
&= \frac{\omega}{2\pi} \frac{|\overline{V}^+|^2}{Z_0} \sin 2\beta d \int_{t=0}^{2\pi/\omega} \sin 2(\omega t + \theta)\,dt \\
&= 0
\end{aligned}
$$

Standing-wave patterns   From (7.6a) and (7.6b) or (7.7a) and (7.7b), or from Figs. 7.3(a) and 7.3(b), we find that the amplitudes of the sinusoidal time variations of the line voltage and line current as functions of distance along the line are

$$
\left| \overline{V}(d) \right| = 2\left| \overline{V}^+ \right| \left| \sin \beta d \right| = 2\left| \overline{V}^+ \right| \left| \sin \frac{2\pi}{\lambda} d \right| \tag{7.8a}
$$

$$
\left| \overline{I}(d) \right| = \frac{2\left| \overline{V}^+ \right|}{Z_0} \left| \cos \beta d \right| = \frac{2\left| \overline{V}^+ \right|}{Z_0} \left| \cos \frac{2\pi}{\lambda} d \right| \tag{7.8b}
$$

Sketches of these quantities versus d are shown in Fig. 7.4. These are known as the *standing-wave patterns*. They are the patterns of line voltage and line current one would obtain by connecting an ac voltmeter between the conductors of the line and an ac ammeter in series with one of the conductors of the line and observing their readings at various points along the line. Alternatively, one can sample the electric and magnetic fields by means of probes. Standing-wave

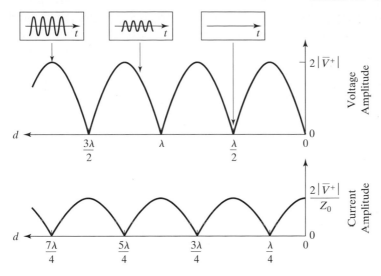

FIGURE 7.4 Standing-wave patterns for voltage and current on a short-circuited line. The insets show time variations of the voltage at points along the line.

patterns should not be misinterpreted as the voltage and current remaining constant with time at a given point. On the other hand, the voltage and current at every point on the line vary sinusoidally with time, as shown in the insets of Fig. 7.4, with the amplitudes of these sinusoidal variations equal to the magnitudes indicated by the standing-wave patterns. Since the distance between successive nodes of voltage or current is equal to $\lambda/2$, a measurement of this distance provides the knowledge of the wavelength. Furthermore, if the phase velocity in the line is known, the frequency of the source can be computed, and vice versa, since $v_p = \lambda f$.

Natural oscillations

Since there is no power flow across a voltage node or a current node of the standing-wave patterns, a constant amount of total energy is locked up in every $\lambda/4$ section between two such adjacent nodes with exchange of energy taking place between the electric and magnetic fields. Thus, once the line is excited by applying a source of energy, then each $\lambda/4$ section of the line between the voltage and current nodes acts as a resonator entirely independent of the remainder of the line. In fact, the $\lambda/4$ section can be removed from the line by cutting it, that is, open circuiting it, at the current node and short circuiting it at the voltage node, and still be made to maintain forever the oscillations of voltage and current. Such oscillations are called *natural oscillations*. Similarly, sections of lengths equal to multiples of $\lambda/4$ can be removed by always cutting the line at current nodes and short circuiting it at voltage nodes, without disturbing the oscillations.

For a fixed physical length of the line, its electrical length, that is, its length in terms of wavelength, depends on the frequency. Thus, a line of length equal to one-quarter wavelength at one frequency behaves as a line of length equal to a different multiple of a wavelength at a different frequency. Let us now consider a line of length l, one end of which is open-circuited and the other end short-circuited, and assume that some energy is stored in this line. Suppose we now pose the question: "What are all the possible standing-wave patterns on this line?" To answer this, we note that the voltage across the short circuit must always be zero, and, hence, the current there must have maximum amplitude. Similarly, the current at the open-circuited end must always be zero, and, hence, the voltage there must have maximum amplitude. We also know that the standing-wave patterns are sinusoidal with the distance between successive nodes corresponding to a half sine wave. Thus, the least possible variation is a quarter cycle of a sine waveform. This corresponds to a wavelength, say, λ_1, equal to $4l$, and the corresponding standing-wave patterns are shown in Fig. 7.5(a).

It is not possible to have a standing-wave pattern for which the wavelength is greater than $4l$ since then the pattern on the line of length l will be less than a quarter cycle of a sine wave. On the other hand, it is possible to have a pattern for which the wavelength is less than $4l$ as long as the conditions of zero voltage (maximum current) at the short circuit and zero current (maximum voltage) at the open circuit are satisfied. Obviously, the next largest wavelength λ_2, less than λ_1, for which this condition is satisfied corresponds to the patterns shown in Fig. 7.5(b). For these patterns, $l = 3\lambda_2/4$, or $\lambda_2 = 4l/3$. The next largest wavelength, λ_3, less than λ_2, corresponds to the patterns shown in Fig. 7.5(c). For these patterns, $l = 5\lambda_3/4$, or $\lambda_3 = 4l/5$.

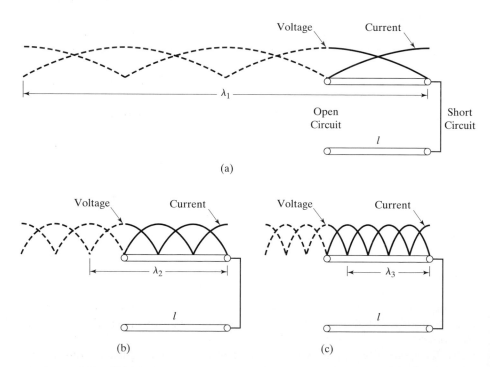

FIGURE 7.5 Standing-wave patterns corresponding to (a) one-quarter cycle, (b) three-quarters cycle, and (c) five-quarters cycle of a sine wave for the voltage and current amplitude distributions for a line of length l open-circuited at one end and short-circuited at the other end.

We can continue in this manner and see that any standing-wave pattern for which the length of the line is an odd multiple of one-quarter wavelength, that is,

$$l = \frac{(2n - 1)\lambda_n}{4} \qquad n = 1, 2, 3, \ldots \tag{7.9}$$

is a valid standing-wave pattern. Alternatively, the wavelengths λ_n, corresponding to the valid standing-wave patterns, are given by

$$\lambda_n = \frac{4l}{2n - 1} \qquad n = 1, 2, 3, \ldots \tag{7.10}$$

The corresponding frequencies are

$$f_n = \frac{v_p}{\lambda_n} = \frac{(2n - 1)v_p}{4l} \qquad n = 1, 2, 3, \ldots \tag{7.11}$$

where v_p is the phase velocity. These frequencies are known as the *natural frequencies of oscillation*. The standing-wave patterns are said to correspond to the different natural modes of oscillation. The lowest frequency (corresponding to the longest wavelength) is known as the *fundamental frequency of oscillation*, and the corresponding mode is known as the *fundamental mode*. The quantity n is called the *mode number*. In the most general case of nonsinusoidal voltage and current distributions on the line, the situation corresponds to the superposition of some or all of the infinite number of natural modes.

Considerations similar to those for the line open-circuited at one end and short-circuited at the other end apply to natural oscillations on lines short-circuited at both ends or open-circuited at both ends.

Returning now to the expressions for the phasor line voltage and the phasor line current given by (7.6a) and (7.6b), respectively, we define the ratio of these two quantities as the line impedance $\bar{Z}(d)$ at that point seen looking toward the short circuit. Thus, *Input impedance*

$$\bar{Z}(d) = \frac{\bar{V}(d)}{\bar{I}(d)} = \frac{2j\bar{V}^+ \sin \beta d}{2(\bar{V}^+/Z_0) \cos \beta d} = jZ_0 \tan \beta d \tag{7.12}$$

In particular, the input impedance $\bar{Z}_{in}$ of a short-circuited line of length l is given by

$$\boxed{\bar{Z}_{in} = jZ_0 \tan \beta l = jZ_0 \tan \frac{2\pi f}{v_p} l} \tag{7.13}$$

We note from (7.13) that the input impedance of the short-circuited line is purely reactive. As the frequency is varied from a low value upward, the input reactance changes from inductive to capacitive and back to inductive, and so on, as illustrated in Fig. 7.6. The input reactance is zero for values of frequency equal to multiples of $v_p/2l$. These are the frequencies for which l is equal to multiples of $\lambda/2$, so that the line voltage is zero at the input and hence the input sees a short circuit. The input reactance is infinity for values of frequency equal to odd multiples

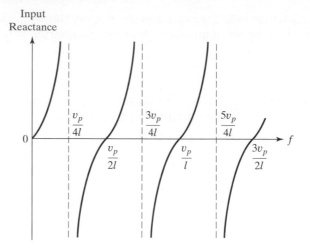

FIGURE 7.6   Variation of the input reactance of a short-circuited transmission line with frequency.

of $v_p/4l$. These are the frequencies for which l is equal to odd multiples of $\lambda/4$, so that the line current is zero at the input and hence the input sees an open circuit.

These properties of the input impedance of a short-circuited line (and, similarly, of an open-circuited line) have several applications. We shall here discuss two of these applications.

Location of
short circuit

1. *Determination of the location of a short circuit (or open circuit) in a line.* The principle behind this lies in the fact that as the frequency of a generator connected to the input of a short-circuited (or open-circuited) line is varied continuously upward, the current drawn from it undergoes alternatively maxima and minima corresponding to zero input reactance and infinite input reactance conditions, respectively. Since the difference between a pair of consecutive frequencies for which the input reactance values are zero and infinity is $v_p/4l$, as can be seen from Fig. 7.6, it follows that the difference between successive frequencies for which the currents drawn from the generator are maxima and minima is $v_p/4l$. As a numerical example, if for an air–dielectric line, it is found that as the frequency is varied from 50 MHz upward, the current reaches a minimum for 50.01 MHz and then a maximum for 50.04 MHz, then the distance l of the short circuit from the generator is given by

$$\frac{v_p}{4l} = (50.04 - 50.01) \times 10^6 = 0.03 \times 10^6 = 3 \times 10^4$$

Since $v_p = 3 \times 10^8$ m/s, it follows that

$$l = \frac{3 \times 10^8}{4 \times 3 \times 10^4} = 2500 \text{ m} = 2.5 \text{ km}$$

Alternatively, if the length l is known, we can compute v_p for the dielectric of the line, from which the permittivity of the dielectric can be found, provided that the value of μ (usually equal to μ_0) is known.

2. *Construction of resonant circuits at microwave frequencies.* The princi-ple behind this lies in the fact that the input reactance of a short-circuited line of a given length can be inductive or capacitive, depending on the frequency, and, hence, two short-circuited lines connected together form a resonant system. To obtain the characteristic equation for the resonant frequencies of such a system, let us consider the system shown in Fig. 7.7, which is made up of two short-circuited line sections of characteristic impedances Z_{01} and Z_{02}, lengths l_1 and l_2, and phase velocities v_{p1} and v_{p2}. Denoting the voltages and currents just to the left and just to the right of the junction to be $\overline{V}_1$ and $\overline{I}_1$ and $\overline{V}_2$ and $\overline{I}_2$, respec-tively, as shown in the figure, we write the boundary conditions at the junction as

$$\overline{V}_1 = \overline{V}_2 \tag{7.14a}$$

$$\overline{I}_1 + \overline{I}_2 = 0 \tag{7.14b}$$

Combining the two, we have

$$\frac{\overline{I}_1}{\overline{V}_1} + \frac{\overline{I}_2}{\overline{V}_2} = 0$$

or

$$\overline{Y}_1 + \overline{Y}_2 = 0 \tag{7.15}$$

where $\overline{Y}_1$ and $\overline{Y}_2$ are the input admittances of the sections to the left and to the right, respectively, of the junction and seen looking toward the short circuits. Equation (7.15) is the condition for resonance of the system. To express it in terms of the line parameters, we note that

$$\overline{Y}_1 = \frac{1}{\overline{Z}_1} = \frac{1}{jZ_{01} \tan \beta_1 l_1} = \frac{1}{jZ_{01} \tan (2\pi f/v_{p1})l_1} \tag{7.16a}$$

$$\overline{Y}_2 = \frac{1}{\overline{Z}_2} = \frac{1}{jZ_{02} \tan \beta_2 l_2} = \frac{1}{jZ_{02} \tan (2\pi f/v_{p2})l_2} \tag{7.16b}$$

Substituting (7.16a) and (7.16b) into (7.15) and simplifying, we obtain the char-acteristic equation for the resonant frequencies to be

$$Z_{01} \tan \frac{2\pi f}{v_{p1}} l_1 + Z_{02} \tan \frac{2\pi f}{v_{p2}} l_2 = 0 \tag{7.17}$$

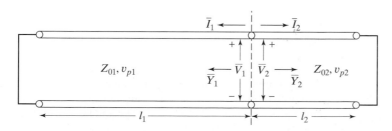

FIGURE 7.7 Resonant system formed by connecting together two short-circuited line sections.

We shall illustrate the computation of the resonant frequencies by means of an example.

Example 7.1

For the system of Fig. 7.7, let us assume $Z_{01} = 2Z_{02} = 60\ \Omega$, $l_1 = 5$ cm, $l_2 = 2$ cm, and $v_{p1} = v_{p2} = c/2$, and obtain the four lowest resonant frequencies of the system.

Substituting the numerical values of the parameters into the characteristic equation (7.17), we obtain

$$\tan \frac{0.2\pi f}{c} + \frac{1}{2} \tan \frac{0.08\pi f}{c} = 0$$

This equation is of the form

$$\tan kx + m \tan x = 0$$

where $k = 2.5$, $m = 0.5$, and $x = 0.08\pi f/c$. In general, an equation of this type can be solved by plotting $\tan kx$ and $-m \tan x$ to scale versus x and finding the points of intersection. Alternatively, a programmable calculator or a computer can be used. Thus, the first four solutions are x equal to $0.3165\pi, 0.5563\pi, 0.8353\pi$, and 1.1647π.

From the values of x obtained from the computer program, we obtain the lowest four resonant frequencies to be 1.1869×10^9, 2.0861×10^9, 3.1324×10^9, and 4.3676×10^9 Hz, or 1.1869, $2.0861, 3.1324$, and 4.3676 GHz.

K7.1. Phasor line voltage and line current; General solutions; Short-circuited line; Complete standing waves; Standing-wave patterns; Natural oscillations; Input impedance; Resonant systems.

D7.1. For each of the following characteristics of standing waves on a lossless short-circuited line, find the frequency of the source exciting the line: **(a)** the distance between successive nodes of voltage amplitude is 50 cm and the dielectric is air; **(b)** the distance between successive nodes of current amplitude is 50 cm and the dielectric is nonmagnetic with $\varepsilon = 9\varepsilon_0$; and **(c)** the distance between successive nodes of instantaneous power flow is 50 cm and the dielectric is air.
Ans. **(a)** 300 MHz; **(b)** 100 MHz; **(c)** 150 MHz.

D7.2. A lossless coaxial cable of characteristic impedance 50 Ω and having a nonmagnetic $(\mu = \mu_0)$, perfect dielectric of permittivity $\varepsilon = 2.25\varepsilon_0$ is short-circuited at the far end. Find the minimum length of the line for which the input impedance is equal to the impedance of each of the following at $f = 100$ MHz: **(a)** an inductor of value equal to 0.5 μH; **(b)** an inductor of value equal to the inductance per unit length of the line; and **(c)** an inductor of value equal to the inductance of the line.
Ans. **(a)** 44.98 cm; **(b)** 40.19 cm; **(c)** 143.03 cm.

D7.3. A lossless transmission line of length $l = 5$ m, characteristic impedance $Z_0 = 100\ \Omega$, and having a nonmagnetic $(\mu = \mu_0)$, perfect dielectric is short-circuited at its far end. A variable frequency voltage source in series with an internal impedance $\overline{Z}_g$ is connected at its input and the line voltage and line current at the input terminals are monitored as the source frequency is varied. It is found that the voltage reaches a maximum amplitude of 10 V at 157.5 MHz and then the current reaches a maximum amplitude of 0.2 A at 165 MHz. Find the following: **(a)** the maximum amplitude of the current in the standing-wave pattern on the line at 157.5 MHz; **(b)** the maximum amplitude of the voltage in the standing-wave pattern on the line at 165 MHz; **(c)** the magnitude of $\overline{Z}_g$; and **(d)** the permittivity of the dielectric of the line.

Ans.   **(a)** 0.1 A;   **(b)** 20 V;   **(c)** 50 Ω;   **(d)** $4\varepsilon_0$.

7.2   LINE TERMINATED BY ARBITRARY LOAD

We devoted the preceding section to the short-circuited line. In this section, we consider a line terminated by an arbitrary load impedance $\overline{Z}_R$, as shown in Fig. 7.8. Then starting with the general solutions for the complex line voltage and line current given by

$$\overline{V}(d) = \overline{V}^+ e^{j\beta d} + \overline{V}^- e^{-j\beta d} \tag{7.18a}$$

$$\overline{I}(d) = \frac{1}{Z_0}(\overline{V}^+ e^{j\beta d} - \overline{V}^- e^{-j\beta d}) \tag{7.18b}$$

and using the boundary condition at $d = 0$, given by

$$\overline{V}(0) = \overline{Z}_R \overline{I}(0) \tag{7.19}$$

we obtain

$$\overline{V}^+ + \overline{V}^- = \frac{\overline{Z}_R}{Z_0}(\overline{V}^+ - \overline{V}^-)$$

or

$$\overline{V}^- = \overline{V}^+ \frac{\overline{Z}_R - Z_0}{\overline{Z}_R + Z_0}$$

Thus, the ratio of $\overline{V}^-$, the reflected wave voltage at the load, to $\overline{V}^+$, the incident wave voltage at the load, that is, the voltage reflection coefficient at the load, denoted by $\overline{\Gamma}_R$, is given by

$$\boxed{\overline{\Gamma}_R = \frac{\overline{V}^-}{\overline{V}^+} = \frac{\overline{Z}_R - Z_0}{\overline{Z}_R + Z_0}} \tag{7.20}$$

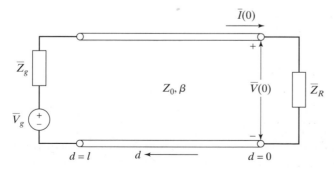

FIGURE 7.8   Line terminated by a complex load impedance.

The solutions for $\overline{V}(d)$ and $\overline{I}(d)$ can then be written as

$$\overline{V}(d) = \overline{V}^+ e^{j\beta d} + \overline{\Gamma}_R \overline{V}^+ e^{-j\beta d} \tag{7.21a}$$

$$\overline{I}(d) = \frac{1}{Z_0}(\overline{V}^+ e^{j\beta d} - \overline{\Gamma}_R \overline{V}^+ e^{-j\beta d}) \tag{7.21b}$$

Generalized reflection coefficient   We now define the generalized voltage reflection coefficient, $\overline{\Gamma}(d)$, that is, the voltage reflection coefficient at any value of d, as the ratio of the reflected wave voltage to the incident wave voltage at that value of d. From (7.21a), we see that

$$\boxed{\overline{\Gamma}(d) = \frac{\overline{\Gamma}_R \overline{V}^+ e^{-j\beta d}}{\overline{V}^+ e^{j\beta d}} = \overline{\Gamma}_R e^{-j2\beta d}} \tag{7.22}$$

so that

$$|\overline{\Gamma}(d)| = |\overline{\Gamma}_R| |e^{-j2\beta d}| = |\overline{\Gamma}_R| \tag{7.23a}$$

and

$$\underline{/\overline{\Gamma}(d)} = \underline{/\overline{\Gamma}_R} + \underline{/e^{-j2\beta d}} = \theta - 2\beta d \tag{7.23b}$$

where θ is the phase angle of $\overline{\Gamma}_R$. Thus, the magnitude of the generalized reflection coefficient remains constant along the line and equal to its value at the load, whereas the phase angle varies linearly with d. In terms of $\overline{\Gamma}(d)$, we can write the solutions for $\overline{V}(d)$ and $\overline{I}(d)$ as

$$\begin{aligned} \overline{V}(d) &= \overline{V}^+ e^{j\beta d}(1 + \overline{\Gamma}_R e^{-j2\beta d}) \\ &= \overline{V}^+ e^{j\beta d}[1 + \overline{\Gamma}(d)] \end{aligned} \tag{7.24a}$$

$$\begin{aligned} \overline{I}(d) &= \frac{\overline{V}^+}{Z_0} e^{j\beta d}(1 - \overline{\Gamma}_R e^{-j2\beta d}) \\ &= \frac{\overline{V}^+}{Z_0} e^{j\beta d}[1 - \overline{\Gamma}(d)] \end{aligned} \tag{7.24b}$$

To study the standing-wave patterns corresponding to (7.24a) and (7.24b), we look at the magnitudes of $\overline{V}(d)$ and $\overline{I}(d)$. These are given by

$$\begin{aligned} |\overline{V}(d)| &= |\overline{V}^+| |e^{j\beta d}| |1 + \overline{\Gamma}(d)| \\ &= |\overline{V}^+| |1 + \overline{\Gamma}_R e^{-j2\beta d}| \end{aligned} \tag{7.25a}$$

$$\begin{aligned} |\overline{I}(d)| &= \frac{|\overline{V}^+|}{Z_0} |e^{j\beta d}| |1 - \overline{\Gamma}(d)| \\ &= \frac{|\overline{V}^+|}{Z_0} |1 - \overline{\Gamma}_R e^{-j2\beta d}| \end{aligned} \tag{7.25b}$$

To sketch $|\overline{V}(d)|$ and $|\overline{I}(d)|$, it is sufficient if we consider the quantities $|1 + \overline{\Gamma}_R e^{-j2\beta d}|$ and $|1 - \overline{\Gamma}_R e^{-j2\beta d}|$ since $|\overline{V}^+|$ is simply a constant, determined by the boundary condition at the source end. Each of these quantities consists of two complex numbers, one of which is a constant equal to $(1 + j0)$ and the other of which has a constant magnitude $|\overline{\Gamma}_R|$ but a variable phase angle $(\theta - 2\beta d)$.

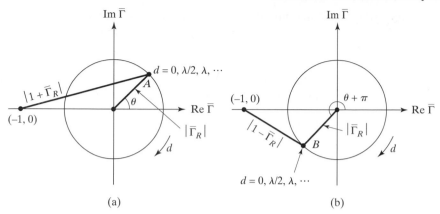

FIGURE 7.9 $\overline{\Gamma}$ plane diagrams for sketching the voltage and current standing-wave patterns for the system of Fig. 7.8.

To evaluate $\left|1 + \overline{\Gamma}_R e^{-j2\beta d}\right|$ and $\left|1 - \overline{\Gamma}_R e^{-j2\beta d}\right|$, we make use of the constructions in the complex $\overline{\Gamma}$-plane, as shown in Fig. 7.9(a) and (b), respectively. In both diagrams, we draw circles with centers at the origin and having radii equal to $\left|\overline{\Gamma}_R\right|$. For $d = 0$, the complex number $\overline{\Gamma}_R e^{-j2\beta d}$ is equal to $\overline{\Gamma}_R$ or $\left|\overline{\Gamma}_R\right|e^{j\theta}$, which is represented by point A in Fig. 7.9(a). To add $(1 + j0)$ and $\overline{\Gamma}_R$, we simply draw a line from the point $(-1, 0)$ to the point A. The length of this line gives $\left|1 + \overline{\Gamma}_R\right|$, which is proportional to the amplitude of the voltage at $d = 0$. As d increases, point A, representing $\overline{\Gamma}_R e^{-j2\beta d}$, moves around the circle in the clockwise direction. The line joining $(-1, 0)$ to the point A whose length is $\left|1 + \overline{\Gamma}_R e^{-j2\beta d}\right|$ executes the motion of a crank. To subtract $\overline{\Gamma}_R$ from $(1 + j0)$, we locate point B in Fig. 7.9(b), which is diametrically opposite to point A in Fig. 7.9(a), and draw a line from $(-1, 0)$ to point B. The length of the line gives $\left|1 - \overline{\Gamma}_R\right|$, which is proportional to the amplitude of the current at $d = 0$. As d increases, B moves around the circle in the clockwise direction following the movement of A in Fig. 7.9(a). The line joining $(-1, 0)$ to the point B whose length is $\left|1 - \overline{\Gamma}_R e^{-j2\beta d}\right|$ executes the motion of a crank. From these constructions and assuming $-\pi \le \theta < \pi$, we note the following facts:

1. Point A lies along the positive real axis and point B lies along the negative real axis for $(\theta - 2\beta d) = 0, -2\pi, -4\pi, -6\pi, \ldots$, or $d = (\lambda/4\pi)(\theta + 2n\pi)$, where $n = 0, 1, 2, 3, \ldots$. Hence, at these values of d, the voltage amplitude is maximum and equal to $\left|\overline{V}^+\right|(1 + \left|\overline{\Gamma}_R\right|)$, whereas the current amplitude is minimum and equal to $(\left|\overline{V}^+\right|/Z_0)(1 - \left|\overline{\Gamma}_R\right|)$. The voltage and current are in phase.

2. Point A lies along the negative real axis and point B lies along the positive real axis for $(\theta - 2\beta d) = -\pi, -3\pi, -5\pi, -7\pi, \ldots$, or $d = (\lambda/4\pi)[\theta + (2n - 1)\pi]$, where $n = 1, 2, 3, 4, \ldots$. Hence, at these values of d, the voltage amplitude is minimum and equal to $\left|\overline{V}^+\right|(1 - \left|\overline{\Gamma}_R\right|)$, whereas the current amplitude is maximum and equal to $(\left|\overline{V}^+\right|/Z_0)(1 + \left|\overline{\Gamma}_R\right|)$. The voltage and current are in phase.

3. Between maxima and minima, the voltage and current vary in accordance with the lengths of the line joining $(-1, 0)$ to the points A and B, respectively, as they move around the circles. These variations are not sinusoidal with distance. The variations near the minima are sharper than are those near the maxima, and, hence, the minima can be located more accurately than can the maxima. Also, the voltage and current are not in phase.

Standing-
wave
parameters

From the preceding discussion, we now sketch the standing-wave patterns for the line voltage and current, as shown in Fig. 7.10. These patterns correspond to partial standing waves, as compared to complete standing waves in the case of the short-circuited line. There are three parameters associated with the standing-wave patterns as follows.

1. *The standing-wave ratio, abbreviated as SWR.* This is the ratio of the maximum voltage amplitude V_{max} to the minimum voltage amplitude V_{min} in the standing-wave pattern. Thus

$$\text{SWR} = \frac{V_{max}}{V_{min}} = \frac{|\overline{V}^+|(1 + |\overline{\Gamma}_R|)}{|\overline{V}^+|(1 - |\overline{\Gamma}_R|)} = \frac{1 + |\overline{\Gamma}_R|}{1 - |\overline{\Gamma}_R|} \qquad (7.26)$$

Note also that SWR is equal to the ratio of the maximum current amplitude I_{max} to the minimum current amplitude I_{min} in the standing-wave

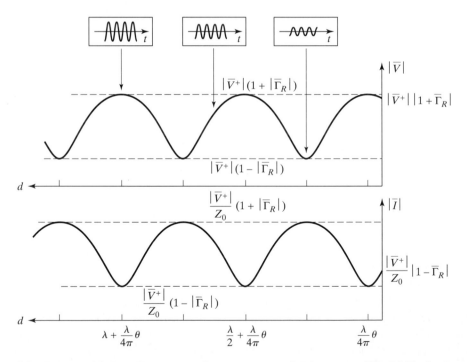

FIGURE 7.10   Voltage and current standing wave patterns for the system of Fig. 7.8. The insets show time variations of voltage at points along the line.

pattern, since

$$\frac{I_{max}}{I_{min}} = \frac{(|\overline{V}^+|/Z_0)(1 + |\overline{\Gamma}_R|)}{(|\overline{V}^+|/Z_0)(1 - |\overline{\Gamma}_R|)} = \frac{1 + |\overline{\Gamma}_R|}{1 - |\overline{\Gamma}_R|}$$

The SWR is a measure of standing waves on the line. It is an easily measurable parameter. We note the following special cases:

a. For $\overline{\Gamma}_R = 0$, SWR = 1 and the standing-wave pattern is simply a line representing constant amplitude. This is the case for a semi-infinitely long line or for a line terminated by its characteristic impedance.

b. For $|\overline{\Gamma}_R| = 1$, SWR = ∞ and the standing-wave pattern possesses perfect nulls. This is the case for complete standing waves.

2. *The distance of the first voltage minimum from the load, denoted by* d_{min}. The voltage minimum nearest to the load occurs when the phase angle of $\overline{\Gamma}(d) = \overline{\Gamma}_R e^{-j2\beta d}$ is equal to $-\pi$, that is, for $(\theta - 2\beta d)$ equal to $-\pi$. Thus,

$$\theta - 2\beta d_{min} = -\pi \tag{7.27}$$

or

$$\boxed{d_{min} = \frac{\theta + \pi}{2\beta} = \frac{\lambda}{4\pi}(\theta + \pi)} \tag{7.28}$$

where $-\pi \le \theta < \pi$. If $\theta = 0$, which occurs when $\overline{Z}_R$ is purely real and greater than Z_0, $d_{min} = \lambda/4$ and a voltage maximum exists right at the load. If $\theta = -\pi$, which occurs when $\overline{Z}_R$ is purely real and less than Z_0, $d_{min} = 0$ and a voltage minimum exists right at the load.

3. *The wavelength* λ. Since the distance between successive voltage minima is equal to $\lambda/2$, the wavelength is twice the distance between successive voltage minima.

For a numerical example involving a complex $\overline{Z}_R$, let us consider $\overline{Z}_R = (15 - j20)\ \Omega$ and $Z_0 = 50\ \Omega$. Then

$$\overline{\Gamma}_R = \frac{\overline{Z}_R - Z_0}{\overline{Z}_R + Z_0} = \frac{(15 - j20) - 50}{(15 - j20) + 50}$$

$$= \frac{-7 - j4}{13 - j4} = \frac{8.06\ \underline{/-150.26°}}{13.60\ \underline{/-17.10°}}$$

$$= 0.593\ \underline{/-133.16°}$$

$$= 0.593 e^{-j0.74\pi}$$

$$SWR = \frac{1 + |\overline{\Gamma}_R|}{1 - |\overline{\Gamma}_R|} = \frac{1 + 0.593}{1 - 0.593} = 3.914$$

$$d_{min} = \frac{\lambda}{4\pi}(\theta + \pi) = \frac{\lambda}{4\pi}(-0.74\pi + \pi)$$

$$= 0.065\lambda$$

Determination of unknown load impedance

Conversely, to the computation of standing-wave parameters for a given load impedance, an unknown load impedance can be determined from standing-wave measurements on a line of known characteristic impedance. An application in practice is the determination of the input impedance of an antenna by making standing-wave measurements on the line feeding the antenna. To outline the basis, we note that by rearranging (7.26) and (7.28), we obtain

$$|\overline{\Gamma}_R| = \frac{SWR - 1}{SWR + 1} \qquad (7.29)$$

and

$$\theta = \frac{4\pi d_{min}}{\lambda} - \pi \qquad (7.30)$$

Thus, the measurement of SWR, d_{min}, and λ provides both the magnitude and phase angle of $\overline{\Gamma}_R$. Then, since from (7.20)

$$\overline{Z}_R = Z_0 \frac{1 + \overline{\Gamma}_R}{1 - \overline{\Gamma}_R} \qquad (7.31)$$

we can compute the value of $\overline{Z}_R$.

Slotted-line measurements

A traditional method of performing standing-wave measurements in the laboratory is by using a *slotted line*. The slotted line is essentially a rigid coaxial line with air dielectric and having a length of about 1 meter (or at least a half-wavelength long). The center conductor is supported by dielectric inserts. A narrow longitudinal slot is cut in the outer conductor, as shown in Fig. 7.11(a). The width of the slot is so small that it has negligible influence on the current flow on the outer conductor, and, hence, on the field configurations between the conductors. A probe of small length, shown in Fig. 7.11(b), intercepts a portion

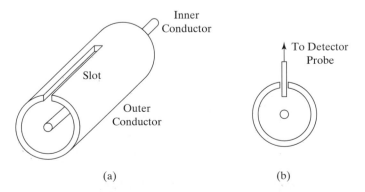

(a)   (b)

FIGURE 7.11　(a) Slotted line. (b) Cross-sectional view of the slotted-line illustrating the probe arrangement.

of the electric field between the inner and outer conductors, and, hence, a small voltage proportional to the line voltage at the probe's location is developed between the probe and the outer conductor. The signal frequency voltage thus developed is detected by some sort of detector, and the resulting output is used as an indicator of the amplitude of the line voltage at the probe's location. The amount of energy picked up by the probe is small enough not to disturb appreciably the fields within the line. The probe and the associated detector components are mounted on a carriage arranged to slide mechanically along the longitudinal slot. As the probe is moved along the slot, the detector indication provides a measure of the variation of the voltage as a function of position on the line. Since the SWR is the ratio of V_{max} to V_{min}, the quantity of interest is the ratio of the two readings rather than the absolute values of the readings themselves. Therefore, absolute calibration of the detector is not required, provided that the detector response is linear in the range of voltages to be measured.

Since it is not always possible to measure the distances of the standing-wave pattern minima from the location of the load, the following procedure is employed. First, the line is terminated by a short circuit in the place of the load. One of the nulls in the resulting standing-wave pattern is taken as the reference point, as shown in Fig. 7.12(a). This establishes that the location of the load is an integral multiple of half-wavelengths from the reference point. Next, the short circuit is removed and the load is connected. The voltage minimum then shifts away from the reference point, as shown in Fig. 7.12(b). By measuring this shift, either away from the load or toward the load, the value of d_{min} can be established. If the shift d_a away from the load is measured, then we can see from Fig. 7.12(b) that d_{min} is simply equal to d_a. On the other hand, if the shift d_t toward the load is measured, then d_{min} is equal to $\lambda/2 - d_t$, where $\lambda/2$ is given by the distance between consecutive nulls either in the case of short circuit or with the unknown load as the termination.

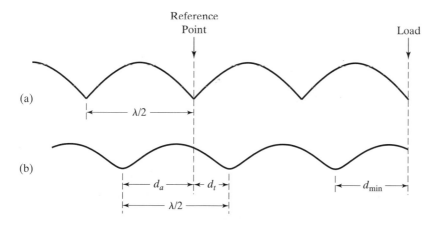

FIGURE 7.12   For illustrating the procedure employed for the determination of d_{min}, the distance of the first voltage minimum of the standing-wave pattern from the load, by making measurements away from the load.

We shall illustrate the computation of $\overline{Z}_R$ from standing-wave measurements by means of an example.

Example 7.2

Let us assume that measurements performed on a slotted line of characteristic imped-ance $Z_0 = 50\ \Omega$ provided the following data. First, with the short circuit as the termina-tion, voltage minima were found to be 20 cm apart. Next, with one of the minima marked as the reference point and the short circuit replaced by the unknown load, the SWR was found to be 3.0 and a voltage minimum was found to be at 5.80 cm from the reference point on the side toward the load. We wish to compute the value of the unknown load im-pedance.

From the value of the SWR, we obtain by using (7.29)

$$|\overline{\Gamma}_R| = \frac{3 - 1}{3 + 1} = 0.5$$

Since the distance between successive voltage minima is 20 cm, $\lambda/2$ is equal to 20 cm, or λ is equal to 40 cm. Since the voltage minimum shifted toward the load from the reference point, $d_{\min}$ is equal to $\lambda/2$ minus the shift, or $20 - 5.8 = 14.2$ cm. Then, from (7.30), we get

$$\theta = \frac{4\pi}{40} \times 14.2 - \pi = 0.42\pi$$

Thus

$$\overline{\Gamma}_R = 0.5e^{j0.42\pi}$$

Finally, using (7.31), we compute the value of the load impedance to be

$$\overline{Z}_R = 50\ \frac{1 + 0.5e^{j0.42\pi}}{1 - 0.5e^{j0.42\pi}}$$

$$= 50\ \frac{1.1243 + j0.4843}{0.8757 - j0.4843}$$

$$= 50\ \frac{1.2242\underline{/23.303°}}{1.0007\underline{/-28.945°}}$$

$$= 61.17\underline{/52.248°}$$

$$= (37.45 + j48.365)\ \Omega$$

Line impedance   Returning now to the solutions for the complex line voltage and current given by (7.24a) and (7.24b), respectively, we find that the line impedance $\overline{Z}(d)$, that is, the impedance at any value of d seen looking toward the load, is given by

$$\boxed{\begin{aligned} \overline{Z}(d) &= \frac{\overline{V}(d)}{\overline{I}(d)} = \frac{\overline{V}^+e^{j\beta d}[1 + \overline{\Gamma}(d)]}{(\overline{V}^+/Z_0)e^{j\beta d}[1 - \overline{\Gamma}(d)]} \\ &= Z_0\ \frac{1 + \overline{\Gamma}(d)}{1 - \overline{\Gamma}(d)} \end{aligned}} \tag{7.32}$$

The following properties of the line impedance are of interest:

1. At the location of a voltage maximum of the standing-wave pattern, $1 + \overline{\Gamma}(d)$ and $1 - \overline{\Gamma}(d)$ are purely real and equal to their maximum and minimum magnitudes $1 + |\overline{\Gamma}_R|$ and $1 - |\overline{\Gamma}_R|$, respectively. Hence, $\overline{Z}(d)$ is purely real and maximum, say, R_{max}, equal to $Z_0[(1 + |\overline{\Gamma}_R|)/(1 - |\overline{\Gamma}_R|)]$, or Z_0 (SWR).

2. At the location of a voltage minimum of the standing-wave pattern, $1 + \overline{\Gamma}(d)$ and $1 - \overline{\Gamma}(d)$ are purely real and equal to their minimum and maximum magnitudes $1 - |\overline{\Gamma}_R|$ and $1 + |\overline{\Gamma}_R|$, respectively. Hence, $\overline{Z}(d)$ is purely real and minimum, say, R_{min}, equal to $Z_0[(1 - |\overline{\Gamma}_R|)/(1 + |\overline{\Gamma}_R|)]$, or $Z_0/(\text{SWR})$.

3. Between voltage maxima and minima, $1 + \overline{\Gamma}(d)$ and $1 - \overline{\Gamma}(d)$ are both complex and out of phase. Hence, $\overline{Z}(d)$ is complex, with magnitude lying between $Z_0(\text{SWR})$ and $Z_0/(\text{SWR})$.

4. Since $\overline{\Gamma}(d \pm n\lambda/2) = \overline{\Gamma}(d)e^{\mp j2\beta n\lambda/2} = \overline{\Gamma}(d)e^{\mp j2n\pi} = \overline{\Gamma}(d), n = 1, 2, 3, \ldots$, $\Gamma(d)$ repeats at intervals of $\lambda/2$, and, hence, $\overline{Z}(d)$ repeats at intervals of $\lambda/2$.

5. The product of the line impedances at two values of d separated by $\lambda/4$ is given by

$$[\overline{Z}(d)]\left[\overline{Z}\left(d \pm \frac{\lambda}{4}\right)\right] = \left[Z_0\frac{1 + \overline{\Gamma}(d)}{1 - \overline{\Gamma}(d)}\right]\left[Z_0\frac{1 + \overline{\Gamma}(d \pm \lambda/4)}{1 - \overline{\Gamma}(d \pm \lambda/4)}\right]$$

$$= Z_0^2\left[\frac{1 + \overline{\Gamma}(d)}{1 - \overline{\Gamma}(d)}\right]\left[\frac{1 + \overline{\Gamma}(d)e^{\mp j2\beta\lambda/4}}{1 - \overline{\Gamma}(d)e^{\mp j2\beta\lambda/4}}\right]$$

$$= Z_0^2\left[\frac{1 + \overline{\Gamma}(d)}{1 - \overline{\Gamma}(d)}\right]\left[\frac{1 + \overline{\Gamma}(d)e^{\mp j\pi}}{1 - \overline{\Gamma}(d)e^{\mp j\pi}}\right]$$

$$= Z_0^2\left[\frac{1 + \overline{\Gamma}(d)}{1 - \overline{\Gamma}(d)}\right]\left[\frac{1 - \overline{\Gamma}(d)}{1 + \overline{\Gamma}(d)}\right]$$

or

$$\boxed{[\overline{Z}(d)]\left[\overline{Z}\left(d \pm \frac{\lambda}{4}\right)\right] = Z_0^2} \qquad (7.33)$$

This is a useful property, as we shall learn in the following section.

For a line of length l, as in Fig. 7.8, the input impedance is given from (7.32) by

Input impedance

$$\boxed{\overline{Z}_{in} = \overline{Z}(l) = Z_0\frac{1 + \overline{\Gamma}(l)}{1 - \overline{\Gamma}(l)}} \qquad (7.34)$$

The input impedance is a useful parameter since for a given generator voltage and internal impedance, the power flow down the line can be computed by considering the line voltage and current at any value of d, since the line is lossless; in particular, it is convenient to do this at the input end of the line from input impedance considerations. We shall illustrate this by means of an example.

Example 7.3

Let us consider the system shown in Fig. 7.13, and find the time-average power delivered to the load from input impedance considerations.

We proceed with the solution in the following step-by-step manner:

a. Compute the reflection coefficient at the load.

$$\overline{\Gamma}_R = \frac{\overline{Z}_R - Z_0}{\overline{Z}_R + Z_0} = \frac{(30 + j40) - 50}{(30 + j40) + 50} = 0.5\underline{/90°}$$

b. Compute the reflection coefficient $\overline{\Gamma}(l)$ at the input end $d = l$.

$$\overline{\Gamma}(l) = \overline{\Gamma}_R e^{-j2\beta l}$$

$$= 0.5\underline{/90°} \times e^{-j(4\pi/\lambda)(0.725\lambda)}$$

$$= 0.5\underline{/90°} \times 1\underline{/-162°}$$

$$= 0.5\underline{/-72°}$$

c. Compute the input impedance.

$$\overline{Z}_{in} = \overline{Z}(l) = Z_0 \frac{1 + \overline{\Gamma}(l)}{1 - \overline{\Gamma}(l)}$$

$$= 50 \frac{1 + 0.5\underline{/-72°}}{1 - 0.5\underline{/-72°}} = 50 \frac{1 + (0.1545 - j0.4755)}{1 - (0.1545 - j0.4755)}$$

$$= 50 \frac{1.2486\underline{/-22.385°}}{0.970\underline{/29.353°}} = 64.361\underline{/-51.738°}$$

$$= (39.86 - j50.54) \ \Omega$$

FIGURE 7.13   Transmission-line system for illustrating the computation of power flow from input impedance considerations.

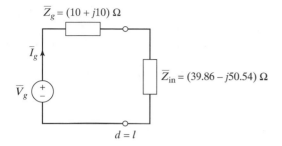

$\overline{Z}_g = (10 + j10)\ \Omega$

$\overline{Z}_{in} = (39.86 - j50.54)\ \Omega$

$d = l$

FIGURE 7.14 Equivalent circuit at the input end $d = l$ for the system of Fig. 7.13.

d. We now have the equivalent circuit at the input, as shown in Fig. 7.14, from which we compute the current $\overline{I}_g = \overline{I}(l)$, drawn from the generator. Thus, we obtain

$$\overline{I}(l) = \overline{I}_g = \frac{\overline{V}_g}{\overline{Z}_g + \overline{Z}_{in}} = \frac{100\underline{/0°}}{(10 + j10) + (39.86 - j50.54)}$$

$$= \frac{100\underline{/0°}}{49.86 - j40.54} = \frac{100\underline{/0°}}{64.261\underline{/-39.114°}}$$

$$= 1.5562\underline{/39.114°}\ A$$

e. The voltage across the input impedance is then given by

$$\overline{V}(l) = \overline{Z}_{in}\overline{I}(l)$$

$$= 64.361\underline{/-51.738°} \times 1.5562\underline{/39.114°}$$

$$= 100.159\underline{/-12.624°}\ V$$

f. Finally, the time-average power delivered to the input and hence to the load is given by

$$\langle P \rangle = \tfrac{1}{2}\,\mathrm{Re}\,[\overline{V}(l)\overline{I}^*(l)]$$

$$= \tfrac{1}{2}\,\mathrm{Re}\,[100.159\underline{/-12.624°} \times 1.5562\underline{/-39.114°}]$$

$$= \tfrac{1}{2} \times 100.159 \times 1.5562 \times \cos 51.738°$$

$$= 48.26\ W$$

Returning to (7.32), we now define the normalized line impedance $\overline{z}(d)$ as the ratio of the line impedance to the line characteristic impedance. Thus, *Normalized impedance and admittance*

$$\boxed{\overline{z}(d) = \frac{\overline{Z}(d)}{Z_0} = \frac{1 + \overline{\Gamma}(d)}{1 - \overline{\Gamma}(d)}} \qquad (7.35)$$

Conversely,

$$\boxed{\overline{\Gamma}(d) = \frac{\overline{z}(d) - 1}{\overline{z}(d) + 1}} \qquad (7.36)$$

Finally, the line admittance is given by

$$\overline{Y}(d) = \frac{1}{\overline{Z}(d)} = \frac{1}{Z_0}\frac{1 - \overline{\Gamma}(d)}{1 + \overline{\Gamma}(d)}$$

or

$$\overline{Y}(d) = Y_0 \frac{1 - \overline{\Gamma}(d)}{1 + \overline{\Gamma}(d)} \tag{7.37}$$

where $Y_0 = 1/Z_0$ is the characteristic admittance of the line. The normalized line admittance is

$$\boxed{\overline{y}(d) = \frac{\overline{Y}(d)}{Y_0} = \frac{1 - \overline{\Gamma}(d)}{1 + \overline{\Gamma}(d)}} \tag{7.38}$$

and, conversely,

$$\boxed{\overline{\Gamma}(d) = \frac{1 - \overline{y}(d)}{1 + \overline{y}(d)}} \tag{7.39}$$

We shall use these relationships in the following sections.

K7.2. Arbitrary load; Generalized reflection coefficient; Partial standing waves; Standing-wave ratio; Standing-wave parameters; Standing-wave measurements; Line impedance; Power flow; Normalized line impedance; Normalized line admittance.

D7.4. A line of characteristic impedance 60 Ω is terminated by a load consisting of the series combination of $R = 30$ Ω, $L = 1$ μH, and $C = 100$ pF. Find the values of SWR and d_{min} for each of the following radian frequencies of the source: **(a)** $\omega = 10^8$; **(b)** $\omega = 2 \times 10^8$; and **(c)** $\omega = 0.8 \times 10^8$.
Ans. **(a)** 2, 0;   **(b)** 14.94, 0.309λ;   **(c)** 3.324, 0.115λ.

D7.5. Standing-wave measurements are performed on a line of characteristic impedance 60 Ω terminated by a load $\overline{Z}_R$. For each of the following sets of standing-wave data, find $\overline{Z}_R$: **(a)** SWR = 1.5, a voltage minimum right at the load; **(b)** SWR = 3.0, two successive voltage minima at 3 cm and 9 cm from the load; and **(c)** SWR = 2.0, two successive voltage minima at 3 cm and 7 cm from the load.
Ans. **(a)** $(40 + j0)$ Ω;   **(b)** $(180 + j0)$ Ω;   **(c)** $(48 + j36)$ Ω.

D7.6. An air-dielectric line of characteristic impedance $Z_0 = 75$ Ω is terminated by a load impedance $(45 + j60)$ Ω. Find the input impedance of the line for each of the following pairs of values of the frequency f and the length l of the line: **(a)** $f = 15$ MHz, $l = 5$ m; **(b)** $f = 50$ MHz, $l = 3$ m; and **(c)** $f = 37.5$ MHz, $l = 5$ m.
Ans. **(a)** $(45 - j60)$ Ω;   **(b)** $(45 + j60)$ Ω;   **(c)** $(225 + j0)$ Ω.

7.3   TRANSMISSION-LINE MATCHING

In the preceding section, we discussed standing waves on a line terminated by an arbitrary load. In the presence of standing waves, that is, when the load impedance is not equal to the characteristic impedance, it follows from (7.34) that the input impedance of the line will vary with frequency since the electrical length of the line and, hence, $\overline{\Gamma}(l) = \overline{\Gamma}_R e^{-j2\beta l}$ changes. This sensitivity to frequency increases with the electrical length of the line. To show this, let the length of the line be $l = n\lambda$. If the frequency is changed by an amount Δf, then the change in

n is given by

$$\Delta n = \Delta\left(\frac{l}{\lambda}\right) = \Delta\left(\frac{lf}{v_p}\right) = \frac{l}{v_p}\Delta f = \frac{n\lambda}{v_p}\Delta f = n\frac{\Delta f}{f} \qquad (7.40)$$

Thus, Δn, the change in the number of wavelengths corresponding to the line length, is proportional to n. The variation of the input impedance with frequency puts a limitation on the performance of a transmission-line system from the point of view of communication. For this and other reasons pertaining to power flow, it is desirable to eliminate standing waves on the line by connecting a *matching* device near the load such that the line views an effective impedance equal to its own characteristic impedance on the generator side of the matching device as shown in Fig. 7.15. The matching device should not at the same time absorb any power. It should be noted that *matching,* as referred to here, is not related to maximum power transfer since the condition for maximum power transfer is that the line input impedance must be the complex conjugate of the generator internal impedance. In the following, we discuss three techniques of matching.

A. Quarter-Wave Transformer Matching

The quarter-wave transformer, or QWT, matching technique makes use of a section of length $\lambda/4$ of a line of characteristic impedance Z_q different from that of the main line, as shown in Fig. 7.16. The principle is based on the property of line impedance given by (7.33). With reference to the notation of Fig. 7.16, we first note that to achieve a match, $\overline{Z}_1$ must be equal to Z_0. Then since from (7.33) $\overline{Z}_1\overline{Z}_2 = Z_q^2$, $\overline{Z}_2 = Z_q^2/\overline{Z}_1 = Z_q^2/Z_0$ must be purely real. We recall from the discussion of line impedance in Section 7.2 that the line impedance is purely real at locations of voltage maxima and minima of the standing-wave pattern. Therefore, within the first half-wavelength from the load, there are two solutions for d_q and hence for Z_q.

Quarter-wave transformer matching

If we choose a voltage minimum for the first solution, then from (7.28)

$$\boxed{d_q^{(1)} = \frac{\lambda}{4\pi}(\theta + \pi)} \qquad (7.41)$$

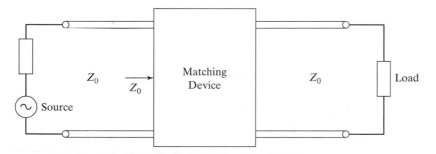

FIGURE 7.15 For illustrating the principle behind transmission-line *macthing.*

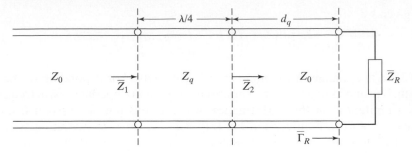

FIGURE 7.16   For illustrating the quarter-wave transformer matching technique.

where θ is the phase angle of $\overline{\Gamma}_R$, and the superscript (1) refers to solution 1. The value of the line impedance is $Z_0 (1 - |\overline{\Gamma}_R|)/(1 + |\overline{\Gamma}_R|)$. Hence, the value of Z_q is given by

$$Z_0 \cdot Z_0 \frac{1 - |\overline{\Gamma}_R|}{1 + |\overline{\Gamma}_R|} = Z_q^2$$

or

$$Z_q^{(1)} = Z_0 \sqrt{\frac{1 - |\overline{\Gamma}_R|}{1 + |\overline{\Gamma}_R|}} \tag{7.42}$$

For the second solution, the value of d_q corresponds to the location of a voltage maximum that occurs at $\pm\lambda/4$ from the location of the voltage minimum. Thus,

$$d_q^{(2)} = d_q^{(1)} \pm \frac{\lambda}{4} \tag{7.43}$$

whichever is positive and less than $\lambda/2$. The corresponding line impedance is $Z_0 (1 + |\overline{\Gamma}_R|)/(1 - |\overline{\Gamma}_R|)$, so that

$$Z_q^{(2)} = Z_0 \sqrt{\frac{1 + |\overline{\Gamma}_R|}{1 - |\overline{\Gamma}_R|}} \tag{7.44}$$

B. Single-Stub Matching

Single-stub matching

Another technique of transmission-line matching known as *stub matching* consists of connecting small sections of short-circuited lines (stubs) of appropriate lengths in parallel with the line, at appropriate distances from the load. In the single-stub matching technique, one stub is used and a match is achieved by

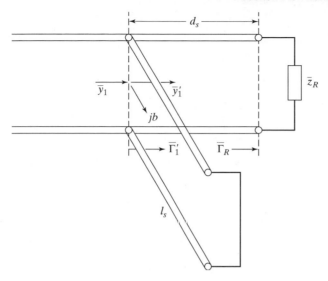

FIGURE 7.17   For illustrating the single-stub matching technique.

varying the location of the stub and the length of the stub. We shall assume the characteristic impedance of the stub to be the same as that of the line and use the notation shown in Fig. 7.17, in which $\bar{z}_R$ is the normalized load impedance, $\bar{y}_1$ and $\bar{y}_1'$ are the normalized line admittances just to the left and just to the right, respectively, of the stub, and b is the normalized input susceptance of the stub. The solution to the single-stub matching problem then consists of finding the values of d_s and l_s for a given value of $\bar{z}_R$ and hence of $\bar{\Gamma}_R$.

First, we observe that to achieve a match, $\bar{y}_1$ must be equal to $(1 + j0)$. Then proceeding to the right of the stub, we can write the following steps:

$$\bar{y}_1' = 1 - jb \tag{7.45a}$$

$$\bar{\Gamma}_1' = \frac{1 - \bar{y}_1'}{1 + \bar{y}_1'} = \frac{jb}{2 - jb} \tag{7.45b}$$

$$\bar{\Gamma}_R = \bar{\Gamma}_1' e^{j2\beta d_s} = \frac{jb}{2 - jb} e^{j2\beta d_s}$$

$$= \frac{|b|}{\sqrt{4 + b^2}} e^{j(\pm \pi/2 + \tan^{-1}b/2 + 2\beta d_s + 2n\pi)} \qquad \text{for} \quad b \gtrless 0 \tag{7.45c}$$

where n is an integer (positive or negative). Thus,

$$|\bar{\Gamma}_R| = \frac{|b|}{\sqrt{4 + b^2}} \tag{7.46a}$$

$$\theta = \pm \frac{\pi}{2} + \tan^{-1}\frac{b}{2} + 2\beta d_s + 2n\pi \qquad \text{for} \quad b \gtrless 0 \tag{7.46b}$$

so that

$$b = \pm \frac{2|\overline{\Gamma}_R|}{\sqrt{1 - |\overline{\Gamma}_R|^2}} \tag{7.47}$$

$$d_s = \frac{\lambda}{4\pi} \left(\theta \mp \frac{\pi}{2} - \tan^{-1} \frac{b}{2} - 2n\pi \right) \qquad \text{for} \quad b \gtrless 0 \tag{7.48}$$

Thus, two solutions are possible for b as given by (7.47) and the corresponding solutions for d_s are given by (7.48), where the integer value for n is chosen such that $0 \le d_s < \lambda/2$. Finally, to find the solutions for the stub length, we note from (7.13) that the normalized input impedance of a short-circuited line of length l_s is $j \tan \beta l_s$, so that

$$\frac{1}{jb} = j \tan \beta l_s$$

$$\tan \beta l_s = -\frac{1}{b}$$

$$l_s = \begin{cases} \dfrac{\lambda}{2\pi} \left[\tan^{-1}\left(-\dfrac{1}{b} \right) \right] + \dfrac{\lambda}{2} & \text{for} \quad b > 0 \\[2ex] \dfrac{\lambda}{2\pi} \left[\tan^{-1}\left(-\dfrac{1}{b} \right) \right] & \text{for} \quad b < 0 \end{cases} \tag{7.49}$$

C. Double-Stub Matching

Double-stub matching

In the single-stub matching technique, it is necessary to vary the distance between the stub and the load as well as the length of the stub to achieve a match for different loads or for different frequencies. This can be inconvenient for some arrangements of lines. When two stubs are used, it is possible to fix their locations and achieve a match for a wide range of loads by adjusting the lengths of the stubs. To discuss the principle behind this *double-stub matching* technique, we make use of the notation shown in Fig. 7.18, in which all admittances and susceptances are normalized quantities with respect to the characteristic admittance of the line. The solution to the double-stub matching problem then consists of finding the values of l_1 and l_2 for a given set of values of $\overline{z}_R$ (and, hence, of $\overline{\Gamma}_R$), d_1, and d_{12}.

First, we observe that to achieve a match, $\overline{y}_2$ must be equal to $(1 + j0)$. Then proceeding to the right in a step-by-step manner, we obtain an expression for $\overline{y}'_1$ in terms of $b_1, b_2,$ and d_{12} as follows:

$$\overline{y}'_2 = \overline{y}_2 - jb_2 = 1 - jb_2 \tag{7.50a}$$

$$\overline{\Gamma}'_2 = \frac{1 - \overline{y}'_2}{1 + \overline{y}'_2} = \frac{jb_2}{2 - jb_2} \tag{7.50b}$$

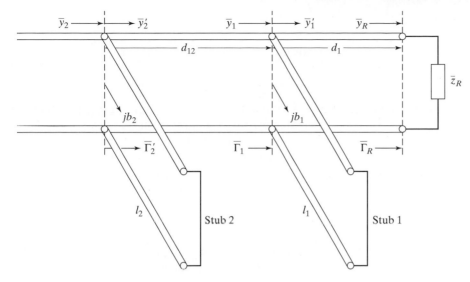

FIGURE 7.18. For illustrating the double-stub matching technique.

$$\overline{\Gamma}_1 = \overline{\Gamma}_2' \, e^{j2\beta d_{12}} = \frac{jb_2}{2 - jb_2} \, e^{j2\beta d_{12}} \tag{7.50c}$$

$$\begin{aligned}
\overline{y}_1 &= \frac{1 - \overline{\Gamma}_1}{1 + \overline{\Gamma}_1} \\
&= \frac{4 - j(4b_2 \cos 2\beta d_{12} - 2b_2^2 \sin 2\beta d_{12})}{4 - 4b_2 \sin 2\beta d_{12} + 4b_2^2 \sin^2 \beta d_{12}}
\end{aligned} \tag{7.50d}$$

$$\begin{aligned}
\overline{y}_1' &= \overline{y}_1 - jb_1 \\
&= \frac{1}{1 - b_2 \sin 2\beta d_{12} + b_2^2 \sin^2 \beta d_{12}} \\
&\quad + j\left(\frac{b_2^2 \sin 2\beta d_{12} - 2b_2 \cos 2\beta d_{12}}{2 - 2b_2 \sin 2\beta d_{12} + 2b_2^2 \sin^2 \beta d_{12}} - b_1 \right)
\end{aligned} \tag{7.50e}$$

For given values of $\overline{z}_R$ and d_1, $\overline{y}_1'$ can be computed in the usual manner, and the real and imaginary parts can be equated to the real and imaginary parts, respectively, on the right side of (7.50e). Noting that b_1 does not appear in the real part expression, we can first compute b_2 by solving the equation for the real parts. Thus, letting the real part of $\overline{y}_1'$ as computed from $\overline{z}_R$ and d_1 to be g', we have

$$\frac{1}{1 - b_2 \sin 2\beta d_{12} + b_2^2 \sin^2 \beta d_{12}} = g' \tag{7.51}$$

Rearranging and solving for b_2, we obtain

$$b_2 = \frac{\sin 2\beta d_{12} \pm \sqrt{\sin^2 2\beta d_{12} - 4(1 - 1/g') \sin^2 \beta d_{12}}}{2 \sin^2 \beta d_{12}}$$

or

$$b_2 = \frac{\cos \beta d_{12} \pm \sqrt{1/g' - \sin^2 \beta d_{12}}}{\sin \beta d_{12}} \qquad (7.52)$$

We now see that a solution does not exist for b_2 if $g' > 1/\sin^2 \beta d_{12}$, and, hence, it is not possible to achieve a match for loads that result in real part of $\bar{y}_1'$ greater than $1/\sin^2 \beta d_{12}$. A simple way to get around this problem is to increase d_1 by $\lambda/4$ (see Problem 7.24). Assuming that the condition $g' < 1/\sin^2 \beta d_{12}$ is achieved, we then compute two possible values for b_2 as given by (7.52). From the equation for the imaginary parts of $\bar{y}_1'$, the corresponding values of b_1 are then given by

$$b_1 = \frac{b_2^2 \sin 2\beta d_{12} - 2b_2 \cos 2\beta d_{12}}{2 - 2b_2 \sin 2\beta d_{12} + 2b_2^2 \sin^2 \beta d_{12}} - b' \qquad (7.53)$$

where b' is the imaginary part of $\bar{y}_1'$ as computed from $\bar{z}_R$ and d_1. Finally, the lengths of the two stubs are computed from b_1 and b_2 as in the case of the single-stub matching technique.

To consider a numerical example for the solution of all three types of matching techniques, let $R_L = 30\ \Omega$, $X_L = -40\ \Omega$, and for the double-stub matching case, $d_1 = 0$ and $d_{12} = 0.375\lambda$. Then the solutions obtained by using the appropriate equations for the three techniques are listed in Table 7.1.

Note that $d_{12} = 0.375\lambda$ is an odd multiple of $\lambda/8$. Values of odd multiples of $\lambda/8$ are commonly used for d_{12}. Also, if the specified value of Z_R is such that $g' > 1/\sin^2 \beta d_{12}$, then the value of d_1 is increased by $\lambda/4$ and the double-stub matching is continued.

Bandwidth   For any transmission-line matched system, the match is disturbed as the frequency is varied from that at which the various electrical lengths and distances are equal to the computed values for achieving the match. For example, in the QWT matched system, the electrical length of the QWT departs from one-quarter wavelength as the frequency is varied from that at which the match is achieved, and the system is no longer matched even if the load does not vary with frequency. A plot of the SWR in the main line to the left of the QWT versus

TABLE 7.1   Solutions for Transmission-Line Matching Example

Technique	Solution Number	Solution
QWT	1	$d_q = 0.125\lambda$, $Z_q = 28.86751\ \Omega$
QWT	2	$d_q = 0.375\lambda$, $Z_q = 86.60254\ \Omega$
Single stub	1	$d_s = 0.20833\lambda$, $l_s = 0.38641\lambda$
Single stub	2	$d_s = 0.04167\lambda$, $l_s = 0.11359\lambda$
Double stub	1	$l_1 = 0.13483\lambda$, $l_2 = 0.32726\lambda$
Double stub	2	$l_1 = 0.05614\lambda$, $l_2 = 0.05996\lambda$

frequency is typically of the shape shown in Fig. 7.19, where f_0 is the design frequency at which the system is matched, and hence the frequency at which the SWR is unity. One can then specify a tolerable value of SWR, say, S, so that there exists an acceptable bandwidth of operation, $f_2 - f_1$. Similar considerations apply to the single-stub and double-stub matched systems.

To discuss a procedure by means of which the SWR versus frequency curve can be computed for all three types of matching techniques discussed, let us consider a transmission-line system having n discontinuities, as shown for $n = 2$ in Fig. 7.20. At each discontinuity, there can exist a stub and a change in characteristic impedance. We shall consider a specification of zero for the length of the stub to mean no stub is present instead of a stub of zero length. This does not result in a conflict since for any matched system using short-circuited stubs, values of zero cannot be obtained for stub lengths since then SWR would be

SWR versus frequency computation

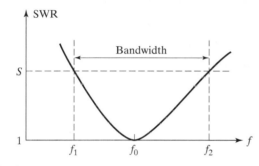

FIGURE 7.19   The SWR versus frequency curve illustrating the bandwidth between the two frequencies, f_1 and f_2, on either side of the design frequency f_0, at which the SWR is a specified value, $S\ (> 1)$.

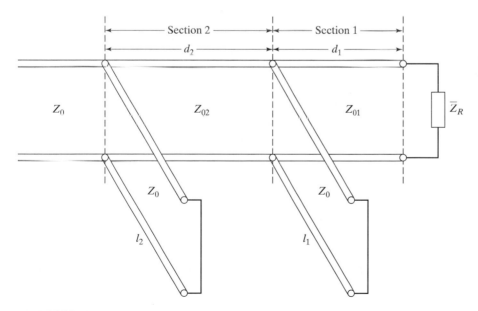

FIGURE 7.20   Transmission-line system for computing the SWR versus frequency curve for QWT, single-stub, and double-stub matched systems.

TABLE 7.2   Values of Parameters for Using the System of Fig. 7.20 for Three Different Cases

System	n	Z_{01}	d_1	l_1	Z_{02}	d_2	l_2
QWT	2	Z_0	d_q	0^a	Z_q	$1/4$	0^a
Single stub	1	Z_0	d_s	l_s	—	—	—
Double stub	2	Z_0	d_1	l_1	Z_0	d_{12}	l_2

^{a}Value of zero means no stub present.

infinity. With this understanding, Fig. 7.20 can be used to represent all three types of matching systems by specifying values for the various parameters, as shown in Table 7.2.

Then to compute the SWR in the main line at a given frequency, we first note that since $\lambda \propto 1/f$, the electrical length of a line section or of a stub is proportional to f. Thus, at a frequency f, the electrical length is equal to f/f_0 times its value at f_0. For a given f/f_0, the procedure consists of starting at the load and computing in succession the line admittance to the left of the stub at each discontinuity from a knowledge of the line admittance at the output of the line section to the right of that stub, until the line admittance to the left of the last discontinuity is found and used to compute the required SWR. In carrying out this procedure, we observe the following:

1. To compute the normalized admittance, say, $\bar{y}_i$, at the input (left) end of a line section of length l from the normalized admittance, say, $\bar{y}_o$, at the output (right) end of that section, we use the formula

$$\bar{y}_i = \frac{1 - \overline{\Gamma}_i}{1 + \overline{\Gamma}_i} = \frac{1 - \overline{\Gamma}_o e^{-j2\beta l}}{1 + \overline{\Gamma}_o e^{-j2\beta l}}$$

$$= \frac{1 - [(1 - \bar{y}_o)/(1 + \bar{y}_o)]e^{-j2\beta l}}{1 + [(1 - \bar{y}_o)/(1 + \bar{y}_o)]e^{-j2\beta l}}$$

or

$$\boxed{\bar{y}_i = \frac{j \sin \beta l + \bar{y}_o \cos \beta l}{\cos \beta l + j\bar{y}_o \sin \beta l}} \tag{7.54}$$

where $\overline{\Gamma}_i$ and $\overline{\Gamma}_o$ are the reflection coefficients at the input and output ends, respectively.

2. To compute the line admittance to the left of a stub, we add the input admittance of the stub to the line admittance to the right of the stub.

The computation of SWR versus f/f_0 can be done by using a computer program. For values of the input parameters pertinent to the first of the two solutions for the double-stub matching case in Table 7.1, the computed values of SWR are listed in Table 7.3. The frequency variation of Z_R is taken into account by assuming Z_R to be the series combination of a single resistor and a single reactive element.

TABLE 7.3   Computed Values of
SWR Versus Frequency

f/f_0	SWR
0.90	1.9249
0.92	1.7124
0.94	1.5117
0.96	1.325
0.98	1.1543
1.00	1.0006
1.02	1.1583
1.04	1.3459
1.06	1.5663
1.08	1.8236
1.10	2.1216

K7.3. Matching; Quarter-wave transformer; Single stub; Double stub; Bandwidth.

D7.7. For a line of characteristic impedance 75 Ω, find the location nearest to the load and the characteristic impedance of a quarter-wave transformer required to achieve a match for each of the following values of $\overline{\Gamma}_R$: **(a)** $1/9$; **(b)** $-j0.5$; and **(c)** $j1/3$.
   Ans. **(a)** 83.85 Ω, 0;   **(b)** 43.30 Ω, 0.125λ;   **(c)** 106.07 Ω, 0.125λ.

D7.8. For each of the following values of $\overline{Z}_R$ terminating a line of characteristic impedance 60 Ω, find the lowest value of d_s and the corresponding smallest value of the length l_s of a single short-circuited stub of characteristic impedance 60 Ω required to achieve a match between the line and the load: **(a)** $\overline{Z}_R = 30$ Ω and **(b)** $\overline{Z}_R = (12 - j24)$ Ω.
   Ans. **(a)** 0.098λ, 0.348λ;   **(b)** 0, 0.074λ.

D7.9. For each of the following sets of values of d_1, d_{12}, and $\overline{z}_R$, associated with the double-stub matching technique, determine whether or not it is possible to achieve a match between the line and the load: **(a)** $d_1 = 0$, $d_{12} = 3\lambda/8$, $\overline{z}_R = 0.3 + j0.4$;   **(b)** $d_1 = \lambda/8$, $d_{12} = 3\lambda/8$, $\overline{z}_R = 0.5$;   and **(c)** $d_1 = \lambda/4$, $d_{12} = 5\lambda/8$, $z_R = 2.5 - j5.0$.
   Ans. **(a)** Yes;   **(b)** yes;   **(c)** no.

7.4   THE SMITH CHART: 1. BASIC PROCEDURES

In the preceding section, we considered transmission-line matching techniques and computer solutions of matching problems. In this section, we discuss some basic procedures using the Smith chart. Introduced in 1939 by P. H. Smith,[1] the Smith chart continues to be a popular graphical aid in the solution of transmission-line problems, including simulation on personal computers.[2]

---

[1]P. H. Smith, "Transmission-Line Calculator," *Electronics,* January 1939, pp. 29–31

[2]See, for example, M. Felton, "Moving the Smith Chart to a Low-Cost Computer," *Microwave Journal,* October 1983, pp. 131–133, and N. N. Rao, "PC-Assisted Instruction of Introductory Elec tromagnetics," *IEEE Transactions on Education,* February 1990, pp. 51–59.

Construction   The *Smith chart* is a transformation from the complex $\bar{Z}$-plane (or $\bar{Y}$-plane) to the complex $\bar{\Gamma}$-plane. To discuss the basis behind the construction of the Smith chart, we begin with the relationship for the reflection coefficient in terms of the normalized line impedance as given by

$$\bar{\Gamma}(d) = \frac{\bar{z}(d) - 1}{\bar{z}(d) + 1} \tag{7.55}$$

Letting $\bar{z}(d) = r + jx$, we have

$$\bar{\Gamma}(d) = \frac{r + jx - 1}{r + jx + 1} = \frac{(r - 1) + jx}{(r + 1) + jx}$$

and

$$|\bar{\Gamma}(d)| = \left[\frac{(r - 1)^2 + x^2}{(r + 1)^2 + x^2}\right]^{1/2} \leq 1$$

for positive values of r. Thus, for the passive line impedances, the reflection coefficient lies inside or on the circle of unit radius in the $\bar{\Gamma}$-plane. We will hereafter call this circle the *unit circle*. Conversely, each point inside or on the unit circle represents a possible value of reflection coefficient corresponding to a unique value of passive normalized line impedance. Hence, all possible values of passive normalized line impedances can be mapped onto the region bounded by the unit circle.

To determine how the normalized line impedance values are mapped onto the region bounded by the unit circle, we note that

$$\bar{\Gamma} = \frac{r + jx - 1}{r + jx + 1} = \frac{r^2 - 1 + x^2}{(r + 1)^2 + x^2} + j\frac{2x}{(r + 1)^2 + x^2}$$

so that

$$\mathrm{Re}(\bar{\Gamma}) = \frac{r^2 - 1 + x^2}{(r + 1)^2 + x^2} \tag{7.56a}$$

$$\mathrm{Im}(\bar{\Gamma}) = \frac{2x}{(r + 1)^2 + x^2} \tag{7.56b}$$

Let us now discuss different cases:

1. $\bar{z}$ is purely real; that is, $x = 0$. Then

$$\mathrm{Re}(\bar{\Gamma}) = \frac{r - 1}{r + 1} \qquad \text{and} \qquad \mathrm{Im}(\bar{\Gamma}) = 0$$

Purely real values of $\bar{z}$ are represented by points on the real axis. For example, $r = 0, \frac{1}{3}, 1, 3,$ and ∞ are represented by $\bar{\Gamma} = -1, -\frac{1}{2}, 0, \frac{1}{2}$ and 1, respectively, as shown in Fig. 7.21(a).

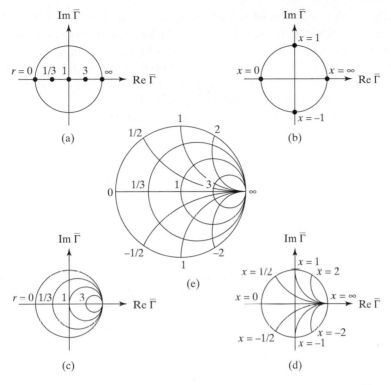

FIGURE 7.21 Development of the Smith chart by transformation from $\bar{z}$ to $\bar{\Gamma}$.

2. $\bar{z}$ is purely imaginary; that is, $r = 0$. Then

$$\left| \bar{\Gamma} - \left| \frac{x^2 - 1}{x^2 + 1} + j \frac{2x}{x^2 + 1} \right| \right| = 1$$

and

$$\underline{/\bar{\Gamma}} = \tan^{-1} \frac{2x}{x^2 - 1}$$

Purely imaginary values of $\bar{z}$ are represented by points on the unit circle. For example, $x = 0$, 1, ∞, -1, and $-\infty$ are represented by $\bar{\Gamma} = 1\underline{/\pi}$, $1\underline{/\pi/2}$, $1\underline{/0°}$, $1\underline{/-\pi/2}$, and $1\underline{/2\pi}$, respectively, as shown in Fig. 7.21(b).

3. $\bar{z}$ is complex, but its real part is constant. Then

$$\left[\mathrm{Re}(\bar{\Gamma}) - \frac{r}{r+1} \right]^2 + [\mathrm{Im}(\bar{\Gamma})]^2$$

$$= \left[\frac{r^2 - 1 + x^2}{(r+1)^2 + x^2} - \frac{r}{r+1} \right]^2 + \left[\frac{2x}{(r+1)^2 + x^2} \right]^2 = \left(\frac{1}{r+1} \right)^2$$

This is the equation of a circle with center at $\text{Re}(\overline{\Gamma}) = r/(r + 1)$ and $\text{Im}(\overline{\Gamma}) = 0$ and radius equal to $1/(r + 1)$. Thus, loci of constant r are circles in the $\overline{\Gamma}$-plane with centers at $[r/(r + 1), 0]$ and radii $1/(r + 1)$. For example, for $r = 0, \frac{1}{3}, 1, 3$, and ∞, the centers of the circles are $(0, 0), (\frac{1}{4}, 0), (\frac{1}{2}, 0), (\frac{3}{4}, 0)$, and $(1, 0)$, respectively, and the radii are $1, \frac{3}{4}, \frac{1}{2}, \frac{1}{4}$, and 0, respectively. These circles are shown in Fig. 7.21(c).

4. $\overline{z}$ is complex, but its imaginary part is constant. Then

$$[\text{Re}(\overline{\Gamma}) - 1]^2 + \left[\text{Im}(\overline{\Gamma}) - \frac{1}{x}\right]^2$$

$$= \left[\frac{r^2 - 1 + x^2}{(r + 1)^2 + x^2} - 1\right]^2 + \left[\frac{2x}{(r + 1)^2 + x^2} - \frac{1}{x}\right]^2 = \left(\frac{1}{x}\right)^2$$

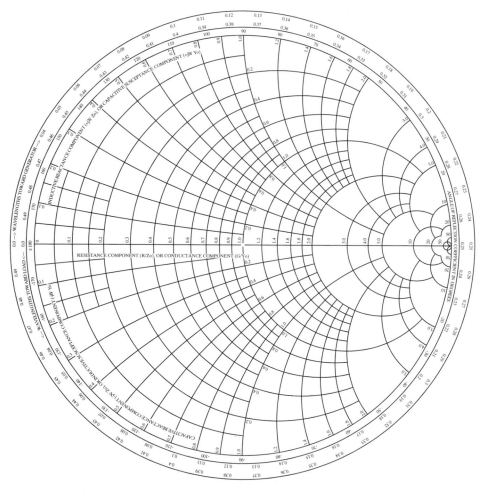

FIGURE 7.22   The Smith chart. (Copyrighted by and reproduced with the permission of Kay Elemetrics Corp., Pine Brook, N.J.)

This is the equation of a circle with center at $\text{Re}(\overline{\Gamma}) = 1$ and $\text{Im}(\overline{\Gamma}) = 1/x$ and radius equal to $1/|x|$. Thus, loci of constant x are circles in the $\overline{\Gamma}$-plane with centers at $(1, 1/x)$ and radii equal to $1/|x|$. For example, for $x = 0$, $\pm\frac{1}{2}, \pm1, \pm2$, and $\pm\infty$, the centers of the circles are $(1, \infty), (1, \pm2), (1, \pm1)$, $(1, \pm\frac{1}{2})$, and $(1, 0)$, respectively, and the radii are $\infty, 2, 1, \frac{1}{2}$, and 0, respectively. Portions of these circles that fall inside the unit circle are shown in Fig. 7.21(d). Portions that fall outside the unit circle represent active impedances.

Combining Figs. 7.21(c) and (d), we obtain the chart of Fig. 7.21(e). In a commercially available form shown in Fig. 7.22, the Smith chart contains circles of constant r and constant x for very small increments of r and x, respectively, so that interpolation between the contours can be carried out accurately. We now consider an example to illustrate some basic procedures using the Smith chart.

Example 7.4

A transmission line of characteristic impedance 50 Ω is terminated by a load impedance $\overline{Z}_R = (15 - j20)$ Ω. It is desired to find the following quantities by using the Smith chart. *Some basic procedures*

1. Reflection coefficient at the load
2. SWR on the line
3. Distance of the first voltage minimum of the standing-wave pattern from the load
4. Line impedance at $d = 0.05\lambda$
5. Line admittance at $d = 0.05\lambda$
6. Location nearest to the load at which the real part of the line admittance is equal to the line characteristic admittance

We proceed with the solution of the problem in the following step-by-step manner with reference to Fig. 7.23.

a. Find the normalized load impedance.

$$\overline{z}_R = \frac{\overline{Z}_R}{Z_0} = \frac{15 - j20}{50} = 0.3 - j0.4$$

b. Locate the normalized load impedance on the Smith chart at the intersection of the 0.3 constant normalized resistance circle and -0.4 constant normalized reactance circle (point A).

c. Locating point A actually amounts to computing the reflection coefficient at the load since the Smith chart is a transformation in the $\overline{\Gamma}$-plane. The magnitude of the reflection coefficient is the distance from the center (O) of the Smith chart (origin of the $\overline{\Gamma}$-plane) to the point A based on a radius of unity for the outermost circle. For this example, $|\overline{\Gamma}_R| = 0.6$. The phase angle of $\overline{\Gamma}_R$ is the angle measured from the horizontal axis to the right of O (positive real axis in the $\overline{\Gamma}$-plane) to the line OA in the counterclockwise direction. This angle is indicated on the chart along its circumference. For this example, $\underline{/\overline{\Gamma}_R} = 227°$. Thus,

$$\overline{\Gamma}_R = 0.6e^{j1.261\pi}$$

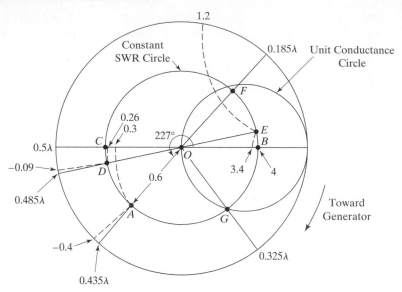

FIGURE 7.23   For illustrating the various procedures to be followed in using the Smith chart.

d. To find the SWR, we recall that at the location of a voltage maximum, the line impedance is purely real and given by

$$R_{\max} = Z_0(\text{SWR}) \tag{7.57}$$

Thus, the normalized value of $R_{\max}$ is equal to the SWR. We therefore move along the line to the location of the voltage maximum, which involves going around the constant $|\overline{\Gamma}|$ circle to the point on the positive real axis. To do this on the Smith chart, we draw a circle passing through A and with center at O. This circle is known as the *constant SWR circle* since for points on this circle, $|\overline{\Gamma}|$ and, hence, SWR $= (1 + |\overline{\Gamma}|)/(1 - |\overline{\Gamma}|)$ is a constant. Impedance values along this circle are normalized line impedances as seen moving along the line. In particular, since point B (the intersection of the constant SWR circle with the horizontal axis to the right of O) corresponds to voltage maximum, the normalized impedance value at point B, which is purely real and maximum, is equal to the SWR. Thus, for this example, SWR $= 4$.

e. Just as point B represents the position of a voltage maximum on the line, point C (intersection of the constant SWR circle with the horizontal axis to the left of O, i.e., the negative real axis of the $\overline{\Gamma}$-plane) represents the location of a voltage minimum. Hence, to find the distance of the first voltage minimum from the load, we move along the constant SWR circle starting at point A (load impedance) toward the generator (clockwise direction on the chart) to reach point C. Distance moved along the constant SWR circle in this process can be determined by recognizing that one complete revolution around the chart ($\overline{\Gamma}$-plane diagram) constitutes movement on the line by 0.5λ. However, it is not necessary to compute in this manner since distance scales in terms of λ are provided along the periphery of the chart for movement in both directions. For this example, the distance from the load to the first voltage minimum = $(0.5 - 0.435)\lambda = 0.065\lambda$. Conversely, if the SWR and the location of the voltage minimum are specified, we can find the load impedance by following the foregoing procedures in reverse.

f. To find the line impedance at $d = 0.05\lambda$, we start at point A and move along the constant SWR circle toward the generator (in the clockwise direction) by a distance of 0.05λ to reach point D. This step is equivalent to finding the reflection coefficient at $d = 0.05\lambda$ knowing the reflection coefficient at $d = 0$ and then computing the normalized line impedance by using (7.35). Thus, from the coordinates corresponding to point D, the normalized line impedance at $d = 0.05\lambda$ is $(0.26 - j0.09)$, and hence the line impedance at $d = 0.05\lambda$ is $50(0.26 - j0.09)$ or $(13 - j4.5)\ \Omega$.

g. To find the line admittance at $d = 0.05\lambda$, we recall that

$$[\overline{Z}(d)]\left[\overline{Z}\left(d + \frac{\lambda}{4}\right)\right] = Z_0^2$$

so that

$$[\overline{z}(d)]\left[\overline{z}\left(d + \frac{\lambda}{4}\right)\right] = 1$$

or

$$\overline{y}(d) = \overline{z}\left(d + \frac{\lambda}{4}\right) \tag{7.58}$$

Thus, the normalized line admittance at point D is the same as the normalized line impedance at a distance $\lambda/4$ from it. Hence, to find $\overline{y}(0.05\lambda)$, we start at point D and move along the constant SWR circle by a distance $\lambda/4$ to reach point E (we note that this point is diametrically opposite to point D) and read its coordinates. This gives $y(0.05\lambda) = (3.4 + j1.2)$. We then have $\overline{Y}(0.05\lambda) - y(0.05\lambda) \times Y_0 = (3.4 + j1.2) \times 1/50 = (0.068 + j0.024)$ S.

h. Relationship (7.58) permits us to use the Smith chart as an admittance chart instead of as an impedance chart. In other words, if we want to find the normalized line admittance $\overline{y}(Q)$ at a point Q on the line, knowing the normalized line admittance $\overline{y}(P)$ at another point P on the line, we can simply locate $\overline{y}(P)$ by entering the chart at coordinates equal to its real and imaginary parts and then moving along the constant SWR circle by the amount of the distance from P to Q in the proper direction to obtain the coordinates equal to the real and imaginary parts of $\overline{y}(Q)$. Thus, it is not necessary first to locate $\overline{z}(P)$ diametrically opposite to $\overline{y}(P)$ on the constant SWR circle, then move along the constant SWR circle to locate $\overline{z}(Q)$, and then find $\overline{y}(Q)$ diametrically opposite to $\overline{z}(Q)$. To find the location nearest to the load at which the real part of the line admittance is equal to the line characteristic admittance, we first locate $\overline{y}(0)$ at point F diametrically opposite to point A which corresponds to $\overline{z}(0)$. We then move along the constant SWR circle toward the generator to reach point G on the circle corresponding to constant real part equal to unity (we call this circle the *unit conductance circle*). Distance moved from F to G is read off the chart as $(0.325 - 0.185)\lambda = 0.14\lambda$. This is the distance closest to the load at which the real part of the normalized line admittance is equal to unity and, hence, the real part of the line admittance is equal to line characteristic admittance.

K7.4. $\overline{\Gamma}$-plane; Unit circle; Transformation from $\overline{z}$ (or $\overline{y}$) to $\overline{\Gamma}$; Smith chart; Constant SWR circle; Unit conductance circle.

D7.10. Find the values of $\overline{\Gamma}$ in polar form onto which the following normalized impedances are mapped: **(a)** $0.25 + j0$; **(b)** $0 - j0.5$; **(c)** $3 + j3$; **(d)** $-1 + j2$.
Ans. **(a)** $0.6\underline{/180°}$; **(b)** $1\underline{/233.13°}$; **(c)** $0.721\underline{/19.44°}$; **(d)** $1.414\underline{/45°}$.

D7.11. Find the following using the Smith chart: **(a)** the normalized input impedance of a line of length 0.1λ and terminated by a normalized load impedance $(2 + j1)$; **(b)** the normalized input admittance of a short-circuited stub of length 0.17λ; and **(c)** the shortest length of an open-circuited stub having the normalized input admittance $j0.4$.
Ans. **(a)** $1.4 - j1.1$; **(b)** $-j0.55$; **(c)** 0.06λ.

7.5   THE SMITH CHART: 2. APPLICATIONS

In the preceding section, we introduced the Smith chart and discussed some basic procedures. In this section, we first consider by means of examples graphical solutions of transmission-line matching problems using the Smith chart and then discuss further applications.

Example 7.5

Single-stub matching solution

Let us consider a transmission line of characteristic impedance $Z_0 = 50\ \Omega$ terminated by a load impedance $\overline{Z}_R = (30 - j40)\ \Omega$ and illustrate the solution of the single-stub matching problem by using the Smith chart, assuming Z_0 of the stub to be $50\ \Omega$.

With reference to the notation in Fig. 7.17, we recall that to achieve a match, the stub must be located at a point on the line at which the real part of the normalized line admittance is equal to unity; the imaginary part of the line admittance at that point is then canceled by appropriately choosing the length of the stub. Hence, we proceed with the solution in the following step-by-step manner with reference to Fig. 7.24.

a. Find the normalized load impedance.

$$\overline{z}_R = \frac{\overline{Z}_R}{Z_0} = \frac{30 - j40}{50} = 0.6 - j0.8$$

Locate the normalized load impedance on the Smith chart at point A.

b. Draw the constant SWR circle passing through point A. This is the locus of the normalized line impedance as well as the normalized line admittance. Starting at point A, go around the constant SWR circle by half a revolution to reach point B diametrically opposite to point A. Point B corresponds to the normalized load admittance.

c. Starting at point B, go around the constant SWR circle toward the generator until point C on the unit conductance circle is reached. This point corresponds to the normalized line admittance having the real part equal to unity, and, hence, it corresponds to the location of the stub. The distance moved from point B to point C (not from point A to point C) is equal to the distance from the load at which the stub must be located. Thus, the location of the stub from the load = $(0.1665 - 0.125)\lambda = 0.0415\lambda$.

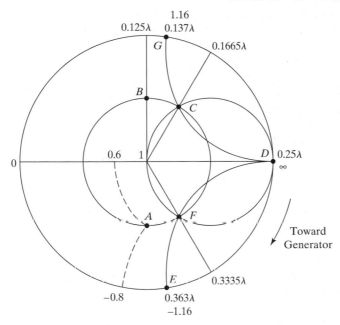

FIGURE 7.24 Solution of single-stub matching problem by using the Smith chart.

d. Read off the Smith chart the normalized susceptance value corresponding to point C. This value is 1.16 and it is the imaginary part of the normalized line admittance at the location of the stub. The imaginary part of the line admittance is equal to $1.16 \times Y_0 = (1.16/50)$ S. The input susceptance of the stub must therefore be equal to $-(1.16/50)$ S.

e. This step consists of finding the length of a short-circuited stub having an input susceptance equal to $-(1.16/50)$ S. We can use the Smith chart for this purpose since this simply consists of finding the distance between two points on a line (the stub in this case) at which the admittances (purely imaginary in this case) are known. Thus, since the short circuit corresponds to a susceptance of infinity, we start at point D and move toward the generator along the constant SWR circle through D (the outermost circle) to reach point E corresponding to $-j1.16$, which is the input admittance of the stub normalized with respect to its own characteristic admittance. The distance moved from D to E is the required length of the stub. Thus, length of the short-circuited stub $= (0.363 - 0.25)\lambda = 0.113\lambda$.

f. The results obtained for the location and the length of the stub agree with one of the solutions found analytically in Section 7.3. The second solution can be obtained by noting that in step (c), we can go around the constant SWR circle from point B until point F on the unit conductance circle is reached, instead of stopping at point C. The stub location for this solution is $(0.3335 - 0.125)\lambda = 0.2085\lambda$. The required input susceptance of the stub is $(1.16/50)$ S. The length of the stub is the distance from point D to point G in the clockwise direction. This is $(0.137 + 0.25)\lambda = 0.387\lambda$. These values are the same as the second solution obtained in Section 7.3.

Example 7.6

Double-stub marching solution

For the line of characteristic impedance $Z_0 = 50\ \Omega$ and load impedance $\bar{Z}_R = (30 - j40)\ \Omega$ of Example 7.5, it is desired to solve the double-stub matching problem by using the Smith chart and assuming Z_0 of both stubs to be $50\ \Omega$, the first stub to be located at the load, and distance between stubs equal to 0.375λ.

With reference to the notation of Fig. 7.18, we first note that to achieve a match, $\bar{y}_2' = 1 - jb_2$ must fall on the unit conductance circle. Now since $\bar{y}_2'$ and $\bar{y}_1$ correspond to locations at the end points of the line section between the stubs, for a given $\bar{y}_1, \bar{y}_2'$ can be obtained by drawing the constant SWR circle through $\bar{y}_1$ and going toward the generator (clockwise direction) by the distance d_{12} from $\bar{y}_1$. Conversely, to obtain $\bar{y}_1$ for a given $\bar{y}_2'$, we start at $\bar{y}_2'$ and go toward the load (counterclockwise direction) by the distance d_{12} along the constant SWR circle. Hence, for $\bar{y}_2'$ to fall on the unit conductance circle, $\bar{y}_1$ must fall on a circle that is obtained by pivoting the unit conductance circle at the center of the Smith chart (point O) and rotating it toward the load by the distance d_{12}, as shown in Fig. 7.25 for $d_{12} = 3\lambda/8$. We shall call this circle the *auxiliary circle*. Thus, the auxiliary circle is the *locus of $\bar{y}_1$ for possible match*.

The matching procedure consists of first locating $\bar{y}_R$ on the Smith chart and then moving along the constant SWR circle through $\bar{y}_R$ toward the generator by the distance d_1 between the load and the first stub, thereby locating $\bar{y}_1'$. The right amount of susceptance is then added to $\bar{y}_1'$ to reach a point on the auxiliary circle. This point corresponds to $\bar{y}_1$ and determines a new constant SWR circle. By going along this new constant SWR

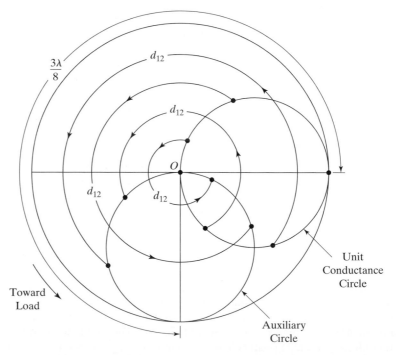

FIGURE 7.25   Rotation of the unit conductance circle by $d_{12}(= 3\lambda/8)$ toward the load about O for illustrating the construction of the auxiliary circle, that is, the locus of $\bar{y}_1$ for possible match for the double-stub matching arrangement of Fig. 7.18.

circle toward the generator by the distance d_{12}, $\bar{y}_2'$ is located on the unit conductance circle. The amount of susceptance added to $\bar{y}_1'$ is the required normalized input susceptance of the first stub, whereas the negative of the imaginary part of $\bar{y}_2'$ is the required normalized input susceptance of the second stub.

Considering now the numerical values of $\bar{z}_R = (30 - j40)/50 = (0.6 - j0.8)$, $d_1 = 0$, and $d_{12} = 0.375\lambda$, we proceed with the solution in the following step-by-step manner with reference to Fig. 7.26.

a. Locate $\bar{z}_R = (0.6 - j0.8)$ at point A and draw the constant SWR circle through A.

b. Locate point B on the constant SWR circle and diametrically opposite to point A. This point corresponds to $\bar{y}_R$. Since d_1 is equal to zero, it also corresponds to $\bar{y}_1'$. If d_1 is not equal to zero, then $\bar{y}_1'$ has to be located by going along the constant SWR circle toward the generator by the distance d_1 from point B.

c. Draw the auxiliary circle, which is the circle obtained by pivoting the unit conductance circle at the center of the chart and rotating it by the distance $d_{12} - 0.375\lambda$ toward the load.

d. This step consists of adding the right amount of susceptance to $\bar{y}_1'$ to get to a point on the auxiliary circle. Hence, starting at point B, go along the constant conductance circle to reach point C on the auxiliary circle. This point corresponds to $\bar{y}_1$. The required normalized input susceptance of the first stub can now be found by noting that $\bar{y}_1 = \bar{y}_1' + jb_1$, and, hence,

$$jb_1 = \bar{y}_1 - \bar{y}_1' = (0.6 - j0.09) - (0.6 + j0.8) = -j0.89$$

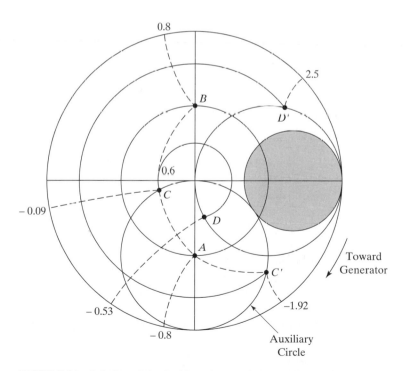

FIGURE 7.26 Solution of the double-stub matching problem by using the Smith chart

e. Starting at point C, go along the constant SWR circle through C toward the generator by $d_{12} = 0.375\lambda$ to reach point D on the unit conductance circle. This point corresponds to $\bar{y}'_2$. Note that the SWR on the portion of the line between the stubs is different from the SWR to the right of the first stub because of the discontinuity introduced by the stub. The required normalized input susceptance of the second stub can now be found by reading the imaginary part of $\bar{y}'_2$ and taking its negative. Thus,

$$jb_2 = -j[\mathrm{Im}(\bar{y}'_2)] = j0.53$$

f. This step consists of finding the lengths of the two stubs having the normalized input susceptances found in steps (d) and (e), by using the procedure discussed in Example 7.5. Thus, we obtain

$$l_1, \text{length of first stub} = (0.385 - 0.25)\lambda = 0.135\lambda$$

$$l_2, \text{length of second stub} = (0.077 + 0.25)\lambda = 0.327\lambda$$

which agree with one of the solutions found analytically in Section 7.3.

g. Finally, the second solution can be obtained by going from point B to point C' on the auxiliary circle and then to point D' on the unit conductance circle, and computing jb_1 and jb_2 as in steps (d) and (e). Thus, we obtain

$$jb_1 = (0.6 - j1.92) - (0.6 + j0.8) = -j2.72$$

$$jb_2 = -j2.5$$

giving us

$$l_1 = (0.306 - 0.25)\lambda = 0.056\lambda$$

$$l_2 = (0.31 - 0.25)\lambda = 0.06\lambda$$

These values are the same as the second solution obtained in Section 7.3.

Before proceeding further, we recall from Section 7.3 that in the double-stub matching technique, it is not possible to achieve a match for loads that result in real part of $\bar{y}'_1$ greater than $1/\sin^2 \beta d_{12}$. For $d_{12} = 3\lambda/8, 1/\sin^2 \beta d_{12} = 2$, and a match cannot be achieved if the real part of $\bar{y}'_1$ is greater than 2. This is easily evident from the Smith chart construction in Fig. 7.26, since if point B falls inside the shaded region (real part > 2), it is not possible to reach a point on the auxiliary circle by moving on the constant conductance circle through B. The shaded region is therefore called the *forbidden region of* $\bar{y}'_1$ *for possible match*. As pointed out in Section 7.3, a solution to the problem is to increase d_1 by $\lambda/4$. This effectively rotates the forbidden region by $180°$ about the center of the chart, thereby making possible a match.

Transformation across a discontinuity

To illustrate the application of the Smith chart further, we shall now discuss a very useful property of the reflection coefficient and hence of the Smith chart. This has to do with the transformation of the reflection coefficient from one side of a discontinuity to the other side of the discontinuity. Let us, for example, consider the system shown in Fig. 7.27, which consists of a junction between two lines of characteristic admittances Y_{01} and Y_{02} and in addition, an admittance $\bar{Y}_d$ connected across the junction. If $\bar{Y}_d = 0$, then the system reduces to a simple

FIGURE 7.27 Transmmission-line system for deriving the transformation of $\overline{\Gamma}$ across a discontinuity.

junction between two lines. If $Y_{01} = Y_{02}$, then the system reduces to an admittance discontinuity in the same line.

Let $\overline{y}_1 = \overline{Y}_1/Y_{01}$ and $\overline{y}_2 = \overline{Y}_2/Y_{02}$ be the normalized admittances to the left and to the right, respectively, of the junction, and let the corresponding reflection coefficients be $\overline{\Gamma}_1$ and $\overline{\Gamma}_2$, respectively, as shown in Fig. 7.27. Then, since $\overline{Y}_1 = \overline{Y}_2 + \overline{Y}_d$, we have

$$
\begin{aligned}
\overline{y}_1 &= \frac{\overline{Y}_1}{Y_{01}} = \frac{\overline{Y}_2}{Y_{01}} + \frac{\overline{Y}_d}{Y_{01}} \\
&= \frac{Y_{02}}{Y_{01}}\left(\frac{\overline{Y}_2}{Y_{02}} + \frac{\overline{Y}_d}{Y_{02}}\right) \\
&= a(\overline{y}_2 + \overline{y}_d)
\end{aligned}
\tag{7.59}
$$

where $a = Y_{02}/Y_{01}$ is the ratio of the characteristic admittances of the two lines and $\overline{y}_d = \overline{Y}_d/Y_{02}$ is the normalized value of $\overline{Y}_d$ with respect to Y_{02}. Substituting for $\overline{y}_1$ and $\overline{y}_2$ in (7.59) in terms of $\overline{\Gamma}_1$ and $\overline{\Gamma}_2$, respectively, we have

$$
\frac{1 - \overline{\Gamma}_1}{1 + \overline{\Gamma}_1} = a\left(\frac{1 - \overline{\Gamma}_2}{1 + \overline{\Gamma}_2} + \overline{y}_d\right)
\tag{7.60}
$$

Rearranging (7.60), we obtain

$$
\overline{\Gamma}_1 = \frac{(1 + a - a\overline{y}_d)\overline{\Gamma}_2 + (1 - a - a\overline{y}_d)}{(1 - a + a\overline{y}_d)\overline{\Gamma}_2 + (1 + a + a\overline{y}_d)}
\tag{7.61}
$$

Equation (7.61) is of the form of the so-called *bilinear transformation,* between two complex planes, a property of which is that circles in one plane are transformed into circles in the second plane. Consequently, loci of $\overline{\Gamma}_2$, which are circles in the $\overline{\Gamma}$-plane, are mapped on to loci of $\overline{\Gamma}_1$, which are also circles in the $\overline{\Gamma}$-plane, and vice versa. Since the Smith chart is a transformation (also bilinear) from $\overline{z}$ or $\overline{y}$ to $\overline{\Gamma}$, this means that loci of $\overline{y}_2$, which are circles are mapped on to loci of $\overline{y}_1$, which are also circles. Since a circle is defined completely by three points, it is therefore sufficient if we use any three points on the locus of $\overline{y}_2$ and find the corresponding three points for $\overline{y}_1$. By locating the center at the intersection of the perpendicular bisectors of lines joining any two pairs of these three points, we can then draw the circle passing through these points, that is, the locus of $\bar{y}_1$. Although we have demonstrated this property by considering the discontinuity of the form shown in Fig. 7.27, it can be shown that the property holds for the case of any linear, passive, bilateral network serving as the discontinuity. We shall now consider an example.

Example 7.7

Let us consider the system shown in Fig. 7.28, in which a line is terminated by a normalized admittance $\bar{y}_R = (0.6 + j0.8)$ and a normalized susceptance of value $b = 0.8$ connected between the two conductors of the line forms the discontinuity. We wish to find the locus of the normalized admittance $\bar{y}_1$ to the left of the discontinuity as the susceptance slides along the line, and then determine the location, nearest to the load, of the susceptance for which the SWR to the left of it is minized.

To construct the locus of $\bar{y}_1$, we first locate $\bar{y}_R = (0.6 + j0.8)$ on the Smith chart at point A and draw the constant SWR circle passing through A, as shown in Fig. 7.29. This circle is the locus of $\bar{y}_2$, the normalized admittance just to the right of the discontinuity as the distance between the load and the discontinuity is varied, that is, as the susceptance slides along the line. We then choose any three points on the locus of $\bar{y}_2$ and locate the corresponding three points for $\bar{y}_1 = \bar{y}_2 + j0.8$. Here, we choose the points A, B, and C. Following the constant conductance circles through these points by the amount of normalized susceptance $+0.8$, we obtain the points D, E, and F, respectively. We then draw the circle passing through these points to obtain the locus of $\bar{y}_1$.

Proceeding further, we note that each point on the locus of $\bar{y}_1$ corresponds to a value of SWR to the left of the susceptance, obtained by following the constant SWR circle through that point to the r value at the V_{max} point. In particular, it can be seen that minimum SWR is achieved to the left of the susceptance for $\bar{y}_1$ lying at point G, which is the closest point to the center of the chart, and the minimum SWR value is 1.35. The distance from the load at which the susceptance must be connected to achieve this minimum SWR can be found by locating the $\bar{y}_2$ corresponding to the $\bar{y}_1$ at G by following the constant conductance circle through G by the amount -0.8 to reach point H. The distance from point A to point H toward the generator is the required distance. It is equal to $(0.346 - 0.125)\lambda$, or 0.221λ.

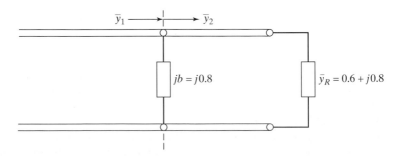

FIGURE 7.28   Transmission-line system in which a susceptance of fixed value sliding along the line forms a discontinuity.

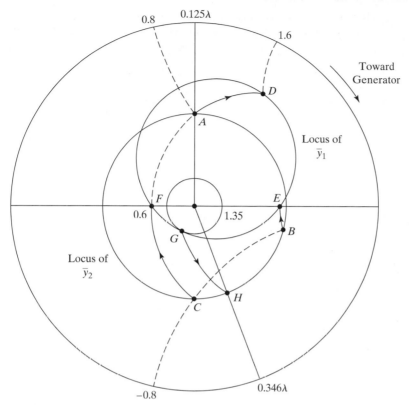

FIGURE 7.29 Construction of the locus of $\bar{y}_1$ for the system of Fig. 7.28 as the susceptance b slides along the line and determination of the minimum SWR that can be achieved to the left of the susceptance and the location of the susceptance to achieve the minimum SWR.

Together with the basic procedures discussed in the previous section, the methods that we have discussed in this section can be extended to solve many other problems using the Smith chart. We include some of these in the problems.

K7.5. Single stub matching; Double-stub matching; Auxiliary circle; Forbidden region of $\bar{y}_1'$ for possible match; Transformation of $\bar{\Gamma}$ across a discontinuity.

D7.12. A line of characteristic impedance 100 Ω is terminated by a load of impedance $(50 + j65)\ \Omega$. Find the following using the Smith chart: **(a)** the SWR on the line; **(b)** the minimum SWR that can be achieved on the line by connecting a stub in parallel with the line at the load; and **(c)** the minimum SWR that can be achieved on the line by connecting a stub in series with the line at the load. *Ans.* **(a)** 3.0; **(b)** 1.33; **(c)** 2.0.

D7.13. A line of characteristic impedance 50 Ω is terminated by a load of impedance $(100 + j100)\ \Omega$. Find the following by using the Smith chart: **(a)** the minimum distance at which a reactance of value 50 Ω must be connected in parallel with the line to minimize the SWR to the left of the reactance and the minimum

SWR achieved; **(b)** the minimum length of a line section of characteristic impedance 100 Ω between the main line and the load required to minimize the SWR on the main line and the minimum SWR achieved; and **(c)** the characteristic impedance of a λ/4 section of line inserted between the main line and the load to minimize the SWR on the main line and the minimum SWR achieved.

Ans.   **(a)** 0.204λ, 1.63;   **(b)** 0.338λ, 1.30;   **(c)** 83.7 Ω, 2.42.

7.6   THE LOSSY LINE

Distributed equivalent circuit

Thus far, we have been concerned with lossless lines. We learned in Section 6.3 that the distributed equivalent circuit for a lossless line consists of series inductors and shunt capacitors, representing energy storage in magnetic and electric fields, respectively. A lossy line is characterized by imperfect but good conductors and imperfect dielectric giving rise to power dissipation, thereby modifying the distributed equivalent circuit. The power dissipation in the conductors is taken into account by a resistance in series with the inductor, whereas the power dissipation in the dielectric is taken into account by a conductance in parallel with the capacitor. In addition, the magnetic field inside the conductors is taken into account by an additional inductance in the series branch. Thus, the distributed equivalent circuit for the lossy line is as shown in Fig. 7.30, where $\mathcal{L}$ includes the additional inductance just mentioned. Note that the notation $\mathcal{R}$ for the series resistance and $\mathcal{G}$ for the shunt conductance is not to be confused to mean that $\mathcal{G}$ is the reciprocal of $\mathcal{R}$.

Transmission-line equations and solution

To discuss wave propagation on a lossy line, we first obtain the transmission-line equations by applying Kirchhoff's voltage and current laws to the circuit of Fig. 7.30. Thus, we have

$$V(z + \Delta z, t) - V(z, t) = -\mathcal{R}\,\Delta z\,I(z, t) - \mathcal{L}\,\Delta z\,\frac{\partial I(z, t)}{\partial t} \tag{7.62a}$$

$$I(z + \Delta z, t) - I(z, t) = -\mathcal{G}\,\Delta z\,V(z, t) - \mathcal{C}\,\Delta z\,\frac{\partial V(z, t)}{\partial t} \tag{7.62b}$$

FIGURE 7.30   Distributed equivalent circuit for a lossy transmission line.

Dividing both sides of (7.62a) and (7.62b) by Δz and letting $\Delta z \to 0$, we obtain the transmission-line equations

$$\frac{\partial V(z, t)}{\partial z} = -\mathcal{R}I(z, t) - \mathcal{L}\frac{\partial I(z, t)}{\partial t} \qquad (7.63a)$$

$$\frac{\partial I(z, t)}{\partial z} = -\mathcal{G}V(z, t) - \mathcal{C}\frac{\partial V(z, t)}{\partial t} \qquad (7.63b)$$

The corresponding equations in terms of phasor voltage and current are

$$\frac{\partial \overline{V}(z)}{\partial z} = -\mathcal{R}\overline{I}(z) - j\omega\mathcal{L}\overline{I}(z)$$

$$\frac{\partial \overline{I}(z)}{\partial z} = -\mathcal{G}\overline{V}(z) - j\omega\mathcal{C}\overline{V}(z)$$

or

$$\frac{\partial \overline{V}}{\partial z} = -(\mathcal{R} + j\omega\mathcal{L})\overline{I} \qquad (7.64a)$$

$$\frac{\partial \overline{I}}{\partial z} = -(\mathcal{G} + j\omega\mathcal{C})\overline{V} \qquad (7.64b)$$

where $\overline{V}$ and $\overline{I}$ are understood to be functions of z.

Combining the two transmission-line equations (7.64a) and (7.64b) by eliminating $\overline{I}$, we obtain the wave equation

$$\frac{\partial^2 \overline{V}}{\partial z^2} = -(\mathcal{R} + J\omega\mathcal{L})\frac{\partial \overline{I}}{\partial z}$$

$$= (\mathcal{R} + j\omega\mathcal{L})(\mathcal{G} + j\omega\mathcal{C})V$$

or

$$\frac{\partial^2 \overline{V}}{\partial z^2} = \overline{\gamma}^2\overline{V} \qquad (7.65)$$

where

$$\overline{\gamma} = \alpha + j\beta$$
$$= \sqrt{(\mathcal{R} + j\omega\mathcal{L})(\mathcal{G} + j\omega\mathcal{C})} \qquad (7.66)$$

The solution for $\overline{V}(z)$ is given by

$$\overline{V}(z) = \overline{A}e^{-\overline{\gamma}z} + \overline{B}e^{\overline{\gamma}z} \qquad (7.67)$$

where $\overline{A} = Ae^{j\theta}$ and $\overline{B} = Be^{j\phi}$ are arbitrary constants. It then follows that

$$V(z, t) = \text{Re}[Ae^{j\theta}e^{-\alpha z}e^{-j\beta z}e^{j\omega t} + Be^{j\phi}e^{\alpha z}e^{j\beta z}e^{j\omega t}]$$
$$= Ae^{-\alpha z}\cos(\omega t - \beta z + \theta) + Be^{\alpha z}\cos(\omega t + \beta z + \phi)$$

Noting that the first and second terms on the right side correspond to waves propagating in the $+z$-and $-z$-directions, respectively, we write (7.67) as

$$\boxed{\overline{V}(z) = \overline{V}^+e^{-\overline{\gamma}z} + \overline{V}^-e^{\overline{\gamma}z}} \qquad (7.68)$$

where the superscripts $+$ and $-$ denote $(+)$ and $(-)$ waves, respectively. The quantity β, which is the imaginary part of $\overline{\gamma}$ is, of course, the phase constant, that is, the rate of change of phase with z for a fixed time, for either wave. The quantity α, which is the real part of $\overline{\gamma}$, is the attenuation constant, denoting that the waves get attenuated by the factor e^{α} per unit distance as they propagate in their respective directions. Thus, the quantity $\overline{\gamma}(= \alpha + j\beta)$ is the propagation constant associated with the wave. We recall that the units of α are nepers per meter. Proceding further, we obtain the corresponding solution for the phasor line current by substituting (7.68) into one of the transmission-line equations. Thus, using (7.64a), we obtain

$$\overline{I}(z) = -\frac{1}{\mathcal{R} + j\omega\mathcal{L}}\frac{\partial\overline{V}}{\partial z}$$
$$= -\frac{1}{\mathcal{R} + j\omega\mathcal{L}}[-\overline{\gamma}\,\overline{V}^+e^{-\overline{\gamma}z} + \overline{\gamma}\overline{V}^-e^{\overline{\gamma}z}]$$

or

$$\boxed{\overline{I}(z) = \frac{1}{\overline{Z}_0}(\overline{V}^+e^{-\overline{\gamma}z} - \overline{V}^-e^{\overline{\gamma}z})} \qquad (7.69)$$

where

$$\boxed{\overline{Z}_0 = \sqrt{\frac{\mathcal{R} + j\omega\mathcal{L}}{\mathcal{G} + j\omega\mathcal{C}}}} \qquad (7.70)$$

is the characteristic impedance of the line, which is now complex.

Low-loss line   Equations (7.68) and (7.69) are the general solutions for the phasor line voltage and current, respectively, with the associated propagation constant and characteristic impedance given by (7.66) and (7.70), respectively. Although it is possible to obtain explicit expressions for α and β, as well as for the real and imaginary parts of $\overline{Z}_0$ in terms of $\omega, \mathcal{R}, \mathcal{L}, \mathcal{G}$ and $\mathcal{C}$, such expressions are often not meaningful since $\mathcal{R}, \mathcal{L}, \mathcal{G}$, and $\mathcal{C}$ are themselves functions of frequency. Hence, in practice, these quantities are obtained from experimental determination of characteristic impedance and propagation constant. However, for the special case of the low-loss line, that is, for $\omega\mathcal{L} \gg \mathcal{R}$ and $\omega\mathcal{C} \gg \mathcal{G}$, we have

$$\bar{\gamma} = \sqrt{j\omega\mathcal{L}\left(1 + \frac{\mathcal{R}}{j\omega\mathcal{L}}\right)j\omega\mathcal{C}\left(1 + \frac{\mathcal{G}}{j\omega\mathcal{C}}\right)}$$

$$\approx j\omega\sqrt{\mathcal{L}\mathcal{C}}\sqrt{1 + \frac{\mathcal{R}}{j\omega\mathcal{L}} + \frac{\mathcal{G}}{j\omega\mathcal{C}}}$$

$$\approx j\omega\sqrt{\mathcal{L}\mathcal{C}}\left[1 + \frac{1}{2}\left(\frac{\mathcal{R}}{j\omega\mathcal{L}} + \frac{\mathcal{G}}{j\omega\mathcal{C}}\right)\right]$$

$$\approx \frac{1}{2}\left(\mathcal{R}\sqrt{\frac{\mathcal{C}}{\mathcal{L}}} + \mathcal{G}\sqrt{\frac{\mathcal{L}}{\mathcal{C}}}\right) + j\omega\sqrt{\mathcal{L}\mathcal{C}}$$

so that

$$\alpha \approx \frac{1}{2}\left(\mathcal{R}\sqrt{\frac{\mathcal{C}}{\mathcal{L}}} + \mathcal{G}\sqrt{\frac{\mathcal{L}}{\mathcal{C}}}\right) \tag{7.71a}$$

$$\beta \approx \omega\sqrt{\mathcal{L}\mathcal{C}} \tag{7.71b}$$

$$v_p = \frac{\omega}{\beta} \approx \frac{1}{\sqrt{\mathcal{L}\mathcal{C}}} \tag{7.71c}$$

Similarly,

$$\bar{Z}_0 = \sqrt{\frac{j\omega\mathcal{L}(1 + \mathcal{R}/j\omega\mathcal{L})}{j\omega\mathcal{C}(1 + \mathcal{G}/j\omega\mathcal{C})}}$$

$$\approx \sqrt{\frac{\mathcal{L}}{\mathcal{C}}}\sqrt{\left(1 + \frac{\mathcal{R}}{j\omega\mathcal{L}}\right)\left(1 - \frac{\mathcal{G}}{j\omega\mathcal{C}}\right)}$$

$$\approx \sqrt{\frac{\mathcal{L}}{\mathcal{C}}}\sqrt{\left(1 + \frac{\mathcal{R}}{j\omega\mathcal{L}} - \frac{\mathcal{G}}{j\omega\mathcal{C}}\right)} \tag{7.71d}$$

$$\approx \sqrt{\frac{\mathcal{L}}{\mathcal{C}}}\left[1 + \frac{1}{2}\left(\frac{\mathcal{R}}{j\omega\mathcal{L}} - \frac{\mathcal{G}}{j\omega\mathcal{C}}\right)\right]$$

$$\approx \sqrt{\frac{\mathcal{L}}{\mathcal{C}}}$$

Thus, for the low-loss line, the expressions for β and $\bar{Z}_0$ are essentially the same as those for a lossless line. Note that the low-loss conditions $\omega\mathcal{L} \gg \mathcal{R}$ and $\omega\mathcal{C} \gg \mathcal{G}$ are valid for very high frequencies or for very small values of $\mathcal{R}$ and $\mathcal{G}$ at lower frequencies.

As already pointed out, for the general case, it is more convenient to determine experimentally the values of $\bar{Z}_0$ and $\bar{\gamma}$ than it is to compute them analytically. The experimental technique is based on the measurements of the input impedance of the line for two values of load impedance. To obtain the expression for the input impedance, we first write the general solutions for the phasor line voltage and current given by (7.68) and (7.69), respectively, in terms of the

Experimental determination of $\bar{Z}_0$ and $\bar{\gamma}$

distance variable d, measured from the load toward the generator, as opposed to z, measured from the generator toward the load. Thus, we have

$$\overline{V}(d) = \overline{V}^+ e^{\overline{\gamma}d} + \overline{V}^- e^{-\overline{\gamma}d} \tag{7.72a}$$

$$\overline{I}(d) = \frac{1}{\overline{Z}_0}(\overline{V}^+ e^{\overline{\gamma}d} - \overline{V}^- e^{-\overline{\gamma}d}) \tag{7.72b}$$

or

$$\overline{V}(d) = \overline{V}^+ e^{\overline{\gamma}d}[1 + \overline{\Gamma}(d)] \tag{7.73a}$$

$$\overline{I}(d) = \frac{\overline{V}^+}{\overline{Z}_0} e^{\overline{\gamma}d}[1 - \overline{\Gamma}(d)] \tag{7.73b}$$

where

$$\overline{\Gamma}(d) = \frac{\overline{V}^-(d)}{\overline{V}^+(d)} = \frac{\overline{V}^- e^{-\overline{\gamma}d}}{\overline{V}^+ e^{\overline{\gamma}d}}$$
$$= \overline{\Gamma}_R e^{-2\overline{\gamma}d} = \overline{\Gamma}_R e^{-2\alpha d} e^{-j2\beta d} \tag{7.74}$$

is the voltage reflection coefficient at any value of d, and $\overline{\Gamma}_R$ is the voltage reflection coefficient at the load.

The line impedance is given by

$$\overline{Z}(d) = \frac{\overline{V}(d)}{\overline{I}(d)} = \overline{Z}_0 \frac{1 + \overline{\Gamma}(d)}{1 - \overline{\Gamma}(d)}$$
$$= \overline{Z}_0 \frac{1 + \overline{\Gamma}_R e^{-2\overline{\gamma}d}}{1 - \overline{\Gamma}_R e^{-2\overline{\gamma}d}} \tag{7.75}$$

The input impedance of a line of length l terminated by a load impedance $\overline{Z}_R$, as shown in Fig. 7.31, is then given in terms of $\overline{Z}_R$ by

$$\overline{Z}_{in} = \overline{Z}(l) = \overline{Z}_0 \frac{1 + \overline{\Gamma}_R e^{-2\overline{\gamma}l}}{1 - \overline{\Gamma}_R e^{-2\overline{\gamma}l}}$$
$$= \overline{Z}_0 \frac{1 + [(\overline{Z}_R - \overline{Z}_0)/(\overline{Z}_R + \overline{Z}_0)]e^{-2\overline{\gamma}l}}{1 - [(\overline{Z}_R - \overline{Z}_0)/(\overline{Z}_R + \overline{Z}_0)]e^{-2\overline{\gamma}l}}$$
$$= \overline{Z}_0 \frac{(\overline{Z}_R + \overline{Z}_0) + (\overline{Z}_R - \overline{Z}_0)e^{-2\overline{\gamma}l}}{(\overline{Z}_R + \overline{Z}_0) - (\overline{Z}_R - \overline{Z}_0)e^{-2\overline{\gamma}l}}$$
$$= \overline{Z}_0 \frac{\overline{Z}_R \cosh \overline{\gamma}l + \overline{Z}_0 \sinh \overline{\gamma}l}{\overline{Z}_R \sinh \overline{\gamma}l + \overline{Z}_0 \cosh \overline{\gamma}l}$$

or

$$\overline{Z}_{in} = \overline{Z}_0 \frac{\overline{Z}_R + \overline{Z}_0 \tanh \overline{\gamma}l}{\overline{Z}_R \tanh \overline{\gamma}l + \overline{Z}_0} \tag{7.76}$$

FIGURE 7.31 Lossy line of length l terminated by $\overline{Z}_R$.

Let us now consider two values of $\overline{Z}_R$; in particular, $\overline{Z}_R = 0$ and $\overline{Z}_R = \infty$, corresponding to a short circuit and an open circuit, respectively. Then denoting the corresponding input impedances to be $\overline{Z}_{in}^s$ and $\overline{Z}_{in}^o$, respectively, we have from (7.76),

$$\overline{Z}_{in}^s = \overline{Z}_0 \tanh \overline{\gamma} l \qquad (7.77a)$$

$$\overline{Z}_{in}^o = \overline{Z}_0 \coth \overline{\gamma} l \qquad (7.77b)$$

from which we obtain

$$\overline{Z}_0 = \sqrt{\overline{Z}_{in}^s \overline{Z}_{in}^o} \qquad (7.78)$$

and

$$\tanh \overline{\gamma} l = \sqrt{\frac{\overline{Z}_{in}^s}{\overline{Z}_{in}^o}} \qquad (7.79)$$

To illustrate the computation of $\overline{Z}_0$ and $\overline{\gamma}$ by means of a numerical example, let us assume that at a certain frequency, measurements indicated

$$\overline{Z}_{in}^s = (30 - j40) \ \Omega$$
$$\overline{Z}_{in}^o = (30 + j40) \ \Omega$$

Then from (7.78),

$$\overline{Z}_0 = \sqrt{(30 - j40)(30 + j40)} = 50 \ \Omega$$

From (7.79),

$$\tanh \overline{\gamma} l = \sqrt{\frac{30 - j40}{30 + j40}} = \sqrt{\frac{50 \underline{/-53.13°}}{50 \underline{/53.13°}}}$$
$$= 1 \underline{/-53.13°} = 0.6 - j0.8$$
$$\overline{\gamma} l = \tanh^{-1}(0.6 - j0.8)$$

Using the identity

$$\tanh^{-1} x = \frac{1}{2} \ln \frac{1 + x}{1 - x}$$

we then have

$$\overline{\gamma}l = \frac{1}{2}\ln\frac{1.6 - j0.8}{0.4 + j0.8} = \frac{1}{2}\ln\frac{1.789\underline{/-26.565°}}{0.894\underline{/63.435°}}$$

$$= \frac{1}{2}\ln 2\underline{/-90°} = \frac{1}{2}\ln\left[2e^{j(2n\pi - \pi/2)}\right]$$

$$= \frac{1}{2}\left[\ln 2 + j(2n\pi - \pi/2)\right]$$

$$= 0.3466 + j(n\pi - \pi/4) \quad n = 0, 1, 2, \ldots$$

Thus,

$$\alpha l = 0.3466$$

$$\alpha = 0.3466/l$$

whereas

$$\beta l = n\pi - \pi/4 \quad n = 1, 2, \ldots$$

where $n = 0$ is ruled out since it gives negative value for β. Note that β can only be determined to within $n\pi$. However, if the approximate value of β is known, then the correct value of n and hence of β can be determined.

In practice, since a perfect open-circuited termination can often be difficult to achieve, it may be desirable to consider the second value of $\overline{Z}_R$ to be arbitrary instead of being equal to ∞. Denoting the corresponding input impedance to be $\overline{Z}_{in}$, we then have from (7.76) and (7.77a)

$$\overline{Z}_{in} = \overline{Z}_0^2\frac{\overline{Z}_R + \overline{Z}_{in}^s}{\overline{Z}_R\,\overline{Z}_{in}^s + \overline{Z}_0^2} \tag{7.80}$$

and hence

$$\overline{Z}_0 = \sqrt{\frac{\overline{Z}_R\,\overline{Z}_{in}^s\,\overline{Z}_{in}}{\overline{Z}_R + \overline{Z}_{in}^s - \overline{Z}_{in}}} \tag{7.81}$$

Knowing the value of $\overline{Z}_0$ from (7.81), we can then compute the value of $\overline{\gamma}$ by using (7.77a).

Power flow   We shall conclude this section with a discussion of power flow down the line. From (7.73a) and (7.73b), the time-average power flow down the line is given by

$$\langle P \rangle = \frac{1}{2}\text{Re}[\overline{V}(d)\overline{I}^*(d)]$$

$$= \frac{1}{2}\text{Re}\left\{\overline{V}^+e^{\overline{\gamma}d}[1 + \overline{\Gamma}(d)]\frac{(\overline{V}^+)^*}{\overline{Z}_0^*}e^{\overline{\gamma}^*d}[1 - \overline{\Gamma}^*(d)]\right\}$$

$$= \frac{1}{2}\text{Re}\left\{\frac{|\overline{V}^+|^2}{\overline{Z}_0^*}e^{2\alpha d}[1 - |\overline{\Gamma}(d)|^2 + \overline{\Gamma}(d) - \overline{\Gamma}^*(d)]\right\}$$

$$= \frac{1}{2}\text{Re}\{|\overline{V}^+|^2\overline{Y}_0^*e^{2\alpha d}[1 - |\overline{\Gamma}(d)|^2 + j2\,\text{Im}\,\overline{\Gamma}(d)]\}$$

or

$$\langle P \rangle = \tfrac{1}{2}|\overline{V}^+|^2 \, e^{2\alpha d} \{G_0[1 - |\overline{\Gamma}(d)|^2] + 2B_0 \, \text{Im} \, \overline{\Gamma}(d)\} \qquad (7.82)$$

where

$$\overline{Y}_0 = \frac{1}{\overline{Z}_0} = G_0 + jB_0$$

is the characteristic admittance of the line. For a given source voltage and impedance, we can compute the value of $|\overline{V}^+|$ from line impedance and power flow considerations at the input end of the line and use that value for further computations. We shall illustrate this by means of an example.

Example 7.8

Let us consider the low-loss line system shown in Fig. 7.32, and compute the time-average power delivered to the input of the line, the time-average power delivered to the load, and the time-average power dissipated in the line.

We proceed with the solution in a step-by-step manner as follows:

a. The reflection coefficient at the load end is given by

$$\overline{\Gamma}_R = \frac{\overline{Z}_R - \overline{Z}_0}{\overline{Z}_R + \overline{Z}_0} = \frac{150 - 50}{150 + 50} = 0.5$$

b. Noting that α is specified in nepers per wavelength, we obtain the reflection coefficient at the input end $d = l$ as

$$\overline{\Gamma}(l) = \overline{\Gamma}_R e^{-2\gamma l} = \overline{\Gamma}_R e^{-2\alpha l} \, e^{-j2\beta l}$$

$$= 0.5 e^{-0.204} e^{-j40.8\pi}$$

$$= 0.4077 \underline{/-144°}$$

FIGURE 7.32 Lossy transmission-line system for illustrating the computation of power flow at the two ends of the line and the power dissipated in the line.

c. The input impedance of the line is given by

$$\overline{Z}_{in} = \overline{Z}(l) = Z_0 \frac{1 + \overline{\Gamma}(l)}{1 - \overline{\Gamma}(l)}$$

$$= 50 \frac{1 + 0.4077\underline{/-144°}}{1 - 0.4077\underline{/-144°}} = 50 \frac{1 + (-0.33 - j0.24)}{1 - (-0.33 - j0.24)}$$

$$= 50 \frac{0.7117\underline{/-19.708°}}{1.3515\underline{/10.229°}} = 26.33\underline{/-29.937°}$$

$$= (22.817 - j13.140) \ \Omega$$

d. The current $\overline{I}_g = \overline{I}(l)$ drawn from the voltage generator can be obtained as

$$\overline{I}(l) = \frac{\overline{V}_g}{\overline{Z}_g + \overline{Z}_{in}} = \frac{100\underline{/0°}}{(30 - j40) + (22.817 - j13.140)}$$

$$= \frac{100\underline{/0°}}{52.817 - j53.140} = \frac{100\underline{/0°}}{74.923\underline{/-45.175°}}$$

$$= 1.3347\underline{/45.175°} \ A$$

e. The voltage at the input end of the line is given by

$$\overline{V}(l) = \overline{Z}_{in}\overline{I}(l)$$

$$= 26.33\underline{/-29.937°} \times 1.3347\underline{/45.175°}$$

$$= 35.143\underline{/15.238°} \ V$$

f. The time-average power flow at the input end of the line is given by

$$\langle P(l)\rangle = \tfrac{1}{2} \operatorname{Re}[\overline{V}(l)\overline{I}*(l)]$$

$$= \tfrac{1}{2} \operatorname{Re}[35.143\underline{/15.238°} \times 1.3347\underline{/-45.175°}]$$

$$= \tfrac{1}{2} \times 35.143 \times 1.3347 \times \cos 29.937°$$

$$= 20.32 \ W$$

g. Noting that $B_0 = 0$, we then obtain the value of $|\overline{V}^+|$ by applying (7.82) to $d = l$. Thus,

$$|\overline{V}^+| = \sqrt{\frac{2\langle P(l)\rangle e^{-2\alpha l}}{G_0[1 - |\overline{\Gamma}(l)|^2]}}$$

$$= \sqrt{\frac{2 \times 20.32 \times e^{-0.204}}{0.02(1 - 0.4077^2)}}$$

$$= 44.58 \ V$$

h. The time-average power delivered to the load is then given by

$$\langle P(0)\rangle = \tfrac{1}{2}|\overline{V}^+|^2 G_0(1 - |\overline{\Gamma}_R|^2)$$

$$= \tfrac{1}{2} \times 44.58^2 \times 0.02(1 - 0.25)$$

$$= 14.91 \ W$$

i. Finally, the time-average power dissipated in the line is

$$\langle P_d\rangle = \langle P(l)\rangle - \langle P(0)\rangle$$

$$= 20.32 - 14.91$$

$$= 5.41 \ W$$

K7.6. Distributed equivalent circuit; Transmission-line equations; Complex propagation constant; Complex characteristic impedance; Input impedance; $\overline{Z}_0$ and $\overline{\gamma}$ from input impedance considerations; Power flow; Power dissipation.

D7.14. For a lossy of length $l = 16.3\lambda$ and characterized by $Z_0 = 60\ \Omega$ and $\alpha = 0.02$ Np/λ, find the input impedance for each of the following values of $\overline{Z}_R$:
(a) $\overline{Z}_R = 0$; (b) $\overline{Z}_R = \infty$; and (c) $\overline{Z}_R = (36 + j0)\ \Omega$.
Ans. (a) $(102.04 - j85.77)\ \Omega$; (b) $(20.67 + j17.38)\ \Omega$;
(c) $(73.17 - j11.39)\ \Omega$.

D7.15. A lossy line of length $l = 10\lambda$ and characterized by $\overline{Z}_0 = 100\ \Omega$ and $\alpha = 10^{-2}$ Np/λ is terminated by a load impedance $\overline{Z}_R$. If a power of 10 W is to be delivered to the load, determine how much power should be delivered to the input terminals of the line for each of the following values of $\overline{Z}_R$: (a) $\overline{Z}_R = 100\ \Omega$; (b) $\overline{Z}_R = 20\ \Omega$; and (c) $\overline{Z}_R = 300\ \Omega$.
Ans. (a) 12.214 W; (b) 15.436 W; (c) 13.556 W.

7.7 SUMMARY

In this chapter, we began our study of sinusoidal steady-state analysis of lossless transmission lines by expressing the general solutions for the phasor line voltage and line current in terms of the distance variable d measured from the load toward the source. These solutions are

$$\overline{V}(d) = \overline{V}^+ e^{j\beta d} + \overline{V}^- e^{-j\beta d}$$

$$\overline{I}(d) = \frac{1}{Z_0}(\overline{V}^+ e^{j\beta d} - \overline{V}^- e^{-j\beta d})$$

By applying these general solutions to the case of a line short-circuited at the far end and obtaining the particular solutions for that case, we discussed the standing-wave phenomenon resulting from the complete reflection of waves by the short circuit. We introduced the concept of a standing-wave pattern and discussed the phenomenon of natural oscillations. We examined the frequency behavior of the input impedance of a short-circuited line of length l, given by

$$\overline{Z}_{in} = jZ_0 \tan \beta l$$

and illustrated (1) its application in a technique for locating short circuit in a line and (2) the computation of resonant frequencies for a system formed by connecting together short-circuited line sections.

Next we considered the general case of a line terminated by an arbitrary load $\overline{Z}_R$ and introduced the concept of the generalized voltage reflection coefficient, as the ratio of the phasor reflected wave voltage at any value of d to the phasor incident wave voltage at that value of d. It is given by

$$\overline{\Gamma}(d) = \overline{\Gamma}_R e^{-j2\beta d}$$

where

$$\overline{\Gamma}_R = |\overline{\Gamma}_R| e^{j\theta} = \frac{\overline{Z}_R - Z_0}{\overline{Z}_R + Z_0}$$

is the voltage reflection coefficient at the load. We then expressed the solutions for the line voltage and line current in terms of $\overline{\Gamma}(d)$ and discussed the construction of standing-wave patterns from the solutions. We learned that together with the property that distance between successive voltage minima of the standing-wave patterns is $\lambda/2$, the quantities

$$\text{SWR} = \frac{1 + |\overline{\Gamma}_R|}{1 - |\overline{\Gamma}_R|}$$

and

$$d_{\min} = \frac{\lambda}{4\pi}(\theta + \pi)$$

constitute an important set of parameters associated with the standing waves. The SWR, which is the ratio of the maximum voltage amplitude to the minimum voltage amplitude in the standing-wave pattern, and $d_{\min}$, which is the distance of the first voltage minimum of the standing-wave pattern from the load, are easily measurable quantities. We then defined the ratio of the complex line voltage to the complex line current at a given value of d to be the line impedance $\overline{Z}(d)$, given by

$$\overline{Z}(d) = Z_0 \frac{1 + \overline{\Gamma}(d)}{1 - \overline{\Gamma}(d)}$$

and discussed its several properties as well as the computation of power flow along the line from considerations of input impedance of the line.

We then turned our attention to the topic of transmission-line matching, which consists of eliminating standing waves by connecting a matching device near the load such that the line views an effective impedance equal to its own characteristic impedance, on the generator side of the matching device. We discussed the need for matching and three techniques of matching: (1) quarter-wave transformer, (2) single stub, and (3) double stub. The quarter-wave transformer technique is based on a property of the line impedance that

$$[\overline{Z}(d)]\left[\overline{Z}\left(d + \frac{\lambda}{4}\right)\right] = Z_0^2$$

whereas the stub-matching techniques make use of the property that the input impedance of a lossless line short-circuited (or open-circuited) at the far end is purely reactive. We also discussed the departure of SWR from unity as the frequency is varied from that at which the match is achieved and illustrated a procedure for computation of the SWR versus frequency.

Next we introduced the Smith chart, a popular graphical aid in the solution of transmission-line problems. We learned that the Smith chart is based on the transformation from the $\overline{z}$-plane to the $\overline{\Gamma}$-plane in accordance with the relationship

$$\overline{\Gamma}(d) = \frac{\overline{z}(d) - 1}{\overline{z}(d) + 1}$$

where

$$\overline{z}(d) = \frac{\overline{Z}(d)}{Z_0}$$

is the normalized line impedance. We discussed the construction of the Smith chart, some basic procedures, and the solution of transmission-line matching problems. We also discussed a useful property associated with the transformation of the reflection coefficient across a discontinuity and illustrated its application by means of an example.

Finally, we extended our analysis of lossless lines briefly to lossy lines, with the discussion of (1) the distributed equivalent circuit, (2) computation of characteristic impedance and propagation constant from input impedance measurements, and (3) computation of power flow at the generator and load ends of the line and power dissipated in the line.

REVIEW QUESTIONS

Q7.1. Discuss the general solutions for the line voltage and line current in terms of the distance variable d in the sinusoidal steady state.

Q7.2. State the boundary condition at a short circuit on a line. For an open-circuited line, what is the boundary condition to be satisfied at the open circuit?

Q7.3. What is a standing wave? How do complete standing waves arise? Discuss their characteristics.

Q7.4. What is a standing-wave pattern? Discuss the voltage and current standing-wave patterns for a short-circuited line.

Q7.5. Explain the phenomenon of natural oscillations and the determination of natural frequencies of oscillation by means of an example.

Q7.6. Discuss the variation with frequency of the input reactance of a short-circuited line and its application in the determination of the location of a short circuit.

Q7.7. Outline the method of computation of resonant frequencies of a system formed by connecting together two short-circuited line sections.

Q7.8. How is the generalized voltage reflection coefficient defined? Discuss its variation along the line.

Q7.9. Discuss the sketching of standing-wave patterns for line voltage and current on a line terminated by an arbitrary load.

Q7.10. Define the standing-wave ratio (SWR). What are the standing-wave ratios for **(a)** a semi-infinitely long line; **(b)** a short-circuited line; **(c)** an open-circuited line; and **(d)** a line terminated by its characteristic impedance?

Q7.11. Discuss the slotted-line technique for performing standing-wave measurements on a line and the determination of an unknown load impedance from the standing-wave measurements.

Q7.12. How is line impedance defined? Summarize the several properties of the line impedance.

Q7.13. Outline the procedure for the determination of time-average power flow down a line from input impedance considerations.

Q7.14. Define normalized line impedance and normalized line admittance. How are they related to the voltage reflection coefficient?

Q7.15. Discuss the reasons for transmission-line matching and the principle behind matching.

Q7.16. Which property of line impedance forms the basis for the quarter-wave transformer (QWT) technique of transmission-line matching? Outline the solution for the QWT matching problem.

Q7.17. What is a stub? Outline the solution for the single-stub matching problem.

Q7.18. Outline the solution for the double-stub matching problem.

Q7.19. Discuss the bandwidth associated with a transmission-line matched system and the procedure for obtaining the SWR in the main line versus frequency.

Q7.20. What is the basis behind the construction of the Smith chart? Briefly discuss the mapping of the normalized line impedances onto the $\overline{\Gamma}$-plane.

Q7.21. Why is a circle with its center at the center of the Smith chart known as a constant SWR circle? Where on the circle is the corresponding SWR value marked?

Q7.22. Using the Smith chart, how do you find the normalized line admittance at a point on the line, given the normalized line impedance at that point?

Q7.23. Briefly describe the solution to the single-stub matching problem by using the Smith chart.

Q7.24. Briefly describe the solution to the double-stub matching problem by using the Smith chart.

Q7.25. Discuss the forbidden region of $\overline{y}'_1$ for possible match associated with the double-stub matching technique.

Q7.26. Discuss the transformation of the reflection coefficient from one side of a transmission-line discontinuity to the other side of the discontinuity and an application of the property associated with this transformation.

Q7.27. Discuss the modification of the distributed equivalent circuit for the lossless line case to the lossy line case.

Q7.28. What are the conditions under which a lossy line can be classified as a low-loss line? Compare the propagation parameters of the low-loss line with those for the lossless line.

Q7.29. Discuss the computation of $\overline{Z}_0$ and $\overline{\gamma}$ for a lossy line from a knowledge of the input impedances of the line with short-circuit and open-circuit terminations.

Q7.30. Briefly outline the procedure for the computation of time-average power flow at the input and the load ends of a lossy line and, hence, the time-average power dissipated in the line.

PROBLEMS

Section 7.1

P7.1. For a line open-circuited at the far end, as shown in Fig. 7.33, obtain the solutions for the phasor line voltage and current and sketch the voltage and current standing-wave patterns, as in Fig. 7.4.

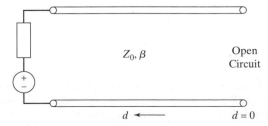

$d \longleftarrow \qquad d = 0$ **FIGURE 7.33**   For Problem P7.1.

P7.2. In the system shown in Fig. 7.34, the line is open-circuited at the far end, the source voltage is

$$V_g(t) = V_0 \cos \pi f_0 t \cos 3\pi f_0 t$$

and $l = \lambda/4$ at $f = f_0$. Find the root-mean-square (rms) values of the line voltage and line current at values of d/l equal to $0, \frac{1}{3}, \frac{1}{2}$, and 1. (*Note:* The rms value of the sum of the voltages of two harmonically related frequencies is equal to the square root of the sum of the squares of the rms values of the individual voltages.)

FIGURE 7.34   For Problem P7.2.

P7.3. In the system shown in Fig. 7.35(a), the source voltage is periodic, as shown in Fig. 7.35(b). Find the reading of the ammeter A if it reads root-mean-square values.

(a)

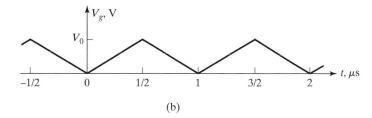

(b)

FIGURE 7.35   For Problem P7.3.

P7.4. In the system shown in Fig. 7.36, a nonmagnetic ($\mu = \mu_0$), lossless material medium is sandwiched between two parallel, perfect conductors. For uniform plane waves bouncing back and forth normal to the conductors, find the following: **(a)** the minimum value of l for which the natural frequency of oscillation is 2.5 GHz if the medium is a perfect dielectric of permittivity $2.25\varepsilon_0$, and **(b)** the

FIGURE 7.36 For Problem P7.4.

expression for the lowest natural frequency of oscillation if the medium is a plasma, which can be thought of as equivalent to a perfect dielectric of permittivity $\varepsilon_0 (1 - f_N^2/f^2)$, where f_N, known as the plasma frequency, is a constant and f is the wave frequency.

P7.5. A ring transmission line is formed as shown in Fig. 7.37(a) by connecting the ends a and a' of the conductors of a line of length l [shown in Fig. 7.37(b)] to the ends b and b', respectively, of the same conductors. Find the natural frequencies of oscillation of the system.

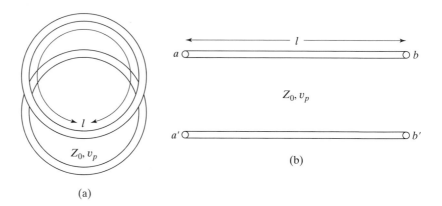

FIGURE 7.37 For Problem P7.5.

P7.6. Repeat Problem P7.5 for a twisted-ring transmission line formed by connecting the ends a and a' of the conductors [see Fig. 7.37(b)] to the ends b' and b, respectively.

P7.7. Show that for $f \ll v_p/2\pi l$ the input impedance of a short-circuited line of length l and phase velocity v_p is essentially that of a single inductor of value $\mathcal{L}l$, where $\mathcal{L}$ is the inductance per unit length of the line. Assuming that the criterion $f \ll v_p/2\pi l$ is satisfied for frequencies $f \leq 0.1 v_p/2\pi l$, compute the maximum length of an air-dielectric short-circuited line for which the input impedance is approximately that of an inductance equal to the total inductance of the line for $f = 100$ MHz.

P7.8. In the example involving the location of a short circuit in a line, solve for the distance of the short circuit from the generator by considering the standing-wave patterns for the two frequencies of interest and deducing the number of wavelengths at one of the two frequencies.

P7.9. A transmission line of characteristic impedance $Z_0 = 100 \ \Omega$, phase velocity $v_p = 2 \times 10^8$ m/s, and length $l = 20$ cm is short-circuited at one end and terminated by an inductor of value 0.1 μH at the other end. Find the three lowest resonant frequencies of the system.

P7.10. The arrangement shown in Fig. 7.38 is that of a parallel-plate resonator made up of two dielectric slabs sandwiched between perfect conductors and in which uniform plane waves bounce back and forth normal to the conductors. **(a)** Show that the resonant frequencies of the system are given by the roots of the characteristic equation

$$\tan \omega \sqrt{\mu_0 \varepsilon_1} \, t + \sqrt{\frac{\varepsilon_1}{\varepsilon_2}} \tan \omega \sqrt{\mu_0 \varepsilon_2} (l - t) = 0$$

(b) Find the five lowest resonant frequencies if $t = l/2, l = 5$ cm, $\varepsilon_1 = 4\varepsilon_0$, and $\varepsilon_2 = 16\varepsilon_0$.

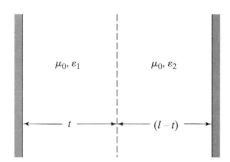

μ_0, ε_1 μ_0, ε_2

$\longleftarrow t \longrightarrow | \longleftarrow (l - t) \longrightarrow$

FIGURE 7.38   For Problem P7.10.

Section 7.2

P7.11. For a line of characteristic impedance Z_0 terminated by a purely reactive load jX, show that the SWR is equal to infinity and the value of d_{min} is $(\lambda/2\pi)[\pi - \tan^{-1}(X/Z_0)]$ for $X > 0$ and $(\lambda/2\pi) \tan^{-1}(|X|/Z_0)$ for $X < 0$.

P7.12. A slotted coaxial line of characteristic impedance 75 Ω was used to measure an unknown load impedance. First, the receiving end of the line was short-circuited. The voltage minima were found to be 20 cm apart. One of the minima was marked as the reference point. Next, the unknown impedance was connected to the receiving end of the line. The SWR was found to be 3.0 and a voltage minimum was found to be 6 cm from the reference point toward the load. Find the value of the unknown load impedance.

P7.13. In the system shown in Fig. 7.39, assume uniform plane waves of frequency f incident normally onto the interface from medium 1. **(a)** Find the SWR in

Medium 1	Medium 2	Medium 3
Free Space	Dielectric	Free Space
μ_0, ε_0	$\mu_0, 2.25\varepsilon_0$	μ_0, ε_0

$\longleftarrow l \longrightarrow$

FIGURE 7.39   For Problem P7.13.

medium 1 for $f = 10^9$ Hz if $l = 5$ cm. **(b)** Find the three lowest values of f for which complete transmission occurs if $l = 5$ cm. **(c)** Find the three lowest values of l for which complete transmission occurs for $f = 10^9$ Hz.

P7.14. For the system shown in Fig. 7.40, find the lowest value of l for which no reflection occurs for a uniform plane wave having the electric field

$$\mathbf{E} = E_0 \cos 4\pi \times 10^9 t \cos \pi \times 10^9 t \, \mathbf{a}_x$$

at $z = 0$ normally incident on the interface from medium 1.

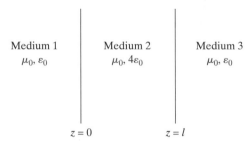

FIGURE 7.40 For Problem P7.14.

P7.15. For uniform plane waves of frequency f incident normally onto the interface from medium 1 in the system shown in Fig. 7.41, find the fraction of the incident power transmitted into medium 3 for each of the following values of f: **(a)** 3000 MHz; **(b)** 6000 MHz; and **(c)** 1500 MHz.

FIGURE 7.41 For Problem P7.15.

P7.16. The arrangement shown in Fig. 7.42 is that of a parallel-plate resonator made up of two plane perfect conductors coated with a dielectric and in which uniform

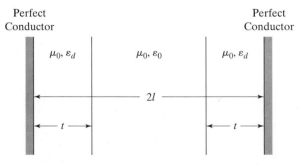

FIGURE 7.42 For Problem P7.16.

plane waves bounce back and forth normal to the plates. **(a)** Show that the characteristic equation for the resonant frequency is given by

$$\tan \omega \sqrt{\mu_0 \varepsilon_0} \, (l - t) \tan \omega \sqrt{\mu_0 \varepsilon_0} t = \sqrt{\frac{\varepsilon_d}{\varepsilon_0}}$$

(b) Find the value of the lowest resonant frequency for $t = l/2 = 5$ cm and $\varepsilon_d = 4\varepsilon_0$.

P7.17. For the system shown in Fig. 7.43, find the input impedance of the line and the time-average power delivered to the load.

FIGURE 7.43   For Problem P7.17.

P7.18. In the system in Fig. 7.44, find: **(a)** the value of the load impedance $\overline{Z}_R$ that enables maximum power transfer from the generator to the load and **(b)** the power transferred to the load for the value found in (a). (*Hint:* Apply maximum power transfer theorem at $d = l$.)

FIGURE 7.44   For Problem P7.18.

P7.19. The ring transmission line of Fig. 7.37(a) is excited by connecting a voltage source $V_g(t) = V_0 \cos \omega t$ across the conductors at some location on the line [such as aa' (or bb') in Fig. 7.37(b)]. Find the impedance viewed by the voltage source.

P7.20. In the system shown in Fig. 7.45, find the values of the reactance X and the characteristic impedance Z_{02} of line 2 for which the power delivered to the load $\overline{Z}_R$ is a maximum.

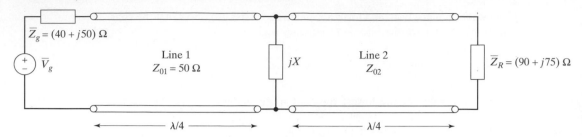

$\overline{Z}_g = (40 + j50)\ \Omega$

$\overline{V}_g$

Line 1
$Z_{01} = 50\ \Omega$

jX

Line 2
Z_{02}

$\overline{Z}_R = (90 + j75)\ \Omega$

$\lambda/4$

$\lambda/4$

FIGURE 7.45   For Problem P7.20.

Section 7.3

P7.21. In the arrangement shown in Fig. 7.46, a quarter-wave dielectric coating is employed to eliminate reflections of uniform plane waves of frequency 1000 MHz incident normally from free space onto a dielectric of permittivity $4\varepsilon_0$. Assuming that $\mu = \mu_0$, find the thickness in centimeters and the permittivity of the dielectric coating.

$\mu_0,\ \varepsilon_0$

$\mu_0,\ 4\varepsilon_0$

Dielectric
Coating
$\mu_0,\ \varepsilon$

1000 MHz

$\lambda/4$

FIGURE 7.46   For Problem P7.21.

P7.22. In the system shown in Fig. 7.47, the $\lambda/4$ section of characteristic impedance 50 Ω is used to minimize the SWR to the left of the section. Find analytically the minimum value of d_1 that minimizes the SWR and the minimum value of the SWR.

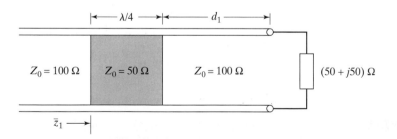

$\lambda/4$ d_1

$Z_0 = 100\ \Omega$ $Z_0 = 50\ \Omega$ $Z_0 = 100\ \Omega$ $(50 + j50)\ \Omega$

$\overline{z}_1$

FIGURE 7.47   For Problem P7.22.

P7.23. Figure 7.48 shows an arrangement, known as the alternated-line transformer, for achieving a matched interconnection between two lines of different characteristic impedances Z_{01} and Z_{02}. It consists of two sections of the same characteristic impedances as those of the lines to be matched but alternated, as shown in the

FIGURE 7.48   For Problem P7.23.

figure. The electrical lengths of the two sections are equal. Show that to achieve a match, the required electrical length of each section is

$$\frac{l}{\lambda} = \frac{1}{2\pi} \tan^{-1} \sqrt{\frac{n}{n^2 + n + 1}}$$

where $n = Z_{02}/Z_{01}$.

P7.24. We learned that in the double-stub matching technique, a solution does not exist for b_2 if $g' > 1/\sin^2 \beta d_{12}$—see (7.52) and associated discussion. Show that one means of resolving this problem is by increasing d_1 by $\lambda/4$.

P7.25. Figure 7.49 shows a hybrid arrangement of a parallel short-circuited stub and a series short-circuited stub connected at a fixed distance d_1 from the load in order to achieve a match between the line and the load. With reference to the notation shown in the figure, show that in order to achieve a match, the required input

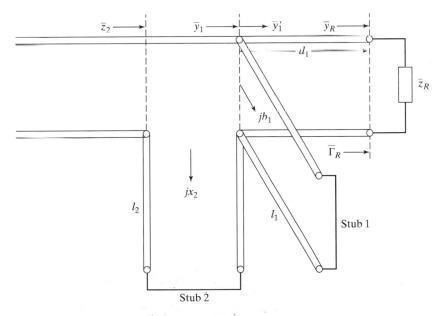

FIGURE 7.49   For Problem P7.25.

reactance of the series stub x_2 is given by

$$x_2 = \sqrt{\frac{1 - g'}{g'}}$$

and the required input susceptance of the parallel stub b_1 is given by

$$b_1 = \frac{x_2}{1 + x_2^2} - b'$$

where $(g' + jb')$ is equal to $\bar{y}_i'$. Discuss the condition for which a solution does not exist for a fixed value of d_1, and a remedy to get around the problem.

P7.26. In the arrangement shown in Fig. 7.50, a quarter-wave transformer is employed to eliminate reflections of uniform plane waves of frequency 2500 MHz incident normally from the free-space side. **(a)** Find analytically the bandwidth between frequencies on either side of 2500 MHz at which the SWR in free space is 2.0. **(b)** What is the maximum SWR in free space as the frequency is varied on either side of 2500 MHz?

Free Space
μ_0, ε_0

Dielectric
$\mu_0, 4\varepsilon_0$

Dielectric
$\mu_0, 16\varepsilon_0$

2500 MHz

← 1.5 cm →

FIGURE 7.50   For Problem P7.26.

Section 7.4

P7.27. The transformation

$$\overline{\Gamma} = \frac{\bar{z} - 1}{\bar{z} + 1}$$

which forms the basis for the construction of the Smith chart maps circles in the complex $\bar{z}$-plane onto circles in the complex $\overline{\Gamma}$-plane. For the circle in the $\bar{z}$-plane given by $(r - 2)^2 + x^2 = 1$, find the equation for the circle in the $\overline{\Gamma}$-plane. (*Hint:* Consider three points on the circle in the $\bar{z}$-plane, find the corresponding three points in the $\overline{\Gamma}$-plane, and then find the equation.)

P7.28. Using the inverse of the procedure suggested in Problem P7.27, find the equation of the circle in the $\bar{z}$-plane that maps onto the circle in the $\overline{\Gamma}$-plane given by $(\text{Re } \overline{\Gamma} - 0.25)^2 + (\text{Im } \overline{\Gamma})^2 = 0.0625$.

P7.29. For a transmission line of characteristic impedance 50 Ω, terminated by a load impedance $(100 + j50)$ Ω, find the following quantities by using the Smith chart: **(a)** reflection coefficient at the load; **(b)** SWR on the line; **(c)** the distance of the first voltage minimum of the standing-wave pattern from the load; **(d)** the line impedance at $d = 0.15\lambda$; **(e)** the line admittance at $d = 0.15\lambda$; and **(f)** the location nearest to the load at which the real part of the line admittance is equal to the line characteristic admittance.

P7.30. In the arrangement shown in Fig. 7.51, uniform plane waves of frequency 2500 MHz are incident normally from medium 1 onto the interface between medium 1 and medium 2. Using the Smith chart, find the SWR in: **(a)** medium 3; **(b)** medium 2; and **(c)** medium 1.

FIGURE 7.51   For Problem P7.30.

Section 7.5

P7.31. Standing-wave measurements on a line of characteristic impedance 60 Ω indicate an SWR of 4.0 and a voltage minimum at a distance of 0.4λ from the load. Determine the value of the load impedance by using the Smith chart.

P7.32. A transmission line of characteristic impedance 60 Ω is terminated by a certain load impedance. It is found that the SWR on the line is equal to 4.0 and that the first voltage minimum of the standing-wave pattern is located to be at 0.2λ from the load. Using the Smith chart, determine the location nearest to the load and the length of a short-circuited stub of characteristic impedance 60 Ω connected in parallel with the line required to achieve a match between the line and the load.

P7.33. Standing-wave measurements on a line of characteristic impedance 60 Ω indicate SWR on the line to be 4.0 and the location of the first voltage minimum of the standing-wave pattern to be 0.1λ from the load. Assuming that $d_1 = 0.05\lambda$ and $d_{12} = 0.375\lambda$ and using the Smith chart, find the lengths of the two short-circuited stubs of characteristic impedance 60 Ω required to achieve a match between the line and the load.

P7.34. It is proposed to match a transmission line of characteristic impedance 100 Ω to a load impedance $(20 - j100)$ Ω by using a double-stub arrangement with spacing between stubs, d_{12}, equal to $5\lambda/8$. Determine the forbidden range of values of d_1 within the first half-wavelength to achieve the match using the Smith chart.

P7.35. Solve Problem P7.22 by using the Smith chart.

P7.36. In the system shown in Fig. 7.52, two line sections, each of length λ/4 and characteristic impedance 50 Ω, are employed. By using the Smith chart, find the locations of the two λ/4 sections, that is, the values of l_1 and l_2 to achieve a match between the 100-Ω line and the load. Use the notation shown in the figure.

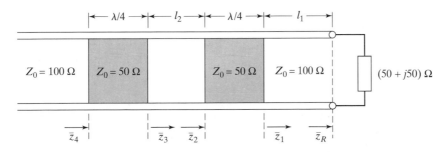

FIGURE 7.52   For Problem P7.36.

P7.37. Solve Problem P7.26 by using the Smith chart.

P7.38. Consider $Z_{01} = 50$ Ω and $Z_{02} = 100$ Ω for the alternated transformer arrangement of Problem P7.23. By using the Smith chart, obtain the minimum value of l/λ for achieving the match and show that it agrees with the solution given in Problem P7.23.

P7.39 For the hybrid parallel-series stub matching arrangement of Problem P7.25, illustrate the solution by using the Smith chart by considering a line of characteristic impedance 50 Ω terminated by a load of $(40 + j40)$ Ω. Assume $d_1 = \lambda/8$ and the characteristic impedance of the stubs to be 50 Ω.

P7.40. In the system shown in Fig. 7.53, it is desired to transfer the maximum possible power from the source to the load. By using the Smith chart, find, if possible, the location and the length of a short-circuited stub of characteristic impedance 100 Ω connected in parallel with the line that will enable this to be achieved.

FIGURE 7.53   For Problem P7.40.

Section 7.6

P7.41. For a lossy line having the parameters $\mathcal{R} = 0.03$ Ω/m, $\mathcal{L} = 1.0$ μH/m, $\mathcal{G} = 3 \times 10^{-9}$ S/m, and $\mathcal{C} = 50$ pF/m, compute the values of $\bar{Z}_0$ and $\bar{\gamma}$ for $f = 10$ kHz.

P7.42. Assume that for a lossy line $\mathcal{R}/\mathcal{L} = \mathcal{G}/\mathcal{C}$ (Heaviside's distortionless line). Show that the expressions for v_p and $\bar{Z}_0$ are then the same as those for the lossless line and that α is equal to $\sqrt{\mathcal{R}\mathcal{G}}$.

P7.43. The input impedance of a lossy line of length 50 m is measured at a frequency of 100 MHz for two cases: with the output short-circuited, it is $(10 + j49)$ Ω, and with the output open circuited, it is $(10 - j49)$ Ω. Find: **(a)** the characteristic impedance of the line; **(b)** the attenuation constant of the line; and **(c)** the phase velocity in the line, assuming its approximate value to be 1.75×10^8 m/s.

P7.44. For the lossy transmission-line system shown in Fig. 7.54, find: **(a)** the time-average power flow at the input end of the line; **(b)** the time-average power delivered to the load; and **(c)** the time-average power dissipated in the line.

FIGURE 7.54   For Problem P7.44.

P7.45. In the arrangement shown in Fig. 7.55, uniform plane waves are incident normally onto a coating of good conductor material of conductivity σ and thickness l on a perfect dielectric slab of thickness $\lambda/4$ and backed by a perfect conductor. Show that no reflection occurs from the coating if $|\bar{\gamma}_c l| \ll 1$ and $\sigma = 1/\eta_0 l$, where $\bar{\gamma}_C$ is the propagation constant in the good conductor material.

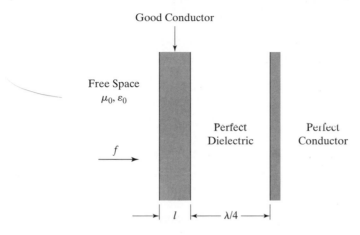

FIGURE 7.55 For Problem P7.45.

REVIEW PROBLEMS

R7.1. In the system shown in Fig. 7.56(a), the source current is periodic, as shown in Fig. 7.56(b). Find the rms value of the current through the short-circuit.

FIGURE 7.56 For Problem R7.1.

R7.2. **(a)** For the system shown in Fig. 7.57, find the value of L for which the system is resonant at $f = 10^9$ Hz. **(b)** What is the next resonant frequency of the system, greater than 10^9 Hz? **(c)** If L is increased by a small amount $\delta\%$, by what percent should l be changed such that the system remains resonant at $f = 10^9$ Hz? **(d)** Repeat part (c) for the resonant frequency found in part (b).

$$Z_0 = 50 \ \Omega$$
$$v_p = 3 \times 10^8 \text{ m/s}$$

$$l = 10 \text{ cm}$$

FIGURE 7.57   For Problem R7.2.

R7.3. In the arrangement shown in Fig. 7.58, a uniform plane wave having the electric field

$$\mathbf{E}_i = E_0 \cos (45\pi \times 10^8 t - 15\pi z) \cos (15\pi \times 10^8 t - 5\pi z) \, \mathbf{a}_x \text{ V/m}$$

is incident on the interface at $z = 0$. Find the fraction of the incident time-average power reflected back into medium 1 and the fraction transmitted into medium 3.

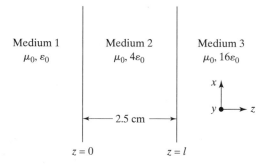

Medium 1   Medium 2   Medium 3
μ_0, ε_0 $\mu_0, 4\varepsilon_0$ $\mu_0, 16\varepsilon_0$

2.5 cm

$z = 0$ $z = l$ FIGURE 7.58   For Problem R7.3.

R7.4. In the system shown in Fig. 7.59, find **(a)** the time-average power delivered to the resistor R_2 and **(b)** the time-average power delivered to the resistor R_3.

R7.5. In the system shown in Fig. 7.60, find the value of the reactance X and the minimum value of the line length l for which the time-average power delivered to the resistor R_L is a maximum. What are the values of this power and the SWR on the line?

R7.6. Standing-wave measurements on a line of characteristic impedance 100 Ω indicate an SWR of 2.80 and a voltage minimum at a distance of 0.1λ from the load. By using the Smith chart, determine the value of the load impedance. It is desired to minimize the SWR on the line by connecting a line section at the load. By using the Smith chart find the minimum required length of the line section and the minimum achievable SWR for each of the following line sections: **(a)** a short-circuited stub of characteristic impedance 100 Ω connected in parallel with the load; **(b)** a short-circuited stub of characteristic impedance100 Ω connected in series with the load; and **(c)** a line section of characteristic impedance 50 Ω inserted between the main line and the load.

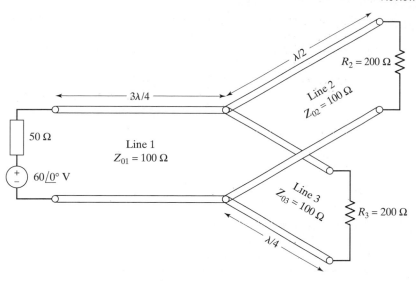

FIGURE 7.59   For Problem R7.4.

FIGURE 7.60   For Problem R7.5.

R7.7. It is desired to achieve a match between a line of characteristic impedance 100 Ω to a load of $(120 - j160)$ Ω by employing the hybrid parallel-series stub matching arrangement of Problem P7.25. Determine the forbidden range of values of d within the first half-wavelength from the load.

R7.8. A lossy line of length $l = 10\lambda$, and characterized by $\overline{Z}_0 = 100$ Ω and $\alpha = 10^{-2}$ Np/λ is terminated by a load impedance $\overline{Z}_R$. If a power of 10 watts is to be delivered to the load, determine how much power should be delivered to the input terminals of the line for each of the following cases: **(a)** $\overline{Z}_R = 100$ Ω, matched termination, and **(b)** $\overline{Z}_R = 20$ Ω.

Topics in Electric- and Magnetic-Field Systems

In Chapter 6, we introduced the distributed-circuit concept in which the circuit parameters inductance and capacitance are distributed uniformly and overlappingly along a transmission line. The phenomenon on a transmission line is one of wave propagation resulting from the complete solution of Maxwell's equations, as governed by the interdependence of time-varying electric and magnetic fields. Thus, for the time-varying case, a given physical structure is strictly speaking a transmission line. However, as already discussed briefly in Section 4.1, by neglecting certain terms in one or both of Maxwell's curl equations, a physical arrangement can be represented by a lumped-circuit equivalent for low frequencies. These approximations are known as *quasistatic approximations*. In this chapter, we introduce this topic.

We begin the chapter by illustrating the determination of the low-frequency terminal behavior of a physical structure via a quasistatic extension of the static field existing in the structure when the frequency of the source driving the structure is zero. The quasistatic extension consists of starting with a time-varying field having the same spatial characteristics as that of the static field and obtaining the field solutions containing terms up to and including the first power in ω, the radian frequency. We shall learn that the quasistatic approximation leads to electric- and magnetic-field systems, that is, systems in which either the electric field or the magnetic field is predominant. The concepts of electric- and magnetic-field systems are important in the study of electromechanics. We shall also consider magnetic circuits, an important class of magnetic-field systems, and illustrate the phenomena of electromechanical energy conversion.

8.1 QUASISTATIC FIELD ANALYSIS

To introduce the quasistatic field approach, we consider the case of an inductor, *Quasistatic* as represented by the structure shown in Fig. 8.1(a), in which an arrangement of *field analysis* two parallel-plane conductors joined at one end by another conducting sheet is *for an* excited by a current source at the other end. We neglect fringing of the fields by *inductor* assuming that the spacing d between the plates is very small compared to the dimensions of the plates or that the structure is part of a structure of much larger extent in the y- and z-directions. For a constant-current source of value I_0 driving the structure at the end $z = -l$, as shown in the figure, such that the surface current densities on the two plates are given by

$$\mathbf{J}_S = \begin{cases} \dfrac{I_0}{w}\mathbf{a}_z & \text{for} \quad x = 0 \\[2ex] -\dfrac{I_0}{w}\mathbf{a}_z & \text{for} \quad x = d \end{cases} \tag{8.1}$$

the medium between the plates is characterized by a uniform y-directed magnetic field, as shown by the cross-sectional view in Fig. 8.1(b). From the boundary condition for the tangential magnetic-field intensity at the surface of a perfect

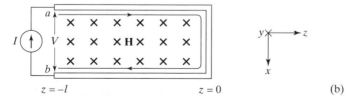

FIGURE 8.1   (a) Parallel-plate structure short-circuited at one end and driven by a current source at the other end. (b) Magnetic field between the plates for a constant-current source.

conductor, the magnitude of this field is I_0/w. Thus, we obtain the static magnetic-field intensity between the plates to be

$$\mathbf{H} = \frac{I_0}{w}\mathbf{a}_y \qquad \text{for} \quad 0 < x < d \tag{8.2}$$

The field is zero outside the plates.

The corresponding magnetic flux density is given by

$$\mathbf{B} = \mu\mathbf{H} = \frac{\mu I_0}{w}\mathbf{a}_y \qquad \text{for} \quad 0 < x < d \tag{8.3}$$

The magnetic flux ψ linking the current is simply the flux crossing the cross-sectional plane of the structure. Since $\mathbf{B}$ is uniform in the cross-sectional plane and normal to it,

$$\psi = B_y(dl) = \frac{\mu dl}{w} I_0 \tag{8.4}$$

The ratio of this magnetic flux to the current, that is, the inductance of the structure, is given by

$$\boxed{L = \frac{\psi}{I_0} = \frac{\mu dl}{w}} \tag{8.5}$$

To discuss the quasistatic behavior of the structure, we now let the current source be varying sinusoidally with time at a frequency ω and assume that the magnetic field between the plates varies accordingly. Thus, for

$$I(t) = I_0 \cos \omega t \tag{8.6}$$

we have

$$\mathbf{H}_0 = \frac{I_0}{w} \cos \omega t \, \mathbf{a}_y \tag{8.7}$$

where the subscript 0 denotes that the field is of the zeroth power in ω. In terms of phasor notation, we have

$$\boxed{\begin{aligned} \bar{I} &= I_0 \\ \bar{H}_{y0} &= \frac{I_0}{w} \end{aligned}} \qquad \begin{aligned} &(8.8) \\ &(8.9) \end{aligned}$$

The time-varying magnetic field (8.7) gives rise to an electric field in accordance with Maxwell's curl equation for $\mathbf{E}$. Expansion of the curl equation for the case under consideration gives

$$\frac{\partial E_x}{\partial z} = -\frac{\partial B_{y0}}{\partial t} = -\mu \frac{\partial H_{y0}}{\partial t}$$

or, in phasor form,

$$\frac{\partial \overline{E}_x}{\partial z} = -j\omega\mu\overline{H}_{y0} \tag{8.10}$$

Substituting for $\overline{H}_{y0}$ from (8.9), we have

$$\frac{\partial \overline{E}_x}{\partial z} = -j\omega\mu\frac{I_0}{w}$$

or

$$\overline{E}_x = -j\omega\mu\frac{I_0}{w}z + \overline{C} \tag{8.11}$$

The constant $\overline{C}$ is, however, equal to zero since $[\overline{E}_x]_{z=0} = 0$ to satisfy the boundary condition of zero tangential electric field on the perfect conductor surface. Thus, we obtain the quasistatic electric field in the structure to be

$$\boxed{\overline{E}_{x1} = -j\omega\frac{\mu z}{w}I_0} \tag{8.12}$$

where the subscript 1 denotes that the field is of the first power in ω. The value of this field at the input of the structure is given by

$$\boxed{[\overline{E}_{x1}]_{z=-l} = j\omega\mu l\frac{I_0}{w}} \tag{8.13}$$

The voltage developed across the current source is now given by

$$\overline{V} = \int_a^b [\overline{E}_{x1}]_{z=-l}\, dx$$

$$= j\omega\frac{\mu dl}{w}I_0$$

or

$$\boxed{\overline{V} = j\omega L I_0} \tag{8.14}$$

Thus, the quasistatic extension of the static field in the structure of Fig. 8.1 illustrates that its input behavior for low frequencies is essentially that of a single inductor of value, the same as that found from static-field considerations.

We shall now determine the condition under which the quasistatic approximation is valid, that is, the condition under which the field of the first power in ω is the predominant part of the total field. To do this, we proceed in the following manner. The electric field $\overline{E}_{x1}$ gives rise to a magnetic field in accordance with Maxwell's curl equation for **H**, which for the case under consideration is

Validity of quasistatic approximation for an inductor

given by

$$\frac{\partial H_y}{\partial z} = -\varepsilon \frac{\partial E_x}{\partial t}$$

or in phasor form by

$$\frac{\partial \overline{H}_y}{\partial z} = -j\omega\varepsilon\overline{E}_x \tag{8.15}$$

Substituting $\overline{E}_{x1}$ from (8.12) for $\overline{E}_x$ in (8.15), we have

$$\frac{\partial \overline{H}_y}{\partial z} = -\omega^2\mu\varepsilon\frac{z}{w}I_0$$

or

$$\overline{H}_{y2} = -\frac{\omega^2\mu\varepsilon z^2}{2w}I_0 + \overline{C}'' \tag{8.16}$$

where the subscript 2 denotes that the field is of power 2 in ω. The constant $\overline{C}''$ can be evaluated by noting that at $z = -l$, $\overline{H}_{y2}$ must be zero since $\overline{H}_{y0}$ by itself satisfies the boundary condition $\mathbf{J}_s = \mathbf{a}_n \times \mathbf{H}$. Thus, we get

$$\overline{H}_{y2} = -\frac{\omega^2\mu\varepsilon(z^2 - l^2)}{2w}I_0 \tag{8.17}$$

This magnetic field gives rise to an electric field in accordance with Maxwell's curl equation for $\mathbf{E}$. Hence, we have

$$\frac{\partial \overline{E}_x}{\partial z} = -j\omega\mu\overline{H}_{y2}$$

$$= \frac{j\omega^3\mu^2\varepsilon(z^2 - l^2)}{2w}I_0$$

or

$$\overline{E}_{x3} = \frac{j\omega^3\mu^2\varepsilon(z^3 - 3l^2 z)}{6w}I_0 + \overline{C}''' \tag{8.18}$$

where the subscript 3 denotes that the field is of power 3 in ω. The constant $\overline{C}'''$ has to be equal to zero to satisfy the boundary condition of zero tangential electric field on the conductor surface $z = 0$. Thus, we obtain

$$\overline{E}_{x3} = \frac{j\omega^3\mu^2\varepsilon(z^3 - 3l^2 z)}{6w}I_0 \tag{8.19}$$

and, hence,

$$[\overline{E}_{x3}]_{z=-l} = j\frac{\omega^3\mu^2\varepsilon l^3}{3}\frac{I_0}{w} \tag{8.20}$$

Continuing in this manner, we would obtain

$$[\overline{E}_{x5}]_{z=-l} = j\frac{2\omega^5\mu^3\varepsilon^2 l^5}{15}\frac{I_0}{w} \tag{8.21}$$

and so on. The total electric field at $z = -l$ can then be written as

$$[\bar{E}_x]_{z=-l} = [\bar{E}_{x1}]_{z=-l} + [\bar{E}_{x3}]_{z=-l} + [\bar{E}_{x5}]_{z=-l} + \cdots$$

$$= j\omega\mu l \frac{I_0}{w} + j\frac{\omega^3\mu^2\varepsilon l^2}{3}\frac{I_0}{w} + j\frac{2\omega^5\mu^3\varepsilon^2 l^3}{15}\frac{I_0}{w} + \cdots$$

$$= j\sqrt{\frac{\mu}{\varepsilon}}\frac{I_0}{w}\left[\omega\sqrt{\mu\varepsilon}l + \frac{1}{3}(\omega\sqrt{\mu\varepsilon}l)^3 + \frac{2}{15}(\omega\sqrt{\mu\varepsilon}l)^5 + \cdots\right]$$

or

$$[\bar{E}_x]_{z=-l} = j\sqrt{\frac{\mu}{\varepsilon}}\frac{I_0}{w}\tan\omega\sqrt{\mu\varepsilon}l \qquad (8.22)$$

From (8.22), it can be seen that for $\omega\sqrt{\mu\varepsilon}l \ll 1$,

$$[\bar{E}_x]_{z=-l} \approx j\sqrt{\frac{\mu}{\varepsilon}}\frac{I_0}{w}\omega\sqrt{\mu\varepsilon}l$$

$$= j\omega\mu l\frac{I_0}{w}$$

which is the same as $[\bar{E}_{x1}]_{z=-l}$. Thus, the condition under which the quasistatic approximation is valid is

$$\omega\sqrt{\mu\varepsilon}l \ll 1$$

or

$$f \ll \frac{1}{2\pi\sqrt{\mu\varepsilon}l} \qquad (8.23)$$

For frequencies beyond which (8.23) is valid, the input behavior of the structure of Fig. 8.1 is no longer essentially that of a single inductor.

    To further investigate the condition for the quasistatic approximation, we recognize that (8.23) can be written as

$$l \ll \frac{v_p}{2\pi f}$$

or,

$$l \ll \frac{\lambda}{2\pi} \qquad (8.24)$$

where $\lambda = v_p/f$ is the wavelength corresponding to f in the dielectric region between the plates. Thus, (8.24) tells us that the length of the structure must be very small compared to the wavelength.

    The criterion (8.24) is a general condition for the quasistatic approximation for the input behavior of a physical structure. Physical structures can be classified as electric-field systems and magnetic-field systems, depending on whether the *Electric- and magnetic-field systems* electric field or the magnetic field is predominant. Quasistatic electric- and magnetic-field systems are particularly important in electromechanics. The structure of Fig. 8.1 is a magnetic-field system, since for the static case the only field present between the plates is the magnetic field and the quasistatic magnetic field has the same spatial dependence as that of the static magnetic field. The quasistatic magnetic field gives rise to a time-varying electric field, but the corresponding displacement current is so small that its effect in adding to the quasistatic magnetic field is negligible and hence it can be omitted from Maxwell's curl equation for **H**. Thus, for a quasistatic magnetic-field system, we have

$$\nabla \times \mathbf{H} = \mathbf{J} \qquad (8.25a)$$

$$\nabla \times \mathbf{E} = -\frac{\partial \mathbf{B}}{\partial t} \qquad (8.25b)$$

$$\nabla \cdot \mathbf{B} = 0 \qquad (8.25c)$$

Likewise, if the structure of Fig. 8.1 is open-circuited at $z = 0$ and driven by a voltage source at $z = -l$, the only field present between the plates in the static case would be the electric field and the quasistatic electric field would have the same spatial dependence as that of the static electric field. The system would then be an electric-field system. The quasistatic electric field would give rise to a time-varying magnetic field, but the corresponding value of $\partial \mathbf{B}/\partial t$ would be so small that its effect in adding to the quasistatic electric field would be negligible and, hence, it can be omitted from Maxwell's curl equation for **E**. Thus, for a quasistatic electric-field system, we have

$$\nabla \times \mathbf{E} = \mathbf{0} \qquad (8.26a)$$

$$\nabla \times \mathbf{H} = \mathbf{J} + \frac{\partial \mathbf{D}}{\partial t} \qquad (8.26b)$$

$$\nabla \cdot \mathbf{D} = \rho \qquad (8.26c)$$

When the medium between the plates is conductive, a conduction current flows between the plates in accordance with $\mathbf{J} = \mathbf{J}_c = \sigma\mathbf{E}$ and the analysis for low-frequency input behavior results in both electric and magnetic fields of the first order in ω. We shall illustrate this by means of an example.

Example 8.1

Low-frequency behavior of a resistor

Let us consider the case of two parallel perfectly conducting plates separated by a lossy medium characterized by conductivity σ, permittivity ε, and permeability μ, and driven by a voltage source at one end, as shown in Fig. 8.2(a). We wish to determine its low-frequency behavior by using the quasistatic-field approach.

Assuming the voltage source to be a constant-voltage source, we first obtain the static electric field in the medium between the plates to be

$$\mathbf{E} = \frac{V_0}{d}\,\mathbf{a}_x$$

FIGURE 8.2 (a) Parallel-plate structure with lossy medium between the plates and driven by a voltage source. (b) Electric and magnetic fields between the plates for a constant-voltage source.

following the procedure of Example 6.2. The conduction current density in the medium is then given by

$$\mathbf{J}_c = \sigma \mathbf{E} = -\frac{\sigma V_0}{d}\mathbf{a}_x$$

The conduction current gives rise to a static magnetic field in accordance with Maxwell's curl equation for $\mathbf{H}$ given for static fields by

$$\nabla \times \mathbf{H} = \mathbf{J}_c = \sigma \mathbf{E}$$

For the case under consideration, this reduces to

$$\frac{\partial H_y}{\partial z} = -\sigma E_x = -\frac{\sigma V_0}{d}$$

giving us

$$H_y = -\frac{\sigma V_0 z}{d} + C_1$$

The constant C_1 is, however, equal to zero since $[H_y]_{z=0} = 0$ in view of the boundary condition that the surface current density on the plates must be zero at $z = 0$. Thus, the static magnetic field in the medium between the plates is given by

$$\mathbf{H} = -\frac{\sigma V_0 z}{d}\mathbf{a}_y$$

The static electric- and magnetic-field distributions are shown by the cross-sectional view of the structure in Fig. 8.2(b).

To determine the quasistatic behavior of the structure, we now let the voltage source be varying sinusoidally with time at a frequency ω and assume that the electric and magnetic fields vary with time accordingly. Thus, for

$$V = V_0 \cos \omega t$$

we have

$$\mathbf{E}_0 = \frac{V_0}{d} \cos \omega t \, \mathbf{a}_x \tag{8.27a}$$

$$\mathbf{H}_0 = -\frac{\sigma V_0 z}{d} \cos \omega t \, \mathbf{a}_y \tag{8.27b}$$

where the subscript 0 denotes that the fields are of the zeroth power in ω. In terms of phasor notation, we have for $\overline{V} = V_0$,

$$\overline{E}_{x0} = \frac{V_0}{d} \tag{8.28a}$$

$$\overline{H}_{y0} = -\frac{\sigma V_0 z}{d} \tag{8.28b}$$

The time-varying electric field (8.27a) gives rise to a magnetic field in accordance with

$$\nabla \times \mathbf{H} = \frac{\partial \mathbf{D}_0}{\partial t} = \varepsilon \frac{\partial \mathbf{E}_0}{\partial t}$$

and the time-varying magnetic field (8.27b) gives rise to an electric field in accordance with

$$\nabla \times \mathbf{E} = -\frac{\partial \mathbf{B}_0}{\partial t} = -\mu \frac{\partial \mathbf{H}_0}{\partial t}$$

For the case under consideration and using phasor notation, these equations reduce to

$$\frac{\partial \overline{H}_y}{\partial z} = -j\omega\varepsilon\overline{E}_{x0} = -j\omega \frac{\varepsilon V_0}{d}$$

$$\frac{\partial \overline{E}_x}{\partial z} = -j\omega\mu\overline{H}_{y0} = j\omega \frac{\mu\sigma V_0 z}{d}$$

giving us

$$\overline{H}_{y1} = -j\omega \frac{\varepsilon V_0 z}{d} + \overline{C}_2$$

$$\overline{E}_{x1} = j\omega \frac{\mu\sigma V_0 z^2}{2d} + \overline{C}_3$$

where the subscript 1 denotes that the fields are of the first power in ω. The constant $\overline{C}_2$ is, however, equal to zero in view of the boundary condition that the surface current density on the plates must be zero at $z = 0$. To evaluate the constant $\overline{C}_3$, we note that $[\overline{E}_{x1}]_{z=-l} = 0$ since the boundary condition at the source end, that is,

$$\overline{V} = \int_a^b [\overline{E}_x]_{z=-l} \, dx$$

is satisfied by $\overline{E}_{x0}$ alone. Thus, we have

$$j\omega \frac{\mu\sigma V_0(-l)^2}{2d} + \overline{C}_3 = 0$$

or

$$\overline{C}_3 = -j\omega \frac{\mu\sigma V_0 l^2}{2d}$$

Substituting for $\overline{C}_3$ and $\overline{C}_2$ in the expressions for $\overline{E}_{x1}$ and $\overline{H}_{y1}$, respectively, we get

$$\overline{E}_{x1} = j\omega \frac{\mu\sigma V_0(z^2 - l^2)}{2d} \tag{8.29a}$$

$$\overline{H}_{y1} = -j\omega \frac{\varepsilon V_0 z}{d}$$

The result for $\overline{H}_{y1}$ is, however, not complete since $\overline{E}_{x1}$ gives rise to a conduction current of density proportional to ω, which in turn provides an additional contribution to $\overline{H}_{y1}$. Denoting this contribution to be $\overline{H}_{y1}^c$, we have

$$\frac{\partial \overline{H}_{y1}^c}{\partial z} = -\sigma \overline{E}_{x1} = -j\omega \frac{\mu\sigma^2 V_0(z^2 - l^2)}{2d}$$

$$\overline{H}_{y1}^c = -j\omega \frac{\mu\sigma^2 V_0(z^3 - 3zl^2)}{6d} + \overline{C}_4$$

The constant $\overline{C}_4$ is zero for the same reason that $\overline{C}_2$ is zero. Hence, setting $\overline{C}_4$ equal to zero and adding the resulting expression for $\overline{H}_{y1}^c$ to the right side of the expression for $\overline{H}_{y1}$, we obtain the complete expression for $\overline{H}_{y1}$ as

$$\overline{H}_{y1} = -j\omega \frac{\varepsilon V_0 z}{d} - j\omega \frac{\mu\sigma^2 V_0(z^3 - 3zl^2)}{6d} \tag{8.29b}$$

The total field components correct to the first power in ω are then given by

$$\overline{E}_x = \overline{E}_{x0} + \overline{E}_{x1}$$

$$= \frac{V_0}{d} + j\omega \frac{\mu\sigma V_0(z^2 - l^2)}{2d} \tag{8.30a}$$

$$\overline{H}_y = \overline{H}_{y0} + \overline{H}_{y1}$$

$$= -\frac{\sigma V_0 z}{d} - j\omega \frac{\varepsilon V_0 z}{d} - j\omega \frac{\mu\sigma^2 V_0(z^3 - 3zl^2)}{6d} \tag{8.30b}$$

The current drawn from the voltage source is

$$\overline{I} = w[\overline{H}_y]_{z=-l}$$

$$= \left(\frac{\sigma wl}{d} + j\omega \frac{\varepsilon wl}{d} - j\omega \frac{\mu\sigma^2 wl^3}{3d} \right) \overline{V} \tag{8.31}$$

Finally, the input admittance of the structure is given by

$$\overline{Y} = \frac{\overline{I}}{\overline{V}} = j\omega \frac{\varepsilon wl}{d} + \frac{\sigma wl}{d} \left(1 - j\omega \frac{\mu\sigma l^2}{3} \right)$$

$$\approx j\omega \frac{\varepsilon wl}{d} + \frac{1}{\dfrac{d}{\sigma wl}\left(1 + j\omega \dfrac{\mu\sigma l^2}{3} \right)} \tag{8.32}$$

FIGURE 8.3   Equivalent circuit for the low-frequency input behavior of the structure of Fig. 8.2.

where we have approximated $[1 - j\omega(\mu\sigma l^2/3)]$ by $[1 + j\omega(\mu\sigma l^2/3)]^{-1}$. Proceeding further, we have

$$\overline{Y} = j\omega \frac{\varepsilon w l}{d} + \cfrac{1}{\cfrac{d}{\sigma w l} + j\omega \cfrac{\mu d l}{3w}}$$

$$= j\omega C + \frac{1}{R + (j\omega L/3)}$$

(8.33)

where $C = \varepsilon w l/d$ is the capacitance of the structure if the material is a perfect dielectric, $R = d/\sigma w l$ is the dc resistance (reciprocal of the conductance) of the structure, and $L = \mu d l/w$ is the inductance of the structure if the material is lossless and the two plates are short-circuited at $z = 0$. The equivalent circuit corresponding to (8.33) consists of capacitance C in parallel with the series combination of resistance R and inductance $L/3$, as shown in Fig. 8.3. Thus, the low-frequency input behavior of the structure of Fig. 8.2 (which acts like a pure resistor at dc) can be represented by the circuit Fig. 8.3, with the understanding of the approximation used in (8.32).

Note that for $\sigma = 0$, (8.31) reduces to

$$\overline{I} = j\omega \frac{\varepsilon w l}{d} \overline{V}$$

$$= j\omega C \overline{V}$$

and the input behavior of the structure is essentially that of a single capacitor of the same value as that found from static-field considerations.

Equivalent circuits beyond quasistatic approximation   Sometimes, it is of interest to consider equivalent circuit representation for the input behavior of a structure for frequencies beyond the quasistatic approximation. For an example, let us consider frequencies slightly beyond those for which the quasistatic approximation is valid for the structure of Fig. 8.1. Then

$$[\overline{E}_x]_{z = -l} = [\overline{E}_{x1}]_{z = -l} + [\overline{E}_{x3}]_{z = -l}$$

and from (8.13) and (8.20), we have

$$[\overline{E}_x]_{z = -l} \approx j\omega\mu l \frac{\overline{I}_0}{w} + j\frac{\omega^3 \mu^2 \varepsilon l^2}{3} \frac{\overline{I}_0}{w}$$

(8.34)

and the voltage developed across the current source is given by

$$\overline{V} = \int_a^b [\overline{E}_x]_{z=-l}\, dx$$

$$\approx j\omega \frac{\mu dl}{w}\overline{I}_0 + j\frac{\omega^3 \mu^2 \varepsilon dl^2}{3}\frac{\overline{I}_0}{w}$$

$$= j\omega \frac{\mu dl}{w}\overline{I}_0 \left[1 + \frac{1}{3}\omega^2\left(\frac{\mu dl}{w}\right)\left(\frac{\varepsilon wl}{d}\right)\right] \qquad (8.35)$$

$$= j\omega L\overline{I}_0 \left[1 + \frac{1}{3}\omega^2 LC\right]$$

where $C = \varepsilon wl/d$ is the capacitance of the structure with the end $z = 0$ open-circuited and from static-field considerations. Rearranging (8.35), we get

$$\overline{I}_0 = \frac{\overline{V}}{j\omega L\left(1 + \frac{1}{3}\omega^2 LC\right)}$$

$$\approx \frac{\overline{V}}{j\omega L}\left(1 - \frac{1}{3}\omega^2 LC\right) \qquad (8.36)$$

$$= \overline{V}\left(\frac{1}{j\omega L} + j\omega \frac{C}{3}\right)$$

Thus, the equivalent circuit consists of the parallel combination of L and $\frac{1}{3}C$, as shown in Fig. 8.4.

K8.1. Quasistatic approximation; Condition for quasistatic approximation; Electric- and magnetic-field systems; Low-frequency terminal behavior; Inductor; Resistor.

D8.1. For the structure of Fig. 8.1, assume that $l = 10\,\text{cm}$, $d = 1\,\text{cm}$, $w = 10\,\text{cm}$, and that the medium between the conductors is free space. Assuming that the condition for quasistatic approximation given by (8.23) is valid for $f < 1/20\pi\sqrt{\mu\varepsilon}l$, find the following: **(a)** the maximum frequency for which the input behavior of the structure is essentially that of a single inductor; **(b)** the value of this inductor; and **(c)** the ratio of the amplitude of the electric field at the input if the structure behaves exactly like a single inductor to the amplitude of the actual electric field at the input for the frequency found in (a).

Ans. **(a)** 47.746 MHz; **(b)** $4\pi \times 10^{-9}$ H; **(c)** 0.9967.

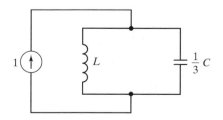

FIGURE 8.4 Equivalent circuit for the input behavior of the structure of Fig 8.1 for frequencies slightly beyond those for which the quasistatic approximation is valid.

8.2   MAGNETIC CIRCUITS

Toroidal conductor versus toroidal magnetic core

Let us consider the two structures shown in Fig. 8.5. The structure of Fig. 8.5(a) is a toroidal conductor of uniform conductivity σ and has a cross-sectional area A and mean circumference l. There is an infinitesimal gap a–b across which a potential difference of V_0 volts is maintained by connecting an appropriate voltage source. Because of the potential difference, an electric field is established in the toroid and a conduction current I_c results from the higher-potential surface a to the lower-potential surface b as shown in the figure. The structure of Fig. 8.5(b) is a toroidal magnetic core of uniform permeability μ and has a cross-sectional area A and mean circumference l. A current I A is passed through a filamentary wire of N turns wound around the toroid by connecting an appropriate current source. Because of the current through the winding, a magnetic field is established in the toroid and a magnetic flux ψ results in the direction of advance of a right-hand screw as it is turned in the sense of the current.

Since the conduction current cannot leak into the free space surrounding the conductor, it is confined entirely to the conductor. On the other hand, the magnetic flux can leak into the free space surrounding the magnetic core and, hence, is not confined completely to the core. However, let us consider the case for which $\mu \gg \mu_0$. Applying the boundary conditions at the boundary between a magnetic material of $\mu \gg \mu_0$ and free space, as shown in Fig. 8.6, we have

$$B_1 \sin \alpha_1 = B_2 \sin \alpha_2$$

$$H_1 \cos \alpha_1 = H_2 \cos \alpha_2$$

or

$$\frac{B_1}{H_1} \tan \alpha_1 = \frac{B_2}{H_2} \tan \alpha_2$$

$$\frac{\tan \alpha_1}{\tan \alpha_2} = \frac{\mu_2}{\mu_1} \ll 1$$

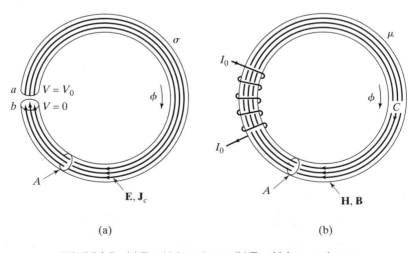

(a)   (b)

FIGURE 8.5   (a) Toroidal conductor. (b) Toroidal magnetic core.

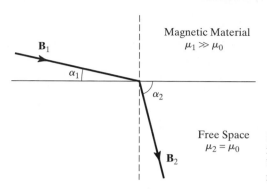

$\mathbf{B}_1$

α_1

Magnetic Material
$\mu_1 \gg \mu_0$

α_2

Free Space
$\mu_2 = \mu_0$

$\mathbf{B}_2$

FIGURE 8.6   Lines of magnetic flux density at the boundary between free space and a magnetic material of $\mu \gg \mu_0$.

Thus, $\alpha_1 \ll \alpha_2$, and

$$\frac{B_2}{B_1} = \frac{\sin \alpha_1}{\sin \alpha_2} \ll 1$$

For example, if the values of μ_1 and α_2 and $1000\mu_0$ and 89°, respectively, then $\alpha_1 = 3°16'$ and $\sin \alpha_1 / \sin \alpha_2 = 0.057$. We can assume for all practical purposes that the magnetic flux is confined to the magnetic core just as the conduction current is confined to the conductor. The structure of Fig. 8.5(b) is then known as a *magnetic circuit* similar to the *electric circuit* of Fig. 8.5(a). Magnetic circuits are encountered in applications involving electromechanical systems, typical examples of which are electromagnets, transformers, and rotating machines.

For the toroidal conductor of Fig. 8.5(a), we have

$$\int_a^b \mathbf{E} \cdot d\mathbf{l} = V_0 \tag{8.37}$$

Assuming E_ϕ to be uniform over the cross-sectional area and equal to its value E_m at the mean radius of the toroid and proceeding, we obtain

$$lE_m = V_0$$

$$E_m = \frac{V_0}{l}$$

$$J_c = \sigma E_m = \frac{\sigma V_0}{l}$$

$$I_c = J_c A = \frac{\sigma V_0 A}{l}$$

Thus, the resistance of the circuit is given by

$$R = \frac{V_0}{I_0} = \frac{l}{\sigma A} \tag{8.38}$$

Similarly, for the toroidal magnetic core of Fig. 8.5(b),

$$\oint_C \mathbf{H} \cdot d\mathbf{l} = NI_0 \tag{8.39}$$

Assuming H_ϕ to be uniform over the cross-sectional area and equal to its value H_m at the mean radius of the toroid, we obtain

$$lH_m = NI_0$$

$$H_m = \frac{NI_0}{l}$$

$$B_m = \mu H_m = \frac{\mu NI_0}{l}$$

$$\psi = B_m A = \frac{\mu NI_0 A}{l}$$

Reluctance defined

We now define the *reluctance* of the magnetic circuit, denoted by the symbol $\mathcal{R}$, as the ratio of the ampere turns NI_0 applied to the magnetic circuit to the magnetic flux ψ. Thus,

$$\boxed{\mathcal{R} = \frac{NI_0}{\psi} = \frac{l}{\mu A}} \qquad (8.40)$$

The reluctance of the magnetic circuit is analogous to the resistance of an electric circuit and has the units of ampere-turns per weber (A-t/Wb). In fact, the complete analogy between the toroidal conductor and the toroidal magnetic core can be seen as follows.

$$V_0 \leftrightarrow NI_0$$
$$\mathbf{E} \leftrightarrow \mathbf{H}$$
$$\mathbf{J}_c \leftrightarrow \mathbf{B}$$
$$\sigma \leftrightarrow \mu$$
$$I_c \leftrightarrow \psi$$
$$R \leftrightarrow \mathcal{R}$$

The equivalent-circuit representations of the two arrangements are shown in Figs. 8.7(a) and (b), respectively. We note from (8.40) that for a given magnetic material, the reluctance appears to be purely a function of the dimensions of the circuit. This is, however, not true since for the ferromagnetic materials used for

(a)   (b)

FIGURE 8.7   Equivalent-circuit representations for the structures of Figs. 8.5(a) and (b), respectively.

FIGURE 8.8. *B* versus *H* curve for annealed sheet steel.

the cores, μ is a function of the magnetic flux density in the material, as we learned in Section 2.6.

As a numerical example of computations involving the magnetic circuit of Fig. 8.5(b), let us consider a core of cross-sectional area 2 cm^2 and mean circumference 20 cm. Let the material of the core be annealed sheet steel for which the *B* versus *H* relationship is shown by the curve of Fig. 8.8. Then to establish a magnetic flux of 3×10^{-4} Wb in the core, the mean flux density must be $(3 \times 10^{-4})/(2 \times 10^{-4})$, or 1.5 Wb/m^2. From Fig. 8.8, the corresponding value of *H* is 1000 A/m. The number of ampere-turns required to establish the flux is then equal to $1000 \times 20 \times 10^{-2}$, or 200, and the reluctance of the core is $200/(3 \times 10^{-4})$, or $(2/3) \times 10^6$ A-t/Wb. We shall now consider a more detailed example.

Example 8.2

A magnetic circuit containing three legs and with an air gap in the right leg is shown in Fig. 8.9(a). A filamentary wire of *N* turns carrying current *I* is wound around the center leg. The core material is annealed sheet steel, for which the *B* versus *H* relationship is shown in Fig. 8.8. The dimensions of the magnetic circuit are

Magnetic circuit with three legs and an air gap

$$A_1 = A_3 = 3 \text{ cm}^2 \qquad A_2 = 6 \text{ cm}^2$$
$$l_1 = l_3 - 20 \text{ cm} \qquad l_2 - 10 \text{ cm} \qquad l_g = 0.2 \text{ mm}$$

Let us determine the value of *NI* required to establish a magnetic flux of 4×10^{-4} Wb in the air gap.

The current in the winding establishes a magnetic flux in the center leg that divides between the right and left legs. Fringing of the flux occurs in the air gap, as shown in Fig. 8.9(b). This is taken into account by using an effective cross section larger than the actual cross section, as shown in Fig. 8.9(c). Using subscripts 1, 2, 3, and *g* for the quantities associated with the left, center, and right legs, and the air gap, respectively, we can write

$$\psi_3 = \psi_g$$
$$\psi_2 = \psi_1 + \psi_3$$

Also, applying Ampère's circuital law to the right and left loops of the magnetic circuit, we obtain, respectively,

$$NI = H_2 l_2 + H_3 l_3 + H_g l_g$$
$$NI = H_2 l_2 + H_1 l_1$$

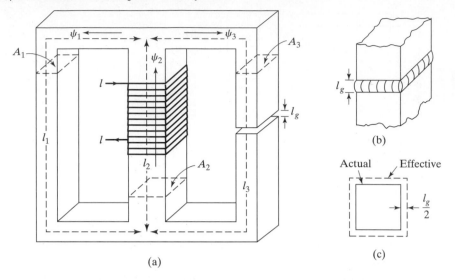

FIGURE 8.9. (a) Magnetic circuit. (b) Fringing of magnetic flux in the air gap of the magnetic circuit. (c) Effective and actual cross sections for the air gap.

It follows from these two equations that

$$H_1 l_1 = H_3 l_3 + H_g l_g$$

which can also be written directly from a consideration of the outer loop of the magnetic circuit.

Noting from Fig. 8.9(c) that the effective cross section of the air gap is $(\sqrt{3} + l_g)^2 = 3.07 \text{ cm}^2$, we find the required magnetic flux density in the air gap to be

$$B_g = \frac{\psi_g}{(A_g)_{\text{eff}}} = \frac{4 \times 10^{-4}}{3.07 \times 10^{-4}} = 1.303 \text{ Wb/m}^2$$

The magnetic-field intensity in the air gap is

$$H_g = \frac{B_g}{\mu_0} = \frac{1.303}{4\pi \times 10^{-7}} = 0.1037 \times 10^7 \text{ A/m}$$

The flux density in leg 3 is

$$B_3 = \frac{\psi_3}{A_3} = \frac{\psi_g}{A_3} = \frac{4 \times 10^{-4}}{3 \times 10^{-4}} = 1.333 \text{ Wb/m}^2$$

From Fig. 8.8, the value of H_3 is 475 A/m.

Knowing the values of H_g and H_3, we then obtain

$$H_1 l_1 = H_3 l_3 + H_g l_g$$
$$= 475 \times 0.2 + 0.1037 \times 10^7 \times 0.2 \times 10^{-3}$$
$$= 302.4 \text{ A}$$
$$H_1 = \frac{302.4}{0.2} = 1512 \text{ A/m}$$

FIGURE 8.10. Equivalent circuit for the magnetic circuit of Fig. 8.9.

From Fig. 8.8, the value of B_1 is 1.56 Wb/m^2, and, hence, the flux in leg 1 is

$$\psi_1 = B_1 A_1 = 1.56 \times 3 \times 10^{-4} = 4.68 \times 10^{-4} \text{ Wb}$$

Thus,

$$\psi_2 = \psi_1 + \psi_3$$
$$= 4.68 \times 10^{-4} + 4 \times 10^{-4} = 8.68 \times 10^{-4} \text{ Wb}$$
$$B_2 = \frac{\psi_2}{A_2} = \frac{8.68 \times 10^{-4}}{6 \times 10^{-4}} = 1.447 \text{ Wb/m}^2$$

From Fig. 8.8, the value of H_2 is 750 A/m. Finally, we obtain the required number of ampere-turns to be

$$NI = H_2 l_2 + H_1 l_1$$
$$= 750 \times 0.1 + 302.4$$
$$= 377.4$$

Note that the equivalent circuit corresponding to the magnetic circuit is as shown in Fig. 8.10, where the reluctances are given by

$$\mathcal{R}_1 = \frac{l_1}{\mu_1 A_1} = \frac{H_1 l_1}{B_1 A_1} = \frac{1512 \times 0.2}{1.56 \times 3 \times 10^{-4}} = 646,154 \text{ A-t/Wb}$$

$$\mathcal{R}_2 = \frac{l_2}{\mu_2 A_2} = \frac{H_2 l_2}{B_2 A_2} = \frac{750 \times 0.1}{1.447 \times 6 \times 10^{-4}} = 86,386 \text{ A-t/Wb}$$

$$\mathcal{R}_3 = \frac{l_3}{\mu_3 A_3} = \frac{H_3 l_3}{B_3 A_3} = \frac{475 \times 0.2}{1.333 \times 3 \times 10^{-4}} = 237,559 \text{ A-t/Wb}$$

$$\mathcal{R}_g = \frac{l_g}{\mu_0 (A_g)_{\text{eff}}} = \frac{0.2 \times 10^{-3}}{4\pi \times 10^{-7} \times 3.07 \times 10^{-4}} = 518,420 \text{ A-t/Wb}$$

K8.2. Magnetic circuit; Analogy with electric circuit; Reluctance; Air gap.

D8.2. Assume that the portion of B versus H curve of Fig. 8.8 in the range $1500 \leq H \leq 3000$ can be approximated by the straight line

$$B = 1.5 + 5 \times 10^{-5} H$$

For a toroidal magnetic circuit made of annealed sheet steel, find the reluctance for each of the following cases: **(a)** $A = 4$ cm^2, $l = 30$ cm, $H = 1800$ A/m; **(b)** $A = 2$ cm^2, $l = 20$ cm, $NI = 500$ A-t; and **(c)** $A = 5$ cm^2, $l = 25$ cm, $\psi = 8 \times 10^{-4}$ Wb. *Ans.* **(a)** 849,057 A-t/Wb; **(b)** 1,538,462 A-t/Wb; **(c)** 625,000 A-t/Wb.

8.3   ELECTROMECHANICAL ENERGY CONVERSION

Parallel-plate capacitor with a movable plate

Let us consider a parallel-plate capacitor with one plate fixed and the other plate free to move, as shown by a cross-sectional view in Fig. 8.11. If we assume a positive charge Q on the movable plate and a negative charge $-Q$ on the fixed plate, resulting from the application of a voltage V between the plates, then a force $\mathbf{F}_e$ directed toward the fixed plate is exerted on the movable plate. If this force is allowed to produce a displacement of the movable plate, mechanical work results thereby converting electrical energy in the system into mechanical energy. Conversely, an externally applied mechanical force can be made to act on the movable plate so as to increase the stored electrical energy in the system. Thus, energy can be converted from electrical to mechanical or vice versa. A familiar example of the former is in the case of an electrical motor, whereas that of the later is in the case of an electrical generator. To determine the amount of energy converted from one form to another, we first need to know how to compute the force $\mathbf{F}_e$. In this section, we illustrate this computation and discuss the determination of energy converted from one form to another.

Computation of mechanical force of electric origin

The computation of the mechanical force $\mathbf{F}_e$ of electric origin follows from considerations of energy balance associated with the electromechanical system. The energy balance can be expressed as

$$
\boxed{\begin{array}{c}\text{Mechanical}\\\text{energy}\\\text{input}\end{array}} + \boxed{\begin{array}{c}\text{Electrical}\\\text{energy}\\\text{input}\end{array}} = \boxed{\begin{array}{c}\text{Increase}\\\text{in stored}\\\text{mechanical}\\\text{energy}\end{array}} + \boxed{\begin{array}{c}\text{Increase}\\\text{in stored}\\\text{electrical}\\\text{energy}\end{array}} + \boxed{\begin{array}{c}\text{Energy}\\\text{dissipated}\end{array}} \qquad (8.41)
$$

For simplicity, we shall consider the system to be lossless so that the last term on the right side of (8.41) is zero. In using (8.41) to find $\mathbf{F}_e$, we shall apply to the movable element of the system an external force equal to $-\mathbf{F}_e$ and displace the

FIGURE 8.11   Parallel-plate capacitor with a movable plate, depicting the force F_e on the movable plate.

element by an infinitesimal distance in the direction of the external force, so that no change in stored mechanical energy occurs. This eliminates the first term on the right side of (8.41). Thus, with reference to the system of Fig. 8.11, we have

$$-F_{ex}\, dx + VI\, dt = dW_e \tag{8.42}$$

where dx is the displacement of the movable plate, I is the current drawn from the voltage source, and W_e is the electric stored energy in the capacitor. Substituting $I = dQ/dt$ from the law of conservation of charge, we obtain

$$-F_{ex}\, dx + V\, dQ = dW_e$$

or

$$\boxed{F_{ex} = -\frac{dW_e}{dx} + V\frac{dQ}{dx}} \tag{8.43}$$

To proceed further, we shall neglect fringing of the electric field at the edges of the capacitor plates so that the charges on the plates and the electric field between the plates are uniformly distributed. Then if A is the area of each plate, we can write the following:

$$W_e = \frac{1}{2}\varepsilon_0 E^2 Ax = \frac{1}{2}\varepsilon_0\left(\frac{V}{x}\right)^2 Ax$$

$$= \frac{\varepsilon_0 V^2 A}{2x}$$

$$\frac{dW_e}{dx} = -\frac{\varepsilon_0 V^2 A}{2x^2}$$

$$Q = CV = \frac{\varepsilon_0 A V}{x}$$

$$\frac{dQ}{dx} = \frac{\varepsilon_0 A V}{x^2}$$

Thus, we obtain

$$F_{ex} = \frac{\varepsilon_0 V^2 A}{2x^2} - \frac{\varepsilon_0 A V^2}{x^2} \tag{8.44}$$

$$= -\frac{1}{2}\frac{\varepsilon_0 A V^2}{x^2}$$

Note that in this procedure, V was held constant since the voltage source was kept connected to the capacitor plates in the process of displacing the plate. If on the other hand, the voltage source is not connected to the capacitor plates in the process of displacing the plate, then Q remains constant, and we can write the following:

$$\frac{dQ}{dx} = 0$$

$$\frac{dW_e}{dx} = \frac{d}{dx}\left(\frac{1}{2}\varepsilon_0 E^2\,Ax\right)$$

$$= \frac{d}{dx}\left[\frac{1}{2}\,\varepsilon_0\left(\frac{Q}{A\varepsilon_0}\right)^2 Ax\right]$$

$$= \frac{1}{2}\frac{Q^2}{A\varepsilon_0}$$

$$F_{ex} = -\frac{1}{2}\frac{Q^2}{A\varepsilon_0} \tag{8.45}$$

The results obtained for F_{ex} in (8.44) and (8.45) appear to be different, but they are not. This can be seen by expressing (8.44) in terms of Q or by expressing (8.45) in terms of V. Choosing the first option, we can write (8.44) as

$$F_{ex} = -\frac{1}{2}\frac{\varepsilon_0 A V^2}{x^2}$$

$$= -\frac{1}{2}\,\varepsilon_0 A E^2$$

$$= -\frac{1}{2}\,\varepsilon_0 A\left(\frac{Q}{A\varepsilon_0}\right)^2$$

$$= -\frac{1}{2}\frac{Q^2}{A\varepsilon_0}$$

which is the same as that given by (8.45). This is to be expected since Q and V are not independent of each other; they are related through the capacitance of the capacitor. Thus, the force $\mathbf{F}_e$ is given by

$$\boxed{\mathbf{F}_e = -\frac{1}{2}\frac{\varepsilon_0 A V^2}{x^2}\,\mathbf{a}_x = -\frac{1}{2}\frac{Q^2}{A\varepsilon_0}\,\mathbf{a}_x} \tag{8.46}$$

We shall now illustrate by means of an example the application of the result we obtained for $\mathbf{F}_e$ in the computation of energy converted from electrical to mechanical, or vice versa, in the energy conversion process.

Example 8.3

Energy conversion computation

Assume that in the parallel-plate capacitor of Fig. 8.11, a source of mechanical force $\mathbf{F}$ is applied to the movable plate such that $\mathbf{F}$ is always maintained equal to $-\mathbf{F}_e$. By appropriately varying V and $\mathbf{F}$, the system is made to traverse the closed cycle in the Q–x-plane, shown in Fig. 8.12. We wish to calculate the energy converted per cycle and determine whether the conversion is from electrical to mechanical or vice versa.

Since the system is made to traverse a closed cycle in the Q–x-plane, there is no change in the electrical stored energy from the initial state to the final state. Hence, the

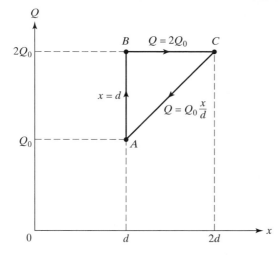

FIGURE 8.12 Closed cycle traversed by the capacitor system of Fig. 8.11.

sum of the mechanical and electrical energy inputs to the system must be zero, or the electrical energy output is equal to the mechanical energy input. The mechanical energy input is given by

$$W_{\substack{\text{mechanical} \\ \text{input}}} = \oint_{ABCA} F_x \, dx = -\oint_{ABCA} F_{ex} \, dx$$

$$= -\int_A^B F_{ex} \, dx - \int_B^C F_{ex} \, dx - \int_C^A F_{ex} \, dx$$

From A to B, x remains constant; hence, $\int_A^B F_{ex} \, dx$ is zero. From (8.45),

$$F_{ex} = \begin{cases} -\dfrac{2Q_0^2}{A\varepsilon_0} & \text{from } B \text{ to } C \\[2mm] -\dfrac{Q_0^2 x^2}{2A\varepsilon_0 d^2} & \text{from } C \text{ to } A \end{cases}$$

Hence,

$$W_{\substack{\text{mechanical} \\ \text{input}}} = \int_{x=d}^{2d} \frac{2Q_0^2}{A\varepsilon_0} \, dx + \int_{x-2d}^{d} \frac{Q_0^2 x^2}{2A\varepsilon_0 d^2} \, dx$$

$$= \frac{2Q_0^2 d}{\varepsilon_0 A} - \frac{7Q_0^2 d}{6\varepsilon_0 A}$$

$$= \frac{5}{6} \frac{Q_0^2 d}{\varepsilon_0 A}$$

Thus, an amount of energy equal to $5Q_0^2 d / 6\varepsilon_0 A$ is converted from mechanical to electrical form.

We have thus far considered an electric-field electromechanical system, *Electromagnet* that is, one in which conversion takes place between energy stored in an electric field and mechanical energy. For an example of a magnetic-field electromechanical system, that is, one in which conversion takes place between energy stored in a magnetic field and mechanical energy, let us consider the arrangement shown in Fig. 8.13 which is the cross section of an electromagnet. When current is passed

FIGURE 8.13 Electromagnet.

through the coil, the armature is pulled upward to close the air gap. The mechanical force $\mathbf{F}_e$ of electric origin can once again be found from energy balance.

In the case of the parallel-plate capacitor of Fig. 8.13, we found $\mathbf{F}_e$ in two ways: by keeping the voltage across the plates constant and by keeping the charge on the plates constant. We found that the two approaches resulted in equivalent expressions for the force. In the present case, we can find $\mathbf{F}_e$ by keeping the current I in the exciting coil to be a constant or by keeping the magnetic flux ψ in the core (and, hence, in the air gap) to be a constant. The two approaches should result in equivalent expressions for $\mathbf{F}_e$. We shall, therefore, take advantage of this to simplify the task of finding $\mathbf{F}_e$ by keeping ψ constant, since then no voltage is induced in the coil and, hence, the electrical energy input term in (8.41) can be set to zero. Also, we shall once again assume a lossless system and apply to the armature an external force equal to $-\mathbf{F}_e$ and displace it by an infinitesimal distance in the direction of the external force. Thus, we obtain

$$-F_{ex}\, dx = dW_m$$

$$F_{ex} = -\frac{dW_m}{dx}$$

where W_m is the magnetic stored energy in the system.

Neglecting fringing of flux across the air gap and noting that the displacement of the armature changes only the magnetic energy stored in the air gap, we write the following:

$$H_{\text{gap}} = \frac{\psi}{A\mu_0}$$

$$(W_m)_{\text{gap}} = 2\left[\frac{1}{2}\mu_0(H_{\text{gap}})^2 Ax\right]$$

$$= \mu_0\left(\frac{\psi}{A\mu_0}\right)^2 Ax$$

$$= \frac{\psi^2 x}{A\mu_0}$$

where A is the cross-sectional area of each gap, and the factor 2 takes into account two gaps. Proceeding further, we have

$$\frac{dW_m}{dx} = \frac{d}{dx}[(W_m)_{\text{gap}}]$$

$$= \frac{\psi^2}{A\mu_0}$$

$$F_{ex} = -\frac{\psi^2}{A\mu_0}$$

$$\boxed{\mathbf{F}_e = -\frac{\psi^2}{A\mu_0}\mathbf{a}_x} \tag{8.47}$$

The expression for $\mathbf{F}_e$ in terms of the current I in the coil that would result from considerations of constant I may now be found by simply expressing ψ in (8.47) in terms of I. Thus, if we assume for simplicity that the permeability of the magnetic core material is so high that

$$H_{\text{core}}l_{\text{core}} \ll H_{\text{gap}}l_{\text{gap}}$$

where l_{core} and l_{gap} arc the lengths of the core and air gap, respectively, then

$$NI \approx 2H_{\text{gap}}x$$

$$H_{\text{gap}} \approx \frac{NI}{2x}$$

$$B_{\text{gap}} \approx \frac{\mu_0 NI}{2x}$$

$$\psi \sim \frac{\mu_0 NIA}{2x}$$

$$\boxed{\mathbf{F}_e \approx -\frac{\mu_0 N^2 I^2 A}{4x^2}\mathbf{a}_x} \tag{8.48}$$

Finally, the computation of energy converted from electrical to mechanical, or vice versa, in a magnetic-field electromechanical system can be performed in a manner similar to that illustrated in Example 8.3 for an electric-field system.

K8.3. Mechanical force of electric origin; Energy conversion; Parallel-plate capacitor with movable plate; Electromagnet.

D8.4. For the parallel-plate capacitor of Fig. 8.13, assume $V = 10$ V, $x = 1$ cm, and $A = 0.01$ m^2 and compute $\mathbf{F}_e$ for each of the following cases: (a) the dielectric between the plates is free space; (b) the dielectric between the plates is a material of permittivity $4\varepsilon_0$; and (c) the lower half of the region between the plates is a dielectric of permittivity $4\varepsilon_0$, whereas the upper half is free space.
Ans. (a) $-5000\varepsilon_0\mathbf{a}_x$ N; (b) $-20,000\varepsilon_0\mathbf{a}_x$, N; (c) $-12,800\varepsilon_0\mathbf{a}_x$ N.

8.4 SUMMARY

In this chapter, we introduced the quasistatic extension of the static field as a means of obtaining the low-frequency behavior of a physical structure. The quasistatic-field approach involves starting with a time-varying field having the same spatial characteristics as the static field and then obtaining field solutions containing terms up to and including the first power in frequency by using Maxwell's curl equations for time-varying fields. We learned that the quasistatic approximation leads to electric- and magnetic-field systems. We illustrated the quasistatic-field analysis by considering two examples, one of them involving a lossy medium.

We then discussed the magnetic circuit, which is essentially an arrangement of closed paths for magnetic flux to flow around just as current in electric circuits. The closed paths are provided by ferromagnetic cores which because of their high permeability relative to that of the surrounding medium confine the flux almost entirely to within the core regions. We illustrated the analysis of magnetic circuits by considering two examples, one of them including an air gap in one of the legs.

Finally, we studied the topic of electromechanical energy conversion. By considering examples of a parallel-plate capacitor with one movable plate, and an electromagnet, we discussed the determination of mechanical forces of electric origin. We also illustrated energy conversion computation for the parallel-plate capacitor example.

REVIEW QUESTIONS

Q8.1. What is meant by the quasistatic extension of the static field in a physical structure?

Q8.2. Outline the steps involved in the quasistatic extension of the static field in a parallel-plate structure short-circuited at one end.

Q8.3. Discuss the derivation of the condition for the validity of the quasistatic approximation for the parallel-plate structure short-circuited at one end.

Q8.4. Discuss the general condition for the quasistatic approximation of a physical structure.

Q8.5. Discuss the classification of physical structures as electric- and magnetic-field systems.

Q8.6. Discuss the low-frequency behavior of a parallel-plate structure with a lossy medium between the plates.

Q8.7. Discuss the quasistatic behavior of the structure of Fig. 8.2 for $\sigma \approx 0$.

Q8.8. What is a magnetic circuit? Why is the magnetic flux in a magnetic circuit confined almost entirely to the core?

Q8.9. Define the reluctance of a magnetic circuit. What is the analogous electric circuit quantity? Why is the reluctance for a given set of dimensions of a magnetic circuit not a constant?

Q8.10. Discuss the complete analogy between a magnetic circuit and an electric circuit using the example of the toroidal magnetic core versus the toroidal conductor.

Q8.11. How is the fringing of the magnetic flux in an air gap in a magnetic circuit taken into account?

Q8.12. Discuss by means of an example the analysis of a magnetic circuit with three legs and its equivalent-circuit representation.

Q8.13. Discuss by means of an example the phenomenon of electromechanical energy conversion.

Q8.14. Outline the computation of mechanical force of electric origin from considerations of energy balance associated with an electromechanical system.

Q8.15. Discuss by means of an example the computation of energy converted from electrical to mechanical, or vice versa, in an electromechanical system.

PROBLEMS

Section 8.1

P8.1. For the structure of Fig. 8.1, assume that $l = 10$ cm, $d = 5$ mm, and $w = 5$ cm and free space for the medium between the plates. **(a)** For a current source $I(t) = 1 \cos 10^6 \pi t$ A, find the voltage developed across the source. **(b)** Repeat part (a) for $I(t) = 1 \cos 10^9 \pi t$ A.

P8.2. For the structure of Fig. 8.2 with $\sigma = 0$, show that the input behavior for frequencies slightly beyond those for which the quasistatic approximation is valid is equivalent to the series combination of $C (= \varepsilon wl/d)$ and $\frac{1}{3}L$, where $L = \mu \, dl/w$ is the inductance of the structure obtained from static-field considerations with the two plates joined by another conductor at $z = 0$, as in Fig. 8.1.

P8.3. Find the conditions under which the quasistatic input behavior of the structure of Fig. 8.2 is essentially equivalent to that of: **(a)** a single resistor; **(b)** a capacitor $C (= \varepsilon wl/d)$ in parallel with a resistor; and **(c)** a resistor in series with an inductor.

P8.4. For the structure of Fig. 8.1, assume that the medium has nonzero conductivity σ. **(a)** Show that the input behavior correct to the first power in ω is the same as if σ were zero. **(b)** Investigate the input behavior correct to the second power in ω and obtain the equivalent circuit.

P8.5. For the structure of Fig. 8.1, obtain the equivalent circuit for the input behavior for frequencies for which the fields up to and including the fifth-order terms in ω are significant.

P8.6. For the structure of Fig. 8.1, show that, under the quasistatic approximation, the rate at which energy flows into the volume of the structure as obtained by surface integration of the Poynting vector over the surface bounding the volume is $d[\frac{1}{2} LI^2(t)]/dt$.

Section 8.2

P8.7. A toroidal magnetic core has the dimensions $A = 5$ cm^2 and $l = 20$ cm. **(a)** If it is found that for NI equal to 200 A-t, a magnetic flux ψ equal to 8×10^{-4} Wb is established in the core, find the permeability μ of the core material. **(b)** If now an air gap of width $l_g = 0.1$ mm is introduced, find the new value of NI required to maintain the flux of 8×10^{-4} Wb, neglecting fringing of flux in the air gap.

P8.8. For the magnetic circuit of Fig. 8.9, assume the air gap to be in the center leg. Find the NI required to establish a magnetic flux of 9×10^{-4} Wb in the air gap.

P8.9. For the magnetic circuit of Fig. 8.9, assume that there is an air gap of length 0.2 mm in the left leg in addition to that in the right leg. Find the NI required to establish a magnetic flux of 4×10^{-4} Wb in the air gap in the right leg.

P8.10. For the magnetic circuit of Fig. 8.9, assume that there is no air gap. Find the magnetic flux established in the center leg for an applied NI equal to 180 A-t.

P8.11. For the magnetic circuit of Fig. 8.9, assume that there is no air gap and that $A_1 = 5$ cm^2 with all other dimensions remaining as specified in Example 8.2. Find the magnetic flux density in the center leg for an applied NI equal to 150 A-t.

Section 8.3

P8.12. In Fig. 8.14, a dielectric slab of permittivity ε sliding between the plates of a parallel-plate capacitor experiences a mechanical force $\mathbf{F}_e$ of electrical origin. Assuming width w for the plates normal to the page and neglecting fringing of fields at the edges of the plates, find the expression for $\mathbf{F}_e$.

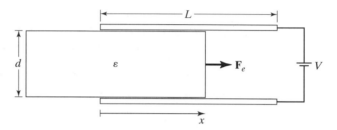

FIGURE 8.14   For Problem P8.12.

P8.13. In Fig. 8.15, a dielectric material of permittivity ε sliding freely in a cylindrical capacitor experiences a mechanical force $\mathbf{F}_e$ of electrical origin in the axial direction. Show that

$$\mathbf{F}_e = \frac{V_0^2 \pi (\varepsilon - \varepsilon_0)}{\ln (b/a)} \mathbf{a}_x$$

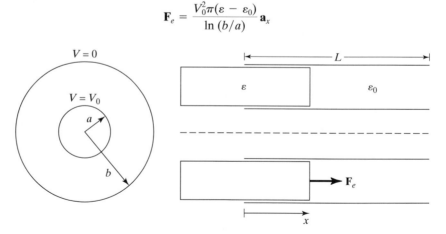

FIGURE 8.15   For Problem P8.13.

P8.14. Assume that in Example 8.3, the parallel-plate capacitor system of Fig. 8.11 is made to traverse the closed cycle in the V-x plane shown in Fig. 8.16 instead of the closed cycle in the Q-x plane shown in Fig. 8.12. Calculate the energy converted per cycle and determine whether the conversion is from mechanical to electrical or vice versa.

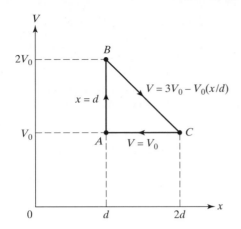

FIGURE 8.16 For Problem P8.14.

P8.15. Figure 8.17(a) shows a magnetic-field electromechanical device in which the magnetic core is free to slide inside a long air-core solenoidal coil. The solenoid has length l, radius a, and number of turns per meter N, and carries a current I. The magnetic core has length $b < l$, radius a, and permeability $\mu \gg \mu_0$, and extends a distance x into the solenoid. **(a)** Neglect fringing of the field and find the mechanical force $\mathbf{F}_e$ of electric origin on the core. Plot F_{ex} versus x. **(b)** Assume that the device is made to traverse the closed path in the I-x plane, as shown in Fig. 8.17(b). Find the energy converted per cycle and determine whether it is from mechanical to electrical or vice versa.

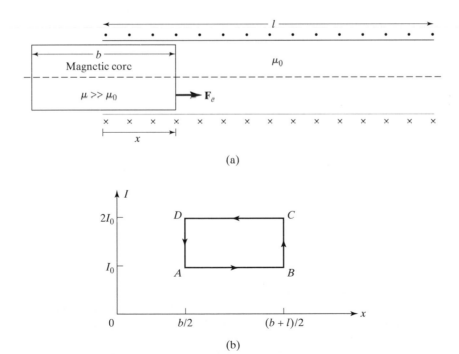

FIGURE 8.17 For Problem P8.15.

REVIEW PROBLEMS

R8.1.   An air-dielectric coaxial cable of inner radius $a = 1$ cm, outer radius $b = 2$ cm, and length $l = 1$ m is short-circuited at one end. Obtain the equivalent circuit for the input behavior of the structure for frequencies slightly beyond those for which the quasistatic approximation is valid. Compute the resonant frequency of the equivalent circuit and comment on its value compared to those for which the circuit is valid.

R8.2.   For the magnetic circuit shown in Fig. 8.18, the dimensions of the legs are $A_1 = A_3 = 2$ cm², $A_2 = 3$ cm², $l_1 = l_3 = 30$ cm, and $l_2 = 10$ cm. The permeability of the core material can be assumed to be $1000\mu_0$. **(a)** Draw the equivalent electric circuit. **(b)** For $N_1 I_1 = 200$ A-t and $N_2 I_2 = 100$ A-t, find the magnetic flux in each leg.

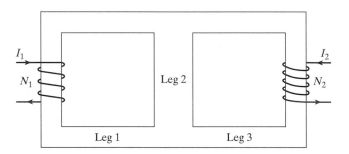

FIGURE 8.18   For Problem R8.2.

R8.3.   In the system shown in Fig. 8.19, the mass M is set in motion in the following manner: (1) the mass is brought to rest at the equilibrium position $x = x_0$ with no charge on the capacitor plates; (2) the mass is constrained to that position and the capacitor plates are charged to $\pm Q$ as shown; and (3) the mass is released, thereby permitting frictionless motion. Obtain the differential equation for the motion of M and find the solution.

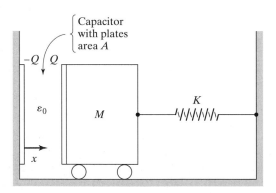

FIGURE 8.19   For Problem R8.3.

Guided Wave Principles for Electronics and Optoelectronics

In Chapter 6, we introduced the transmission line and studied propagation and bouncing of waves along a line, applicable to digital electronics. We devoted Chapter 7 to sinusoidal steady-state analysis of waves on transmission lines. We learned that wave propagation along a transmission line is characterized by the waves sliding parallel to its conductors, with electric and magnetic fields entirely transverse to the direction of propagation, with the waves being known as transverse electromagnetic (TEM) waves.

Another kind of waveguiding mechanism is one in which the waves bounce obliquely between the parallel planes as they progress along the structure, resulting in transverse electric (TE) and transverse magnetic (TM) waves. The arrangement is then familiarly referred to as a waveguide, although the transmission line is also a waveguide. To continue our study of guided waves for electronics, we introduce in this chapter the principles of TE and TM waves, supported by plane conductors as in the case of the parallel-plate transmission line, as well as those supported by a plane dielectric slab, without the conductors. The latter arrangement is particularly applicable to optoelectronics.

Thus, we first consider the parallel-plate waveguide, consisting of two parallel plane conductors. To do this, we make use of the superposition of two uniform plane waves propagating at an angle to each other. Hence, we begin the chapter with a discussion of uniform plane wave propagation in an arbitrary direction relative to the coordinate axes.

9.1 UNIFORM PLANE WAVE PROPAGATION IN AN ARBITRARY DIRECTION

Two dimensions   In Chapter 5, we introduced the uniform plane wave propagating in the z-direction by considering an infinite plane current sheet lying in the xy-plane. If the current sheet lies in a plane making an angle to the xy-plane, the uniform plane wave would then propagate in a direction different from the z-direction. Thus, let us first consider the two-dimensional case of a uniform plane wave propagating in a perfect dielectric medium in the z'-direction making an angle θ with the negative x-axis, as shown in Fig. 9.1. Let the electric field of the wave be entirely in the y-direction. The magnetic field would then be directed as shown in the figure so that $\mathbf{E} \times \mathbf{H}$ points in the z'-direction.

We can write the expression for the electric field of the wave as

$$\mathbf{E} = E_0 \cos (\omega t - \beta z') \, \mathbf{a}_y \tag{9.1}$$

where $\beta = \omega \sqrt{\mu \varepsilon}$ is the phase constant, that is, the rate of change of phase with distance along the z'-direction for a fixed value of time. From the construction of Fig. 9.2(a), we, however, have

$$z' = -x \cos \theta + z \sin \theta \tag{9.2}$$

so that

$$
\begin{aligned}
\mathbf{E} &= E_0 \cos \left[\omega t - \beta(-x \cos \theta + z \sin \theta) \right] \mathbf{a}_y \\
&= E_0 \cos \left[\omega t - (-\beta \cos \theta)x - (\beta \sin \theta)z \right] \mathbf{a}_y \\
&= E_0 \cos (\omega t - \beta_x x - \beta_z z) \, \mathbf{a}_y
\end{aligned}
\tag{9.3}
$$

where $\beta_x = -\beta \cos \theta$ and $\beta_z = \beta \sin \theta$ are the phase constants in the positive x- and positive z-directions, respectively.

We note that $|\beta_x|$ and $|\beta_z|$ are less than β, the phase constant along the direction of propagation of the wave. This can also be seen from Fig. 9.1 in which two constant phase surfaces are shown by dashed lines passing through the

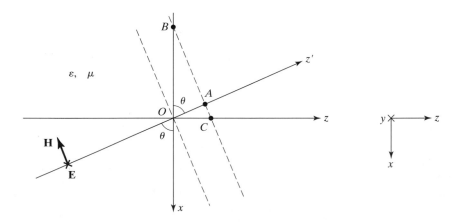

FIGURE 9.1   Uniform plane wave propagating in the z'-direction lying in the xz-plane and making an angle θ with the negative x-axis.

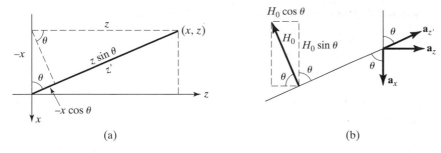

FIGURE 9.2   Constructions pertinent to the formulation of the expressions for the fields of the uniform plane wave of Fig. 9.1.

points O and A on the z'-axis. Since the distance along the x-direction between the two constant phase surfaces, that is, the distance OB is equal to $OA/\cos\theta$, the rate of change of phase with distance along the x-direction is equal to

$$\beta\frac{OA}{OB} = \frac{\beta(OA)}{OA/\cos\theta} = \beta\cos\theta$$

The minus sign for β_x simply signifies the fact that insofar as the x-axis is concerned, the wave is progressing in the negative x-direction. Similarly, since the distance along the z-direction between the two constant phase surfaces, that is, the distance OC is equal to $OA/\sin\theta$, the rate of change of phase with distance along the z-direction is equal to

$$\beta\frac{OA}{OC} = \frac{\beta(OA)}{OA/\sin\theta} = \beta\sin\theta$$

Since the wave is progressing along the positive z-direction, β_z is positive. We further note that

$$\beta_x^2 + \beta_z^2 = (-\beta\cos\theta)^2 + (\beta\sin\theta)^2 = \beta^2 \tag{9.4}$$

and that

$$-\cos\theta\,\mathbf{a}_x + \sin\theta\,\mathbf{a}_z = \mathbf{a}_{z'} \tag{9.5}$$

where $\mathbf{a}_{z'}$ is the unit vector directed along the z'-direction, as shown in Fig. 9.2(b). Thus, the vector

$$\boldsymbol{\beta} = (-\beta\cos\theta)\mathbf{a}_x + (\beta\sin\theta)\mathbf{a}_z = \beta_x\mathbf{a}_x + \beta_z\mathbf{a}_z \tag{9.6}$$

defines completely the direction of propagation and the phase constant along the direction of propagation. Hence, the vector $\boldsymbol{\beta}$ is known as the *propagation vector.*

The expression for the magnetic field of the wave can be written as

$$\mathbf{H} = \mathbf{H}_0\cos(\omega t - \beta z') \tag{9.7}$$

where

$$|\mathbf{H}_0| - \frac{E_0}{\sqrt{\mu/\varepsilon}} = \frac{E_0}{\eta} \tag{9.8}$$

since the ratio of the electric field intensity to the magnetic field intensity of a uniform plane wave is equal to the intrinsic impedance of the medium. From the construction in Fig. 9.2(b), we observe that

$$\mathbf{H}_0 = H_0(-\sin \theta \, \mathbf{a}_x - \cos \theta \, \mathbf{a}_z) \tag{9.9}$$

Thus, using (9.9) and substituting for z' from (9.2), we obtain

$$\begin{aligned}
\mathbf{H} &= H_0(-\sin \theta \, \mathbf{a}_x - \cos \theta \, \mathbf{a}_z) \cos \left[\omega t - \beta(-x \cos \theta + z \sin \theta)\right] \\
&= -\frac{E_0}{\eta}(\sin \theta \, \mathbf{a}_x + \cos \theta \, \mathbf{a}_z) \cos \left[\omega t - \beta_x x - \beta_z z\right]
\end{aligned} \tag{9.10}$$

Generalization to three dimensions   Generalizing the foregoing treatment to the case of a uniform plane wave propagating in a completely arbitrary direction in three dimensions, as shown in Fig. 9.3, and characterized by phase constants β_x, β_y, and β_z in the x-, y-, and z-directions, respectively, we can write the expression for the electric field as

$$\begin{aligned}
\mathbf{E} &= \mathbf{E}_0 \cos (\omega t - \beta_x x - \beta_y y - \beta_z z + \phi_0) \\
&= \mathbf{E}_0 \cos \left[\omega t - (\beta_x \mathbf{a}_x + \beta_y \mathbf{a}_y + \beta_z \mathbf{a}_z) \cdot (x\mathbf{a}_x + y\mathbf{a}_y + z\mathbf{a}_z) + \phi_0\right] \\
&= \mathbf{E}_0 \cos (\omega t - \boldsymbol{\beta} \cdot \mathbf{r} + \phi_0)
\end{aligned} \tag{9.11}$$

where

$$\boldsymbol{\beta} = \beta_x \mathbf{a}_x + \beta_y \mathbf{a}_y + \beta_z \mathbf{a}_z \tag{9.12}$$

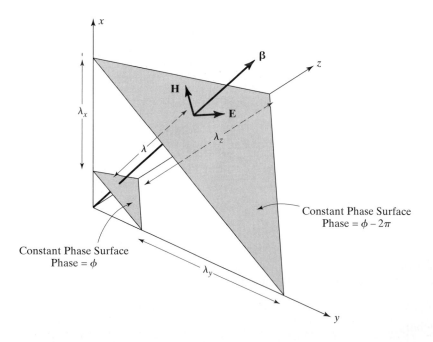

FIGURE 9.3   The various quantities associated with a uniform plane wave propagating in an arbitrary direction.

is the propagation vector,

$$\mathbf{r} = x\mathbf{a}_x + y\mathbf{a}_y + z\mathbf{a}_z \qquad (9.13)$$

is the position vector, and ϕ_0 is the phase at the origin at $t = 0$. We recall that the position vector is the vector drawn from the origin to the point (x, y, z) and hence has components x, y, and z along the x-, y-, and z-axes, respectively. The expression for the magnetic field of the wave is then given by

$$\mathbf{H} = \mathbf{H}_0 \cos (\omega t - \boldsymbol{\beta} \cdot \mathbf{r} + \phi_0) \qquad (9.14)$$

where

$$|\mathbf{H}_0| - \frac{|\mathbf{E}_0|}{\eta} \qquad (9.15)$$

Since $\mathbf{E}$, $\mathbf{H}$, and the direction of propagation are mutually perpendicular to each other, it follows that

$$\mathbf{E}_0 \cdot \boldsymbol{\beta} = 0 \qquad (9.16a)$$

$$\mathbf{H}_0 \cdot \boldsymbol{\beta} = 0 \qquad (9.16b)$$

$$\mathbf{E}_0 \cdot \mathbf{H}_0 = 0 \qquad (9.16c)$$

In particular, $\mathbf{E} \times \mathbf{H}$ should be directed along the propagation vector $\boldsymbol{\beta}$, as illustrated in Fig. 9.3, so that $\boldsymbol{\beta} \times \mathbf{E}_0$ is directed along $\mathbf{H}_0$. We can therefore combine the facts (9.16) and (9.15) to obtain

$$\mathbf{H}_0 = \frac{\mathbf{a}_\beta \times \mathbf{E}_0}{\eta} = \frac{\mathbf{a}_\beta \times \mathbf{E}_0}{\sqrt{\mu/\varepsilon}} = \frac{\omega\sqrt{\mu\varepsilon} \, \mathbf{a}_\beta \times \mathbf{E}_0}{\omega\mu}$$

$$= \frac{\beta \mathbf{a}_\beta \times \mathbf{E}_0}{\omega\mu} = \frac{\boldsymbol{\beta} \times \mathbf{E}_0}{\omega\mu} \qquad (9.17)$$

where $\mathbf{a}_\beta$ is the unit vector along $\boldsymbol{\beta}$. Thus,

$$\mathbf{H} = \frac{1}{\omega\mu} \boldsymbol{\beta} \times \mathbf{E} \qquad (9.18)$$

Returning to Fig. 9.3, we can define several quantities pertinent to the uniform plane wave propagation in an arbitrary direction. The apparent wavelengths $\lambda_x, \lambda_y,$ and λ_z along the coordinate axes $x, y,$ and z, respectively, are the distances measured along those respective axes between two consecutive constant

Apparent wavelengths and phase velocities

phase surfaces between which the phase difference is 2π, as shown in the figure, at a fixed time. From the interpretations of β_x, β_y, and β_z as being the phase constants along the x-, y-, and z-axes, respectively, we have

$$\lambda_x = \frac{2\pi}{\beta_x} \tag{9.19a}$$

$$\lambda_y = \frac{2\pi}{\beta_y} \tag{9.19b}$$

$$\lambda_z = \frac{2\pi}{\beta_z} \tag{9.19c}$$

We note that the wavelength λ along the direction of propagation is related to λ_x, λ_y, and λ_z in the manner

$$\frac{1}{\lambda^2} = \frac{1}{(2\pi/\beta)^2} = \frac{\beta^2}{4\pi^2} = \frac{\beta_x^2 + \beta_y^2 + \beta_z^2}{4\pi^2}$$
$$= \frac{1}{\lambda_x^2} + \frac{1}{\lambda_y^2} + \frac{1}{\lambda_z^2} \tag{9.20}$$

The apparent phase velocities v_{px}, v_{py}, and v_{pz} along the x-, y-, and z-axes, respectively, are the velocities with which the phase of the wave progresses with time along the respective axes. Thus,

$$v_{px} = \frac{\omega}{\beta_x} \tag{9.21a}$$

$$v_{py} = \frac{\omega}{\beta_y} \tag{9.21b}$$

$$v_{pz} = \frac{\omega}{\beta_z} \tag{9.21c}$$

The phase velocity v_p along the direction of propagation is related to v_{px}, v_{py}, and v_{pz} in the manner

$$\frac{1}{v_p^2} = \frac{1}{(\omega/\beta)^2} = \frac{\beta^2}{\omega^2} = \frac{\beta_x^2 + \beta_y^2 + \beta_z^2}{\omega^2}$$
$$= \frac{1}{v_{px}^2} + \frac{1}{v_{py}^2} + \frac{1}{v_{pz}^2} \tag{9.22}$$

The apparent wavelengths and phase velocities along the coordinate axes are greater than the actual wavelength and phase velocity, respectively, along the direction of propagation of the wave. This fact can be understood physically by considering, for example, water waves in an ocean striking the shore at an angle. The distance along the shoreline between two successive crests is greater than the distance between the same two crests measured along a line normal to the orientation of the crests. Also, to keep pace with a particular crest an observer has to run faster along the shoreline than in a direction normal to the orientation of the crests. We shall now consider an example.

Example 9.1

Let us consider a 30-MHz uniform plane wave propagating in free space and given by the electric field vector

$$\mathbf{E} = 5(\mathbf{a}_x + \sqrt{3}\mathbf{a}_y) \cos [6\pi \times 10^7 t - 0.05\pi(3x - \sqrt{3}y + 2z)] \text{ V/m}$$

We wish to verify the properties and find the magnetic field vector $\mathbf{H}$ and other parameters associated with the wave.

Comparing the given expression for $\mathbf{E}$ with the general expression (9.11), we have

$$\mathbf{E}_0 = 5(\mathbf{a}_x + \sqrt{3}\mathbf{a}_y)$$
$$\boldsymbol{\beta} \cdot \mathbf{r} = 0.05\pi(3x - \sqrt{3}y + 2z)$$
$$= 0.05\pi(3\mathbf{a}_x - \sqrt{3}\mathbf{a}_y + 2\mathbf{a}_z) \cdot (x\mathbf{a}_x + y\mathbf{a}_y + z\mathbf{a}_z)$$
$$\boldsymbol{\beta} = 0.05\pi(3\mathbf{a}_x - \sqrt{3}\mathbf{a}_y + 2\mathbf{a}_z)$$
$$\boldsymbol{\beta} \cdot \mathbf{E}_0 = 0.05\pi(3\mathbf{a}_x - \sqrt{3}\mathbf{a}_y + 2\mathbf{a}_z) \cdot 5(\mathbf{a}_x + \sqrt{3}\mathbf{a}_y)$$
$$= 0.25\pi(3 - 3) = 0$$

Hence, (9.16a) is satisfied; $\mathbf{E}_0$ is perpendicular to $\boldsymbol{\beta}$.

$$\beta = |\boldsymbol{\beta}| = 0.05\pi|3\mathbf{a}_x - \sqrt{3}\mathbf{a}_y + 2\mathbf{a}_z| = 0.05\pi\sqrt{9 + 3 + 4} = 0.2\pi$$

$$\lambda = \frac{2\pi}{\beta} = \frac{2\pi}{0.2\pi} = 10 \text{ m}$$

This does correspond to a frequency of $(3 \times 10^8)/10$ Hz, or 30 MHz, in free space. The direction of propagation is along the unit vector

$$\mathbf{a}_\beta = \frac{\boldsymbol{\beta}}{|\boldsymbol{\beta}|} = \frac{3\mathbf{a}_x - \sqrt{3}\mathbf{a}_y + 2\mathbf{a}_z}{\sqrt{9 + 3 + 4}} = \frac{3}{4}\mathbf{a}_x - \frac{\sqrt{3}}{4}\mathbf{a}_y + \frac{1}{2}\mathbf{a}_z$$

From (9.17),

$$\mathbf{H}_0 = \frac{1}{\omega\mu_0} \boldsymbol{\beta} \times \mathbf{E}_0$$

$$= \frac{0.05\pi \times 5}{6\pi \times 10^7 \times 4\pi \times 10^{-7}} (3\mathbf{a}_x - \sqrt{3}\mathbf{a}_y + 2\mathbf{a}_z) \times (\mathbf{a}_x + \sqrt{3}\mathbf{a}_y)$$

$$= \frac{1}{96\pi} \begin{vmatrix} \mathbf{a}_x & \mathbf{a}_y & \mathbf{a}_z \\ 3 & -\sqrt{3} & 2 \\ 1 & \sqrt{3} & 0 \end{vmatrix}$$

$$= \frac{1}{48\pi} (-\sqrt{3}\mathbf{a}_x + \mathbf{a}_y + 2\sqrt{3}\mathbf{a}_z)$$

Thus,

$$\mathbf{H} = \frac{1}{48\pi}(-\sqrt{3}\mathbf{a}_x + \mathbf{a}_y + 2\sqrt{3}\mathbf{a}_z) \cos [6\pi \times 10^7 t - 0.05\pi(3x - \sqrt{3}y + 2z)] \text{ A/m}$$

To verify the expression for **H** just derived, we note that

$$\mathbf{H}_0 \cdot \boldsymbol{\beta} = \left[\frac{1}{48\pi} (-\sqrt{3}\mathbf{a}_x + \mathbf{a}_y + 2\sqrt{3}\mathbf{a}_z) \right] \cdot [0.05\pi(3\mathbf{a}_x - \sqrt{3}\mathbf{a}_y + 2\mathbf{a}_z)]$$

$$= \frac{0.05}{48} (-3\sqrt{3} - \sqrt{3} + 4\sqrt{3}) = 0$$

$$\mathbf{E}_0 \cdot \mathbf{H}_0 = 5(\mathbf{a}_x + \sqrt{3}\mathbf{a}_y) \cdot \frac{1}{48\pi} (-\sqrt{3}\mathbf{a}_x + \mathbf{a}_y + 2\sqrt{3}\mathbf{a}_z)$$

$$= \frac{5}{48\pi} (-\sqrt{3} + \sqrt{3}) = 0$$

$$\frac{|\mathbf{E}_0|}{|\mathbf{H}_0|} = \frac{5|\mathbf{a}_x + \sqrt{3}\mathbf{a}_y|}{(1/48\pi)|-\sqrt{3}\mathbf{a}_x + \mathbf{a}_y + 2\sqrt{3}\mathbf{a}_z|} = \frac{5\sqrt{1 + 3}}{(1/48\pi)\sqrt{3 + 1 + 12}}$$

$$= \frac{10}{1/12\pi} = 120\pi = \eta_0$$

Hence, (9.16b), (9.16c), and (9.15) are satisfied.

Proceeding further, we find that

$$\beta_x = 0.05\pi \times 3 = 0.15\pi$$

$$\beta_y = -0.05\pi \times \sqrt{3} = -0.05\sqrt{3}\pi$$

$$\beta_z = 0.05\pi \times 2 = 0.1\pi$$

We then obtain

$$\lambda_x = \frac{2\pi}{\beta_x} = \frac{2\pi}{0.15\pi} = \frac{40}{3} \text{ m} = 13.333 \text{ m}$$

$$\lambda_y = \frac{2\pi}{|\beta_y|} = \frac{2\pi}{0.05\sqrt{3}\pi} = \frac{40}{\sqrt{3}} \text{ m} = 23.094 \text{ m}$$

$$\lambda_z = \frac{2\pi}{\beta_z} = \frac{2\pi}{0.1\pi} = 20 \text{ m}$$

$$v_{px} = \frac{\omega}{\beta_x} = \frac{6\pi \times 10^7}{0.15\pi} = 4 \times 10^8 \text{ m/s}$$

$$v_{py} = \frac{\omega}{|\beta_y|} = \frac{6\pi \times 10^7}{0.05\sqrt{3}\pi} = 4\sqrt{3} \times 10^8 \text{ m/s} = 6.928 \times 10^8 \text{ m/s}$$

$$v_{pz} = \frac{\omega}{\beta_z} = \frac{6\pi \times 10^7}{0.1\pi} = 6 \times 10^8 \text{ m/s}$$

Finally, to verify (9.20) and (9.22), we note that

$$\frac{1}{\lambda_x^2} + \frac{1}{\lambda_y^2} + \frac{1}{\lambda_z^2} = \frac{1}{(40/3)^2} + \frac{1}{(40/\sqrt{3})^2} + \frac{1}{20^2}$$

$$= \frac{9}{1600} + \frac{3}{1600} + \frac{4}{1600} = \frac{1}{100} = \frac{1}{10^2} = \frac{1}{\lambda^2}$$

and

$$\frac{1}{v_{px}^2} + \frac{1}{v_{py}^2} + \frac{1}{v_{pz}^2} = \frac{1}{(4 \times 10^8)^2} + \frac{1}{(4\sqrt{3} \times 10^8)^2} + \frac{1}{(6 \times 10^8)^2}$$

$$= \frac{1}{16 \times 10^{16}} + \frac{1}{48 \times 10^{16}} + \frac{1}{36 \times 10^{16}}$$

$$= \frac{1}{9 \times 10^{16}} = \frac{1}{(3 \times 10^8)^2} = \frac{1}{v_p^2}$$

K9.1. Uniform plane wave; Propagation in an arbitrary direction; Propagation vector; Apparent wavelengths; Apparent phase velocities.

D9.1. For each of the following cases of a uniform plane wave propagating in free space, find the frequency f: **(a)** wavelength along the direction of propagation of the wave is 2 m; **(b)** the propagation vector is $\pi(1.2\mathbf{a}_x + 0.9\mathbf{a}_y)$ rad/m; and **(c)** the apparent wavelengths along three mutually perpendicular directions are 1 m, 1 m, and 2 m.

Ans. **(a)** 150 MHz; **(b)** 225 MHz; **(c)** 450 MHz.

D9.2. For a uniform plane wave of frequency 150 MHz propagating away from the origin into the first octant in a nonmagnetic ($\mu = \mu_0$), perfect dielectric medium of $\varepsilon = 2\varepsilon_0$, the apparent wavelengths along the x- and y-directions are found to be $2\frac{1}{2}$ m and $3\frac{1}{3}$ m, respectively. Find **(a)** the phase constant along the x-direction; **(b)** the apparent wavelength along the z-direction; **(c)** the apparent phase velocity along the direction of the unit vector $\frac{1}{13}(3\mathbf{a}_x - 4\mathbf{a}_y + 12\mathbf{a}_z)$; and **(d)** the equation of the plane if the source of the wave is an infinite plane sheet of uniform current density passing through the origin.

Ans. **(a)** 0.8π rad/m; **(b)** 2 m; **(c)** 3.25×10^8 m/s; **(d)** $4x + 3y + 5z = 0$.

9.2 TE AND TM WAVES IN A PARALLEL-PLATE WAVEGUIDE

TE waves

In the preceding section, we introduced uniform plane wave propagation in an arbitrary direction. Let us now consider the superposition of two uniform plane waves propagating symmetrically with respect to the z-axis as shown in Fig. 9.4 and having the electric fields entirely in the y-direction as given by

$$\mathbf{E}_1 = -\frac{E_0}{2} \cos(\omega t - \boldsymbol{\beta}_1 \cdot \mathbf{r}) \mathbf{a}_y$$

$$= -\frac{E_0}{2} \cos(\omega t + \beta x \cos\theta - \beta z \sin\theta) \mathbf{a}_y \qquad (9.23a)$$

$$\mathbf{E}_2 = \frac{E_0}{2} \cos(\omega t - \boldsymbol{\beta}_2 \cdot \mathbf{r}) \mathbf{a}_y$$

$$= \frac{E_0}{2} \cos(\omega t - \beta x \cos\theta - \beta z \sin\theta) \mathbf{a}_y \qquad (9.23b)$$

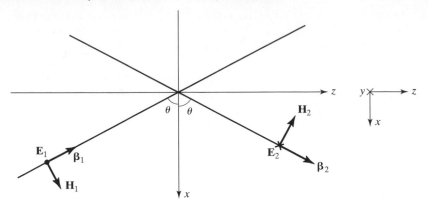

FIGURE 9.4   Superposition of two uniform plane waves propagating symmetrically with respect to the z-axis.

where $\beta = \omega\sqrt{\mu\varepsilon}$, with ε and μ being the permittivity and the permeability, respectively, of the medium. The corresponding magnetic fields are given by

$$\mathbf{H}_1 = \frac{E_0}{2\eta} (\sin\theta\, \mathbf{a}_x + \cos\theta\, \mathbf{a}_z) \cos(\omega t + \beta x \cos\theta - \beta z \sin\theta) \qquad (9.24a)$$

$$\mathbf{H}_2 = \frac{E_0}{2\eta} (-\sin\theta\, \mathbf{a}_x + \cos\theta\, \mathbf{a}_z) \cos(\omega t - \beta x \cos\theta - \beta z \sin\theta) \qquad (9.24b)$$

where $\eta = \sqrt{\mu/\varepsilon}$. The electric and magnetic fields of the superposition of the two waves are given by

$$
\begin{aligned}
\mathbf{E} &= \mathbf{E}_1 + \mathbf{E}_2 \\
&= -\frac{E_0}{2} [\cos(\omega t - \beta z \sin\theta + \beta x \cos\theta) \\
&\qquad - \cos(\omega t - \beta z \sin\theta - \beta x \cos\theta)]\, \mathbf{a}_y \\
&= E_0 \sin(\beta x \cos\theta) \sin(\omega t - \beta z \sin\theta)\, \mathbf{a}_y
\end{aligned}
\qquad (9.25a)
$$

$$
\begin{aligned}
\mathbf{H} &= \mathbf{H}_1 + \mathbf{H}_2 \\
&= \frac{E_0}{2\eta} \sin\theta\, [\cos(\omega t - \beta z \sin\theta + \beta x \cos\theta) \\
&\qquad - \cos(\omega t - \beta z \sin\theta - \beta x \cos\theta)]\, \mathbf{a}_x \\
&\quad + \frac{E_0}{2\eta} \cos\theta\, [\cos(\omega t - \beta z \sin\theta + \beta x \cos\theta) \\
&\qquad + \cos(\omega t - \beta z \sin\theta - \beta x \cos\theta)]\, \mathbf{a}_z \\
&= -\frac{E_0}{\eta} \sin\theta \sin(\beta x \cos\theta) \sin(\omega t - \beta z \sin\theta)\, \mathbf{a}_x \\
&\quad + \frac{E_0}{\eta} \cos\theta \cos(\beta x \cos\theta) \cos(\omega t - \beta z \sin\theta)\, \mathbf{a}_z
\end{aligned}
\qquad (9.25b)
$$

In view of the factors $\sin(\beta x \cos\theta)$ and $\cos(\beta x \cos\theta)$ for the x-dependence and the factors $\sin(\omega t - \beta z \sin\theta)$ and $\cos(\omega t - \beta z \sin\theta)$ for the z-dependence, the composite fields have standing wave character in the x-direction and traveling wave character in the z-direction. Thus, we have standing waves in the x-direction moving bodily in the z-direction, as illustrated in Fig. 9.5, by considering the electric field for two different times. In fact, we find that the

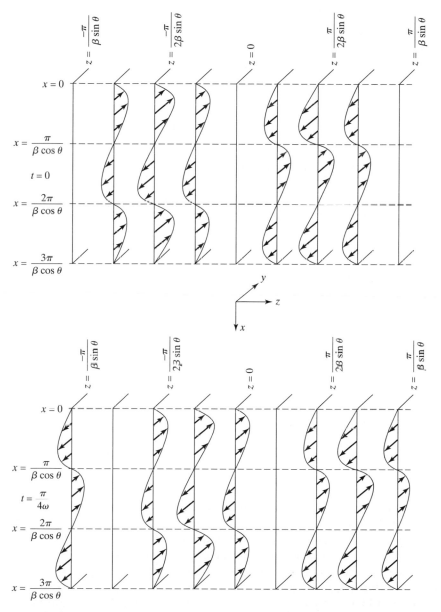

FIGURE 9.5   Standing waves in the x-direction moving bodily in the z-direction.

Poynting vector is given by

$$
\begin{aligned}
\mathbf{P} = \mathbf{E} \times \mathbf{H} &= E_y \mathbf{a}_y \times (H_x \mathbf{a}_x + H_z \mathbf{a}_z) \\
&= -E_y H_x \mathbf{a}_z + E_y H_z \mathbf{a}_x \\
&= \frac{E_0^2}{\eta} \sin \theta \sin^2 (\beta x \cos \theta) \sin^2 (\omega t - \beta z \sin \theta)\, \mathbf{a}_z \qquad (9.26) \\
&\quad + \frac{E_0^2}{4\eta} \cos \theta \sin (2\beta x \cos \theta) \sin 2(\omega t - \beta z \sin \theta)\, \mathbf{a}_x
\end{aligned}
$$

The time-average Poynting vector is given by

$$
\begin{aligned}
\langle \mathbf{P} \rangle &= \frac{E_0^2}{\eta} \sin \theta \sin^2 (\beta x \cos \theta) \langle \sin^2 (\omega t - \beta z \sin \theta) \rangle\, \mathbf{a}_z \\
&\quad + \frac{E_0^2}{4\eta} \cos \theta \sin (2\beta x \cos \theta) \langle \sin 2(\omega t - \beta z \sin \theta) \rangle\, \mathbf{a}_x \qquad (9.27) \\
&= \frac{E_0^2}{2\eta} \sin \theta \sin^2 (\beta x \cos \theta)\, \mathbf{a}_z
\end{aligned}
$$

Thus, the time-average power flow is entirely in the z-direction, thereby verifying our interpretation of the field expressions. Since the composite electric field is directed entirely transverse to the z-direction, that is, the direction of time-average power flow, whereas the composite magnetic field is not, the composite wave is known as the *transverse electric,* or TE, wave.

From the expressions for the fields for the TE wave given by (9.25a) and (9.25b), we note that the electric field is zero for sin ($\beta x \cos \theta$) equal to zero, or

$$
\beta x \cos \theta = \pm m\pi, \qquad m = 0, 1, 2, 3, \ldots
$$

$$
x = \pm \frac{m\pi}{\beta \cos \theta} = \pm \frac{m\lambda}{2 \cos \theta}, \qquad m = 0, 1, 2, 3, \ldots \qquad (9.28)
$$

where

$$
\lambda = \frac{2\pi}{\beta} = \frac{2\pi}{\omega \sqrt{\mu \varepsilon}} = \frac{1}{f \sqrt{\mu \varepsilon}}
$$

Thus, if we place perfectly conducting sheets in these planes, the waves will propagate undisturbed, that is, as though the sheets were not present since the boundary condition that the tangential component of the electric field be zero on the surface of a perfect conductor is satisfied in these planes. The boundary condition that the normal component of the magnetic field be zero on the surface of a perfect conductor is also satisfied since H_x is zero in these planes.

Parallel-plate
waveguide

If we consider any two adjacent sheets, the situation is actually one of uniform plane waves bouncing obliquely between the sheets, as illustrated in Fig. 9.6 for two sheets in the planes $x = 0$ and $x = \lambda/(2 \cos \theta)$, thereby guiding the wave and hence the energy in the z-direction, parallel to the plates. Thus, we have a *parallel-plate metallic waveguide,* as compared to the parallel-plate transmission line in which the uniform plane wave slides parallel to the plates. We

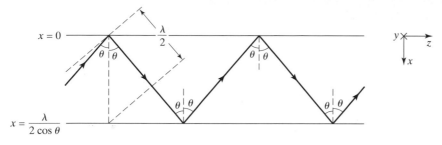

FIGURE 9.6   Uniform plane waves bouncing obliquely between two parallel plane perfectly conducting sheets.

note from the constant phase surfaces of the obliquely bouncing wave shown in Fig. 9.6 that $\lambda/(2 \cos \theta)$ is simply one-half of the apparent wavelength of that wave in the x-direction, that is, normal to the plates. Thus, the fields have one-half apparent wavelength in the x-direction. If we place the perfectly conducting sheets in the planes $x = 0$ and $x = m\lambda/(2 \cos \theta)$, the fields will then have m number of one-half apparent wavelengths in the x-direction between the plates. The fields have no variations in the y-direction. Thus, the fields are said to correspond to $\text{TE}_{m,0}$ *modes,* where the subscript m refers to the x-direction, denoting m number of one-half apparent wavelengths in that direction and the subscript 0 refers to the y-direction, denoting zero number of one-half apparent wavelengths in that direction.

   Let us now consider a parallel-plate waveguide with perfectly conducting plates situated in the planes $x = 0$ and $x = a$, that is, having a fixed spacing a between them, as shown in Fig. 9.7(a). Then, for $\text{TE}_{m,0}$ waves guided by the plates,   *Cutoff phenomenon*

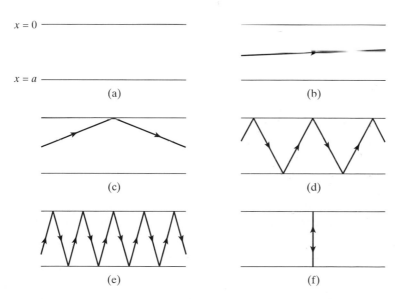

FIGURE 9.7   For illustrating the phenomenon of cutoff in a parallel-plate waveguide.

we have from (9.28),

$$a = \frac{m\lambda}{2 \cos \theta}$$

or

$$\cos \theta = \frac{m\lambda}{2a} = \frac{m}{2a} \frac{1}{f\sqrt{\mu\varepsilon}} \tag{9.29}$$

Thus, waves of different wavelengths (or frequencies) bounce obliquely between the plates at different values of the angle θ. For very small wavelengths (very high frequencies), $m\lambda/2a$ is small, $\cos \theta \approx 0$, $\theta \approx 90°$, and the waves simply slide between the plates as in the case of the transmission line, as shown in Fig. 9.7(b). As λ increases (f decreases), $m\lambda/2a$ increases, θ decreases, and the waves bounce more and more obliquely, as shown in Figs. 9.7(c)–(e), until λ becomes equal to $2a/m$ for which $\cos \theta = 1$, $\theta = 0°$, and the waves simply bounce back and forth normally to the plates, as shown in Fig. 9.7(f), without any feeling of being guided parallel to the plates. For $\lambda > 2a/m$, $m\lambda/2a > 1$, $\cos \theta > 1$, and θ has no real solution, indicating that propagation does not occur for these wavelengths in the waveguide mode. This condition is known as the *cutoff* condition.

The cutoff wavelength, denoted by the symbol λ_c, is given by

$$\boxed{\lambda_c = \frac{2a}{m}} \tag{9.30}$$

This is simply the wavelength for which the spacing a is equal to m number of one-half wavelengths. Propagation of a particular mode is possible only if λ is less than the value of λ_c for that mode. The cutoff frequency is given by

$$\boxed{f_c = \frac{m}{2a\sqrt{\mu\varepsilon}}} \tag{9.31}$$

Propagation of a particular mode is possible only if f is greater than the value of f_c for that mode. Consequently, waves of a given frequency f can propagate in all modes for which the cutoff wavelengths are greater than the wavelength or the cutoff frequencies are less than the frequency. Note that since the cutoff phenomenon corresponds to the waves bouncing back and forth normal to the plates, that is, transverse to the guide axis, the expressions for the cutoff wavelength and frequency can be obtained directly from considerations of transverse resonance.

Substituting λ_c for $2a/m$ in (9.29), we have

$$\cos \theta = \frac{\lambda}{\lambda_c} = \frac{f_c}{f} \tag{9.32a}$$

$$\sin \theta = \sqrt{1 - \cos^2 \theta} = \sqrt{1 - \left(\frac{\lambda}{\lambda_c}\right)^2} = \sqrt{1 - \left(\frac{f_c}{f}\right)^2}. \tag{9.32b}$$

$$\beta \cos \theta = \frac{2\pi}{\lambda} \frac{\lambda}{\lambda_c} = \frac{2\pi}{\lambda_c} = \frac{m\pi}{a} \tag{9.32c}$$

$$\beta \sin \theta = \frac{2\pi}{\lambda} \sqrt{1 - \left(\frac{\lambda}{\lambda_c}\right)^2} \tag{9.32d}$$

We see from (9.32d) that the phase constant along the z-direction, that is, $\beta_z(=\beta \sin \theta)$, is real for $\lambda < \lambda_c$ and imaginary for $\lambda > \lambda_c$. Since

$$\cos (\omega t \mp j|\beta_z|z) = \text{Re } e^{j(\omega t \mp j|\beta_z|z)}$$
$$= \text{Re}(e^{\pm |\beta_z|z} e^{j\omega t})$$
$$= e^{\pm |\beta_z|z} \cos \omega t$$

an imaginary value of the phase constant does not correspond to wave propagation. This once again explains the cutoff phenomenon. We now define the guide wavelength, λ_g, to be the wavelength in the z-direction, that is, along the guide. This is given by

$$\lambda_g = \frac{2\pi}{\beta_z} = \frac{2\pi}{\beta \sin \theta} = \frac{\lambda}{\sqrt{1 - (\lambda/\lambda_c)^2}} = \frac{\lambda}{\sqrt{1 - (f_c/f)^2}} \tag{9.33}$$

This is simply the apparent wavelength, in the z-direction, of the obliquely bouncing uniform plane waves. The phase velocity along the guide axis, which is simply the apparent phase velocity, in the z-direction, of the obliquely bouncing uniform plane waves, is

$$v_{pz} = \frac{\omega}{\beta_z} = \frac{\omega}{\beta \sin \theta} = \frac{v_p}{\sqrt{1 - (\lambda/\lambda_c)^2}} = \frac{v_p}{\sqrt{1 - (f_c/f)^2}} \tag{9.34}$$

Finally, substituting (9.32a)–(9.32d) in the field expressions (9.25a) and (9.25b), we obtain

Field expressions for TE$_{m,0}$ modes

$$\mathbf{E} = E_0 \sin \left(\frac{m\pi x}{a}\right) \sin (\omega t - \beta_z z) \mathbf{a}_y \tag{9.35a}$$

$$\mathbf{H} = -\frac{E_0}{\eta} \frac{\lambda}{\lambda_g} \sin \left(\frac{m\pi x}{a}\right) \sin (\omega t - \beta_z z) \mathbf{a}_x$$

$$+ \frac{E_0}{\eta} \frac{\lambda}{\lambda_c} \cos \left(\frac{m\pi x}{a}\right) \cos (\omega t - \beta_z z) \mathbf{a}_z \tag{9.35b}$$

These expressions for the TE$_{m,0}$ mode fields in the parallel-plate waveguide do not contain the angle θ. They clearly indicate the standing-wave character of the fields in the x-direction, having m one-half sinusoidal variations between the plates. We shall now consider an example.

Example 9.2

Let us assume the spacing a between the plates of an air-dielectric parallel-plate waveguide to be 5 cm and investigate the propagating $TE_{m,0}$ modes for $f = 10,000$ MHz.

From (9.30), the cutoff wavelengths for $TE_{m,0}$ modes are given by

$$\lambda_c = \frac{2a}{m} = \frac{10}{m} \text{ cm} = \frac{0.1}{m} \text{ m}$$

This result is independent of the dielectric between the plates. Since the medium between the plates is free space, the cutoff frequencies for the $TE_{m,0}$ modes are

$$f_c = \frac{3 \times 10^8}{\lambda_c} = \frac{3 \times 10^8}{0.1/m} = 3m \times 10^9 \text{ Hz}$$

For $f = 10,000$ MHz $= 10^{10}$ Hz, the propagating modes are $TE_{1,0}(f_c = 3 \times 10^9$ Hz), $TE_{2,0}(f_c = 6 \times 10^9$ Hz), and $TE_{3,0}(f_c = 9 \times 10^9$ Hz).

For each propagating mode, we can find θ, λ_g, and v_{pz} by using (9.32a), (9.33), and (9.34), respectively. Values of these quantities are listed in the following:

Mode	λ_c (cm)	f_c (MHz)	θ (deg)	λ_g (cm)	v_{pz} (m/s)
$TE_{1,0}$	10	3000	72.54	3.145	3.145×10^8
$TE_{2,0}$	5	6000	53.13	3.75	3.75×10^8
$TE_{3,0}$	3.33	9000	25.84	6.882	6.882×10^8

TM waves

We have thus far considered transverse electric or TE waves in a parallel-plate waveguide. In a similar manner, it is possible to have propagation of transverse magnetic or TM waves, so called because the magnetic field is directed entirely transverse to the direction of time-average power flow whereas the electric field is not. The field expressions for TM waves can be obtained by starting with two uniform plane waves having their magnetic fields entirely in the y-direction and proceeding in a manner similar to the development of TE waves. We shall however not pursue that approach. Instead, we shall, by analogy with (9.35a), write the expression for the magnetic field of the TM wave and then derive the electric field by using one of Maxwell's curl equations.

Field expressions for $TM_{m,0}$ modes

Thus, assuming the guide to be made up of parallel plates in the $x = 0$ and $x = a$ planes, and writing the expression for the magnetic field of the $TM_{m,0}$ wave and using

$$\nabla \times \mathbf{H} = \frac{\partial \mathbf{D}}{\partial t}$$

we obtain the fields for the TM modes to be

$$\mathbf{H} = H_0 \cos\left(\frac{m\pi x}{a}\right) \sin(\omega t - \beta_z z)\, \mathbf{a}_y \qquad (9.36a)$$

$$\mathbf{E} = \frac{\lambda}{\lambda_g}\, \eta H_0 \cos\left(\frac{m\pi x}{a}\right) \sin(\omega t - \beta_z z)\, \mathbf{a}_x$$
$$+ \frac{\lambda}{\lambda_c}\, \eta H_0 \sin\left(\frac{m\pi x}{a}\right) \cos(\omega t - \beta_z z)\, \mathbf{a}_z \qquad (9.36b)$$

Note that the x-variation of H_y is cosinusoidal, which leads to sinusoidal variation for E_z so that the boundary condition of zero tangential electric field is satisfied on the two plates. The parameters λ_c and λ_g in (9.36a) and (9.36b) and the
other parameters f_c and v_{pz} for the TM modes are the same as those for the TE
modes, given by (9.30), (9.33), (9.31), and (9.34), respectively.

Let us now consider reflection and transmission at a dielectric discontinu- *Parallel-plate*
ity in a parallel-plate guide, as shown in Fig. 9.8. If a TE or TM wave is incident *waveguide*
on the junction from section 1, then it will set up a reflected wave into section 1 *discontinuity*
and a transmitted wave into section 2, provided that mode propagates in that
section. The fields corresponding to these incident, reflected, and transmitted
waves must satisfy the boundary conditions at the dielectric discontinuity.

Considering first TE waves and denoting the incident, reflected, and transmitted wave fields by the subscripts i, r, and t, respectively, we have from the continuity of the tangential component of $\mathbf{E}$ at a dielectric discontinuity,

$$E_{yi} + E_{yr} = E_{yt} \qquad \text{at } z = 0 \qquad (9.37a)$$

and from the continuity of the tangential component of $\mathbf{H}$ at a dielectric discontinuity,

$$H_{xi} + H_{xr} = H_{xt} \qquad \text{at } z = 0 \qquad (9.37b)$$

We now define the guide characteristic impedance, η_{g1}, of section 1 as

$$\eta_{g1} = \frac{E_{yi}}{-H_{xi}} \qquad (9.38)$$

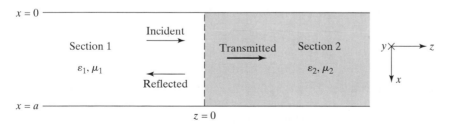

FIGURE 9.8   For consideration of reflection and transmission at a dielectric discontinuity in
a parallel-plate waveguide.

Recognizing that $\mathbf{a}_y \times (-\mathbf{a}_x) = \mathbf{a}_z$, we note that η_{g1} is simply the ratio of the transverse components of the electric and magnetic fields of the $TE_{m,0}$ wave that give rise to time-average power flow down the guide. From (9.35a) and (9.35b) applied to section 1, we have

$$\eta_{g1} = \eta_1 \frac{\lambda_{g1}}{\lambda_1} = \frac{\eta_1}{\sqrt{1 - (\lambda_1/\lambda_c)^2}} = \frac{\eta_1}{\sqrt{1 - (f_{c1}/f)^2}} \tag{9.39}$$

The guide characteristic impedance is analogous to the characteristic impedance of a transmission line, if we recognize that E_{yi} and $-H_{xi}$ are analogous to V^+ and I^+, respectively. In terms of the reflected wave fields, it then follows that

$$\eta_{g1} = -\left(\frac{E_{yr}}{-H_{xr}}\right) = \frac{E_{yr}}{H_{xr}} \tag{9.40}$$

This result can also be seen from the fact that for the reflected wave, the power flow is in the negative z-direction, and since $\mathbf{a}_y \times \mathbf{a}_x = -\mathbf{a}_z$, η_{g1} is equal to E_{yr}/H_{xr}. For the transmitted wave fields, we have

$$\frac{E_{yt}}{-H_{xt}} = \eta_{g2} \tag{9.41}$$

where

$$\eta_{g2} = \eta_2 \frac{\lambda_{g2}}{\lambda_2} = \frac{\eta_2}{\sqrt{1 - (\lambda_2/\lambda_c)^2}} = \frac{\eta_2}{\sqrt{1 - (f_{c2}/f)^2}} \tag{9.42}$$

is the guide characteristic impedance of section 2.

Using (9.38), (9.40), and (9.41), (9.37b) can be written as

$$\frac{E_{yi}}{\eta_{g1}} - \frac{E_{yr}}{\eta_{g1}} = \frac{E_{yt}}{\eta_{g2}} \tag{9.43}$$

Solving (9.37a) and (9.43), we get

$$E_{yi}\left(1 - \frac{\eta_{g2}}{\eta_{g1}}\right) + E_{yr}\left(1 + \frac{\eta_{g2}}{\eta_{g1}}\right) = 0$$

or the reflection coefficient at the junction is given by

$$\boxed{\Gamma = \frac{E_{yr}}{E_{yi}} = \frac{\eta_{g2} - \eta_{g1}}{\eta_{g2} + \eta_{g1}}} \tag{9.44}$$

This expression for the reflection coefficient is the same as that for the voltage reflection coefficient at the load of a lossless transmission line of characteristic impedance η_{g1} terminated by a resistive load η_{g2}. It is also the same as the voltage reflection coefficient at the junction between two transmission lines 1 and 2 having the characteristic impedances η_{g1} and η_{g2}, respectively, as shown in Fig. 9.9, where line 2 is infinitely long and hence its input impedance is equal

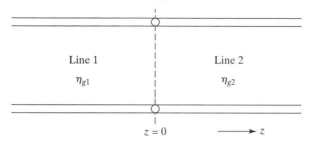

FIGURE 9.9 Transmission-line equivalent of parallel-plate waveguide discontinuity.

to η_{g2}. Thus, insofar as reflection and transmission at the discontinuity are concerned, each waveguide section can be replaced by a transmission line of characteristic impedance equal to the guide characteristic impedance given for the TE modes by

$$[\eta_g]_{TE} = \frac{\eta}{\sqrt{1 - (f_c/f)^2}} \tag{9.45}$$

It should be noted that unlike the characteristic impedance of a lossless line, which is a constant independent of frequency, the guide characteristic impedance of the lossless waveguide is a function of the frequency and the mode of propagation. Before considering TM modes, it should be pointed out that the power reflection coefficient is Γ^2 so that the reflected power is Γ^2 times the incident power and the transmitted power into section 2 is $(1 - \Gamma^2)$ times the incident power.

Turning now to TM waves, we observe from (9.36a) and (9.36b) that the ratio of the transverse electric field component F_x to the transverse magnetic field component H_y, which together are responsible for time-average power flow in the z-direction, is equal to $\eta\lambda/\lambda_g$, and hence the guide characteristic impedance for TM waves is given by

$$[\eta_g]_{TM} = \eta\sqrt{1 - (f_c/f)^2} \tag{9.46}$$

Thus, the transmission-line equivalent for reflection and transmission of TM waves at the waveguide discontinuity is the same as in Fig. 9.9 except that η_{g1} and η_{g2} follow from (9.46). We shall now consider an example.

Example 9.3

Let us consider the parallel-plate waveguide discontinuity shown in Fig. 9.10. We wish to find the power reflection coefficients for $TE_{1,0}$ and $TM_{1,0}$ waves of frequency $f = 5000$ MHz incident on the junction from the free-space side.

FIGURE 9.10   For illustrating the computation of reflection and transmission coefficients at a parallel-plate waveguide discontinuity.

For the $TE_{1,0}$ mode or for the $TM_{1,0}$ mode, $\lambda_c = 2a = 10$ cm, independent of the dielectric. For $f = 5000$ MHz,

$$\lambda_1 = \text{wavelength on the free space side} = \frac{3 \times 10^8}{5 \times 10^9} = 6 \text{ cm}$$

$$\lambda_2 = \text{wavelength on the dielectric side} = \frac{3 \times 10^8}{\sqrt{9} \times 5 \times 10^9} = \frac{6}{3} = 2 \text{ cm}$$

Since $\lambda < \lambda_c$ in both sections, $TE_{1,0}$ and $TM_{1,0}$ modes propagate in both sections. Thus, for the $TE_{1,0}$ mode.

$$\eta_{g1} = \frac{\eta_1}{\sqrt{1 - (\lambda_1/\lambda_c)^2}} = \frac{120\pi}{\sqrt{1 - (6/10)^2}} = 471.24 \ \Omega$$

$$\eta_{g2} = \frac{\eta_2}{\sqrt{1 - (\lambda_2/\lambda_c)^2}} = \frac{120\pi/\sqrt{9}}{\sqrt{1 - (2/10)^2}} = \frac{40\pi}{\sqrt{1 - 0.04}} = 128.25 \ \Omega$$

$$\Gamma^2 = \left(\frac{\eta_{g2} - \eta_{g1}}{\eta_{g2} + \eta_{g1}}\right)^2 = \left(\frac{128.25 - 471.24}{128.25 + 471.24}\right)^2 = (-0.572)^2 = 0.327$$

For the $TM_{1,0}$ mode,

$$\eta_{g1} = \eta_1 \sqrt{1 - (\lambda_1/\lambda_c)^2} = 301.59 \ \Omega$$

$$\eta_{g2} = \eta_2 \sqrt{1 - (\lambda_2/\lambda_c)^2} = 123.12 \ \Omega$$

$$\Gamma^2 = \left(\frac{\eta_{g2} - \eta_{g1}}{\eta_{g2} + \eta_{g1}}\right)^2 = \left(\frac{123.12 - 301.59}{123.12 + 301.59}\right)^2 = (-0.42)^2 = 0.176$$

We have in this section introduced the principle of metallic waveguides by considering the parallel-plate waveguide. In practice, however, metallic waveguides are generally made up of a single conductor having rectangular or circular cross section. We shall defer the consideration of rectangular metallic waveguides to Section 10.1 and discuss in Section 9.3 the important phenomenon of dispersion, characteristic of propagation in parallel-plate as well as rectangular and circular waveguides and leading to the concept of group velocity.

Earth-ionosphere waveguide

But first we shall conclude this section with a brief description of a naturally occurring waveguide, although of spherical geometry. This is the Earth-ionosphere waveguide. The ionosphere is a region of the upper atmosphere extending from approximately 50 km to more than 1000 km above Earth. In this region, the constituent gases are ionized, mostly because of ultraviolet radiation from the Sun, thereby resulting in the production of positive ions and electrons that are free to move under the influence of the fields of a wave incident on the medium. The positive ions are, however, heavy compared to the electrons, and hence they are relatively immobile. The electron motion produces a current that influences the wave propagation. The electron density in the ionosphere exists in several layers known as the D, E, and F layers in which the ionization changes with the hour of the day, the season, and the sunspot cycle. However, for the purpose of our discussion, it is sufficient to assume that the electron density increases continuously from zero at the lower boundary, reaching a peak at some height, typically lying between 250 and 350 km, and then decreases continuously, as shown in Fig. 9.11(a). The wave propagation is influenced by the electrons in such a manner that waves of very low frequencies are reflected at the base. As the frequency is increased, the waves penetrate deeper into the region but still return to Earth after reflection. When their frequency exceeds a certain value, typically between 20 and 40 MHz depending on the angle of incidence, they penetrate through the maximum of the layer and hence do not return to Earth. Thus, for frequencies in the VLF range and lower, the lower boundary of the ionosphere and Earth form a waveguide, thereby permitting waveguide mode of propagation, as shown in Fig. 9.11(b).

K9.2. Transverse electric wave; Transverse magnetic wave; Parallel-plate waveguide; Cutoff frequency; Cutoff wavelength; Guide wavelength; Guide characteristic impedance.

D9.3. The dimension a of an air-dielectric parallel-plate waveguide is 3 cm. Find the values of θ and λ_g for each of the following cases: **(a)** $f = 6000$ MHz, $TE_{1,0}$ mode; **(b)** $f = 15,000$ MHz, $TE_{1,0}$ mode; and **(c)** $f = 15,000$ MHz, $TE_{2,0}$ mode.
Ans. **(a)** 33.56°, 9.045 cm; **(b)** 70.53°, 2.121 cm; **(c)** 48.19°, 2.683 cm.

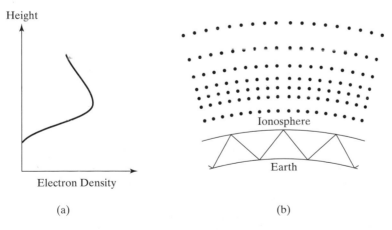

(a)

(b)

FIGURE 9.11 **(a)** Variation of electron density with height for a simplified ionosphere. **(b)** Depiction of waveguide mode of propagation in the Earth-ionosphere waveguide.

D9.4. TE waves are excited in an air-dielectric parallel-plate waveguide having the plates in the $x = 0$ and $x = 5$ cm planes by setting up at its input $z = 0$ a field distribution having

$$\mathbf{E} = 40 \sin^3 20\pi x \sin 2 \times 10^{10} \pi t \, \mathbf{a}_y \text{ V/m}$$

Noting that the electric field of a propagating $TE_{m,0}$ mode is of the form given by (9.35a), find E_0 for each of the following modes: **(a)** $TE_{1,0}$; **(b)** $TE_{2,0}$; and **(c)** $TE_{3,0}$.
Ans. **(a)** 30 V/m; **(b)** 0 V/m; **(c)** −10 V/m.

D9.5. For a parallel-plate waveguide of spacing $a = 3$ cm and filled with a dielectric of $\varepsilon = 6.25\varepsilon_0$ and $\mu = \mu_0$, find the values of the guide characteristic impedance for each of the following cases: **(a)** $TE_{1,0}$ mode of $f = 3000$ MHz; **(b)** $TM_{1,0}$ mode of $f = 3000$ MHz; and **(c)** $TE_{1,0}$ mode of $f = 6000$ MHz.
Ans. **(a)** 202.3 Ω; **(b)** 112.4 Ω; **(c)** 159.94 Ω.

9.3 DISPERSION AND GROUP VELOCITY

In the preceding section, we learned that for the propagating range of frequencies, the phase velocity and the wavelength along the axis of the parallel-plate waveguide are given by

$$v_{pz} = \frac{v_p}{\sqrt{1 - (f_c/f)^2}} \tag{9.47}$$

and

$$\lambda_g = \frac{\lambda}{\sqrt{1 - (f_c/f)^2}} \tag{9.48}$$

where $v_p = 1/\sqrt{\mu\varepsilon}$, $\lambda = v_p/f = 1/f\sqrt{\mu\varepsilon}$, and f_c is the cutoff frequency. We note that for a particular mode, the phase velocity of propagation along the guide axis varies with the frequency. As a consequence of this characteristic of the guided wave propagation, the field patterns of the different frequency components of a signal comprising a band of frequencies do not maintain the same phase relationships as they propagate down the guide. This phenomenon is known as *dispersion,* so called after the phenomenon of dispersion of colors by a prism.

To discuss dispersion, let us consider a simple example of two infinitely long trains A and B traveling in parallel, one below the other, with each train made up of boxcars of identical size and having wavy tops, as shown in Fig. 9.12. Let the spacings between the peaks (centers) of successive boxcars be 50 m and 90 m, and let the speeds of the trains be 20 m/s and 30 m/s, for trains A and B, respectively. Let the peaks of the cars numbered 0 for the two trains be aligned at time $t = 0$, as shown in Fig. 9.12(a). Now, as time progresses, the two peaks get out of alignment as shown, for example, for $t = 1$ s in Fig. 9.12(b), since train B is traveling faster than train A. But at the same time, the gap between the peaks of cars numbered −1 decreases. This continues until at $t = 4$ s, the peak of car −1 of train A having moved by a distance of 80 m aligns with the peak of car −1 of train B, which will have moved by a distance of 120 m, as shown in Fig. 9.12(c). For an observer following the movement of the two trains as a group, the group

FIGURE 9.12   For illustrating the concept of group velocity.

appears to have moved by a distance of 30 m although the individual trains will have moved by 80 m and 120 m, respectively. Thus, we can talk of a *group velocity*, that is, the velocity with which the group as a whole is moving. In this case, the group velocity is (30 m)/(4 s) or 7.5 m/s.

The situation in the case of the guided wave propagation of two different frequencies in the parallel-plate waveguide is exactly similar to the two-train example just discussed. The distance between the peaks of two successive cars is analogous to the guide wavelength, and the speed of the train is analogous to the phase velocity along the guide axis. Thus, let us consider the field patterns corresponding to two waves of frequencies f_A and f_B propagating in the same mode, having guide wavelengths λ_{gA} and λ_{gB}, and phase velocities along the guide axis v_{pzA} and v_{pzB}, respectively, as shown, for example, for the electric field of the $\text{TE}_{1,0}$ mode in Fig. 9.13. Let the positive peaks numbered 0 of the two patterns be aligned at $t = 0$, as shown in Fig. 9.13(a). As the individual waves travel with their respective phase velocities along the guide, these two peaks get out of alignment, but some time later, say, Δt, the positive peaks numbered -1 will align at some distance, say, Δz, from the location of the alignment of the 0 peaks, as shown in Fig. 9.13(b). Since the -1th peak of wave A will have traveled a

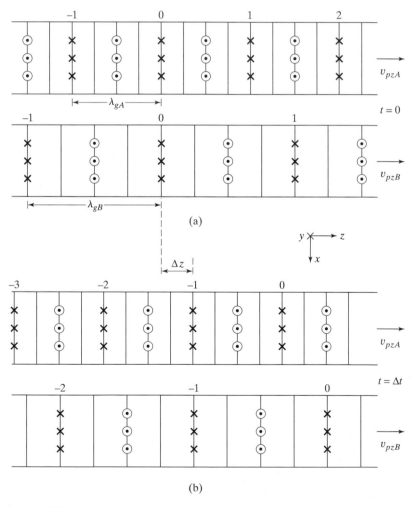

FIGURE 9.13   For illustrating the concept of group velocity for guided wave propagation.

distance $\lambda_{gA} + \Delta z$ with a phase velocity v_{pzA} and the -1th peak of wave B will have traveled a distance $\lambda_{gB} + \Delta z$ with a phase velocity v_{pzB} in this time Δt, we have

$$\lambda_{gA} + \Delta z = v_{pzA} \, \Delta t \tag{9.49a}$$

$$\lambda_{gB} + \Delta z = v_{pzB} \, \Delta t \tag{9.49b}$$

Solving (9.49a) and (9.49b) for Δt and Δz, we obtain

$$\Delta t = \frac{\lambda_{gA} - \lambda_{gB}}{v_{pzA} - v_{pzB}} \tag{9.50a}$$

and

$$\Delta z = \frac{\lambda_{gA} v_{pzB} - \lambda_{gB} v_{pzA}}{v_{pzA} - v_{pzB}} \tag{9.50b}$$

The group velocity, v_g, is then given by

$$v_g = \frac{\Delta z}{\Delta t} = \frac{\lambda_{gA} v_{pzB} - \lambda_{gB} v_{pzA}}{\lambda_{gA} - \lambda_{gB}} = \frac{\lambda_{gA} \lambda_{gB} f_B - \lambda_{gB} \lambda_{gA} f_A}{\lambda_{gA} \lambda_{gB}[(1/\lambda_{gB}) - (1/\lambda_{gA})]}$$

$$= \frac{f_B - f_A}{(1/\lambda_{gB}) - (1/\lambda_{gA})}$$

or

$$\boxed{v_g = \frac{\omega_B - \omega_A}{\beta_{zB} - \beta_{zA}}} \tag{9.51}$$

where β_{zA} and β_{zB} are the phase constants along the guide axis, corresponding to f_A and f_B, respectively. Thus, the group velocity of a signal comprised of two frequencies is the ratio of the difference between the two radian frequencies to the difference between the corresponding phase constants along the guide axis.

If we now have a signal comprised of a number of frequencies, then a value *Dispersion* of group velocity can be obtained for each pair of these frequencies in accor- *diagram* dance with (9.51). In general, these values of group velocity will all be different. In fact, this is the case for wave propagation in the parallel-plate guide, as can be seen from Fig. 9.14, which is a plot of ω versus β_z corresponding to the parallel-plate guide for which

$$\beta_z = \frac{2\pi}{\lambda_g} = \frac{2\pi}{\lambda} \sqrt{1 - \left(\frac{\lambda}{\lambda_c}\right)^2} = \omega\sqrt{\mu\varepsilon} \sqrt{1 - \left(\frac{f_c}{f}\right)^2}. \tag{9.52}$$

Such a plot is known as the ω–β_z *diagram* or *dispersion diagram*. Note that the dispersion diagram begins at $\omega = \omega_c$ on the ω-axis, since for $\omega < \omega_c$, propagation does not occur. The phase velocity along the guide axis given for a particular frequency by

$$\boxed{v_{pz} = \frac{\omega}{\beta_z}} \tag{9.53}$$

FIGURE 9.14   Dispersion diagram for the parallel-plate waveguide.

is equal to the slope of the line drawn from the origin to the point, on the dispersion curve, corresponding to that frequency, as shown in the figure for the three frequencies ω_1, ω_2, and ω_3. The group velocity for a particular pair of frequencies is given by the slope of the line joining the two points, on the curve, corresponding to the two frequencies as shown in the figure for the two pairs ω_1, ω_2 and ω_2, ω_3. Since the curve is nonlinear, it can be seen that the two group velocities are not equal. We cannot then attribute a particular value of group velocity for the group of the three frequencies ω_1, ω_2, and ω_3.

If, however, the three frequencies are very close, as in the case of a narrowband signal, it is meaningful to assign a group velocity to the entire group having a value equal to the slope of the tangent to the dispersion curve at the center frequency. Thus, the group velocity corresponding to a narrow band of frequencies centered around a predominant frequency ω is given by

$$v_g = \frac{d\omega}{d\beta_z} \tag{9.54}$$

For the parallel-plate waveguide under consideration, we have from (9.52)

$$
\begin{aligned}
\frac{d\beta_z}{d\omega} &= \sqrt{\mu\varepsilon}\sqrt{1 - \left(\frac{f_c}{f}\right)^2} + \omega\sqrt{\mu\varepsilon} \cdot \frac{1}{2}\left(1 - \frac{f_c^2}{f^2}\right)^{-1/2}\frac{f_c^2}{\pi f^3} \\
&= \sqrt{\mu\varepsilon}\left(1 - \frac{f_c^2}{f^2} + \frac{\omega}{2\pi}\frac{f_c^2}{f^3}\right)\left(1 - \frac{f_c^2}{f^2}\right)^{-1/2} \\
&= \sqrt{\mu\varepsilon}\left(1 - \frac{f_c^2}{f^2}\right)^{-1/2}
\end{aligned}
$$

and hence, from (9.54),

$$v_g = \frac{d\omega}{d\beta_z} = \frac{1}{\sqrt{\mu\varepsilon}}\sqrt{1 - \frac{f_c^2}{f^2}} = v_p\sqrt{1 - \left(\frac{f_c}{f}\right)^2} \tag{9.55}$$

From (9.47) and (9.55), we note that $v_{pz} > v_p$, $v_g < v_p$, and

$$v_{pz}v_g = v_p^2$$

For a numerical example, let us consider the air-dielectric parallel-plate waveguide of spacing $a = 5$ cm and a narrow-band signal of center frequency $f = 10,000$ MHz propagating in the $\text{TE}_{1,0}$ mode. Then from Example 9.2, $f_c = 3000$ MHz, and from (9.55),

$$v_g = 3 \times 10^8 \sqrt{1 - (3/10)^2}$$
$$= 2.862 \times 10^8 \text{ m/s}$$

as compared to $v_{pz} = 3.145 \times 10^8$ m/s found in Example 9.2.

Amplitude-modulated signal

An example of a narrow-band signal is an amplitude-modulated signal, having a carrier frequency ω modulated by a low-frequency $\Delta\omega \lll \omega$ as given by

$$E_x(t) = E_{x0}(1 + m \cos \Delta\omega \cdot t) \cos \omega t \qquad (9.56)$$

where m is the percentage modulation. Such a signal is actually equivalent to a superposition of unmodulated signals of three frequencies $\omega - \Delta\omega$, ω, and $\omega + \Delta\omega$, as can be seen by expanding the right side of (9.56). Thus,

$$E_x(t) = E_{x0} \cos \omega t + mE_{x0} \cos \omega t \cos \Delta\omega \cdot t$$
$$= E_{x0} \cos \omega t + \frac{mE_{x0}}{2} [\cos (\omega - \Delta\omega)t + \cos (\omega + \Delta\omega)t] \qquad (9.57)$$

The frequencies $\omega - \Delta\omega$ and $\omega + \Delta\omega$ are the side frequencies. When the amplitude-modulated signal propagates in a dispersive channel, such as the parallel-plate waveguide under consideration, the different frequency components undergo phase changes in accordance with their respective phase constants. Thus, if $\beta_z - \Delta\beta_z$, β_z, and $\beta_z + \Delta\beta_z$ are the phase constants corresponding to $\omega - \Delta\omega$, ω, and $\omega + \Delta\omega$, respectively, assuming linearity of the dispersion curve within the narrow band, the amplitude-modulated wave is given by

$$E_x(z, t) = E_{x0} \cos (\omega t - \beta_z z)$$
$$+ \frac{mE_{x0}}{2} \{\cos [(\omega - \Delta\omega)t - (\beta_z - \Delta\beta_z)z]$$
$$+ \cos [(\omega + \Delta\omega)t - (\beta_z + \Delta\beta_z)z]\}$$
$$= E_{x0} \cos (\omega t - \beta_z z) \qquad (9.58)$$
$$+ \frac{mE_{x0}}{2} \{\cos [(\omega t - \beta_z z) - (\Delta\omega \cdot t - \Delta\beta_z \cdot z)]$$
$$+ \cos [(\omega t - \beta_z z) + (\Delta\omega \cdot t - \Delta\beta_z \cdot z)]\}$$
$$= E_{x0} \cos (\omega t - \beta_z z) + mE_{x0} \cos (\omega t - \beta_z z) \cos (\Delta\omega \cdot t - \Delta\beta_z \cdot z)$$
$$= E_{x0}[1 + m \cos (\Delta\omega \cdot t - \Delta\beta_z \cdot z)] \cos (\omega t - \beta_z z)$$

This indicates that although the carrier-frequency phase changes in accordance with the phase constant β_z, the modulation envelope and hence the information travels with the group velocity $\Delta\omega/\Delta\beta_z$, as shown in Fig. 9.15. In view of this and

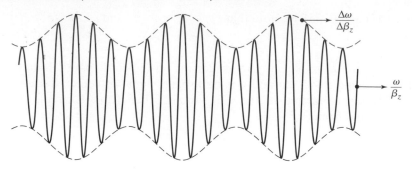

FIGURE 9.15   For illustrating that the modulation envelope travels with the group velocity.

since v_g is less than v_p, the fact that v_{pz} is greater than v_p is not a violation of the theory of relativity. Since it is always necessary to use some modulation technique to convey information from one point to another, the information always takes more time to reach from one point to another in a dispersive channel than in the corresponding nondispersive medium. For further understanding of the concept of group velocity, the reader is advised to view a movie narrated by Van Duzer.[1]

K9.3.   Dispersion; Group velocity; Dispersion diagram; Narrow-band signal.

D9.6.   The ω–β_z curve for a dispersive channel can be approximated by

$$\omega = \omega_0 + k\beta_z^2$$

in the vicinity of $\omega = 1.5\omega_0$, where k is a positive constant. Find **(a)** the phase velocity for a signal of $\omega = 1.4\omega_0$; **(b)** the phase velocity for a signal of $\omega = 1.6\omega_0$; **(c)** the group velocity for a signal composed of two radian frequencies $1.4\omega_0$ and $1.6\omega_0$; and **(d)** the group velocity for a narrow-band signal having the center radian frequency $1.5\omega_0$.
Ans.   **(a)** $2.2136\sqrt{k\omega_0}$;   **(b)** $2.0656\sqrt{k\omega_0}$;   **(c)** $1.4071\sqrt{k\omega_0}$;
(d) $1.4142\sqrt{k\omega_0}$.

9.4   REFLECTION AND REFRACTION OF PLANE WAVES

Let us now consider a uniform plane wave incident obliquely on a plane boundary between two different perfect dielectric media at an angle of incidence θ_i to the normal to the boundary, as shown in Fig. 9.16. To satisfy the boundary conditions at the interface between the two media, a reflected wave and a transmitted wave will be set up. Let θ_r be the angle of reflection and θ_t be the angle of transmission. Then without writing the expressions for the fields, we can find the relationship among θ_i, θ_r, and θ_t by noting that for the incident, reflected, and transmitted waves to be in step at the boundary, their apparent phase velocities

---

[1]T. Van Duzer, *Wave Velocities, Dispersion and the Omega–Beta Diagram* (Newton, MA: Educational Development Center).

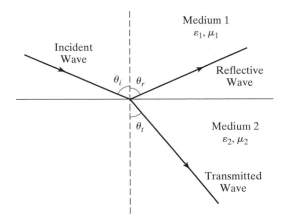

FIGURE 9.16 Reflection and transmission of an obliquely incident uniform plane wave on a plane boundary between two different perfect dielectric media.

parallel to the boundary must be equal; that is,

$$\frac{v_{p1}}{\sin \theta_i} - \frac{v_{p1}}{\sin \theta_r} = \frac{v_{p2}}{\sin \theta_t} \tag{9.59}$$

where $v_{p1}\,(= 1/\sqrt{\mu_1 \varepsilon_1})$ and $v_{p2}\,(= 1/\sqrt{\mu_2 \varepsilon_2})$ are the phase velocities along the directions of propagation of the waves in medium 1 and medium 2, respectively. From (9.59), we have

Laws of reflection and refraction

$$\sin \theta_r = \sin \theta_i \tag{9.60a}$$

$$\sin \theta_t = \frac{v_{p2}}{v_{p1}} \sin \theta_i = \sqrt{\frac{\mu_1 \varepsilon_1}{\mu_2 \varepsilon_2}} \sin \theta_i \tag{9.60b}$$

or

$$\theta_r = \theta_i \tag{9.61a}$$

$$\theta_t = \sin^{-1}\left(\sqrt{\frac{\mu_1 \varepsilon_1}{\mu_2 \varepsilon_2}} \sin \theta_i\right) \tag{9.61b}$$

Equation (9.61a) is known as the *law of reflection* and (9.61b) is known as the *law of refraction,* or *Snell's law.* Snell's law is commonly cast in terms of the refractive index, denoted by the symbol n and defined as the ratio of the velocity of light in free space to the phase velocity in the medium. Thus, if $n_1\,(= c/v_{p1})$ and $n_2\,(= c/v_{p2})$ are the (phase) refractive indices for media 1 and 2, respectively, then

$$\theta_t = \sin^{-1}\left(\frac{n_1}{n_2} \sin \theta_i\right) \tag{9.62}$$

For two dielectrics having $\mu_1 = \mu_2 = \mu_0$, which is usually the case, (9.62) reduces to

$$\theta_t = \sin^{-1}\left(\sqrt{\frac{\varepsilon_1}{\varepsilon_2}}\,\sin\theta_i\right) \tag{9.63}$$

We shall now consider the derivation of the expressions for the reflection and transmission coefficients at the boundary. To do this, we distinguish between two cases: (1) the electric field vector of the wave linearly polarized parallel to the interface and (2) the magnetic field vector of the wave linearly polarized parallel to the interface. The law of reflection and Snell's law hold for both cases since they result from the fact that the apparent phase velocities of the incident, reflected, and transmitted waves parallel to the boundary must be equal.

Perpendicular polarization   The geometry pertinent to the case of the electric field vector parallel to the interface is shown in Fig. 9.17 in which the interface is assumed to be in the $x = 0$ plane, and the subscripts i, r, and t associated with the field symbols denote incident, reflected, and transmitted waves, respectively. The plane of incidence, that is, the plane containing the normal to the interface and the propagation vectors, is assumed to be in the xz-plane, so that the electric field vectors are entirely in the y-direction. The corresponding magnetic field vectors are then as shown in the figure so as to be consistent with the condition that $\mathbf{E}$, $\mathbf{H}$, and $\boldsymbol{\beta}$ form a right-handed mutually orthogonal set of vectors. Since the electric field vectors are perpendicular to the plane of incidence, this case is also said to correspond to perpendicular polarization. The angle of incidence is assumed to be θ_1. From the law of reflection (9.61a), the angle of reflection is then also θ_1. The angle of transmission, assumed to be θ_2, is related to θ_1 by Snell's law, given by (9.61b).

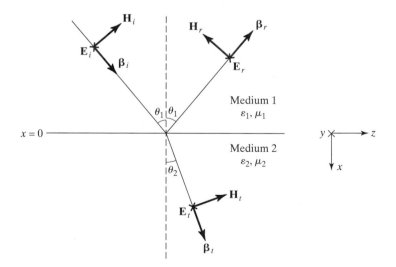

FIGURE 9.17   For obtaining the reflection and transmission coefficients for an obliquely incident uniform plane wave on a dielectric interface with its electric field perpendicular to the plane of incidence.

The boundary conditions to be satisfied at the interface $x = 0$ are that (1) the tangential component of the electric field intensity be continuous and (2) the tangential component of the magnetic field intensity be continuous. Thus, we have at the interface $x = 0$

$$E_{yi} + E_{yr} = E_{yt} \tag{9.64a}$$

$$H_{zi} + H_{zr} = H_{zt} \tag{9.64b}$$

Expressing the quantities in (9.64a) and (9.64b) in terms of the total fields, we obtain

$$E_i + E_r = E_t \tag{9.65a}$$

$$H_i \cos \theta_1 - H_r \cos \theta_1 = H_t \cos \theta_2 \tag{9.65b}$$

We also know from one of the properties of uniform plane waves that

$$\frac{E_i}{H_i} = \frac{E_r}{H_r} = \eta_1 = \sqrt{\frac{\mu_1}{\varepsilon_1}} \tag{9.66a}$$

$$\frac{E_t}{H_t} = \eta_2 = \sqrt{\frac{\mu_2}{\varepsilon_2}} \tag{9.66b}$$

Substituting (9.66a) and (9.66b) into (9.65b) and rearranging, we get

$$E_i - E_r = E_t \frac{\eta_1 \cos \theta_2}{\eta_2 \cos \theta_1} \tag{9.67}$$

Solving (9.65a) and (9.67) for E_i and E_r, we have

$$E_i = \frac{E_t}{2} \left(1 + \frac{\eta_1 \cos \theta_2}{\eta_2 \cos \theta_1} \right) \tag{9.68a}$$

$$E_r = \frac{E_t}{2} \left(1 - \frac{\eta_1 \cos \theta_2}{\eta_2 \cos \theta_1} \right) \tag{9.68b}$$

We now define the reflection coefficient $\Gamma_\perp$ and the transmission coefficient $\tau_\perp$ as

$$\Gamma_\perp = \frac{E_r}{E_i} = \frac{E_{yr}}{E_{yi}} \tag{9.69a}$$

$$\tau_\perp = \frac{E_t}{E_i} = \frac{E_{yt}}{E_{yi}} \tag{9.69b}$$

where the subscript $\perp$ refers to perpendicular polarization. From (9.68a) and (9.68b), we then obtain

$$\Gamma_\perp = \frac{\eta_2 \cos \theta_1 - \eta_1 \cos \theta_2}{\eta_2 \cos \theta_1 + \eta_1 \cos \theta_2} \tag{9.70a}$$

$$\tau_\perp = \frac{2\eta_2 \cos \theta_1}{\eta_2 \cos \theta_1 + \eta_1 \cos \theta_2} \tag{9.70b}$$

Equations (9.70a) and (9.70b) are known as the Fresnel reflection and transmission coefficients, respectively, for perpendicular polarization.

Parallel
polarization

Before we discuss the result given by (9.70a) and (9.70b), we shall derive the corresponding expressions for the case in which the magnetic field of the wave is parallel to the interface. The geometry pertinent to this case is shown in Fig. 9.18. Here again the plane of incidence is chosen to be the xz-plane, so that the magnetic field vectors are entirely in the y-direction. The corresponding electric field vectors are then as shown in the figure so as to be consistent with the condition that $\mathbf{E}$, $\mathbf{H}$, and $\boldsymbol{\beta}$ form a right-handed mutually orthogonal set of vectors. Since the electric field vectors are parallel to the plane of incidence, this case is also said to correspond to parallel polarization.

Once again the boundary conditions to be satisfied at the interface $x = 0$ are that (1) the tangential component of the electric field intensity be continuous and (2) the tangential component of the magnetic field intensity be continuous. Thus, we have at the interface $x = 0$,

$$E_{zi} + E_{zr} = E_{zt} \tag{9.71a}$$

$$H_{yi} + H_{yr} = H_{yt} \tag{9.71b}$$

Expressing the quantities in (9.71a) and (9.71b) in terms of the total fields and also using (9.66a) and (9.66b), we obtain

$$E_i - E_r = E_t \frac{\cos \theta_2}{\cos \theta_1} \tag{9.72a}$$

$$E_i + E_r = E_t \frac{\eta_1}{\eta_2} \tag{9.72b}$$

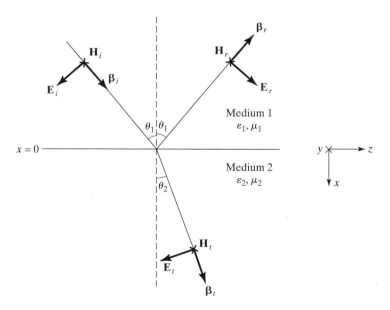

FIGURE 9.18   For obtaining the reflection and transmission coefficients for an obliquely incident uniform plane wave on a dielectric interface with its electric field parallel to the plane of incidence.

Solving (9.72a) and (9.72b) for E_i and E_r, we have

$$E_i = \frac{E_t}{2}\left(\frac{\eta_1}{\eta_2} + \frac{\cos\theta_2}{\cos\theta_1}\right) \tag{9.73a}$$

$$E_r = \frac{E_t}{2}\left(\frac{\eta_1}{\eta_2} - \frac{\cos\theta_2}{\cos\theta_1}\right) \tag{9.73b}$$

We now define the reflection coefficient $\Gamma_\|$ and the transmission coefficient $\tau_\|$ as

$$\Gamma_\| = -\frac{E_r}{E_i} \tag{9.74a}$$

$$\tau_\| = \frac{F_t}{E_i} \tag{9.74b}$$

where the subscript $\|$ refers to parallel polarization. From (9.73a) and (9.73b), we then obtain

$$\Gamma_\| = \frac{\eta_2\cos\theta_2 - \eta_1\cos\theta_1}{\eta_2\cos\theta_2 + \eta_1\cos\theta_1} \tag{9.75a}$$

$$\tau_\| = \frac{2\eta_2\cos\theta_1}{\eta_2\cos\theta_2 + \eta_1\cos\theta_1} \tag{9.75b}$$

Note from (9.74a) and (9.74b) that

$$\frac{E_{zr}}{E_{zi}} = \frac{E_r\cos\theta_1}{-E_i\cos\theta_1} = -\frac{E_r}{E_i} = \Gamma_\| \tag{9.76a}$$

$$\frac{E_{zt}}{E_{zi}} = \frac{-E_t\cos\theta_2}{-E_i\cos\theta_1} = \tau_\|\frac{\cos\theta_2}{\cos\theta_1} \tag{9.76b}$$

Equations (9.75a) and (9.75b) are known as the Fresnel reflection and transmission coefficients, respectively, for parallel polarization.

We shall now discuss the results given by (9.70a), (9.70b), (9.75a), and (9.75b) for the reflection and transmission coefficients for the two cases:

1. For $\theta_1 = 0$, that is, for the case of normal incidence of the uniform plane wave upon the interface, $\theta_2 = 0$ and

$$\Gamma_\perp = \frac{\eta_2 - \eta_1}{\eta_2 + \eta_1}, \qquad \Gamma_\| = \frac{\eta_2 - \eta_1}{\eta_2 + \eta_1}$$

$$\tau_\perp = \frac{2\eta_2}{\eta_2 + \eta_1}, \qquad \tau_\| = \frac{2\eta_2}{\eta_2 + \eta_1}$$

Thus, the reflection coefficients as well as the transmission coefficients for the two cases become equal as they should, since for normal incidence there is no difference between the two polarizations except for rotation by 90° parallel to the interface.

Total internal
reflection

2. $\Gamma_\perp = 1$ and $\Gamma_\parallel = -1$ if $\cos \theta_2 = 0$; that is,

$$\sqrt{1 - \sin^2 \theta_2} = \sqrt{1 - \frac{\mu_1 \varepsilon_1}{\mu_2 \varepsilon_2} \sin^2 \theta_1} = 0$$

or

$$\sin \theta_1 = \sqrt{\frac{\mu_2 \varepsilon_2}{\mu_1 \varepsilon_1}} \tag{9.77}$$

where we have used Snell's law given by (9.61b) to express $\sin \theta_2$ in terms of $\sin \theta_1$. If we assume $\mu_2 = \mu_1 = \mu_0$ as is usually the case, (9.77) has real solutions for θ_1 for $\varepsilon_2 < \varepsilon_1$. Thus, for $\varepsilon_2 < \varepsilon_1$, that is, for transmission from a dielectric medium of higher permittivity into a dielectric medium of lower permittivity, there is a critical angle of incidence θ_c given by

$$\boxed{\theta_c = \sin^{-1} \sqrt{\frac{\varepsilon_2}{\varepsilon_1}}} \tag{9.78}$$

for which θ_2 is equal to 90° and $|\Gamma_\perp|$ and $|\Gamma_\parallel| = 1$. For $\theta_1 > \theta_c$, $\sin \theta_2$ becomes greater than 1, $\cos \theta_2$ becomes imaginary, and $\Gamma_\perp$ and $\Gamma_\parallel$ become complex, but with their magnitudes equal to unity, and *total internal reflection* occurs; that is, the time-average power of incident wave is entirely reflected, the boundary condition being satisfied by an evanescent field in medium 2. To explain the evanescent nature, we note with reference to the geometry of Fig. 9.17 or Fig. 9.18 that

$$\beta_{x2}^2 + \beta_{z2}^2 = \beta_t^2 = \omega^2 \mu_2 \varepsilon_2$$

or

$$\beta_{x2}^2 = \omega^2 \mu_2 \varepsilon_2 - \beta_{z2}^2$$

For $\theta_1 = \theta_c$, $\beta_{z2} = \beta_{z1} = \omega^2 \mu_1 \varepsilon_1 \sin^2 \theta_c = \omega^2 \mu_2 \varepsilon_2$, and $\beta_{x2}^2 = 0$. Therefore, for $\theta_1 > \theta_c$, $\beta_{z2} = \beta_{z1} = \omega^2 \mu_1 \varepsilon_1 \sin^2 \theta_1 > \omega^2 \mu_2 \varepsilon_2$, and $\beta_{x2}^2 < 0$. Thus β_{x2} should be replaced by $-j\alpha_{x2}$, corresponding to exponential decay of field in the x-direction without a propagating wave character. The phenomenon of total internal reflection is the fundamental principle of optical waveguides, since if we have a dielectric slab of permittivity ε_1 sandwiched between two dielectric media of permittivity $\varepsilon_2 < \varepsilon_1$, then by launching waves at an angle of incidence greater than the critical angle, it is possible to achieve guided wave propagation within the slab, as we shall learn in the next section.

3. $\Gamma_\perp = 0$ for $\eta_2 \cos \theta_1 = \eta_1 \cos \theta_2$; that is, for

$$\eta_2 \sqrt{1 - \sin^2 \theta_1} = \eta_1 \sqrt{1 - \frac{\mu_1 \varepsilon_1}{\mu_2 \varepsilon_2} \sin^2 \theta_1}$$

or

$$\sin^2 \theta_1 = \frac{\eta_2^2 - \eta_1^2}{\eta_2^2 - \eta_1^2(\mu_1 \varepsilon_1/\mu_2 \varepsilon_2)} = \mu_2 \frac{\mu_2 - \mu_1(\varepsilon_2/\varepsilon_1)}{\mu_2^2 - \mu_1^2} \qquad (9.79)$$

For the usual case of transmission between two dielectric materials, that is, for $\mu_2 = \mu_1$ and $\varepsilon_2 \neq \varepsilon_1$, this equation has no real solution for θ_1, and hence there is no angle of incidence for which the reflection coefficient is zero for the case of perpendicular polarization.

4. $\Gamma_\parallel = 0$ for $\eta_2 \cos \theta_2 = \eta_1 \cos \theta_1$; that is, for

$$\eta_2 \sqrt{1 - \frac{\mu_1 \varepsilon_1}{\mu_2 \varepsilon_2} \sin^2 \theta_1} = \eta_1 \sqrt{1 - \sin^2 \theta_1}$$

or

$$\sin^2 \theta_1 = \frac{\eta_2^2 - \eta_1^2}{\eta_2^2(\mu_1 \varepsilon_1/\mu_2 \varepsilon_2) - \eta_1^2} = \varepsilon_2 \frac{(\mu_2/\mu_1)\varepsilon_1 - \varepsilon_2}{\varepsilon_1^2 - \varepsilon_2^2} \qquad (9.80)$$

If we assume $\mu_2 = \mu_1$, this equation reduces to

$$\sin^2 \theta_1 = \frac{\varepsilon_2}{\varepsilon_1 + \varepsilon_2}$$

which then gives

$$\cos^2 \theta_1 = 1 - \sin^2 \theta_1 = \frac{\varepsilon_1}{\varepsilon_1 + \varepsilon_2}$$

and

$$\tan \theta_1 = \sqrt{\frac{\varepsilon_2}{\varepsilon_1}}$$

Thus, there exists a value of the angle of incidence θ_p, given by

$$\theta_p = \tan^{-1} \sqrt{\frac{\varepsilon_2}{\varepsilon_1}} \qquad (9.81)$$

for which the reflection coefficient is zero, and hence there is complete transmission for the case of parallel polarization.

5. In view of cases 3 and 4, for an elliptically polarized wave incident on the interface at the angle θ_p, the reflected wave will be linearly polarized perpendicular to the plane of incidence. For this reason, the angle θ_p is known as the *polarizing angle*. It is also known as the *Brewster angle*. The phenomenon associated with the Brewster angle has several applications. An example is in gas lasers in which the discharge tube lying between the mirrors of a Fabry–Perot resonator is sealed by glass windows placed at the Brewster angle, as shown in Fig. 9.19, to minimize reflections from the ends of the tube so that the laser behavior is governed by the mirrors external to the tube.

Brewster angle

FIGURE 9.19   For illustrating the application of the Brewster angle effect in gas lasers.

We shall now consider an example.

Example 9.4

A uniform plane wave having the electric field

$$\mathbf{E}_i = E_0 \left(\frac{\sqrt{3}}{2} \mathbf{a}_x - \frac{1}{2} \mathbf{a}_z \right) \cos [6\pi \times 10^9 t - 10\pi (x + \sqrt{3}z)]$$

is incident on the interface between free space and a dielectric medium of $\varepsilon = 1.5\varepsilon_0$ and $\mu = \mu_0$, as shown in Fig. 9.20. We wish to obtain the expressions for the electric fields of the reflected and transmitted waves.

First, we note from the given $\mathbf{E}_i$ that the propagation vector of the incident wave is given by

$$\boldsymbol{\beta}_i = 10\pi (\mathbf{a}_x + \sqrt{3}\mathbf{a}_z) = 20\pi \left(\frac{1}{2} \mathbf{a}_x + \frac{\sqrt{3}}{2} \mathbf{a}_z \right)$$

the direction of which is consistent with the angle of incidence of 60°. We also note that the electric field vector (which is perpendicular to $\boldsymbol{\beta}_i$) is entirely in the plane of incidence. Thus the situation corresponds to one of parallel polarization, as in Fig. 9.18.

To obtain the required fields, we first find, by using (9.63) and with reference to the notation of Fig. 9.18, that

$$\sin \theta_2 = \sqrt{\frac{\varepsilon_0}{1.5\varepsilon_0}} \sin 60° = \frac{1}{\sqrt{2}}$$

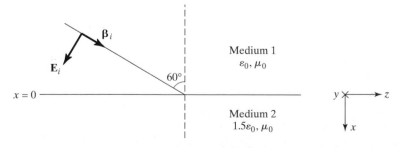

FIGURE 9.20   For Example 9.4.

or $\theta_2 = 45°$. Then from (9.75a)–(9.75b) and (9.76a)–(9.76b), we have

$$\Gamma_{\parallel} = \frac{(\eta_0/\sqrt{1.5})\cos 45° - \eta_0 \cos 60°}{(\eta_0/\sqrt{1.5})\cos 45° + \eta_0 \cos 60°}$$

$$= \frac{2 - \sqrt{3}}{2 + \sqrt{3}} = 0.072$$

$$\tau_{\parallel} = \frac{2(\eta_0/\sqrt{1.5})\cos 60°}{(\eta_0/\sqrt{1.5})\cos 45° + \eta_0 \cos 60°}$$

$$= \frac{2\sqrt{2}}{2 + \sqrt{3}} = 0.758$$

$$\frac{E_r}{E_i} = -0.072$$

$$\frac{E_t}{E_i} = 0.758$$

Finally, noting with the aid of Fig. 9.21 that

$$\boldsymbol{\beta}_r = 20\pi\left(-\frac{1}{2}\mathbf{a}_x + \frac{\sqrt{3}}{2}\mathbf{a}_z\right) = 10\pi(-\mathbf{a}_x + \sqrt{3}\,\mathbf{a}_z)$$

and

$$\boldsymbol{\beta}_t = 20\pi\sqrt{1.5}\left(\frac{1}{\sqrt{2}}\mathbf{a}_x + \frac{1}{\sqrt{2}}\mathbf{a}_z\right) = 10\sqrt{3}\pi(\mathbf{a}_x + \mathbf{a}_z)$$

we write the expressions for the reflected and transmitted wave fields to be

$$\mathbf{E}_r = -0.072 E_0\left(\frac{\sqrt{3}}{2}\mathbf{a}_x + \frac{1}{2}\mathbf{a}_z\right)\cos\left[6\pi \times 10^9 t + 10\pi(x - \sqrt{3}z)\right]$$

and

$$\mathbf{E}_t = 0.758 E_0\left(\frac{1}{\sqrt{2}}\mathbf{a}_x - \frac{1}{\sqrt{2}}\mathbf{a}_z\right)\cos\left[6\pi \times 10^9 t - 10\sqrt{3}\pi(x + z)\right]$$

Note that for $x = 0$, $E_{ri} + E_{rr} = E_{zt}$ and $E_{ri} + E_{rr} = 1.5E_{xt}$, so that the fields do indeed satisfy the boundary conditions.

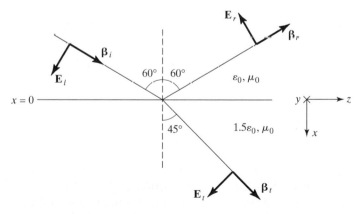

FIGURE 9.21.   For writing the expressions for the reflected and transmitted wave electric fields for Example 9.4.

K9.4. Oblique incidence of uniform plane waves; Plane interface; Law of reflection; Snell's law; Perpendicular and parallel polarizations; Total internal reflection; Brewster angle.

D9.7. Consider a plane boundary between medium 1 ($\varepsilon = \varepsilon_1, \mu = \mu_0$) and medium 2 ($\varepsilon = \varepsilon_2, \mu = \mu_0$). Find the value of $\varepsilon_2/\varepsilon_1$ for each of the following cases of uniform plane waves incident on the boundary from medium 1. **(a)** Total internal reflection occurs for $\theta_i \geq 60°$; **(b)** the reflection coefficient for parallel polarization is zero for $\theta_i = 60°$; and **(c)** the critical angle of incidence for total internal reflection is the same as the Brewster angle for incidence from medium 2.
Ans.   **(a)** 0.75;   **(b)** 3;   **(c)** 0.618.

D9.8. In Figs. 9.17 and 9.18, assume that $\varepsilon_1 = 3\varepsilon_0, \varepsilon_2 = 9\varepsilon_0, \mu_1 = \mu_2 = \mu_0$, and $\theta_i = 45°$. Find **(a)** E_r/E_i and E_t/E_i for the case of perpendicular polarization (Fig. 9.17) and **(b)** E_r/E_i and E_t/E_i for the case of parallel polarization (Fig. 9.18).
Ans.   **(a)** $-0.382, 0.618$;   **(b)** $0.146, 0.662$.

9.5   DIELECTRIC SLAB GUIDE

In the preceding section, we learned that for a wave incident obliquely from a dielectric medium of permittivity ε_1 onto another dielectric medium of permittivity $\varepsilon_2 < \varepsilon_1$, total internal reflection occurs for angles of incidence θ_i exceeding the critical angle θ_c given by

$$\theta_c = \sin^{-1}\sqrt{\frac{\varepsilon_2}{\varepsilon_1}} \tag{9.82}$$

where it is assumed that $\mu = \mu_0$ everywhere. In this section, we shall consider the dielectric slab waveguide, which forms the basis for thin-film waveguides, used extensively in integrated optics.

A.   Wave-Bounce Approach

Description   The dielectric slab waveguide consists of a dielectric slab of permittivity ε_1, sandwiched between two dielectric media of permittivities less than ε_1. For simplicity, we shall consider the symmetric waveguide, that is, one for which the permittivities of the dielectrics on either side of the slab are the same and equal to ε_2, as shown in Fig. 9.22. Then by launching waves at an angle of incidence $\theta_i > \theta_c$, where θ_c is given by (9.82), it is possible to achieve guided wave propagation within the slab, as shown in the figure. For a given thickness d of the slab and for a given frequency of the waves, there are only discrete values of θ_i for which the guiding can take place. In other words, guiding of a wave of a given frequency is not ensured simply because the condition for total internal reflection is met.

Self-consistency condition for guidance   The allowed values of θ_i are dictated by the self-consistency condition, which can be explained with the aid of the construction in Fig. 9.23, as follows. If we consider a point A on a given wavefront designated 1 and follow that wavefront as it moves to position 1' passing through point B, reflects at the interface $x = d/2$ giving rise to wavefront designated 2, then moves to position 2' passing through point C, reflects at the interface $x = -d/2$ giving rise to wavefront designated 3, and

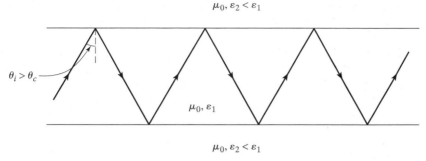

FIGURE 9.22 Total internal reflection in a dielectric slab waveguide.

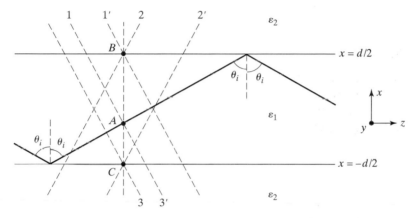

FIGURE 9.23 For explaining the self-consistency condition for waveguiding in a dielectric slab guide.

moves to position $3'$ passing through A, then the total phase shift undergone must be equal to an integer multiple of 2π. If λ_0 is the wavelength in free space corresponding to the wave frequency, the self-consistency condition is given by

$$\frac{2\pi\sqrt{\varepsilon_{r1}}}{\lambda_0}(AB\cos\theta_i) + \underline{/\Gamma_B} + \frac{2\pi\sqrt{\varepsilon_{r1}}}{\lambda_0}(BC\cos\theta_i)$$

$$+ \underline{/\Gamma_A} + \frac{2\pi\sqrt{\varepsilon_{r1}}}{\lambda_0}(CA\cos\theta_i) = 2m\pi, \qquad m = 0, 1, 2, \ldots \qquad (9.83)$$

where $\overline{\Gamma}_A$ and $\overline{\Gamma}_B$ are the reflection coefficients at the interfaces $x = -d/2$ and $x = d/2$, respectively, and $\varepsilon_{r1} = \varepsilon_1/\varepsilon_0$. We recall that under conditions of total internal reflection, the reflection coefficients (9.70a) and (9.75a) become complex with their magnitudes equal to unity. For the symmetric waveguide, $\overline{\Gamma}_A = \overline{\Gamma}_B$. Thus, substituting $\overline{\Gamma}$ for $\overline{\Gamma}_A$ and $\overline{\Gamma}_B$ and $2d$ for $(AB + BC + CA)$, we write (9.83) as

$$\frac{4\pi d\sqrt{\varepsilon_{r1}}}{\lambda_0}\cos\theta_i + 2\underline{/\overline{\Gamma}} = 2m\pi, \qquad m = 0, 1, 2, \ldots$$

or

$$\boxed{\frac{2\pi d \sqrt{\varepsilon_{r1}}}{\lambda_0} \cos \theta_i + \underline{/\overline{\Gamma}} = m\pi, \qquad m = 0, 1, 2, \ldots}$$ (9.84)

Characteristic
equation for
TE modes
and solution

To proceed further, we need to distinguish between the cases of perpendicular and parallel polarizations as defined in the preceding section, since the reflection coefficients for the two cases are different. We shall here consider only the case of perpendicular polarization. The situation then corresponds to TE modes since the electric field has no longitudinal or z-component. Thus, substituting

$$\cos \theta_1 = \cos \theta_i$$

and

$$\begin{aligned}
\cos \theta_2 &= \sqrt{1 - \sin^2 \theta_2} \\
&= j\sqrt{\sin^2 \theta_2 - 1} \\
&= j\sqrt{\frac{\varepsilon_1}{\varepsilon_2} \sin^2 \theta_i - 1}
\end{aligned}$$

in (9.70a), we obtain

$$\overline{\Gamma}_\perp = \frac{\eta_2 \cos \theta_i - j\eta_1 \sqrt{(\varepsilon_1/\varepsilon_2) \sin^2 \theta_i - 1}}{\eta_2 \cos \theta_i + j\eta_1 \sqrt{(\varepsilon_1/\varepsilon_2) \sin^2 \theta_i - 1}}$$ (9.85)

so that

$$\begin{aligned}
\underline{/\overline{\Gamma}_\perp} &= -2 \tan^{-1} \frac{\eta_1 \sqrt{(\varepsilon_1/\varepsilon_2) \sin^2 \theta_i - 1}}{\eta_2 \cos \theta_i} \\
&= -2 \tan^{-1} \frac{\sqrt{\sin^2 \theta_i - (\varepsilon_2/\varepsilon_1)}}{\cos \theta_i}
\end{aligned}$$ (9.86)

Substituting (9.86) into (9.84), we then obtain

$$\frac{2\pi d \sqrt{\varepsilon_{r1}}}{\lambda_0} \cos \theta_i - 2 \tan^{-1} \frac{\sqrt{\sin^2 \theta_i - (\varepsilon_2/\varepsilon_1)}}{\cos \theta_i} = m\pi, \qquad m = 0, 1, 2, \ldots$$

or

$$\tan \left(\frac{\pi d \sqrt{\varepsilon_{r1}}}{\lambda_0} \cos \theta_i - \frac{m\pi}{2} \right) = \frac{\sqrt{\sin^2 \theta_i - (\varepsilon_2/\varepsilon_1)}}{\cos \theta_i}, \qquad m = 0, 1, 2, \ldots$$

or

$$\boxed{\tan \left[f(\theta_i) \right] = \begin{cases} g(\theta_i), & m = 0, 2, 4, \ldots \\ -\dfrac{1}{g(\theta_i)}, & m = 1, 3, 5, \ldots \end{cases}}$$ (9.87)

where

$$f(\theta_i) = \frac{\pi d \sqrt{\varepsilon_{r1}}}{\lambda_0} \cos \theta_i \tag{9.88a}$$

$$g(\theta_i) = \frac{\sqrt{\sin^2 \theta_i - (\varepsilon_2/\varepsilon_1)}}{\cos \theta_i} \tag{9.88b}$$

Equation (9.87) is the characteristic equation for the guiding of TE waves in the dielectric slab. For given values of $\varepsilon_1, \varepsilon_2, d,$ and λ_0, the solutions for θ_i can be obtained by plotting the two sides of (9.87) versus θ_i and finding the points of intersection. The nature of this construction is shown in Fig. 9.24. Each solution corresponds to one mode. It can be seen from (9.88a) and Fig. 9.24 that for a given set of values of ε_1 and ε_2, fewer solutions are obtained for θ_i as the ratio (d/λ_0) becomes smaller, since the number of branches of the plot of $\tan[f(\theta_i)]$ between $\theta_i = \pi/2$ and $\theta_i = \theta_c$ become fewer. It can also be seen that there is always one solution even for arbitrarily low values of (d/λ_0), that is, for large values of λ_0 or low frequencies, for a given d.

Alternative to the graphical solution, we can use a computer to solve (9.87) for the allowed values of θ_i for specified values of $\varepsilon_{r1}, \varepsilon_{r2}, d,$ and λ_0. Computed values of θ_i for values of $\varepsilon_{r1} = 4, \varepsilon_{r2} = 1, d = 10$ mm, and $\lambda_0 = 5$ mm are listed in Table 9.1.

Returning now to Fig. 9.24, we designate the modes associated with the solutions as TE_m modes, where $m = 0, 1, 2$, correspond to the values of m on the plot. We note from the plot that the solution for a given TE_m mode for $m > 1$

Cutoff frequencies

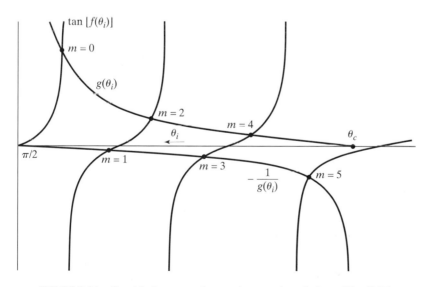

FIGURE 9.24 Graphical construction pertinent to the solution of Eq. (9.87).

TABLE 9.1   Allowed Values of θ_i for
Dielectric Slab Guide Example

m	θ_i (deg)
0	83.42783
1	76.77756
2	69.96263
3	62.87805
4	55.38428
5	47.28283
6	38.30225

does not exist if $f(\theta_c) < m\pi/2$. Therefore, the cutoff condition is given by

$$\frac{\pi d\sqrt{\varepsilon_{r1}}}{\lambda_0} \cos\theta_c < \frac{m\pi}{2}$$

$$\frac{\pi d\sqrt{\varepsilon_{r1}}}{\lambda_0}\sqrt{1 - \frac{\varepsilon_2}{\varepsilon_1}} < \frac{m\pi}{2}$$

$$\lambda_0 > \frac{2d\sqrt{\varepsilon_{r1} - \varepsilon_{r2}}}{m} \tag{9.89}$$

where we have used (9.82). The cutoff frequency is given by

$$f_c = \frac{c}{\lambda_0} = \frac{mc}{2d\sqrt{\varepsilon_{r1} - \varepsilon_{r2}}}$$

The fundamental mode, TE_0, has no cutoff frequency. Thus,

$$\boxed{f_c = \frac{mc}{2d\sqrt{\varepsilon_{r1} - \varepsilon_{r2}}}, \qquad m = 0, 1, 2, \ldots} \tag{9.90}$$

Example 9.5

*Finding
propagating
modes*

For the symmetric dielectric slab waveguide of Fig. 9.23, let $\varepsilon_1 = 2.56\varepsilon_0$, $\varepsilon_2 = \varepsilon_0$, and $d = 10\lambda_0$. We wish to find the number of TE modes that can propagate by guidance in the slab.

From (9.90),

$$f_c = \frac{mc}{20\lambda_0\sqrt{2.56 - 1}}$$

$$= \frac{mf}{24.98}, \qquad m = 0, 1, 2, \ldots$$

Thus, for $m > 24$, $f_c > f$ and the modes are cut off. Therefore, the number of propagating TE modes is 25, corresponding to $m = 0, 1, 2, \ldots, 24$.

The entire discussion for guided waves in the dielectric slab guide can be repeated for TM modes by using $\overline{\Gamma}_{\parallel}$ in the place of $\overline{\Gamma}_{\perp}$ in (9.84) to derive the characteristic equation for guidance. We shall include the derivation as Problem 9.23.

TM modes

B. Wave-Field Approach

A formal approach to the investigation of guided modes in the dielectric slab involves the derivation of the field expressions. This is done by recognizing with reference to the geometry in Fig. 9.23 that (a) in the slab the fields have standing wave character in the x-direction and traveling wave character in the z-direction; (b) outside the slab, the fields are evanescent, that is, they decay exponentially away from it in the x-direction and have traveling wave character in the z-direction; and (c) from symmetry considerations, the fields should be even or odd with respect to x.

Field behavior for guided modes

Let us first consider even TE modes, that is, modes with the transverse field components having even symmetry with respect to x. Then we write the expression for the (only) electric field component $\overline{E}_y$ to be

Field expressions for even TE modes

$$
\overline{E}_y = \begin{cases} \overline{A} \cos \beta_{x1} x \, e^{-j\beta_z z} & \text{for} \quad |x| < d/2 \\ \overline{B} e^{-\alpha_{x2} x} \, e^{-j\beta_z z} & \text{for} \quad x > d/2 \\ \overline{B} e^{\alpha_{x2} x} \, e^{-j\beta_z z} & \text{for} \quad x < -d/2 \end{cases} \tag{9.91}
$$

where $\overline{A}$ and $\overline{B}$ are constants. Note that subscripts 1 and 2 denote regions of permittivities ε_1 and ε_2, respectively, and that the phase constant β_z does not have a subscript 1 or 2, since it must be the same in all three regions in view of the requirement that the fields be in phase at the boundaries $x = \pm d/2$ for all z. Continuity of $\overline{E}_y$ at $x = \pm d/2$ further requires that

$$
\overline{A} \cos \beta_{x1} \frac{d}{2} = \overline{B} e^{-\alpha_{x2} d/2}
$$

so that

$$
\overline{B} = \overline{A} e^{\alpha_{x2} d/2} \cos \beta_{x1} \frac{d}{2} \tag{9.92}
$$

and hence

$$
\overline{E}_y = \begin{cases} \overline{A} \cos \beta_{x1} x \, e^{-j\beta_z z} & \text{for} \quad |x| < d/2 \\ \overline{A} \cos \beta_{x1} \dfrac{d}{2} e^{-\alpha_{x2}(x - d/2)} e^{-j\beta_z z} & \text{for} \quad x > d/2 \\ \overline{A} \cos \beta_{x1} \dfrac{d}{2} e^{\alpha_{x2}(x + d/2)} e^{-j\beta_z z} & \text{for} \quad x < -d/2 \end{cases} \tag{9.93}
$$

To obtain the corresponding magnetic field components, we use the phasor forms of (4.12a)–(4.12c) with the understanding that $(\partial/\partial t) \to j\omega$, $(\partial/\partial y) = 0$, $(\partial/\partial z) \to -j\beta_z$ and $E_x = E_z = 0$:

$$
j\beta_z \overline{E}_y = -j\omega \overline{B}_x \tag{9.94a}
$$

$$
\frac{\partial \overline{E}_y}{\partial x} = -j\omega \overline{B}_z \tag{9.94b}
$$

Thus

$$\overline{H}_x = -\frac{\beta_z}{\omega\mu_0}\,\overline{E}_y \tag{9.95}$$

$$\overline{H}_z = \begin{cases} -\dfrac{j\beta_{x1}}{\omega\mu_0}\,\overline{A}\sin\beta_{x1}x\,e^{-j\beta_z z} & \text{for } |x| < d/2 \\[2ex] -\dfrac{j\alpha_{x2}}{\omega\mu_0}\,\overline{A}\cos\beta_{x1}\dfrac{d}{2}\,e^{-\alpha_{x2}(x-d/2)}e^{-j\beta_z z} & \text{for } x > d/2 \\[2ex] \dfrac{j\alpha_{x2}}{\omega\mu_0}\,\overline{A}\cos\beta_{x1}\dfrac{d}{2}\,e^{\alpha_{x2}(x+d/2)}e^{-j\beta_z z} & \text{for } x < -d/2 \end{cases} \tag{9.96}$$

Now, continuity of $\overline{H}_z$ at $x = \pm d/2$ requires that

$$\tan\beta_{x1}\frac{d}{2} = \frac{\alpha_{x2}}{\beta_{x1}} \tag{9.97}$$

We also know that $\beta_{x1}, \beta_z,$ and α_{x2} are not independent since together the field components must also satisfy the component equations of (4.22) in phasor form with $\mathbf{J} = \mathbf{0}$:

$$-j\beta_z\overline{H}_x - \frac{\partial\overline{H}_z}{\partial x} = \begin{cases} j\omega\varepsilon_1\overline{E}_y & \text{for } |x| < d/2 \\ j\omega\varepsilon_2\overline{E}_y & \text{for } x > d/2 \\ j\omega\varepsilon_2\overline{E}_y & \text{for } x < -d/2 \end{cases} \tag{9.98}$$

Substitution of (9.93), (9.95), and (9.96) gives us

$$\beta_{x1}^2 + \beta_z^2 = \omega^2\mu_0\varepsilon_1 \tag{9.99a}$$

$$-\alpha_{x2}^2 + \beta_z^2 = \omega^2\mu_0\varepsilon_2 \tag{9.99b}$$

or

$$\frac{\alpha_{x2}}{\beta_{x1}} = \sqrt{\frac{\omega^2\mu_0(\varepsilon_1 - \varepsilon_2)}{\beta_{x1}^2} - 1} \tag{9.100}$$

Combining (9.97) and (9.100), we obtain the characteristic equation for guidance to be

$$\tan\left(\beta_{x1}\frac{d}{2}\right) = \sqrt{\frac{\omega^2\mu_0(\varepsilon_1 - \varepsilon_2)}{\beta_{x1}^2} - 1}$$

$$\tan\left(\frac{\beta_1 d}{2}\cos\theta_i\right) = \sqrt{\frac{\omega^2\mu_0(\varepsilon_1 - \varepsilon_2)}{\omega^2\mu_0\varepsilon_1\cos^2\theta_i} - 1}$$

$$\tan\left(\frac{\pi d\sqrt{\varepsilon_{r1}}}{\lambda_0}\cos\theta_i\right) = \frac{\sqrt{\sin^2\theta_i - (\varepsilon_2/\varepsilon_1)}}{\cos\theta_i}$$

$$\boxed{\tan[f(\theta_i)] = g(\theta_i)} \tag{9.101}$$

which is the same as (9.87) for $m = 0, 2, 4, \ldots$.

Proceeding further, we can interpret the mode number m in terms of the field variations with x in the following manner. For a given value of m, we observe from Fig. 9.24 that $m\pi/2 < \beta_{x1}d/2 \; [= f(\theta_i)] < (m+1)\pi/2$. Thus, from $x = 0$ to $x = \pm d/2$, $\cos \beta_{x1}x$ varies from $\cos 0°$ to some value between $\cos m\pi/2$ and $\cos (m+1)\pi/2$. Near cutoff, $\beta_{x1}d/2 \rightarrow m\pi/2$, $\tan (\beta_{x1}d/2) \rightarrow 0$, and $\alpha_{x2} \rightarrow 0$. The variation of E_y with x is as illustrated in Fig. 9.25(a) for $m = 2$. At high frequencies far from cutoff, $\beta_{x1}d/2 \rightarrow (m+1)\pi/2$, $\tan \beta_{x1}d/2 \rightarrow \infty$, and $\alpha_{x2} \rightarrow \infty$. The variation of E_y with x is as illustrated in Fig. 9.25(b). Figure 9.25(c) illustrates the situation intermediate to those near cutoff and far from cutoff. Thus, within the thickness of the slab, the behavior of the field components varies from m half-sine variations near cutoff toward $(m+1)$ half-sine variations far from cutoff, with the evanescence outside the dielectric slab dictated by $\alpha_{x2} \rightarrow 0$ near cutoff toward $\alpha_{x2} \rightarrow \infty$ far from cutoff.

The field expressions for the odd TE modes, that is, modes with the transverse field components having odd symmetry with respect to x, can be obtained by writing the expression for $\overline{E}_y$ to be

Field expressions for odd TE modes

$$\overline{E}_y = \begin{cases} \overline{C}\sin\beta_{x1}x \, e^{-j\beta_z z} & \text{for} \quad |x| < d/2 \\ \overline{D}e^{-\alpha_{x2}x}e^{-j\beta_z z} & \text{for} \quad x > d/2 \\ -\overline{D}e^{\alpha_{x2}x}e^{-j\beta_z z} & \text{for} \quad x < -d/2 \end{cases} \qquad (9.102)$$

where $\overline{C}$ and $\overline{D}$ are constants and proceeding in a manner similar to that for the even modes. We shall omit the details and write down the final results:

$$\overline{E}_y = \begin{cases} \overline{C}\sin\beta_{x1}x \, e^{-j\beta_z z} & \text{for} \quad |x| < d/2 \\ \overline{C}\sin\beta_{x1}\dfrac{d}{2} e^{-\alpha_{x2}(x-d/2)} e^{-j\beta_z z} & \text{for} \quad x > d/2 \\ -\overline{C}\sin\beta_{x1}\dfrac{d}{2} e^{\alpha_{x2}(x+d/2)} e^{-j\beta_z z} & \text{for} \quad x < -d/2 \end{cases} \qquad (9.103)$$

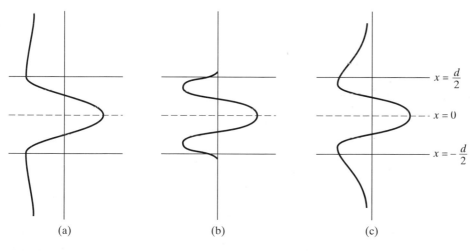

(a)   (b)   (c)

$x = \dfrac{d}{2}$

$x = 0$

$x = -\dfrac{d}{2}$

FIGURE 9.25 Variations of E_y with x for the TE_2 mode in the symmetric dielectric slab waveguide for (a) near cutoff, (b) far from cutoff; and (c) intermediate to (a) and (b).

$$\overline{H}_x = -\frac{\beta_z}{\omega \mu_0} \overline{E}_y \tag{9.104}$$

$$\overline{H}_z = \begin{cases} \dfrac{j\beta_{x1}}{\omega \mu_0} \overline{C} \cos \beta_{x1} x \; e^{-j\beta_z z} & \text{for} \quad |x| < d/2 \\[2ex] -\dfrac{j\alpha_{x2}}{\omega \mu_0} \overline{C} \sin \beta_{x1}\dfrac{d}{2} e^{-\alpha_{x2}(x-d/2)} e^{-j\beta_z z} & \text{for} \quad x > d/2 \\[2ex] -\dfrac{j\alpha_{x2}}{\omega \mu_0} \overline{C} \sin \beta_{x1}\dfrac{d}{2} e^{\alpha_{x2}(x+d/2)} e^{-j\beta_z z} & \text{for} \quad x < -d/2 \end{cases} \tag{9.105}$$

Continuity of $\overline{H}_z$ at $x = \pm d/2$ requires that

$$\cot \beta_{x1}\frac{d}{2} = -\frac{\alpha_{x2}}{\beta_{x1}} \tag{9.106}$$

$$= -\sqrt{\frac{\omega^2 \mu_0 (\varepsilon_1 - \varepsilon_2)}{\beta_{x1}^2} - 1}$$

where we have used (9.100). Thus, the characteristic equation for guidance is

$$\boxed{\tan f(\theta_i) = -\frac{1}{g(\theta_i)}} \tag{9.107}$$

which is the same as (9.87) for $m = 1, 3, 5, \ldots$.

Proceeding further, we observe from Fig. 9.24 that for a given value of m, $m\pi/2 < \beta_{x1} d/2 \, [= f(\theta_i)] < (m + 1)\pi/2$. Thus, from $x = 0$ to $x = \pm d/2$, $\sin \beta_{x1} x$ varies from $\sin 0°$ to some value between $\sin m\pi/2$ and $\sin (m + 1)\pi/2$. Near cutoff, $\beta_{x1} d/2 \to m\pi/2$, $\cot (\beta_{x1} d/2) \to 0$, and $\alpha_{x2} \to 0$. At high frequencies far from cutoff, $\beta_{x1} d/2 \to (m + 1)\pi/2$, $\cot (\beta_{x1} d/2) \to \infty$, and $\alpha_{x2} \to \infty$. Thus, the variation of E_y with x for $m = 1$ is illustrated in Fig. 9.26 for three situations:

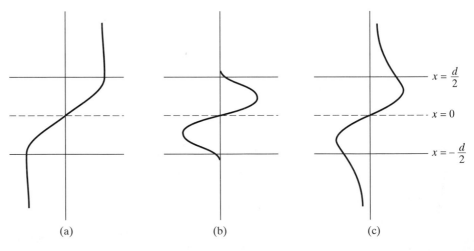

(a)   (b)   (c)

FIGURE 9.26   Variations of E_y with x for TE$_1$ mode in the symmetric dielectric slab waveguide for (a) near cutoff; (b) far from cutoff; and (c) intermediate to (a) and (b).

(a) near cutoff; (b) far from cutoff; and (c) intermediate to (a) and (b). As in the case of the even modes, the behavior of the field components varies from m half-sine variations near cutoff toward $(m + 1)$ half-sine variations far from cutoff.

Let us now investigate the time-average power flow down the symmetric *Power flow* slab waveguide for TE modes. First, we write the complex Poynting vector associated with the TE mode fields as given by

$$\overline{\mathbf{P}} = \tfrac{1}{2}\overline{\mathbf{E}} \times \overline{\mathbf{H}}^*$$
$$= \tfrac{1}{2}(\overline{E}_y \overline{H}_z^* \mathbf{a}_x - \overline{E}_y \overline{H}_x^* \mathbf{a}_z) \tag{9.108}$$

Then noting from (9.93), (9.95), and (9.96) that $\overline{E}_y \overline{H}_x^*$ is real, whereas as $\overline{E}_y \overline{H}_z^*$ is imaginary, we obtain the time-average Poynting vector as given by

$$\langle \mathbf{P} \rangle = \operatorname{Re} \overline{\mathbf{P}}$$

$$= \frac{\beta_z}{2\omega\mu_0} |\overline{E}_y|^2 \mathbf{a}_z$$

$$= \frac{\beta_z |\overline{A}|^2}{2\omega\mu_0} \begin{cases} \cos^2 \beta_{x1} x \, \mathbf{a}_z & \text{for } |x| < d/2 \\[2mm] \cos^2 \beta_{x1} \dfrac{d}{2} e^{-2\alpha_{x2}(x-d/2)} \mathbf{a}_z & \text{for } x > d/2 \\[2mm] \cos^2 \beta_{x1} \dfrac{d}{2} e^{2\alpha_{x2}(x+d/2)} \mathbf{a}_z & \text{for } x < -d/2 \end{cases} \tag{9.109}$$

where we have used the even mode field expression. For the odd modes, the $\cos^2$ terms will be replaced by $\sin^2$ terms and the final result will be the same. The time-average power flow along the guide per unit length in the y-direction (because of the independence of the fields with y) is then given by

$$\langle P \rangle = \int_{y=0}^{1} \int_{x=-\infty}^{\infty} \langle \mathbf{P} \rangle \cdot dx\, dy\, \mathbf{a}_z$$

$$= \frac{\beta_z |\overline{A}|^2}{\omega\mu_0} \left[\int_0^{d/2} \cos^2 \beta_{x1} x \, dx + \int_{d/2}^{\infty} \cos^2 \beta_{x1} \frac{d}{2} e^{-2\alpha_{x2}(x-d/2)} dx \right] \tag{9.110}$$

$$= \frac{\beta_z |\overline{A}|^2}{\omega\mu_0} \left[\frac{d}{4} + \frac{\sin \beta_{x1} d}{4\beta_{x1}} + \frac{\cos^2 \beta_{x1} d/2}{2\alpha_{x2}} \right]$$

Using (9.97) and substituting

$$\sin \beta_{x1} d = 2 \sin \beta_{x1} \frac{d}{2} \cos \beta_{x1} \frac{d}{2}$$

$$= \frac{2\beta_{x1}}{\alpha_{x2}} \sin^2 \beta_{x1} \frac{d}{2}$$

We obtain

$$\langle P \rangle = \frac{\beta_z |\overline{A}|^2}{\omega\mu_0} \left(\frac{d}{4} + \frac{1}{2\alpha_{x2}} \right)$$
$$- \frac{\beta_z |\overline{A}|^2}{4\omega\mu_0} \left(d + \frac{2}{\alpha_{x2}} \right) \tag{9.111}$$

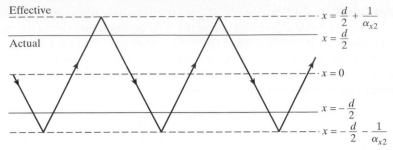

FIGURE 9.27   For illustrating the effective boundaries for waveguiding along the symmetric dielectric slab guide.

Besides giving the expression for the time-average power flow along the guide, (9.111) leads to the definition of fictitious effective boundaries at $x = \pm d_{\text{eff}}/2$, where

$$d_{\text{eff}} = d + \frac{2}{\alpha_{x2}} \tag{9.112}$$

as shown in Fig. 9.27. The physical interpretation of the placement of these effective boundaries relates to the phase shift that the waves experience upon being total internally reflected at the actual boundaries.

Example 9.6

For the values of $\varepsilon_{r1}, \varepsilon_{r2}, d,$ and λ_0 used for Table 9.1, it is desired to find d_{eff} for the first three modes.

From (9.99b),

$$\alpha_{x2}^2 = \beta_z^2 - \omega^2 \mu_0 \varepsilon_2$$
$$= \omega^2 \mu_0 \varepsilon_1 \sin^2 \theta_i - \omega^2 \mu_0 \varepsilon_2$$
$$= \omega^2 \mu_0 \varepsilon_0 (\varepsilon_{r1} \sin^2 \theta_i - \varepsilon_{r2})$$
$$\alpha_{x2} = \frac{2\pi}{\lambda_0} \sqrt{\varepsilon_{r1} \sin^2 \theta_i - \varepsilon_{r2}}$$

Substituting $\lambda_0 = 5$ mm, $\varepsilon_{r1} = 4$, and $\varepsilon_{r2} = 1$, we have

$$\alpha_{x2} = \frac{2\pi}{5 \times 10^{-3}} \sqrt{4 \sin^2 \theta_i - 1}$$

From Table 9.1, we can then compute α_{x2} and hence d_{eff} for the first three modes, as listed in Table 9.2.

TABLE 9.2   For Example 9.6

Mode	θ_i (deg)	$\alpha_{x2}(\text{m}^{-1})$	$d_{\text{eff}}(\text{mm})$
TE_0	83.42783	2157.47	10.927
TE_1	76.77756	2099.27	10.953
TE_2	69.96263	1998.97	11.001

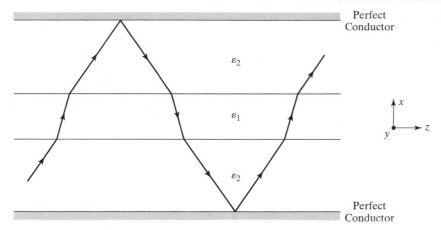

FIGURE 9.28 For explaining the mechanism pertinent to radiation modes in a dielectric slab guide.

The entire solution for the field components can be repeated for TM modes *TM modes* by starting with the expression for the (only) magnetic field component $\overline{H}_y$ and proceeding in a manner similar to that used for the TE mode case. We shall not however pursue these details here but include them as Problems P9.25 and P9.26.

We have thus far discussed modes that are guided within the slab. Another *Radiation* type of mode that is possible for the dielectric slab guide is that for which the *modes* field variations with x are sinusoidal not only in the slab but also outside it. The situation can be visualized by locating perfectly conducting plates on either side of the slab and parallel to it as in Fig. 9.28 and displacing the conductors to infinity, thereby obtaining the slab waveguide in the limiting case. Waves that are incident from medium 1 on the interface at angles of incidence less than the critical angle for total internal reflection are transmitted into medium 2 and are reflected at the conductor, giving rise to ray paths such as the one shown. The modes established when the associated self-consistency condition is satisfied are known as the radiation modes. These modes are important in the coupling of energy in and out of the dielectric slab.

K9.5. Dielectric slab waveguide; Guiding by total internal reflection; Self-consistency condition; Characteristic equation for guidance; Propagating modes; Derivation of field expressions; Characteristic equation for guidance; Mode behavior; Power flow; Radiation modes.

D9.9. For a symmetric dielectric slab waveguide, $\varepsilon_1 = 2.25\varepsilon_0$ and $\varepsilon_2 = \varepsilon_0$. Find the following for TE modes: **(a)** the lowest value of d/λ_0 for which an allowed value of θ_i is 60°; **(b)** the lowest value of d/λ_0 for which an allowed value of θ_i is 75°; and **(c)** the second lowest value of d/λ_0 for which an allowed value of θ_i is 75°.
Ans. **(a)** 0.3545; **(b)** 0.9972; **(c)** 2.2852.

D9.10. For a symmetric dielectric slab waveguide, $\varepsilon_{r1} = 4$ and $\varepsilon_{r2} = 1$. Find the value of d_{eff}/d for the TE$_0$ mode for each of the following values of d/λ_0: **(a)** 2; **(b)** 5; and **(c)** 0.5.
Ans. **(a)** 1.0927; **(b)** 1.0368; **(c)** 1.4047.

9.6   RAY TRACING AND GRADED-INDEX GUIDE

For the dielectric slab waveguide of the preceding section, the permittivity undergoes an abrupt discontinuity from a uniform value of ε_1 in the slab to a uniform value of ε_2 on either side of the slab. When the permittivity varies within the thickness of the slab, the arrangement is known as a graded-index guide, as compared to the step-index guide of the previous section, where the word "index" refers to the refractive index n ($= c/v_p = \sqrt{\varepsilon_r}$ for a nonmagnetic dielectric). To extend our discussion of guided wave propagation to a graded-index slab waveguide, we first introduce the general topic of geometrical optics and ray tracing.

Geometrical optics approximation explained
  Geometrical optics is that branch of optics that allows us to study wave phenomena by tracing "rays," which are paths normal to the wavefronts, from the local application of the laws of reflection and refraction (Snell's law). Whenever the wavefront extends and is uniform over many wavelengths and when the boundaries are large compared to the wavelength, ray tracing can be usefully employed. Also, as long as the radii of curvature are large in comparison to the wavelength, the boundaries as well as the wavefronts can be nonplanar. In fact, we have already made use of geometrical optics concepts to introduce the cut-off phenomenon in a parallel-plate waveguide in Section 9.2 and to derive the characteristic equation for guidance in the dielectric slab guide in Section 9.5. In all of these cases, the media were uniform and the boundaries plane and abrupt so that the ray paths were all straight lines. For nonuniform media, the ray paths become curved.

Ray tracing formulation
  To formulate the ray tracing procedure, let us consider the arrangement shown in Fig. 9.29 in which a medium of continuously varying refractive index $n(x)$ is approximated by a series of plane slabs of uniform refractive indices n_1, $n_2, n_3, \ldots$. Let a wave be incident from the medium of refractive index n_0 at an angle θ_0 from the vertical. Then assuming $n_0 < n_1 < n_2 < n_3 < \ldots$, the ray path bends more and more away from the vertical in accordance with Snell's law

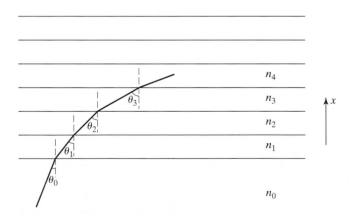

FIGURE 9.29   The bending of ray paths in a series of plane dielectric slabs of uniform refractive indices.

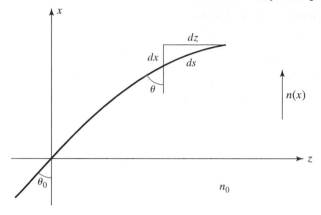

FIGURE 9.30 For the formulation of ray tracing for continuous variation of refractive index.

applied at the interfaces

$$n_0 \sin \theta_0 = n_1 \sin \theta_1 = n_2 \sin \theta_2 = \cdots \qquad (9.113)$$

with the path in each layer being a straight line. In the limit that the thickness of each layer goes to zero, the refractive index varies continuously with x and the ray path becomes curved. To trace the path of the ray, let us consider a differential segment ds along the ray path, having the components dx and dz in the x- and z-directions, respectively, as shown in Fig. 9.30. Then

$$\frac{dz}{dx} = \tan \theta = \frac{\sin \theta}{\sqrt{1 - \sin^2 \theta}} \qquad (9.114)$$

From (9.113), $n \sin \theta = n_0 \sin \theta_0$, or $\sin \theta = (n_0/n) \sin \theta_0$. Substituting in (9.114), we obtain

$$\frac{dz}{dx} = \frac{n_0 \sin \theta_0}{\sqrt{n^2 - n_0^2 \sin^2 \theta_0}} \qquad (9.115)$$

For a given refractive index profile $n(x)$, the solution of (9.115) gives the ray trajectory in the xz-plane. In general, the solution has to be carried out numerically. For certain functions for $n(x)$, an analytical solution is possible. We shall illustrate by means of an example.

Example 9.7

Let us consider a variation of refractive index as given by

$$n^2 = n_0^2 (1 - \alpha x)$$

where $\alpha > 0$. Note that for a nonmagnetic dielectric medium, this corresponds to a linear

Ray path for linear profile of permittivity

profile of permittivity. We wish to find the ray trajectory in the medium in the xz-plane for a wave entering the medium at $x = 0$ and $z = 0$ at an angle θ_0 from the vertical (x-direction).

From (9.115), we have

$$\frac{dz}{dx} = \frac{n_0 \sin \theta_0}{\sqrt{n_0^2 - n_0^2 \alpha x - n_0^2 \sin^2 \theta_0}}$$

$$= \frac{\sin \theta_0}{\sqrt{\cos^2 \theta_0 - \alpha x}}$$

The ray trajectory is given by

$$z(x) = \int_0^x \frac{\sin \theta_0}{\sqrt{\cos^2 \theta_0 - \alpha x}} \, dx$$

$$= -\sin \theta_0 \left[\frac{2}{\alpha} \sqrt{\cos^2 \theta_0 - \alpha x} \right]_0^x$$

$$= -\frac{2 \sin \theta_0}{\alpha} [\sqrt{\cos^2 \theta_0 - \alpha x} - \cos \theta_0]$$

Rearranging, we have

$$\cos^2 \theta_0 - \alpha x = \left(-\frac{\alpha z}{2 \sin \theta_0} + \cos \theta_0 \right)^2$$

or

$$x = \frac{\cos^2 \theta_0}{\alpha} - \left(\frac{\cos \theta_0}{\sqrt{\alpha}} - \frac{\sqrt{\alpha}}{2 \sin \theta_0} z \right)^2$$

Thus, the ray trajectory is parabolic, with the parabola having its apex at $x = (\cos^2 \theta_0)/\alpha$ and $z = (\sin 2\theta_0)/\alpha$. Note that at $x = (\cos^2 \theta_0)/\alpha$, $n^2 = n_0^2(1 - \cos^2 \theta_0)$ or $n = n_0 \sin \theta_0$, and $\theta = 90°$, consistent with the solution obtained.

Paraxial ray approximation   When the ray trajectories are nearly along the propagation axis, the rays are known as paraxial rays. For paraxial rays, the angle δ ($= 90° - \theta$) that the ray makes with the propagation axis, which here is the z-axis as shown in Fig. 9.31, is small such that approximations of $\sin \delta \approx \delta$ and $\cos \delta \approx 1$ can be used.

FIGURE 9.31   Geometry pertinent to the paraxial ray approximation.

Example 9.8

An important example of a graded refractive index profile is given by

$$n^2 = n_0^2(1 - \alpha^2 x^2)$$

(9.116)

Paraxial rays in parabolic index profile

For α sufficiently small that $\alpha x \ll 1$, which is usually the case,

$$n(x) = n_0\sqrt{1 - \alpha^2 x^2} \approx n_0(1 - \tfrac{1}{2}\alpha^2 x^2)$$

(9.117)

corresponds to a parabolic variation. We wish to investigate paraxial rays for this profile.

Let us consider a ray making an angle δ_0 with the z-axis at the point $x = 0$. Then from (9.115),

$$\frac{dz}{dx} = \frac{n_0 \sin \theta_0}{\sqrt{n^2 - n_0^2 \sin^2 \theta_0}} = \frac{n_0 \cos \delta_0}{\sqrt{n^2 - n_0^2 \cos^2 \delta_0}}$$

$$= \frac{n_0 \cos \delta_0}{\sqrt{n_0^2(1 - \alpha^2 x^2) - n_0^2 \cos^2 \delta_0}}$$

$$= \frac{\cos \delta_0}{\sqrt{\sin^2 \delta_0 - \alpha^2 x^2}}$$

$$\approx \frac{1}{\delta_0^2 - \alpha^2 x^2}$$

$$z = \int_0^x \frac{dx}{\sqrt{\delta_0^2 - \alpha^2 x^2}}$$

$$= \left[\frac{1}{\alpha} \sin^{-1}\left(\frac{\alpha}{\delta_0}x\right)\right]_0^x$$

$$= \frac{1}{\alpha} \sin^{-1} \frac{\alpha x}{\delta_0}$$

or

$$x = \frac{\delta_0}{\alpha} \sin \alpha z$$

(9.118)

The ray oscillates about the axis with a period of $2\pi/\alpha$, known as the *pitch*, independent of δ_0, and a peak amplitude δ_0/α, as shown in Fig. 9.32 for a few values of δ_0.

FIGURE 9.32   Paraxial rays in a parabolic index profile.

Guidance condition for graded-index guide

We may now discuss wave guidance in a graded-index guide. To do this, let us consider, for simplicity, the symmetric refractive index profile of the shape shown in Fig. 9.33(a). Then for those waves that are total internally reflected within $-d/2 < x < d/2$, a sketch of the ray path can be drawn, as in Fig. 9.33(b), with apex points at $x = \pm x_a$, where $x_a < d/2$. From Snell's law, $n(\pm x_a) = n_1(x) \sin \theta(x)$, since $\theta = 90°$ at $x = \pm x_a$. Using the same reasoning as for writing (9.83), we can write the self-consistency condition for guidance to be

$$\int_{x=-x_a}^{x_a} \frac{2\pi}{\lambda_0} \sqrt{\varepsilon_{r1}(x)} \cos \theta(x) \, dx \; + \; \underline{/[\overline{\Gamma}]}_{x=x_a} + \int_{x=x_a}^{-x_a} \frac{2\pi}{\lambda_0} \sqrt{\varepsilon_{r1}(x)} \cos \theta(x) \, dx$$

$$+ \; \underline{/[\overline{\Gamma}]}_{x=-x_a} = 2m\pi, \qquad m = 0, 1, 2, \ldots \qquad (9.119)$$

In view of symmetry, (9.119) reduces to

$$\frac{4\pi}{\lambda_0} \int_{x=0}^{x_a} \sqrt{\varepsilon_{r1}(x)} \cos \theta(x) \, dx \; + \; \underline{/\overline{\Gamma}_a} = m\pi, \qquad m = 0, 1, 2, \ldots \qquad (9.120)$$

where $\overline{\Gamma}_a = [\overline{\Gamma}]_{x=x_a}$.

To proceed further, we consider the TE case and find $\underline{/\overline{\Gamma}_a}$ by using (9.86). We first write $\overline{\Gamma}_a$ as

$$\overline{\Gamma}_a = \frac{\eta_2' \cos \theta_i - j\eta_1' \sqrt{(\varepsilon_1'/\varepsilon_2') \sin^2 \theta_i - 1}}{\eta_2' \cos \theta_i + j\eta_1' \sqrt{(\varepsilon_1'/\varepsilon_2') \sin^2 \theta_i - 1}}$$

$$= \frac{\sqrt{\varepsilon_1'} \cos \theta_i - j\sqrt{\varepsilon_1' \sin^2 \theta_i - \varepsilon_2'}}{\sqrt{\varepsilon_1'} \cos \theta_i + j\sqrt{\varepsilon_1' \sin^2 \theta_i - \varepsilon_2'}} \qquad (9.121)$$

where we have inserted primes so as not to confuse with the notation of Fig. 9.33.

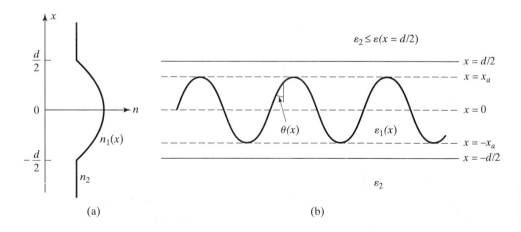

(a)   (b)

FIGURE 9.33   (a) Refractive index profile for a symmetrical graded-index guide. (b) Ray path within the graded-index region.

Now we note that for the situation under consideration,

$$\varepsilon_1' = [\varepsilon]_{x=x_a - \Delta x} = \varepsilon(x_a-)$$

$$\varepsilon_2' = [\varepsilon]_{x=x_a + \Delta x} = \varepsilon(x_a+)$$

$$\varepsilon_1' \sin^2 \theta_i = [\varepsilon]_{x=x_a} \sin^2 90° = \varepsilon(x_a)$$

$$\sqrt{\varepsilon_1'} \cos \theta_i = \sqrt{\varepsilon_1' - \varepsilon_1' \sin^2 \theta_i} = \sqrt{\varepsilon(x_a-) - \varepsilon(x_a)}$$

so that

$$\overline{\Gamma}_a = \frac{\sqrt{\varepsilon(x_a-) - \varepsilon(x_a)} - j\sqrt{\varepsilon(x_a) - \varepsilon(x_a+)}}{\sqrt{\varepsilon(x_a-) - \varepsilon(x_a)} + j\sqrt{\varepsilon(x_a) - \varepsilon(x_a+)}} \qquad (9.122)$$

In view of the continuous variation of ε_1, we now have to take the limit of the right side of (9.122) as x_a- and x_a+ tend to x_a. We note however that this results in a situation of zero divided by zero. To avoid this, we write that in the vicinity of $x = x_a$,

$$\varepsilon(x) \approx \varepsilon(x_a) + (x - x_a)\frac{d\varepsilon}{dx} = \varepsilon(x_a) + \Delta x\frac{d\varepsilon}{dx}$$

Thus,

$$\varepsilon(x_a^-) - \varepsilon(x_a) \approx \varepsilon(x_a) - \Delta x\frac{d\varepsilon}{dx} - \varepsilon(x_a) = -\Delta x\frac{d\varepsilon}{dx}$$

$$\varepsilon(x_a) - \varepsilon(x_a^+) \approx \varepsilon(x_a) - \varepsilon(x_a) - \Delta x\frac{d\varepsilon}{dx} = -\Delta x\frac{d\varepsilon}{dx}$$

$$\overline{\Gamma}_a = \lim_{\Delta x \to 0} \frac{\sqrt{-\Delta x\frac{d\varepsilon}{dx}} - j\sqrt{-\Delta x\frac{d\varepsilon}{dx}}}{\sqrt{-\Delta x\frac{d\varepsilon}{dx}} + j\sqrt{-\Delta x\frac{d\varepsilon}{dx}}} \qquad (9.123)$$

$$= \frac{1 - j}{1 + j} = 1\underline{/-\pi/2}$$

so that $\underline{/\Gamma_a} = -\pi/2$. The same result can be shown to hold for the TM case, which makes use of $\Gamma_\parallel$ (see Problem P9.31).

Finally, substituting $\underline{/\Gamma_a} = -\pi/2$ and also

$$\sqrt{\varepsilon_{r1}(x)} \cos \theta = \sqrt{\varepsilon_{r1}(x) - \varepsilon_{r1}(x)\sin^2 \theta} \qquad (9.124)$$

$$= \sqrt{\varepsilon_{r1}(x) - \varepsilon_{r1}(x_a)}$$

into (9.120), we obtain the characteristic equation for guidance to be

$$\boxed{\frac{4\pi}{\lambda_0} \int_{x=0}^{x_a} \sqrt{\varepsilon_{r1}(x) - \varepsilon_{r1}(x_a)}\, dx = \left(m + \frac{1}{2}\right)\pi, \qquad m = 0, 1, 2, \ldots} \qquad (9.125)$$

or in terms of refractive index

$$\frac{4\pi}{\lambda_0} \int_{x=0}^{x_a} \sqrt{n_1^2(x) - n_1^2(x_a)} \, dx = \left(m + \frac{1}{2}\right)\pi, \qquad m = 0, 1, 2, \ldots \qquad (9.126)$$

As in the case of the step-index guide, each value of m corresponds to a mode. For a given value of m and for a given profile $n_1(x)$, (9.126) must in general be solved numerically. There are however certain refractive index profiles that permit analytical solution. An example is in order.

Example 9.9

Modes in parabolic-index guide

Let us consider the refractive index profile of Example 9.8 given by

$$n_1^2(x) = n_0^2 [1 - \alpha^2 x^2] \qquad (9.127)$$

where α is such that $[n_1]_{x=\pm d/2} \geq n_2$, and investigate guided waves in the slab.

Substituting for $n_1^2(x)$ in (9.126), we have

$$\frac{4\pi}{\lambda_0} \int_{x=0}^{x_a} \sqrt{n_0^2(1 - \alpha^2 x^2) - n_0^2(1 - \alpha^2 x_a^2)} \, dx = \left(m + \frac{1}{2}\right)\pi, \qquad m = 0, 1, 2, \ldots$$

$$\frac{4\pi n_0 \alpha}{\lambda_0} \int_{x=0}^{x_a} \sqrt{x_a^2 - x^2} \, dx = \left(m + \frac{1}{2}\right)\pi, \qquad m = 0, 1, 2, \ldots$$

$$\frac{2\pi n_0 \alpha}{\lambda_0} \left[x\sqrt{x_a^2 - x^2} + x_a^2 \sin^{-1}\frac{x}{x_a}\right]_{x=0}^{x_a} = \left(m + \frac{1}{2}\right)\pi, \qquad m = 0, 1, 2, \ldots$$

$$x_a^2 = \frac{(2m + 1)\lambda_0}{2\pi n_0 \alpha}, \qquad m = 0, 1, 2, \ldots \qquad (9.128)$$

The value of x_a increases with the mode number m, as can also be seen in general from (9.126). Recall that $x_a < d/2$. For paraxial modes, the ray trajectories are given by

$$x = x_a \sin \alpha z$$
$$= \sqrt{\frac{(2m + 1)\lambda_0}{2\pi n_0 \alpha}} \sin \alpha z \qquad (9.129)$$

In contrast with the modes of the step-index guide in which the ray paths for all modes extend to the boundaries of the slab with varying values of the pitch, for these modes the ray paths possess amplitudes increasing with m but with a fixed pitch.

The numerical solution of (9.126) involves an iterative procedure. For a given refractive index profile and specified values of d and λ_0, the iterative procedure consists of starting with $m = 0$ and computing x_a that satisfies (9.126).

To do this, x_a is set equal to $d/2$ and the integral on the left side is evaluated numerically. If this results in a value of less than $\pi/2$ for the left side of (9.126), then it means that a solution does not exist for any value of m and the computation is terminated. If the value is greater than $\pi/2$, then an interval-bisection procedure is used iteratively, beginning with the interval from 0 to $d/2$, until a value of x_a that satisfies (9.126) to a desired accuracy is found. The value of m is then increased in steps of unity and the computation repeated for each value of m, beginning with the search interval extending from the solution for x_a found for the previous value of m to $d/2$. The entire computation is terminated when a value of m is reached for which the left side of (9.126) yields a value of less than $(m + \frac{1}{2})\pi$.

Returning now to the result of (9.129), we consider its consequence in intermodal dispersion, the type of dispersion resulting from different travel times of rays corresponding to different modes. Because rays of higher modes travel farther but with greater velocities (lower refractive index), the travel times of the different rays are nearly equalized, thereby almost eliminating intermodal dispersion. To discuss in quantitative terms, we note that the phase constant along the guide axis is given by

Intermodal dispersion

$$\begin{aligned}
\beta_z &= \omega \sqrt{\mu[\varepsilon]}_{x=x_a} \\
&= \omega \sqrt{\mu_0 \varepsilon_0 [n_1]}_{x=x_a} \\
&= \frac{\omega}{c} n_0 \sqrt{1 - \alpha^2 x_a^2} \\
&\approx \frac{\omega n_0}{c} \left(1 - \frac{1}{2}\alpha^2 x_a^2\right) \qquad \text{for} \quad \alpha^2 x_a^2 \ll 1 \\
&= \frac{\omega n_0}{c} \left[1 - \frac{(2m + 1)\lambda_0 \alpha}{4\pi n_0}\right] \\
&= \frac{\omega n_0}{c} - \frac{(2m + 1)\alpha}{2}
\end{aligned}$$

$$\frac{d\beta_z}{d\omega} = \frac{n_0}{c}$$

Thus, the group velocity along the guide axis is given by

$$v_{gz} = \frac{d\omega}{d\beta_z} = \frac{c}{n_0}$$

independent of m.

K9.6. Ray tracing; Paraxial rays; Parabolic index profile; Graded-index guide; Intermodal dispersion.

D9.11. In Fig. 9.34, a medium of continuously varying refractive index is approximated by a series of plane slabs of uniform refractive indices. If $n_0 = 1$ and $\theta_0 = 60°$, find the following: **(a)** θ_1 if $n_1 = 1.5$; **(b)** n_3 if $\theta_3 = 30°$; and **(c)** θ_r.
Ans. **(a)** 35.26°; **(b)** $\sqrt{3}$; **(c)** 30°.

FIGURE 9.34   For Problem D9.11.

D9.12.   In Example 9.9, let $n_0 = 1.5, \alpha = 0.05/\lambda_0$, and $d = 10\lambda_0$. Compute the following: **(a)** x_a for $m = 0$; **(b)** the maximum value of m; and **(c)** the paraxial angle δ_0 for the value of m computed in (b).
Ans.   **(a)** $1.4567\lambda_0$;   **(b)** 5;   **(c)** 13.88°.

9.7   SUMMARY

In this chapter, we studied the principles of waveguides. To introduce the wave-guiding phenomenon, we first learned how to write the expressions for the electric and magnetic fields of a uniform plane wave propagating in an arbitrary direction with respect to the coordinate axes. These expressions are given by

$$\mathbf{E} = \mathbf{E}_0 \cos (\omega t - \boldsymbol{\beta} \cdot \mathbf{r} + \phi_0)$$
$$\mathbf{H} = \mathbf{H}_0 \cos (\omega t - \boldsymbol{\beta} \cdot \mathbf{r} + \phi_0)$$

where $\boldsymbol{\beta}$ and $\mathbf{r}$ are the propagation and position vectors given by

$$\boldsymbol{\beta} = \beta_x \mathbf{a}_x + \beta_y \mathbf{a}_y + \beta_z \mathbf{a}_z$$
$$\mathbf{r} = x \mathbf{a}_x + y \mathbf{a}_y + z \mathbf{a}_z$$

and ϕ_0 is the phase of the wave at the origin at $t = 0$. The magnitude of $\boldsymbol{\beta}$ is equal to $\omega\sqrt{\mu\varepsilon}$, the phase constant along the direction of propagation of the wave. The direction of $\boldsymbol{\beta}$ is the direction of propagation of the wave. We learned that

$$\mathbf{E}_0 \cdot \boldsymbol{\beta} = 0$$
$$\mathbf{H}_0 \cdot \boldsymbol{\beta} = 0$$
$$\mathbf{E}_0 \cdot \mathbf{H}_0 = 0$$

that is, $\mathbf{E}_0, \mathbf{H}_0$, and $\boldsymbol{\beta}$ are mutually perpendicular, and that

$$\frac{|\mathbf{E}_0|}{|\mathbf{H}_0|} = \eta = \sqrt{\frac{\mu}{\varepsilon}}$$

Also, since $\mathbf{E} \times \mathbf{H}$ should be directed along the propagation vector $\boldsymbol{\beta}$, it then follows that

$$\mathbf{H} = \frac{1}{\omega\mu} \boldsymbol{\beta} \times \mathbf{E}$$

The quantities $\beta_x, \beta_y,$ and β_z are the phase constants along the x-, y-, and z-axes, respectively. The apparent wavelengths and the apparent phase velocities along the coordinate axes are given, respectively, by

$$\lambda_i = \frac{2\pi}{\beta_i}, \qquad i = x, y, z$$

$$v_{pi} = \frac{\omega}{\beta_i}, \qquad i = x, y, z$$

By considering the superposition of two uniform plane waves having only y-components of electric fields and propagating at an angle to each other and placing perfect conductors in two constant-x planes such that the boundary condition of zero tangential electric field is satisfied, we introduced the parallel-plate waveguide. We learned that the composite wave is a transverse electric, or TE, wave since the electric field is entirely transverse to the direction of time-average power flow, that is, the guide axis, but the magnetic field is not. In terms of the uniform plane wave propagation, the phenomenon is one of waves bouncing obliquely between the conductors as they progress down the guide. For a fixed spacing a between the conductors of the guide, waves of different frequencies bounce obliquely at different angles such that the spacing a is equal to an integer, say, m number of one-half apparent wavelengths normal to the plates and hence the fields have m number of one-half sinusoidal variations normal to the plates. These are said to correspond to $TE_{m,0}$ modes, where the subscript 0 implies no variations of the fields in the direction parallel to the plates and transverse to the guide axis. When the frequency is such that the spacing a is equal to m one-half wavelengths, the waves bounce normally to the plates without the feeling of being guided along the axis, thereby leading to the cutoff condition. Thus, the cutoff wavelengths corresponding to $TE_{m,0}$ modes are given by

$$\lambda_c = \frac{2a}{m}$$

and the cutoff frequencies are given by

$$f_c = \frac{v_p}{\lambda_c} = \frac{m}{2a\sqrt{\mu\varepsilon}}$$

A given frequency signal can propagate in all modes for which $\lambda < \lambda_c$ or $f > f_c$. For the propagating range of frequencies, the wavelength along the guide axis, that is, the guide wavelength, and the phase velocity along the guide axis are given, respectively, by

$$\lambda_g = \frac{\lambda}{\sqrt{1 - (\lambda/\lambda_c)^2}} = \frac{\lambda}{\sqrt{1 - (f_c/f)^2}}$$

$$v_{pz} = \frac{v_p}{\sqrt{1 - (\lambda/\lambda_c)^2}} = \frac{v_p}{\sqrt{1 - (f_c/f)^2}}$$

As compared to TE modes, the transverse magnetic, or TM, modes have their magnetic fields entirely transverse to the direction of time-average power flow. They are obtained by considering two uniform plane waves having only y-components of magnetic fields and propagating at an angle to each other and placing two perfect conductors in two constant-x planes. The expressions for the propagation parameters λ_c, f_c, λ_g, and v_{pz} for the TM modes are the same as those for the TE modes.

We discussed the solution of problems involving reflection and transmission at a discontinuity in a waveguide by using the transmission-line analogy. This consists of replacing each section of the waveguide by a transmission line whose characteristic impedance is equal to the guide characteristic impedance and then computing the reflection and transmission coefficients as in the transmission-line case. The guide characteristic impedance, η_g, is the ratio of a transverse electric-field component to the corresponding transverse magnetic-field component, which together with the electric-field component gives rise to time-average power flow down the guide. It is given for TE modes by

$$[\eta_g]_{\text{TE}} = \frac{\eta}{\sqrt{1 - (\lambda/\lambda_c)^2}} = \frac{\eta}{\sqrt{1 - (f_c/f)^2}}$$

and for TM modes by

$$[\eta_g]_{\text{TM}} = \eta \sqrt{1 - \left(\frac{\lambda}{\lambda_c}\right)^2} = \eta \sqrt{1 - \left(\frac{f_c}{f}\right)^2}$$

We then discussed the phenomenon of dispersion arising from the frequency dependence of the phase velocity along the guide axis, and we introduced the concept of group velocity. Group velocity is the velocity with which the envelope of a narrow-band modulated signal travels in the dispersive channel, and hence it is the velocity with which the information is transmitted. It is given by

$$v_g = \frac{d\omega}{d\beta_z}$$

where β_z is the phase constant along the guide axis. On the dispersion diagram or the ω–β_z diagram, the group velocity is equal to the slope of the tangent to the dispersion curve at the center frequency of the narrow-band signal. For the parallel-plate waveguide, it is given by

$$v_g = v_p \sqrt{1 - \left(\frac{f_c}{f}\right)^2}$$

To extend the discussion of waveguides to integrated optics, we then considered oblique incidence of a uniform plane wave on the boundary between two perfect dielectric media. We derived the *laws of reflection and refraction,* given, respectively, by

$$\theta_r = \theta_i$$

$$\theta_t = \sin^{-1}\left(\sqrt{\frac{\mu_1 \varepsilon_1}{\mu_2 \varepsilon_2}} \sin \theta_i\right)$$

where θ_i, θ_r, and θ_t are the angles of incidence, reflection, and transmission, respectively, of a uniform plane wave incident from medium 1 (ε_1, μ_1) onto medium 2 (ε_2, μ_2). The law of refraction is also known as *Snell's law*. We then derived the expressions for the reflection and transmission coefficients for the cases of perpendicular and parallel polarizations. An examination of these expressions revealed the following, under the assumption of $\mu_1 = \mu_2$: (1) for incidence from a medium of higher permittivity onto one of lower permittivity, there is a critical angle of incidence given by

$$\theta_c = \sin^{-1} \sqrt{\frac{\varepsilon_2}{\varepsilon_1}}$$

beyond which total internal reflection occurs, and (2) for the case of parallel polarization, there is an angle of incidence, known as the Brewster angle and given by

$$\theta_p = \tan^{-1} \sqrt{\frac{\varepsilon_2}{\varepsilon_1}}$$

for which the reflection coefficient is zero.

Next, we introduced the dielectric slab waveguide, consisting of a dielectric slab of permittivity ε_1, sandwiched between two dielectric media of permittivities $\varepsilon_2 < \varepsilon_1$. We learned that by launching waves at an angle of incidence θ_i greater than the critical angle for total internal reflection, it is possible to achieve guided wave propagation within the slab. For a given frequency, several modes are possible corresponding to values of θ_i that satisfy the self-consistency condition associated with the bouncing waves. We derived the characteristic equation for computing these values of θ_i for the TE case and discussed its solution. The modes are designated TE_m modes and their cutoff frequencies are given by

$$f_c = \frac{mc}{2d\sqrt{\varepsilon_{r1} - \varepsilon_{r2}}}, \qquad m = 0, 1, 2, \ldots$$

where d is the thickness of the slab. The fundamental mode, TE_0, has no cutoff frequency. We also discussed the guided modes by using the approach of deriving the field expressions, based on the behavior that (a) in the transverse (x-) direction, the fields have standing wave character inside the slab and are evanescent outside the slab, and (b) in the longitudinal (z-) direction, the fields have traveling wave character both inside and outside the slab. Dividing the modes into even and odd modes with respect to x from symmetry considerations, we derived the field expressions for the TE modes and (a) obtained the associated characteristic equation for guidance to be the same as that obtained from the wave-bounce approach, (b) discussed the field behavior from near cutoff to far from cutoff, and (c) investigated power flow down the guide.

To extend the treatment of dielectric waveguide to one of graded-index guide, that is, one in which the refractive index varies within the thickness of the slab, we first introduced the topic of ray tracing, making use of the geometrical optics concept. The ray tracing procedure involves the application of Snell's law in conjunction with the geometry associated with the problem. Although in general the solution has to be carried out numerically, for certain functions for the refractive index variation analytical solutions are possible, as illustrated by considering (a) a linear profile of permittivity, and (b) paraxial rays, that is, rays that make small angles to the propagation axis, in a parabolic index profile. For the latter case, we found that the ray oscillates about the axis with a pitch independent of the angle of takeoff from the axis at the starting point. We then considered the graded-index guide having a symmetric refractive index profile and derived the condition for guidance. Applying the guidance condition to investigate modes in a parabolic index guide, we showed that for paraxial rays in the parabolic index guide intermodal dispersion is nearly eliminated.

REVIEW QUESTIONS

Q9.1. What is the propagation vector? Interpret the significance of its magnitude and direction.

Q9.2. Discuss how the phase constants along the coordinate axes are less than the phase constant along the direction of propagation of a uniform plane wave propagating in an arbitrary direction.

Q9.3. Write the expressions for the electric and magnetic fields of a uniform plane wave propagating in an arbitrary direction and list all the conditions to be satisfied by the electric field, magnetic field, and propagation vectors.

Q9.4. What are apparent wavelengths? Why are they longer than the wavelength along the direction of propagation?

Q9.5. What are apparent phase velocities? Why are they greater than the phase velocity along the direction of propagation?

Q9.6. Discuss how the superposition of two uniform plane waves propagating at an angle to each other gives rise to a composite wave consisting of standing waves traveling bodily transverse to the standing waves.

Q9.7. What is a transverse electric wave? Discuss the reasoning behind the nomenclature $TE_{m,0}$ modes.

Q9.8. Compare the phenomenon of guiding of uniform plane waves in a parallel-plate waveguide with that in a parallel-plate transmission line.

Q9.9. Discuss how the cutoff condition arises in a parallel-plate waveguide. Explain the relationship between the cutoff wavelength and the spacing between the plates of a parallel-plate waveguide based on the phenomenon at cutoff.

Q9.10. Is the cutoff wavelength dependent on the dielectric in the waveguide? Is the cutoff frequency dependent on the dielectric in the waveguide?

Q9.11. What is guide wavelength?

Q9.12. Provide a physical explanation for the frequency dependence of the phase velocity along the guide axis.

Q9.13. What is a transverse magnetic wave? Compare and contrast TE and TM waves in a parallel-plate waveguide.

Q9.14. How is guide characteristic impedance defined? Discuss guide characteristic impedance for both TE and TM modes.

Q9.15. Discuss the use of the transmission-line analogy for solving problems involving reflection and transmission at a waveguide discontinuity.

Q9.16. Why are the reflection and transmission coefficients for a given mode at a loss-less waveguide discontinuity dependent on frequency whereas the reflection and transmission coefficients at the junction of two lossless lines are independent of frequency?

Q9.17. Discuss the phenomenon of guiding of waves in the Earth-ionosphere waveguide.

Q9.18. Discuss the phenomenon of dispersion.

Q9.19. Discuss the concept of group velocity with the aid of an example.

Q9.20. What is a dispersion diagram? Explain how the phase and group velocities can be determined from a dispersion diagram.

Q9.21. When is it meaningful to attribute a group velocity to a signal comprised of more than two frequencies? Why?

Q9.22. Discuss the propagation of a narrow-band amplitude-modulated signal in a dispersive channel.

Q9.23. Discuss the condition required to be satisfied by the incident, reflected, and transmitted waves at the interface between two dielectric media.

Q9.24. What is Snell's law?

Q9.25. What is meant by the plane of incidence? Distinguish between the two different linear polarizations pertinent to the derivation of the reflection and transmission coefficients for oblique incidence on a dielectric interface.

Q9.26. Briefly discuss the determination of the Fresnel reflection and transmission coefficients for an obliquely incident wave on a dielectric interface.

Q9.27. What is total internal reflection? Discuss the nature of the reflection coefficient and the manner in which the boundary condition is satisfied for an angle of incidence greater than the critical angle for total internal reflection?

Q9.28. What is the Brewster angle? What is the polarization of the reflected wave for an elliptically polarized wave incident on a dielectric interface at the Brewster angle? Discuss an application of the Brewster angle effect.

Q9.29. Discuss the principle of optical waveguides by considering the dielectric slab waveguide.

Q9.30. Explain the self-consistency condition for waveguiding in a dielectric slab waveguide.

Q9.31. Discuss the dependence of the number of propagating modes in a dielectric slab waveguide on the ratio of the thickness d of the dielectric slab to the wavelength λ_0.

Q9.32. Considering TE modes in a dielectric slab guide, specify the fundamental mode and discuss the associated cutoff condition.

Q9.33. Outline the considerations that come into play in deriving the field expressions for the modes in a dielectric slab guide.

Q9.34. Discuss the mode designation for a dielectric slab guide with reference to the field variations in the guide. Further discuss the behavior of the even and odd modes as the situation changes from near cutoff to far from cutoff.

Q9.35. Discuss the concept of effective boundary for waveguiding along a dielectric slab guide.

Q9.36. Explain radiation modes with reference to a dielectric slab guide.

Q9.37. What is geometrical optics approximation? Under what conditions is it valid?

Q9.38. Outline the formulation of the procedure for ray tracing in a plane-stratified medium of continuously varying refractive index.

Q9.39. What is paraxial ray approximation? Discuss paraxial rays in a medium of parabolic index profile.

Q9.40. Outline the derivation of the self-consistency condition for wave guidance in a graded-index dielectric slab guide.

Q9.41. Compare and contrast the ray trajectories associated with modes in a graded-index guide with those associated with modes in a step-index guide.

Q9.42. What is intermodal dispersion? Why is it minimized for the case of paraxial rays in a parabolic index guide?

PROBLEMS

Section 9.1

P9.1. The electric field of a uniform plane wave propagating in a perfect dielectric medium having $\varepsilon = 9\varepsilon_0$ and $\mu = \mu_0$ is given by

$$\mathbf{E} = 10(2\mathbf{a}_x + \mathbf{a}_y - 2\mathbf{a}_z) \cos [3\pi \times 10^7 t - 0.1\pi(2x - 2y + z)]$$

Find: **(a)** the frequency; **(b)** the direction of propagation; **(c)** the wavelength along the direction of propagation; **(d)** the apparent wavelengths along the x-, y-, and z-axes; and **(e)** the apparent phase velocities along the x-, y-, and z-axes.

P9.2. Given

$$\mathbf{E} = 10(4\mathbf{a}_x + 5\mathbf{a}_y - 3\mathbf{a}_z) \cos [3\pi \times 10^7 t - 0.02\pi(3x + 4z)]$$

(a) Determine if the given $\mathbf{E}$ represents the electric field of a uniform plane wave propagating in free space. **(b)** If the answer is "yes," find the corresponding magnetic-field vector $\mathbf{H}$.

P9.3. Given

$$\mathbf{E} = (\mathbf{a}_x + 2\mathbf{a}_y + \sqrt{3}\mathbf{a}_z) \cos [15\pi \times 10^6 t - 0.05\pi(\sqrt{3}x - z)]$$

$$\mathbf{H} = \frac{1}{60\pi} (\mathbf{a}_x - 2\mathbf{a}_y + \sqrt{3}\mathbf{a}_z) \cos [15\pi \times 10^6 t - 0.05\pi(\sqrt{3}x - z)]$$

(a) Perform all necessary tests and determine if these fields represent a uniform plane wave propagating in a perfect dielectric medium. **(b)** If the answer is "yes," find the permittivity and the permeability of the medium.

P9.4. The apparent phase velocities of a uniform plane wave propagating in a perfect dielectric medium are measured in three directions as follows: 4×10^8 m/s along the x-direction, 2×10^8 m/s along the direction of the unit vector $\frac{1}{5}(4\mathbf{a}_x + 3\mathbf{a}_y)$, and 12×10^8 m/s along the direction of the unit vector $\frac{1}{3}(\mathbf{a}_x - 2\mathbf{a}_y + 2\mathbf{a}_z)$. Find the direction of propagation of the wave and the phase velocity along the direction of the unit vector $\frac{1}{3}(2\mathbf{a}_x + 2\mathbf{a}_y - \mathbf{a}_z)$.

P9.5. The electric field of a uniform plane wave propagating in free space is given in phasor form by

$$\overline{\mathbf{E}} = 10(\mathbf{a}_x + j0.4\mathbf{a}_y + j0.3\mathbf{a}_z)e^{j(0.6y - 0.8z)}$$

(a) Determine the frequency of the wave. **(b)** What is the direction of propagation? **(c)** Obtain the associated magnetic field in phasor form. **(d)** Discuss the polarization of the wave. **(e)** Find the time-average power flow per unit area normal to the direction of propagation.

Section 9.2

P9.6. The dimension a of a parallel-plate waveguide with a dielectric of $\varepsilon = 4\varepsilon_0$ and $\mu = \mu_0$ is 3 cm. Determine the propagating modes for a wave of frequency 6000 MHz. For each propagating mode, find $\lambda_c, f_c, \theta, \lambda_g,$ and v_{pz}.

P9.7. TE modes are excited in an air-dielectric parallel-plate waveguide having the plates in the $x = 0$ and $x = 5$ cm planes by setting up at its input $z = 0$ a field distribution such that

$$\mathbf{E} = E_0 \sin^3 20\pi x \cos 5\pi \times 10^9 t \cos 15\pi \times 10^9 t \, \mathbf{a}_y$$

Find the expression for the electric field of the propagating wave.

P9.8. TM mode is excited in a parallel-plate waveguide filled with a dielectric of $\varepsilon = 4\varepsilon_0$ and $\mu = \mu_0$ and having the plates in the $x = 0$ and $x = 5$ cm planes by setting up at its input $z = 0$ the magnetic field distribution

$$\mathbf{H} = H_0 \cos 40\pi x \sin 8\pi \times 10^9 t \, \mathbf{a}_y$$

Find the expressions for the electric and magnetic fields of the propagating wave.

P9.9. For the parallel-plate waveguide discontinuity of Example 9.3, find the power reflection coefficient for $f = 7500$ MHz for each of the following cases: **(a)** TEM mode; **(b)** $\mathrm{TE}_{1,0}$ mode; and **(c)** $\mathrm{TM}_{1,0}$ mode.

P9.10. The left half (section 1) of a parallel-plate waveguide of dimensions $a = 5$ cm is filled with a dielectric of $\varepsilon = 9\varepsilon_0$ and $\mu = \mu_0$. The right half (section 2) is filled with a dielectric of $\varepsilon = 16\varepsilon_0$ and $\mu = \mu_0$. For waves of frequency 2500 MHz incident on the discontinuity from the left, find the power reflection coefficient for each of the following cases: **(a)** $\mathrm{TE}_{1,0}$ mode; **(b)** $\mathrm{TE}_{2,0}$ mode; **(c)** $\mathrm{TM}_{1,0}$ mode; and **(d)** $\mathrm{TM}_{2,0}$ mode.

Section 9.3

P9.11. The ω–β curve for a dispersive channel can be approximated by

$$\frac{1}{\omega^2} = \frac{1}{\omega_0^2} + \frac{k^2}{\beta^2}$$

in the vicinity of $\omega = 0.5\omega_0$, where k is a constant. Find the following: **(a)** the phase velocity for a signal of $\omega = 0.5\omega_0$; **(b)** the group velocity for a signal composed of two frequencies $0.4\omega_0$ and $0.6\omega_0$; and **(c)** the group velocity for a narrow-band signal having the center frequency $0.5\omega_0$.

P9.12. For a parallel-plate waveguide of dimension $a = 2.5$ cm and having a perfect dielectric of $\varepsilon = 9\varepsilon_0$ and $\mu = \mu_0$, find the group velocity for: **(a)** a signal composed of the two frequencies $f_1 = 2500$ MHz and $f_2 = 3000$ MHz; and **(b)** a narrow-band signal having the center frequency 2500 MHz.

P9.13. Find the group velocity of propagation of a TE wave in a parallel-plate waveguide filled with a perfect dielectric of $\varepsilon = 2.25\varepsilon_0$ and $\mu = \mu_0$ and having the plates in the $x = 0$ and $x = 10$ cm planes for each of the following cases of electric field distribution at its input $z = 0$:

(a) $E_0 \sin 10\pi x \cos 10^9 \pi t \cos 4 \times 10^9 \pi t \, \mathbf{a}_y$

(b) $E_0 \sin 10\pi x (1 + 0.5 \cos 10^4 \pi t) \cos 4 \times 10^9 \pi t \, \mathbf{a}_y$

P9.14. By considering the parallel-plate waveguide, show that a point on the obliquely bouncing wavefront, traveling with the phase velocity along the oblique direction, progresses parallel to the guide axis with the group velocity.

Section 9.4

P9.15. For the case of two nonmagnetic ($\mu = \mu_0$), perfect dielectric media, show that the Fresnel coefficients for perpendicular polarization (9.70a) and (9.70b) reduce to

$$\Gamma_\perp = \frac{\sin(\theta_2 - \theta_1)}{\sin(\theta_2 + \theta_1)} \quad \text{and} \quad \tau_\perp = \frac{2\cos\theta_1 \sin\theta_2}{\sin(\theta_2 + \theta_1)}$$

respectively, and the Fresnel coefficients for parallel polarization (9.75a) and (9.75b) reduce to

$$\Gamma_\parallel = \frac{\tan(\theta_2 - \theta_1)}{\tan(\theta_2 + \theta_1)} \quad \text{and} \quad \tau_\parallel = \frac{2\cos\theta_1 \sin\theta_2}{\sin(\theta_1 + \theta_2)\cos(\theta_1 - \theta_2)}$$

respectively.

P9.16. In Example 9.4, assume that

$$E_i = E_0(\mathbf{a}_x - \mathbf{a}_z)\cos[6\pi \times 10^8 t - \sqrt{2}\pi(x + z)]$$

and the angle of incidence is $45°$. Obtain the expressions for the electric fields of the reflected and transmitted waves.

P9.17. Repeat Problem P9.16 for

$$\mathbf{E}_i = E_0\mathbf{a}_y \cos[6\pi \times 10^8 t - \sqrt{2}\pi(x + z)]$$

P9.18. In Example 9.4, assume that the permittivity ε_2 of medium 2 is unknown and that

$$\mathbf{E}_i = E_0\left(\frac{\sqrt{3}}{2}\mathbf{a}_x - \frac{1}{2}\mathbf{a}_z\right)\cos[6\pi \times 10^9 t - 10\pi(x + \sqrt{3}z)]$$
$$+ E_0\mathbf{a}_y \sin[6\pi \times 10^9 t - 10\pi(x + \sqrt{3}z)]$$

(a) Find the value of ε_2 for which the reflected wave is linearly polarized.

(b) For the value of ε_2 found in (a), find the expressions for the reflected and transmitted wave electric fields.

P9.19. For oblique incidence of a uniform plane wave on a dielectric interface, show that the Fresnel reflection and transmission coefficients are consistent with the condition that for power flow normal to the interface, the sum of the reflected power and the transmitted power be equal to the incident power, for each of the two cases: **(a)** perpendicular polarization and **(b)** parallel polarization.

Section 9.5

P9.20. A thin-film waveguide employed in integrated optics consists of a substrate on which a thin film of refractive index (c/v_p) greater than that of the substrate is deposited. The medium above the film is air. For relative permittivities of the substrate and the film equal to 2.25 and 2.4, respectively, find the minimum bouncing angle of total internally reflected waves in the film. Assume $\mu = \mu_0$ for both substrate and film.

P9.21. For a symmetric dielectric slab waveguide, $\varepsilon_1 = 2.25\varepsilon_0$ and $\varepsilon_2 = \varepsilon_0$. **(a)** Find the number of propagating TE modes for $d/\lambda_0 = 10$. **(b)** Find the maximum value of d/λ_0 for which the waveguide supports only one TE mode.

P9.22. Design a symmetric dielectric slab waveguide, with $\varepsilon_{r1} = 2.25$ and $\varepsilon_{r2} = 2.13$, by finding the value of d/λ_0 such that the TE_1 mode operates at 20% above its cut-off frequency.

P9.23. Consider the derivation of the characteristic equation for guiding of waves in the symmetric dielectric slab waveguide for the case of parallel polarization, which corresponds to TM modes. Noting that in Fig. 9.18, $H_r/H_i = E_r/E_i = -\Gamma_\parallel$, where $\Gamma_\parallel$ is given by (9.75a), show that the characteristic equation is given by

$$\tan[f(\theta_i)] = \begin{cases} g(\theta_i), & m = 0, 2, 4, \ldots \\ -\dfrac{1}{g(\theta_i)}, & m = 1, 3, 5, \ldots \end{cases}$$

where

$$f(\theta_i) = \frac{\pi d \sqrt{\varepsilon_{r1}}}{\lambda_0} \cos\theta_i$$

$$g(\theta_i) = \frac{\sqrt{\sin^2\theta_i - (\varepsilon_2/\varepsilon_1)}}{(\varepsilon_2/\varepsilon_1)\cos\theta_i}$$

P9.24. The ratio of the power associated with the region of the slab ($|x| < d/2$) to the total power ($-\infty < x < \infty$) is known as the *power confinement factor*. Show that for a given frequency, the power confinement factor is the highest for the dominant mode. Find the power confinement factors for the three modes in Example 9.6.

P9.25. Beginning with an expression for $\overline{H}_y$ analogous to that of $\overline{E}_y$ for even TE modes given by (9.91), derive the field expressions for even TM modes for the symmetric dielectric slab waveguide and obtain the characteristic equation for guidance.

P9.26. Repeat Problem P9.25 for odd TM modes, beginning with an expression for $\overline{H}_y$ analogous to that of $\overline{E}_y$ for odd TE modes given by (9.102).

P9.27. For the symmetric dielectric slab guide, show that the ω versus β_z, or the dispersion relation for the TE_m mode, is given by

$$\tan^2\left(\frac{d}{2}\sqrt{\omega^2\mu_0\varepsilon_1 - \beta_z^2} - \frac{m\pi}{2}\right) = \frac{\beta_z^2 - \omega^2\mu_0\varepsilon_2}{\omega^2\mu_0\varepsilon_1 - \beta_z^2}$$

Then find the value of d/λ_0 for which the phase velocity along the guide axis v_{pz} is equal to $\sqrt{v_{p1}v_{p2}}$, where $v_{p1} = 1/\sqrt{\mu_0\varepsilon_1}$ and $v_{p2} = 1/\sqrt{\mu_0\varepsilon_2}$, for the fundamental mode, if $\varepsilon_{r1} = 4.0$ and $\varepsilon_{r2} = 2.25$. Note that since $\tan^2$ is always positive, $\omega^2\mu_0\varepsilon_2 < \beta_z^2 < \omega^2\mu_0\varepsilon_1$, or $v_{p2} > v_{pz} > v_{p1}$, as it should be.

Section 9.6

P9.28. The laws of geometrical optics can be derived from Fermat's principle, which states that the optical path length $\int_A^B n\,ds$ of a ray of light from point A to point B is an extremum, so that the variation in the optical path length, $\delta\int_A^B n\,ds$, is equal to zero. Derive from this property the laws of reflection and refraction for oblique incidence on a plane boundary between two different perfect dielectric media.

P9.29. Consider ray tracing in spherical geometry, as shown in Fig. 9.35, in which n is a function of r, the radial distance from the center of a spherical interface

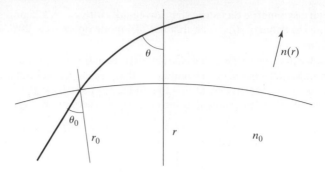

FIGURE 9.35   For Problem P9.29.

(radius r_0). A ray is incident on the interface at an angle θ_0 from a medium of uniform refractive index n_0. Show that the ray path for $r > r_0$ is governed by a modified form of Snell's law given by

$$nr \sin \theta = \text{constant} = n_0 r_0 \sin \theta_0$$

and hence by the solution of the differential equation

$$\frac{d\theta}{dr} = \frac{n_0 r_0 \sin \theta}{r\sqrt{n^2 r^2 - n_0^2 r_0^2 \sin^2 \theta_0}}$$

P9.30. In Example 9.8, assume

$$n(x) = n_0(1 - \alpha|x|)$$

Obtain the solution for $z(x)$ for paraxial rays and find the approximate peak amplitude and the approximate pitch in terms of δ_0.

P9.31. Show that the guidance condition for the graded-index guide for the TM case is the same as that given by (9.125) or (9.126) for the TE case by showing that $\overline{\Gamma}_a$ for the TM case is $1 \underline{/-\pi/2}$. Note that $\overline{\Gamma}$ to be used for the TM case is $-\Gamma_\parallel$ (see Problem P9.23), where $\Gamma_\parallel$ is given by (9.75a).

P9.32. In Example 9.9, assume

$$n_1^2(x) = n_0^2(1 - \alpha|x|)$$

Obtain the expression for x_a for the mth mode.

REVIEW PROBLEMS

R9.1. The apparent phase velocities of a uniform plane wave of frequency 10^9 Hz propagating in a nonmagnetic ($\mu = \mu_0$), perfect dielectric medium are given for three directions as follows: 2.25×10^8 m/s along the direction of the unit vector $\frac{1}{3}(2\mathbf{a}_x + \mathbf{a}_y + 2\mathbf{a}_z)$; 4.5×10^8 m/s along the direction of the unit vector $\frac{1}{3}(2\mathbf{a}_x - 2\mathbf{a}_y + \mathbf{a}_z)$; and 18×10^8 m/s along the direction of the unit vector $\frac{1}{3}(2\mathbf{a}_x + \mathbf{a}_y - 2\mathbf{a}_z)$. Find **(a)** the permittivity of the medium and **(b)** the apparent wavelength along the direction of the unit vector $\frac{1}{3}(2\mathbf{a}_x + 2\mathbf{a}_y - \mathbf{a}_z)$.

R9.2. The left half ($z < 0$) of a parallel-plate waveguide having the plates in the $x = 0$ and $x = 2.5$ cm planes is filled with a dielectric of $\varepsilon = 4\varepsilon_0$ and $\mu = \mu_0$, whereas the right half ($z > 0$) is filled with a dielectric of $\varepsilon = 2.25\varepsilon_0$ and $\mu = \mu_0$. A TM wave having the magnetic field at $z = 0$ given by

$$\mathbf{H}_i = H_0 \cos^3 40\pi x \cos 3\pi \times 10^{10} t \, \mathbf{a}_y$$

is incident on the discontinuity $z = 0$ from the left. Find the expressions for the electric and magnetic fields as functions of z and t for the incident and reflected waves for $z < 0$ and the transmitted wave for $z > 0$.

R9.3. The ω–β curve for a certain dispersive channel is such that at a frequency $\omega = \omega_0$, the group and phase velocities are equal, and the group velocity is proportional to ω in a certain frequency interval around $\omega = \omega_0$. Find the ratio v_g/v_p in that frequency interval.

R9.4. Region 1 ($x < 0$) is free space, whereas region 2 ($x > 0$) is a perfect dielectric of $\varepsilon = 16\varepsilon_0$ and $\mu = \mu_0$. For an elliptically polarized uniform plane wave incident from free space onto the boundary $x = 0$, it is found that the reflected wave is linearly polarized with electric-field amplitude E_1 and the transmitted wave is circularly polarized with electric-field magnitude E_2. **(a)** Find E_2/E_1. **(b)** Find the axial ratio of the polarization ellipse for the incident wave.

R9.5. **Design Problem.** Design a symmetric dielectric slab waveguide with $\varepsilon_{r1} = 2.25$ and $\varepsilon_{r2} = 1.00$ by finding the value of d/λ_0 such that θ_i for the TM_0 mode is 88°. Then find the value of θ_i for the TE_0 mode and the number of remaining propagating TE and TM modes.

R9.6. For a graded-index guide with the refractive index profile given by

$$n_1^2(x) = n_0^2(1 - \alpha|x|)$$

obtain the solution for $z(x)$ for paraxial rays and find the approximate amplitude and approximate pitch in terms of δ_0. Then find the approximate value of δ_0 in terms of mode number.

C H A P T E R 1 0

Several Topics for Electronics and Photonics

In the previous chapter, we introduced the principles of guided waves and learned that the mechanism of waveguiding is one in which the waves bounce obliquely between parallel planes as they progress along the structure. We studied transverse electric (TE) and transverse magnetic (TM) waves, supported by plane conductors as well those supported by a plane dielectric slab. Thus, we restricted our study of guided waves to one-dimensional structures. In this chapter, we extend the treatment to two dimensions.

As in the previous chapter, we consider metallic waveguides, that is, those in which the TE and TM waves propagate in the medium between and parallel to the metallic boundaries of the guide, as well as optical fibers in which the supporting structure for the guided waves is a cylindrical dielectric rod surrounded by a dielectric cladding and without the presence of metallic boundaries.

We shall begin the chapter by introducing rectangular metallic waveguides, first based on a geometrical extension of the parallel-plate waveguide arrangement and then by solving Maxwell's equations in two dimensions. We make use of the latter approach to study cylindrical metallic waveguides and later optical fibers. We shall also consider losses in metallic waveguides and resonators and extend our discussion of dispersion in Chapter 9 to study the phenomenon of pulse broadening, which is of particular importance in photonic systems. Finally, we shall consider the topics of interference and diffraction, and wave propagation in an anisotropic medium.

10.1 RECTANGULAR METALLIC WAVEGUIDE AND CAVITY RESONATOR

TE modes in rectangular waveguide

To introduce the rectangular metallic waveguide, we begin with TE modes in a parallel-plate waveguide. We recall that the parallel-plate waveguide is made up of two perfectly conducting sheets in the planes $x = 0$ and $x = a$ and that the electric field of the $TE_{m,0}$ mode has only a y-component with m number of one-half sinusoidal variations in the x-direction and no variations in the y-direction. If we now introduce two perfectly conducting sheets in two constant-y planes, say, $y = 0$ and $y = b$, the field distribution will remain unaltered since the electric field is entirely normal to the plates, and hence the boundary condition of zero tangential electric field is satisfied for both sheets. We then have the rectangular waveguide, a metallic pipe with rectangular cross section in the xy-plane, as shown in Fig. 10.1.

Since the $TE_{m,0}$ mode field expressions derived for the parallel-plate waveguide satisfy the boundary conditions for the rectangular waveguide, those expressions as well as the entire discussion of the parallel-plate waveguide case hold also for $TE_{m,0}$ mode propagation in the rectangular waveguide case. We learned that the $TE_{m,0}$ modes can be interpreted as due to uniform plane waves having electric field in the y-direction and bouncing obliquely between the conducting walls $x = 0$ and $x = a$, and with the associated cutoff condition characterized by bouncing of the waves back and forth normally to these walls, as shown in Fig. 10.2(a). For the cutoff condition, the dimension a is equal to m number of one-half wavelengths such that

$$[\lambda_c]_{TE_{m,0}} = \frac{2a}{m} \tag{10.1}$$

In a similar manner, we can have uniform plane waves having electric field in the x-direction and bouncing obliquely between the walls $y = 0$ and $y = b$, and with the associated cutoff condition characterized by bouncing of the waves back and forth normally to these walls, as shown in Fig. 10.2(b), thereby resulting in $TE_{0,n}$ modes having no variations in the x-direction and n number of one-half sinusoidal variations in the y-direction. For the cutoff condition, the dimension b is equal to n number of one-half wavelengths such that

$$[\lambda_c]_{TE_{0,n}} = \frac{2b}{n} \tag{10.2}$$

FIGURE 10.1 Rectangular waveguide.

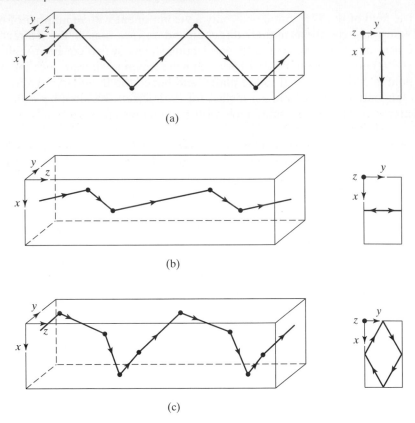

FIGURE 10.2   Propagation and cutoff of (a) $\mathrm{TE}_{m,0}$, (b) $\mathrm{TE}_{0,n}$, and (c) $\mathrm{TE}_{m,n}$ modes in a rectangular waveguide.

We can even have $\mathrm{TE}_{m,n}$ modes having m number of one-half sinusoidal variations in the x-direction and n number of one-half sinusoidal variations in the y-direction due to uniform plane waves having both x- and y-components of the electric field and bouncing obliquely between all four walls of the guide and with the associated cutoff condition characterized by bouncing of the waves back and forth obliquely between the four walls as shown, for example, in Fig. 10.2(c). For the cutoff condition, the dimension a must be equal to m number of one-half apparent wavelengths in the x-direction, and the dimension b must be equal to n number of one-half apparent wavelengths in the y-direction such that

$$\frac{1}{[\lambda_c]_{\mathrm{TE}_{m,n}}^2} = \frac{1}{(2a/m)^2} + \frac{1}{(2b/n)^2}$$

or

$$[\lambda_c]_{\mathrm{TE}_{m,n}} = \frac{1}{\sqrt{(m/2a)^2 + (n/2b)^2}} \qquad (10.3)$$

The discussion thus far of $\text{TE}_{m,n}$ modes in a rectangular waveguide has been based on qualitative reasoning. We shall now derive the field expressions for the TE modes. To do this, we shall first show, by making use of the expansions for the Maxwell's curl equations in Cartesian coordinates, that all transverse (x and y) field components are derivable from the longitudinal field component H_z. It is convenient to use the phasor forms of the field components and differential equations. Since all components of the fields are then dependent on t and z in the manner $e^{j[\omega t - \beta_z z]}$, we can replace $\partial/\partial t$ by $j\omega$ and $\partial/\partial z$ by $-j\beta_z$. Furthermore, $E_z = 0$ in view of TE modes and J_x, J_y, and J_z are all zero since the medium inside the waveguide is a perfect dielectric. Thus, the phasor forms of (4.12a)–(4.12c) and of the component equations of (4.22) pertinent to the discussion here are

Derivation of field expressions for TE modes

$$j\beta_z \overline{E}_y = -j\omega\mu\overline{H}_x \quad (10.4a)$$

$$-j\beta_z \overline{E}_x = -j\omega\mu\overline{H}_y \quad (10.4b)$$

$$\frac{\partial \overline{E}_y}{\partial x} - \frac{\partial \overline{E}_x}{\partial y} = -j\omega\mu\overline{H}_z \quad (10.4c)$$

$$\frac{\partial \overline{H}_z}{\partial y} + j\beta_z\overline{H}_y = j\omega\varepsilon\overline{E}_x \quad (10.4d)$$

$$-j\beta_z\overline{H}_x - \frac{\partial \overline{H}_z}{\partial x} = j\omega\varepsilon\overline{E}_y \quad (10.4e)$$

$$\frac{\partial \overline{H}_y}{\partial x} - \frac{\partial \overline{H}_x}{\partial y} = 0 \quad (10.4f)$$

Solving (10.4a), (10.4b), (10.4d), and (10.4e), for $\overline{E}_x, \overline{E}_y, \overline{H}_x$, and $\overline{H}_y$ in terms of $\overline{H}_z$, we obtain

$$\overline{E}_x = \frac{j\omega\mu}{\beta_z^2 - \beta^2}\frac{\partial \overline{H}_z}{\partial y} \quad (10.5a)$$

$$\overline{E}_y = -\frac{j\omega\mu}{\beta_z^2 - \beta^2}\frac{\partial \overline{H}_z}{\partial x} \quad (10.5b)$$

$$\overline{H}_x = j\frac{\beta_z}{\beta_z^2 - \beta^2}\frac{\partial \overline{H}_z}{\partial x} \quad (10.5c)$$

$$\overline{H}_y = j\frac{\beta_z}{\beta_z^2 - \beta^2}\frac{\partial \overline{H}_z}{\partial y} \quad (10.5d)$$

Furthermore, by substituting (10.5a) and (10.5b) into (10.4c) and rearranging, we obtain a differential equation for $\overline{H}_z$ as given by

$$\frac{\partial^2 \overline{H}_z}{\partial x^2} + \frac{\partial^2 \overline{H}_z}{\partial y^2} - (\beta_z^2 - \beta^2)\overline{H}_z = 0 \quad (10.6)$$

Recall that $\beta = \omega\sqrt{\mu\varepsilon}$, so that $\beta^2 = \omega^2\mu\varepsilon$.

To solve (10.6) for $\overline{H}_z$, we make use of the *separation of variables* technique. This consists of assuming that the required function of the two variables x and y is the product of two functions, one of which is a function of x only and the second is a function of y only. Thus, denoting these functions to be X and Y,

Separation of variables technique

we have

$$\boxed{\overline{H}_z(x, y, z) = \overline{X}(x)\overline{Y}(y)e^{-j\beta_z z}} \tag{10.7}$$

Substituting (10.7) into (10.6), we then obtain

$$\overline{X}''(x)\overline{Y}(y) + \overline{X}(x)\overline{Y}''(y) - (\beta_z^2 - \beta^2)\overline{X}(x)\overline{Y}(y) = 0$$

or

$$\frac{\overline{X}''}{\overline{X}} + \frac{\overline{Y}''}{\overline{Y}} = \beta_z^2 - \beta^2 \tag{10.8}$$

where the primes denote differentiation with respect to the respective variables. Equation (10.8) says that a function of x only plus a function of y only is equal to a constant. For this to be satisfied, both functions must be equal to constants. Hence, we write

$$\frac{\overline{X}''}{\overline{X}} = -\beta_x^2, \text{ a constant} \tag{10.9a}$$

and

$$\frac{\overline{Y}''}{\overline{Y}} = -\beta_y^2, \text{ a constant} \tag{10.9b}$$

or

$$\frac{d^2\overline{X}}{dx^2} = -\beta_x^2\overline{X} \tag{10.10a}$$

and

$$\frac{d^2\overline{Y}}{dy^2} = -\beta_y^2\overline{Y} \tag{10.10b}$$

We have thus obtained two ordinary differential equations involving separately the two variables x and y; hence, the technique is known as the *separation of variables* technique.

The solutions to (10.10a) and (10.10b) are given by

$$\overline{X}(x) = \overline{A}_1 e^{j\beta_x x} + \overline{A}_2 e^{-j\beta_x x}$$

$$\overline{Y}(y) = \overline{B}_1 e^{j\beta_y y} + \overline{B}_2 e^{-j\beta_y y}$$

so that

$$\overline{H}_z = (\overline{A}_1 e^{j\beta_x x} + \overline{A}_2 e^{-j\beta_x x})(\overline{B}_1 e^{j\beta_y y} + \overline{B}_2 e^{-j\beta_y y})e^{-j\beta_z z} \tag{10.11}$$

where $\overline{A}_1, \overline{A}_2, \overline{B}_1$, and $\overline{B}_2$ are constants. We also note from substitution of (10.9a) and (10.9b) into (10.8) that

$$-\beta_x^2 - \beta_y^2 = \beta_z^2 - \beta^2$$

or

$$\beta_z^2 = \beta^2 - \beta_x^2 - \beta_y^2 \tag{10.12}$$

Now, to determine the constants in (10.11), we make use of the boundary conditions that require that the tangential components of the electric-field intensity on all four walls of the guide be zero. Thus, we have

$$\overline{E}_x = 0 \quad \text{for} \quad y = 0, 0 < x < a$$

$$\overline{E}_x = 0 \quad \text{for} \quad y = b, 0 < x < a$$

$$\overline{E}_y = 0 \quad \text{for} \quad x = 0, 0 < y < b$$

$$\overline{E}_y = 0 \quad \text{for} \quad x = a, 0 < y < b$$

To apply these boundary conditions to (10.11), we have to translate them into boundary conditions involving $\overline{H}_z$. From (10.5a) and (10.5b), these are

$$\frac{\partial \overline{H}_z}{\partial y} = 0 \quad \text{for} \quad y = 0, 0 < x < a \tag{10.13a}$$

$$\frac{\partial \overline{H}_z}{\partial y} = 0 \quad \text{for} \quad y = b, 0 < x < a \tag{10.13b}$$

$$\frac{\partial \overline{H}_z}{\partial x} = 0 \quad \text{for} \quad x = 0, 0 < y < b \tag{10.13c}$$

$$\frac{\partial \overline{H}_z}{\partial x} = 0 \quad \text{for} \quad x = a, 0 < y < b \tag{10.13d}$$

Using (10.13c) and (10.13a) in conjunction with (10.11), we get

$$\overline{A}_1 - \overline{A}_2 = 0 \quad \text{or} \quad \overline{A}_2 = \overline{A}_1$$

$$\overline{B}_1 - \overline{B}_2 = 0 \quad \text{or} \quad \overline{B}_2 = \overline{B}_1$$

which then simplifies (10.11) to

$$\overline{H}_z = \overline{A} \cos \beta_x x \cos \beta_y y \, e^{-j\beta_z z} \tag{10.14}$$

where $\overline{A}$ is a constant. Using the remaining two boundary conditions (10.13d) and (10.13b), we then obtain

$$\sin \beta_x a = 0 \quad \text{or} \quad \beta_x = \frac{m\pi}{a}, \quad m = 0, 1, 2, \ldots \tag{10.15a}$$

$$\sin \beta_y b = 0 \quad \text{or} \quad \beta_y = \frac{n\pi}{b}, \quad n = 0, 1, 2, \ldots \tag{10.15b}$$

Thus, the solution for $\overline{H}_z$ for the $\text{TE}_{m,n}$ mode is given by

$$\boxed{\overline{H}_z = \overline{A} \cos \frac{m\pi x}{a} \cos \frac{n\pi y}{b} e^{-j\beta_z z}} \tag{10.16}$$

We also note by substituting (10.15a) and (10.15b) in (10.12) that

$$\beta_z^2 = \beta^2 - \left(\frac{m\pi}{a}\right)^2 - \left(\frac{n\pi}{b}\right)^2 \tag{10.17}$$

For propagation to occur, the exponent β_z in (10.16) must be real. Hence, the cutoff condition is given by

$$\omega^2 \mu\varepsilon - \left(\frac{m\pi}{a}\right)^2 - \left(\frac{n\pi}{b}\right)^2 = 0 \tag{10.18}$$

or the cutoff frequency is given by

$$f_c = \frac{1}{\sqrt{\mu\varepsilon}} \sqrt{\left(\frac{m}{2a}\right)^2 + \left(\frac{n}{2b}\right)^2} \tag{10.19}$$

and the cutoff wavelength is

$$\lambda_c = \frac{1}{\sqrt{\mu\varepsilon}f_c}$$
$$= \frac{1}{\sqrt{(m/2a)^2 + (n/2b)^2}} \tag{10.20}$$

which is the same as given by (10.3). Now, from (10.17) and (10.20), we have

$$\beta_z^2 - \beta^2 = -\left(\frac{m\pi}{a}\right)^2 - \left(\frac{n\pi}{b}\right)^2$$
$$= -(2\pi)^2 \left[\left(\frac{m}{2a}\right)^2 + \left(\frac{n}{2b}\right)^2\right] \tag{10.21}$$
$$= -\left(\frac{2\pi}{\lambda_c}\right)^2$$

Substituting (10.16) and (10.21) into (10.5a)–(10.5d), we obtain the expressions for the transverse field components:

$$\overline{E}_x = j\frac{\omega\mu\lambda_c^2}{4\pi^2}\frac{n\pi}{b}\overline{A}\cos\frac{m\pi x}{a}\sin\frac{n\pi y}{b}e^{-j\beta_z z} \tag{10.22a}$$

$$\overline{E}_y = -j\frac{\omega\mu\lambda_c^2}{4\pi^2}\frac{m\pi}{a}\overline{A}\sin\frac{m\pi x}{a}\cos\frac{n\pi y}{b}e^{-j\beta_z z} \tag{10.22b}$$

$$\overline{H}_x = j\frac{\lambda_c^2}{2\pi\lambda_g}\frac{m\pi}{a}\overline{A}\sin\frac{m\pi x}{a}\cos\frac{n\pi y}{b}e^{-j\beta_z z} \tag{10.22c}$$

$$\overline{H}_y = j\frac{\lambda_c^2}{2\pi\lambda_g}\frac{n\pi}{b}\overline{A}\cos\frac{m\pi x}{a}\sin\frac{n\pi y}{b}e^{-j\beta_z z} \tag{10.22d}$$

Note that the sine terms in these field expressions satisfy the boundary conditions of zero tangential electric field and zero normal magnetic field at the walls of the waveguide. It can also be seen that if both m and n are equal to zero, then all transverse field components go to zero. Therefore, *for TE modes, either m or n can be zero, but both m and n cannot be zero.*

The entire procedure for the derivation of the field expressions can be repeated for TM waves by starting with the longitudinal field component $\overline{E}_z$. We shall however not pursue the derivation here but instead present the final expressions. This is done in Table 10.1 together with the TE mode field expressions. The upper signs of the $\mp$ and $\pm$ signs in these expressions refer to waves propagating in the $+z$-direction, whereas the lower signs refer to waves propagating in the $-z$-direction. Note from the expression for $\overline{E}_z$ in Table 10.1 that the x- and y-variations of $\overline{E}_z$ are sinusoidal so that $\overline{E}_z$ goes to zero on all four walls of the waveguide. This arises because $\overline{E}_z$, being longitudinal, is tangential to all four walls, and the boundary conditions require that the tangential components of **E** be zero on the walls. It can also be seen that if either m or n is equal to zero, then $\overline{E}_z = 0$. Therefore, *for TM modes both m and n must be nonzero.* Also listed in

TM modes in rectangular waveguide

TABLE 10.1 Field Expressions and Associated Parameters for TE and TM Modes in a Rectangular Waveguide

Transverse electric (TE) waves	Transverse magnetic (TM) waves
Field Expressions: $(m, n = 0, 1, 2, ..., \text{but not both zero})$	Field Expressions: $(m, n = 1, 2, 3, ...)$
$\overline{E}_z = 0$	$\overline{H}_z = 0$
$\overline{H}_z = \overline{A} \cos \dfrac{m\pi x}{a} \cos \dfrac{n\pi y}{b} e^{\mp j\beta_z z}$	$\overline{E}_z = \overline{A} \sin \dfrac{m\pi x}{a} \sin \dfrac{n\pi y}{b} e^{\mp j\beta_z z}$
$\overline{E}_x = j \dfrac{\lambda_c^2}{4\pi^2} \omega\mu \dfrac{n\pi}{b} \overline{A} \cos \dfrac{m\pi x}{a} \sin \dfrac{n\pi y}{b} e^{\mp j\beta_z z}$	$\overline{E}_x = \mp j \dfrac{\lambda_c^2}{2\pi\lambda_g} \dfrac{m\pi}{a} \overline{A} \cos \dfrac{m\pi x}{a} \sin \dfrac{n\pi y}{b} e^{\mp j\beta_z z}$
$\overline{E}_y = -j \dfrac{\lambda_c^2}{4\pi^2} \omega\mu \dfrac{m\pi}{a} \overline{A} \sin \dfrac{m\pi x}{a} \cos \dfrac{n\pi y}{b} e^{\mp j\beta_z z}$	$\overline{E}_y = \mp j \dfrac{\lambda_c^2}{2\pi\lambda_g} \dfrac{n\pi}{b} \overline{A} \sin \dfrac{m\pi x}{a} \cos \dfrac{n\pi y}{b} e^{\mp j\beta_z z}$
$\overline{H}_x = \mp \dfrac{\overline{E}_y}{\eta_g}$	$\overline{H}_x = \mp \dfrac{\overline{E}_y}{\eta_g}$
$\overline{H}_y = \pm \dfrac{\overline{E}_x}{\eta_g}$	$\overline{H}_y = \pm \dfrac{\overline{E}_x}{\eta_g}$
$f_c = \dfrac{1}{2\sqrt{\mu\varepsilon}} \sqrt{\left(\dfrac{m}{a}\right)^2 + \left(\dfrac{n}{b}\right)^2}$	$f_c = \dfrac{1}{2\sqrt{\mu\varepsilon}} \sqrt{\left(\dfrac{m}{a}\right)^2 + \left(\dfrac{n}{b}\right)^2}$
$\lambda_c = \dfrac{2}{\sqrt{(m/a)^2 + (n/b)^2}}$	$\lambda_c = \dfrac{2}{\sqrt{(m/a)^2 + (n/b)^2}}$
$\lambda_g = \dfrac{\lambda}{\sqrt{1 - (\lambda/\lambda_c)^2}} = \dfrac{\lambda}{\sqrt{1 - (f_c/f)^2}}$	$\lambda_g = \dfrac{\lambda}{\sqrt{1 - (\lambda/\lambda_c)^2}} = \dfrac{\lambda}{\sqrt{1 - (f_c/f)^2}}$
$v_{pz} = \dfrac{1}{\sqrt{\mu\varepsilon}\sqrt{1 - (f_c/f)^2}} = \dfrac{1}{\sqrt{\mu\varepsilon}\sqrt{1 - (\lambda/\lambda_c)^2}}$	$v_{pz} = \dfrac{1}{\sqrt{\mu\varepsilon}\sqrt{1 - (f_c/f)^2}} = \dfrac{1}{\sqrt{\mu\varepsilon}\sqrt{1 - (\lambda/\lambda_c)^2}}$
$\eta_g = \dfrac{\sqrt{\mu/\varepsilon}}{\sqrt{1 - (f_c/f)^2}} = \dfrac{\sqrt{\mu/\varepsilon}}{\sqrt{1 - (\lambda/\lambda_c)^2}}$	$\eta_g = \sqrt{\dfrac{\mu}{\varepsilon}} \sqrt{1 - \left(\dfrac{f_c}{f}\right)^2} = \sqrt{\dfrac{\mu}{\varepsilon}} \sqrt{1 - \left(\dfrac{\lambda}{\lambda_c}\right)^2}$

Table 10.1 are the expressions for the cutoff frequency f_c, the cutoff wavelength λ_c, the guide wavelength λ_g, the phase velocity v_{pz} along the guide axis, and the guide characteristic impedance η_g, all of which have the same interpretations as the corresponding quantities for the parallel-plate waveguide case.

Dominant
mode

The foregoing discussion of the modes of propagation in a rectangular waveguide points out that a signal of given frequency can propagate in several modes, namely, all modes for which the cutoff frequencies are less than the signal frequency or the cutoff wavelengths are greater than the signal wavelength. Waveguides are, however, designed so that only one mode, the mode with the lowest cutoff frequency (or the largest cutoff wavelength), propagates. This is known as the *dominant mode*. From Table 10.1, we can see that the dominant mode is the $TE_{1,0}$ mode or the $TE_{0,1}$ mode, depending on whether the dimension a or the dimension b is the larger of the two. By convention, the larger dimension is designated to be a, and hence the $TE_{1,0}$ mode is the dominant mode. We shall now consider an example.

Example 10.1

Finding
propagating
modes

It is desired to determine the lowest four cutoff frequencies referred to the cutoff frequency of the dominant mode for three cases of rectangular waveguide dimensions: $b/a = 1$, $b/a = \frac{1}{2}$, and $b/a = \frac{1}{3}$. Given $a = 3$ cm, it is then desired to find the propagating mode(s) for $f = 9000$ MHz for each of the three cases.

From Table 10.1, the expression for the cutoff wavelength for a $TE_{m,n}$ mode, where $m = 0, 1, 2, 3, \ldots$ and $n = 0, 1, 2, 3, \ldots$ but not both m and n equal to zero, and for a $TM_{m,n}$ mode where $m = 1, 2, 3, \ldots$ and $n = 1, 2, 3, \ldots$ is given by

$$\lambda_c = \frac{1}{\sqrt{(m/2a)^2 + (n/2b)^2}}$$

The corresponding expression for the cutoff frequency is

$$f_c = \frac{v_p}{\lambda_c} = \frac{1}{\sqrt{\mu\varepsilon}} \sqrt{\left(\frac{m}{2a}\right)^2 + \left(\frac{n}{2b}\right)^2}$$

$$= \frac{1}{2a\sqrt{\mu\varepsilon}} \sqrt{m^2 + \left(n\frac{a}{b}\right)^2}$$

The cutoff frequency of the dominant mode $TE_{1,0}$ is $1/2a\sqrt{\mu\varepsilon}$. Hence,

$$\frac{f_c}{[f_c]_{TE_{1,0}}} = \sqrt{m^2 + \left(n\frac{a}{b}\right)^2}$$

By assigning different pairs of values for m and n, the lowest four values of $f_c/[f_c]_{TE_{1,0}}$ can be computed for each of the three specified values of b/a. These computed values and the corresponding modes are shown in Fig. 10.3.

For $a = 3$ cm, and assuming free space for the dielectric in the waveguide,

$$[f_c]_{TE_{1,0}} = \frac{1}{2a\sqrt{\mu\varepsilon}} = \frac{3 \times 10^8}{2 \times 0.03} = 5000 \text{ MHz}$$

FIGURE 10.3 Lowest four cutoff frequencies referred to the cutoff frequency of the dominant mode for three cases of rectangular waveguide dimensions.

Hence, for a signal of frequency $f = 9000$ MHz, all the modes for which $f_c/[f_c]_{TE_{1,0}}$ is less than 1.8 propagate. From Fig. 10.3, these are

$$TE_{1,0}, TE_{0,1}, TM_{1,1}, TE_{1,1} \quad \text{for} \quad b/a = 1$$

$$TE_{1,0} \quad \text{for} \quad b/a = \tfrac{1}{2}$$

$$TE_{1,0} \quad \text{for} \quad b/a - \tfrac{1}{3}$$

It can be seen from Fig. 10.3 that for $b/a < \tfrac{1}{2}$, the second lowest cutoff frequency which corresponds to that of the $TE_{2,0}$ mode is twice the cutoff frequency of the dominant mode $TE_{1,0}$. For this reason, the dimension b of a rectangular waveguide is generally chosen to be less than or equal to $a/2$ in order to achieve single-mode transmission over a complete octave (factor of 2) range of frequencies.

As in the case of the parallel-plate waveguide, reflection and transmission at discontinuities in rectangular waveguides can be studied by using the transmission-line analogy. We shall illustrate this by means of an example.

Transmission-line analogy

Example 10.2

A rectangular waveguide extending in the z-direction and having the dimensions $a = 4$ cm and $b = 2$ cm has a dielectric discontinuity at $z = 0$, as shown in Fig. 10.4. For $TE_{1,0}$ waves of frequency $f = 5000$ MHz incident from section 1, we wish to find (a) the transmission-line equivalent and (b) the length and the permittivity of a quarter-wave section required to achieve a match between the two sections.

FIGURE 10.4   Rectangular waveguide discontinuity.

a. First, we note that for the $TE_{1,0}$ mode, $\lambda_c = 2a = 8$ cm for both sections. For $f = 5000$ MHz, the wavelength in free space is $\lambda_1 = 6$ cm and the wavelength in a dielectric of permittivity $9\varepsilon_0$ is $\lambda_2 = 2$ cm. Since λ_1 and λ_2 are both less than λ_c, the $TE_{1,0}$ mode propagates in both sections. Denoting the guide parameters associated with sections 1 and 2 by subscripts 1 and 2, respectively, we then obtain

$$\eta_{g1} = \frac{\eta_1}{\sqrt{1 - (\lambda_1/\lambda_c)^2}} = \frac{377}{\sqrt{1 - (6/8)^2}} = 570 \ \Omega$$

$$\eta_{g2} = \frac{\eta_2}{\sqrt{1 - (\lambda_2/\lambda_c)^2}} = \frac{377/3}{\sqrt{1 - (2/8)^2}} = 129.8 \ \Omega$$

Thus, the transmission-line equivalent is as shown in Fig. 10.5.

b. The characteristic impedance of a quarter-wave section required to achieve a match between line 1 and line 2 must be equal to $\sqrt{\eta_{g1}\eta_{g2}}$. Denoting the parameters associated with the quarter-wave section by subscript 3, we then have

$$\eta_{g3} = \frac{\eta_3}{\sqrt{1 - (\lambda_3/\lambda_c)^2}} = \sqrt{\eta_{g1}\eta_{g2}}$$

or

$$\frac{\eta_1\sqrt{\varepsilon_0/\varepsilon_3}}{\sqrt{1 - (\lambda_1/\lambda_c)^2(\varepsilon_0/\varepsilon_3)}} = \sqrt{\eta_{g1}\eta_{g2}}$$

$$\frac{\varepsilon_0/\varepsilon_3}{1 - (6/8)^2(\varepsilon_0/\varepsilon_3)} = \frac{570 \times 129.8}{(377)^2} = 0.5205$$

solving which we obtain $\varepsilon_3 = 2.484\varepsilon_0$. To find the length of the quarter-wave section, we compute

$$\lambda_{g3} = \frac{\lambda_3}{\sqrt{1 - (\lambda_3/\lambda_c)^2}} = \frac{\lambda_1\sqrt{\varepsilon_0/\varepsilon_3}}{\sqrt{1 - (\lambda_1/\lambda_c)^2(\varepsilon_0/\varepsilon_3)}}$$

$$= \frac{6 \times 0.6345}{\sqrt{1 - (9/16) \times 0.4026}} = 4.33 \text{ cm}$$

Thus, the length of the quarter-wave section is $\lambda_{g3}/4$, or 1.0825 cm.

$\eta_{g1} = 570 \ \Omega$ $\eta_{g2} = 129.8 \ \Omega$

FIGURE 10.5   Transmission-line equivalent for the rectangular waveguide discontinuity of Fig. 10.4 for $TE_{1,0}$ waves of frequency 5000 MHz.

Let us now consider guided waves of equal amplitude propagating in the positive z- and negative z-directions in a rectangular waveguide. This can be achieved by terminating the guide by a perfectly conducting sheet in a constant-z plane, that is, a transverse plane of the guide. Due to perfect reflection from the sheet, the fields will then be characterized by standing wave nature along the guide axis, that is, in the z-direction, in addition to the standing wave nature in the x- and y-directions. The standing wave pattern along the guide axis will have nulls of transverse electric field on the terminating sheet and in planes parallel to it at distances of integer multiples of $\lambda_g/2$ from that sheet. Placing of perfect conductors in these planes will not disturb the fields since the boundary condition of a zero tangential electric field is satisfied in those planes.

Conversely, if we place two perfectly conducting sheets in two constant-z planes separated by a distance d, then, for the boundary conditions to be satisfied, d must be equal to an integer multiple of $\lambda_g/2$. We then have a rectangular box of dimensions a, b, and d in the x-, y-, and z-directions, respectively, as shown in Fig. 10.6. Such a structure is known as a *cavity resonator* and is the counterpart of the low-frequency lumped parameter resonant circuit at microwave frequencies, since it supports oscillations at frequencies for which the foregoing condition, that is,

$$d = l\,\frac{\lambda_g}{2}, \qquad l = 1, 2, 3, \ldots \tag{10.23}$$

is satisfied. Substituting for λ_g in (10.23) from Table 10.1 and rearranging, we obtain

$$\frac{2d}{l} = \frac{\lambda}{\sqrt{1 - (\lambda/\lambda_c)^2}}$$

or

$$\frac{1}{\lambda^2} - \frac{1}{\lambda_c^2} = \left(\frac{l}{2d}\right)^2$$

which upon substitution for λ_c gives

$$\frac{1}{\lambda^2} = \left(\frac{m}{2a}\right)^2 + \left(\frac{n}{2b}\right)^2 + \left(\frac{l}{2d}\right)^2$$

$$\lambda = \frac{1}{\sqrt{(m/2a)^2 + (n/2b)^2 + (l/2d)^2}} \tag{10.24}$$

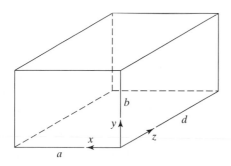

FIGURE 10.6   Rectangular cavity resonator.

The expression for the frequencies of oscillation is thus given by

$$
f_{osc} = \frac{v_p}{\lambda} = \frac{1}{\sqrt{\mu\varepsilon}} \sqrt{\left(\frac{m}{2a}\right)^2 + \left(\frac{n}{2b}\right)^2 + \left(\frac{l}{2d}\right)^2}
\qquad (10.25)
$$

The modes are designated by three subscripts in the manner $TE_{m,n,l}$ and $TM_{m,n,l}$. Since m, n, and l can assume combinations of integer values, an infinite number of frequencies of oscillation are possible for a given set of dimensions of the cavity resonator. Also, a given frequency of oscillation may correspond to more than one mode. We recall that for TE modes $m, n = 0, 1, 2, \ldots$, but not both zero, whereas for TM modes $m, n = 1, 2, 3, \ldots$. For both TE and TM modes $l = 1, 2, 3, \ldots$, as given in (10.23). In addition TM modes at cutoff ($\beta_z = 0$, $\lambda_g = \infty$ and $\eta_g = 0$) satisfy the boundary conditions since then $\overline{E}_x$ and $\overline{E}_y$ both go to zero. Hence, for TM modes $l = 0$ is allowed. We shall now consider an example.

Example 10.3

The dimensions of a rectangular cavity resonator with air dielectric are $a = 4$ cm, $b = 2$ cm, and $d = 4$ cm. It is desired to determine the three lowest frequencies of oscillation and specify the mode(s) of oscillation, transverse with respect to the z-direction, for each frequency.

By substituting $\mu = \mu_0$, $\varepsilon = \varepsilon_0$, and the given dimensions for $a, b,$ and d in (10.25), we obtain

$$
f_{osc} = 3 \times 10^8 \sqrt{\left(\frac{m}{0.08}\right)^2 + \left(\frac{n}{0.04}\right)^2 + \left(\frac{l}{0.08}\right)^2}
$$

$$
= 3750\sqrt{m^2 + 4n^2 + l^2} \ \text{MHz}
$$

By assigning combinations of integer values for m, n, and l and keeping in mind the restrictions on these values as discussed, we obtain the three lowest frequencies of oscillation and the corresponding modes to be

$3750 \times \sqrt{2} = 5303$ MHz   for $TE_{1,0,1}$ mode

$3750 \times \sqrt{5} = 8385$ MHz   for $TE_{0,1,1}, TE_{2,0,1}, TE_{1,0,2},$ and $TM_{1,1,0}$ modes

$3750 \times \sqrt{6} = 9186$ MHz   for $TE_{1,1,1}$ and $TM_{1,1,1}$ modes

---

K10.1.   Rectangular waveguide; TE and TM modes; Dominant mode; Cavity resonator; Frequency of oscillation.

D10.1.   A generator of fundamental frequency 2000 MHz and rich in harmonics excites a rectangular waveguide. Find all frequencies that propagate in only TE modes for each of the following cases: **(a)** $a = 5$ cm, $b = 2.5$ cm, $\varepsilon = \varepsilon_0$; **(b)** $a = 4.5$ cm, $b = 1.5$ cm, $\varepsilon = 4\varepsilon_0$; and **(c)** $a = 6$ cm, $b = 6$ cm, $\varepsilon = \varepsilon_0$. Assume that $\mu = \mu_0$ for all cases.
Ans.   **(a)** 4000, 6000 MHz;   **(b)** 2000, 4000 MHz;   **(c)** None.

D10.2.   For the rectangular waveguide discontinuity of Fig. 10.4, find the power reflection coefficient for incidence from section 1 for each of the following cases:

(a) $TE_{1,0}$ wave of frequency $f = 10{,}000$ MHz; **(b)** $TE_{1,1}$ wave of frequency $f = 10{,}000$ MHz; and **(c)** $TM_{1,1}$ wave of frequency $f = 10{,}000$ MHz.
Ans.   **(a)** 0.2756;   **(b)** 0.4649;   **(c)** 0.0676.

D10.3.   The frequencies of oscillation for an air-dielectric rectangular cavity resonator of dimensions a, b, and d, in the x-, y-, and z-directions, respectively, are given for three modes as follows:

$$f_{osc} = 3000\sqrt{5} \text{ MHz for } TE_{1,0,1} \text{ mode}$$

$$f_{osc} = 3000\sqrt{26} \text{ MHz for } TE_{0,1,1} \text{ mode}$$

$$f_{osc} = 3000\sqrt{30} \text{ MHz for } TM_{1,1,1} \text{ mode}$$

Find the values of a, b, and d in cm.
Ans.   2.5 cm, 1 cm, 5 cm.

10.2   CYLINDRICAL METALLIC WAVEGUIDE AND CAVITY RESONATOR

Thus far in this chapter, we have been concerned with the guiding of waves between metallic boundaries involving rectangular geometries. We shall now extend the treatment to cylindrical geometry by considering the case of a cylindrical waveguide, which is simply a hollow tube of circular cross section of a radius a and extending along the z-direction, as shown in Fig. 10.7. Thus, for TM waves in a cylindrical waveguide,

$$H_z = 0 \qquad E_z \neq 0$$

and all other field components can be expressed in terms of E_z, whereas for TE waves,

$$E_z = 0 \qquad H_z \neq 0$$

and all other field components can be expressed in terms of H_z.

To derive the field expressions for TM and TE modes in a cylindrical waveguide, we begin with Maxwell's curl equations in cylindrical coordinates and proceed similarly to the rectangular waveguide case, using the phasor forms of

Solution of Maxwell's curl equations in cylindrical coordinates

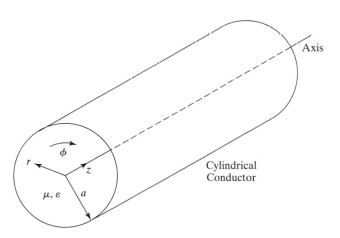

FIGURE 10.7   A cylindrical waveguide extending in the z-direction.

the field components in the manner $e^{j[\omega t \mp \beta_z z]}$, where the upper and lower signs represent wave propagation in the $+z$- and $-z$-directions, respectively. Thus, replacing $\partial/\partial t$ by $j\omega$ and $\partial/\partial z$ by $\mp j\beta_z$ in the expansions for the Maxwell's curl equations in cylindrical coordinates, we have for a perfect dielectric medium inside the waveguide,

$$\frac{1}{r}\frac{\partial \overline{E}_z}{\partial \phi} \pm j\beta_z \overline{E}_\phi = -j\omega\mu \overline{H}_r \tag{10.26a}$$

$$\mp j\beta_z \overline{E}_r - \frac{\partial \overline{E}_z}{\partial r} = -j\omega\mu \overline{H}_\phi \tag{10.26b}$$

$$\frac{1}{r}\frac{\partial}{\partial r}(rE_\phi) - \frac{1}{r}\frac{\partial \overline{E}_r}{\partial \phi} = -j\omega\mu \overline{H}_z \tag{10.26c}$$

$$\frac{1}{r}\frac{\partial \overline{H}_z}{\partial \phi} \pm j\beta_z \overline{H}_\phi = j\omega\varepsilon \overline{E}_r \tag{10.26d}$$

$$\mp j\beta_z \overline{H}_r - \frac{\partial \overline{H}_z}{\partial r} = j\omega\varepsilon \overline{E}_\phi \tag{10.26e}$$

$$\frac{1}{r}\frac{\partial}{\partial r}(r\overline{H}_\phi) - \frac{1}{r}\frac{\partial \overline{H}_r}{\partial \phi} = j\omega\varepsilon \overline{E}_z \tag{10.26f}$$

Solving (10.26a), (10.26b), (10.26d), and (10.26e), we obtain

$$\overline{E}_r = \frac{1}{\beta_z^2 - \beta^2}\left[\pm j\beta_z \frac{\partial \overline{E}_z}{\partial r} + \frac{j\omega\mu}{r}\frac{\partial \overline{H}_z}{\partial \phi}\right] \tag{10.27a}$$

$$\overline{E}_\phi = -\frac{1}{\beta_z^2 - \beta^2}\left[\mp j\frac{\beta_z}{r}\frac{\partial \overline{E}_z}{\partial \phi} + j\omega\mu \frac{\partial \overline{H}_z}{\partial r}\right] \tag{10.27b}$$

$$\overline{H}_r = -\frac{1}{\beta_z^2 - \beta^2}\left[\frac{j\omega\varepsilon}{r}\frac{\partial \overline{E}_z}{\partial \phi} \mp j\beta_z \frac{\partial \overline{H}_z}{\partial r}\right] \tag{10.27c}$$

$$\overline{H}_\phi = \frac{1}{\beta_z^2 - \beta^2}\left[j\omega\varepsilon \frac{\partial \overline{E}_z}{\partial r} \pm j\frac{\beta_z}{r}\frac{\partial \overline{H}_z}{\partial \phi}\right] \tag{10.27d}$$

thereby expressing the transverse field components in terms of the longitudinal field components. Recall that $\beta^2 = \omega^2\mu\varepsilon$.

Now, setting $\overline{H}_z = 0$ and substituting for $\overline{H}_r$ and $\overline{H}_\phi$ in (10.26f), we obtain the differential equation for $\overline{E}_z$ for the TM-mode case

$$\frac{1}{r}\frac{\partial}{\partial r}\left(r\frac{\partial \overline{E}_z}{\partial r}\right) + \frac{1}{r^2}\frac{\partial^2 \overline{E}_z}{\partial \phi^2} - (\beta_z^2 - \beta^2)\overline{E}_z = 0 \tag{10.28a}$$

Similarly, setting $\overline{E}_z = 0$ and substituting for $\overline{E}_r$ and $\overline{E}_\phi$ in (10.26c), we obtain the differential equation for $\overline{H}_z$ for the TE-mode case

$$\frac{1}{r}\frac{\partial}{\partial r}\left[r\frac{\partial \overline{H}_z}{\partial r}\right] + \frac{1}{r^2}\frac{\partial^2 \overline{H}_z}{\partial \phi^2} - (\beta_z^2 - \beta^2)\overline{H}_z = 0 \tag{10.28b}$$

In view of the similarity of (10.28a) and (10.28b), we let $\overline{\psi}$ stand for $\overline{E}_z$ in the case of TM waves and $\overline{H}_z$ in the case of TE waves and consider the solution of the differential equation

$$\frac{1}{r}\frac{\partial}{\partial r}\left(r\frac{\partial\overline{\psi}}{\partial r}\right) + \frac{1}{r^2}\frac{\partial^2\overline{\psi}}{\partial\phi^2} - (\beta_z^2 - \beta^2)\overline{\psi} = 0$$

or

$$\boxed{\frac{1}{r}\frac{\partial}{\partial r}\left[r\frac{\partial\overline{\psi}}{\partial r}\right] + \frac{1}{r^2}\frac{\partial^2\overline{\psi}}{\partial\phi^2} - \beta_c^2\overline{\psi} = 0} \qquad (10.29)$$

where

$$\boxed{\beta_c^2 = \beta^2 - \beta_z^2} \qquad (10.30)$$

To solve this equation, we make use of the separation of variables technique as in Section 10.1. Thus, letting

$$\overline{\psi} = \overline{R}(r)\overline{\Phi}(\phi)e^{\mp j\beta_z z} \qquad (10.31)$$

and substituting into (10.29), we have

$$\overline{\Phi}\frac{1}{r}\frac{\partial}{\partial r}\left(r\frac{\partial\overline{R}}{\partial r}\right) + \frac{\overline{R}}{r^2}\frac{\partial^2\overline{\Phi}}{\partial\phi^2} + \beta_c^2\overline{R}\,\overline{\Phi} = 0$$

or

$$\frac{r}{\overline{R}}(\overline{R}' + r\overline{R}'') + \beta_c^2 r^2 = -\frac{\overline{\Phi}''}{\overline{\Phi}} = n^2,\text{ a constant} \qquad (10.32)$$

where the primes denote differentiation with respect to the respective variables, thereby obtaining the two separate differential equations

$$\overline{\Phi}'' + n^2\overline{\Phi} = 0 \qquad (10.33a)$$

$$\overline{R}'' + \frac{1}{r}\overline{R}' + \left(\beta_c^2 - \frac{n^2}{r^2}\right)\overline{R} = 0 \qquad (10.33b)$$

The solution for (10.33a) is of the form

Bessel functions

$$\overline{\Phi}(\phi) = \overline{A}_n\cos n\phi + \overline{B}_n\sin n\phi \qquad (10.34a)$$

where $\overline{A}_n$ and $\overline{B}_n$ are constants. Equation (10.33b) is a Bessel's equation that has the solution

$$\overline{R}(r) = \overline{C}_n J_n(\beta_c r) + \overline{D}_n N_n(\beta_c r) \qquad (10.34b)$$

where $\overline{C}_n$ and $\overline{D}_n$ are constants, and

$$\boxed{\begin{aligned} J_n(\beta_c r) &= \text{Bessel function of the first kind of order } n \text{ and argument } \beta_c r \\ N_n(\beta_c r) &= \text{Bessel function of the second kind of order } n \text{ and argument } \beta_c r \end{aligned}}$$

The variations of $J_n(x)$ and $N_n(x)$ with their argument x are shown for a few values of n in Figs. 10.8(a) and (b), respectively. The function $N_n(x)$ has the property that $N_n(0) \to \infty$ for all orders. Since the fields must remain finite inside the guide, which includes $r = 0$ (the z-axis), it follows that $\overline{D}_n$ must be zero. Hence, the solution for $R(r)$ pertinent to the cylindrical waveguide problem is given by

$$R(r) = \overline{C}_n J_n(\beta_c r)$$

(a)

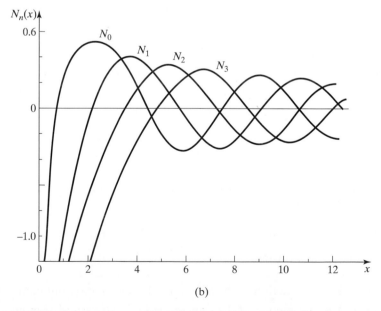

(b)

FIGURE 10.8   Variations of (a) $J_n(x)$ and (b) $N_n(x)$, with the argument x for a few values of n.

The solution for $\overline{\psi}$ is thus given by

$$\overline{\psi} = J_n(\beta_c r)(\overline{A}_n \cos n\phi + \overline{B}_n \sin n\phi)e^{\mp j\beta_z z} \qquad (10.35)$$

where we have absorbed $\overline{C}_n$ into $\overline{A}_n$ and $\overline{B}_n$. This is the general solution for $\overline{\psi}$, which we use for $\overline{E}_z$ in the case of TM waves and the $\overline{H}_z$ in the case of TE waves to obtain their particular solutions satisfying the boundary conditions.

For TM modes, $\overline{H}_z = 0$ and $\overline{\psi} = \overline{E}_z$. Hence, _TM modes_

$$\boxed{\overline{E}_z = J_n(\beta_c r)(\overline{A}_n \cos n\phi + \overline{B}_n \sin n\phi)e^{\mp j\beta_z z}} \qquad (10.36)$$

Since for a cylindrical waveguide, $0 \le \phi \le 2\pi$ and $\overline{E}_z$ must be a single-valued function, the solution must be periodic in ϕ with a period 2π. This requires that n be an integer. Thus,

$$n = 0, 1, 2, 3, \ldots$$

The boundary condition at $r = a$ requires that the tangential component of $\mathbf{E}$ be zero. Since $\overline{E}_z$ is tangential, it follows that $[\overline{E}_z]_{r=a} = 0$ for all ϕ, giving us

$$0 = J_n(\beta_c a)(\overline{A}_n \cos n\phi + \overline{B}_n \sin n\phi)e^{\mp j\beta_z z}$$

or

$$\boxed{J_n(\beta_c a) - 0} \qquad (10.37)$$

Note that this also makes the remaining tangential component $\overline{E}_\phi$ equal to zero at $r = a$ since from (10.27b), $\overline{E}_\phi \propto \partial \overline{E}_z / \partial \phi$. Equation (10.37) tells us that only certain values of $\beta_c a$ and hence of β_c are allowed. These are the values for which the Bessel function goes to zero, that is, they are the roots of the equation

$$J_n(x) = 0 \qquad (10.38)$$

For a given value of n, there are an infinite number of roots as can be seen from Fig. 10.8(a). Denoting the number of the root to be m, we list in Table 10.2 the lowest three ($m = 1, 2, 3$) nonvanishing roots for the first five values of n ($n = 0, 1, 2, 3, 4$).

With the understanding that the values of $\beta_c a$ are given by those in Table 10.2, we can obtain the expressions for the transverse-field components

TABLE 10.2 Roots of $J_n(x)$

m \ n	0	1	2	3	4
1	2.40	3.83	5.14	6.38	7.59
2	5.52	7.02	8.42	9.76	11.06
3	8.65	10.17	11.62	13.02	14.37

by noting that $\overline{H}_z = 0$ and substituting (10.36) into (10.27a)–(10.27d):

$$\overline{E}_r = \mp j \frac{\beta_z}{\beta_c} J_n'(\beta_c r)(\overline{A}_n \cos n\phi + \overline{B}_n \sin n\phi)e^{\mp j\beta_z z} \qquad (10.39a)$$

$$\overline{E}_\phi = \pm j \frac{\beta_z n}{\beta_c^2 r} J_n(\beta_c r)(\overline{A}_n \sin n\phi - \overline{B}_n \cos n\phi)e^{\mp j\beta_z z} \qquad (10.39b)$$

$$\overline{H}_r = \mp \frac{\omega\varepsilon}{\beta_z} \overline{E}_\phi \qquad (10.39c)$$

$$\overline{H}_\phi = \pm \frac{\omega\varepsilon}{\beta_z} \overline{E}_r \qquad (10.39d)$$

In (10.39a), the prime associated with J_n denotes differentiation with respect to the argument $\beta_c r$.

The allowed values of $\beta_c a$ may be written as $(\beta_c a)_{n,m}$, where the first subscript refers to the order of the Bessel function and the second subscript denotes the mth root of the nth order Bessel function. The corresponding modes are designated as $TM_{n,m}$ modes. In terms of the field configurations, the first subscript refers to the number of complete sinusoidal variations of the field components in the ϕ direction. For example, $n = 0$ means that the field components have no variations in ϕ, $n = 1$ means that they have one sinusoidal variation in ϕ, and so on. The second subscript refers to the number of quasi half-cycle variations (except that in the case of $n = 0$, the first one is a one-quarter cycle) of the field components in the r direction in accordance with the behavior of the Bessel functions, as shown in Fig. 10.8(a), and their derivatives, depending upon the component.

TE modes   For TE modes, $\overline{E}_z = 0$ and $\overline{\psi} = \overline{H}_z$. Hence,

$$\overline{H}_z = J_n(\beta_c r)(\overline{A}_n \cos n\phi + \overline{B}_n \sin n\phi)e^{\mp j\beta_z z} \qquad (10.40)$$

Again, n must be an integer as in the case of TM modes since the solution must be periodic in ϕ with period 2π in order that $\overline{H}_z$ is single-valued. Thus,

$$n = 0, 1, 2, 3, \ldots$$

The boundary condition at $r = a$ requires that the tangential component of $\overline{\mathbf{E}}$ be zero. Since $\overline{E}_z \equiv 0$ for TE modes, we need to consider $\overline{E}_\phi$, which is the remaining tangential component of $\overline{\mathbf{E}}$. From (10.27b), we note that $\overline{E}_\phi \propto \partial\overline{H}_z/\partial r$. Thus, we have

$$\left[\frac{\partial\overline{H}_z}{\partial r} \right]_{r=a} = 0$$

or

$$0 = J_n'(\beta_c a)(A_n \cos n\phi + B_n \sin n\phi)e^{\mp j\beta_z z}$$

where the prime associated with J_n denotes the derivative of the Bessel function. It follows that

$$\boxed{J_n'(\beta_c a) = 0} \tag{10.41}$$

Equation (10.41) tells us that again in this case, only certain values of $\beta_c a$ and hence of β_c are allowed. These are the values for which the derivative of the Bessel function goes to zero, that is, they are the roots of the equation

$$J_n'(x) = 0 \tag{10.42}$$

It can be seen by visualizing the derivatives of the graphs of the Bessel functions in Fig. 10.8(a) that for a given value of n, there are infinite number of roots for (10.42), corresponding to the points at which the slopes of the Bessel functions are zero. Denoting the number of the root to be m, we list in Table 10.3 the lowest three $(m = 1, 2, 3)$ nonvanishing roots for the first five values of $n(n = 0, 1, 2, 3, 4)$.

With the understanding that the values of $\beta_c a$ are given by those in Table 10.3, we can obtain the expressions for the transverse field components by noting that $\bar{E}_z = 0$ and substituting (10.40) into (10.27a)–(10.27b):

$$\bar{E}_r = j\frac{\omega\mu n}{\beta_c^2 r} J_n(\beta_c r)(\bar{A}_n \sin n\phi - \bar{B}_n \cos n\phi)e^{\mp j\beta_z z} \tag{10.43a}$$

$$\bar{E}_\phi = j\frac{\omega\mu}{\beta_c} J_n'(\beta_c r)(\bar{A}_n \cos n\phi + \bar{B}_n \sin n\phi)e^{\mp j\beta_z z} \tag{10.43b}$$

$$\bar{H}_r = \mp\frac{\beta_z}{\omega\mu}\bar{E}_\phi \tag{10.43c}$$

$$\bar{H}_\phi = \pm\frac{\beta_z}{\omega\mu}\bar{E}_r \tag{10.43d}$$

The allowed values of $\beta_c a$ may once again be written as $(\beta_c a)_{n,m}$ and the modes designated as $TE_{n,m}$ modes. As in the case of $TM_{n,m}$ modes, the first subscript refers to the number of complete sinusoidal variations of the field components in the ϕ direction. The second subscript refers to the number of quasi-half-cycle variations (except that for $n \ne 0$, the first one is a one-quarter cycle) of the field components in the r direction.

TABLE 10.3   Roots of $J_n'(x)$

m \ n	0	1	2	3	4
1	3.83	1.84	3.05	4.20	5.32
2	7.02	5.33	6.71	8.02	9.28
3	10.17	8.54	9.97	11.35	12.68

Let $(\beta_c)_{n,m}$ be the values of β_c for the (n,m)th mode found from the roots of $J_n(\beta_c a)$ for TM modes and of $J'_n(\beta_c a)$ for TE modes. Then recalling that

$$\beta_c^2 = \beta^2 - \beta_z^2 = \omega^2 \mu \varepsilon - \beta_z^2$$

and that the cutoff condition occurs for β_z equal to zero, we note that the cutoff frequencies are given by

$$\omega^2 \mu \varepsilon = (\beta_c)_{n,m}^2$$

$$(f_c)_{n,m} = \frac{(\beta_c)_{n,m}}{2\pi \sqrt{\mu\varepsilon}} \qquad (10.44)$$

and the cutoff wavelengths are given by

$$(\lambda_c)_{n,m} = \frac{1}{\sqrt{\mu\varepsilon}(f_c)_{n,m}}$$
$$= \frac{2\pi}{(\beta_c)_{n,m}} \qquad (10.45)$$

Proceeding further, for the propagating range of frequencies for a given mode, we have from (10.30),

$$\beta_z^2 = \beta^2 - \beta_c^2$$
$$= \left(\frac{2\pi}{\lambda}\right)^2 \left[1 - \left(\frac{f_c}{f}\right)^2\right]$$
$$= \left(\frac{2\pi}{\lambda}\right)^2 \left[1 - \left(\frac{\lambda}{\lambda_c}\right)^2\right]$$

or

$$\lambda_g = \frac{\lambda}{\sqrt{1 - (f_c/f)^2}} = \frac{\lambda}{\sqrt{1 - (\lambda/\lambda_c)^2}} \qquad (10.46)$$

The phase velocity along the guide axis is given by

$$v_{pz} = \lambda_g f = \frac{1}{\sqrt{\mu\varepsilon}} \frac{1}{\sqrt{1 - (f_c/f)^2}} = \frac{1}{\sqrt{\mu\varepsilon}} \frac{1}{\sqrt{1 - (\lambda/\lambda_c)^2}} \qquad (10.47)$$

Finally, by taking the ratios of appropriate pairs of transverse electric- and magnetic-field components given by (10.39a)–(10.39d) for TM modes and (10.43a)–(10.43d) for TE modes, we obtain the guide characteristic impedances to be

$$[\eta_g]_{\text{TM}} = \frac{\overline{E}_r}{\pm \overline{H}_\phi} = \frac{\overline{E}_\phi}{\mp \overline{H}_r} = \frac{\beta_z}{\omega\varepsilon} = \frac{2\pi}{\lambda_g \omega\varepsilon} = \sqrt{\frac{\mu}{\varepsilon}} \frac{\lambda}{\lambda_g}$$
$$= \sqrt{\frac{\mu}{\varepsilon}} \sqrt{1 - \left(\frac{f_c}{f}\right)^2} = \sqrt{\frac{\mu}{\varepsilon}} \sqrt{1 - \left(\frac{\lambda}{\lambda_c}\right)^2} \qquad (10.48a)$$

and

$$[\eta_g]_{TE} = \frac{\overline{E}_r}{\pm \overline{H}_\phi} = \frac{\overline{E}_\phi}{\mp \overline{H}_r} = \frac{\omega\mu}{\beta_z} = \frac{\omega\mu\lambda_g}{2\pi} = \sqrt{\frac{\mu}{\varepsilon}} \frac{\lambda_g}{\lambda}$$

$$= \frac{\sqrt{\mu/\varepsilon}}{\sqrt{1 - (f_c/f)^2}} = \frac{\sqrt{\mu/\varepsilon}}{\sqrt{1 - (\lambda/\lambda_c)^2}} \qquad (10.48b)$$

Note that the expressions (10.46)–(10.48b) are the same as the corresponding expressions for the rectangular waveguide case. We shall now consider an example.

Example 10.4

The radius of an air-dielectric cylindrical waveguide is given by $a = 5$ cm. It is desired to find the propagating modes and their characteristics for a signal of frequency $f = 3$ GHz. *Finding propagating modes*

From Tables 10.2 and 10.3, we note that the lowest root is 1.84, corresponding to the $TE_{1,1}$ mode. Thus, $TE_{1,1}$ mode is the dominant mode and its cutoff frequency is given by

$$[f_c]_{TF\,1,1} = \frac{[(\beta_c a)_{1,1}]/a}{2\pi\sqrt{\mu_0\varepsilon_0}} = \frac{1.84}{5 \times 10^{-2}} \times \frac{3 \times 10^8}{2\pi}$$

$$= 1.757 \text{ GHz}$$

In fact, by choosing the roots in increasing order of value from Tables 10.2 and 10.3 and dividing them by a and substituting in (10.44), we can find the cutoff frequencies in the increasing order. The four lowest cutoff frequencies and the corresponding modes found in this manner are listed in Table 10.4. For any given frequency f, propagation occurs in all modes for which $f > f_c$. Thus, for a signal of frequency 3 GHz, the propagating modes are $TE_{1,1}$, $TM_{0,1}$, and $TE_{2,1}$. The corresponding values of λ_c, λ_g, v_{pz}, and η_g, computed by using (10.45)–(10.48b), are also listed in Table 10.4.

TABLE 10.4   The Four Lowest Cutoff Frequencies and the Corresponding Modes for an Air-Dielectric Cylindrical Waveguide of Radius $a = 5$ cm and the Parameters for the Propagating Modes for $f = 3$ GHz

$\beta_c a$	Mode(s)	f_c (GHz)	λ_c (cm)	λ_g (cm)	v_{pz} (m/s)	η_g (ohms)
1.84	$TE_{1,1}$	1.757	17.074	12.337	3.701×10^8	465.10
2.40	$TM_{0,1}$	2.292	13.090	15.500	4.650×10^8	243.24
3.05	$TE_{2,1}$	2.913	10.300	41.827	12.548×10^8	1576.84
3.83	$TE_{0,1}, TM_{1,1}$	3.657	8.203	—	—	—

As in the case of the rectangular waveguide, by placing two perfectly conducting sheets in two constant-z planes separated by a distance d, we can have a cylindrical cavity resonator that supports oscillations at frequencies for which *Cylindrical cavity resonator*

$$d = l\frac{\lambda_g}{2}, \qquad l = 1, 2, 3, \ldots \qquad (10.49)$$

Substituting for λ_g from (10.46), we have

$$\frac{2d}{l} = \frac{\lambda}{\sqrt{1 - (\lambda/\lambda_c)^2}}$$

$$\frac{1}{\lambda^2} = \frac{1}{\lambda_c^2} + \left(\frac{l}{2d}\right)^2$$

$$\boxed{f_{osc} = \frac{v_p}{\lambda} = \frac{1}{\sqrt{\mu\varepsilon}} \sqrt{\frac{1}{\lambda_c^2} + \left(\frac{l}{2d}\right)^2}} \qquad (10.50)$$

The modes are designated $TE_{n,m,l}$ and $TM_{n,m,l}$, where n and m are allowed values as discussed earlier. For both TE and TM modes, $l = 1, 2, 3, \ldots$, as given by (10.49). In addition, as for the rectangular waveguide case, TM modes at cutoff ($\beta_z = 0, \lambda_g = \infty$ and $\eta_g = 0$) satisfy the boundary conditions since both $\overline{E}_r$ and $\overline{E}_\phi$ go to zero. Hence, for TM modes $l = 0$ is allowed. Let us consider an example.

Example 10.5

A cylindrical cavity resonator is formed by placing two perfectly conducting sheets 5 cm apart in the cross-sectional planes of the cylindrical waveguide of Example 10.4, so that $d = 5$ cm. It is desired to find the four lowest frequencies of oscillation and identify the mode(s) of oscillation for each frequency.

By substituting $\mu = \mu_0$, $\varepsilon = \varepsilon_0$, and $d = 5$ cm, in (10.50), we obtain

$$f_{osc} = 3 \times 10^8 \sqrt{\frac{1}{\lambda_c^2} + \left(\frac{l}{0.1}\right)^2}$$

$$= 3000 \sqrt{\left(\frac{0.1}{\lambda_c}\right)^2 + l^2} \text{ MHz}$$

By using the results of Table 10.4 and assigning values to l, as discussed, we obtain the four lowest frequencies of oscillation and the corresponding modes to be

$$3000 \times \frac{10}{13.09} = 2292 \text{ MHz for } TM_{0,1,0} \text{ mode}$$

$$3000 \sqrt{\left(\frac{10}{17.074}\right)^2 + 1} = 3477 \text{ MHz for } TE_{1,1,1} \text{ mode}$$

$$3000 \times \frac{10}{8.203} = 3657 \text{ MHz for } TM_{1,1,0} \text{ mode}$$

$$3000 \sqrt{\left(\frac{10}{13.09}\right)^2 + 1} = 3775 \text{ MHz for } TM_{0,1,1} \text{ mode}$$

K10.2.   Cylindrical waveguide; Bessel functions; TM and TE modes; Cylindrical cavity resonator.

D10.4.   An air-dielectric waveguide with the z-axis as its axis has a semicircular cross section of radius $a = 5$ cm, as shown in Fig. 10.9. Find the mode with the lowest

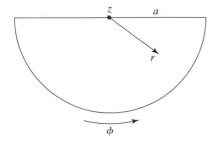

FIGURE 10.9 For Problem D10.4.

cutoff frequency and the corresponding value of the cutoff frequency for **(a)** TE
waves and **(b)** TM waves.
Ans. **(a)** $TE_{1,1}$, 1.757 GHz; **(b)** $TM_{1,1}$, 3.657 GHz.

D10.5. The resonant frequencies for the $TM_{0,1,0}$ mode and the $TE_{1,1,1}$ mode of an air-
dielectric cylindrical cavity resonator are both known to be 3000 MHz. Find the
values of the dimensions a and d of the resonator.
Ans. 3.82 cm; 7.79 cm.

10.3 LOSSES IN METALLIC WAVEGUIDES AND RESONATORS

In this section, we shall extend our discussion of waveguides and resonators to *Loss in*
consider the effects of lossy materials. Power dissipation in the imperfect dielec- *dielectric*
tric of a guide results in loss that follows simply from the attenuation constant for
the case of a uniform plane wave propagating in the dielectric. If for the purpose
of illustration, we consider the TE or TM wave in a parallel-plate waveguide,
then we know that progress of the composite TE or TM wave along the guide by
a distance d involves travel of the component uniform plane waves obliquely to
the plates by a distance $d/\sqrt{1 - (f_c/f)^2}$. Thus, if α_d' is the attenuation constant
for uniform plane wave propagation in the dielectric, then the attenuation constant α_d for the TE or TM wave along the guide axis is $\alpha_d'/\sqrt{1 - (f_c/f)^2}$ and the
attenuation $e^{\mp\alpha_d z}$ is equal to $e^{\mp[\alpha_d'/\sqrt{1-(f_c/f)^2}]z}$. From Section 5.4, we recall that
for a slightly imperfect dielectric ($\sigma/\omega\varepsilon \ll 1$),

$$\alpha_d' \approx \frac{\sigma}{2}\sqrt{\frac{\mu}{\varepsilon}}\left(1 - \frac{\sigma^2}{8\omega^2\varepsilon^2}\right) \approx \frac{\sigma}{2}\sqrt{\frac{\mu}{\varepsilon}}$$

Unlike the case of the imperfect dielectric, attenuation of the wave due to *Basis for*
power flow into the imperfect conductors of a guide as the wave propagates *analysis of*
down the guide involves an elaborate treatment. Since the conductors are only *loss in*
slightly imperfect ($\sigma/\omega\varepsilon \gg 1$), the procedure is based on considering the situa- *conductors*
tion as though a plane wave having the same magnetic field components as
those given by the appropriate tangential magnetic field components on that
wall for the perfect conductor case propagates normally into the conductor and
then computing the power flow into the wall (assumed to be of infinite depth in
view of the rapid attenuation of fields as they propagate into a good conductor).
Now, for a tangential magnetic field $\overline{\mathbf{H}}_t$ on a given wall, the electric-field vector

of a uniform plane wave propagating into the wall (designated to be in the direction $\mathbf{a}_n$) is $\overline{\eta}_c \overline{\mathbf{H}}_t \times \mathbf{a}_n$, where $\overline{\eta}_c$ is the intrinsic impedance of the conductor. The complex Poynting vector is

$$
\begin{aligned}
\overline{\mathbf{P}} &= \tfrac{1}{2} \overline{\mathbf{E}} \times \overline{\mathbf{H}}^* = \tfrac{1}{2} \overline{\eta}_c (\overline{\mathbf{H}}_t \times \mathbf{a}_n) \times \overline{\mathbf{H}}_t^* \\
&= \tfrac{1}{2} \overline{\eta}_c [\mathbf{a}_n(\overline{\mathbf{H}}_t \cdot \overline{\mathbf{H}}_t^*) - \overline{\mathbf{H}}_t(\mathbf{a}_n \cdot \overline{\mathbf{H}}_t^*)] \\
&= \tfrac{1}{2} \overline{\eta}_c \overline{\mathbf{H}}_t \cdot \overline{\mathbf{H}}_t^* \mathbf{a}_n
\end{aligned}
\tag{10.51}
$$

The time-average power flowing into the conductor of conductivity σ for a length Δz along the guide is given by

$$
\begin{aligned}
\Delta \langle P_d \rangle &= \int_l (\mathrm{Re}\, \overline{\mathbf{P}}) \cdot dl\, \Delta z\, \mathbf{a}_n \\
&= \int_l \mathrm{Re}\left(\frac{1}{2} \overline{\eta}_c \overline{\mathbf{H}}_t \cdot \overline{\mathbf{H}}_t^*\right) dl\, \Delta z \\
&= \frac{\Delta z}{2\sigma\delta} \int_l \overline{\mathbf{H}}_t \cdot \overline{\mathbf{H}}_t^*\, dl
\end{aligned}
\tag{10.52}
$$

where $\delta (= 1/\sqrt{\pi f \mu \sigma})$ is the skin depth at the frequency of operation f, dl is the differential length element along the transverse dimension, and $\int_l$ denotes integration performed along the transverse dimension. We shall illustrate the application of (10.52) by means of an example.

Example 10.6

Attenuation constant for $TE_{1,0}$ mode in a rectangular waveguide

Let us consider the propagation of $TE_{1,0}$ mode in a rectangular waveguide and obtain the expression for the attenuation constant α_c associated with it due to imperfect but good conductors comprising the walls of the guide.

To obtain the attenuation constant α_c, we note that since for a given mode the fields are attenuated in the manner $e^{-\alpha_c z}$ where the z-direction is assumed to be the guide axis, the time-average power flow $\langle P_f \rangle$ down the guide varies in the manner $e^{-2\alpha_c z}$. The time-average power dissipated over an infinitesimal distance Δz at any value of z along the guide is then given by

$$
\begin{aligned}
\Delta \langle P_d \rangle &= \frac{-\partial \langle P_f \rangle}{\partial z} \Delta z \\
&= 2\alpha_c \langle P_f \rangle\, \Delta z
\end{aligned}
$$

so that

$$
\boxed{\alpha_c = \frac{1}{2\langle P_f \rangle} \frac{\Delta \langle P_d \rangle}{\Delta z}}
\tag{10.53}
$$

Thus, the attenuation constant is one-half the ratio of the time-average power dissipated per unit distance at any value of z along the guide to the time-average power flow down the guide at that value of z. In fact, this is a general result applicable for any lossy traveling wave. The procedure for computing α_c for a given mode therefore consists of evaluating $\Delta \langle P_d \rangle$ and $\langle P_f \rangle$ for that mode.

To find $\Delta\langle P_d\rangle$, we consider the different walls of the waveguide separately. For each wall, we compute the time-average power flowing into the conductor over a surface made up of distance Δz along the guide axis and the entire transverse dimension of that wall by using (10.52). To proceed further, we substitute $m = 1$ and $n = 0$ in the TE mode field expressions given in Table 10.1, and considering the (+) wave only, we obtain the $TE_{1,0}$ mode field components in a lossless waveguide to be

$$\overline{E}_z = \overline{E}_x = \overline{H}_y = 0 \tag{10.54a}$$

$$\overline{H}_z = \overline{A}\,\cos\frac{\pi x}{a}e^{-j\beta_z z} \tag{10.54b}$$

$$\overline{E}_y = -j\frac{\lambda_c^2}{4\pi^2}\,\omega\mu\,\frac{\pi}{a}\,\overline{A}\,\sin\frac{\pi x}{a}e^{-j\beta_z z} \tag{10.54c}$$

$$\overline{H}_x = -\frac{\overline{E}_y}{\eta_g} = j\frac{2a}{\lambda_g}\,\overline{A}\,\sin\frac{\pi x}{a}e^{-j\beta_z z} \tag{10.54d}$$

where a and b are the dimensions of the guide in the x- and y-directions, respectively, as shown in Fig. 10.10. For the lossy case, the field components are multiplied by $e^{-\alpha_c z}$. Since we compute the quantities $\Delta\langle P_d\rangle$ and $\langle P_f\rangle$ at some particular value of z, say, z_0, we can absorb the factors $e^{-\alpha_c z_0}$ and $e^{-j\beta_z z_0}$ into the constant $\overline{A}$. Also, each nonzero tangential component of magnetic field on a given wall will be accompanied by a tangential electric field perpendicular to it so as to produce power flow into the conductor. Since some of these tangential electric-field components are longitudinal, the mode is no longer exactly TE mode. However, these components are very small in magnitude; hence, the mode is almost a TE mode.

We shall now consider the different walls and compute the corresponding values of $\Delta\langle P_d\rangle$ with the aid of Fig. 10.10.

RIGHT SIDE WALL ($x = 0$)

$$\overline{\mathbf{H}}_t = \overline{A}\,\mathbf{a}_z$$
$$\overline{\mathbf{H}}_t \cdot \overline{\mathbf{H}}_t^* = |A|^2$$
$$\Delta\langle P_d\rangle = \frac{\Delta z}{2\sigma\delta}\int_{y=0}^{b}|\overline{A}|^2\,dy$$
$$= \frac{|\overline{A}|^2 b}{2\sigma\delta}\,\Delta z \tag{10.55a}$$

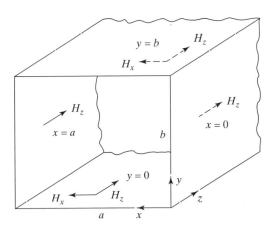

FIGURE 10.10   Rectangular waveguide with imperfectly conducting walls and showing the tangential magnetic field components on the walls for the $TE_{1,0}$ mode.

LEFT SIDE WALL ($x = a$)
Same as for the right side wall.

$$\Delta\langle P_d\rangle = \frac{|\overline{A}|^2 b}{2\sigma\delta}\,\Delta z \tag{10.55b}$$

BOTTOM WALL ($y = 0$)

$$\overline{\mathbf{H}}_t = \overline{A}\left(j\frac{2a}{\lambda_g}\sin\frac{\pi x}{a}\,\mathbf{a}_x + \cos\frac{\pi x}{a}\,\mathbf{a}_z\right)$$

$$\overline{\mathbf{H}}_t \cdot \overline{\mathbf{H}}_t^* = |\overline{A}|^2\left(\frac{4a^2}{\lambda_g^2}\sin^2\frac{\pi x}{a} + \cos^2\frac{\pi x}{a}\right)$$

$$\Delta\langle P_d\rangle = \frac{\Delta z}{2\sigma\delta}\int_{x=0}^{a}|\overline{A}|^2\left(\frac{4a^2}{\lambda_g^2}\sin^2\frac{\pi x}{a} + \cos^2\frac{\pi x}{a}\right)dx$$

$$= \frac{|\overline{A}|^2\,\Delta z}{4\sigma\delta}\left(\frac{4a^3}{\lambda_g^2} + a\right) \tag{10.55c}$$

TOP WALL ($y = b$)
Same as for the bottom wall.

$$\Delta\langle P_d\rangle = \frac{|\overline{A}|^2\,\Delta z}{4\sigma\delta}\left(\frac{4a^3}{\lambda_g^2} + a\right) \tag{10.55d}$$

Adding up the contributions from all four walls, we obtain the total time-average power dissipated over an infinitesimal length Δz along the guide to be

$$\Delta\langle P_d\rangle = \frac{|\overline{A}|^2}{2\sigma\delta}\left(\frac{4a^3}{\lambda_g^2} + a + 2b\right)\Delta z$$

$$= \frac{|\overline{A}|^2}{2\sigma\delta}\left[\frac{4a^3}{\lambda^2}\left(1 - \frac{\lambda^2}{4a^2}\right) + a + 2b\right]\Delta z \tag{10.56}$$

$$= \frac{|\overline{A}|^2}{\sigma\delta}\left(\frac{2a^3}{\lambda^2} + b\right)\Delta z$$

Now, to find the time-average power transmitted down the guide, we note that the time-average Poynting vector is given by

$$\langle\mathbf{P}\rangle = \tfrac{1}{2}\,\mathrm{Re}\,\overline{\mathbf{E}} \times \overline{\mathbf{H}}^*$$

$$= -\frac{1}{2\eta_g}\overline{E}_y\overline{E}_y^*\,\mathbf{a}_z$$

$$= \frac{1}{2\eta_g}\frac{\lambda_c^4}{16\pi^4}\omega^2\mu^2\frac{\pi^2}{a^2}|\overline{A}|^2\sin^2\frac{\pi x}{a}\,\mathbf{a}_z \tag{10.57}$$

$$= \frac{\omega^2\mu^2 a^2}{2\eta_g\pi^2}|\overline{A}|^2\sin^2\frac{\pi x}{a}\,\mathbf{a}_z$$

The time-average power transmitted down the guide is then given by

$$\langle P_f\rangle = \int_{y=0}^{b}\int_{x=0}^{a}\langle\mathbf{P}\rangle \cdot dx\,dy\,\mathbf{a}_z$$

$$= \frac{\omega^2\mu^2 a^2}{2\eta_g\pi^2}|\overline{A}|^2\int_{y=0}^{b}\int_{x=0}^{a}\sin^2\frac{\pi x}{a}\,dx\,dy \tag{10.58}$$

$$= \frac{\omega^2\mu^2 a^2 b}{4\eta_g\pi^2}|\overline{A}|^2$$

Finally, the attenuation constant is given by

$$
\alpha_c = \frac{\Delta\langle P_d\rangle}{2\langle P_f\rangle\,\Delta z}
$$

$$
= \frac{|\overline{A}|^2}{2\sigma\delta}\left(\frac{2a^3}{\lambda^2} + b\right)\frac{4\eta_g\pi^2}{\omega^2\mu^2a^3b|\overline{A}|^2}
$$

$$
= \frac{1}{\sigma\delta}\left(1 + \frac{b\lambda^2}{2a^3}\right)\frac{\eta\varepsilon}{\mu b\sqrt{1 - (f_c/f)^2}}
$$

$$
= \frac{1}{\sigma\delta}\left[1 + \frac{2b}{a}\left(\frac{\lambda}{\lambda_c}\right)^2\right]\frac{1}{b\eta\sqrt{1 - (f_c/f)^2}}
$$

or

$$
\alpha_c = \frac{1}{\sigma\delta b\eta\sqrt{1 - (f_c/f)^2}}\left[1 + \frac{2b}{a}\left(\frac{f_c}{f}\right)^2\right] \tag{10.59}
$$

Note that for $f \to f_c$, $\alpha_c \to \infty$. For $f \gg f_c$, $\alpha_c \approx 1/\sigma\delta b\eta \propto \sqrt{f}$, so that for $f \to \infty$, $\alpha_c \to \infty$. Thus, as f varies from f_c to infinity, α_c varies from infinity to some minimum value and then increases to infinity. The minimum value of α_c occurs for

$$
\left(\frac{f}{f_c}\right)^2 = \left(1.5 + \frac{3b}{a}\right) + \sqrt{\left(1.5 + \frac{3b}{a}\right)^2 - \frac{2b}{a}} \tag{10.60}
$$

For example, for $b/a = 1/2$, the minimum value occurs for $f/f_c \approx 2.4142$.

To proceed further, let us now consider the walls of a cavity resonator to be imperfect but good conductors. Then we can talk of the quality or Q factor of the resonator and derive the expression for it. The Q factor, which is a measure of the frequency selectivity of the resonator, is defined as

Q factor of a resonator

$$
Q = 2\pi\,\frac{\text{energy stored in the resonator}}{\text{energy dissipated per cycle}}
$$

$$
= 2\pi f\,\frac{\text{energy stored in the resonator}}{\text{time average power dissipated}} \tag{10.61}
$$

Since the conductors are good conductors, the power dissipated in them can be computed by using analysis as in Example 10.6 for the waveguide case. As for the energy stored in the cavity, it is distributed between the electric and magnetic fields at any arbitrary instant of time. But there are particular values of time at which the electric field is maximum and the magnetic field is zero, and vice versa. At these values of time, the entire energy is stored in one of the two fields. This can be taken advantage of to obtain the stored energy. We shall illustrate the determination of the Q factor by means of an example.

Example 10.7

Q factor for TE$_{1,0,1}$ *mode of a rectangular cavity resonator*

Let us consider the TE$_{1,0,1}$ mode of oscillation in the rectangular cavity resonator of Fig. 10.6 and obtain the expression for the Q factor associated with it due to imperfect but good conductors comprising the walls of the resonator.

First, we obtain the expressions for TE$_{1,0,1}$ mode field components by superimposing the $(+)$ and $(-)$ wave field components for the TE$_{1,0}$ waves from Table 10.1 and satisfying the boundary conditions of zero tangential electric fields at the ends $z = 0$ and $z = d$. Thus, we have

$$\overline{E}_z = \overline{E}_x = \overline{H}_y = 0 \tag{10.62a}$$

$$\overline{H}_z = \cos \frac{\pi x}{a} [\overline{A}_1 e^{-j\beta_z z} + \overline{A}_2 e^{j\beta_z z}] \tag{10.62b}$$

$$\overline{E}_y = -j \frac{\lambda_c^2}{4\pi^2} \omega\mu \frac{\pi}{a} \sin \frac{\pi x}{a} [\overline{A}_1 e^{-j\beta_z z} + \overline{A}_2 e^{j\beta_z z}] \tag{10.62c}$$

$$\overline{H}_x = j \frac{1}{\eta_g} \frac{\lambda_c^2}{4\pi^2} \omega\mu \frac{\pi}{a} \sin \frac{\pi x}{a} [\overline{A}_1 e^{-j\beta_z z} - \overline{A}_2 e^{j\beta_z z}] \tag{10.62d}$$

$$[\overline{E}_y]_{z=0} = [\overline{E}_y]_{z=d} = 0 \tag{10.63}$$

so that

$$\overline{A}_1 + \overline{A}_2 = 0 \quad \text{or} \quad \overline{A}_2 = -\overline{A}_1 \tag{10.64a}$$

$$\sin \beta_z d = 0 \quad \text{or} \quad \beta_z = \pi/d \tag{10.64b}$$

giving us the required field components

$$\overline{E}_z = \overline{E}_x = \overline{H}_y = 0 \tag{10.65a}$$

$$\overline{H}_z = \overline{A} \cos \frac{\pi x}{a} \sin \frac{\pi z}{d} \tag{10.65b}$$

$$\overline{E}_y = -j\overline{A} \frac{\omega\mu a}{\pi} \sin \frac{\pi x}{a} \sin \frac{\pi z}{d} \tag{10.65c}$$

$$\overline{H}_x = -\overline{A} \frac{a}{d} \sin \frac{\pi x}{a} \cos \frac{\pi z}{d} \tag{10.65d}$$

where $\overline{A} = -2j\overline{A}_1$ and we have also substituted $\lambda_c = 2a$ and $\omega\mu/\eta_g = 2\pi/\lambda_g = \pi/d$.

To find the energy stored in the cavity, we shall make use of the electric field. Noting that the amplitude of the only electric field component E_y, which is the value of E_y at the instant of time the magnetic field throughout the cavity is zero, is given by

$$|\overline{E}_y| = |\overline{A}| \frac{\omega\mu a}{\pi} \sin \frac{\pi x}{a} \sin \frac{\pi z}{d} \tag{10.66}$$

and integrating the energy density throughout the volume of the cavity, we obtain the energy stored in the cavity to be

$$W_{\text{stored}} = \int_{x=0}^{a} \int_{y=0}^{b} \int_{z=0}^{d} \frac{1}{2}\varepsilon |\overline{E}_y|^2 \, dv$$

$$= \frac{1}{2}\varepsilon |\overline{A}|^2 \frac{\omega^2 \mu^2 a^2}{\pi^2} \int_{x=0}^{a} \int_{y=0}^{b} \int_{z=0}^{d} \sin^2 \frac{\pi x}{a} \sin^2 \frac{\pi z}{d} \, dx \, dy \, dz \qquad (10.67)$$

$$= |\overline{A}|^2 \frac{\omega^2 \mu^2 \varepsilon}{8\pi^2} a^3 b d$$

To find the time-average power dissipated in the walls of the cavity, we note from the application of (10.51) that for a given wall, the time-average power dissipated is

$$\langle P_d \rangle = \int_S (\text{Re } \overline{\mathbf{P}}) \cdot dS \, \mathbf{a}_n$$

$$= \int_S \text{Re} \left(\frac{1}{2} \overline{\eta}_c \, \overline{\mathbf{H}}_t \cdot \overline{\mathbf{H}}_t^* \right) dS \qquad (10.68)$$

$$= \frac{1}{2\sigma\delta} \int_S \overline{\mathbf{H}}_t \cdot \overline{\mathbf{H}}_t^* \, dS$$

where S is the surface of the wall. Applying this result to the different walls of the cavity, we compute the corresponding values of $\langle P_d \rangle$ with the aid of Fig. 10.11.

RIGHT SIDE WALL ($x = 0$)

$$\overline{\mathbf{H}}_t = \overline{A} \sin \frac{\pi z}{d} \mathbf{a}_z$$

$$\overline{\mathbf{H}}_t \cdot \overline{\mathbf{H}}_t^* = |\overline{A}|^2 \sin^2 \frac{\pi z}{d}$$

$$\langle P_d \rangle = \frac{1}{2\sigma\delta} \int_{y=0}^{b} \int_{z=0}^{d} \overline{\mathbf{H}}_t \cdot \overline{\mathbf{H}}_t^* \, dy \, dz$$

$$= \frac{|\overline{A}|^2 \, bd}{4\sigma\delta} \qquad (10.69a)$$

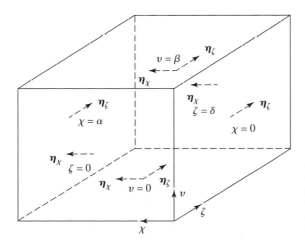

FIGURE 10.11   Rectangular cavity resonator with imperfectly conducting walls and showing the tangential magnetic field components on the walls for the $TE_{1,0,1}$ mode.

LEFT SIDE WALL ($x = a$)
Same as for right side wall.

$$\langle P_d \rangle = \frac{|\overline{A}|^2 \, bd}{4\sigma\delta} \tag{10.69b}$$

BOTTOM WALL ($y = 0$)

$$\overline{\mathbf{H}}_t = \overline{A}\left[-\frac{a}{d}\sin\frac{\pi x}{a}\cos\frac{\pi z}{d}\,\mathbf{a}_x + \cos\frac{\pi x}{a}\sin\frac{\pi z}{d}\,\mathbf{a}_z\right]$$

$$\overline{\mathbf{H}}_t \cdot \overline{\mathbf{H}}_t^* = |\overline{A}|^2\left[\left(\frac{a}{d}\right)^2\sin^2\frac{\pi x}{a}\cos^2\frac{\pi z}{d} + \cos^2\frac{\pi x}{a}\sin^2\frac{\pi z}{d}\right]$$

$$\langle P_d \rangle = \frac{1}{2\sigma\delta}\int_{x=0}^{a}\int_{z=0}^{d}\overline{\mathbf{H}}_t \cdot \overline{\mathbf{H}}_t^*\, dx\, dz$$

$$= \frac{|\overline{A}|^2}{8\sigma\delta}\left(\frac{a^3}{d} + ad\right) \tag{10.69c}$$

TOP WALL ($y = b$)
Same as for bottom wall.

$$\langle P_d \rangle = \frac{|\overline{A}|^2}{8\sigma\delta}\left(\frac{a^3}{d} + ad\right) \tag{10.69d}$$

FRONT WALL ($z = 0$)

$$\overline{\mathbf{H}}_t = -\overline{A}\,\frac{a}{d}\sin\frac{\pi x}{a}\,\mathbf{a}_x$$

$$\overline{\mathbf{H}}_t \cdot \overline{\mathbf{H}}_t^* = |\overline{A}|^2\left(\frac{a}{d}\right)^2\sin^2\frac{\pi x}{a}$$

$$\langle P_d \rangle = \frac{1}{2\sigma\delta}\int_{x=0}^{a}\int_{y=0}^{b}\overline{\mathbf{H}}_t \cdot \overline{\mathbf{H}}_t^*\, dx\, dy$$

$$= \frac{|\overline{A}|^2}{4\sigma\delta}\frac{a^3 b}{d^2} \tag{10.69e}$$

BACK WALL ($z = d$)
Same as for front wall.

$$\langle P_d \rangle = \frac{|\overline{A}|^2}{4\sigma\delta}\frac{a^3 b}{d^2} \tag{10.69f}$$

Adding up the contributions from all six walls, we obtain the total time-average power dissipated to be

$$\langle P_d \rangle = \frac{|\overline{A}|^2}{2\sigma\delta}\left(bd + \frac{a^3}{2d} + \frac{ad}{2} + \frac{a^3 b}{d^2}\right) \tag{10.70}$$

substituting (10.67) and (10.70) into (10.61), we obtain

$$Q = \omega\,\frac{\omega^2\mu\varepsilon a^3 bd/8\pi^2}{\dfrac{1}{2\sigma\delta}\left(bd + \dfrac{a^3}{2d} + \dfrac{ad}{2} + \dfrac{a^3 b}{d^2}\right)}$$

$$= \frac{\omega^3\mu^2\varepsilon\sigma\delta}{4\pi^2}\,\frac{2a^3 bd^3}{2bd^3 + a^3 d + ad^3 + 2a^3 b} \tag{10.71}$$

From (10.25), the resonant frequency ω for the $TE_{1,0,1}$ mode is however given by

$$\omega = \frac{\pi}{\sqrt{\mu\varepsilon}}\left[\left(\frac{1}{a}\right)^2 + \left(\frac{1}{d}\right)^2\right]^{1/2} \tag{10.72}$$

Thus, (10.71) reduces to

$$Q = \frac{\pi\sigma\delta\eta}{2}\frac{b(a^2 + d^2)^{3/2}}{ad(a^2 + d^2) + 2b(a^3 + d^3)} \tag{10.73}$$

For a numerical example, for the $TE_{1,0,1}$ mode of Example 10.3 and with the walls of the cavity made of copper ($\sigma = 5.80 \times 10^7$ S/m), $\delta = 9.075 \times 10^{-7}$ m and the value of Q is about 11020.

K10.3 Attenuation constant; $TE_{1,0}$ mode in a rectangular waveguide; Q factor; $TF_{1,0,1}$ mode in a rectangular cavity resonator.

D10.6 For each of the following cases of $TE_{1,0}$ waves propagating in a rectangular waveguide with copper walls, find the frequency of operation for which the attenuation constant α_c is a minimum and the minimum value of α_c: **(a)** $a = 3$ cm, $b = 1.5$ cm, and air-dielectric and **(b)** $a = b = 3$ cm, and dielectric of $\varepsilon = 4\varepsilon_0$ and $\mu = \mu_0$.
Ans. **(a)** 12.0711 GHz, 0.00653; **(b)** 7.4045 GHz, 0.0047.

D10.7 Find Q for $TE_{1,0,1}$ mode for each of the following cases of a rectangular cavity resonator with copper walls: **(a)** $a = b = d = 5$ cm, air-dielectric; **(b)** $a = 2.5$ cm, $b = 2$ cm, $d = 5$ cm, air-dielectric; **(c)** $a = b = d = 5$ cm; dielectric with $\varepsilon = 2.25\varepsilon_0$ and $\mu = \mu_0$.
Ans. **(a)** 16,434; **(b)** 10,160; **(c)** 13,417.

10.4 OPTICAL FIBER

Thus far in this chapter, we have discussed the guiding of waves between metal-*Description* lic boundaries. We now turn our attention to the optical fiber. An optical fiber, so called because of its filamentary appearance, consists typically of a core and a cladding, having circular cross sections, as shown in Fig. 10.12(a). The core is made up of a material of permittivity greater than that of the cladding so that a critical angle exists for waves inside the core incident on the interface between the core and the cladding, and hence waveguiding is made possible in the core by total internal reflection. The phenomenon of guiding may be visualized by considering a longitudinal cross section of the fiber through its axis, shown in

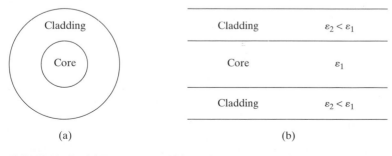

FIGURE 10.12 (a) Transverse and (b) longitudinal cross sections of an optical fiber.

Fig. 10.12(b), and comparing it with that of the slab waveguide, shown in Fig. 9.22. Whereas this situation corresponds to meridional rays, skewed rays whose paths lie in planes offset from the fiber axis also explain the guiding mechanism. Although the cladding is not essential for the purpose of waveguiding in the core since the permittivity of the core material is greater than that of free space, the cladding serves two useful purposes: (1) It avoids scattering and field distortion by the supporting structure of the fiber since the field decays exponentially outside the core and hence is negligible outside the cladding; (2) it allows a single-mode propagation for a larger value of the radius of the core than permitted in the absence of the cladding.

To simplify analysis of waveguiding in an optical fiber, we shall consider the cladding region to extend to infinity, so that the geometry is one of a cylindrical dielectric rod of permittivity greater than that of the surrounding medium, as shown in Fig. 10.13. In addition, we shall consider the permittivity of the core to be uniform, thereby corresponding to the case of a step-index fiber. To carry out the field analysis, we make use of our previous experience with the cylindrical metallic waveguide and the dielectric slab waveguide. First, we know that the transverse field components can be expressed in terms of the longitudinal field components E_z and H_z. We shall, however, find that the modes do not all separate into TE and TM modes and that the situation leads to the so-called hybrid modes. Second, we know in analogy with the case of the slab guide that radially decaying fields can be expected outside the core.

We recall from Section 10.2 that the differential equation to be satisfied by the fields for the cylindrical geometry is that given by (10.29)

$$\frac{1}{r}\frac{\partial}{\partial r}\left(r\frac{\partial\overline{\psi}}{\partial r}\right) + \frac{1}{r^2}\frac{\partial^2\overline{\psi}}{\partial\phi^2} + \beta_c^2\,\overline{\psi} = 0 \tag{10.74}$$

where $\overline{\psi}$ stands for $\overline{E}_z$ or $\overline{H}_z$ and

$$\beta_c^2 = \omega^2\mu\varepsilon - \left(\frac{2\pi}{\lambda_g}\right)^2 = \beta^2 - \beta_z^2 \tag{10.75}$$

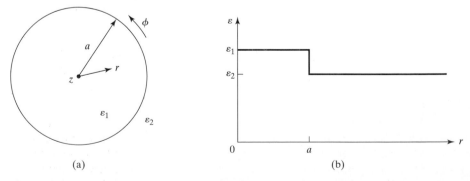

(a)   (b)

FIGURE 10.13   (a) Cross section of a cylindrical dielectric rod surrounded by a cladding region extending to infinity. (b) Permittivity profile for the arrangement of (a).

We also learned that (10.74) is separable into a differential equation involving r only and one involving ϕ only. The solution for the r-variation is a superposition of the Bessel functions $J_l(\beta_c r)$ and $N_l(\beta_c r)$, whereas the solutions for the ϕ-variation are sinusoidal ($\cos l\phi$ and $\sin l\phi$). The sinusoidal variations in ϕ require l to be an integer. The behaviors of the Bessel functions for real arguments are shown in Fig. 10.8.

Inside the core, $\beta = \beta_1 = \omega\sqrt{\mu\varepsilon_1}$, $\beta_{c1}^2 = \beta_1^2 - \beta_z^2 > 0$, and β_c is real. The solutions for the r-variation are J_l and N_l with real arguments. But since N_l has the property that $N_l(0) \to \infty$ for all orders, we rule it out. Thus,   *Field solution in the core*

$$\overline{\psi}(r, \phi, z) = \overline{A}J_l(\beta_{c1}r)\begin{Bmatrix}\cos l\phi \\ \sin l\phi\end{Bmatrix}e^{-j\beta_z z} \qquad \text{for} \quad r < a \qquad (10.76)$$

where we consider only the $(+)$ wave, for simplicity. Defining $u = \beta_{c1}a$, we can write the longitudinal components and then the transverse field components by using (10.27a)–(10.27d) as follows:

$$\overline{E}_{z1} = \overline{A}J_l\left(u\frac{r}{a}\right)\begin{Bmatrix}\cos l\phi \\ \sin l\phi\end{Bmatrix}e^{-j\beta_z z} \qquad (10.77\text{a})$$

$$\overline{H}_{z1} = \overline{B}J_l\left(u\frac{r}{a}\right)\begin{Bmatrix}\sin l\phi \\ \cos l\phi\end{Bmatrix}e^{-j\beta_z z} \qquad (10.77\text{b})$$

$$\overline{E}_{r1} = \left[-\frac{j\beta_z a}{u}\overline{A}J_l'\left(u\frac{r}{a}\right) + \frac{j\omega\mu a^2 l}{u^2 r}\overline{B}J_l\left(u\frac{r}{a}\right)\right]\begin{Bmatrix}\cos l\phi \\ \sin l\phi\end{Bmatrix}e^{-j\beta_z z} \qquad (10.77\text{c})$$

$$\overline{H}_{r1} = \left[\mp\frac{j\omega\varepsilon_1 a^2 l}{u^2 r}\overline{A}J_l\left(u\frac{r}{a}\right) - \frac{j\beta_z a}{u}\overline{B}J_l'\left(u\frac{r}{a}\right)\right]\begin{Bmatrix}\sin l\phi \\ \cos l\phi\end{Bmatrix}e^{-j\beta_z z} \qquad (10.77\text{d})$$

$$\overline{E}_{\phi1} = \left[\pm\frac{j\beta_z a^2 l}{u^2 r}\overline{A}J_l\left(u\frac{r}{a}\right) + \frac{j\omega\mu a}{u}\overline{B}J_l'\left(u\frac{r}{a}\right)\right]\begin{Bmatrix}\sin l\phi \\ \cos l\phi\end{Bmatrix}e^{-j\beta_z z} \qquad (10.77\text{e})$$

$$\overline{H}_{\phi1} = \left[-\frac{j\omega\varepsilon_1 a}{u}\overline{A}J_l'\left(u\frac{r}{a}\right) \mp \frac{j\beta_z a^2 l}{u^2 r}\overline{B}J_l\left(u\frac{r}{a}\right)\right]\begin{Bmatrix}\cos l\phi \\ \sin l\phi\end{Bmatrix}e^{-j\beta_z z} \qquad (10.77\text{f})$$

In (10.77a)–(10.77f), the upper functions and signs go together and the lower functions and signs go together. Note that when $\overline{E}_{z1} \propto \cos l\phi$, $\overline{H}_{z1} \propto \sin l\phi$. This is because when a given transverse component is proportional to $\overline{E}_z$ (or $\overline{H}_z$), it is proportional to $\partial\overline{H}_z/\partial\phi$ (or $\partial\overline{E}_z/\partial\phi$).

In the cladding, $\beta = \beta_2 = \omega\sqrt{\mu\varepsilon_2}$, $\beta_{c2}^2 = \beta_2^2 - \beta_z^2 < 0$, and β_c is imaginary.   *Field solution in the cladding*   The solutions for the r-variations are J_l and N_l with imaginary arguments jwr/a, where $w^2 = -\beta_{c2}^2a^2$. It is common practice then to represent the solution in terms of the modified Bessel functions I_l and K_l with real arguments wr/a. The behaviors of these functions for a few values of l are shown in Fig. 10.14. Since I_l has the property that $I_l(\infty) \to \infty$ for all orders, we rule it out. We can thus write

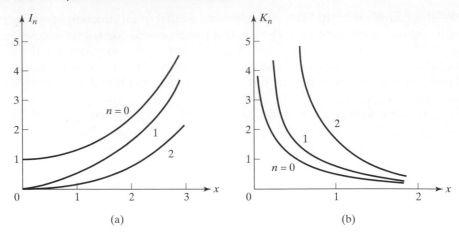

FIGURE 10.14 Behaviors of the modified Bessel functions (a) $I_n(x)$ and (b) $K_n(x)$.

the solutions for the field components as follows:

$$\overline{E}_{z2} = \overline{C}K_l\left(w\,\frac{r}{a}\right)\begin{Bmatrix}\cos l\phi\\\sin l\phi\end{Bmatrix}e^{-j\beta_z z} \tag{10.78a}$$

$$\overline{H}_{z2} = \overline{D}K_l\left(w\,\frac{r}{a}\right)\begin{Bmatrix}\sin l\phi\\\cos l\phi\end{Bmatrix}e^{-j\beta_z z} \tag{10.78b}$$

$$\overline{E}_{r2} = \left[\frac{j\beta_z a}{w}\,\overline{C}K_l'\left(w\,\frac{r}{a}\right) \pm \frac{j\omega\mu a^2 l}{w^2 r}\,\overline{D}K_l\left(w\,\frac{r}{a}\right)\right]\begin{Bmatrix}\cos l\phi\\\sin l\phi\end{Bmatrix}e^{-j\beta_z z} \tag{10.78c}$$

$$\overline{H}_{r2} = \left[\mp\frac{j\omega\varepsilon_2 a^2 l}{w^2 r}\,\overline{C}K_l\left(w\,\frac{r}{a}\right) + \frac{j\beta_z a}{w}\,\overline{D}K_l'\left(w\,\frac{r}{a}\right)\right]\begin{Bmatrix}\sin l\phi\\\cos l\phi\end{Bmatrix}e^{-j\beta_z z} \tag{10.78d}$$

$$\overline{E}_{\phi2} = \left[\mp\frac{j\beta_z a^2 l}{w^2 r}\,\overline{C}K_l\left(w\,\frac{r}{a}\right) - \frac{j\omega\mu a}{w}\,\overline{D}K_l'\left(w\,\frac{r}{a}\right)\right]\begin{Bmatrix}\sin l\phi\\\cos l\phi\end{Bmatrix}e^{-j\beta_z z} \tag{10.78e}$$

$$\overline{H}_{\phi2} = \left[\frac{j\omega\varepsilon_2 a}{w}\,\overline{C}K_l'\left(w\,\frac{r}{a}\right) \pm \frac{j\beta_z a^2 l}{w^2 r}\,\overline{D}K_l\left(w\,\frac{r}{a}\right)\right]\begin{Bmatrix}\cos l\phi\\\sin l\phi\end{Bmatrix}e^{-j\beta_z z} \tag{10.78f}$$

Guidance condition   To obtain the guidance condition, we need to satisfy the boundary conditions for the field components. We now see that for $l = 0$ the set of equations separates into two groups $\overline{E}_{z1,2}$, $\overline{E}_{r1,2}$, and $\overline{H}_{\phi1,2}$ involving the constants $\overline{A}$ and $\overline{C}$ and $\overline{H}_{z1,2}$, $\overline{H}_{r1,2}$, and $\overline{E}_{\phi1,2}$ involving the constants $\overline{B}$ and $\overline{D}$. The first group corresponds to the TM case and the characteristic equation obtained by setting $\overline{E}_{z1}/\overline{H}_{\phi1} = \overline{E}_{z2}/\overline{H}_{\phi2}$ at $r = a$ is

$$\boxed{\frac{J_1(u)}{J_0(u)} = -\frac{\varepsilon_2 u}{\varepsilon_1 w}\frac{K_1(w)}{K_0(w)}} \tag{10.79}$$

where we have used the property that $J_0'(u) = -J_1(u)$ and $K_0'(w) = -K_1(w)$. The second group corresponds to the TE case and the characteristic equation obtained by setting $\bar{E}_{\phi 1}/\bar{H}_{z1} = \bar{E}_{\phi 2}/\bar{H}_{z2}$ at $r = a$ is

$$\frac{J_1(u)}{J_0(u)} = -\frac{u}{w}\frac{K_1(w)}{K_0(w)} \tag{10.80}$$

For $l \neq 0$, the boundary conditions cannot be satisfied by the two separate groups and hence the fields can no longer be separated into TM and TE modes but are known as *hybrid modes*. The characteristic equation obtained by setting $\bar{E}_{z1} = \bar{E}_{z2}, \bar{H}_{z1} = \bar{H}_{z2}, \bar{E}_{z1}/\bar{H}_{\phi 1} = \bar{E}_{z2}/\bar{H}_{\phi 2}$, and $\bar{E}_{\phi 1}/\bar{H}_{z1} = \bar{E}_{\phi 2}/\bar{H}_{z2}$ at $r = a$ is then given by (see Problem P10.20)

$$\left[\frac{J_l'(u)}{uJ_l(u)} + \frac{K_l'(w)}{wK_l(w)}\right]\left[\frac{\beta_1^2 J_l'(u)}{uJ_l(u)} + \frac{\beta_2^2 K_l'(w)}{wK_l(w)}\right]$$
$$= \frac{\beta_z^2 l^2 V^4}{u^4 w^4} = l^2\left(\frac{1}{u^2} + \frac{1}{w^2}\right)\left(\frac{\beta_1^2}{u^2} + \frac{\beta_2^2}{w^2}\right) \tag{10.81}$$

where

$$V^2 = u^2 + w^2 = (\beta_1^2 - \beta_2^2)a^2 \tag{10.82}$$

Note that

$$V = \sqrt{\beta_1^2 - \beta_2^2}\, a = \frac{2\pi a}{\lambda_0} = \sqrt{\varepsilon_{r1} - \varepsilon_{r2}}$$
$$= \frac{2\pi a}{\lambda_0}\sqrt{n_1^2 - n_2^2} = \frac{2\pi u}{\lambda_0}(NA) \tag{10.83}$$

where

$$NA = \sqrt{n_1^2 - n_2^2} \tag{10.84}$$

is known as the *numerical aperture*, an important parameter of physical signifi-cance (see Problem P10.22).

Determination of the characteristics of the modes for a given value of l re-quires the solution of (10.79) or (10.80) for $l = 0$ (TM and TE modes) and (10.81) for $l \neq 0$ (hybrid modes), all with the constraint (10.82), from a knowledge of the values of $\omega, \varepsilon_1, \varepsilon_2$, and a. It is convenient to replace the derivatives of the Bessel functions by the Bessel functions themselves, using the recursive formulas

$$J_{l-1}(x) + J_{l+1}(x) = \frac{2l}{x}J_l(x) \tag{10.85a}$$

$$J_{l-1}(x) - J_{l+1}(x) = 2J_l'(x) \tag{10.85b}$$

$$K_{l-1}(x) + K_{l+1}(x) = -2K_l'(x) \tag{10.85c}$$

$$K_{l-1}(x) - K_{l+1}(x) = \frac{2l}{x}K_l(x) \tag{10.85d}$$

and express (10.81) in the manner (see Problem P10.23)

$$(J^+ + K^+)(\beta_1^2 J^- - \beta_2^2 K^-) + (J^- - K^-)(\beta_1^2 J^+ + \beta_2^2 K^+) = 0 \qquad (10.86)$$

where

$$J^\pm = \frac{J_{l\pm1}(u)}{uJ_l(u)} \qquad \text{and} \qquad K^\pm = \frac{K_{l\pm1}(w)}{wK_l(w)} \qquad (10.87)$$

Hybrid modes
   Let us first consider the hybrid modes. Then a method of solution consists of plotting (10.82) and (10.86) in the uw-plane and looking for the points of intersection. To do this, we note that (10.82) is simply the equation of a circle in the uw-plane, whereas (10.86) results in many branches. The approximate plot of Fig. 10.15 shows the nature of the first three branches for $l = 1$ and the circle for $V = 2$. The mode designations $HE_{l,m}$ and $EH_{l,m}$ for the hybrid modes denote, although hybrid, H_z is predominant for HE modes and E_z is predominant for EH modes. The subscript m denotes the number of the solution for the particular value of l. It can be seen that for the situation shown in Fig. 10.15 $HE_{1,2}$ and $EH_{1,1}$ are cut off, whereas a solution exists for $HE_{1,1}$. In fact, since the branch for $HE_{1,1}$ originates at $(0,0)$, that mode has no cutoff. Thus, $HE_{1,1}$ mode is the dominant mode. As ω increases, V increases, and more and more points of intersection with the V-circle occur, corresponding to higher-order modes.

Single-mode operation
   The modes with the lowest nonzero cutoff frequency are the TM and TE modes governed by the characteristic equations (10.79) and (10.80), respectively. Therefore, to determine the condition for single-mode operation, we consider these equations. Since the cutoff occurs for $w = 0$ (condition of fields extending to infinity in medium 2) and $K_1/K_0 = \infty$ for $w = 0$, we get $J_1(u) = 0$ or $u = 2.405$ for both modes. Also for $w = 0$, $V = \sqrt{u^2 + w^2} = u$. Thus, for single-mode operation, V must be less than 2.405.

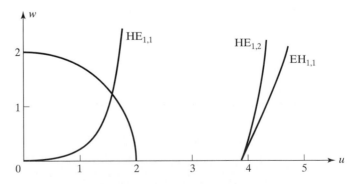

FIGURE 10.15   For illustrating the nature of solution of the characteristic equation for guided modes in an optical fiber.

Example 10.8

A fiber with core and cladding refractive indices $n_1 = 1.45$ and $n_2 = 1.44$, respectively, has a core radius of $a = 25\ \mu m$. We wish to find the minimum free-space wavelength, λ_0, for single-mode operation.

For $n_1 = 1.45$ and $n_2 = 1.44$,

$$NA = \sqrt{(1.45)^2 - (1.44)^2} = 0.17$$

Thus, λ_0 is given by

$$V = \frac{2\pi a}{\lambda_0}(NA) < 2.405$$

$$\lambda_0 > \frac{2\pi a(NA)}{2.405}$$

$$= \frac{2\pi \times 25 \times 0.17}{2.405}\ \mu m$$

$$= 11.1033\ \mu m$$

For most practical fibers, the refractive indices of the core and the cladding are nearly equal $(\varepsilon_1 \approx \varepsilon_2)$, corresponding to weak guidance. The rays are then paraxial and the longitudinal components of the fields are much smaller than the transverse components, so that the waves are almost TEM, the simplest such waves being linearly polarized along two orthogonal axes. The associated modes are designated as $LP_{l,m}$ modes, formed in general by superpositions of the exact modes. The $LP_{0,1}$ mode, which corresponds to just the $HE_{1,1}$ mode, is the dominant mode with no cutoff. Thus, in terms of LP modes, single-mode operation refers to the $LP_{0,1}$ mode and it is ensured for $V < 2.405$.

Weak guidance and LP modes

K10.4. Core; Cladding; Field solutions; Hybrid modes; $HE_{1,1}$ mode; LP modes.

D10.8. A fiber has core radius $a = 10\ \mu m$. Find the following for single-mode propagation at $\lambda_0 = 10\ \mu m$; **(a)** maximum allowable value of the numerical aperture; **(b)** minimum allowable value of the cladding refractive index n_2, if the core refractive index $n_1 - 1.5$, and **(c)** maximum allowable value of the core refractive index n_1, if the difference between the core and cladding refractive indices is 0.04.

Ans. **(a)** 0.3828; **(b)** 1.4503; **(c)** 1.8513.

10.5 PULSE BROADENING IN DISPERSIVE MEDIUM

In Section 9.3, we introduced the phenomenon of dispersion resulting from the variation of phase velocity of a wave with frequency. We learned that for a signal comprising a band of frequencies, the field patterns of the different frequency components do not maintain the same phase relationships as they travel down the propagation medium, and hence they disperse. For the case of a narrow-band signal, it is then meaningful to associate a group velocity, which determines the travel time of the signal. In this section, we consider another consequence of dispersion, which is the broadening of a pulse as it travels down the propagation medium. Although this characteristic is pertinent to any dispersive medium, it is particularly important for propagation of pulses in an optical fiber, because of its predominant contribution to the limitations in signal bandwidths that can be transmitted over long distances.

A heuristic explanation of pulse broadening can be provided by considering the dispersion diagram of Fig. 9.14. The slope of the curve at a point on the dispersion curve represents the group velocity at the frequency corresponding to that point. If we consider two frequencies ω_1 and ω_2 ($> \omega_1$), then we note that the group velocity at the higher frequency is greater than that at the lower frequency. Thus, a positive group velocity differential exists between the two frequencies. Now, we know that a signal having a pulse envelope has a certain bandwidth associated with it. The narrower the pulse, the greater is the bandwidth. For a given distance, the lower frequencies undergo a greater group delay than for the higher frequencies, or the higher frequencies "run away" from the lower frequencies, thereby causing loss of bandwidth and hence broadening of the pulse. In fact, the wider the bandwidth of the signal, the greater is the group delay spread within the bandwidth, so that the narrower the pulse envelope of the signal, the more pronounced is the pulse broadening.

Signal with Gaussian envelope

For a quantitative discussion of pulse broadening due to dispersion, we shall consider the propagation of an amplitude-modulated signal consisting of a carrier frequency signal modulated by a pulse having a Gaussian distribution with time, that is, a carrier frequency signal with a Gaussian envelope. Thus, let the electric field at the input ($z = 0$) of the dispersive channel be

$$E(0, t) = E_0 \, e^{-(t/T_0)^2} \cos \omega_0 t \tag{10.88}$$

where ω_0 is the carrier frequency, so that the phasor electric field is

$$\boxed{\overline{E}(0, t) = E_0 \, e^{-(t/T_0)^2} \, e^{j\omega_0 t}} \tag{10.89}$$

In (10.89), the function $e^{-(t/T_0)^2}$ corresponds to Gaussian distribution with $1/e$ half-width equal to T_0, as shown in Fig. 10.16(a).

Analysis for signal with Gaussian envelope

To analyze the propagation of the pulse, we need to use the Fourier technique, which consists of following the Fourier spectrum of (10.89) as it propagates down the medium and examine the inverse Fourier transform at the distant location. The Fourier transform of (10.89) is given by

$$\overline{E}(0, \omega) = \int_{-\infty}^{\infty} \overline{E}(0, t) \, e^{-j\omega t} \, dt$$

$$= E_0 \int_{-\infty}^{\infty} e^{-(t/T_0)^2} \, e^{-j(\omega - \omega_0)t} \, dt$$

$$= 2E_0 \int_{0}^{\infty} e^{-(t/T_0)^2} \cos(\omega - \omega_0)t \, dt$$

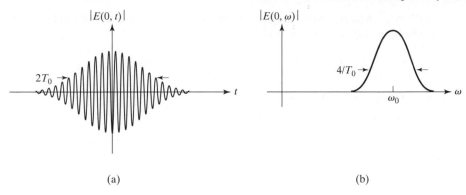

FIGURE 10.16 (a) Amplitude-modulated signal with Gaussian envelope, and (b) its Fourier spectrum.

or

$$E(0, \omega) = \sqrt{\pi}\, E_0 T_0\, e^{-(1/4)(\omega - \omega_0)^2 T_0^2} \qquad (10.90)$$

Note that the Fourier transform is also Gaussian, with half-width $2/T_0$ and centered at the carrier frequency ω_0, as shown in Fig. 10.16(b).

In traveling by a distance z from the input, each frequency component of the Fourier spectrum undergoes a phase change $e^{-j\beta_z z}$, where it is understood that β_z, the phase constant along the direction of propagation, is a function of ω. Assuming lossless medium, the Fourier transform of the signal at that location is then given by

$$\overline{E}(z, \omega) = \overline{E}(0, \omega)\, e^{-j\beta_z z} \qquad (10.91)$$

Taking the inverse Fourier transform of (10.91), we obtain the propagated signal to be

$$\overline{E}(z, t) = \frac{1}{2\pi} \int_{-\infty}^{\infty} \overline{E}(z, \omega)\, e^{j\omega t}\, d\omega$$

$$= \frac{1}{2\pi} \int_{-\infty}^{\infty} \overline{E}(0, \omega)\, e^{-j\beta_z z}\, e^{j\omega t}\, d\omega \qquad (10.92)$$

To evaluate (10.92), we consider the pulse to be narrow-band and β_z to be a slowly varying function of ω, so that we can expand β_z in Taylor's series about the carrier frequency ω_0 and keep the first three terms:

$$\beta_z(\omega) = \beta_z(\omega_0) + (\omega - \omega_0)\frac{d\beta_z}{d\omega}\bigg|_{\omega_0} + \frac{1}{2}(\omega - \omega_0)^2 \frac{d^2\beta_z}{d\omega^2}\bigg|_{\omega_0}$$

$$= \beta_z^{(0)} + (\omega - \omega_0)\beta_z^{(1)} + \frac{1}{2}(\omega - \omega_0)^2\, \beta_z^{(2)} \qquad (10.93)$$

where $\beta_z^{(0)}$, $\beta_z^{(1)}$, and $\beta_z^{(2)}$ are β_z, $d\beta_z/d\omega$, and $d^2\beta_z/d\omega^2$, respectively, evaluated at $\omega = \omega_0$. Substituting (10.93) into (10.92), we have

$$\bar{E}(z, t) = \frac{1}{2\pi} \int_{-\infty}^{\infty} \bar{E}(0, \omega) \, e^{-j\beta_z^{(0)}z - j(\omega - \omega_0)\beta_z^{(1)}z - j(1/2)(\omega - \omega_0)^2 \beta_z^{(2)}z} e^{j\omega t} \, d\omega$$

$$= \frac{1}{2\pi} \int_{-\infty}^{\infty} \bar{E}(0, \omega') \, e^{-j\beta_z^{(0)}z + j\omega'(t - t_0) - j(1/2)\omega'^2 \beta_z^{(2)}z} e^{j\omega_0 t} \, d\omega' \qquad (10.94)$$

where

$$\boxed{t_0 = z\,\beta_z^{(1)} = \frac{z}{[v_g]_{\omega_0}} = \text{group delay at } \omega_0} \qquad (10.95)$$

and $\omega' = \omega - \omega_0$. Proceeding further, we obtain

$$\bar{E}(z, t) = \frac{E_0 T_0}{2\sqrt{\pi}} e^{j[\omega_0 t - \beta_z^{(0)}z]}$$

$$\cdot \int_{-\infty}^{\infty} e^{-(1/4)\omega'^2 T_0^2} e^{j\omega'(t - t_0) - j(1/2)\omega'^2 \beta_z^{(2)}z} \, d\omega'$$

$$= \frac{E_0 T_0}{2\sqrt{\pi}} e^{j[\omega_0 t - \beta_z^{(0)}z]}$$

$$\cdot \int_{-\infty}^{\infty} e^{-(1/4)\omega'^2[T_0^2 + j2\beta_z^{(2)}z]} e^{j\omega'(t - t_0)} \, d\omega'$$

$$= \frac{E_0 T_0}{2\sqrt{\pi}} e^{j[\omega_0 t - \beta_z^{(0)}z]} \int_{-\infty}^{\infty} e^{-(1/4)\omega'^2 T_0^2[1 + j(S/T_0)]} e^{j\omega'(t - t_0)} \, d\omega'$$

or

$$\boxed{\bar{E}(z, t) = \frac{E_0 e^{j[\omega_0 t - \beta_z^{(0)}z]}}{\sqrt{1 + j(S/T_0)}} e^{-\{(t - t_0)^2[1 - j(S/T_0)]\}/(T_0^2 + S^2)}} \qquad (10.96)$$

where

$$\boxed{\begin{aligned} S &= \frac{2}{T_0} \beta_z^{(2)} z = \frac{2}{T_0} \frac{\partial t_0}{\partial \omega} \\ &= \text{half-bandwidth of the pulse at } z = 0 \text{ times} \\ &\quad \text{group delay per unit frequency} \\ &= \text{group delay for the entire half-bandwidth} \end{aligned}} \qquad (10.97)$$

Pulse broadening parameter   The result that we have obtained for the transmitted pulse tells us that the pulse envelope is $e^{-(t - t_0)^2/(T_0^2 + S^2)}$. Thus, it reaches a peak at $t = t_0$, which is simply the group delay at the carrier frequency ω_0, and its half-width is broadened

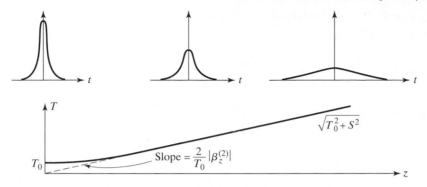

FIGURE 10.17 Broadening of the Gaussian envelope of a signal as a function of distance in a dispersive medium.

from the initial value T_0 to $T(z) = \sqrt{T_0^2 + S^2}$. The quantity S, which has the physical meaning given by (10.97), is known as the "pulse broadening parameter." It is of interest to discuss this further since it governs the pulse spread. In fact, for $|S| \gg T_0$, the pulse half-width is approximately equal to $|S|$. From (10.97), it can be seen that the value of S is influenced by three factors: (1) the bandwidth, $2/T_0$, of the input pulse; (2) dispersion, through the factor $\beta_z^{(2)}$; and (3) the distance, z, traveled by the signal. Figure 10.17 depicts the pulse broadening as a function of distance. Note from the plot of T versus z that for large distances the half-width of the pulse increases approximately linearly with distance at the rate of $|S|/z$, or $(2/T_0)|\beta_z^{(2)}|$, which is inversely proportional to the initial half-width T_0.

Example 10.9

Consider the propagation of a signal with Gaussian pulse envelope of half-width T_0 in a dispersive medium having the dispersion relation

$$\omega = \omega_p\left(1 + \frac{a^2\beta_z^2}{2}\right)$$

where ω_p and a are constants. It is desired to investigate the broadening of the transmitted pulse.

For the given dispersion relation,

$$\frac{d\omega}{d\beta_z} = \omega_p a^2 \beta_z$$

$$\frac{d^2\omega}{d\beta_z^2} = \omega_p a^2$$

From (10.97), the pulse broadening parameter is given by

$$S = \frac{2}{T_0}\omega_p^2 a^2 z$$

and the half-width of the transmitted pulse is

$$\sqrt{T_0^2 + S^2} = \sqrt{T_0^2 + \left(\frac{2}{T_0}\,\omega_p^2 a^2 z\right)^2}$$

$$= T_0\sqrt{1 + \left(\frac{2}{T_0^2}\,\omega_p^2 a^2 z\right)^2}$$

Note that for large distances from the input, the half-width is approximately equal to $(2/T_0)\omega_p^2\,az$.

Dispersion in optical waveguides

Having discussed the phenomenon of pulse spreading due to dispersion, we shall now turn to causes of dispersion in optical waveguides. In optical waveguides, dispersion is influenced by three factors: (1) waveguide dispersion, (2) intermodal dispersion, and (3) material dispersion. Waveguide dispersion is due to the fact that the phase velocity of propagation of the wave along the guide axis is a function of the wave frequency. This is the same as in metallic waveguides, which led to the introduction of the topic of dispersion in Section 9.3. Intermodal dispersion results from different travel times of rays corresponding to different modes, as discussed in Section 9.6. Material dispersion results from the frequency dependence of the refractive index of optical materials. We shall here elaborate on this.

Material dispersion

For a plane wave propagating in a uniform material medium having the refractive index n, the phase constant is given by

$$\boxed{\beta = \frac{\omega}{v_p} = \frac{\omega n}{c}} \tag{10.98}$$

To investigate pulse broadening due to material dispersion, we need to evaluate $\beta^{(2)} = d^2\beta/d\omega^2$. First, we note that

$$\frac{d\beta}{d\omega} = \frac{1}{c}\left(n + \omega\frac{dn}{d\omega}\right)$$

Since it is common practice to deal with n as a function of free-space wavelength λ_0, we make use of the relationship

$$f\lambda_0 = c$$

and obtain

$$\frac{dn}{d\omega} = \frac{dn}{d\lambda_0}\cdot\frac{d\lambda_0}{d\omega} = \frac{dn}{d\lambda_0}\left(-\frac{2\pi c}{\omega^2}\right)$$

so that

$$\frac{d\beta}{d\omega} = \frac{1}{c}\left(n - \lambda_0\frac{dn}{d\lambda_0}\right)$$

Note that the group velocity is

$$v_g = \frac{d\omega}{d\beta} = \frac{c}{n - \lambda_0(dn/d\lambda_0)} \qquad (10.99)$$

The quantity $[n - \lambda_0(dn/d\lambda_0)]$, denoted N, is known as the *group refractive index*.

Proceeding further, we obtain

$$\frac{d^2\beta}{d\omega^2} = \frac{d}{d\lambda_0}\left(\frac{d\beta}{d\omega}\right) \cdot \frac{d\lambda_0}{d\omega}$$

$$= -\frac{\lambda_0}{c}\frac{d^2n}{d\lambda_0^2}\frac{d\lambda_0}{d\omega} \qquad (10.100)$$

The pulse broadening parameter is given by

$$S = \text{half-bandwidth of the pluse} \times \beta^{(2)}z$$

$$= -\frac{\lambda_0}{c}\frac{d^2n}{d\lambda_0^2}\frac{d\lambda_0}{d\omega}(\Delta\omega)z \qquad (10.101)$$

$$= -\left(\frac{\lambda_0}{c}\frac{d^2n}{d\lambda_0^2}\right)(\Delta\lambda_0)z$$

where $\Delta\omega$ is the half bandwidth of the pulse and $\Delta\lambda_0 = (d\lambda_0/d\omega)\,\Delta\omega$ is the half-spectral width, that is, the half-bandwidth in the wavelength domain.

The quantity $[(\lambda_0/c)(d^2n/d\lambda_0^2)]$ is known as the *dispersion coefficient,* denoted D_λ, usually given in units of ps/km-nm, where the pulse broadening is measured in picoseconds, the length of the medium in kilometers, and the source spectral width in nanometers. The wavelength dependences of n, N, and D_λ for fused silica are shown plotted in Fig. 10.18. For $\lambda_0 < 1.312\ \mu m$, $D_\lambda < 0$ and the dispersion is normal. For $\lambda_0 > 1.312\ \mu m$, $D_\lambda > 0$ and the dispersion is anomalous. For $\lambda_0 \approx 1.312\ \mu m$, D_λ is zero and dispersion is negligible.

K10.5. Signal with Gaussian envelope; Pulse broadening parameter; Dispersion in optical guides; Material dispersion; Dispersion coefficient.

D10.9. The phase refractive index of an optical material used in a single-mode fiber can be approximated as

$$n(\lambda_0) = 1.58 + \frac{1.46 \times 10^{-14}}{\lambda_0^2}$$

Obtain the following for λ_0 equal to 0.5 μm: **(a)** the phase refractive index; **(b)** the group refractive index; and **(c)** the dispersion coefficient.
Ans. **(a)** 1.6384; **(b)** 1.7552; **(c)** 2,336 ps/km-nm.

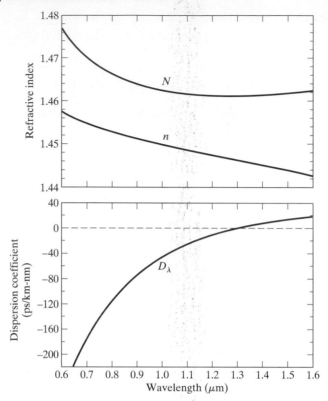

FIGURE 10.18   Wavelength dependences of n, N, and D_λ for fused silica.

10.6   INTERFERENCE AND DIFFRACTION

Interference explained   In this section, we turn our attention to two related topics that are based on the superposition of waves. When two or more waves are superimposed, the resulting distribution of intensity is in general not merely the sum of the distributions of the intensities of the individual waves; instead, it varies between maxima, which exceed the sum of the individual intensities, and minima, which may go to zero. This phenomenon is called interference. In the terminology of light, intensity means the time-average power crossing a unit area perpendicular to the direction of power flow, that is, the time-average Poynting vector, which is proportional to the square of the amplitude of the electric field. Thus, denoting intensity by the symbol I, we have

$$I = k\langle \mathbf{E} \cdot \mathbf{E} \rangle = k\langle E^2 \rangle \tag{10.102}$$

where k is the constant of proportionality, dependent on the medium. For free space, k is equal to $1/2\eta_0$.

Let us now consider two uniform plane waves whose electric fields at a point P are $\mathbf{E}_1$ and $\mathbf{E}_2$, respectively. Then the intensities of the individual waves are

$$I_1 = k\langle E_1^2 \rangle \quad \text{and} \quad I_2 = k\langle E_2^2 \rangle \qquad (10.103)$$

The intensity of the superposition of the two waves is

$$\begin{aligned} I &= k\langle (\mathbf{E}_1 + \mathbf{E}_2) \cdot (\mathbf{E}_1 + \mathbf{E}_2) \rangle \\ &= k\langle E_1^2 \rangle + k\langle E_2^2 \rangle + 2k\langle \mathbf{E}_1 \cdot \mathbf{E}_2 \rangle \qquad (10.104) \\ &= I_1 + I_2 + 2k\langle E_1 E_2 \cos \alpha \rangle \end{aligned}$$

where α is the angle between the directions of E_1 and E_2. The third term on the right side is seen to be the interference term. Depending on the sign of this term, it can be seen that the intensity of the composite wave can be greater than or less than the sum of the individual intensities. It is equal to the sum of the two intensities only if $\cos \alpha$ is equal to zero, which occurs when $\mathbf{E}_1$ and $\mathbf{E}_2$ are perpendicular. Thus, two waves whose fields are polarized perpendicular to each other do not produce interference.

An experimental demonstration of interference of light was first performed by Thomas Young. Young's experiment consisted of light from a monochromatic point source S_0 behind a screen containing two pinholes separated by a distance of the order of a millimeter passing through the pinholes, thereby giving rise to two light beams creating an interference pattern on an observation screen, located at a distance R of the order of several meters beyond the pinholes, as shown in Fig. 10.19.

Two-beam interference

The two waves generated at the pinholes are spherical waves and in phase. At a point P on the observation screen, the waves can be approximated as plane waves. The interference phenomenon can therefore be analyzed by

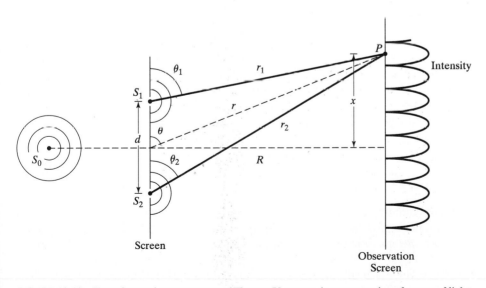

FIGURE 10.19 Experimental arrangement of Thomas Young to demonstrate interference of light.

superimposing two traveling waves propagating at the angles θ_1 and θ_2 to the vertical. Because of the large value of R, the difference between these angles is small so that the two lines S_1P and S_2P can be considered to be parallel and $(r_2 - r_1) \approx d \cos \theta$. As the point P moves vertically on the observation screen, the path difference $d \cos \theta$ varies and hence the phase difference between the two waves varies, creating the interference pattern.

Assuming equal amplitudes and zero phase angle at the slits, the individual electric fields of the two beams can be written as

$$E_1 = \frac{A}{r_1} \cos (\omega t - \beta_{r1}) \tag{10.105a}$$

$$E_2 = \frac{A}{r_2} \cos (\omega t - \beta_{r2}) \tag{10.105b}$$

Setting $r_1 \approx r_2 \approx r$ in the amplitude factors, we obtain the total electric field to be

$$E = E_1 + E_2 \approx \frac{A}{r} [\cos (\omega t - \beta r_1) + \cos (\omega t - \beta r_2)] \tag{10.106}$$

The intensity of the beam on the screen is therefore given by

$$
\begin{aligned}
I &= k \langle E^2 \rangle \\
&= \frac{kA^2}{r^2} \langle \cos^2 (\omega t - \beta r_1) + \cos^2 (\omega t - \beta r_2) \\
&\quad + 2 \cos (\omega t - \beta r_1) \cos (\omega t - \beta r_2) \rangle \\
&= \frac{kA^2}{r^2} \left[\frac{1}{2} + \frac{1}{2} + \cos \beta (r_2 - r_1) \right] \\
&= \frac{2kA^2}{r^2} \cos^2 \frac{\beta (r_2 - r_1)}{2} \\
&= \frac{2kA^2}{r^2} \cos^2 \frac{\beta d \cos \theta}{2}
\end{aligned}
\tag{10.107}
$$

Thus, the intensity varies between zero and $2kA^2/r^2$. In terms of the distance x, we can write $\cos \theta = x/r = x/\sqrt{R^2 + x^2}$. Furthermore, for $x \ll R$, we can set $\sqrt{R^2 + x^2} \approx R$, so that we obtain

$$\boxed{I \approx \frac{2kA^2}{R^2} \cos^2 \frac{\pi dx}{\lambda R}} \tag{10.108}$$

Thus, the maxima of the intensity occur for $x = n\lambda R/d, n = 0, \pm 1, \pm 2, \ldots$, and the minima occur for $x = (2n + 1)\lambda R/d, n = 0, \pm 1, \pm 2, \ldots$. The separation between the maxima and the minima is a constant, equal to $\Delta x = \lambda R/d$. For a numerical example, for $d = 0.1$ mm, $R = 2$ m, and $\lambda = 0.5$ μm, $\Delta x = 1$ cm.

Conversely to the preceding, if λ is unknown, it can be computed from a measurement of Δx and given the values of R and d. For two beams, the

sharpness of the interference fringes is not sufficient to permit accurate measurement of Δx, but by using a large number of beams the sharpness can be increased. We shall not pursue the multiple-source problem here but consider it in
connection with antenna arrays in Section 11.4. We shall however include here
an example of a similar phenomenon resulting form multiple reflections and
transmissions of waves from a single source.

Example 10.10

Let us consider a uniform plane wave of electric field $\bar{E}_i$ and wavelength λ_0, incident at an
angle θ_i on a plane dielectric slab of thickness d and refractive index n, as shown in
Fig. 10.20. For simplicity, we shall consider the medium on either side of the dielectric
slab to be free space. We wish to investigate the interference resulting from the (infinite)
number of waves produced by reflections and transmissions at the two interfaces, a few
of which are shown in Fig. 10.20.

*Multiple-
beam
interference*

   With reference to the notation shown in Fig. 10.20, and denoting Γ and τ to be the
reflection and transmission coefficients, respectively, for incidence from free space on to
the dielectric and Γ' and τ' to be the reflection and transmission coefficients, respectively,
for incidence from the dielectric on to free space, we can write the expressions for the
successively reflected and transmitted wave electric fields as follows:

$$\bar{E}_{r1} = \bar{E}_i \, \Gamma \qquad\qquad (10.109\text{a})$$

$$\bar{E}_{r2} = \bar{E}_i \tau \Gamma' \tau' e^{-j\delta} \qquad\qquad (10.109\text{b})$$

$$\bar{E}_{r3} = \bar{E}_i \tau (\Gamma')^3 \tau' e^{-j2\delta} \qquad\qquad (10.109\text{c})$$

$$\bar{E}_{t1} = \bar{E}_i \tau \tau' \qquad\qquad (10.109\text{d})$$

$$\bar{E}_{t2} = \bar{E}_i \tau (\Gamma')^2 \, \tau' e^{-j\delta} \qquad\qquad (10.109\text{e})$$

$$\bar{E}_{t3} = \bar{E}_i \tau (\Gamma')^4 \tau' e^{-j2\delta} \qquad\qquad (10.109\text{f})$$

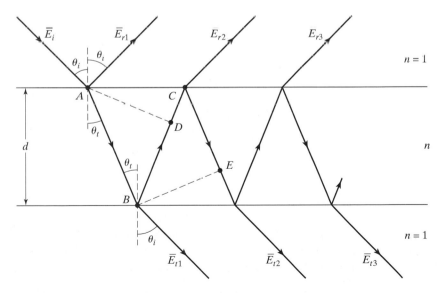

FIGURE 10.20 Multiple reflections and transmissions for plane wave incidence on a plane
dielectric slab.

where

$$
\begin{aligned}
\delta &= \frac{2\pi n}{\lambda_0}(AB + BD) = \frac{2\pi n}{\lambda_0}(BC + CE) \\[2mm]
&= \frac{2\pi n}{\lambda_0}\frac{d}{\cos\theta_t}(1 + \cos 2\theta_t) \\[2mm]
&= \frac{4\pi n d}{\lambda_0}\cos\theta_t
\end{aligned}
\tag{10.110}
$$

is the additional phase shift undergone by successive reflected (or transmitted) waves. Summing up all the reflected wave fields, we obtain the total reflected wave field to be

$$
\begin{aligned}
\overline{E}_r &= \overline{E}_i\{\Gamma + \tau\tau'\overline{\Gamma}'e^{-j\delta}[1 + (\Gamma')^2 e^{-j\delta} + \cdots]\} \\[2mm]
&= \overline{E}_i\left[\Gamma + \frac{\tau\tau'\Gamma'e^{-j\delta}}{1 - (\Gamma')^2 e^{-j\delta}}\right]
\end{aligned}
\tag{10.111a}
$$

Similarly, the total transmitted wave field is given by

$$
\begin{aligned}
\overline{E}_t &= \overline{E}_i\tau\tau'[1 + (\Gamma')^2 e^{-j\delta} + \cdots] \\[2mm]
&= \overline{E}_i\left[\frac{\tau\tau'}{1 - (\Gamma')^2 e^{-j\delta}}\right]
\end{aligned}
\tag{10.111b}
$$

Although the specific expressions for $\Gamma, \Gamma', \tau,$ and τ' depend on the polarization of $\mathbf{E}_i$ and are given in Section 9.4 by the Fresnel coefficients, we can write, regardless of polarization (see Problem P10.28), that

$$
\Gamma' = -\Gamma
\tag{10.112a}
$$

and

$$
\tau\tau' - \Gamma^2 = 1
\tag{10.112b}
$$

Substituting (10.112a) and (10.112b) into (10.111a) and (10.111b), we obtain

$$
\overline{E}_r = \overline{E}_i\left[\frac{\Gamma(1 - e^{-j\delta})}{1 - \Gamma^2 e^{-j\delta}}\right]
\tag{10.113a}
$$

$$
\overline{E}_t = \overline{E}_i\frac{1 - \Gamma^2}{1 - \Gamma^2 e^{-j\delta}}
\tag{10.113b}
$$

The fractions of the incident intensity that are reflected and transmitted are given, respectively, by

$$
\frac{I_r}{I_i} = \frac{|\overline{E}_r|^2}{|\overline{E}_i|^2} = \frac{4F^2 \sin^2(\delta/2)}{\pi^2 + 4F^2 \sin^2(\delta/2)}
\tag{10.114a}
$$

$$
\frac{I_t}{I_i} = \frac{|\overline{E}_t|^2}{|\overline{E}_i|^2} = \frac{\pi^2}{\pi^2 + 4F^2 \sin^2(\delta/2)}
\tag{10.114b}
$$

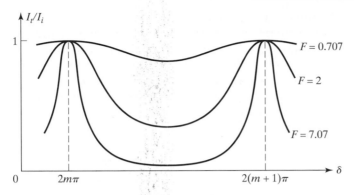

FIGURE 10.21   Plots of the transmission characteristics of the arrangement of Fig. 10.20 for several values of F.

where

$$F = \frac{\pi |\Gamma|}{1 - \Gamma^2} \qquad (10.115)$$

is called the finesse. Note that the sum of right sides of (10.114a) and (10.114b) is equal to unity, as it should be.

The transmission characteristic of the arrangement of Fig. 10.20, a model for the Fabry–Perot etalon, or interferometer, can now be discussed with the aid of the plot of the right side of (10.114b) versus δ, which is shown in Fig. 10.21, for several values of F and hence Γ. It can be seen that maximum transmission of unity occurs for

Fabry–Perot etalon

$$\delta = \frac{4\pi nd}{\lambda_0} \cos \theta_t = 2m\pi, \qquad m = 1, 2, 3, \ldots \qquad (10.116)$$

with the sharpness of the maxima increasing with F. For given values of λ_0, d, and n, the plot can be thought of as variation of I_t/I_0 with θ_t (and hence θ_i), thereby corresponding to the interference pattern. For fixed values of d, n, and θ_t, the peaks in Fig. 10.21 correspond to two adjacent frequencies at which I_t/I_i is unity. From (10.116), this frequency separation can be seen to be equal to $c/(2nd \cos \theta_t)$. Also, for given values of d and n, different values of λ_0 (and hence f) give rise to interference patterns of different periodicities, thereby allowing resolution of closely spaced frequencies for high values of F (see Problem P10.29).

When an object is placed between a source of light and an observation screen, the shadow on the screen contains a fine structure of interference fringes in the vicinity of the boundary separating the dark shadow from the rest of the brightly illuminated screen, as compared to a simple sharp boundary between the dark and bright regions. This phenomenon, which occurs due to the *bending* of a portion of the beam, is known as *diffraction*. Just as interference is a manifestation of the superposition of light beams, diffraction is also a manifestation of the superposition of light beams. Interference usually applies to the interaction of only a few beams with one another, whereas diffraction usually pertains

Diffraction explained

to the superposition of a large number, even a continuous distribution of beams, although the distinction is not sharp. The phenomenon of diffraction is in contrast to the principle of geometrical optics, which has to do with light traveling in straight lines except for bending by reflection and refraction, and which is strictly valid under certain conditions: (a) The dimensions of the object in the path of light are very large compared to the wavelength and (b) the region of importance is not close to the boundary of the shadow.

Huygens–
Fresnel
principle

Although the exact treatment of the phenomenon of diffraction involves solution as a boundary value problem and is very difficult, it can be studied in approximate but sufficiently accurate terms by using the Huygens–Fresnel principle, as long as the distance from the diffracting object to the point of observation is more than about ten wavelengths. To explain this principle, let us consider a plane monochromatic wave incident normally on a screen in the xy-plane with an aperture cut into it, as shown in Fig. 10.22. Then, according to this principle, the incident wave may be thought of as giving rise to secondary (spherical) waves emanating from every point in the aperture and that interfere with one another to the right of the screen. The scalar field at a point P is approximately given by

$$\overline{E}(P) \approx \frac{j\beta}{2\pi} \int_S \frac{\overline{E}(x', y', 0)}{R} e^{-j\beta R} \, dS \qquad (10.117)$$

where S is the area of the aperture, and $\overline{E}(x', y', 0)$ is the scalar field in the aperture. We shall illustrate the application of (10.117) by means of an example.

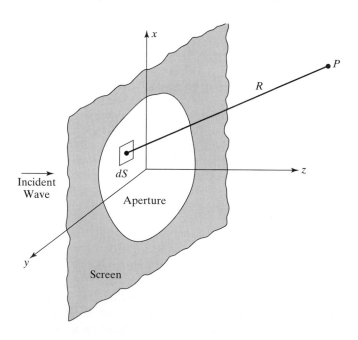

FIGURE 10.22   Geometry pertinent to diffraction by an aperture on a screen.

Example 10.11

Let us assume that the aperture of Fig. 10.22 is a circular hole of radius a having its center at the origin and illuminated by a uniform plane wave of electric-field intensity E_0 at the aperture, as shown in Fig. 10.23. We wish to investigate the diffracted field along the z-axis.   *Diffraction by a circular aperture*

Applying (10.117) to the geometry in Fig. 10.23 and noting the circular symmetry about the z-axis, we obtain

$$\overline{E}(0, 0, z) = \frac{j\beta}{2\pi} \int_{r=0}^{a} \int_{\phi=0}^{2\pi} \frac{E_0}{\sqrt{r^2 + z^2}} e^{-j\beta \sqrt{r^2+z^2}} \, r \, dr \, d\phi \qquad (10.118)$$

Making the change of variable $R = \sqrt{r^2 + z^2}$, and hence $dR = (r \, dr)/R$, we obtain

$$\overline{E}(0, 0, z) = j\beta E_0 \int_{z}^{\sqrt{a^2+z^2}} e^{-j\beta R} \, dR$$

$$= -E_0 e^{-j\beta \sqrt{a^2+z^2}} + E_0 e^{-j\beta z} \qquad (10.119)$$

Proceeding further, the intensity is given by

$$I = \frac{1}{2\eta_0} |\overline{E}|^2 = \frac{1}{2\eta_0} \overline{E}\,\overline{E}^*$$

$$= \frac{E_0^2}{2\eta_0} [2 - 2\cos\beta(\sqrt{a^2 + z^2} - z)] \qquad (10.120)$$

$$- \frac{E_0^2}{\eta_0} \sin^2 \frac{\beta(\sqrt{a^2 + z^2} - z)}{2}$$

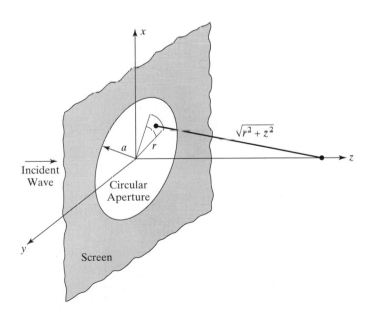

FIGURE 10.23 Geometry pertinent to diffraction by a circular aperture illuminated by a plane wave.

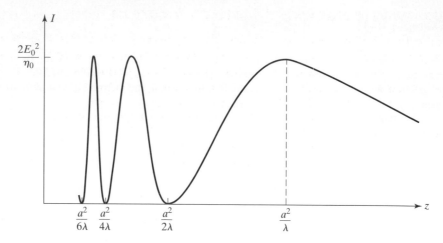

FIGURE 10.24   Variation of the intensity along the z-axis for the arrangement of Fig. 10.23.

For $z \gg a$, (10.120) reduces to

$$
\boxed{
\begin{aligned}
I &\approx \frac{2E_0^2}{\eta_0} \sin^2 \frac{\beta a^2}{4z} \\
&= \frac{2E_0^2}{\eta_0} \sin^2 \frac{\pi a^2}{2\lambda z}
\end{aligned}
}
\tag{10.121}
$$

Fresnel vs. Fraunhofer diffraction   A sketch of the result given by (10.121) is shown in Fig. 10.24. It can be seen that for $z < a^2/\lambda$, the intensity fluctuates between maxima of $2E_0^2/\eta_0$ and minima of zero, corresponding to constructive and destructive interference, respectively, of the spherical waves. The situation is said to correspond to Fresnel diffraction. For $z > a^2/\lambda$, the intensity decreases monotonically. For $z \gg a^2/\lambda$, the diffraction is known as Fraunhofer diffraction and the evaluation of the integral in (10.117) becomes easier, because waves arriving at P from the aperture approach plane waves, thereby permitting the simplifying plane wave approximation in the integrand. We shall however not pursue the topic here but defer the consideration to Section 11.5, where the determination of the far field due to an aperture distribution is identical to that of the solution for Fraunhofer diffraction. In practice, the boundary between the Fresnel and Fraunhofer diffraction regions is taken to be $2D^2/\lambda$, where D is the diameter of the circular aperture, or in the case of a noncircular aperture it is its maximum linear dimension (see Problem P10.30).

K10.6.   Intensity; Two-beam interference; Multiple-beam interference; Finesse; Fabry–Perot etalon; Diffraction; Huygens–Fresnel principle; Circular aperture; Fresnel diffraction; Fraunhofer diffraction.

D10.10.   For a uniform plane wave incident from free space onto a plane dielectric slab of thickness $d = 5.1\lambda_0$ and refractive index $n = 4$ at an angle $\theta_i = 60°$, find F and I_t/I_i for each of the two polarizations: **(a)** perpendicular and **(b)** parallel. *Ans.* **(a)** 6.034, 0.2113; **(b)** 1.226, 0.8665.

10.7   WAVE PROPAGATION IN ANISOTROPIC MEDIUM

In Section 2.5, we learned that for certain dielectric materials known as *anisotropic dielectric materials,* **D** is not in general parallel to **E** and the relationship between **D** and **E** is expressed by means of a permittivity tensor consisting of a 3×3 matrix. Let us consider an anisotropic dielectric medium characterized by the simple **D** to **E** relationship given by

Linear characteristic polarizations

$$\begin{bmatrix} D_x \\ D_y \\ D_z \end{bmatrix} = \begin{bmatrix} \varepsilon_x & 0 & 0 \\ 0 & \varepsilon_y & 0 \\ 0 & 0 & \varepsilon_z \end{bmatrix} \begin{bmatrix} E_x \\ E_y \\ E_z \end{bmatrix} \tag{10.122}$$

and having the permeability μ_0. It is easy to see that the characteristic polarizations for this case are all linear directed along the coordinate axes and having the effective permittivities ε_x, ε_y, and ε_z for the x-, y-, and z-directed polarizations, respectively. These axes are then known as the principal axes. Let us consider a uniform plane wave propagating along one of the principal axes, say, the z-direction. The wave will then generally contain both x- and y-components of the fields. It can be decomposed into two waves, one having an x-directed electric field and the other having a y-directed electric field. These component waves travel individually in the anisotropic medium as though it is isotropic but with different phase velocities since the effective permittivities are different. In view of this, the phase relationship between the two waves, and hence the polarization of the composite wave, changes with distance along the direction of propagation. Also, when they encounter a discontinuity, the component waves undergo reflection and transmission by different amounts. We shall illustrate by means of an example.

Example 10.12

Let us consider a uniform plane wave of frequency 1500 MHz incident from free space ($z < 0$) normally onto an anisotropic perfect dielectric medium ($z > 0$), characterized by the permittivity matrix

Propagation along a principal axis

$$[\varepsilon] = \varepsilon_0 \begin{bmatrix} 4 & 0 & 0 \\ 0 & 9 & 0 \\ 0 & 0 & 4 \end{bmatrix}$$

and $\mu = \mu_0$. We wish to discuss the reflected and transmitted waves for several cases of incident waves.

Case 1. The incident wave has only an x-component of **E** as given by

$$\mathbf{E}_i = E_0 \cos (3 \times 10^9 \pi t - 10\pi z)\, \mathbf{a}_x$$

Then the effective permittivity of the anisotropic medium is $4\varepsilon_0$, and from (5.118) and

(5.119), $\overline{\Gamma} = -\frac{1}{3}$ and $\overline{\tau} = \frac{2}{3}$. The reflected and transmitted wave electric fields are

$$\mathbf{E}_r = -\frac{E_0}{3}\cos(3 \times 10^9\pi t + 10\pi z)\,\mathbf{a}_x$$

$$\mathbf{E}_t = \frac{2E_0}{3}\cos(3 \times 10^9\pi t - 20\pi z)\,\mathbf{a}_x$$

where we have made use of the fact that for the transmitted wave, the phase constant is $\omega\sqrt{\mu_0 \cdot 4\varepsilon_0} = 2\omega\sqrt{\mu_0\varepsilon_0} = 2 \times 10\pi = 20\pi$.

Case 2. The incident wave has only a y-component of **E** as given by

$$\mathbf{E}_i = E_0\cos(3 \times 10^9\,\pi t - 10\pi z)\,\mathbf{a}_y$$

Then the effective permittivity of the anisotropic medium is $9\varepsilon_0$, and from (5.118) and (5.119), $\overline{\Gamma} = -\frac{1}{2}$ and $\overline{\tau} = \frac{1}{2}$. The reflected and transmitted wave electric fields are

$$\mathbf{E}_r = -\frac{E_0}{2}\cos(3 \times 10^9\pi t + 10\pi z)\,\mathbf{a}_y$$

$$\mathbf{E}_t = \frac{E_0}{2}\cos(3 \times 10^9\pi t - 30\pi z)\,\mathbf{a}_y$$

where we have made use of the fact that for the transmitted wave, the phase constant is $\omega\sqrt{\mu_0 \cdot 9\varepsilon_0} = 3\omega\sqrt{\mu_0\varepsilon_0} = 3 \times 10\pi = 30\pi$.

Case 3. The incident wave has both x- and y-components of **E** and is linearly polarized, as given by

$$\mathbf{E}_i = E_1\cos(3 \times 10^9\pi t - 10\pi z)\,\mathbf{a}_x + E_2\cos(3 \times 10^9\pi t - 10\pi z)\,\mathbf{a}_y$$

Then from superposition of cases 1 and 2, the reflected and transmitted wave electric fields are given by

$$\mathbf{E}_r = -\frac{E_1}{3}\cos(3 \times 10^9\pi t + 10\pi z)\,\mathbf{a}_x - \frac{E_2}{2}\cos(3 \times 10^9\pi t + 10\pi z)\,\mathbf{a}_y$$

$$\mathbf{E}_t = \frac{2E_1}{3}\cos(3 \times 10^9\pi t - 20\pi z)\,\mathbf{a}_x + \frac{E_2}{2}\cos(3 \times 10^9\pi t - 30\pi z)\,\mathbf{a}_y$$

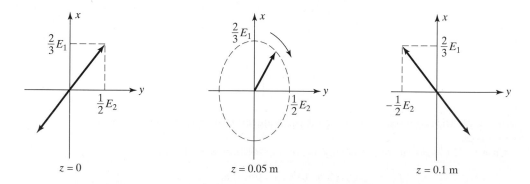

FIGURE 10.25   Change in polarization versus z of the transmitted wave electric field of Example 10.12.

Note that $\mathbf{E}_r$ is linearly polarized, although along a direction making an angle to that of the direction of polarization of $\mathbf{E}_i$. The polarization of $\mathbf{E}_t$, on the other hand, varies with z since the phase difference between the x- and y-components of the electric field is $\Delta\phi = 10\pi z$. As the transmitted wave propagates in the z-direction, $\Delta\phi$ changes from zero at $z = 0$ to $\pi/2$ at $z = 0.05$ m to π at $z = 0.1$ m, and so on. Thus, the polarization changes from linear at $z = 0$ to elliptical for $z > 0$, becoming linear again at $z = 0.1$ m, but rotated by an angle as shown in Fig. 10.25, and so on.

The simple form of permittivity tensor given by (10.122) can be realized for certain anisotropic crystals by an appropriate choice of the coordinate system. If the permittivities ε_x, ε_y, and ε_z are all different, then the crystal is said to be *biaxial.* If two of the three are equal, then it is said to be *uniaxial.*

To generalize the observation in Example 10.12, the phase difference *Wave plates* between the x- and y-components of $\mathbf{E}$ in the anisotropic medium can be expressed as

$$
\begin{aligned}
\Delta\phi &= (\omega\sqrt{\mu_0\varepsilon_x} - \omega\sqrt{\mu_0\varepsilon_y})z \\
&= \frac{\omega}{c}(\sqrt{\varepsilon_{rx}} - \sqrt{\varepsilon_{ry}})z \\
&= \frac{2\pi}{\lambda_0}(n_x - n_y)z
\end{aligned}
\tag{10.123}
$$

where λ_0 is the free-space wavelength, and n_x and n_y are the refractive indices. The result given by (10.123) is the basis behind *wave plates* or *retardation plates.* The word *retardation* refers to the fact that the phase of one of the two components lags that of the second component. If $\Delta\phi = 2m\pi$, where m is an integer, the plate is called a *full-wave plate.* If reflections from the surfaces of the plate are considered to be negligible, as is usually the case, then it can be seen that the state of polarization of the wave at the output plane of the plate is the same as that at the input plane. For $\Delta\phi = (2m + 1)\pi$ and $(2m + 1)\pi/2$, the arrangement corresponds to *half-wave plate* and *quarter-wave plate,* respectively. A half-wave plate results in a rotation of the direction of linear polarization, as illustrated in Fig. 10.25, which corresponds to $\Delta\phi = \pi$. Note that the direction has shifted by twice the angle that it initially makes with the x- (or y-) direction. A quarter-wave plate can transform a linearly polarized wave into a circularly polarized wave.

For a different example of an anisotropic medium than that characterized *Circular* by (10.122), let us consider the $\mathbf{D}$ to $\mathbf{E}$ relationship of the form *characteristic polarizations*

$$
\overline{\mathbf{D}} = \varepsilon\overline{\mathbf{E}} + j\varepsilon_0\gamma\overline{\mathbf{B}} \times \overline{\mathbf{E}}
\tag{10.124}
$$

which is exhibited by certain materials when placed in a static magnetic field $\mathbf{B}$, where γ is a constant depending on the material. For a uniform plane wave propagating in the z-direction, and $\mathbf{B} = B_0\mathbf{a}_z$, we have

$$\overline{D}_x = \varepsilon\overline{E}_x - j\varepsilon_0\gamma B_0\overline{E}_y$$

$$\overline{D}_y = j\varepsilon_0\gamma B_0\overline{E}_x + \varepsilon\overline{E}_y$$

To find the characteristic polarizations, we set $\overline{D}_x/\overline{D}_y = \overline{E}_x/\overline{E}_y$. Thus,

$$\frac{\varepsilon\overline{E}_x - j\varepsilon_0\gamma B_0\overline{E}_y}{j\varepsilon_0\gamma B_0\overline{E}_x + \varepsilon\overline{E}_y} = \frac{\overline{E}_x}{\overline{E}_y}$$

Solving for $\overline{E}_x/\overline{E}_y$, we get

$$\boxed{\frac{\overline{E}_x}{\overline{E}_y} = \pm j}$$

This result corresponds to equal amplitudes of E_x and E_y and phase differences of $\pm 90°$. Thus, the characteristic polarizations are both circular, rotating in opposite senses as viewed along the z-direction.

The effective permittivities of the medium corresponding to the characteristic polarizations are

$$\frac{\overline{D}_x}{\overline{E}_x} = \frac{\varepsilon\overline{E}_x - j\varepsilon_0\gamma B_0\overline{E}_y}{\overline{E}_x}$$

$$= \varepsilon - j\varepsilon_0\gamma B_0\frac{\overline{E}_y}{\overline{E}_x} \qquad (10.125)$$

$$= \varepsilon \mp \varepsilon_0\gamma B_0 \qquad \text{for} \quad \frac{\overline{E}_x}{\overline{E}_y} = \pm j$$

The phase constants associated with the propagation of the characteristic waves are

$$\boxed{\beta_\pm = \omega\sqrt{\mu(\varepsilon \mp \varepsilon_0\gamma B_0)}} \qquad (10.126)$$

where the subscripts $+$ and $-$ refer to $\overline{E}_x/\overline{E}_y = +j$ and $\overline{E}_x/\overline{E}_y = -j$, respectively.

Let us now consider the electric field of the wave to be linearly polarized in the x-direction at $z = 0$, that is,

$$\mathbf{E}(0) = E_0\cos\omega t\,\mathbf{a}_x \qquad (10.127)$$

Then we can express (10.127) as the superposition of two circularly polarized fields having opposite senses of rotation in the xy-plane in the manner

$$\mathbf{E}(0) = \left(\frac{E_0}{2}\cos\omega t\,\mathbf{a}_x + \frac{E_0}{2}\sin\omega t\,\mathbf{a}_y\right) + \left(\frac{E_0}{2}\cos\omega t\,\mathbf{a}_x - \frac{E_0}{2}\sin\omega t\,\mathbf{a}_y\right) \quad (10.128)$$

The circularly polarized field inside the first pair of parentheses on the right side of (10.128) corresponds to

$$\frac{\overline{E}_x}{\overline{E}_y} = \frac{E_0/2}{-jE_0/2} = +j$$

whereas that inside the second pair of parentheses corresponds to

$$\frac{\overline{E}_x}{\overline{E}_y} = \frac{E_0/2}{jE_0/2} = -j$$

Assuming propagation in the positive z-direction, the field at an arbitrary value of z is then given by

$$
\begin{aligned}
\mathbf{E}(z) &= \left[\frac{E_0}{2} \cos \left(\omega t - \beta_+ z \right) \mathbf{a}_x + \frac{E_0}{2} \sin \left(\omega t - \beta_+ z \right) \mathbf{a}_y \right] \\
&\quad + \left[\frac{E_0}{2} \cos \left(\omega t - \beta_- z \right) \mathbf{a}_x - \frac{E_0}{2} \sin \left(\omega t - \beta_- z \right) \mathbf{a}_y \right] \\
&= \left[\frac{E_0}{2} \cos \left(\omega t - \frac{\beta_+ + \beta_-}{2} z - \frac{\beta_+ - \beta_-}{2} z \right) \mathbf{a}_x \right. \\
&\quad \left. + \frac{E_0}{2} \sin \left(\omega t - \frac{\beta_+ + \beta_-}{2} z - \frac{\beta_+ - \beta_-}{2} z \right) \mathbf{a}_y \right] \\
&\quad + \left[\frac{E_0}{2} \cos \left(\omega t - \frac{\beta_+ + \beta_-}{2} z + \frac{\beta_+ - \beta_-}{2} z \right) \mathbf{a}_x \right. \\
&\quad \left. - \frac{E_0}{2} \sin \left(\omega t - \frac{\beta_+ + \beta_-}{2} z + \frac{\beta_+ - \beta_-}{2} z \right) \mathbf{a}_y \right] \\
&= \left[E_0 \cos \left(\frac{\beta_- - \beta_+}{2} z \right) \mathbf{a}_x \right. \\
&\quad \left. + E_0 \sin \left(\frac{\beta_- - \beta_+}{2} z \right) \mathbf{a}_y \right] \cos \left(\omega t - \frac{\beta_+ + \beta_-}{2} z \right)
\end{aligned}
$$

(10.129)

The result given by (10.129) indicates that the x- and y-components of the field are in phase at any given value of z. Hence, the field is linearly polarized for all values of z. The direction of polarization is, however, a function of z since $\tan^{-1} (E_y/E_x)$, the angle made by the field vector with the x-axis, is $[(\beta_- - \beta_+)/2]z$. Thus, the direction of polarization rotates linearly with z at a rate of $(\beta_- - \beta_+)/2$. This phenomenon is known as *Faraday rotation* and is illustrated with the aid of the sketches in Fig. 10.26. The sketches in any given column correspond to a fixed value of z whereas the sketches in a given row correspond to a fixed value of t. At $z = 0$, the field is linearly polarized in the x-direction and is the superposition of two counterrotating circularly polarized fields, as shown by the time series of sketches in the first column. If the medium is isotropic, the two counterrotating circularly polarized fields undergo the same amount of phase lag with z and the field remains linearly polarized in the x-direction, as shown by the dashed lines in the second and third columns. For the case of the anisotropic medium, the two circularly polarized fields undergo

Faraday rotation

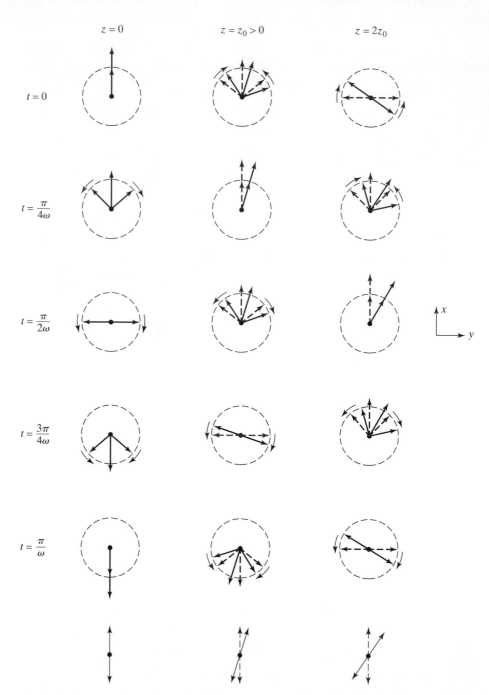

FIGURE 10.26   For illustrating the phenomenon of Faraday rotation.

different amounts of phase lag with z. Hence, their superposition results in a linear polarization making an angle with the x-direction and increasing linearly with z, as shown by the solid lines in the second and third columns.

The phenomenon of Faraday rotation that we have just discussed forms the basis for a number of devices. A simple example is illustrated by the magneto-optical switch. The magneto-optical switch is a device for modulating a laser beam by switching an electric current on and off. The electric current generates a magnetic field that rotates the magnetization vector in a magnetic iron-garnet film on a substrate of garnet in the plane of the film through which a light wave passes. When it enters the film, the light wave field is linearly polarized normal to the plane of the film. If the current in the electric circuit is off, the magnetization vector is normal to the direction of propagation of the wave and the wave emerges out of the film without change of polarization, as shown in Fig. 10.27(a). If the current in the electric circuit is on, the magnetization vector is parallel to the direction of propagation of the wave, the light wave undergoes Faraday rotation and emerges out of the film with its polarization rotated by 90°, as shown in Fig. 10.27(b). After it emerges out of the film, the light beam is passed through a polarizer, which has the property of absorbing light of the original polarization but passing through the light of the 90°-rotated polarization. Thus, the beam is made to turn on and off by the switching on and off of the current in the electric circuit. In this manner, any coded message can be made to be carried by the light beam.

Magneto-optical switch

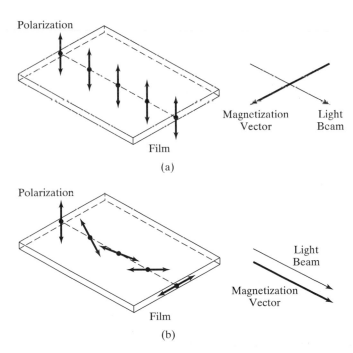

FIGURE 10.27   For illustrating the principle of operation of a magneto-optical switch.

K10.7. Anisotropic dielectric materials; Characteristic polarizations; Wave plates; Faraday rotation.

D10.11. At $\lambda_0 = 0.633$ μm, the refractive indices of mica are given by $n_x = 1.594$ (fast axis) and $n_y = 1.599$ (slow axis). Find the following: **(a)** the minimum thickness of a mica sheet to act as a half-wave plate; **(b)** the number of wavelengths undergone by the wave in the thickness of the plate for the x-polarization; and **(c)** the number of wavelengths undergone by the wave in the thickness of the plate for the y-polarization.

Ans. **(a)** 63.3 μm; **(b)** 159.4; **(c)** 159.9.

10.8   SUMMARY

In this chapter, we extended the treatment of the parallel-plate waveguide to the rectangular waveguide, which is a metallic pipe of rectangular cross section. By considering a rectangular waveguide of cross-sectional dimensions a and b, we discussed transverse electric, or TE, modes, as well as transverse magnetic or TM modes, and learned that although $\text{TE}_{m,n}$ modes can include values of m or n equal to zero, $\text{TM}_{m,n}$ modes require that both m and n be nonzero, where m and n refer to the number of one-half sinusoidal variations of the fields along the dimensions a and b, respectively. The cutoff wavelengths for the $\text{TE}_{m,n}$ or $\text{TM}_{m,n}$ modes are given by

$$\lambda_c = \frac{1}{\sqrt{(m/2a)^2 + (n/2b)^2}}$$

The mode that has the largest cutoff wavelength or the lowest cutoff frequency is the dominant mode, which here is the $\text{TE}_{1,0}$ mode.

By placing perfect conductors in two transverse planes of a rectangular waveguide separated by an integer multiple of one-half the guide wavelength, we introduced the cavity resonator, which is the microwave counterpart of the lumped parameter resonant circuit encountered in low-frequency circuit theory. For a rectangular cavity resonator having dimensions a, b, and d, the frequencies of oscillation for the $\text{TE}_{m,n,l}$ or $\text{TM}_{m,n,l}$ modes are given by

$$f_{osc} = \frac{1}{\sqrt{\mu\varepsilon}} \sqrt{\left(\frac{m}{2a}\right)^2 + \left(\frac{n}{2b}\right)^2 + \left(\frac{l}{2d}\right)^2}$$

where l refers to the number of one-half sinusoidal variations of the fields along the dimension d.

Next we introduced the cylindrical waveguide, which is a metallic pipe of cylindrical cross section. We learned that guided modes in the cylindrical waveguide are characterized by field variations in the radial direction in accordance with Bessel functions and sinusoidal variations in the angular direction. The modes are designated as $\text{TE}_{n,m}$ and $\text{TM}_{n,m}$, where the first subscript refers to the angular variations and the second to the radial variations. The $\text{TE}_{1,1}$ mode is the dominant mode. We also discussed the cylindrical cavity resonator formed by placing perfect conductors in two transverse planes of the guide, as in the case of the rectangular cavity resonator.

We then considered losses in waveguides and resonators and discussed by means of examples the determination of attenuation constant for a propagating mode in a waveguide and the Q factor, a measure of frequency selectivity, for an oscillating mode in a resonator.

Proceeding further, we introduced the optical fiber, which consists typically of a core and a cladding having circular cross sections. By assuming the cladding region to extend to infinity so that the situation corresponds to one of a cylindrical dielectric rod, we carried out the field analysis and learned that in addition to TE ($E_z = 0$) and TM ($H_z = 0$) modes, the fields correspond to the so-called hybrid modes, designated HE and EH. For the hybrid modes, both E_z and H_z are not equal to zero; however, for the HE modes, H_z is predominant, whereas for the EH modes, E_z is predominant. In fact, the dominant mode is the $HE_{1,1}$ mode, having no cutoff. For single-mode operation at a given wavelength λ_0, the condition is given by

$$V = \frac{2\pi a}{\lambda_0} (NA) < 2.405$$

where a is the radius of the core, and NA is the numerical aperture given by

$$NA = \sqrt{n_1^2 - n_2^2}$$

n_1 and n_2 being the refractive indices of the core and the cladding, respectively. For weak guidance ($n_1 \approx n_2$), the modes are designated LP, with $LP_{0,1}$ mode corresponding to the $HE_{1,1}$ mode.

We then extended our discussion of dispersion in the previous chapter to the propagation of a signal with pulse envelope to study the phenomenon of pulse broadening in a dispersive medium. In particular, we considered a signal with a Gaussian pulse envelope and derived the expression for the *pulse broadening parameter,* which governs the pulse spread with distance in the medium. Following a brief discussion of the types of dispersion in optical waveguides, we derived the expression for the pulse broadening parameter for the case of material dispersion and discussed dispersion characteristics due to material dispersion in fused silica.

Next we turned our attention to two related topics, interference and diffraction, which are both based on superposition of waves. Although the distinction is not sharp, interference usually applies to the interaction of only a few light beams with one another, whereas diffraction usually pertains to the superposition of a large number, even a continuous distribution, of beams. We discussed interference by considering (a) the two-beam interference experiment of Thomas Young and (b) multiple-beam interference due to plane wave incidence obliquely on a plane dielectric slab, the latter arrangement constituting a model for the Fabry–Perot etalon or interferometer. For diffraction, we introduced the Huygens–Fresnel principle, according to which each point on a wavefront generates a spherical wave, and illustrated its application by considering the example of a plane wave incident on a circular aperture in a screen. By investigating diffraction along the axis of the aperture, we discussed briefly Fresnel versus Fraunhofer diffraction.

Finally, we discussed the topic of wave propagation in an anisotropic medium. By considering the example of a uniform plane wave incident normally along a principal axis of a uniaxial crystal, thereby resulting in the characteristic polarizations to be linear, we illustrated the principle behind wave plates or retardation plates. By means of another example of an anisotropic medium for which the characteristic polarizations are circular, we introduced Faraday rotation, which is the phenomenon of rotation of the direction of polarization of a linearly polarized wave as it propagates in the medium, and discussed the operation of a magneto-optical switch, a device employing Faraday rotation for modulating a light beam.

REVIEW QUESTIONS

Q10.1. Discuss the nomenclature associated with the modes of propagation in a rectangular waveguide.

Q10.2. Explain the relationship between the cutoff wavelength and the dimensions of a rectangular waveguide based on the phenomenon at cutoff.

Q10.3. Briefly outline the procedure for deriving the expressions for TE mode fields in a rectangular waveguide.

Q10.4. Compare and contrast TE and TM modes in a rectangular waveguide.

Q10.5. What is the dominant mode? Which one of the rectangular waveguide modes is the dominant mode?

Q10.6. Why is the dimension b of a rectangular waveguide generally chosen to be less than or equal to one-half the dimension a?

Q10.7. What is a cavity resonator?

Q10.8. How do the dimensions of a rectangular cavity resonator determine the frequencies of oscillation of the resonator?

Q10.9. Briefly outline the procedure for deriving the expressions for the TE and TM mode fields in a cylindrical waveguide.

Q10.10. Compare and contrast TE and TM modes in a cylindrical waveguide.

Q10.11. Which one of the cylindrical waveguide modes is the dominant mode?

Q10.12. discuss the basis for the computation of power loss associated with slightly imperfect conductors comprising the walls of a waveguide.

Q10.13. Briefly outline the procedure for the determination of the attenuation constant for a propagating mode in a waveguide with slightly imperfect conductors.

Q10.14. How is the Q factor of a resonator defined? Briefly outline the procedure for the determination of the Q factor of a cavity resonator with slightly imperfect conductors.

Q10.15. Provide a brief description of the optical fiber.

Q10.16. Outline the steps involved in obtaining the guidance condition for a wave along a cylindrical dielectric rod surrounded by a cladding region extending to infinity.

Q10.17. What are hybrid modes? Why do they arise for guided waves in an optical fiber but not for those in a cylindrical metallic waveguide?

Q10.18. Discuss the condition for single-mode operation of an optical fiber.

Q10.19. Discuss the heuristic explanation for pulse broadening in a dispersive medium.

Q10.20. Outline the analysis for a signal with a Gaussian envelope for studying pulse broadening in a dispersive medium.

Q10.21. Discuss the *pulse broadening parameter* associated with the propagation of a signal with Gaussian envelope in a dispersive medium.

Q10.22. Briefly discuss the types of dispersion in optical waveguides.

Q10.23. Outline the derivation of the pulse broadening parameter for material dispersion.

Q10.24. Discuss the *dispersion coefficient* associated with material dispersion, by considering the example of fused silica.

Q10.25. What is interference? Under what condition(s) do two waves not produce interference?

Q10.26. Describe Young's two-beam interference experiment.

Q10.27. Discuss the phenomenon of multiple-beam interference resulting from the incidence of a uniform plane wave obliquely on a plane dielectric slab and its application to the Fabry–Perot etalon.

Q10.28. What is diffraction? Compare and contrast the phenomenon of diffraction with the principle of geometrical optics.

Q10.29. Describe the Huygens–Fresnel principle for the solution of a diffraction problem.

Q10.30. Using the example of diffraction along the axis of a circular aperture in a plane screen, discuss Fresnel versus Fraunhofer diffraction.

Q10.31. When does a wave propagate in an anisotropic medium without change in its polarization?

Q10.32. Discuss the principle behind wave plates, providing specific examples.

Q10.33. What is Faraday rotation? When does Faraday rotation take place in an anisotropic medium?

Q10.34. Consult appropriate reference books and list three applications of Faraday rotation.

Q10.35. What is a magneto-optical switch? Discuss its operation.

PROBLEMS

Section 10.1

P10.1. For a rectangular waveguide of dimensions $a = 3.75$ cm and $b = 1.25$ cm, and having a dielectric of $\varepsilon = 6.25\varepsilon_0$ and $\mu = \mu_0$, find all propagating modes for $f = 5000$ MHz, and for each mode, find the values of β_z, λ_g, v_{pz}, and η_g.

P10.2. **Design Problem.** Consider propagation of TM waves of frequency $f = 6000$ MHz in an air-dielectric waveguide of square cross section ($b = a$). Find the range of a for which the $TM_{1,1}$ mode propagates with a 20% safety factor ($f > 1.20f_c$) but also such that f is at least 20% below the f_c of the next higher-order mode.

P10.3. A rectangular waveguide of dimensions $a = 3$ cm and $b = 1.5$ cm has a dielectric discontinuity, as shown in Fig. 10.28. A $TE_{1,0}$ wave of frequency 6000 MHz is incident on the discontinuity from the free-space side. **(a)** Find the SWR in the free-space section. **(b)** Find the length and the permittivity of a quarter-wave section required to achieve a match between the two media. Assume $\mu = \mu_0$ for the quarter-wave section.

FIGURE 10.28   For Problem P10.3.

P10.4. A rectangular waveguide of dimensions $a = 4$ cm and $b = 2$ cm has dielectric discontinuity, as shown in Fig. 10.29. A $TM_{1,1}$ wave of frequency 10,000 MHz is incident on the discontinuity form the free-space side. **(a)** Find the SWR in the free-space section. **(b)** Find the length and the permittivity of a quarter-wave section required to achieve a match between the two media. Assume $\mu = \mu_0$ for the quarter-wave section.

FIGURE 10.29   For Problem P10.4.

P10.5. Consider the use of the alternated transformer arrangement (see Problem P7.23) to achieve a match between two sections of a rectangular waveguide, as shown in Fig. 10.30. Find the minimum values of l_2 and l_3 in centimeters to achieve the desired match for $TE_{1,0}$ mode at a frequency $f = 6000$ MHz.

FIGURE 10.30   For Problem P10.5.

P10.6. A dielectric slab of thickness 4 cm and permittivity $2.25\varepsilon_0$ exists in an air-dielectric rectangular waveguide of dimensions $a = 3$ cm and $b = 1.5$ cm, as shown in Fig. 10.31. Find the lowest frequency for which the dielectric slab is transparent (i.e., allows complete transmission) for $TE_{1,0}$ mode propagation in the waveguide.

FIGURE 10.31   For Problem P10.6.

P10.7. For a rectangular cavity resonator having the dimensions $a = 2.5$ cm, $b = 2$ cm, and $d = 5$ cm, and filled with a dielectric of $\varepsilon = 2.25\varepsilon_0$ and $\mu = \mu_0$, find the five lowest frequencies of oscillation. Identify the mode(s) for each frequency.

P10.8. For an air-dielectric cubical cavity resonator having the dimensions $a = b = d = 5$ cm, find the three lowest frequencies of oscillation. Identify the mode(s) for each frequency.

Section 10.2

P10.9. For a cylindrical waveguide of radius $a = 3$ cm and having a dielectric of $\varepsilon = 2.25\varepsilon_0$ and $\mu = \mu_0$, find the propagating modes for a signal of frequency 5.0 GHz. For each of the propagating modes, find and tabulate the values of f_c, λ_c, λ_g, v_{pz}, and η_g, as in Table 10.4.

P10.10. An air-dielectric waveguide has the cross section shown in Fig. 10.32. For the radius $a = 3$ cm, determine the lowest two cutoff frequencies and identify the corresponding modes.

FIGURE 10.32 For Problem P10.10.

P10.11. Consider the use of the alternated transformer arrangement (see Problem P7.23) to achieve a match between two sections of a cylindrical waveguide, as shown in Fig. 10.33. Find the minimum values of l_2 and l_3 in centimeters to achieve the desired match for $TE_{1,1}$ mode at a frequency $f = 3$ GHz.

FIGURE 10.33 For Problem P10.11.

P10.12. Repeat Problem P10.11 for $TM_{0,1}$ mode.

P10.13. A cylindrical cavity resonator is formed by placing two perfectly conducting sheets 4 cm apart in the cross-sectional planes of the cylindrical waveguide of Problem P10.9. Find the five lowest frequencies of oscillation and identify the mode(s) of oscillation for each frequency.

P10.14. Consider $TE_{n,m}$ modes bouncing between the walls $z = 0$ and $z = 2d$ of a cylindrical cavity resonator of radius a with the end regions $0 < z < t$ and

$(2d - t) < z < 2d$ filled with a nonmagnetic ($\mu = \mu_0$), perfect dielectric of permittivity ε_1, and the region $t < z < (2d - t)$ filled with a nonmagnetic, perfect dielectric of permittivity ε_2. **(a)** Obtain the condition for oscillation. **(b)** Compute the lowest resonant frequency for $a = 5$ cm, $d = 2$ cm, $t = 1$ cm, $\varepsilon_1 = 4\varepsilon_0$, and $\varepsilon_2 = \varepsilon_0$.

Section 10.3

P10.15. For a parallel-plate waveguide with imperfect but good conductors of conductivity σ and spacing a, show that the attenuation constant α_c for TEM wave propagation along the guide is $1/\sigma\delta a\eta$. Compute the value of α_c for $a = 5$ cm, $f = 5000$ MHz, copper plates, and air-dielectric.

P10.16. Repeat Problem P10.15 for TE wave propagation to show that α_c is equal to $2(f_c/f)^2/[\sigma\delta a\eta\sqrt{1 - (f_c/f)^2}]$ and compute the value of α_c for the data specified in that problem and for the $TE_{1,0}$ mode.

P10.17. For a parallel-plate resonator consisting of two infinite, plane, perfectly conducting plates in the $z = 0$ and $z = l$ planes and separated by a perfect dielectric, as shown in Fig. 10.34, the electric and magnetic fields are given by

$$\mathbf{E} = E_0 \sin \frac{n\pi z}{l} \sin \frac{n\pi t}{l\sqrt{\mu\varepsilon}} \mathbf{a}_x$$

$$\mathbf{H} = \frac{E_0}{\sqrt{\mu/\varepsilon}} \cos \frac{n\pi z}{l} \cos \frac{n\pi t}{l\sqrt{\mu\varepsilon}} \mathbf{a}_y$$

where $n = 1, 2, 3, \ldots$. **(a)** Show that the energy stored in the resonator per unit area of the plates is $\frac{1}{4}\delta E_0^2 l$. **(b)** If the plates are made of imperfect but good

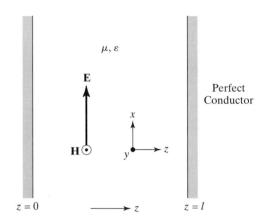

FIGURE 10.34   For Problem P10.17.

conductors, show that the Q of the resonator is $l/2\delta$. **(c)** Compute the value of Q for the fundamental mode of oscillation ($n = 1$), for $l = 1$ cm, assuming air-dielectric and copper plates.

P10.18. For the parallel-plate resonator of Problem P10.17, assume that the dielectric is slightly lossy with conductivity $\sigma_d \ll \omega\varepsilon$. **(a)** Assuming the plates to be perfect conductors, show that the Q of the resonator is given by $Q_1 = \omega\varepsilon/\sigma_d$. **(b)** If in addition to the slightly lossy dielectric, the plates are made up of imperfect but good conductors, show that the Q of the resonator is given by

$$\frac{1}{Q} = \frac{1}{Q_1} + \frac{1}{Q_2}$$

where Q_1 is as given in part (a) and Q_2 is equal to $l/2\delta$, as in Problem P10.17.

P10.19. Obtain the expression for the Q factor for $TM_{m,n,l}$ ($l \neq 0$) mode in a cubical cavity resonator of sides a and show that it is equal to $a/4\delta$.

Section 10.4

P10.20. Supply the missing steps in the derivation of the characteristic equation (10.81) for the case of $l \neq 0$ from the boundary conditions $\overline{E}_{z1} = \overline{E}_{z2}$, $\overline{H}_{z1} = \overline{H}_{z2}$, $\overline{E}_{z1}/\overline{H}_{\phi1} = \overline{E}_{z2}/\overline{H}_{\phi2}$, and $\overline{E}_{\phi1}/\overline{H}_{z1} = \overline{E}_{\phi2}/\overline{H}_{z2}$ at $r = a$.

P10.21. Show that the guidance conditions (10.79)–(10.81) are consistent with the boundary conditions for r-components of the fields given by (10.77c), (10.77d), (10.78c), and (10.78d).

P10.22. Assume that a wave is incident from air onto the core of an optical fiber at an angle θ_a, as shown by the cross-sectional view in Fig. 10.35. Show that the maximum allowable value of θ_a for guiding of the wave in the core by total internal reflection is given by

$$\sin [\theta_a]_{max} = \sqrt{n_1^2 - n_2^2}$$

which is defined to be the numerical aperture (NA) of the optical fiber. This provides the physical interpretation for NA.

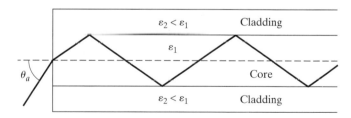

FIGURE 10.35   For Problem P10.22.

P10.23. Supply the missing steps in the derivation of (10.86) from (10.81) by using (10.85a)–(10.85d). [*Hint:* Use (10.85b) and (10.85c) for the left side of (10.81) and (10.85a) and (10.85d) for the right side of (10.81).]

P10.24. For an optical fiber with core and cladding refractive indices $n_1 = 1.50$ and $n_2 = 1.40$, find the maximum value of a/λ_0 for single-mode operation.

Section 10.5

P10.25. With reference to the result given by (10.96), show that there is an optimum pulse half-width for which the transmitted pulse half-width at a given value of z is minimized. Find the expression for this optimum pulse half-width and the corresponding transmitted pulse half-width.

P10.26. A measure of the pulse spread due to intermodal dispersion is provided by the difference in travel times between the longest ray path and the shortest ray path. Show that this difference per unit distance along the fiber is approximately equal to Δ/c, where $\Delta = (n_1 - n_2)$.

P10.27. A signal with a Gaussian pulse envelope of half-width 5 ps propagates in an optical fiber made of fused silica with the characteristics shown in Fig. 10.18. If the wavelength of the signal is 0.8 μm, obtain approximate values for **(a)** the time taken by the signal to propagate 1 km and **(b)** the half-width of the Gaussian pulse envelope of the signal at the distance of 1 km.

Section 10.6

P10.28. Verify that the Fresnel coefficients given in Section 9.4 satisfy (10.112a) and (10.112b) for both cases of the polarization of $\mathbf{E}_i$: **(a)** perpendicular and **(b)** parallel. Further show that (10.112b) is consistent with conservation of power flow normal to the dielectric slab.

P10.29. Assume that the limiting frequency resolution of the Fabry–Perot etalon is defined as the separation between the two frequencies at which I_t/I_i is $\frac{1}{2}$. Then, assuming further that this frequency difference is small compared to the frequency range, $\Delta f = c/(2nd \cos \theta_t)$, between two adjacent peaks of the interference pattern for fixed d and θ_t, show that it is equal to $\Delta f/F$.

P10.30. The boundary between the Fresnel and Fraunhofer diffraction regions is determined by the maximum allowable departure of the phase of the waves emanating from the aperture and arriving at the observation point (and, hence, vice versa) from that of a plane wave. For a maximum allowable departure of $\pi/8$, which is the value used in practice, show by considering the circular aperture that the boundary between the two regions is $2D^2/\lambda$, where D is the diameter of the hole.

Section 10.7

P10.31. For all three cases in Example 10.12, find the expressions for the incident, reflected, and transmitted wave magnetic fields.

P10.32. Show that for plane wave propagation in an anisotropic medium, the angle between $\mathbf{E}$ and $\mathbf{H}$ is not in general equal to $90°$. For the anisotropic medium of Example 10.12, find the angle between $\mathbf{E}$ and $\mathbf{H}$ at $z = 0+$, for $E_2 = E_1 = E_0$ in case 3.

P10.33. Medium 1 ($z < 0$) is free space, whereas medium 2 ($z > 0$) is a nonmagnetic ($\mu = \mu_0$), anisotropic perfect dielectric characterized by

$$[\varepsilon] = \varepsilon_0 \begin{bmatrix} 6.25 & 0 & 0 \\ 0 & 2.25 & 0 \\ 0 & 0 & 6.25 \end{bmatrix}$$

For a uniform plane wave having the electric field

$$\mathbf{E}_i = E_0[\cos (6\pi \times 10^8 t - 2\pi z)\, \mathbf{a}_x + \sin (6\pi \times 10^8 t - 2\pi z)\, \mathbf{a}_y]$$

incident on the interface $z = 0$ from medium 1, find the following: **(a)** the reflected wave electric and magnetic fields; and **(b)** the transmitted wave electric and magnetic fields.

P10.34. Medium 1 $(z < 0)$ is free space, whereas medium 2 $(z > 0)$ is a nonmagnetic $(\mu = \mu_0)$, anisotropic perfect dielectric characterized by

$$[\varepsilon] = \varepsilon_0 \begin{bmatrix} 8 & 2 & 0 \\ 2 & 5 & 0 \\ 0 & 0 & 4 \end{bmatrix}$$

A uniform plane wave having the electric field

$$\mathbf{E}_i = E_0 \cos (6\pi \times 10^9 t - 20\pi z)\, \mathbf{a}_x$$

is incident on the interface $z = 0$ from medium 1. Obtain the reflected and transmitted wave electric and magnetic fields.

P10.35. In Fig. 10.36, medium 2 is a nonmagnetic $(\mu = \mu_0)$, anisotropic perfect dielectric characterized by

$$[\varepsilon] = \varepsilon_0 \begin{bmatrix} 4 & 0 & 0 \\ 0 & 16 & 0 \\ 0 & 0 & 4 \end{bmatrix}$$

A circularly polarized uniform plane wave having the electric field

$$\mathbf{E}_i = E_0[\cos (6\pi \times 10^9 t - 2\pi z)\, \mathbf{a}_x + \sin (6\pi \times 10^9 t - 2\pi z)\, \mathbf{a}_y]$$

FIGURE 10.36   For Problem P10.35.

is incident normally onto medium 2 from medium 1. **(a)** Find the minimum value of L, the thickness of medium 2, for which the reflected wave is linearly polarized. **(b)** Find the reflected wave electric field for the value of L in (a). **(c)** Find the minimum value of L for which there is no reflected wave.

P10.36.   Show that for $\gamma B_0 \ll \varepsilon/\varepsilon_0$, the Faraday rotating power $(\beta_- - \beta_+)/2$ of the medium characterized by (10.124) is approximately equal to $\pi\gamma B_0/(\lambda_0\sqrt{\varepsilon/\varepsilon_0})$.

REVIEW PROBLEMS

R10.1.   Show that for a waveguide of square cross section $(a = b)$, the number of propagating modes at a frequency f many times the cutoff frequency of the dominant mode is approximately equal to $2\pi\mu\varepsilon f^2 a^2$.

R10.2.   A rectangular waveguide system consisting of two sections is shown in Fig. 10.37. For $\mathrm{TE}_{1,0}$ waves of frequency 5000 MHz incident from Section 1 onto the discontinuity, find the SWR in Section 1. It is desired to minimize the SWR in Section 1 by placing a $\lambda_g/4$ section of $\varepsilon = 4\varepsilon_0$ and $\mu = \mu_0$ at a distance d to the left of the discontinuity. **(a)** What is the length of the $\lambda_g/4$ section? **(b)** Using the Smith chart, find the minimum value of d and the value of the minimum SWR to the left of the $\lambda_g/4$ section.

FIGURE 10.37   For Problem R10.2.

R10.3.   Consider an air-dielectric cavity resonator made up by placing conductors in the $z = 0$ and $z = d$ planes of the waveguide of Problem P10.10. Determine the two lowest frequencies of oscillation for $a = 3$ cm and $d = 5$ cm, and identify the corresponding modes.

R10.4.   Repeat Problem P10.15 for TM wave propagation to show that α_c is equal to $2/[\sigma\delta a\eta\sqrt{1 - (f_c/f)^2}]$ and compute the value of α_c for the data specified in that problem and for the $\mathrm{TM}_{1,0}$ mode.

R10.5.   For an optical fiber with core and cladding refractive indices $n_1 = 3.503$ and $n_2 = 3.5$, find the maximum value of the core radius for single-mode operation at $\lambda_0 = 1.2\ \mu\mathrm{m}$.

R10.6.   A light-emitting diode at the wavelength 0.9 μm with spectral width 25 nm excites a single-mode fiber made of fused silica with the characteristics shown in Fig. 10.18. Find the approximate pulse width after broadening due to material dispersion over a distance of 1 km.

R10.7. Consider the arrangement complementary to that in Example 10.11, that is, a circular disk of radius a having its center at the origin and illuminated by a plane wave of electric field intensity E_0. Obtain the expression for the diffracted field along the axis of the disk and show that the intensity is a constant, independent of distance from the center of the disk. (*Hint:* Use the fact that for two complementary screens placed together in the same plane, no aperture exists so that no diffraction results.)

R10.8. In Fig. 10.38, medium 2 is a nonmagnetic ($\mu = \mu_0$), anisotropic perfect dielectric characterized by

$$[\varepsilon] = \varepsilon_0 \begin{bmatrix} 8 & 2 & 0 \\ 2 & 5 & 0 \\ 0 & 0 & 9 \end{bmatrix}$$

For a uniform plane wave with the electric field

$$\mathbf{E}_i = E_0 \mathbf{a}_x \cos (6\pi \times 10^8 t - 2\pi z)$$

incident on medium 2 normally on medium 1, find the following: **(a)** the reflected wave electric and magnetic fields if the thickness l of medium 2 is $\frac{1}{4}$ m and **(b)** the minimum value of l for which medium 2 is transparent for the incident wave.

| Medium 1 | Medium 2 | Medium 3 |
| μ_0, ε_0 | $\mu_0, [\varepsilon]$ | μ_0, ε_0 |

(+)

l

FIGURE 10.38 For Problem R10.8.

CHAPTER 11

Principles of Radiation
and Antennas

In Chapters 5, 6, 7, 9, and 10, we studied the principles and applications of propagation and transmission of electromagnetic waves. The remaining important topic pertinent to electromagnetic wave phenomena is radiation of electromagnetic waves. We have, in fact, touched on the principle of radiation of electromagnetic waves in Chapter 5 when we derived the electromagnetic field due to the infinite plane sheet of time-varying, spatially uniform current density. We learned that the current sheet gives rise to uniform plane waves radiating away from the sheet to either side of it. We pointed out at that time that the infinite plane current sheet is, however, an idealized, hypothetical source. With the experience gained thus far in our study of the elements of engineering electromagnetics, we are now in a position to learn the principles of radiation from physical antennas, which is our goal in this chapter.

We begin the chapter with the derivation of the electromagnetic field due to an elemental wire antenna, known as the *Hertzian dipole*. After studying the radiation characteristics of the Hertzian dipole, we consider the example of a half-wave dipole to illustrate the use of superposition to represent an arbitrary wire antenna as a series of Hertzian dipoles to determine its radiation fields. We also discuss the principles of arrays of physical antennas and the concept of image antennas to take into account ground effects. Next we study radiation from aperture antennas. Finally, we consider briefly the receiving properties of antennas and learn of their reciprocity with the radiating properties.

11.1 HERTZIAN DIPOLE

The *Hertzian dipole* is an elemental antenna consisting of an infinitesimally long piece of wire carrying an alternating current $I(t)$, as shown in Fig. 11.1. To maintain the current flow in the wire, we postulate two point charges $Q_1(t)$ and

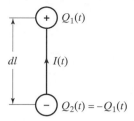

FIGURE 11.1 Hertzian dipole.

$Q_2(t)$ terminating the wire at its two ends, so that the law of conservation of charge is satisfied. Thus, if

$$I(t) = I_0 \cos \omega t \tag{11.1}$$

then

$$\frac{dQ_1}{dt} = I(t) = I_0 \cos \omega t \tag{11.2a}$$

$$\frac{dQ_2}{dt} = -I(t) = -I_0 \cos \omega t \tag{11.2b}$$

and

$$Q_1(t) = \frac{I_0}{\omega} \sin \omega t \tag{11.3a}$$

$$Q_2(t) = -\frac{I_0}{\omega} \sin \omega t = -Q_1(t) \tag{11.3b}$$

The time variations of I, Q_1, and Q_2, given by (11.1), (11.3a) and (11.3b), respectively, are illustrated by the curves and the series of sketches for the dipoles in Fig. 11.2, corresponding to one complete period. The different sizes of the arrows associated with the dipoles denote the different strengths of the current whereas the number of the plus or minus signs is indicative of the strength of the charges.

To determine the electromagnetic field due to the Hertzian dipole, we consider the dipole to be situated at the origin and oriented along the z-axis, in a perfect dielectric medium. We shall use an approach based on the magnetic vector potential and obtain electric and magnetic fields consistent with Maxwell's equations, while fulfilling certain other pertinent requirements. We shall begin with the magnetic vector potential for the static case and then extend it to the time-varying current element. To do this, we recall from Section 4.5 that for a current element of length $d\mathbf{l} = dl \, \mathbf{a}_z$ situated at the origin, as shown in Fig. 11.3 and carrying current I, the magnetic field at a point $P(r, \theta, \phi)$ is given by

$$\mathbf{A} = \frac{\mu I \, d\mathbf{l}}{4\pi r} = \frac{\mu I \, dl}{4\pi r} \mathbf{a}_z \tag{11.4}$$

If the current in the element is now assumed to be time varying in the manner $I = I_0 \cos \omega t$, we might expect the corresponding magnetic vector potential

Retarded potential

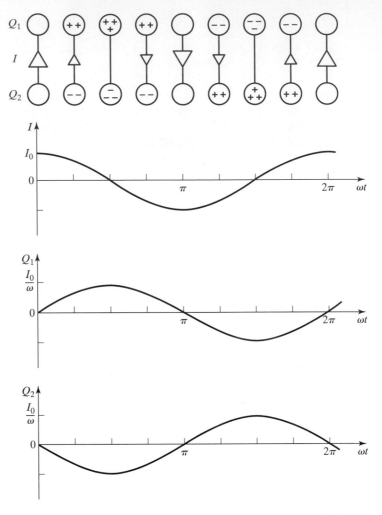

FIGURE 11.2 Time variations of charges and current associated with the Hertzian dipole.

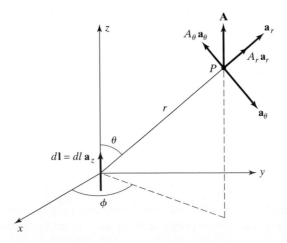

FIGURE 11.3 For finding the magnetic vector potential due to an infinitesimal current element.

to be that in (11.4) with *I* replaced by $I_0 \cos \omega t$. Proceeding in this manner would however lead to fields inconsistent with Maxwell's equations. The reason is that time-varying electric and magnetic fields give rise to wave propagation, according to which the effect of the source current at a given value of time is felt at a distance *r* from the origin after a time delay of r/v_p, where v_p is the velocity of propagation of the wave. Conversely, the effect felt at a distance *r* from the origin at time *t* is due to the current that existed at the origin at an earlier time $(t - r/v_p)$. Thus, for the time-varying current element $I_0 \, dl \cos \omega t \, \mathbf{a}_z$ situated at the origin, the magnetic vector potential is given by

$$
\begin{aligned}
\mathbf{A} &= \frac{\mu I_0 \, dl}{4\pi r} \cos \omega \left(t - \frac{r}{v_p} \right) \mathbf{a}_z \\
&= \frac{\mu I_0 \, dl}{4\pi r} \cos (\omega t - \beta r) \, \mathbf{a}_z
\end{aligned}
\tag{11.5}
$$

where we have replaced ω / v_p by β, the phase constant. The result given by (11.5) is known as the *retarded* magnetic vector potential in view of the phase-lag factor βr contained in it.

To augment the reasoning behind the retarded magnetic vector potential, recall that in Section 4.4, we derived differential equations for the electromagnetic potentials. For the magnetic vector potential, we obtained

$$
\nabla^2 \mathbf{A} - \mu \varepsilon \frac{\partial^2 \mathbf{A}}{\partial t^2} = -\mu \mathbf{J}
\tag{11.6}
$$

which reduces to

$$
\nabla^2 A_z - \mu \varepsilon \frac{\partial^2 A_z}{\partial t^2} = -\mu J_z
\tag{11.7}
$$

for $\mathbf{A} = A_z \mathbf{a}_z$ and $\mathbf{J} = J_z \mathbf{a}_z$. Equation (11.7) has the form of the wave equation, except in three dimensions and with the source term on the right side. Thus, the solution for A_z must be of the form of a traveling wave while reducing to the static field case for no time variations.

Expressing **A** in (11.5) in terms of its components in spherical coordinates, as shown in Fig. 11.3, we obtain

Fields due to Hertzian dipole

$$
\mathbf{A} = \frac{\mu I_0 \, dl \cos (\omega t - \beta r)}{4\pi r} (\cos \theta \, \mathbf{a}_r - \sin \theta \, \mathbf{a}_\theta)
\tag{11.8}
$$

The magnetic field due to the Hertzian dipole is then given by

$$
\mathbf{H} = \frac{\mathbf{B}}{\mu} = \frac{1}{\mu} \nabla \times \mathbf{A}
$$

$$
= \frac{1}{\mu}
\begin{vmatrix}
\dfrac{\mathbf{a}_r}{r^2 \sin \theta} & \dfrac{\mathbf{a}_\theta}{r \sin \theta} & \dfrac{\mathbf{a}_\phi}{r} \\[2mm]
\dfrac{\partial}{\partial r} & \dfrac{\partial}{\partial \theta} & \dfrac{\partial}{\partial \phi} \\[2mm]
A_r & r A_\theta & 0
\end{vmatrix}
= \frac{1}{\mu r} \left[\frac{\partial}{\partial r} (r A_\theta) - \frac{\partial A_r}{\partial \theta} \right] \mathbf{a}_\phi
$$

or

$$\boxed{\mathbf{H} = \frac{I_0 \, dl \sin \theta}{4\pi} \left[\frac{\cos (\omega t - \beta r)}{r^2} - \frac{\beta \sin (\omega t - \beta r)}{r} \right] \mathbf{a}_\phi} \qquad (11.9)$$

Using Maxwell's curl equation for $\mathbf{H}$ with $\mathbf{J}$ set equal to zero in view of perfect dielectric medium, we then have

$$\frac{\partial \mathbf{E}}{\partial t} = \frac{1}{\varepsilon} \nabla \times \mathbf{H}$$

$$= \frac{1}{\varepsilon} \begin{vmatrix} \dfrac{\mathbf{a}_r}{r^2 \sin \theta} & \dfrac{\mathbf{a}_\theta}{r \sin \theta} & \dfrac{\mathbf{a}_\phi}{r} \\[2mm] \dfrac{\partial}{\partial r} & \dfrac{\partial}{\partial \theta} & \dfrac{\partial}{\partial \phi} \\[2mm] 0 & 0 & r \sin \theta \, H_\phi \end{vmatrix}$$

$$= \frac{1}{\varepsilon r^2 \sin \theta} \frac{\partial}{\partial \theta} (r \sin \theta \, H_\phi) \, \mathbf{a}_r - \frac{1}{\varepsilon r \sin \theta} \frac{\partial}{\partial r} (r \sin \theta \, H_\phi) \, \mathbf{a}_\theta$$

or

$$\boxed{\begin{aligned} \mathbf{E} = {}& \frac{2 I_0 \, dl \cos \theta}{4 \pi \varepsilon \omega} \left[\frac{\sin (\omega t - \beta r)}{r^3} + \frac{\beta \cos (\omega t - \beta r)}{r^2} \right] \mathbf{a}_r \\[2mm] & + \frac{I_0 \, dl \sin \theta}{4\pi \, \varepsilon \omega} \left[\frac{\sin (\omega t - \beta r)}{r^3} + \frac{\beta \cos (\omega t - \beta r)}{r^2} \right. \\[2mm] & \qquad \left. - \frac{\beta^2 \sin (\omega t - \beta r)}{r} \right] \mathbf{a}_\theta \end{aligned}} \qquad (11.10)$$

Equations (11.10) and (11.9) represent the electric and magnetic fields, respectively, due to the Hertzian dipole. The following observations are pertinent to these field expressions:

1. They satisfy the two maxwell's curl equations. In fact, we have obtained (11.10) from (11.9) by using the curl equation for $\mathbf{H}$. The reader is urged to verify that (11.9) follows from (11.10) through the curl equation for $\mathbf{E}$.

2. They contain terms involving $1/r^3, 1/r^2$, and $1/r$. Far from the dipole such that $\beta r \gg 1$, the $1/r^3$ and $1/r^2$ terms are negligible compared to the $1/r$ terms so that the fields vary inversely with r. Furthermore, for any value of r, the time-average value of the θ-component of the Poynting vector due to the fields is zero, and the contribution to the time-average value of the r-component is completely from the $1/r$ terms (see Problem P11.2). Thus, the time-average Poynting vector varies proportionately to $1/r^2$ and is directed entirely in the radial direction. This is consistent with the physical requirement that for the time-average power crossing all possible spheri-

cal surfaces centered at the dipole to be the same, the power density must be inversely proportional to r^2, since the surface areas of the spherical surfaces are proportional to the squares of their radii.

3. For $\beta r \ll 1$, the $1/r^3$ terms dominate the $1/r^2$ terms which in turn dominate the $1/r$ terms. Also, $\sin(\omega t - \beta r) \approx (\sin \omega t - \beta r \cos \omega t)$ and $\cos(\omega t - \beta r) \approx (\cos \omega t + \beta r \sin \omega t)$, so that

$$\mathbf{E} \approx \frac{I_0 \, dl \sin \omega t}{4\pi\varepsilon\omega r^3} (2\cos\theta \, \mathbf{a}_r + \sin\theta \, \mathbf{a}_\theta) \tag{11.11}$$

$$\mathbf{H} \approx \frac{I_0 \, dl \cos \omega t}{4\pi r^2} \sin\theta \, \mathbf{a}_\phi \tag{11.12}$$

Equation (11.11) is the same as (4.102) with Q replaced by $(I_0/\omega)\sin\omega t$, that is, $Q_1(t)$ in Fig. 11.1, and d replaced by dl. Equation (11.12) gives the same $\mathbf{B}$ as the magnetic field given by Biot–Savart law applied to a current element $I \, dl \, \mathbf{a}_z$ at the origin and then I replaced by $I_0 \cos \omega t$, that is, $I(t)$ in Fig. 11.1. Thus, electrically close to the dipole, where retardation effects are negligible, the field expressions approach toward the corresponding static field expressions with the static source terms simply replaced by the time-varying source terms.

Example 11.1

Let us consider in free space a Hertzian dipole of length 0.1 m situated at the origin and along the z-axis, carrying the current $10 \cos 2\pi \times 10^7 t$ A. We wish to obtain the electric and magnetic fields at the point $(5, \pi/6, 0)$.

For convenience in computation of the amplitudes and phase angles of the field components, we shall express the field components in phasor form. Thus, replacing $\cos(\omega t - \beta r)$ by $e^{-j\beta r}$ and $\sin(\omega t - \beta r)$ by $-je^{-j\beta r}$, we have

$$\begin{aligned} \bar{E}_r &= \frac{2I_0 \, dl \cos \theta}{4\pi\varepsilon\omega} \left(-\frac{j}{r^3} + \frac{\beta}{r^2}\right) e^{-j\beta r} \\ &= \frac{2\beta^2 \eta I_0 \, dl \cos \theta}{4\pi} \left[-j\frac{1}{(\beta r)^3} + \frac{1}{(\beta r)^2}\right] e^{-j\beta r} \end{aligned} \tag{11.13}$$

$$\begin{aligned} \bar{E}_\theta &= \frac{I_0 \, dl \sin \theta}{4\pi\varepsilon\omega} \left(-\frac{j}{r^3} + \frac{\beta}{r^2} + \frac{j\beta^2}{r}\right) e^{-j\beta r} \\ &= \frac{\beta^2 \eta I_0 \, dl \sin \theta}{4\pi} \left[-j\frac{1}{(\beta r)^3} + \frac{1}{(\beta r)^2} + j\frac{1}{\beta r}\right] e^{-j\beta r} \end{aligned} \tag{11.14}$$

$$\begin{aligned} \bar{H}_\phi &= \frac{I_0 \, dl \sin \theta}{4\pi} \left(\frac{1}{r^2} + \frac{j\beta}{r}\right) e^{-j\beta r} \\ &= \frac{\beta^2 I_0 \, dl \sin \theta}{4\pi} \left[\frac{1}{(\beta r)^2} + j\frac{1}{\beta r}\right] e^{-j\beta r} \end{aligned} \tag{11.15}$$

where $\eta = \sqrt{\mu/\varepsilon}$ is the intrinsic impedance of the medium. Using $I_0 = 10$ A, $dl = 0.1$ m, $f = 10^7$ Hz, $\mu = \mu_0$, $\varepsilon = \varepsilon_0$, $\beta = \pi/15$, $\eta = 120\pi$, $r = 5$ m, and $\theta = \pi/6$, and carrying out the computations, we obtain

$$\overline{E}_r = 2.8739\underline{/-103.679°} \text{ V/m}$$

$$\overline{E}_\theta = 0.6025\underline{/-54.728°} \text{ V/m}$$

$$\overline{H}_\phi = 0.0023\underline{/-13.679°} \text{ A/m}$$

Thus, the required fields are

$$\mathbf{E} = 2.8739 \cos(2\pi \times 10^7 t - 0.576\pi)\, \mathbf{a}_r$$
$$+ 0.6025 \cos(2\pi \times 10^7 t - 0.304\pi)\, \mathbf{a}_\theta \text{ V/m}$$
$$\mathbf{H} = 0.0023 \cos(2\pi \times 10^7 t - 0.076\pi)\, \mathbf{a}_\phi \text{ A/m}$$

K11.1. Hertzian dipole; Retarded magnetic vector potential; Complete electromagnetic field; Behavior far from the dipole ($\beta r \gg 1$); Behavior close to the dipole ($\beta r \ll 1$).

D11.1. Consider a Hertzian dipole of length 0.1λ carrying sinusoidally time-varying current of amplitude 4π A. Find the magnitude of the electric dipole moment for each of the following cases: **(a)** $f = 10$ MHz, medium is free space; **(b)** $f = 100$ kHz, medium is free space; and **(c)** $f = 25$ kHz, medium is seawater ($\sigma = 4$ S/m, $\varepsilon = 80\varepsilon_0$, and $\mu = \mu_0$).
Ans. **(a)** 6×10^{-7} C-m; **(b)** 6×10^{-3} C-m; **(c)** 8×10^{-5} C-m.

D11.2. Three Hertzian dipoles of lengths 1, 1, and 2 m are situated at the origin oriented along the positive x-, y-, and z-axes, respectively, and carrying currents $1 \cos 2\pi \times 10^6 t$, $2 \sin 2\pi \times 10^6 t$, and $2 \cos 2\pi \times 10^6 t$ A, respectively. The medium is free space. Find the following at $(0, 0, 50)$ in Cartesian coordinates: **(a)** E_x; **(b)** E_y; and **(c)** E_z.
Ans. **(a)** $12.051 \cos(2\pi \times 10^6 t + 0.696\pi)$ mV/m; **(b)** $24.102 \cos(2\pi \times 10^6 t + 0.196\pi)$ mV/m; **(c)** $0.1327 \cos(2\pi \times 10^6 t - 0.576\pi)$ V/m.

11.2 RADIATION RESISTANCE AND DIRECTIVITY

Radiation fields

In the preceding section, we derived the expressions for the complete electromagnetic field due to the Hertzian dipole. These expressions look very complicated. Fortunately, it is seldom necessary to work with the complete field expressions because one is often interested in the field far from the dipole that is governed predominantly by the terms involving $1/r$. Thus, from (11.10) and (11.9), we find that for a Hertzian dipole of length dl oriented along the z-axis and carrying current

$$I = I_0 \cos \omega t \tag{11.16}$$

the electric and magnetic fields at values of r *far from the dipole* are given by

$$\mathbf{E} = -\frac{\beta^2 I_0\, dl \sin\theta}{4\pi\, \varepsilon\omega r} \sin(\omega t - \beta r)\, \mathbf{a}_\theta$$

$$= -\frac{\eta\beta I_0\, dl \sin\theta}{4\pi r} \sin(\omega t - \beta r)\, \mathbf{a}_\theta \tag{11.17a}$$

$$\mathbf{H} = -\frac{\beta I_0\, dl \sin\theta}{4\pi r} \sin(\omega t - \beta r)\, \mathbf{a}_\phi \tag{11.17b}$$

These fields are known as the *radiation fields*, since they are the components of the total fields that contribute to the time-average radiated power away from the dipole. Before we discuss the nature of these fields, let us find out quantitatively what we mean by *far from the dipole*. To do this, we look at the expression for the complete magnetic field given by (11.9) and note that the ratio of the amplitudes of the $1/r^2$ and $1/r$ terms is equal to $1/\beta r$. Hence, for $\beta r \gg 1$, the $1/r^2$ term is negligible compared to the $1/r$ term as already pointed out in the previous section. This means that for $r \gg 1/\beta$, or $r \gg \lambda/2\pi$, that is, even at a distance of a few wavelengths from the dipole, the fields are predominantly radiation fields.

Returning now to the expressions for the radiation fields given by (11.17a) and (11.17b), we note that at any given point, (1) the electric field (E_θ), the magnetic field (H_ϕ) and the direction of propagation (r) are mutually perpendicular, and (2) the ratio of E_θ to H_ϕ is equal to η, which are characteristic of uniform plane waves. The phase of the field, however, is uniform over the surfaces $r = $ constant, that is, spherical surfaces centered at the dipole, whereas the amplitude of the field is uniform over surfaces $(\sin\theta)/r = $ constant. Hence, the fields are only locally uniform plane waves, that is, over any small area normal to the r-direction at a given point.

The Poynting vector due to the radiation fields is given by

$$
\begin{aligned}
\mathbf{P} &= \mathbf{E} \times \mathbf{H} \\
&= E_\theta \, \mathbf{a}_\theta \times H_\phi \, \mathbf{a}_\phi = E_\theta H_\phi \, \mathbf{a}_r \\
&= \frac{\eta \beta^2 I_0^2 \, (dl)^2 \sin^2\theta}{16\pi^2 r^2} \sin^2(\omega t - \beta r) \, \mathbf{a}_r
\end{aligned}
\tag{11.18}
$$

By evaluating the surface integral of the Poynting vector over any surface enclosing the dipole, we can find the power flow out of that surface, that is, the power "radiated" by the dipole. For convenience in evaluating the surface integral, we choose the spherical surface of radius r and centered at the dipole, as shown in Fig. 11.4. Thus, noting that the differential surface area on the spherical

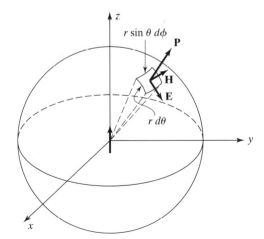

FIGURE 11.4   For computing the power radiated by the Hertzian dipole.

surface is $(r\, d\theta)(r \sin \theta\, d\phi)\mathbf{a}_r$ or $r^2 \sin \theta\, d\theta\, d\phi\, \mathbf{a}_r$, we obtain the instantaneous power radiated to be

$$
\begin{aligned}
P_{\text{rad}} &= \int_{\theta=0}^{\pi} \int_{\phi=0}^{2\pi} \mathbf{P} \cdot r^2 \sin \theta\, d\theta\, d\phi\, \mathbf{a}_r \\
&= \int_{\theta=0}^{\pi} \int_{\phi=0}^{2\pi} \frac{\eta \beta^2 I_0^2\, (dl)^2 \sin^3 \theta}{16\pi^2} \sin^2 (\omega t - \beta r)\, d\theta\, d\phi \\
&= \frac{\eta \beta^2 I_0^2\, (dl)^2}{8\pi} \sin^2 (\omega t - \beta r) \int_{\theta=0}^{\pi} \sin^3 \theta\, d\theta \qquad (11.19) \\
&= \frac{\eta \beta^2 I_0^2\, (dl)^2}{6\pi} \sin^2 (\omega t - \beta r) \\
&= \frac{2\pi \eta I_0^2}{3} \left(\frac{dl}{\lambda}\right)^2 \sin^2 (\omega t - \beta r)
\end{aligned}
$$

The time-average power radiated by the dipole, that is, the average of P_{rad} over one period of the current variation, is

$$
\begin{aligned}
\langle P_{\text{rad}} \rangle &= \frac{2\pi \eta I_0^2}{3} \left(\frac{dl}{\lambda}\right)^2 \langle \sin^2 (\omega t - \beta r) \rangle \\
&= \frac{\pi \eta I_0^2}{3} \left(\frac{dl}{\lambda}\right)^2 \qquad (11.20) \\
&= \frac{1}{2} I_0^2 \left[\frac{2\pi \eta}{3} \left(\frac{dl}{\lambda}\right)^2 \right]
\end{aligned}
$$

Radiation resistance   We now define a quantity known as the *radiation resistance* of the antenna, denoted by the symbol R_{rad}, as the value of a fictitious resistor that dissipates the same amount of time-average power as that radiated by the antenna when a current of the same peak amplitude as that in the antenna is passed through it. Recalling that the average power dissipated in a resistor R when a current $I_0 \cos \omega t$ is passed through it is $\frac{1}{2} I_0^2 R$, we note from (11.20) that the radiation resistance of the Hertzian dipole is

$$
\boxed{R_{\text{rad}} = \frac{2\pi \eta}{3} \left(\frac{dl}{\lambda}\right)^2 \ \Omega} \qquad (11.21)
$$

For free space, $\eta = \eta_0 = 120\pi\ \Omega$, and

$$
\boxed{R_{\text{rad}} = 80\pi^2 \left(\frac{dl}{\lambda}\right)^2 \ \Omega} \qquad (11.22)
$$

As a numerical example, for (dl/λ) equal to 0.01, $R_{\text{rad}} = 80\pi^2 (0.01)^2 = 0.08\ \Omega$. Thus, for a current of peak amplitude 1 A, the time-average radiated power is equal to 0.04 W. This indicates that a Hertzian dipole of length 0.01λ is not a very effective radiator.

We note from (11.21) that the radiation resistance and, hence, the radiated power are proportional to the square of the electrical length, that is, the physical length expressed in terms of wavelength, of the dipole. The result given by (11.21) is, however, valid only for small values of dl/λ since if dl/λ is not small, the amplitude of the current along the antenna can no longer be uniform and its variation must be taken into account in deriving the radiation fields and hence the radiation resistance. We shall do this in the following section for a half-wave dipole, that is, for a dipole of length equal to $\lambda/2$.

Let us now examine the directional characteristics of the radiation from *Radiation* the Hertzian dipole. We note from (11.17a) and (11.17b) that, for a constant r, *pattern* the amplitude of the fields is proportional to $\sin\theta$. Similarly, we note from (11.18) that for a constant r, the power density is proportional to $\sin^2\theta$. Thus, an observer wandering on the surface of an imaginary sphere centered at the dipole views different amplitudes of the fields and of the power density at different points on the surface. The situation is illustrated in Fig. 11.5(a) for the power density by attaching to different points on the spherical surface vectors having lengths proportional to the Poynting vectors at those points. It can be seen that the power density is largest for $\theta = \pi/2$, that is, in the plane normal to the axis of the dipole, and decreases continuously toward the axis of the dipole, becoming zero along the axis.

It is customary to depict the radiation characteristic by means of a *radiation pattern,* as shown in Fig. 11.5(b), which can be imagined to be obtained by shrinking the radius of the spherical surface in Fig. 11.5(a) to zero with the Poynting vectors attached to it and then joining the tips of the Poynting vectors. Thus, the distance from the dipole point to a point on the radiation pattern is proportional to the power density in the direction of that point. Similarly, the radiation pattern for the fields can be drawn as shown in Fig. 11.5(c), based on the $\sin\theta$ dependence of the fields. In view of the independence of the fields from ϕ, the patterns of Figs. 11.5(b) and (c) are valid for any plane containing the axis of the dipole. In fact, the three-dimensional radiation patterns can be imagined to be the figures obtained by revolving these patterns about the dipole axis. For a general case, the radiation may also depend on ϕ, and hence it will be necessary to draw a radiation pattern for the $\theta = \pi/2$ plane. Here, this pattern is merely a circle centered at the dipole.

We now define a parameter known as the *directivity* of the antenna, de- *Directivity* noted by the symbol D, as the ratio of the maximum power density radiated by the antenna to the average power density. To elaborate on the definition of D, imagine that we take the power radiated by the antenna and distribute it equally in all directions by shortening some of the vectors in Fig. 11.5(a) and lengthening the others so that they all have equal lengths. The pattern then becomes nondirectional, and the power density, which is the same in all directions, will be less than the maximum power density of the original pattern. Obviously, the more directional the radiation pattern of an antenna is, the greater is the directivity.

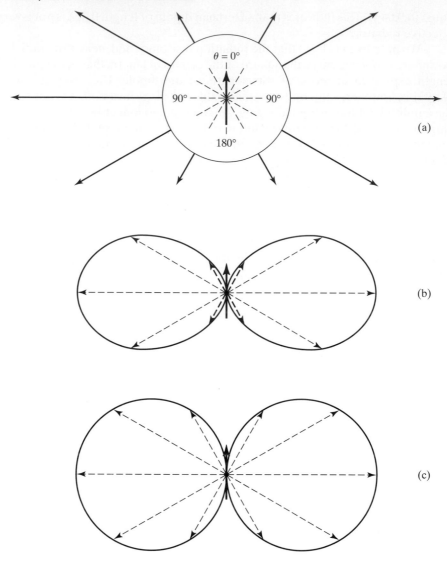

FIGURE 11.5   Directional characteristics of radiation from the Hertzian dipole.

From (11.18), we obtain the maximum power density radiated by the Hertzian dipole to be

$$[P_r]_{max} = \frac{\eta\beta^2 I_0^2 (dl)^2 [\sin^2\theta]_{max}}{16\pi^2 r^2} \sin^2(\omega t - \beta r)$$

$$= \frac{\eta\beta^2 I_0^2 (dl)^2}{16\pi^2 r^2} \sin^2(\omega t - \beta r) \tag{11.23}$$

By dividing the radiated power given by (11.19) by the surface area $4\pi r^2$ of the sphere of radius r, we obtain the average power density to be

$$[P_r]_{av} = \frac{P_{rad}}{4\pi r^2} = \frac{\eta \beta^2 I_0^2 (dl)^2}{24\pi^2 r^2} \sin^2 (\omega t - \beta r) \qquad (11.24)$$

Thus, the directivity of the Hertzian dipole is given by

$$\boxed{D = \frac{[P_r]_{max}}{[P_r]_{av}} = 1.5} \qquad (11.25)$$

To generalize the computation of directivity for an arbitrary radiation pattern, let us consider

$$P_r = \frac{P_0 \sin^2 (\omega t - \beta r)}{r^2} f(\theta, \phi) \qquad (11.26)$$

where P_0 is a constant, and $f(\theta, \phi)$ is the power density pattern. Then

$$[P_r]_{max} = \frac{P_0 \sin^2 (\omega t - \beta r)}{r^2} [f(\theta, \phi)]_{max}$$

$$[P_r]_{av} = \frac{P_{rad}}{4\pi r^2}$$

$$= \frac{1}{4\pi r^2} \int_{\theta=0}^{2\pi} \int_{\phi=0}^{\pi} \frac{P_0 \sin^2 (\omega t - \beta r)}{r^2} f(\theta, \phi)\, \mathbf{a}_r \cdot r^2 \sin \theta \, d\theta \, d\phi \, \mathbf{a}_r$$

$$= \frac{P_0 \sin^2 (\omega t - \beta r)}{4\pi r^2} \int_{\theta=0}^{\pi} \int_{\phi=0}^{2\pi} f(\theta, \phi) \sin \theta \, d\theta \, d\phi$$

$$\boxed{D - 4\pi \frac{[f(\theta, \phi)]_{max}}{\int_{\theta=0}^{\pi} \int_{\phi=0}^{2\pi} f(\theta, \phi) \sin \theta \, d\theta \, d\phi}} \qquad (11.27)$$

Example 11.2

Let us compute the directivity corresponding to the power density pattern function $f(\theta, \phi) = \sin^2 \theta \cos^2 \theta$.
From (11.27),

$$D = 4\pi \frac{[\sin^2 \theta \cos^2 \theta]_{max}}{\int_{\theta=0}^{\pi} \int_{\phi=0}^{2\pi} \sin^3 \theta \cos^2 \theta \, d\theta \, d\phi}$$

$$= 4\pi \frac{[\frac{1}{4} \sin^2 2\theta]_{max}}{2\pi \int_{\theta=0}^{\pi} (\sin^3 \theta - \sin^5 \theta) \, d\theta}$$

$$= \frac{1}{2} \frac{1}{(4/3) - (16/15)}$$

$$= 1\frac{7}{8}$$

The ratio of the power density radiated by the antenna as a function of direction to the average power density is given by $Df(\theta, \phi)$. This quantity is known as the *directive gain of the antenna.* Another useful parameter is the power gain of the antenna, which takes into account the ohmic power losses in the antenna. It is denoted by the symbol G and is proportional to the directive gain, the proportionality factor being the power efficiency of the antenna, which is the ratio of the power radiated by the antenna to the power supplied to it by the source of excitation.

K11.2. Radiation fields; $\beta r \gg 1$; $1/r$ terms; Time-average radiated power; Radiation resistance; Radiation pattern; Power density; Directivity.

D11.3. Three Hertzian dipoles of lengths 1, 2, and 2 m are situated at the origin oriented along the positive x-, y-, and z-axes, respectively, carrying currents $1 \cos 2\pi \times 10^6 t$, $2 \cos 2\pi \times 10^6 t$, and $2 \sin 2\pi \times 10^6 t$ A, respectively. Determine the polarizations (including right-hand or left-hand sense in the case of circular and elliptical) of the radiation field at each of the following points: **(a)** a point on the x-axis; **(b)** a point on the y-axis; and **(c)** a point on the z-axis.
Ans. **(a)** right circular; **(b)** left elliptical; **(c)** linear.

D11.4. Compute the directivity responding to each of the following functions $f(\theta, \phi)$ in (11.27):

(a) $f(\theta, \phi) = \begin{cases} 1 & \text{for} \quad 0 < \theta < \pi/2 \\ 0 & \text{otherwise} \end{cases}$

(b) $f(\theta, \phi) = \begin{cases} \sin^2 \theta & \text{for} \quad 0 < \theta < \pi/2 \\ 0 & \text{otherwise} \end{cases}$

(c) $f(\theta, \phi) = \begin{cases} 1 & \text{for} \quad 0 < \theta < \pi/2 \\ \sin^2 \theta & \text{for} \quad \pi/2 < \theta < \pi \end{cases}$

Ans. **(a)** 2; **(b)** 3; **(c)** 1.2.

11.3   LINEAR ANTENNAS

In the preceding section, we found the radiation fields due to a Hertzian dipole, which is an elemental antenna of infinitesimal length. If we now have an antenna of any length having a specified current distribution, we can divide it into a series of Hertzian dipoles, and by applying superposition, we can find the radiation fields for that antenna. We illustrate this procedure in this section by first considering the half-wave dipole, which is a commonly used form of antenna.

Half-wave dipole

The half-wave dipole is a center-fed, straight-wire antenna of length L equal to $\lambda/2$ and having the current distribution

$$I(z) = I_0 \cos \frac{\pi z}{L} \cos \omega t \qquad \text{for} \quad -L/2 < z < L/2 \qquad (11.28)$$

where the dipole is assumed to be oriented along the z-axis with its center at the origin, as shown in Fig. 11.6(a). As can be seen from Fig. 11.6(a), the amplitude of the current distribution varies cosinusoidally along the antenna with zeros at the ends and maximum at the center. To see how this distribution comes about,

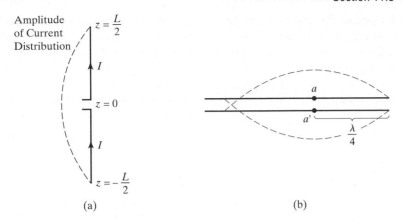

FIGURE 11.6 (a) Half-wave dipole. (b) Open-circuited transmission line for illustrating the evolution of the half-wave dipole.

the half-wave dipole may be imagined to be the evolution of an open-circuited transmission line with the conductors folded perpendicularly to the line at points $\lambda/4$ from the end of the line. The current standing wave pattern for an open-circuited line is shown in Fig. 11.6(b). It consists of zero current at the open circuit and maximum current at $\lambda/4$ from the open circuit, that is, at points a and a'. Hence, it can be seen that when the conductors are folded perpendicularly to the line at a and a', the half-wave dipole shown in Fig. 11.6(a) results.

Now to find the radiation field due to the half-wave dipole, we divide it into a number of Hertzian dipoles, each of length dz', as shown in Fig. 11.7. If we

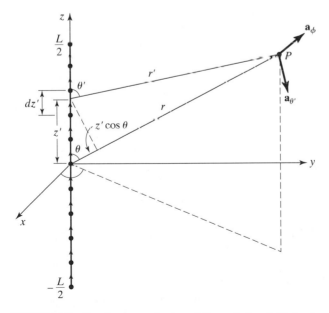

FIGURE 11.7 For the determination of the radiation field due to the half-wave dipole.

consider one of these dipoles situated at distance z' from the origin, then from (11.28), the current in this dipole is $I_0 \cos (\pi z'/L) \cos \omega t$. From (11.17a) and (11.17b), the radiation fields due to this dipole at point P situated at distance r' from it are given by

$$d\mathbf{E} = -\frac{\eta \beta I_0 \cos (\pi z'/L) \, dz' \sin \theta'}{4\pi r'} \sin (\omega t - \beta r') \, \mathbf{a}_{\theta'} \qquad (11.29\text{a})$$

$$d\mathbf{H} = -\frac{\beta I_0 \cos (\pi z'/L) \, dz' \sin \theta'}{4\pi r'} \sin (\omega t - \beta r') \, \mathbf{a}_{\phi} \qquad (11.29\text{b})$$

where θ' is the angle between the z-axis and the line from the current element to the point P and $\mathbf{a}_{\theta'}$ is the unit vector perpendicular to that line, as shown in Fig. 11.7. The fields due to the entire current distribution of the half-wave dipole are then given by

$$\mathbf{E} = \int_{z'=-L/2}^{L/2} d\mathbf{E}$$

$$= -\int_{z'=-L/2}^{L/2} \frac{\eta \beta I_0 \cos (\pi z'/L) \sin \theta' \, dz'}{4\pi r'} \sin (\omega t - \beta r') \, \mathbf{a}_{\theta'} \qquad (11.30\text{a})$$

$$\mathbf{H} = \int_{z'=-L/2}^{L/2} d\mathbf{H}$$

$$= -\int_{z'=-L/2}^{L/2} \frac{\beta I_0 \cos (\pi z'/L) \sin \theta' \, dz'}{4\pi r'} \sin (\omega t - \beta r') \, \mathbf{a}_{\phi} \qquad (11.30\text{b})$$

where r', θ', and $\mathbf{a}_{\theta'}$ are functions of z'.

For radiation fields, r' is at least equal to several wavelengths and hence $\gg L$. We can therefore set $\mathbf{a}_{\theta'} \approx \mathbf{a}_{\theta}$ and $\theta' \approx \theta$ since they do not vary significantly for $-L/2 < z' < L/2$. We can also set $r' \approx r$ in the amplitude factors for the same reason, but for r' in the phase factors, we substitute $r - z' \cos \theta$ since the phase angle in $\sin (\omega t - \beta r') = \sin (\omega t - \pi r'/L)$ can vary appreciably over the range $-L/2 < z' < L/2$. For example, if $L = 2$ m ($\lambda = 4$ m), $\theta = 0$, and $r = 10$, then r' varies from 11 for $z' = -L/2$ to 9 for $z' = L/2$, and $\pi r'/L$ varies from 5.5π for $z' = -L/2$ to 4.5π for $z' = L/2$. Thus, we have

$$\mathbf{E} = E_{\theta} \mathbf{a}_{\theta}$$

where

$$E_{\theta} = -\int_{z'=-L/2}^{L/2} \frac{\eta \beta I_0 \cos (\pi z'/L) \sin \theta}{4\pi r} \sin (\omega t - \beta r + \beta z' \cos \theta) \, dz'$$

$$= -\frac{\eta (\pi/L) I_0 \sin \theta}{4\pi r} \int_{z'=-L/2}^{L/2} \cos \frac{\pi z'}{L} \sin \left(\omega t - \frac{\pi}{L} r + \frac{\pi}{L} z' \cos \theta \right) dz'$$

Evaluating the integral, we obtain

$$E_{\theta} = -\frac{\eta I_0}{2\pi r} \frac{\cos [(\pi/2) \cos \theta]}{\sin \theta} \sin \left(\omega t - \frac{\pi}{L} r \right) \qquad (11.31\text{a})$$

Similarly,

$$\mathbf{H} = H_\phi \mathbf{a}_\phi$$

where

$$H_\phi = -\frac{I_0}{2\pi r} \frac{\cos\left[(\pi/2)\cos\theta\right]}{\sin\theta} \sin\left(\omega t - \frac{\pi}{L}r\right) \tag{11.31b}$$

The Poynting vector due to the radiation fields of the half-wave dipole is given by

$$\mathbf{P} = \mathbf{E} \times \mathbf{H} = E_\theta H_\phi \mathbf{a}_r$$

$$\tag{11.32}$$

$$= \frac{\eta I_0^2}{4\pi^2 r^2} \frac{\cos^2\left[(\pi/2)\cos\theta\right]}{\sin^2\theta} \sin^2\left(\omega t - \frac{\pi}{L}r\right) \mathbf{a}_r$$

The power radiated by the half-wave dipole is given by

$$P_{\text{rad}} = \int_{\theta=0}^{\pi} \int_{\phi=0}^{2\pi} \mathbf{P} \cdot r^2 \sin\theta \, d\theta \, d\phi \, \mathbf{a}_r$$

$$= \int_{\theta=0}^{\pi} \int_{\phi=0}^{2\pi} \frac{\eta I_0^2}{4\pi^2} \frac{\cos^2\left[(\pi/2)\cos\theta\right]}{\sin\theta} \sin^2\left(\omega t - \frac{\pi}{L}r\right) d\theta \, d\phi$$

$$\tag{11.33}$$

$$= \frac{\eta I_0^2}{\pi} \sin^2\left(\omega t - \frac{\pi}{L}r\right) \int_{\theta=0}^{\pi/2} \frac{\cos^2\left[(\pi/2)\cos\theta\right]}{\sin\theta} d\theta$$

$$= \frac{0.609\eta I_0^2}{\pi} \sin^2\left(\omega t - \frac{\pi}{L}r\right)$$

where we have used the result

$$\int_{\theta=0}^{\pi/2} \frac{\cos^2\left[(\pi/2)\cos\theta\right]}{\sin\theta} d\theta = 0.609$$

obtainable by numerical integration. The time-average radiated power is

$$\langle P_{\text{rad}} \rangle = \frac{0.609\eta I_0^2}{\pi} \left\langle \sin^2\left(\omega t - \frac{\pi}{L}r\right) \right\rangle$$

$$\tag{11.34}$$

$$= \frac{1}{2} I_0^2 \left(\frac{0.609\eta}{\pi}\right)$$

Thus, the radiation resistance of the half-wave dipole is

$$R_{\text{rad}} = \frac{0.609\eta}{\pi} \, \Omega \tag{11.35}$$

For free space, $\eta = \eta_0 = 120\pi \, \Omega$, and

$$R_{\text{rad}} = 0.609 \times 120 = 73 \, \Omega \tag{11.36}$$

Turning our attention now to the directional characteristics of the half-wave dipole, we note from (11.31a) and (11.31b) that the radiation pattern for the fields is $\{\cos\,[(\pi/2)\,\cos\,\theta]\}/\sin\,\theta$, whereas for the power density, it is $\{\cos^2\,[(\pi/2)\,\cos\,\theta]\}/\sin^2\,\theta$. These patterns, shown in Fig. 11.8(a) and (b), are slightly more directional than the corresponding patterns for the Hertzian dipole. The directivity of the half-wave dipole may now be found by using (11.27). Thus,

$$D = 4\pi\,\frac{\{\cos^2\,[(\pi/2)\,\cos\,\theta]/\sin^2\,\theta\}_{max}}{\int_{\theta=0}^{\pi}\int_{\phi=0}^{2\pi}\{\cos^2\,[(\pi/2)\,\cos\,\theta]/\sin^2\,\theta\}\,\sin\,\theta\,d\theta\,d\phi}$$

$$= 4\pi\,\frac{1}{2\pi \times 2 \times 0.609}$$

or

$$\boxed{D = 1.642} \tag{11.37}$$

Linear antenna of arbitrary length

For a center-fed linear antenna of length L equal to an arbitrary number of wavelengths, the current distribution can be written as

$$I(z) = \begin{cases} I_0 \sin \beta\left(\dfrac{L}{2} + z\right) \cos \omega t & \text{for} \quad -\dfrac{L}{2} < z < 0 \\[2ex] I_0 \sin \beta\left(\dfrac{L}{2} - z\right) \cos \omega t & \text{for} \quad 0 < z < \dfrac{L}{2} \end{cases} \tag{11.38}$$

where once again the antenna is assumed to be oriented along the z-axis with its center at the origin. Note that the current distribution is such that the amplitude of the current goes to zero at the two ends of the antenna and varies sinusoidally

(a)

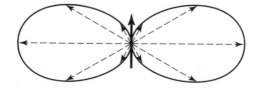

(b)   FIGURE 11.8   Radiation patterns for (a) the fields and (b) the power density due to the half-wave dipole.

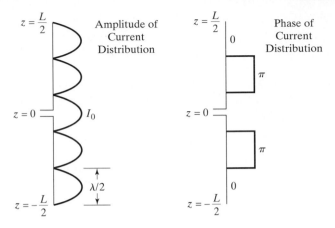

FIGURE 11.9   Variations of amplitude and phase of current distribution along a linear antenna of length $L = 5\lambda/2$.

along the antenna with phase reversals every half wavelength from the ends, as
shown, for example, for $L = 5\lambda/2$ in Fig. 11.9. Note also that for $L = \lambda/2$,
(11.38) reduces to (11.28). Using (11.38) and proceeding in the same manner as
for the half-wave dipole, the components of the radiation fields, the radiation resistance, and the directivity for the linear antenna of arbitrary electrical length
can be obtained. The results are

$$E_\theta = -\frac{\eta I_0}{2\pi r} F(\theta) \sin(\omega t - \beta r) \tag{11.39a}$$

$$H_\phi = -\frac{I_0}{2\pi r} F(\theta) \sin(\omega t - \beta r) \tag{11.39b}$$

$$R_{\text{rad}} = \frac{\eta}{\pi} \int_{\theta=0}^{\pi/2} F^2(\theta) \sin\theta\, d\theta \tag{11.39c}$$

$$D = \frac{[F^2(\theta)]_{\text{máx}}}{\int_{\theta=0}^{\pi/2} F^2(\theta) \sin\theta\, d\theta} \tag{11.39d}$$

where

$$F(\theta) = \frac{\cos[(\beta L/2)\cos\theta] - \cos(\beta L/2)}{\sin\theta} \tag{11.40}$$

is the radiation pattern for the fields. For $L = k\lambda$, (11.40) reduces to

$$F(\theta) = \frac{\cos(k\pi\cos\theta) - \cos(k\pi)}{\sin\theta} \tag{11.41}$$

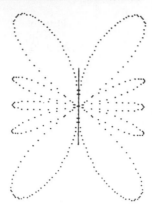

FIGURE 11.10   Computer-generated plot of radiation pattern for a linear antenna of length 2.5λ.

For a specified value of k, the radiation pattern can be obtained by substituting (11.41) for $0 < \theta < \pi$. As an example, Fig. 11.10 shows a computer-generated plot of the radiation pattern for $k = 2.5$. The radiation resistance and directivity can be computed by evaluating numerically the integrals in (11.39c) and (11.39d), respectively. For $k = 2.5$, these are 120.768 and 3.058, respectively.

K11.3.   Half-wave dipole; Radiation fields; Radiation characteristics; Linear antenna; Arbitrary length.

D11.5.   A center-fed linear antenna in free space has the current distribution of the form given by (11.38), where $I_0 = 1$ A. Find the amplitude of E_θ at $r = 100$ m for each of the following cases: **(a)** $L = 2$ m, $f = 75$ MHz, $\theta = 60°$; **(b)** $L = 2$ m, $f = 200$ MHz, $\theta = 60°$; and **(c)** $L = 4$ m, $f = 300$ MHz, $\theta = 30°$.
Ans.   **(a)** 0.49 V/m;   **(b)** 0 V/m;   **(c)** 1.335 V/m.

11.4   ANTENNA ARRAYS

In Section 5.2, we illustrated the principle of an antenna array by considering an array of two parallel, infinite plane, current sheets of uniform densities. We learned that by appropriately choosing the spacing between the current sheets and the amplitudes and phases of the current densities, a desired radiation characteristic can be obtained. The infinite plane current sheet is, however, a hypothetical antenna for which the fields are truly uniform plane waves propagating in the one dimension normal to the sheet. Now that we have gained some knowledge of physical antennas, in this section, we consider arrays of such antennas.

Array of two Hertzian dipoles

The simplest array we can consider consists of two Hertzian dipoles, oriented parallel to the z-axis and situated at points on the x-axis on either side of and equidistant from the origin, as shown in Fig. 11.11. We shall consider the amplitudes of the currents in the two dipoles to be equal, but we shall allow a phase difference α between them. Thus, if $I_1(t)$ and $I_2(t)$ are the currents in the dipoles

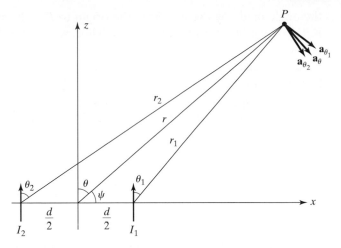

FIGURE 11.11 For computing the radiation field due to an array of two Hertzian dipoles.

situated at $(d/2, 0, 0)$ and $(-d/2, 0, 0)$, respectively, then

$$I_1 = I_0 \cos\left(\omega t + \frac{\alpha}{2}\right) \qquad (11.42a)$$

$$I_2 = I_0 \cos\left(\omega t - \frac{\alpha}{2}\right) \qquad (11.42b)$$

For simplicity, we consider a point P in the xz-plane and compute the radiation field at that point due to the array of the two dipoles. To do this, we note from (11.17a) that the electric field intensities at the point P due to the individual dipoles are given by

$$\mathbf{E}_1 = -\frac{\eta \beta I_0 \, dl \sin \theta_1}{4\pi r_1} \sin\left(\omega t - \beta r_1 + \frac{\alpha}{2}\right) \mathbf{a}_{\theta_1} \qquad (11.43a)$$

$$\mathbf{E}_2 = -\frac{\eta \beta I_0 \, dl \sin \theta_2}{4\pi r_2} \sin\left(\omega t - \beta r_2 - \frac{\alpha}{2}\right) \mathbf{a}_{\theta_2} \qquad (11.43b)$$

where $\theta_1, \theta_2, r_1, r_2, \mathbf{a}_{\theta_1}$, and $\mathbf{a}_{\theta_2}$ are as shown in Fig. 11.11.

For $r \gg d$, that is, for points far from the array, which is the region of interest, we can set $\theta_1 \approx \theta_2 \approx \theta$ and $\mathbf{a}_{\theta_1} \approx \mathbf{a}_{\theta_2} \approx \mathbf{a}_\theta$. Also, we can set $r_1 \approx r_2 \approx r$ in the amplitude factors, but for r_1 and r_2 in the phase factors, we substitute

$$r_1 \approx r - \frac{d}{2} \cos \psi \qquad (11.44a)$$

$$r_2 \approx r + \frac{d}{2} \cos \psi \qquad (11.44b)$$

where ψ is the angle made by the line from the origin to P with the axis of the array, that is, the x-axis, as shown in Fig. 11.11. Thus, we obtain the resultant field to be

$$
\begin{aligned}
\mathbf{E} &= \mathbf{E}_1 + \mathbf{E}_2 \\
&= -\frac{\eta\beta I_0 \, dl \sin\theta}{4\pi r} \left[\sin\left(\omega t - \beta r + \frac{\beta d}{2}\cos\psi + \frac{\alpha}{2}\right) \right. \\
&\qquad\left. + \sin\left(\omega t - \beta r - \frac{\beta d}{2}\cos\psi - \frac{\alpha}{2}\right) \right] \mathbf{a}_\theta \\
&= -\frac{2\eta\beta I_0 \, dl \sin\theta}{4\pi r} \cos\left(\frac{\beta d \cos\psi + \alpha}{2}\right) \sin(\omega t - \beta r)\, \mathbf{a}_\theta
\end{aligned}
\tag{11.45}
$$

Unit, group, and resultant patterns

Comparing (11.45) with the expression for the electric field at P due to a single dipole situated at the origin, we note that the resultant field of the array is simply equal to the single dipole field multiplied by the factor $2\cos\left[(\beta d \cdot \cos\psi + \alpha)/2\right]$, known as the *array factor*. Thus, the radiation pattern of the resultant field is given by the product of $\sin\theta$, which is the radiation pattern of the single dipole field, and $\cos\left[(\beta d \cos\psi + \alpha)/2\right]$, which is the radiation pattern of the array if the antennas were isotropic. We shall call these three patterns the *resultant pattern,* the *unit pattern,* and the *group pattern,* respectively. It is apparent that the group pattern is independent of the nature of the individual antennas as long as they have the same spacing and carry currents having the same relative amplitudes and phase differences. It can also be seen that the group pattern is the same in any plane containing the axis of the array. In other words, the three-dimensional group pattern is simply the pattern obtained by revolving the group pattern in the xz-plane about the x-axis, that is, the axis of the array.

Example 11.3

For the array of two antennas carrying currents having equal amplitudes, let us consider several pairs of d and α and investigate the group patterns.

Case 1: $d = \lambda/2, \alpha = 0$.   The group pattern is

$$
\left| \cos\left(\frac{\beta\lambda}{4}\cos\psi\right) \right| = \cos\left(\frac{\pi}{2}\cos\psi\right)
$$

This is shown in Fig. 11.12(a). It has maxima perpendicular to the axis of the array and nulls along the axis of the array. Such a pattern is known as a *broadside pattern.*

Case 2: $d = \lambda/2, \alpha = \pi$.   The group pattern is

$$
\left| \cos\left(\frac{\beta\lambda}{4}\cos\psi + \frac{\pi}{2}\right) \right| = \left| \sin\left(\frac{\pi}{2}\cos\psi\right) \right|
$$

This is shown in Fig. 11.12(b). It has maxima along the axis of the array and nulls perpendicular to the axis of the array. Such a pattern is known as an *endfire pattern.*

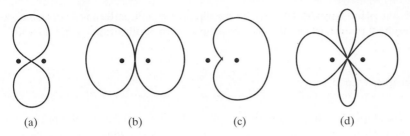

FIGURE 11.12 Group patterns for an array of two antennas carrying currents of equal amplitude for (a) $d = \lambda/2, \alpha = 0$, (b) $d = \lambda/2, \alpha = \pi$, (c) $d = \lambda/4, \alpha = -\pi/2$, and (d) $d = \lambda, \alpha = 0$.

Case 3: $d = \lambda/4, \alpha = -\pi/2$. The group pattern is

$$\left| \cos\left(\frac{\beta\lambda}{8} \cos\psi - \frac{\pi}{4} \right) \right| = \cos\left(\frac{\pi}{4} \cos\psi - \frac{\pi}{4} \right)$$

This is shown in Fig. 11.12(c). It has a maximum along $\psi = 0$ and null along $\psi = \pi$. Again, this is an endfire pattern, but directed to one side. This case is the same as the one considered in Section 5.2.

Case 4: $d = \lambda, \alpha = 0$. The group pattern is

$$\left| \cos\left(\frac{\beta\lambda}{2} \cos\psi \right) \right| = \left| \cos\left(\pi \cos\psi \right) \right|$$

This is shown in Fig. 11.12(d). It has maxima along $\psi = 0°, 90°$, and $180°$ and nulls along $\psi = 60°$ and $120°$.

Proceeding further, we can obtain the resultant pattern for an array of two Hertzian dipoles by multiplying the unit pattern by the group pattern. Thus, recalling that the unit pattern for the Hertzian dipole is $\sin\theta$ in the plane of the dipole and considering values of $\lambda/2$ and 0 for d and α, respectively, for which the group pattern is given in Fig. 11.12(a), we obtain the resultant pattern in the xz-plane, as shown in Fig. 11.13(a). In the xy-plane,

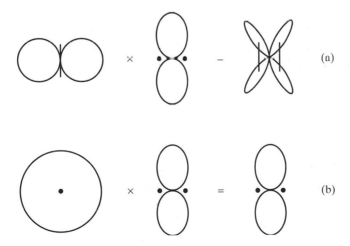

FIGURE 11.13 Determination of the resultant pattern of an antenna array by multiplication of unit and group patterns.

that is, the plane normal to the axis of the dipole, the unit pattern is a circle, and, hence, the resultant pattern is the same as the group pattern, as illustrated in Fig. 11.13(b).

Example 11.4

Pattern multiplication

The procedure of multiplication of the unit and group patterns to obtain the resultant pattern illustrated in Example 11.3 is known as the *pattern multiplication* technique. Let us consider a linear array of four isotropic antennas spaced $\lambda/2$ apart and fed in phase, as shown in Fig. 11.14(a), and obtain the resultant pattern by using the pattern multiplication technique.

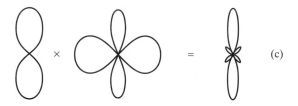

FIGURE 11.14   Determination of the resultant pattern for a linear array of four isotropic antennas.

To obtain the resultant pattern of the four-element array, we replace it by a two-element array of spacing λ, as shown in Fig. 11.14(b), in which each element forms a unit representing a two-element array of spacing $\lambda/2$. The unit pattern is then the pattern shown in Fig. 11.12(a). The group pattern, which is the pattern of two isotropic radiators having $d = \lambda$ and $\alpha = 0$, is the pattern given in Fig. 11.12(d). The resultant pattern of the four-element array is the product of these two patterns, as illustrated in Fig. 11.14(c). If the individual elements of the four-element array are not isotropic, then this pattern becomes the group pattern for the determination of the new resultant pattern.

Uniform linear array of n antennas

Let us now consider a uniform linear array of n antennas of spacing d, as shown in Fig. 11.15. Then assuming currents of equal amplitude I_0 and progressive phase shift α, that is, in the manner $I_0 \cos \omega t, I_0 \cos (\omega t + \alpha), I_0 \cos (\omega t + 2\alpha), \ldots$ for antennas $1, 2, 3, \ldots$, respectively, we can obtain the far field ($r \gg nd$) as follows. If the complex electric field at the point (r_0, ψ) due to element 1 is assumed to be $1e^{-j\beta r_0}$, then the complex electric fields at that point due to elements $2, 3, \ldots$

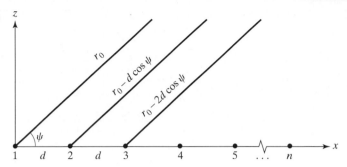

FIGURE 11.15   For obtaining the group pattern for a uniform linear array of n antennas.

are $1e^{j\alpha}e^{j\beta(r_0 - d\cos\psi)}$, $1e^{j2\alpha}e^{j\beta(r_0 - 2d\cos\psi)}$, . . . , so that the field due to the n-element array is

$$
\begin{aligned}
\bar{E}(\psi) &= 1e^{-j\beta r_0} + 1e^{j\alpha}e^{-j\beta(r_0 - d\cos\psi)} \\
&\quad + 1e^{j2\alpha}e^{-j\beta(r_0 - 2d\cos\psi)} + \cdots \\
&\quad + 1e^{j(n-1)\alpha}e^{-j\beta[r_0 - (n-1)d\cos\psi]} \\
&= [1 + e^{j(\beta d\cos\psi + \alpha)} + e^{j2(\beta d\cos\psi + \alpha)} \quad\quad (11.46) \\
&\quad + \cdots + e^{j(n-1)(\beta d\cos\psi + \alpha)}]e^{-j\beta r_0} \\
&= \frac{1 - e^{jn(\beta d\cos\psi + \alpha)}}{1 - e^{j(\beta d\cos\psi + \alpha)}} e^{-j\beta r_0}
\end{aligned}
$$

The magnitude of $\bar{E}$ is given by

$$
\begin{aligned}
|\bar{E}(\psi)| &= \left| \frac{1 - e^{jn(\beta d\cos\psi + \alpha)}}{1 - e^{j(\beta d\cos\psi + \alpha)}} \right| \\
&= \left| \frac{\sin n[(\beta d\cos\psi + \alpha)/2]}{\sin[(\beta d\cos\psi + \alpha)/2]} \right|
\end{aligned}
\quad\quad (11.47)
$$

which has a maximum value of n for $\beta d\cos\psi + \alpha = 0, 2\pi, 4\pi, \ldots$. Thus, the group pattern is

$$
\boxed{F(\psi) = \frac{1}{n} \left| \frac{\sin n[(\beta d\cos\psi + \alpha)/2]}{\sin[(\beta d\cos\psi + \alpha)/2]} \right|}
\quad\quad (11.48)
$$

Note that for $n = 2$, (11.48) reduces to $\cos[(\beta d\cos\psi + \alpha)/2]$, which is the group pattern obtained for the two-element array. The nulls of the pattern occur for $n(\beta d\cos\psi + \alpha) = 2m\pi$, where m is any integer but not equal to $0, n, 2n, \ldots$. For $d = k\lambda$, (11.48) reduces to

$$
F(\psi) = \frac{1}{n} \left| \frac{\sin n(\pi k\cos\psi + \alpha/2)}{\sin(\pi k\cos\psi + \alpha/2)} \right|
\quad\quad (11.49)
$$

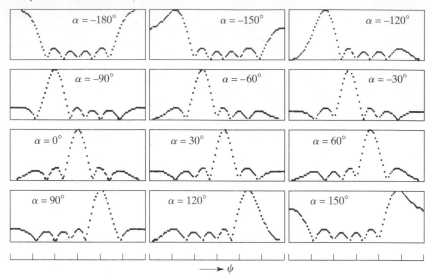

FIGURE 11.16   Plots of group patterns for the uniform linear array of Fig. 11.15 for $n = 6$ and $k = 0.5$. The horizontal scale for ψ for each plot is such that ψ varies for 0 to 180°.

Figure 11.16 shows computer-generated sequence of plots of F versus ψ ($0 \leq \psi \leq 180°$) for values of α ranging from $-180°$ to 150° in steps of 30°, for $n = 6$ and $k = 0.5$. It can be seen that as the value of α is varied, the value of ψ along which the principal maximum of the group pattern occurs varies in a continuous manner, as to be expected.

Principle of phased array

The behavior illustrated in Fig. 11.16 is the basis for the principle of phased arrays. In a phased array, the phase differences between the elements of the array are varied electronically to scan the radiation pattern over a desired angle without having to move the antenna structure mechanically.

Log-periodic dipole array

A type of array that is commonly seen is the *log-periodic dipole array*, which is an example of a broadband array. To discuss briefly, we first note that the directional properties of antennas and antenna arrays depend on their electrical dimensions, that is, the dimensions expressed in terms of the wavelength at the operating frequency. Hence, an antenna of fixed physical dimensions exhibits frequency-dependent characteristics. This very fact suggests that for an antenna to be frequency-independent, its electrical size must remain constant with frequency, and, hence, its physical size should increase proportionately to the wavelength. Alternatively, for an antenna of fixed physical dimensions, the active region, that is, the region responsible for the predominant radiation should vary with frequency, that is, scale itself in such a manner that its electrical size remains the same. An example in which this is the case is the log-periodic dipole array, shown in Fig. 11.17. As the name implies, it employs a number of dipoles. The dipole lengths and the spacings between consecutive dipoles increase along the array by a constant scale factor such that

$$\frac{l_{i+1}}{l_i} = \frac{d_{i+1}}{d_i} = \tau \qquad (11.50)$$

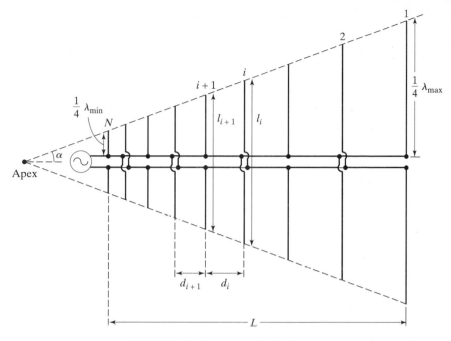

FIGURE 11.17   Log-periodic dipole array.

From the principle of scaling, it is evident that for this structure extending from zero to infinity and energized at the apex, the properties repeat at frequencies given by $\tau^n f$, where n takes integer values. When plotted on a logarithmic scale, these frequencies are equally spaced at intervals of $\log \tau$. It is for this reason that the structure is called *log periodic*.

The log-periodic dipole array is fed by a transmission line, as shown in Fig. 11.17, such that a 180° phase shift is introduced between successive elements in addition to that corresponding to the spacing between the elements. The resulting radiation pattern is directed toward the apex, that is, toward the source. Almost all the radiation takes place from those elements that are in the vicinity of a half wavelength long. The operating band of frequencies is therefore bounded on the low side by frequencies at which the largest elements are approximately a half wavelength long and on the high side by frequencies corresponding to the size of the smallest elements. As the frequency is varied, the radiating, or active, region moves back and forth along the array. Since practically all the input power is radiated by the active region, the larger elements to the right of it are not excited. Furthermore, because the radiation is toward the apex, these larger elements are essentially in a field-free region and hence do not significantly influence the operation. Although the shorter elements to the left of the active region are in the antenna beam, they have small influence on the pattern because of their short lengths, close spacings, and the 180° phase shift.

Thus far, we have considered the antennas to be situated in an unbounded medium, so that the waves radiate in all directions from the antenna without   *Image antennas* giving rise to reflections from any obstacles. In practice, however, we have to consider the effect of the ground even if no other obstacles are present. To do this, it is reasonable to assume that the ground is a perfect conductor and use the concept of image antennas, which together with the actual antennas form arrays.

To introduce this concept, let us consider a Hertzian dipole oriented vertically and located at a height h above a plane, perfect conductor surface, as shown in Fig. 11.18(a). Since no waves can penetrate into the perfect conductor,

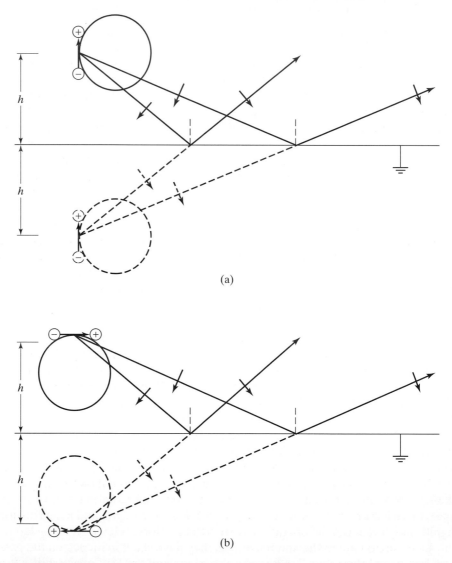

FIGURE 11.18   For illustrating the concept of image antennas. (a) Vertical Hertzian dipole and (b) horizontal Hertzian dipole above a plane perfect conductor surface.

as we learned in Section 5.4, the waves radiated from the dipole onto the conductor give rise to reflected waves, as shown in Fig. 11.18(a) for two directions of incidence. For a given incident wave onto the conductor surface, the angle of reflection is equal to the angle of incidence, as can be seen intuitively from the following reasons: (1) the reflected wave must propagate away from the conductor surface, (2) the apparent wavelengths of the incident and reflected waves parallel to the conductor surface must be equal, and (3) the tangential component of the resultant electric field on the conductor surface must be zero, which also determines the polarity of the reflected wave electric field. Also because of (3), the reflected wave amplitude must equal the incident wave amplitude. If we now produce the directions of propagation of the two reflected waves backward, they meet at a point that is directly beneath the dipole and at the same distance h below the conductor surface as the dipole is above it. Thus, the reflected waves appear to be originating from an antenna, which is the *image* of the actual antenna about the conductor surface. This image antenna must also be a vertical antenna since in order for the boundary condition of zero tangential electric field to be satisfied at all points on the conductor surface, the image antenna must have the same radiation pattern as that of the actual antenna, as shown in Fig. 11.18(a). In particular, the current in the image antenna must be directed in the same sense as that in the actual antenna to be consistent with the polarity of the reflected wave electric field. It can be seen, therefore, that the charges associated with the image dipole have signs opposite to those of the corresponding charges associated with the actual dipole.

A similar reasoning can be applied to the case of a horizontal Hertzian dipole above a perfect conductor surface, as shown in Fig. 11.18(b). Here it can be seen that the current in the image antenna is directed in the opposite sense to that in the actual antenna. This again results in charges associated with the image dipole having signs opposite to those of the corresponding charges associated with the actual dipole. In fact, this is always the case.

From the foregoing discussion, it can be seen that the field due to an antenna in the presence of the conductor is the same as the resultant field of the array formed by the actual antenna and the image antenna. There is, of course, no field inside the conductor. The image antenna is only a virtual antenna that serves to simplify the field determination outside the conductor. The simplification results from the fact that we can use the knowledge gained on antenna arrays to determine the radiation pattern.

For example, for a vertical Hertzian dipole at a height of $\lambda/2$ above the conductor surface, the radiation pattern in the vertical plane is the product of the unit pattern, which is the radiation pattern of the single dipole in the plane of its axis, and the group pattern corresponding to an array of two isotropic radiators spaced λ apart and fed in phase. This multiplication and the resultant pattern are illustrated in Fig. 11.19. The radiation patterns for the case of the horizontal dipole can be obtained in a similar manner.

To discuss another example of the application of the image-antenna concept, we consider the corner reflector, an arrangement of two plane perfect conductors at an angle to each other, as shown by the cross-sectional view in

Corner reflector

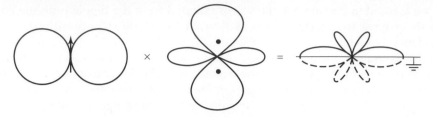

FIGURE 11.19   Determination of radiation pattern in the vertical plane for a vertical Hertzian dipole above a plane perfect conductor surface.

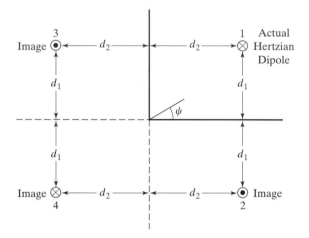

FIGURE 11.20   Application of image-antenna concept to obtain the radiation pattern for a Hertzian dipole in the presence of a corner reflector.

Fig. 11.20 for the case of the 90° angle. We shall assume that each conductor is semi-infinite in extent. For a Hertzian dipole situated parallel to both conductors, the locations and polarities of the images can be obtained to be as shown in the figure. By using the pattern multiplication technique, the radiation pattern in the cross-sectional plane can then be obtained.

For an example, let $d_1 = d_2 = \lambda/4$. Then using the notation in Fig. 11.20, we can consider antennas 1 and 2 as constituting a unit for which the pattern is $|\sin [(\pi/2) \sin \psi]|$, which is that of case 2 in Example 11.3, except that ψ is measured from the line which is perpendicular to the axis of the array. Antennas 3 and 4 constitute a similar unit except for opposite polarity so that the group pattern for the two units is $|\sin [(\pi/2) \cos \psi]|$. Thus, the required radiation pattern is

$$\left| \sin \left(\frac{\pi}{2} \sin \psi \right) \sin \left(\frac{\pi}{2} \cos \psi \right) \right|$$

which is shown plotted in Fig. 11.21.

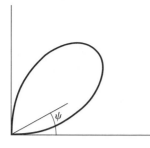

FIGURE 11.21 Radiation pattern in the cross-sectional plane for the case of $d_1 = d_2 = \lambda/4$ in the arrangement of Fig. 11.20.

K11.4. Antenna array; Unit pattern; Group pattern; Resultant pattern; Pattern multiplication; Uniform linear array; Image antenna concept; Corner reflector.

D11.6. For the array of two antennas of Example 11.3, assume that $d = 3\lambda/2$ and $\alpha = \pi/2$. Find the three lowest values of ψ for which the group pattern has nulls. *Ans.* 33.56°, 80.41°, 120°.

D11.7. Obtain the expression for the resultant pattern for each of the following cases of linear array of isotropic antennas: **(a)** three antennas carrying currents with amplitudes in the ratio 1:2:1, spaced λ apart and fed in phase; **(b)** five antennas carrying currents with amplitudes in the ratio 1:2:2:2:1, spaced $\lambda/2$ apart and with progressive phase shift of 180°; and **(c)** five antennas carrying currents in the ratio 1:2:3:2:1, spaced λ apart and fed in phase.
Ans. **(a)** $\cos^2(\pi \cos\psi)$; **(b)** $\sin^2[(\pi/2)\cos\psi]|\cos(\pi\cos\psi)|$;
(c) $[\sin^2(3\pi\cos\psi)]/[9\sin^2(\pi\cos\psi)]$.

D11.8. For the Hertzian dipole in the presence of the corner reflector of Fig. 11.20, let r be the ratio of the radiation field at a point in the cross-sectional plane and along the line extending from the corner through the dipole to the radiation field at the same point in the absence of the corner reflector. Find the value of r for each of the following cases: **(a)** $d_1 = d_2 = \sqrt{2}\lambda$; **(b)** $d_1 = d_2 = \lambda/4\sqrt{2}$; and **(c)** $d_1 = 0.3\lambda, d_2 = 0.4\lambda$.
Ans. **(a)** 0; **(b)** 2; **(c)** 3.275.

11.5 APERTURE ANTENNAS

An important class of antennas, called *aperture antennas,* is one for which the radiation is computed from a knowledge of the field distribution in an aperture instead of from a current distribution associated with the source of radiation, as has been the case thus far. The corner reflector discussed in the previous section is, in the practical case of finite-sized conductors (and, hence, defining an aperture), an example of such an antenna. Besides reflectors such as the corner reflector, other examples of aperture antennas are horns extending from waveguides, slots in conducting enclosures, and lenses. Essentially for an aperture antenna, the primary source, which is elsewhere, sets up the field distribution in the aperture, which in turn is assumed to give rise to secondary waves in accordance with the Huygens–Fresnel principle, introduced in Section 10.6. *Description and examples*

In particular, as mentioned in Section 10.6, the determination of the far field from an aperture antenna is the same as setting up the problem to solve for Fraunhofer diffraction from the aperture. To review briefly, consider a plane monochromatic wave incident normally on a screen in the *xy*-plane, with an *Far-field determination*

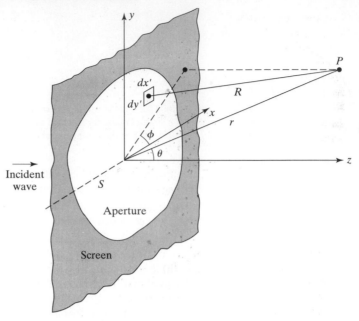

FIGURE 11.22   Geometry pertinent to the determination of the far field for radiation from an aperture antenna.

aperture cut into it, as shown in Fig. 11.22. Then, according to the Huygens–Fresnel principle, the incident wave may be thought of as giving rise to secondary (spherical) waves emanating from every point in the aperture and that interfere with one another to produce the field distribution away from the aperture. The scalar field at a point P is approximately given by

$$\overline{E}(P) \approx \frac{j\beta}{2\pi} \int_S \frac{\overline{E}(x', y', 0)}{R} e^{-j\beta R} \, dS \qquad (11.51)$$

where S is the area of the aperture, and $\overline{E}(x', y', 0)$ is the scalar field in the aperture. For the Fraunhofer approximation, the waves arriving at P approach plane waves, thereby permitting simplification of the integrand in (11.51) by using the plane wave approximation. This consists of assuming that the lines from points in the aperture $(x', y', 0)$ to the observation point $P(x, y, z)$ are all parallel $(x', y' \ll r)$, so that

$$
\begin{aligned}
R &= \sqrt{(x - x')^2 + (y - y')^2 + z^2} \\
&= \sqrt{r^2 - 2xx' - 2yy' + (x')^2 + (y')^2} \\
&= r\left[1 - \frac{2xx'}{r^2} - \frac{2yy'}{r^2} + \left(\frac{x'}{r}\right)^2 + \left(\frac{y'}{r}\right)^2\right]^{1/2} \qquad (11.52) \\
&\approx r\left(1 - \frac{xx'}{r^2} - \frac{yy'}{r^2}\right) \\
&= r - x' \sin\theta \cos\phi - y' \sin\theta \sin\phi
\end{aligned}
$$

For the R in the denominator in the integrand, further approximation can be made as $R \approx r$. Thus, (11.51) reduces to

$$\overline{E}(x, y, z) \approx \frac{j\beta}{2\pi r} e^{-j\beta r} \int_S \overline{E}(x', y', 0)\, e^{j\beta \sin\theta\,(x'\cos\phi + y'\sin\phi)}\, dx'\, dy' \qquad (11.53)$$

Equation (11.53) is the starting point for the determination of the far-field distribution for an aperture antenna. We shall illustrate by means of an example.

Example 11.5

Let us consider a rectangular aperture in the xy-plane and centered at the origin with a uniform field distribution $\overline{\mathbf{E}} = E_0 \mathbf{a}_y$ in it, as shown in Fig. 11.23, and investigate the characteristics of the far field due to it.

Rectangular aperture with uniform excitation

Applying (11.53) to the rectangular aperture, we have at a point $P(r, \theta, \phi)$ far from the aperture

$$\begin{aligned}
\overline{E} &\approx \frac{j\beta e^{-j\beta r}}{2\pi r} \int_{x'=-a/2}^{a/2} \int_{y'=-b/2}^{b/2} E_0 e^{j\beta \sin\theta\,(x'\cos\phi + y'\sin\phi)}\, dx'\, dy' \\
&= \frac{j\beta E_0 e^{-j\beta r}}{2\pi r} \int_{-a/2}^{a/2} e^{jx'\beta \sin\theta\cos\phi}\, dx' \int_{-b/2}^{b/2} e^{jy'\beta\sin\theta\sin\phi}\, dy'
\end{aligned} \qquad (11.54)$$

Evaluating the integrals, we obtain

$$\overline{E} = \frac{j\beta E_0\, abe^{-j\beta r}}{2\pi r} \left(\frac{\sin\psi_1}{\psi_1}\right)\left(\frac{\sin\psi_2}{\psi_2}\right) \qquad (11.55)$$

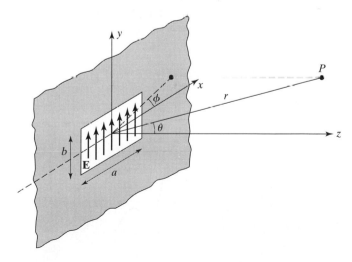

FIGURE 11.23 Rectangular aperture antenna with a uniform field distribution in the aperture.

where

$$\psi_1 = \frac{\beta a \sin \theta \cos \phi}{2} \tag{11.56a}$$

and

$$\psi_2 = \frac{\beta b \sin \theta \sin \phi}{2} \tag{11.56b}$$

Radiation   The quantities of interest in (11.55) are the $(\sin \psi)/\psi$ type of terms, which deter-
characteristics   mine the radiation pattern. To discuss this, we consider the two coordinate planes $\phi = 0$
and $\phi = 90°$ and find from (11.55) that the amplitudes of the fields in these two planes
are given by

$$\begin{aligned}
|\overline{E}|_{\phi=0} &= \frac{\beta E_0 ab}{2\pi r} \left| \frac{\sin \psi_1}{\psi_1} \right|_{\phi=0} \\
&= \frac{\beta E_0 ab}{2\pi r} \left| \frac{\sin [(\beta a \sin \theta)/2]}{(\beta a \sin \theta)/2} \right|
\end{aligned} \tag{11.57a}$$

and

$$\begin{aligned}
|\overline{E}|_{\phi=90°} &= \frac{\beta E_0 ab}{2\pi r} \left| \frac{\sin \psi_2}{\psi_2} \right|_{\phi=90°} \\
&= \frac{\beta E_0 ab}{2\pi r} \left| \frac{\sin [(\beta b \sin \theta)/2]}{(\beta b \sin \theta)/2} \right|
\end{aligned} \tag{11.57b}$$

where we have used the fact that $\lim_{\Delta \to 0} (\sin \Delta)/\Delta$ is equal to 1. Thus, in both planes,
the behavior is the same except for the appearance of the different dimensions a and b in
the $(\sin \psi)/\psi$ factors in (11.57a) and (11.57b), respectively.

To examine this behavior, we consider the plot of $|(\sin \psi)/\psi|$ versus ψ, which is
shown in Fig. 11.24. We note that it indicates a strong central maximum of unity at $\psi = 0$

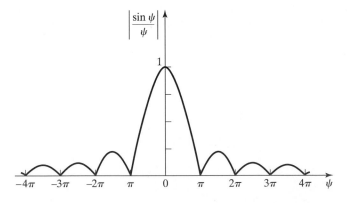

FIGURE 11.24   Variation of $|(\sin \psi)/\psi|$ with ψ, pertinent to the radia-
tion pattern for the rectangular aperture antenna of Fig. 11.23.

and a series of secondary (weaker) maxima on either side of it, with nulls occurring at $|\psi| = m\pi, m = 1, 2, 3, \ldots$. The secondary maxima, which occur at $|\psi| = 1.4303\pi, 2.459\pi, 3.471\pi, \ldots$, are successively less intense, having values $0.2172, 0.1284, 0.0913, \ldots$, respectively. If we consider the fact that the power density is proportional to $|\overline{E}|^2$, then the insignificance of these maxima becomes more evident, since the successive maxima of $|(\sin\psi)/\psi|^2$ are $1, 0.0472, 0.0165, 0.0083, \ldots$. Thus, the quantity of interest is the beam width between the first nulls (BWFN) between which the radiation is concentrated. The BWFN is given by twice the value of θ corresponding to the first null. For the $\phi = 0$ plane, this value is given by

$$\frac{\beta a \sin\theta}{2} = \pi \tag{11.58}$$

For narrow beams, which is the case in practice, $\sin\theta \approx \theta$ in this range, so that (11.58) can be written as

$$\frac{\beta a \theta}{2} \approx \pi \tag{11.59}$$

$$\theta \approx \frac{2\pi}{\beta a} = \frac{\lambda}{a}$$

or

$$\boxed{[\text{BWFN}]_{\phi=0} \approx \frac{2\lambda}{a}} \tag{11.60a}$$

Similarly,

$$\boxed{[\text{BWFN}]_{\phi=90°} \approx \frac{2\lambda}{b}} \tag{11.60b}$$

Finally, we consider the determination of the directivity of the rectangular aperture antenna. To do this, it is convenient to use the basic definition that

$$D = \frac{[P_r]_{\max}}{[P_r]_{\text{av}}} = \frac{\langle P_r\rangle_{\max}}{\langle P_r\rangle_{\text{av}}} = \frac{4\pi r^2\langle P_r\rangle_{\max}}{\langle P_{\text{rad}}\rangle} \tag{11.61}$$

instead of using (11.27), since P_{rad}, the power radiated from the antenna, being the same as that passing through the aperture, is much easier to compute from the aperture field distribution as compared to the evaluation of the integral in (11.27). Thus, in view of the uniform distribution of $\overline{E}(x, y, 0) = E_0\mathbf{a}_y$ in the aperture,

$$\langle P_{\text{rad}}\rangle = \frac{1}{2}\frac{E_0^2}{\eta_0}(ab) \tag{11.62}$$

and from (11.55),

$$\overline{E}_{\max} = \frac{j\beta E_0 abe^{-j\beta r}}{2\pi r}$$

$$\langle P_r\rangle_{\max} = \frac{1}{2}\frac{|\overline{E}_{\max}|^2}{\eta_0}$$

$$= \frac{\beta^2 E_0^2 a^2 b^2}{8\pi^2 r^2\eta_0} \tag{11.63}$$

Substituting (11.62) and (11.63) into (11.61), we obtain

$$D = \frac{\beta^2 ab}{\pi} = \frac{4\pi}{\lambda^2} (ab) \qquad (11.64)$$

This result tells us that the directivity of the rectangular aperture antenna is $4\pi/\lambda^2$ times the physical aperture, ab. Although we have derived it here for the rectangular aperture, it is true for an aperture of any shape with uniform excitation.

K11.5. Aperture antenna; Far field; Rectangular aperture; Uniform excitation; BWFN.

D11.9. For the rectangular aperture antenna of Fig. 11.23, the BWFN in the $\phi = 0$ plane is 0.1 rad and the directivity is 800π. Find the following in degrees: **(a)** the BWFN in the $\phi = 90°$ plane; **(b)** the half-power beamwidth (HPBW), that is, twice the value of θ for which the power density is one-half of the maximum power density in the $\phi = 0$ plane; and **(c)** the beamwidth between the first secondary maxima in the $\phi = 0$ plane.
Ans. **(a)** 11.46; **(b)** 2.54; **(c)** 8.19.

11.6 RECEIVING PROPERTIES

Reciprocity

Thus far, we have considered the radiating, or transmitting, properties of antennas. Fortunately, it is not necessary to repeat all the derivations for the discussion of the receiving properties of antennas since reciprocity dictates that the receiving pattern of an antenna be the same as its transmitting pattern. To illustrate this in simple terms without going through the general proof of reciprocity, let us consider a Hertzian dipole situated at the origin and directed along the z-axis, as shown in Fig. 11.25. We know that the radiation pattern is then given by $\sin \theta$ and that the polarization of the radiated field is such that the electric field is in the plane of the dipole axis.

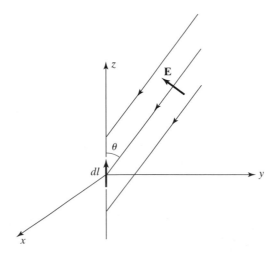

FIGURE 11.25 For investigating the receiving properties of a Hertzian dipole.

To investigate the receiving properties of the Hertzian dipole, we assume that it is situated in the radiation field of a second antenna so that the incoming waves are essentially uniform plane waves. Thus, let us consider a uniform plane wave with its electric field **E** in the plane of the dipole and incident on the dipole at an angle θ with its axis, as shown in Fig. 11.25. Then the component of the incident electric field parallel to the dipole is $E \sin \theta$. Since the dipole is infinitesimal in length, the voltage induced in the dipole, which is the line integral of the electric-field intensity along the length of the dipole, is simply equal to $(E \sin \theta)\, dl$ or to $E\, dl \sin \theta$. This indicates that for a given amplitude of the incident wave field, the induced voltage in the dipole is proportional to $\sin \theta$. Furthermore, for an incident uniform plane wave having its electric field normal to the dipole axis, the voltage induced in the dipole is zero; that is, the dipole does not respond to polarization with electric field normal to the plane of its axis. These properties are reciprocal to the transmitting properties of the dipole. Since an arbitrary antenna can be decomposed into a series of Hertzian dipoles, it then follows that reciprocity holds for an arbitrary antenna. Thus, the receiving pattern of an antenna is the same as its transmitting pattern.

Let us consider the loop antenna, a common type of receiving antenna. A simple form of loop antenna consists of a circular loop of wire with a pair of terminals. We shall orient the circular loop antenna with its axis aligned with the z-axis, as shown in Fig. 11.26, and we shall assume that it is electrically short; that is, its dimensions are small compared to the wavelength of the incident wave, so that the spatial variation of the field over the area of the loop is negligible. For a uniform plane wave incident on the loop, we can find the voltage induced in the loop, that is, the line integral of the electric-field intensity around the loop, by using Faraday's law. Thus, if **H** is the magnetic-field intensity associated with the wave, the magnitude of the induced voltage is given by

Loop antenna

$$|V| = \left| -\frac{d}{dt} \int_{\substack{\text{area of} \\ \text{the loop}}} \mathbf{B} \cdot d\mathbf{S} \right|$$

$$= \left| -\mu \frac{d}{dt} \int_{\substack{\text{area of} \\ \text{the loop}}} \mathbf{H} \cdot dS\, \mathbf{a}_z \right| \tag{11.65}$$

$$= \mu A \left| \frac{\partial H_z}{\partial t} \right|$$

where A is the area of the loop. Hence, the loop does not respond to a wave having its magnetic field entirely parallel to the plane of the loop, that is, normal to the axis of the loop.

For a wave having its magnetic field in the plane of the axis of the loop and incident on the loop at an angle θ with its axis, as shown in Fig. 11.26, $H_z = H \sin \theta$ and, hence, the induced voltage has a magnitude

$$\boxed{ |V| = \mu A \left| \frac{\partial H}{\partial t} \right| \sin \theta } \tag{11.66}$$

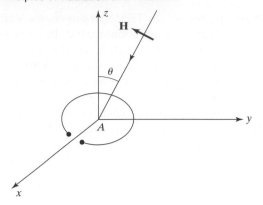

FIGURE 11.26   Circular loop antenna.

Thus, the receiving pattern of the loop antenna is given by sin θ, the same as that of a Hertzian dipole aligned with the axis of the loop antenna. The loop antenna, however, responds best to polarization with the magnetic field in the plane of its axis, whereas the Hertzian dipole responds best to polarization with the electric field in the plane of its axis.

Example 11.6

The directional properties of a receiving antenna can be used to locate the source of an incident signal. To illustrate the principle, as already discussed in Section 3.2, let us consider two vertical loop antennas, numbered 1 and 2, situated on the x-axis at $x = 0$ m and $x = 200$ m, respectively. By rotating the loop antennas about the vertical (z-axis), it is found that no (or minimum) signal is induced in antenna 1 when it is in the xz-plane and in antenna 2 when it is in a plane making an angle of 5° with the axis, as shown by the top view in Fig. 11.27. Let us find the location of the source of the signal.

Since the receiving properties of a loop antenna are such that no signal is induced for a wave arriving along its axis, the source of the signal is located at the intersection of the axes of the two loops when they are oriented so as to receive no (or minimum) signal. From simple geometrical considerations, the source of the signal is therefore located on the y-axis at $y = 200/\tan 5°$, or 2.286 km.

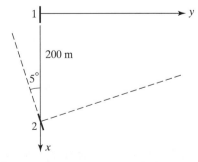

FIGURE 11.27   Top view of two loop antennas used to locate the source of an incident signal.

A useful parameter associated with the receiving properties of an antenna *Effective area*
is the effective area, denoted A_e and defined as the ratio of the time-average
power delivered to a matched load connected to the antenna to the time-average
power density of the appropriately polarized incident wave at the antenna. The
matched condition is achieved when the load impedance is equal to the complex
conjugate of the antenna impedance.

Let us consider the Hertzian dipole and derive the expression for its effective area. First, with reference to the equivalent circuit shown in Fig. 11.28, where
$\overline{V}_{oc}$ is the open-circuit voltage induced between the terminals of the antenna,
$\overline{Z}_A = R_A + jX_A$ is the antenna impedance, and $\overline{Z}_L = \overline{Z}_A^*$ is the load impedance,
we note that the time-average power delivered to the matched load is

$$P_R = \frac{1}{2}\left(\frac{|\overline{V}_{oc}|}{2R_A}\right)^2 R_A = \frac{|\overline{V}_{oc}|^2}{8R_A} \tag{11.67}$$

For a Hertzian dipole of length l, the open-circuit voltage is

$$\overline{V}_{oc} = \overline{E}l \tag{11.68}$$

where $\overline{E}$ is the electric field of an incident wave linearly polarized parallel to the
dipole axis. Substituting (11.68) into (11.67), we get

$$P_R = \frac{|\overline{E}|^2 l^2}{8R_A} \tag{11.69}$$

For a lossless dipole, $R_A = R_{rad} = 80\pi^2(l/\lambda)^2$, so that

$$P_R = \frac{|\overline{E}|^2 \lambda^2}{640\pi^2} \tag{11.70}$$

The time-average power density at the antenna is

$$\frac{|\overline{E}|^2}{2\eta_0} = \frac{|\overline{E}|^2}{240\pi} \tag{11.71}$$

Thus, the effective area is

$$A_e = \frac{|\overline{E}|^2\lambda^2/640\pi^2}{|\overline{E}|^2/240\pi} = \frac{3\lambda^2}{8\pi} \tag{11.72}$$

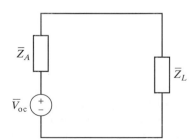

FIGURE 11.28   Equivalent circuit for a
receiving antenna connected to a load.

or

$$A_e = 0.1194\lambda^2 \tag{11.73}$$

In practice, R_A is greater than R_{rad} due to losses in the antenna, and the effective area is less than that given by (11.73). Rewriting (11.72) as

$$A_e = 1.5 \times \frac{\lambda^2}{4\pi}$$

and recalling that the directivity of the Hertzian dipole is 1.5, we observe that

$$A_e = \frac{\lambda^2}{4\pi} D \tag{11.74}$$

Although we have obtained this result for a Hertzian dipole, it can be shown that it holds for any antenna. It is of interest to note from (11.74) and (11.64) that the effective area of a rectangular aperture antenna for uniform field distribution in the aperture is equal to the physical aperture, which is to be expected.

Friis transmission formula   We shall now derive the *Friis transmission formula*, an important equation in making communication link calculations. To do this, let us consider two antennas, one transmitting and the other receiving, separated by a distance d. Let us assume that the antennas are oriented and polarization matched so as to maximize the received signal. Then if P_T is the transmitter power radiated by the transmitting antenna, the power density at the receiving antenna is $(P_T/4\pi d^2)D_T$, where D_T is the directivity of the transmitting antenna. The power received by a matched load connected to the terminals of the receiving antenna is then given by

$$P_R = \frac{P_T D_T}{4\pi d^2} A_{eR} \tag{11.75}$$

where A_{eR} is the effective area of the receiving antenna. Thus, the ratio of P_R to P_T is given by

$$\frac{P_R}{P_T} = \frac{D_T A_{eR}}{4\pi d^2} \tag{11.76}$$

Denoting A_{eT} to be the effective area of the transmitting antenna if it were receiving and using (11.74), we obtain

$$\frac{P_R}{P_T} = \frac{A_{eT} A_{eR}}{\lambda^2 d^2} \tag{11.77}$$

Equation (11.77) is the Friis transmission formula. It gives the maximum value of P_R/P_T for a given d and for a given pair of transmitting and receiving antennas. If the antennas are not oriented to receive the maximum signal, or if a polarization mismatch exists, or if the receiving antenna is not matched to its load,

P_R/P_T would be less than that given by (11.77). Losses in the antennas would also decrease the value of P_R/P_T.

An alternative formula to (11.77) is obtained by substituting for A_{eR} in (11.76) in terms of the directivity D_R of the receiving antenna if it were used for transmitting. Thus, we obtain

$$\boxed{\frac{P_R}{P_T} = \frac{D_T D_R \lambda^2}{16\pi^2 d^2}} \qquad (11.78)$$

K11.6. Receiving pattern; Reciprocity with transmitting pattern; Effective area; Communication link; Friis transmission formula.

D11.10. A communication link in free space uses two linear antennas of equal lengths L, oriented parallel to each other and normal to the line joining their centers. The antennas are separated by a distance $d = 1$ km. Find the maximum value of P_R/P_T for each of the following cases: **(a)** $L = 1$ m, $f = 10$ MHz; **(b)** $L = 1$ m, $f = 150$ MHz; and **(c)** $L = 2$ m, $f = 75$ MHz.
Ans. **(a)** 12.8×10^{-6}; **(b)** 6.8×10^{-8}; **(c)** 27.25×10^{-8}.

11.7 SUMMARY

In this chapter, we studied the principles of antennas. We first introduced the Hertzian dipole, which is an elemental wire antenna, and derived the electromagnetic field due to the Hertzian dipole by using the retarded magnetic vector potential. For a Hertzian dipole of length dl, oriented along the z-axis at the origin and carrying current

$$I(t) = I_0 \cos \omega t$$

we found the complete electromagnetic field to be given by

$$\mathbf{E} = \frac{2I_0 \, dl \cos \theta}{4\pi\varepsilon\omega} \left[\frac{\sin(\omega t - \beta r)}{r^3} + \frac{\beta \cos(\omega t - \beta r)}{r^2} \right] \mathbf{a}_r$$

$$+ \frac{I_0 \, dl \sin \theta}{4\pi\varepsilon\omega} \left[\frac{\sin(\omega t - \beta r)}{r^3} + \frac{\beta \cos(\omega t - \beta r)}{r^2} - \frac{\beta^2 \sin(\omega t - \beta r)}{r} \right] \mathbf{a}_\theta$$

$$\mathbf{H} = \frac{I_0 \, dl \sin \theta}{4\pi} \left[\frac{\cos(\omega t - \beta r)}{r^2} - \frac{\beta \sin(\omega t - \beta r)}{r} \right] \mathbf{a}_\phi$$

where $\beta = \omega\sqrt{\mu\varepsilon}$ is the phase constant.

For $\beta r \gg 1$ or for $r \gg \lambda/2\pi$, the only important terms in the complete field expressions are the $1/r$ terms since the remaining terms are negligible compared to these terms. Thus, for $r \gg \lambda/2\pi$, the Hertzian dipole fields are given by

$$\mathbf{E} = -\frac{\eta\beta I_0 \, dl \sin \theta}{4\pi r} \sin(\omega t - \beta r) \, \mathbf{a}_\theta$$

$$\mathbf{H} = -\frac{\beta I_0 \, dl \sin \theta}{4\pi r} \sin(\omega t - \beta r) \, \mathbf{a}_\phi$$

where $\eta = \sqrt{\mu/\varepsilon}$ is the intrinsic impedance of the medium. These fields, known as the radiation fields, correspond to locally uniform plane waves radiating away from the dipole and, in fact, are the only components of the complete fields contributing to the time-average radiated power. We found the time-average power radiated by the Hertzian dipole to be given by

$$\langle P_{rad} \rangle = \frac{1}{2} I_0^2 \left[\frac{2\pi\eta}{3} \left(\frac{dl}{\lambda} \right)^2 \right]$$

and identified the quantity inside the brackets to be its radiation resistance. The radiation resistance, R_{rad}, of an antenna is the value of a fictitious resistor that will dissipate the same amount of time-average power as that radiated by the antenna when a current of the same peak amplitude as that in the antenna is passed through it. Thus, for the Hertzian dipole,

$$R_{rad} = \frac{2\pi\eta}{3} \left(\frac{dl}{\lambda} \right)^2$$

We then examined the directional characteristics of the radiation fields of the Hertzian dipole as indicated by the factor $\sin\theta$ in the field expressions and hence by the factor $\sin^2\theta$ for the power density. We discussed the radiation patterns and introduced the concept of the directivity of an antenna. The directivity, D, of an antenna is defined as the ratio of the maximum power density radiated by the antenna to the average power density. For the Hertzian dipole,

$$D = 1.5$$

For the general case of a power density pattern $f(\theta, \phi)$, the directivity is given by

$$D = 4\pi \frac{[f(\theta, \phi)]_{max}}{\int_{\theta=0}^{\pi} \int_{\phi=0}^{2\pi} f(\theta, \phi) \sin\theta \, d\theta \, d\phi}$$

As an illustration of obtaining the radiation fields due to a wire antenna of arbitrary length and arbitrary current distribution by representing it as a series of Hertzian dipoles and using superposition, we considered the example of a center-fed half-wave dipole of length $L(= \lambda/2)$, oriented along the z-axis with its center at the origin and having the current distribution given by

$$I(z) = I_0 \cos \frac{\pi z}{L} \cos \omega t \qquad \text{for} \quad -L/2 < z < L/2$$

and found that the radiation fields are

$$\mathbf{E} = -\frac{\eta I_0}{2\pi r} \frac{\cos\left[(\pi/2) \cos\theta\right]}{\sin\theta} \sin\left(\omega t - \frac{\pi}{L} r\right) \mathbf{a}_\theta$$

$$\mathbf{H} = -\frac{I_0}{2\pi r} \frac{\cos\left[(\pi/2) \cos\theta\right]}{\sin\theta} \sin\left(\omega t - \frac{\pi}{L} r\right) \mathbf{a}_\phi$$

From these, we sketched the radiation patterns and computed the radiation resistance and the directivity of the half-wave dipole to be

$$R_{rad} = 73 \ \Omega \quad \text{for free space}$$
$$D = 1.642$$

We then extended the computation of these quantities to the case of a center-fed linear antenna of length equal to an arbitrary number of wavelengths.

We discussed antenna arrays and introduced the technique of obtaining the resultant radiation pattern of an array by multiplication of the unit and the group patterns. For an array of two antennas having the spacing d and fed with currents of equal amplitude but differing in phase by α, we found the group pattern for the fields to be $\left|\cos\left[(\beta d \cos \psi + \alpha)/2\right]\right|$, where ψ is the angle measured from the axis of the array, and we investigated the group patterns for several pairs of values of d and α. For example, for $d = \lambda/2$ and $\alpha = 0$, the pattern corresponds to maximum radiation broadside to the axis of the array, whereas for $d = \lambda/2$ and $\alpha = \pi$, the pattern corresponds to maximum radiation endfire to the axis of the array. We generalized the treatment to a uniform linear array of n antennas and briefly discussed the principle of a broadband array.

To take into account the effect of ground on antennas, we introduced the concept of an image antenna in a perfect conductor and discussed the application of the array techniques in conjunction with the actual and the image antennas to obtain the radiation pattern of the actual antenna in the presence of the ground. As another example of the image-antenna concept, we considered the corner reflector.

Next we discussed the far-field determination for an aperture antenna by recalling that it is equivalent to setting up the problem to solve for Fraunhofer diffraction from the aperture, which consists of using the plane wave approximation. By considering the example of a rectangular aperture with uniform field distribution in it, we illustrated the solution and studied the resulting radiation pattern and its characteristics.

Finally, we discussed receiving properties of antennas. In particular, (1) we discussed the reciprocity between the receiving and radiating properties of an antenna by considering the simple case of a Hertzian dipole, (2) we considered the loop antenna and illustrated the application of its directional properties for locating the source of a radio signal, and (3) we introduced the effective area concept and derived the Friis transmission formula.

REVIEW QUESTIONS

Q11.1. What is a Hertzian dipole? Discuss the time variations of the current and charges associated with the Hertzian dipole.

Q11.2. Discuss the analogy between the magnetic vector potential due to an infinitesimal current element and the electric scalar potential due to a point charge.

Q11.3. To what does the word *retarded* in the terminology *retarded magnetic vector potential* refer? Explain.

Q11.4. Outline the derivation of the electromagnetic field due to the Hertzian dipole.

Q11.5. Discuss the characteristics of the electromagnetic field due to the Hertzian dipole.

Q11.6. What are radiation fields? Why are they important? Discuss their characteristics.

Q11.7. Define the radiation resistance of an antenna.

Q11.8. Why is the expression for the radiation resistance of a Hertzian dipole not valid for a linear antenna of any length?

Q11.9. What is a radiation pattern?

Q11.10. Discuss the radiation pattern for the power density due to the Hertzian dipole.

Q11.11. Define the directivity of an antenna. What is the directivity of a Hertzian dipole?

Q11.12. How do you find the radiation fields due to an antenna of arbitrary length and arbitrary current distribution?

Q11.13. Discuss the evolution of the half-wave dipole from an open-circuited transmission line.

Q11.14. Justify the approximations involved in evaluating the integrals in the determination of the radiation fields due to the half-wave dipole.

Q11.15. What are the values of the radiation resistance and the directivity for a half-wave dipole?

Q11.16. What is an antenna array?

Q11.17. Justify the approximations involved in the determination of the resultant field of an array of two antennas.

Q11.18. What is an array factor? Provide a physical explanation for the array factor.

Q11.19. Discuss the concept of unit and group patterns and their multiplication to obtain the resultant pattern of an array.

Q11.20. Distinguish between broadside and endfire radiation patterns.

Q11.21. Discuss the principle of a phased array.

Q11.22. Discuss the principle of a broadband array using as an example the log-periodic dipole array.

Q11.23. Discuss the concept of an image antenna to find the field of an antenna in the vicinity of a perfect conductor.

Q11.24. What determines the sense of the current flow in an image antenna relative to that in the actual antenna?

Q11.25. How does the concept of an image antenna simplify the determination of the radiation pattern of an antenna above a perfect conductor surface?

Q11.26. Discuss the application of the image-antenna concept to the 90° corner reflector.

Q11.27. Explain the distinguishing feature pertinent to the computation of radiation from an aperture antenna.

Q11.28. Give examples of aperture antennas.

Q11.29. Discuss the determination of the far field for an aperture antenna.

Q11.30. Describe the radiation pattern for the far field of a rectangular aperture antenna with uniform field distribution in the aperture and discuss its characteristics.

Q11.31. Discuss the reciprocity associated with the transmitting and receiving properties of an antenna. Can you think of a situation in which reciprocity does not hold?

Q11.32. What is the receiving pattern of a loop antenna? How should you orient a loop antenna to receive **(a)** a maximum signal and **(b)** a minimum signal?

Q11.33. Discuss the application of the directional receiving properties of a loop antenna in the location of the source of a radio signal.

Q11.34. How is the effective area of a receiving antenna defined?

Q11.35. Outline the derivation of the expression for the effective area of a Hertzian dipole.

Q11.36. Discuss the derivation of the Friis transmission formula.

PROBLEMS

Section 11.1

P11.1. Show that (11.9) and (11.10) satisfy the Maxwell's curl equations for **E**.

P11.2. For the electromagnetic field due to the Hertzian dipole, show that **(a)** the time-average value of the θ-component of the Poynting vector is zero and **(b)** the contribution to the time-average value of the r-component of the Poynting vector is completely from the terms involving $1/r$.

P11.3. Show that the field expressions obtained by replacing ωt in (11.11) and (11.12) by $(\omega t - \beta r)$ do not satisfy Maxwell's curl equations.

P11.4. A Hertzian dipole of length 1 m situated at the origin and oriented along the positive z-direction carries the current $10 \cos 2\pi \times 10^6 t \cos 6\pi \times 10^6 t$ A. Find the root-mean-square values of $E_r, E_\theta,$ and H_ϕ at the point $(10, \pi/3, 0)$. Assume free space for the medium.

Section 11.2

P11.5. Show that the radiation fields given by (11.17a) and (11.17b) do not by themselves satisfy simultaneously the Maxwell's curl equations.

P11.6. Find the value of r at which the amplitude of the radiation field in the θ-component of **E** in (11.10) is equal to the resultant amplitude of the remaining two terms.

P11.7. Find the amplitude I_0 of the current with which a Hertzian dipole of length 0.5 m has to be excited at a frequency of 10 MHz to produce an electric-field intensity of amplitude 1 mV/m at a distance of 1 km broadside to the dipole, in free space. What is the time-average power radiated for the computed value of I_0?

P11.8. The power density pattern for an antenna located at the origin is given by

$$f(\theta, \phi) = \begin{cases} \csc^2 \theta & \text{for} \quad \pi/6 \le \theta \le \pi/2 \\ 0 & \text{otherwise} \end{cases}$$

Find the directivity of the antenna.

P11.9. Find the ratio of the currents in two antennas having directivities D_1 and D_2 and radiation resistances R_{rad1} and R_{rad2} for which the maximum time-average radiated power densities are equal.

P11.10. For the Hertzian dipole of Problem P11.4, calculate the time-average power radiated by the dipole.

Section 11.3

P11.11. For the half-wave dipole of Section 11.3, find the magnetic vector potential for the radiation fields and show that the radiation fields obtained from it are the same as those given by (11.31a) and (11.31b).

P11.12. Find the maximum amplitude I_0 of the current with which a linear dipole of length 15 m has to be excited at a frequency of 10 MHz in order to produce an electric-field intensity of amplitude 1 mV/m at a distance of 1 km broadside to the dipole, in free space. What is the time-average power radiated for the computed value of I_0?

P11.13. Repeat Problem P11.12 for a linear dipole of length 15 m at a frequency of 50 MHz.

P11.14. A short dipole is a center-fed straight-wire antenna having a length small compared to a wavelength. The amplitude of the current distribution can then be approximated as decreasing linearly from a maximum at the center to zero at the ends. Thus, for a short dipole of length L lying along the z-axis between $z = -L/2$ and $z = L/2$, the current distribution is given by

$$I(z) = \begin{cases} I_0\left(1 + \dfrac{2z}{L}\right)\cos\omega t & \text{for} \quad -L/2 < z < 0 \\[2mm] I_0\left(1 - \dfrac{2z}{L}\right)\cos\omega t & \text{for} \quad 0 < z < L/2 \end{cases}$$

(a) Obtain the radiation fields of the short dipole. **(b)** Find the radiation resistance and the directivity of the short dipole.

P11.15. Consider a circular loop antenna of radius a such that the circumference is small compared to the wavelength. Assume the loop antenna to be in the xy-plane with its center at the origin and the loop current to be $I = I_0 \cos\omega t$ in the sense of increasing ϕ. Show that for obtaining the radiation fields, the magnetic vector potential due to the loop antenna is given by

$$\mathbf{A} = \frac{\mu_0 I_0 \pi a^2 \beta \sin\theta}{4\pi r} \sin(\omega t - \beta r)\,\mathbf{a}_\phi$$

where $\beta = \omega/v_p$. Then show that the radiation fields are

$$\mathbf{E} = \frac{\eta I_0 \pi a^2 \beta^2 \sin\theta}{4\pi r} \cos(\omega t - \beta r)\,\mathbf{a}_\phi$$

$$\mathbf{H} = \frac{I_0 \pi a^2 \beta^2 \sin\theta}{4\pi r} \cos(\omega t - \beta r)\,\mathbf{a}_\theta$$

P11.16. Find the radiation resistance and the directivity of the circular loop antenna of Problem P11.15. Compare the dependence of the radiation resistance on the electrical size (circumference/wavelength) to the dependence of the radiation resistance of the Hertzian dipole on its electrical size (length/wavelength).

Section 11.4

P11.17. For the array of two antennas of Example 11.3, find and sketch the group pattern for each of the following cases: **(a)** $d = \lambda, \alpha = \pi/2$; and **(b)** $d = 2\lambda, \alpha = 0$.

P11.18. For the array of two Hertzian dipoles of Fig. 11.11, find and sketch the resultant pattern in the xz-plane for each of the following cases: **(a)** $d = \lambda/2, \alpha = \pi$; and **(b)** $d = \lambda/4, \alpha = -\pi/2$.

P11.19. Use the pattern multiplication technique in reverse to synthesize an array of isotropic elements for the group pattern

$$\cos^2\left(\frac{\pi}{2}\cos\psi\right)\left|\sin\left(\frac{\pi}{2}\cos\psi\right)\right|$$

P11.20. Repeat Problem P11.19 for the group pattern

$$\frac{\cos^2 (6\pi \cos \psi)}{9 \cos^2 (2\pi \cos \psi)}$$

P11.21. For a horizontal half-wave dipole at a height $\lambda/4$ above a plane, perfect conductor surface, find and sketch the radiation pattern in **(a)** the vertical plane perpendicular to the axis of the antenna and **(b)** the vertical plane containing the axis of the antenna.

P11.22. For a vertical antenna of length $\lambda/4$ above a plane, perfect conductor surface, find **(a)** the radiation pattern in the vertical plane and **(b)** the directivity.

P11.23. A Hertzian dipole is situated parallel to one side and perpendicular to the other side of a 90° corner reflector, as shown in Fig. 11.29. Find the expression for the radiation pattern in the plane of the paper as a function of the angle θ and the distances d_1 and d_2.

FIGURE 11.29   For Problem P11.23.

P11.24. A $\lambda/4$ monopole is situated parallel to one side and perpendicular to the other side of a 90° corner reflector, as shown in Fig. 11.30. Find the radiation pattern in the plane of the paper as a function of the angle θ.

FIGURE 11.30   For Problem P11.24.

P11.25. A corner reflector is made up of two semi-infinite, plane, perfect conductors at an angle of 60°, as shown by the cross-sectional view in Fig. 11.31. A Hertzian dipole is situated parallel to the conductors at a distance of $k\lambda$ from the corner along the bisector of the two conductors. Find the ratio of the radiation field at a point broadside to the dipole and along the bisector of the conductors to the radiation field at the same point in the absence of the corner reflector for the following values of k: **(a)** $\frac{1}{4}$; **(b)** $\frac{1}{2}$; and **(c)** 1.

FIGURE 11.31   For Problem P11.25.

Section 11.5

P11.26. For the rectangular aperture antenna of Example 11.5, assume that the field distribution in the aperture is nonuniform as given by

$$\overline{E}(x, y, 0) = E_0 \cos \frac{\pi x}{a} \mathbf{a}_y \qquad \text{for} \quad -a/2 < x < a/2, -b/2 < y < b/2$$

Obtain the expression for the far field and hence the expressions for the following: **(a)** BWFN in the $\phi = 0$ plane; **(b)** BWFN in the $\phi = 90°$ plane; **(c)** HPBW in the $\phi = 0$ plane; and **(d)** the directivity.

P11.27. Repeat Problem P11.26 for

$$\overline{E}(x, y, 0) = E_0 \cos^2 \frac{\pi x}{a} \mathbf{a}_y \qquad \text{for} \quad -a/2 < x < a/2, -b/2 < y < b/2$$

P11.28. Consider a circular aperture of radius a in the xy-plane and centered at the origin. For uniform field distribution $\mathbf{E} = E_0 \mathbf{a}_x$ in the aperture, show that the far-field radiation pattern is in accordance with $J_1(\beta a \sin \theta)/(\beta a \sin \theta)$. Further, given that the first nonzero root of $J_1(x) = 0$ is 3.83, show that in any constant-ϕ-plane, the BWFN is approximately equal to $1.22\lambda/a$.

$$\left[\textit{Note:} \; \frac{1}{2\pi} \int_0^{2\pi} e^{jx \cos \alpha} d\alpha = J_0(x) \int x J_0(x) \, dx = x J_1(x) \right]$$

P11.29. Consider the uniform linear array of n isotropic antennas of Fig. 11.15 for the case of $\alpha = 0$ so that the group pattern is a broadside pattern. Show that for large n and for $nd \gg \lambda$, the radiation pattern is the same as that in one of the coordinate planes ($\phi = 0$ or $\phi = 90°$) for the rectangular aperture antenna with uniform field distribution of Example 11.5, and hence the BWFN is approximately equal to $2\lambda/L$, where L is the length of the array.

Section 11.6

P11.30. An arrangement of two identical Hertzian dipoles situated at the origin and oriented along the x- and y-axes, known as the turnstile antenna, is used for receiving circularly polarized signals arriving along the z-axis. Determine how you would combine the voltages induced in the two dipoles so that the turnstile antenna is responsive to circular polarization rotating in the clockwise sense as viewed by the antenna but not to that of the counterclockwise sense of rotation.

P11.31. A uniform plane wave is incident on an interferometer consisting of an array of two identical antennas with spacing $d = 3\lambda$ at an angle $\psi = 50°$ to the axis of the array, producing a phase difference $\Delta\phi$ between the voltages induced in the two antennas. Find all possible values of $0° < \psi < 180°$ that result in a phase difference equal to $\Delta\phi \pm 2n\pi$, where n is an integer, between the two induced voltages.

P11.32. A communication link at a frequency of 30 MHz uses a half-wave dipole for the transmitting antenna and a small loop for the receiving antenna, involving a distance of 100 km. The antennas are oriented so as to receive maximum signal and the receiving antenna is matched to its load. If the received time-average power is to be 1 μW, find the minimum required value of the maximum amplitude I_0 of the current with which the transmitting antenna has to be excited. Assume the antennas to be lossless.

REVIEW PROBLEMS

R11.1. Two identical current elements are located at the origin, one directed along the positive x-axis and the other directed along the positive z-axis. They carry currents equal in amplitude and 90° out of phase. Find the expression for the locus of all points at which the polarization of the field is circular on the surface of a sphere of radius r, where $\beta r \gg 1$.

R11.2. For the array of two Hertzian dipoles in Fig. 11.11, assume that $d \ll \lambda$ and $\alpha = \pi$. Obtain an approximate expression for the three-dimensional power density pattern $f(\theta, \phi)$ and find the directivity and the radiation resistance.

R11.3. Show that the radiation pattern for a center-fed linear antenna of length equal to an odd-integer number of half-wavelengths, n, obtained by setting $k = n/2$ in (11.41) agrees with the one obtained by considering the antenna as an array of n half-wave dipoles of currents of equal amplitudes and appropriate progressive phase shift.

R11.4. Synthesize an array of isotropic elements for the group pattern

$$\frac{\sin (2\pi \cos \psi)}{\sin [(\pi/2) \cos \psi]} \cos^2 \left(\frac{\pi}{2} \cos \psi\right)$$

R11.5. A Hertzian dipole is situated at a distance d from the corner along the bisector of the two conductors of a 90° corner reflector and oriented normal to the bisector in the cross-sectional plane, as shown in Fig. 11.32. Obtain the expression for the radiation pattern in the cross-sectional plane, as a function of the angle θ.

FIGURE 11.32   For Problem R11.5.

R11.6. Consider the uniform linear array of n isotropic antennas of Fig. 11.15 for the case of $\alpha = -\beta d$, so that the group pattern is an endfire pattern. Show that for large n and for $nd \gg \lambda$, the BWFN is approximately equal to $\sqrt{8\lambda/L}$.

CHAPTER 12

Topics in Numerical Electromagnetics

Having provided the theoretical and analytical treatment on the various elements of engineering electromagnetics in the previous chapters, we turn our attention in this final chapter to an introduction of numerical methods. In particular, we are interested in the application of well-known numerical techniques for solving integral equations and partial differential equations to the numerical solution of Maxwell's equations and equations involving potential.

In this context, we shall consider four methods: (1) the finite-difference method, leading to the solution of Laplace's equation in two dimensions by using appropriate finite-difference approximations to the derivative terms; (2) the method of moments involving the inversion of an integral equation relating the electric potential to charge distribution by approximating the integral as a summation; (3) the finite-element method, also for solving Laplace's equation in two dimensions, but based on the minimization of electric energy expressed as an integral over the region of interest; and (4) the finite-difference time-domain method, for solving the one-dimensional wave equation or the first-order differential equations leading to it by extending the finite-difference approximations to the time derivative terms. We shall also present several examples of applications, including the determination of transmission-line parameters and the time-domain analysis of an initially charged transmission line.

12.1 FINITE-DIFFERENCE METHOD

The finite-difference method is employed for solving differential equations, and it is perhaps the simplest method for that purpose. It consists of replacing the derivative terms in the differential equation by their finite-difference approximations and solving the resulting algebraic equations. To do this, the region of interest is discretized by selecting a set of grid points and the derivatives of the function of interest at each grid point are expressed in terms of the values of the function at a subset of the grid points by using approximations such as the central-difference formulas. The resulting set of algebraic equations are solved for the values of the function at the grid points. We shall illustrate this first in one-dimension.

Thus, let us consider solving the differential equation

$$\frac{d^2f(x)}{dx^2} + f(x) = 0 \qquad (12.1)$$

Solution of one-dimensional differential equation

over the region $0 \leq x \leq 1$, with the boundary conditions specified as $f(0) = 0$ and $f(1) = 1$. Then we divide the region of interest $0 \leq x \leq 1$ into n equal segments, thereby identifying $(n + 1)$ grid points, including the two end points. The situation is illustrated in Fig. 12.1(a) for $n = 4$. Since the values of f at the end points 0 and 4 are specified, we need to find the values at the three interior grid points 1, 2, and 3, and hence we need to obtain a set of three algebraic equations.

Let us consider the kth grid point, where $k - 1, 2, 3$. Then, at that grid point, $x = ka$, where a is the spacing between two adjacent grid points, as shown in Fig. 12.1(b) We can approximate d^2f/dx^2 at this grid point as

$$\left[\frac{d^2f}{dx^2}\right]_k = \left[\frac{d^2f}{dx^2}\right]_{x=ka}$$

$$\approx \frac{1}{a}\left[\left(\frac{df}{dx}\right)_{x=(k-0.5)a} - \left(\frac{df}{dx}\right)_{x=(k+0.5)a}\right] \qquad (12.2)$$

$$\approx \frac{1}{a}\left[\left(\frac{f_{k+1} - f_k}{a}\right) - \left(\frac{f_k - f_{k-1}}{a}\right)\right]$$

$$= \frac{1}{a^2}(f_{k+1} - 2f_k + f_{k-1})$$

(a)

(b)

FIGURE 12.1   For the solution of one-dimensional differential equation using the finite-difference method.

where f_k is the value of f at the kth grid point, that is, at $x = ka$. The right side of (12.2) is the central-difference approximation for the second derivative of f at the grid point k.

Using (12.2) and noting that here a is equal to $\frac{1}{4}$, we can write the finite-difference approximation for the differential equation at the kth grid point as

$$16(f_{k+1} - 2f_k + f_{k-1}) + f_k = 0$$

or

$$16f_{k-1} - 31f_k + 16f_{k+1} = 0 \tag{12.3}$$

Applying this result to the three interior grid points 1, 2, and 3, we obtain the set of three equations

$$16f_0 - 31f_1 + 16f_2 = 0 \tag{12.4a}$$
$$16f_1 - 31f_2 + 16f_3 = 0 \tag{12.4b}$$
$$16f_2 - 31f_3 + 16f_4 = 0 \tag{12.4c}$$

Recognizing that $f_0 = 0$ and $f_4 = 1$, these three equations can be arranged in matrix form as

$$\begin{bmatrix} -31 & 16 & 0 \\ 16 & -31 & 16 \\ 0 & 16 & -31 \end{bmatrix} \begin{bmatrix} f_1 \\ f_2 \\ f_3 \end{bmatrix} = \begin{bmatrix} 0 \\ 0 \\ -16 \end{bmatrix} \tag{12.5}$$

Solving we obtain $f_1 = 0.2943$, $f_2 = 0.5702$, and $f_3 = 0.8109$. An analytical solution reveals that the exact solution for $f(x)$ is $(\sin x)/(\sin 1)$, which gives $f_1 = 0.2940$, $f_2 = 0.5697$, and $f_3 = 0.8101$. Thus, the numerical solution is accurate to the fourth decimal place even for the number of interior grid points as small as 3.

Solution of two-dimensional Laplace's equation

The procedure can be extended to two-dimensional and three-dimensional differential equations. One equation of interest is the Laplace's equation (6.14). We shall consider the two-dimensional Laplace's equation in the Cartesian coordinates x and y, given by

$$\nabla^2 V = \frac{\partial^2 V}{\partial x^2} + \frac{\partial^2 V}{\partial y^2} = 0 \tag{12.6}$$

To introduce the principle behind the numerical solution of (12.6), let us suppose that we know the potentials V_1, V_2, V_3, and V_4 at four points equidistant from a point $P(0, 0, 0)$ and lying on mutually perpendicular axes, x and y, passing through P as shown in Fig. 12.2, and that we wish to find the potential V_0 at P in terms of V_1, V_2, V_3, and V_4. Then we require that

$$[\nabla^2 V]_P = \left[\frac{\partial^2 V}{\partial x^2} + \frac{\partial^2 V}{\partial y^2} \right]_{(0,0,0)} = 0 \tag{12.7}$$

To solve this equation approximately for V_0, we note that

$$\left[\frac{\partial^2 V}{\partial x^2} \right]_{(0,0,0)} = \left[\frac{\partial}{\partial x} \left(\frac{\partial V}{\partial x} \right) \right]_{(0,0,0)}$$

$$\approx \frac{1}{a} \left\{ \left[\frac{\partial V}{\partial x} \right]_{(a/2,0,0)} - \left[\frac{\partial V}{\partial x} \right]_{(-a/2,0,0)} \right\}$$

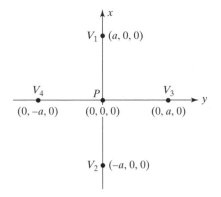

FIGURE 12.2   For illustrating the principle behind the numerical solution of Laplace's equation in two dimensions.

$$\approx \frac{1}{a} \left\{ \frac{[V]_{(a,0,0)} - [V]_{(0,0,0)}}{a} - \frac{[V]_{(0,0,0)} - [V]_{(-a,0,0)}}{a} \right\}$$

$$= \frac{1}{a^2} [(V_1 - V_0) - (V_0 - V_2)]$$

$$= \frac{1}{a^2} (V_1 + V_2 - 2V_0) \tag{12.8a}$$

Similarly,

$$\left[\frac{\partial^2 V}{\partial y^2} \right]_{(0,0,0)} \approx \frac{1}{a^2} (V_3 + V_4 - 2V_0) \tag{12.8b}$$

Substituting (12.8a) and (12.8b) into (12.7) and rearranging, we obtain

$$\boxed{ V_0 \approx \tfrac{1}{4}(V_1 + V_2 + V_3 + V_4) } \tag{12.9}$$

Thus, the potential at P is approximately equal to the average of the potentials at the four equidistant points lying along mutually perpendicular axes through P. The result becomes more and more accurate as the spacing a becomes less and less. Equation (12.9) is the finite-difference approximation to (12.6) and forms the basis for its numerical solution by the finite-difference method. We shall illustrate this by means of an example.

Example 12.1

Let us consider four infinitely long conducting strips of equal widths, situated such that the cross section of the arrangement is a square and held at potentials V_a, V_b, V_i, and V_r, as shown in Fig. 12.3. Note that the corners are insulated so that the plates do not touch. By dividing the area between the conductors into a 6×6 grid of squares, and using (12.9), we wish to find the approximate values of the potentials at the grid points by the finite-difference method.

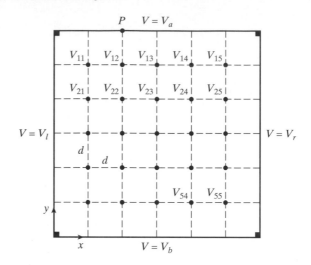

FIGURE 12.3   Cross-sectional view of an arrangement of four infinitely long conducting strips, with the region inside divided into a 6×6 grid of squares.

The solution consists of obtaining a set of values for the potentials at the grid points such that the potential at each grid point is the average of the potentials at the neighboring four grid points to within a specified tolerance. Thus, if we denote the potentials to be $V_{11}, V_{12}, V_{13}, V_{14}, V_{15}, V_{21}, V_{22}, \ldots V_{55}$, and if the specified tolerance is denoted to be Δ, then the values of the potentials must be such that

$$\left| V_{11} - \tfrac{1}{4}(V_a + V_{12} + V_{21} + V_l) \right| < \Delta \tag{12.10a}$$

$$\left| V_{12} - \tfrac{1}{4}(V_a + V_{13} + V_{22} + V_{11}) \right| < \Delta \tag{12.10b}$$

and so on. The simplest technique adaptable to computer solution is to begin with values of zero for all unknown potentials. By traversing the grid in a systematic manner, the average of the four neighboring potentials is computed for each grid point and is used to replace the potential at that grid point if that value differs from the computed average by more than Δ. This procedure is repeated until a final set of values for the unknown potentials consistent with (12.10a), (12.10b), . . . is obtained.

Let us consider some numerical values: $V_a = 100$ V, $V_b = 0$ V, $V_l = 40$ V, $V_r = 0$ V, and $\Delta = 0.01$ V. Then we first set all unknown potentials equal to zero. Beginning at the grid point 11 and traversing the grid rowwise, we replace the zero value for V_{11} by $\tfrac{1}{4}(100 + 40 + 0 + 0)$, or 35 V, then replace the zero value for V_{12} by $\tfrac{1}{4}(100 + 35 + 0 + 0)$, or 33.75 V, and so on. After one traversal is completed, we come back to the grid point 11 and traverse the grid again, replacing the potential value at each grid point by the average of the then existing values of the four neighboring potentials, as necessary. This procedure is repeated until the desired set of values is obtained.

The procedure just discussed can be very conveniently carried out by using a computer program. The final set of values from the run of such a program for $V_a = 100$ V, $V_b = 0$ V, $V_l = 40$ V, $V_r = 0$ V, and $\Delta = 0.01$ V is shown in Fig. 12.4, which also shows the residuals, where a residual at a grid point is the absolute value of the difference between the potential at that grid point and the average of the four neighboring potentials. The

	100	100	100	100	100	
40	65.60	72.71	72.30	65.76	48.10	0
	0.006	0.006	0.004	0.006	0.006	
40	49.69	52.99	50.73	42.68	26.66	0
	0.006	0.007	0.004	0.007	0.006	
40	40.21	38.84	34.95	27.61	15.89	0
	0.004	0.004	0.000	0.004	0.004	
40	32.32	27.23	22.65	16.92	9.29	0
	0.006	0.007	0.004	0.007	0.006	
40	21.86	15.14	11.49	8.19	4.36	0
	0.006	0.006	0.004	0.006	0.006	
	0	0	0	0	0	

ITERATION NO. = 25
SOLUTION COMPLETED
VALUE OF DELTA ACHIEVED <= 7.423401E-03

FIGURE 12.4   Final set of values of potentials and residuals for the arrangement of Fig. 12.3, for V_a = 100 V, V_b = 0 V, V_l = 40 V, V_r = 0 V, and Δ = 0.01 V.

residuals are shown below the potential values. It can be seen that all residuals are less than 0.01 V.

The method we just discussed is known as the *iteration* technique since it involves the iterative process of converging an initially assumed solution to a final one consistent with Laplace's equation in the approximate sense given by (12.9). There are several variations of the iteration technique. For example, by employing an initial guess other than zeros, a faster convergence may be achieved. The end result will, however, still be only to within the specified accuracy. Alternative to the iteration technique, one can write a set of simultaneous equations by applying (12.9) to each grid point and then solve the equations for the unknown potentials, as already illustrated for the one-dimensional case.

Iteration technique

The solution obtained for the potentials at the grid points by any method can be used to plot approximately the equipotential lines by interpolating between grid points. An example of such plotting also by using a computer is shown in Fig. 12.5, which corresponds to that of $V_a = V_l = V_b = 0$ V and $V_r = 100$ V in Fig. 12.3, and an 8×8 grid of squares. Figure 12.5(a) shows the computed potential values at a 4×4 set of grid points (with the remaining grid points omitted for the sake of clarity) and the 25-V equipotential line being plotted. Figure 12.5(b) shows a complete set of equipotential lines from 0 to 100 V in steps of 10 V. Note that in Fig. 12.5(b) the 0-V equipotential line does not follow the boundary at the upper- and lower-left corners. This is because in view of the division of the region into a finite grid of squares (8×8 here), the solution is not influenced by the corner points; that is, the solution for the case of the 0-V conductor following the plotted 0-V line is the same as that for which it follows the original boundary.

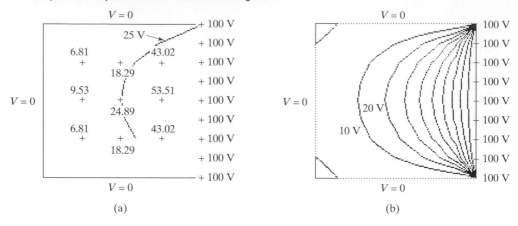

FIGURE 12.5   (a) Plotting of an equipotential line by interpolation between grid points. (b) Set of equipotential lines from 0 to 100 V in steps of 10 V.

Finally, the solution for the potentials can also be used to find approximate electric field intensities at the grid points by using the potential values to obtain approximate values of $\partial V/\partial x$ and $\partial V/\partial y$. For example, in Fig. 12.3, the electric-field intensity at the grid point 12 is given approximately by

$$[\mathbf{E}]_{12} \approx \frac{V_{11} - V_{13}}{2d}\mathbf{a}_x + \frac{V_{22} - V_a}{2d}\mathbf{a}_y$$

where d is the spacing between two adjacent grid points. Similarly, the electric-field intensities at points on the conductors can be found and used to obtain the surface charge densities. For example, the surface charge density at the point P on the conductor of potential V_a and adjacent to the grid point 12 is given approximately by

$$[\rho_S]_P \approx -\mathbf{a}_y \cdot \varepsilon \frac{V_{12} - V_a}{d}\mathbf{a}_y$$

$$= \varepsilon \frac{V_a - V_{12}}{d}$$

where ε is the permittivity of the medium between the conductors.

K12.1. Finite-difference method; Solution of one-dimensional differential equation; Solution of Laplace's equation in two dimensions; Iteration technique.

D12.1. Three infinitely long conductor strips are arranged such that the cross section is an isosceles triangle, as shown in Fig. 12.6. The region between the conductors is divided into a grid of points as shown in the figure, where the spacing between adjacent pairs of points is d. By writing equations consistent with (12.9) for the potentials at the grid points A, B, and C and solving them, find the following: **(a)** the approximate potential at the grid point C; **(b)** the approximate electric field intensity at the grid point C; and **(c)** the approximate surface charge density at the point P, assuming the medium between the conductors to be free space.
Ans.   **(a)** 8 V;   **(b)** $-(5\mathbf{a}_x + 7\mathbf{a}_y)/d$ V/m;   **(c)** $-4\varepsilon_0/d$ C/m^2.

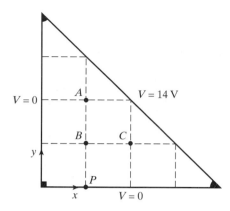

FIGURE 12.6 For Problem D12.1.

12.2 METHOD OF MOMENTS

When the boundaries of the physical arrangement extend to infinity, the finite-difference method cannot be used unless some approximations are made to limit the extent for the grid to be finite. Another numerical technique, known as the *method of moments*, is useful in such situations. The method of moments is commonly used to solve integral equations. An example consists of finding the charge distribution on the conductors held at known constant potentials. Thus, the problem is the inverse of the problem of finding the potential for a known charge distribution. To cast the technique in general terms, let us consider a surface charge distribution $\rho_S(x, y, z)$ on a given surface. Then applying superposition in conjunction with the expression for the potential due to a point charge given by (4.100), the potential due to the charge distribution can be expressed as

$$V(x, y, z) = \frac{1}{4\pi\varepsilon_0} \int_{\substack{\text{surface of} \\ \text{the charge} \\ \text{distribution}}} \frac{\rho_S(x', y', z')}{R} \, dS' \qquad (12.11)$$

where the primes denote source point coordinates. The procedure consists of dividing the surface into a finite number of subsections to approximate the integral in (12.11) by a summation and apply the equation to points on the subsections to obtain a set of linear algebraic equations. The set of equations is then inverted to obtain the desired solution. We shall illustrate the method by means of an example.

Example 12.2

Let us consider a thin, straight wire of length l and radius a ($\ll l$), as shown in Fig. 12.7(a) and held at a potential of 1 V. We wish to obtain the resulting (surface) charge distribution on the wire by the method of moments.

Thin, straight wire held at known potential

The determination of the charge distribution by the method of moments consists of dividing the wire into a number of segments, assuming the charge density in each

FIGURE 12.7   (a) Thin wire divided into five equal segments. (b) For the determination of the potential due to a cylindrical surface charge.

segment to be uniform and setting up and solving a set of algebraic equations. For simplicity of illustration, we shall divide the wire into five equal segments numbered 1 through 5 and having surface charge densities $\rho_{S1}, \rho_{S2}, \ldots, \rho_{S5}$. From considerations of symmetry, there are then only three unknowns, since $\rho_{S4} = \rho_{S2}$ and $\rho_{S5} = \rho_{S1}$. Hence, we need three independent equations.

An equation is obtained by writing the potential at the center point of a given segment to be the superposition of the potentials at that point due to the charges in the five segments. To obtain the contribution due to a segment, we consider the cylindrical surface charge of uniform density ρ_{S0} coaxial with the z-axis and located symmetrically about the origin, as shown in Fig. 12.7(b), and compute the potential due to it at two points: (1) at the origin and (2) at a point $(0, 0, z)$, where $z > d$, using the approximation $a \ll d$. Case 1 is appropriate to finding the potential due to the charge in a given segment in Fig. 12.7(a) at its own center point, whereas case 2 is appropriate to finding the potential due to the charge in a given segment in Fig. 12.7(a) at the center point of another segment.

Dividing the cylindrical surface charge in Fig. 12.7(b) into a number of ring charges, one of which is shown in the figure, and using superposition, we obtain

$$[V]_{(0,0,0)} = \int_{z'=-d}^{d} \int_{\phi=0}^{2\pi} \frac{\rho_{S0}a \, d\phi \, dz'}{4\pi\varepsilon_0 \sqrt{a^2 + (z')^2}}$$

$$= \frac{\rho_{S0}a}{2\varepsilon_0} \{\ln [z' + \sqrt{a^2 + (z')^2}]\}_{z'=-d}^{d}$$

$$= \frac{\rho_{S0}a}{2\varepsilon_0} \ln \frac{d + \sqrt{a^2 + d^2}}{-d + \sqrt{a^2 + d^2}}$$

which for $a \ll d$ reduces to

$$[V]_{(0,0,0)} \approx \frac{\rho_{S0}a}{2\varepsilon_0} \ln \frac{2d}{-d + d(1 + a^2/2d^2)}$$

$$= \frac{\rho_{S0}a}{\varepsilon_0} \ln \frac{2d}{a}$$ (12.12a)

For a point $P(0,0,z)$, where $z > d$, we can consider the cylindrical surface charge to be a line charge of density $2\pi a\rho_{S0}$ and write

$$[V]_P = \int_{z'=-d}^{d} \frac{2\pi a\rho_{S0}\, dz'}{4\pi\varepsilon_0(z - z')}$$

$$= \frac{\rho_{S0}a}{2\varepsilon_0} [-\ln (z - z')]_{z'=-d}^{d}$$ (12.12b)

$$= \frac{\rho_{Su}a}{2\varepsilon_0} \ln \frac{z + d}{z - d}$$

Applying (12.12a) and (12.12b) to write the equation for the potential at the center of segment 1 in Fig. 5.11(a), we obtain

$$\frac{\rho_{S1}a}{\varepsilon_0} \ln \frac{l}{5a} + \frac{\rho_{S2}a}{2\varepsilon_0} \ln 3 + \frac{\rho_{S3}a}{2\varepsilon_0} \ln \frac{5}{3} + \frac{\rho_{S4}a}{2\varepsilon_0} \ln \frac{7}{5} + \frac{\rho_{S5}a}{2\varepsilon_0} \ln \frac{9}{7} = 1$$

or

$$\rho_{S1}\left(2 \ln \frac{l}{5a} + \ln \frac{9}{7}\right) + \rho_{S2}(\ln 3 + \ln 1.4) + \rho_{S3} \ln \frac{5}{3} = \frac{2\varepsilon_0}{a}$$ (12.13)

where we have substituted $\rho_{S5} = \rho_{S1}$ and $\rho_{S4} = \rho_{S2}$. Similarly, writing the equations for the potentials at the center points of segments 2 and 3 and arranging the three equations in matrix form, we get

$$\begin{bmatrix} 2 \ln \dfrac{l}{5a} + \ln \dfrac{9}{7} & \ln 3 + \ln 1.4 & \ln \dfrac{5}{3} \\[2mm] \ln 3 + \ln 1.4 & 2 \ln \dfrac{l}{5a} + \ln \dfrac{5}{3} & \ln 3 \\[2mm] 2 \ln \dfrac{5}{3} & 2 \ln 3 & 2 \ln \dfrac{l}{5a} \end{bmatrix} \begin{bmatrix} \rho_{S1} \\[2mm] \rho_{S2} \\[2mm] \rho_{S3} \end{bmatrix} = \begin{bmatrix} \dfrac{2\varepsilon_0}{a} \\[2mm] \dfrac{2\varepsilon_0}{a} \\[2mm] \dfrac{2\varepsilon_0}{a} \end{bmatrix}$$ (12.14)

By inverting (12.14), the solutions for ρ_{S1}, ρ_{S2}, and ρ_{S3} can be obtained. For a numerical example, if $l = 1$ m and $a = 1$ mm, the values of ρ_{S1}, ρ_{S2}, and ρ_{S3} are $158.38\varepsilon_0$, $145.42\varepsilon_0$, and $143.32\varepsilon_0$, respectively. When a larger number of segments are used, a more accurate solution is obtained for the charge distribution on the wire. For example, the result for $n = 40$ obtained by using a computer program is shown in Fig. 12.8, where the height of the first rectangle is $204\varepsilon_0$ C/m^2.

FIGURE 12.8   Charge distribution along a thin, straight wire of length 1 m and radius 1 mm, and held at a potential of 1 V. The height of the first rectangle is $204\varepsilon_0$ C/m^2.

*Capacitance
of a
parallel-plate
capacitor*

Proceeding further, we recall that in Example 6.2 we discussed the solution of Laplace's equation for the one-dimensional case of two infinite, plane, parallel, perfectly conducting plates, which may be considered an idealization of a parallel-plate capacitor with its plates having large dimensions compared to the spacing between them. We then obtained the expression for the capacitance of the arrangement per unit area of the plates. Because of the idealization, this expression is only approximate for a capacitor with finite-sized plates. It becomes less and less accurate as the size of the plates becomes less and less large for a given spacing between them, since the fringing of the field at the edges of the plates becomes more and more severe. Thus, the problem is that in the nonideal case, the field distribution between the capacitor plates and the charge distribution on the capacitor plates are not uniform, whereas for the ideal case, they are uniform. Hence, it is not in general possible to obtain an analytical expression for the capacitance; one has to resort to numerical of graphical techniques. The method of moments serves as a useful tool for such cases.

For an example, let us consider an arrangement in which the spacing between the plates is a, the dimensions of the plates are $2a \times 3a$, and from symmetry considerations, the upper plate is held at a potential of 1 V and the lower plate is held at a potential of -1 V. For the purposes of illustration of the method, we shall divide each plate into a 2×3 set of squares, as shown in Fig. 12.9, and assume that within each square, the (surface) charge density is uniform. From symmetry considerations, we then have only two unknown charge densities ρ_{s1}, and ρ_{s2}, as shown in the figure. Therefore, it is sufficient to write two independent equations. We shall do this by considering squares 1 and 2 and equating the potentials at the center points of these squares to 1 V.

To write the expression for the potential at the center point of a square due to the charge in a different square, we shall consider that charge to be a point charge at the center of the square. Thus, the potential at point 1 due to the charge in square 4 is $\rho_{s1}a^2/4\pi\varepsilon_0 a$, the potential at point 2 due to the charge in

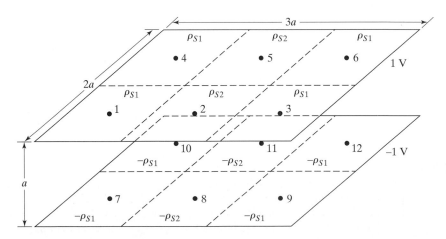

FIGURE 12.9   For finding the capacitance of a parallel-plate capacitor by the method of moments.

square 12 is $-\rho_{S1}a^2/4\pi\varepsilon_0(\sqrt{3}a)$, and so on. To write the expression for the potential at the center point of a square due to the charge in that square, we shall use the result given in Problem P4.28. For example, the potential at point 1 due to the charge in square 1 is $(\rho_{S1}a/\pi\varepsilon_0)\ln(1+\sqrt{2})$. Proceeding in this manner, we obtain the two equations to be

$$\frac{\rho_{S1}a}{\pi\varepsilon_0}\ln(1+\sqrt{2}) + \frac{\rho_{S1}a^2}{4\pi\varepsilon_0}\left(\frac{1}{2a}+\frac{1}{a}+\frac{1}{\sqrt{5a}}-\frac{1}{a}-\frac{1}{\sqrt{5a}}-\frac{1}{\sqrt{2a}}-\frac{1}{\sqrt{6a}}\right)$$

$$+ \frac{\rho_{S2}a^2}{4\pi\varepsilon_0}\left(\frac{1}{a}+\frac{1}{\sqrt{2a}}-\frac{1}{\sqrt{2a}}-\frac{1}{\sqrt{3a}}\right) = 1 \qquad (12.15a)$$

$$\frac{\rho_{S2}a}{\pi\varepsilon_0}\ln(1+\sqrt{2}) + \frac{\rho_{S1}a^2}{4\pi\varepsilon_0}\left(\frac{2}{a}+\frac{2}{\sqrt{2a}}-\frac{2}{\sqrt{2a}}-\frac{2}{\sqrt{3a}}\right)$$

$$+ \frac{\rho_{S2}a^2}{4\pi\varepsilon_0}\left(\frac{1}{a}-\frac{1}{a}-\frac{1}{\sqrt{2a}}\right) = 1 \qquad (12.15b)$$

or

$$2.9101\rho_{S1} + 0.4226\rho_{S2} = \frac{4\pi\varepsilon_0}{a} \qquad (12.16a)$$

$$0.8453\rho_{S1} + 2.8184\rho_{S2} = \frac{4\pi\varepsilon_0}{a} \qquad (12.16b)$$

Solving (12.16a) and (12.16b) for ρ_{S1} and ρ_{S2}, we obtain $\rho_{S1} = 3.8378\varepsilon_0/a$ and $\rho_{S2} = 3.3075\varepsilon_0/a$. The magnitude of charge on either plate is then equal to $(4a^2 \times 3.8378\varepsilon_0/a + 2a^2 \times 3.3075\varepsilon_0/a)$, or $21.9662\varepsilon_0 a$. Finally, noting that the potential difference between the plates is 2 V, the capacitance can be computed to be $10.983\varepsilon_0 a$. A more accurate result can be obtained by dividing each plate into a larger number of squares, but it is instructive to compare the value just obtained with the value of $6\varepsilon_0 a$, which follows from the application of $C = \varepsilon_0 A/d$, where A is the area of the plates, and d is the spacing between the plates.

K12.2. Method of moments; Thin, straight wire held at known potential; Determination of charge distribution; Parallel-plate capacitor; Determination of capacitance.

D12.2. For the problem of Example 12.2, consider that to compute the potential at the center of a given segment due to the charge in another segment, the charge in that segment can be assumed to be a point charge at the center of that segment. Modify the formulation to obtain the new matrix equation in the place of (12.14) and find the values of ρ_{S1}, ρ_{S2}, and ρ_{S3}, for $l = 1$ m and $a = 1$ mm.
Ans. $159.48\varepsilon_0$ C/m²; $147.94\varepsilon_0$ C/m²; $145.77\varepsilon_0$ C/m².

D12.3. Consider a parallel-plate capacitor having square-shaped plates of sides a and spacing a between the plates. Find the following: **(a)** the capacitance of the capacitor if fringing of fields at the edges of the plates is neglected; **(b)** the capacitance by using the method of moments, considering each plate as one square; and **(c)** the capacitance by dividing each plate into a 2 × 2 set of squares and using the method of moments. Assume free space for the dielectric.
Ans **(a)** $\varepsilon_0 a$; **(b)** $2.488\varepsilon_0 a$; **(c)** $2.8367\varepsilon_0 a$.

12.3   DETERMINATION OF TRANSMISSION-LINE PARAMETERS

In this section, we shall illustrate the application of the numerical methods introduced in the previous two sections for the determination of transmission-line parameters, by means of two examples.

Example 12.3

Parallel-strip line

The parallel-strip line is the same as the parallel-plate line (see Fig. 6.12) without the imposition of the approximation $d/w \ll 1$, such that fringing of fields can not be neglected. We wish to find the capacitance per unit length and hence the characteristic impedance of the parallel-strip line embedded in a homogeneous medium (which we shall assume here to be free space) for the case of $d = w$, by using the method of moments.

The procedure for the application of method of moments to find the capacitance per unit length of a parallel-strip line is similar to that used for finding the capacitance of a parallel-plate capacitor in Section 12.2. Thus, let us consider the cross-sectional view of the parallel-strip line and divide each conductor into $2n$ substrips, as shown in Fig. 12.10 for $n = 3$, and assume the charge density in each substrip to be uniform. From symmetry consideration, we can apply a potential of 1 V to one of the conductors and -1 V to the other conductor. Also from symmetry considerations, there are only $n (= 3)$ unknown charge densities to be determined, namely, the charge densities associated with the substrips in one half of one of the conductors. Thus, we need to write a set of $n (= 3)$ independent equations for the $n (= 3)$ unknown charge densities. To do this, we consider pairs of substrips $11', 22', \ldots$, situated opposite to each other and write the expression for the potential difference between the center points of each pair and set that equal to 2 V. The expression for the potential difference between the center points of a given pair is the sum of the contributions to the potential difference from all $2n(= 6)$ pairs. To obtain the contribution from a given pair, we make use of the result given in Problem P4.29 for the potential difference between two points due to an infinitely long strip of uniform surface charge density. For example, let us consider the potential difference between the center points of 1 and $1'$. Then the contribution to it from the pair of substrips 1 and $1'$ is

$$
\frac{\rho_{S1}}{2\pi\varepsilon_0}\left[\frac{w}{6}\ln\frac{(w/12)^2 + d^2}{(w/12)^2} + 4d\tan^{-1}\frac{w}{12d}\right]
$$

whereas the contribution from the pair of substrips 2 and $2'$ is

$$
\frac{\rho_{S2}}{2\pi\varepsilon_0}\left[\frac{w}{4}\ln\frac{(w/4)^2 + d^2}{(w/4)^2} - \frac{w}{12}\ln\frac{(w/12)^2 + d^2}{(w/12)^2} + 2d\left(\tan^{-1}\frac{w}{4d} - \tan^{-1}\frac{w}{12d}\right)\right]
$$

FIGURE 12.10   Division of the conductors of a parallel-strip line into substrips.

Writing contributions in this manner and adding appropriately, we obtain the matrix equation for the three unknown charge densities ρ_{S1}, ρ_{S2}, and ρ_{S3} for the case of $d = w$, that is, $k = 1$, as given by

$$\begin{bmatrix} 1.311 & 0.815 & 0.658 \\ 0.815 & 1.432 & 1.005 \\ 0.658 & 1.005 & 1.779 \end{bmatrix} \begin{bmatrix} \rho_{S1} \\ \rho_{S2} \\ \rho_{S3} \end{bmatrix} = \begin{bmatrix} 4\pi\varepsilon_0/w \\ 4\pi\varepsilon_0/w \\ 4\pi\varepsilon_0/w \end{bmatrix}$$

so that

$$\rho_{S1} = \frac{6.0854\varepsilon_0}{w} \text{ C/m}^2$$

$$\rho_{S2} = \frac{3.2062\varepsilon_0}{w} \text{ C/m}^2$$

$$\rho_{S3} = \frac{3.003\varepsilon_0}{w} \text{ C/m}^2$$

The magnitude of the charge per unit length on either conductor is $(w/3) \times (\rho_{S1} + \rho_{S2} + \rho_{S3})(1) = 4.098\varepsilon_0$ C. Thus, the capacitance per unit length is given by

$$\mathcal{C} = \frac{4.0982\varepsilon_0}{2} = 2.0491\varepsilon_0 \text{ F/m}$$

Finally, the characteristic impedance of the parallel-strip line for the case of $d = w$ is $\sqrt{\mu_0\varepsilon_0}/2.0491\varepsilon_0$, or 183.98 Ω. For larger values of n, the solution can be carried out by using a computer program. For example, for $d = w = 1$ cm, and $n = 10$, the values of $\mathcal{C}$ and Z_0 can be computed to be 18.5252 pF/m and 179.9352 Ω, respectively.

Example 12.4

When the bottom conductor of the microstrip line of Fig. 6.15 is extended so as to surround the top conductor, we get the enclosed-microstrip line, as shown by the cross-sectional view in Fig. 12.11. Here we assume a square cross section for the outer conductor and wish to determine the propagation parameters for the line by using the finite-difference method to find the values of capacitance per unit length with and without the dielectric substrate in place, as required by (6.59a) and (6.59b), in view of the inhomogeneity.

Enclosed-microstrip line

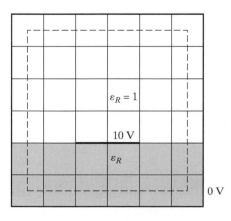

FIGURE 12.11   Division of the region between the conductors of an enclosed-microstrip line into a set of squares.

For purposes of illustration, we divide the region inside the outer conductor into a 6×6 set of squares with the grid points identified as (i, j), where i is the row number (1 to 5 from top to bottom), and j is the column number (1 to 5 from left to right). We place the inner conductor along the line from grid point $(4, 2)$ to grid point $(4, 4)$ so that the region below row 4 is dielectric substrate (relative permittivity ε_R), and the region above row 4 is free space. We further assume the inner conductor to be kept at 10 V and the outer conductor at 0 V, and apply the iteration procedure, illustrated in Example 12.1 to compute the potentials at the grid points not on the conductors. We note, however, that in view of the inhomogeneity when the dielectric substrate is in place, the modified form of (12.9) given by (see Problem P12.4)

$$V_0 \approx \frac{V_1 + \varepsilon_r V_2}{2(1 + \varepsilon_r)} + \frac{V_3 + V_4}{4} \tag{12.17}$$

needs to be used. Thus, the procedure consists of the following steps:

a. With the dielectric substrate in place, find the solution for the potentials at the grid points not on the conductors consistent with (12.17) to within a specified tolerance (Δ) assumed here to be 0.01 V. Find the magnitude of the charge per unit length along the conductors by applying Gauss' law in integral form to a surface having as the cross section the contour that passes through the center points of the squares adjacent to the outer conductor, as shown in Fig. 12.11. Find the capacitance per unit length ($\mathscr{C}$).

b. With the dielectric replaced by free space, repeat step (a) to obtain the capacitance per unit length ($\mathscr{C}_0$).

c. Find Z_0 and v_p by using (6.59a) and (6.59b), respectively.

The solution just outlined can be carried out by using a computer program. The final set of values for the potentials obtained from the run of such a program for the specific arrangement of Fig. 12.11 for $\varepsilon_R = 10$, and the results for $\mathscr{C}$, $\mathscr{C}_0$, Z_0, and v_p are shown in Fig. 12.12. The upper rows of potential values at the interior grid points correspond to

0	0	0	0	0	0	0
0	0.69	1.23	1.43	1.23	0.69	0
	0.69	1.24	1.44	1.24	0.69	
0	1.53	2.84	3.27	2.84	1.53	0
	1.54	2.85	3.28	2.85	1.54	
0	2.58	5.34	5.99	5.34	2.58	0
	2.63	5.36	6.00	5.36	2.63	
0	3.47	10.00	10.00	10.00	3.47	0
	3.64	10.00	10.00	10.00	3.64	
0	1.89	4.10	4.55	4.10	1.89	0
	1.93	4.12	4.55	4.12	1.94	
0	0	0	0	0	0	0

C = 226.4795 PF/m   C0 = 37.85039 PF/m

Z0 = 36.00221 OHMS

VP = 1.226428E+08 M/S

FIGURE 12.12 Final set of values for the potentials and the results for $\mathscr{C}$, $\mathscr{C}_0$, Z_0, and v_p for the enclosed-microstrip line of Fig. 12.11 for $\varepsilon_R = 10$.

the case of the dielectric substrate in place and the lower rows correspond to the case of the dielectric replaced by free space.

K12.3. Parallel-strip line; Method of moments; Enclosed-microstrip line; Finite-difference method.

D12.4. For the parallel-strip line of Fig. 12.10, find the following for $d = w$: **(a)** the contribution to the potential difference between the center points of substrips 2 and 2' from the pair of substrips 2 and 2'; **(b)** the contribution to the potential difference between the center points of substrips 2 and 2' from the pair of substrips 5 and 5'; and **(c)** the contribution to the potential difference between the center points of substrips 1 and 1' from the pair of substrips 6 and 6'.
Ans.   **(a)** $0.1849\rho_{S2}w/\varepsilon_0$;   **(b)** $0.043\rho_{S2}w/\varepsilon_0$;   **(c)** $0.0238\rho_{S1}w/\varepsilon_0$.

12.4   FINITE-ELEMENT METHOD

The finite-element method, a general technique for solving differential equations, was first developed by structural engineers for the analysis of stresses and strains in complex systems. It was not until 1968 that its applications to the solution of electromagnetic-field problems were initiated. Unlike the finite-difference method, which provides solutions at an array of grid points in the region of interest, the finite-element method provides solution over the entire region of interest. Furthermore, it is difficult to apply the finite-difference method to regions having irregularly shaped boundaries, whereas the finite-element method is particularly suitable for such regions. However, the finite-element method in its precise form is elaborate and we shall here present only an introduction by considering the early simple approach.

The basic concept of the finite-element method is that although the be- *Solution*
havior of a function may be complex when viewed over a large region, a simple *procedure*
approximation may be sufficient for a small subregion. The total region is divided into a number of nonoverlapping subregions called finite elements. Within each element, the function of interest is approximated by an algebraic expression, and where the adjoining elements overlap, the algebraic representations must agree to provide continuity of the function. The equations to be solved are derived not directly from the differential equations that govern the function, but from the minimization of an integral-type functional such as the electric energy in the case of the electric potential. The solution procedure in this manner consists essentially of four steps: (1) discretizing the region of interest into the finite elements, (2) deriving the governing equations for the individual finite elements, (3) relating the individual finite elements to the assembly of the elements, and (4) obtaining and solving the system of equations for the potentials. We shall describe these steps in the context of finding the solution for the two-dimensional Laplace's equation in Cartesian coordinates x and y, given by (12.6) and then illustrate by means of an example.

1. *Discretization of region into finite elements.* In two dimensions, the finite elements are usually polygons, the simplest of which are triangles and quadrilaterals. We shall confine our presentation to triangles to keep the analysis simple. Figure 12.13 shows an example in which a region is divided into five triangular

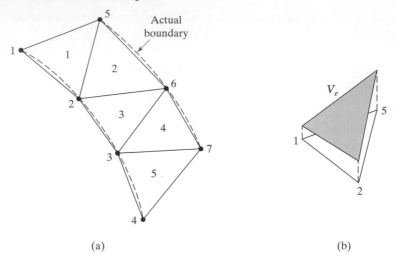

(a)   (b)

FIGURE 12.13   (a) Discretization of a region into triangular finite elements.
(b) Linear variation of potential within a triangular finite element.

elements, with a total of seven nodes. The most common type of expression for
V within an element is a polynomial expansion. For a triangular element, it is
given by

$$V_e(x, y) = a + bx + cy \qquad (12.18)$$

where the subscript e denotes element. Note that this represents linear variation
of potential within the element, as shown, for example, in Fig. 12.13(b) for one
element. Also, this approximation is the same as assuming that the electric field
is uniform within the element, since

$$\mathbf{E} = -\boldsymbol{\nabla} V_e = -(b\mathbf{a}_x + c\mathbf{a}_y) \qquad (12.19)$$

   2. *Equations governing the elements.* Let us consider a typical element
shown in Fig. 12.14. Using (12.18), we can then express the potentials V_{e1}, V_{e2}, and

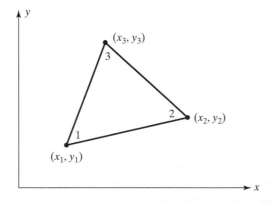

FIGURE 12.14   A typical triangular finite element for setting up
the equations governing the
element.

V_{e3} at nodes 1, 2, and 3, respectively, as

$$V_{e1} = a + bx_1 + cy_1 \tag{12.20a}$$

$$V_{e2} = a + bx_2 + cy_2 \tag{12.20b}$$

$$V_{e3} = a + bx_3 + cy_3 \tag{12.20c}$$

from which we can write

$$
\begin{bmatrix} a \\ b \\ c \end{bmatrix} = \begin{bmatrix} 1 & x_1 & y_1 \\ 1 & x_2 & y_2 \\ 1 & x_3 & y_3 \end{bmatrix}^{1} \begin{bmatrix} V_{e1} \\ V_{e2} \\ V_{e3} \end{bmatrix}
$$

$$
= \frac{1}{2A} \begin{bmatrix} (x_2 y_3 - x_3 y_2) & (x_3 y_1 - x_1 y_3) & (x_1 y_2 - x_2 y_1) \\ (y_2 - y_3) & (y_3 - y_1) & (y_1 - y_2) \\ (x_3 - x_2) & (x_1 - x_3) & (x_2 - x_1) \end{bmatrix} \begin{bmatrix} V_{e1} \\ V_{e2} \\ V_{e3} \end{bmatrix} \tag{12.21}
$$

where

$$
A = \frac{1}{2} \begin{vmatrix} 1 & x_1 & y_1 \\ 1 & x_2 & y_2 \\ 1 & x_3 & y_3 \end{vmatrix}
$$

$$
= \tfrac{1}{2}[(x_2 - x_1)(y_3 - y_1) - (x_3 - x_1)(y_2 - y_1)] \tag{12.22}
$$

is the area of the element. Note that for A to be positive, the nodes need to be numbered counterclockwise as in Fig. 12.14.

Proceeding further, we have, by substituting (12.21) into (12.18)

$$
V_e = \frac{1}{2A} \begin{bmatrix} 1 & x & y \end{bmatrix} \begin{bmatrix} (x_2 y_3 - x_3 y_2) & (x_3 y_1 - x_1 y_3) & (x_1 y_2 - x_2 y_1) \\ (y_2 - y_3) & (y_3 - y_1) & (y_1 - y_2) \\ (x_3 - x_2) & (x_1 - x_3) & (x_2 - x_1) \end{bmatrix} \begin{bmatrix} V_{e1} \\ V_{e2} \\ V_{e3} \end{bmatrix}
$$

$$\tag{12.23}$$

or

$$V_e = \sum_{i=1}^{3} \alpha_i(x, y) V_{ei} \tag{12.24}$$

where

$$\alpha_1 = \frac{1}{2A}[(x_2 y_3 - x_3 y_2) + (y_2 - y_3)x + (x_3 - x_2)y] \tag{12.25a}$$

$$\alpha_2 = \frac{1}{2A}[(x_3 y_1 - x_1 y_3) + (y_3 - y_1)x + (x_1 - x_3)y] \tag{12.25b}$$

$$\alpha_3 = \frac{1}{2A}[(x_1 y_2 - x_2 y_1) + (y_1 - y_2)x + (x_2 - x_1)y] \tag{12.25c}$$

The quantities α_i are called the shape functions.

Functional
for solution
of Laplace's
equation

Now, for the solution of Laplace's equation in two dimensions for the electric potential, the functional to be minimized is the electric energy per unit length normal to the two dimensions, that is,

$$W = \int \frac{1}{2} \varepsilon |\mathbf{E}|^2 \, dS \tag{12.26}$$

over the region of interest. For the element under consideration, this is given by

$$
\begin{aligned}
W_e &= \int_{\substack{\text{area} \\ \text{of } e}} \frac{1}{2} \varepsilon |\mathbf{E}|^2 \, dS \\[2mm]
&= \frac{1}{2} \int_A \varepsilon |\nabla V_e|^2 \, dS \\[2mm]
&= \frac{1}{2} \varepsilon \int_A (\nabla V_e \cdot \nabla V_e) \, dS \\[2mm]
&= \frac{1}{2} \varepsilon \int_A \left(\sum_{i=1}^{3} V_{ei} \nabla \alpha_i \cdot \sum_{i=1}^{3} V_{ei} \nabla \alpha_i \right) dS \\[2mm]
&= \frac{1}{2} \varepsilon \sum_{i=1}^{3} \sum_{j=1}^{3} V_{ei} \left(\int_A \nabla \alpha_i \cdot \nabla \alpha_j \, dS \right) V_{ej}
\end{aligned}
\tag{12.27}
$$

where we have used (12.24). We now define

$$C_{ij}^{(e)} = \int_A (\nabla \alpha_i \cdot \nabla \alpha_j) \, dS \tag{12.28}$$

so that we can write (12.27) as

$$W_e = \tfrac{1}{2} \varepsilon [V_e]^T [C^{(e)}] [V_e] \tag{12.29}$$

where

$$[V_e] = \begin{bmatrix} V_{e1} \\ V_{e2} \\ V_{e3} \end{bmatrix} \tag{12.30a}$$

$$[V_e]^T = [V_{e1} \quad V_{e2} \quad V_{e3}] = \text{transpose of } [V_e] \tag{12.30b}$$

and

$$[C^{(e)}] = \begin{bmatrix} C_{11}^{(e)} & C_{12}^{(e)} & C_{13}^{(e)} \\ C_{21}^{(e)} & C_{22}^{(e)} & C_{23}^{(e)} \\ C_{31}^{(e)} & C_{32}^{(e)} & C_{33}^{(e)} \end{bmatrix} \tag{12.30c}$$

The matrix $[C^{(e)}]$ is known as the *element coefficient matrix*. Substituting (12.25a)–(12.25c) into (12.28) and evaluating, we obtain

$$C_{11}^{(e)} = \frac{1}{4A} [(y_2 - y_3)^2 + (x_2 - x_3)^2] \tag{12.31a}$$

$$C_{12}^{(e)} = \frac{1}{4A} [(y_2 - y_3)(y_3 - y_1) + (x_2 - x_3)(x_3 - x_1)] \tag{12.31b}$$

$$C_{13}^{(e)} = \frac{1}{4A} [(y_2 - y_3)(y_1 - y_2) + (x_2 - x_3)(x_1 - x_2)] \tag{12.31c}$$

$$C_{21}^{(e)} = C_{12}^{(e)} \tag{12.31d}$$

$$C_{22}^{(e)} = \frac{1}{4A} [(y_3 - y_1)^2 + (x_3 - x_1)^2] \tag{12.31e}$$

$$C_{23}^{(e)} = \frac{1}{4A} [(y_3 - y_1)(y_1 - y_2) + (x_3 - x_1)(x_1 - x_2)] \tag{12.31f}$$

$$C_{31}^{(e)} = C_{13}^{(e)} \tag{12.31g}$$

$$C_{32}^{(e)} = C_{23}^{(e)} \tag{12.31h}$$

$$C_{33}^{(e)} = \frac{1}{4A} [(y_1 - y_2)^2 + (x_1 - x_2)^2] \tag{12.31i}$$

Note that $\sum_{i=1}^{3} C_{ij}^{(e)}$, that is, the sum of each row of $[C^{(e)}]$, and $\sum_{j=1}^{3} C_{ij}^{(e)}$, that is, the sum of each column of $[C^{(e)}]$, are zero.

3. *Relating the individual elements to the assembly.* Proceeding further, we consider the assembly of all elements in the region of interest to write the expression for the total energy in the region. This is given by

$$W = \sum_{e=1}^{n} W_e = \frac{1}{2} \varepsilon [V]^t [C] [V] \tag{12.32}$$

where

$$[V] = \begin{bmatrix} V_1 \\ V_2 \\ V_3 \\ \vdots \\ V_n \end{bmatrix} \tag{12.33}$$

is the column matrix of the potentials at the nodes, $[V]^T$ is the transpose of $[V]$, and $[C]$, which is known as the *global coefficient matrix,* is the matrix resulting from the assemblage of the individual element coefficient matrices.

To illustrate the determination of $[C]$, let us consider the assembly of three elements shown in Fig. 12.15. For each element, the node numbers 1, 2, and 3 are indicated inside the triangle in the counterclockwise sense. These are called *lcoal nodes.* The nodes for the assembly, which are called *global nodes,* are indicated

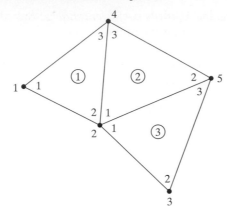

FIGURE 12.15   An assembly of three triangular finite elements, for relating the individual elements to the assembly.

at the vertices outside the assembly. The element numbers are circumscribed by circles. Since there are five global nodes, $[C]$ is a 5×5 matrix as given by

$$[C] = \begin{bmatrix} C_{11} & C_{12} & C_{13} & C_{14} & C_{15} \\ C_{21} & C_{22} & C_{23} & C_{24} & C_{25} \\ C_{31} & C_{32} & C_{33} & C_{34} & C_{35} \\ C_{41} & C_{42} & C_{43} & C_{44} & C_{45} \\ C_{51} & C_{52} & C_{53} & C_{54} & C_{55} \end{bmatrix} \tag{12.34}$$

The global matrix $[C]$ is not to be confused with the element matrices $[C^{(e)}]$. For triangular elements, all element matrices are 3×3, whereas the global matrix has the size $n \times n$.

Since the total energy as expressed in (12.32) is the sum of the energies in the three individual elements, and since the potential distribution must be continuous across the boundaries of pairs of adjacent elements, the elements of $[C]$ are related to the elements of the individual matrices $[C^{(e)}]$. Thus, for example, since global node 1 belongs to element 1 only and is the same as the local node 1,

$$C_{11} = C_{11}^{(1)}$$

Since global node 2 belongs to all three elements and is the same as local node 2 for element 1, local node 1 for element 2, and local node 1 for element 3,

$$C_{22} = C_{22}^{(1)} + C_{11}^{(2)} + C_{11}^{(3)}$$

Similarly,

$$C_{33} = C_{22}^{(3)}$$
$$C_{44} = C_{33}^{(1)} + C_{33}^{(2)}$$
$$C_{55} = C_{22}^{(2)} + C_{33}^{(2)}$$

The global link 12 is the same as the local link 12 for element 1. Hence,

$$C_{12} = C_{12}^{(1)}$$

The global link 23 is the same as the local link 12 for element 3. Hence,

$$C_{23} = C_{12}^{(3)}$$

The global link 24 is common to elements 1 and 2, and is the same as the local link 23 for element 1 and the local link 13 for element 2. Therefore,

$$C_{24} = C_{23}^{(1)} + C_{13}^{(2)}$$

Similarly,

$$C_{25} = C_{12}^{(2)} + C_{13}^{(3)}$$

Since there is no coupling between global nodes 1 and 3,

$$C_{13} = 0$$

In this manner, the elements of the entire matrix $[C]$ can be written as follows:

$$[C] = \begin{bmatrix} C_{11}^{(1)} & C_{12}^{(1)} & 0 & C_{13}^{(1)} & 0 \\ C_{21}^{(1)} & C_{22}^{(1)} + C_{11}^{(2)} + C_{11}^{(3)} & C_{12}^{(3)} & C_{23}^{(1)} + C_{13}^{(2)} & C_{12}^{(3)} + C_{13}^{(3)} \\ 0 & C_{21}^{(3)} & C_{22}^{(3)} & 0 & C_{23}^{(3)} \\ C_{31}^{(1)} & C_{32}^{(1)} + C_{31}^{(2)} & 0 & C_{33}^{(1)} + C_{33}^{(2)} & C_{32}^{(2)} \\ 0 & C_{21}^{(2)} + C_{31}^{(3)} & C_{32}^{(3)} & C_{23}^{(2)} & C_{22}^{(2)} + C_{33}^{(3)} \end{bmatrix}$$

$$(12.35)$$

The global matrix has the following properties: (a) It is symmetric, which can be understood if we recall that the local matrices are all symmetric, and (b) it is singular, that is, the determinant formed by its elements is zero, as we shall discuss later.

4. Equations governing the potentials and solution. Having obtained the elements of the global matrix, we now set the derivatives of the energy given by (12.32) with respect to the node potentials equal to zero, to minimize the functional, that is, the total energy in the region. Thus, *Minimization of functional*

$$\frac{\partial W}{\partial V_k} = 0, \qquad \text{for } k = 1, 2, \ldots, n \tag{12.36}$$

For example, from (12.32) and (12.34), we obtain

$$\frac{\partial W}{\partial V_1} = 2C_{11}V_1 + C_{12}V_2 + C_{13}V_3 + C_{14}V_4$$
$$+ C_{15}V_5 + C_{21}V_2 + C_{31}V_3 + C_{41}V_4 + C_{51}V_5 \tag{12.37}$$
$$= 0$$

which gives

$$C_{11}V_1 + C_{12}V_2 + C_{13}V_3 + C_{14}V_4 + C_{15}V_5 = 0 \tag{12.38}$$

In general,

$$\sum_{i=1}^{n} C_{ki}V_i = 0 \qquad \text{for } k = 1, 2, \ldots, n \tag{12.39}$$

For the case under consideration, $n = 5$, and we get five equations for the five potentials. Noting that the right sides of all five equations are zero, we now observe that the global matrix $[C]$ must be singular in order to have a nontrivial solution. Since $[C]$ is singular, it also means that the solution of (12.39) is not unique. The situation is that for a given problem, the potentials are specified at a subset of the global nodes and, hence, we can only use the subset of (12.39) that is pertinent to the derivatives with respect to the unknown potentials. Thus, if the potentials at nodes 1, 3, and 5 are specified, then we use only those two equations resulting from setting $\partial W / \partial V_2$ and $\partial W / \partial V_4$ equal to zero to solve for V_2 and V_4.

Let us now consider an example.

Example 12.5

An assembly of two finite elements is shown in Fig. 12.16. Global node 3 is kept at 10-V potential, whereas global node 1 is at 0 V. It is desired to find the values of the potentials at global nodes 2 and 4 by using the finite-element method.

We proceed with the solution by executing the four steps as discussed:

Step 1.   The region of interest is already discretized. With reference to the numbering of the elements, local nodes, and global nodes, as in Fig. 12.16, we proceed with the remaining three steps as follows.

Step 2.   Compute the element coefficient matrix for each element. Using (12.22) and (12.31a)–(12.31i) for each of the two elements, we obtain the following values.

ELEMENT 1

$$A = 2$$

$$[C^{(1)}] = \begin{bmatrix} 1 & -1/2 & -1/2 \\ -1/2 & 1/2 & 0 \\ -1/2 & 0 & 1/2 \end{bmatrix}$$

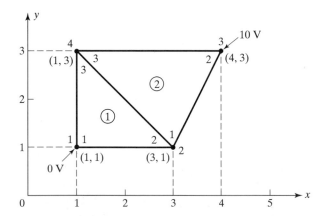

FIGURE 12.16   Assembly of two finite elements for Example 12.5.

ELEMENT 2

$$A = 3$$

$$[C^{(2)}] = \begin{bmatrix} 3/4 & -1/2 & -1/4 \\ -1/2 & 2/3 & -1/6 \\ -1/4 & -1/6 & 5/12 \end{bmatrix}$$

Note that for each of $[C^{(1)}]$ and $[C^{(2)}]$, the sum of each row and the sum of each column are zero.

Step 3. Compute the global coefficient matrix.

$$[C] = \begin{bmatrix} C_{11}^{(1)} & C_{12}^{(1)} & 0 & C_{13}^{(1)} \\ C_{21}^{(1)} & C_{22}^{(1)} + C_{11}^{(2)} & C_{12}^{(2)} & C_{23}^{(1)} + C_{13}^{(2)} \\ 0 & C_{21}^{(2)} & C_{22}^{(2)} & C_{23}^{(2)} \\ C_{31}^{(1)} & C_{32}^{(1)} + C_{31}^{(2)} & C_{32}^{(2)} & C_{33}^{(1)} + C_{33}^{(2)} \end{bmatrix}$$

$$= \begin{bmatrix} 1 & -1/2 & 0 & -1/2 \\ -1/2 & 5/4 & -1/2 & -1/4 \\ 0 & -1/2 & 2/3 & -1/6 \\ -1/2 & -1/4 & -1/6 & 11/12 \end{bmatrix}$$

Step 4. Using

$$\frac{\partial W}{\partial V_2} = 0 \quad \text{and} \quad \frac{\partial W}{\partial V_4} = 0$$

and noting that $V_1 = 0$ and $V_3 = 10$ V, we have

$$\begin{bmatrix} -1/2 & 5/4 & -1/2 & -1/4 \\ -1/2 & -1/4 & -1/6 & 11/12 \end{bmatrix} \begin{bmatrix} 0 \\ V_2 \\ 10 \\ V_4 \end{bmatrix} = \begin{bmatrix} 0 \\ 0 \end{bmatrix}$$

or,

$$5V_2 - V_4 = 20$$

$$-3V_2 + 11V_4 = 20$$

Solving, we obtain

$$V_2 = 4.615 \text{ V}$$

$$V_4 = 3.077 \text{ V}$$

K12.5. Finite-element method; Functional; Minimization of functional; Solution of Laplace's equation in two dimensions; Element coefficient matrices; Global coefficient matrix.

D12.5. For the assembly of four triangular elements in the xy-plane, shown in Fig. 12.17, find the numerical values of the following elements of the global coefficient matrix: **(a)** C_{12}; **(b)** C_{25}; **(c)** C_{31}; **(d)** C_{44}; and **(e)** C_{55}.
Ans. **(a)** 0; **(b)** −2; **(c)** 0; **(d)** 2; **(e)** 5.

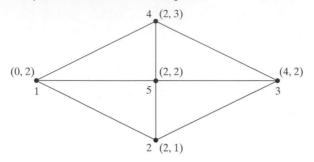

FIGURE 12.17   For Problem D12.5.

12.5   FINITE-DIFFERENCE TIME-DOMAIN METHOD

Solution of one-dimensional wave equation

In section 12.1, we introduced the finite-difference method for solving differential equations. We recall that it consists of replacing the derivatives (with respect to space coordinates) in the differential equations by their finite-difference approximations and solving the resulting algebraic equations. The finite-difference time-domain (FD-TD) method extends this procedure to derivatives involving time variation in addition to the space derivatives, and it is a useful technique for the numerical solution of a wide range of problems in electromagnetics. We shall here include only a very elementary treatment of the topic.

The simplest differential equation involving space and the time variations is the one-dimensional second-order partial differential equation (5.8) given by

$$\frac{\partial^2 E_x(z, t)}{\partial z^2} = \mu_0 \varepsilon_0 \frac{\partial^2 E_x(z, t)}{\partial t^2} \tag{12.40}$$

We already know that the solution to this equation consists of a superposition of traveling waves propagating in the $+z$-direction and the $-z$-direction and, hence, it is known as the *one-dimensional wave equation*. To solve this equation numerically, we discretize the range of interest in z and replace the left side by its central-difference approximation. Similarly, the time interval of interest can be discretized and the derivative on the right side replaced by its central difference approximation. The resulting algebraic equation can be rearranged to express E_x at a given point in a space-time $(z$–$t)$ grid of points in terms of its (previously computed) values at certain other neighboring grid points (see Problem P12.21). The rearranged equation thus permits the progressive computation of $E_x(z, t)$ at the grid points, beginning with its values specified by the boundary conditions and initial conditions pertinent to the problem.

A more illuminating approach, that is also illustrative of the physical phenomenon, emanates from the use of the two first-order coupled partial differential equations (5.7a) and (5.7b), given by

$$\frac{\partial E_x(z, t)}{\partial z} = -\mu_0 \frac{\partial H_y(z, t)}{\partial t} \tag{12.41a}$$

$$\frac{\partial H_y(z, t)}{\partial z} = -\varepsilon_0 \frac{\partial E_x(z, t)}{\partial t} \tag{12.41b}$$

and from which (5.8) was derived. Recall that these equations follow from Maxwell's curl equations for the special case of $\mathbf{E} = E_x(z, t)\mathbf{a}_x$ and $\mathbf{H} = H_y(z, t)\mathbf{a}_y$, and free space for the medium. In the continuous solution of these equations, both E_x and H_y are variables defined at the same point (z, t) in the space-time coordinate system. The starting point in the solution by the FD-TD method is to consider E_x and H_y as variables not at the same point but at alternate points in the space-time grid, as illustrated, for example, in Fig. 12.18.

Note that the arrangement of Fig. 12.18 leaves certain points in the grid unlabelled and surrounded by four labelled points. For example, the point $(1, 2)$ is unlabelled and surrounded by the two points $(0, 2)$ and $(2, 2)$ labelled E_x along the line parallel to the z-axis, and the two points $(1, 1)$ and $(1, 3)$ labelled H_y along the line parallel to the t-axis. We can use this arrangement to express $\partial E_x / \partial z$ and $\partial H_y / \partial t$ at this point in terms of their finite-difference approximations and substitute them in (12.41a). Thus, we have

$$\boxed{\frac{E_x(2, 2) - E_x(0, 2)}{2a} = -\mu_0 \frac{H_y(1, 3) - H_y(1, 1)}{2\delta}} \qquad (12.42a)$$

where a and δ are the spacings between the grid points for the z- and t-variations, respectively. Similarly, the point $(2, 3)$ is unlabelled and surrounded by the two points $(1, 3)$ and $(3, 3)$ labelled H_y along the line parallel to the z-axis, and the two points $(2, 2)$ and $(2, 4)$ labelled E_x along the line parallel to the t-axis. We can use this arrangement to approximate $\partial H_y / \partial z$ and $\partial E_x / \partial t$ at the grid point $(2, 3)$ to write (12.41b) as

$$\boxed{\frac{H_y(3, 3) - H_y(1, 3)}{2a} = -\varepsilon_0 \frac{E_x(2, 4) - E_x(2, 2)}{2\delta}} \qquad (12.42b)$$

Equations (12.42a) and (12.42b) can be used in a "leap-frog" scheme to progress on the grid with the solution. We shall illustrate this by means of an analogous transmission-line example.

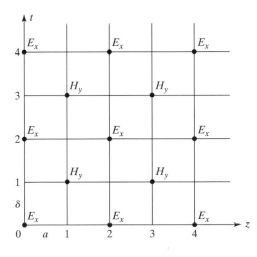

FIGURE 12.18   Space-time grid for the FD-TD solution of the coupled partial differential equations (12.41a) and (12.41b).

Example 12.6

Initially charged transmission line

Figure 12.19(a) shows a lossless transmission line of length $l = 12$ m, characteristic impedance $Z_0 = 100\ \Omega$, and velocity of propagation $v_p = 10^8$ m/s, and short-circuited at both ends. At $t = 0$, the line current is zero everywhere along the line, and the line voltage has the distribution

$$V(z, 0) = 10 \sin \frac{\pi z}{12}\ \text{V}$$

as shown in Fig. 12.19(b). It is desired to apply the FD-TD method to investigate the line voltage and line current for $t > 0$.

The differential equations of interest are the transmission-line equations (6.47a) and (6.47b) given by

$$\frac{\partial V(z, t)}{\partial z} = -\mathscr{L}\,\frac{\partial I(z, t)}{\partial t} \tag{12.43a}$$

$$\frac{\partial I(z, t)}{\partial z} = -\mathscr{C}\,\frac{\partial V(z, t)}{\partial t} \tag{12.43b}$$

which are analogous to (12.41a) and (12.41b), respectively. From the given values of Z_0 and v_p, we obtain

$$\mathscr{L} = \frac{Z_0}{v_p} = \frac{100}{10^8} = 10^{-6}\,\text{H/m}$$

$$\mathscr{C} = \frac{1}{Z_0 v_p} = \frac{1}{100 \times 10^8} = 10^{-10}\,\text{F/m}$$

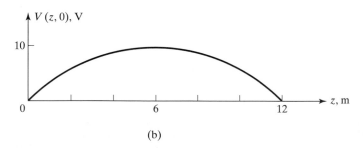

FIGURE 12.19 (a) Lossless transmission line short-circuited at both ends. (b) Voltage distribution on the line for $t = 0$.

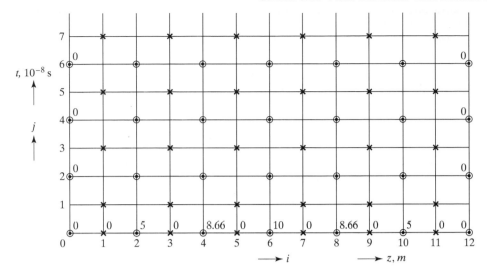

FIGURE 12.20 Space-time grid of points for the FD-TD solution for the line voltage for $t > 0$ for the transmission line of Fig. 12.19. Numbers beside the circled points represent voltage in volts and those beside the crossed points represent current in amperes.

We shall divide the line into 12 equal segments of width 1 m and use a time step of $1/v_p = 10^{-8}$ s, so that the space-time grid is as shown in Fig. 12.20. The grid points at which V is known or to be computed are denoted by circles, and the grid points at which I is known or to be computed are denoted by crosses. Initial values, as specified by the initial distributions of voltage and current, are marked at the grid points on the $t = 0$ line. Boundary values of $V = 0$, as required by short circuits at either end of the line are marked at the grid points on the $z = 0$ ordinate and $z = 12$ ordinate.

Denoting the grid points by (i, j), where i refers to space (z) and j refers to time (t), and applying the finite-difference approximations to the derivatives in (12.43a), we have

$$\frac{V(i + 1, j - 1) - V(i - 1, j - 1)}{2 \times 1} = -10^{-6} \frac{I(i, j) - I(i, j - 2)}{2 \times 10^{-8}}$$

or

$$\boxed{I(i, j) = I(i, j - 2) + \frac{V(i - 1, j - 1) - V(i + 1, j - 1)}{100}} \qquad (12.44)$$

where $i = 1, 3, 5, 7, 9, 11$, and $j = 3, 5, 7, \ldots$. To find the line currents corresponding to $j = 1$, we use the initial values of zero corresponding to $j = 0$ and only one time step. Thus,

$$\boxed{I(i, 1) = \frac{V(i - 1, 0) - V(i + 1, 0)}{200}} \qquad (12.45)$$

Applying the finite-difference approximations to the derivatives in (12.43b), we have

$$\frac{I(i + 1, j - 1) - I(i - 1, j - 1)}{2 \times 1} = -10^{-10} \frac{V(i, j) - V(i, j - 2)}{2 \times 10^{-8}}$$

TABLE 12.1
Progression of FD-TD Solution for the Initially Charged Line of Fig. 12.19

	i / j	0	1	2	3	4	5	6	7	8	9	10	11	12
V, V	0	0	0	5		8.66		10		8.66		5		0
I, A	0	0		0		0		0		0		0		
I, A	1		−0.025		−0.0183		−0.0067		0.0067		0.0183		0.025	
V, V	2	0		4.33		7.50		8.66		7.50		4.33		0
I, A	3		−0.0683		−0.05		−0.0183		0.0183		0.05		0.0683	
V, V	4	0		2.50		4.33		5.00		4.33		2.50		0

The values of *i* and *j* correspond to the space-time grid in Fig. 12.20.

or

$$V(i, j) = V(i, j - 2) + 100[I(i - 1, j - 1) - I(i + 1, j - 1)] \qquad (12.46)$$

We can now proceed with the solution, as shown in Table 12.1. To begin the solution, we use (12.45) to compute the values of *I* corresponding to $j = 1$. Then, we use alternatingly (12.44) and (12.46) to compute values of *V* followed by the values of *I* for successive values of *j*. The resulting solutions are shown in Table 12.1.

Magic time step
    The continuous solution to this problem can be obtained by using the procedures discussed in Section 6.7 for time-domain analysis of transmission lines with initial conditions, or by using the natural oscillations concepts in Section 7.1. This solution is given exactly by

$$V(z, t) = 10 \sin \frac{\pi z}{12} \cos \frac{10^8 \pi t}{12} \text{ V}$$

It can be seen that the computed values of *V* at the grid points agree with the exact analytical solution. In fact, it can be shown that this is the case when the grid points are chosen such that the time increment is equal to the so-called "magic time step,"[1] which is the space segment width divided by the velocity of propagation. The solution may however be insufficient to represent the actual behavior in the continuous region if the discretization in *z* does not correspond to a fraction of a wavelength. A discussion of such considerations as accuracy of the solution and stability of the solution process is beyond the scope here.

---

[1]See A. Taflove, *Computational Electrodynamics; The Finite-Difference Time-Domain Method* (Norwood, MA: Artech House, 1995), p. 38.

K12.5. Finite-difference time-domain method; One-dimensional second-order partial differential equation; First-order coupled partial differential equations; Leap-frog scheme.

D12.6. For the transmission-line problem of Example 12.6, extend the solution beyond the grid points in Table 12.1 to find the following quantities: **(a)** $I(5, 3)$; **(b)** $V(6, 8)$; **(c)** $I(7, 7)$; and **(d)** $V(8, 4)$.
Ans.   **(a)** -0.0683 A;   **(b)** 0 V;   **(c)** 0.025 A;   **(d)** -4.33 V.

12.6 SUMMARY

In this chapter, we introduced four numerical methods for solving problems in electromagnetics: (1) the finite-difference method, (2) the method of moments, (3) the finite-element method, and (4) the finite-difference time-domain method.

The finite-difference method is based on replacing the derivative terms in a differential equation by their finite-difference approximations and solving the resulting algebraic equations. We illustrated the technique by applying it to a one-dimensional differential equation. We then discussed and illustrated by means of an example the numerical solution of Laplace's equation in two dimensions

$$\frac{\partial^2 V}{\partial x^2} + \frac{\partial^2 V}{\partial y^2} = 0 \tag{12.47}$$

The numerical solution is based on the finite-difference approximation to (12.47) that the potential V_0 at a point P in the charge-free region is given by

$$V_0 \approx \tfrac{1}{4}(V_1 + V_2 + V_3 + V_4) \tag{12.48}$$

where V_1, V_2, V_3, and V_4 are the potentials at four equidistant points lying along mutually perpendicular axes through P. By using an iterative technique, a set of values for the potentials at appropriately chosen grid points is obtained such that the potential at each grid point satisfies (12.48) to within a specified tolerance.

We then turned our attention to the method of moments, which is a numerical technique useful for solving a class of problems for which exact analytical solutions are in general not possible. Considering, for example, a surface charge distribution $\rho_S(x, y, z)$ on a given surface, the method of moments technique consists of inverting the integral equation

$$V(x, y, z) = \frac{1}{4\pi\varepsilon_0} \int_{\substack{\text{surface of} \\ \text{the charge} \\ \text{distribution}}} \frac{\rho_S(x', y', z')}{R}\, dS'$$

by approximating the integral as a summation. We illustrated the method of moments technique by means of two examples: (1) finding the charge distribution on a thin, straight wire held at a known potential and (2) finding the capacitance of a parallel-plate capacitor, taking into account fringing of the field at the edges of the plates.

We then applied the method of moments and the finite-difference method for the determination of transmission-line parameters. Specifically, we illustrated the determination of Z_0 and v_p for a parallel-strip line embedded in a homogeneous medium by using the method of moments and for an enclosed-microstrip line by using the finite-difference method.

The finite-element method, a more general technique than the finite-difference method for solving differential equations, is based on the minimization of an integral-type functional such as the electric energy in the case of the electric potential, instead of solving the differential equations directly. The solution procedure consists of (1) discretizing the region of interest into a set of finite elements, (2) deriving the governing equations for the individual finite elements, (3) relating the individual elements to the assembly of the elements, and (4) obtaining and solving the system of equations for the potentials.

We illustrated this procedure for the solution of the two-dimensional Laplace's equation (12.47) by considering triangles for the finite elements. A linear variation is assumed for the potential within each triangle and the *element coefficient matrix* that relates the energy within the triangle to the potentials at the vertices of the triangle is derived. A *global coefficient matrix* that relates the assembly of elements to the individual element is then obtained and used to complete the solution. We considered an example involving an assembly of two triangles with the potentials specified at two of the four global nodes and computed the two unknown potentials to illustrate this procedure.

The finite-difference time-domain method extends the numerical technique to solving differential equations involving time. A simple example is the solution of the one-dimensional wave equation, consisting of replacing the second derivatives with respect to the spatial dimension and time by the corresponding central-difference approximations and applying the resulting algebraic equation to a space-time grid of points. Alternatively, two first-order coupled partial differential equations following from Maxwell's curl equations can be used in a more illuminating manner by applying their approximations to a space-time grid of points in a "leap-frog" scheme. We illustrated this procedure by considering a lossless transmission line initially charged to a voltage and finding the line voltage and current values at points on the line at a later value of time.

REVIEW QUESTIONS

Q12.1. Describe the formulation behind the finite-difference method of solving differential equations.

Q12.2. Outline the procedure for solving a one-dimensional differential equation by the finite-difference method.

Q12.3. Discuss the basis behind the numerical solution of Laplace's equation in two dimensions by the finite-difference method.

Q12.4. Describe the iteration technique for the computer solution of Laplace's equation in two dimensions by the finite-difference method.

Q12.5. How would you apply the iteration technique for the computer solution of Laplace's equation in three dimensions?

Q12.6. Discuss the formulation behind the problem of finding the charge distribution on a conductor of known potential by the method of moments.

Q12.7. Outline by means of an example the procedure for obtaining the charge distribution on a conductor of known potential by the method of moments technique.

Q12.8. Why is the expression for the capacitance of a parallel-plate capacitor obtained by using Laplace's equation in one dimension approximate?

Q12.9. Discuss the determination of the capacitance of a parallel-plate capacitor by the method of moments technique.

Q12.10. Describe the procedure for obtaining Z_0 and v_p for a parallel-strip line embedded in a homogeneous medium by using the method of moments.

Q12.11. Outline the procedure for obtaining Z_0 and v_p for an enclosed-microstrip line by using the finite-difference method.

Q12.12. Describe the basic concept of the finite-element method for solving differential equations and outline the steps involved in its implementation.

Q12.13. Discuss the linear approximation for the potential within a triangular finite element.

Q12.14. Discuss the functional to be minimized for the solution of Laplace's equation in two dimensions by the finite-element method.

Q12.15. Discuss the derivation of the element coefficient matrix for a triangular finite element and its properties.

Q12.16. Describe the determination of the global coefficient matrix for an assembly of triangular elements from the individual element coefficient matrices and discuss its properties.

Q12.17. Describe the solution of the one-dimensional wave equation by the FD-TD method.

Q12.18. Discuss the key to the solution of Maxwell's curl equations for the special case of $\mathbf{E} - E_x(z, t)\mathbf{a}_x$ and $\mathbf{H} = H_y(z, t)\mathbf{a}_y$, and free space for the medium, by the FD-TD method, as compared to the solution of the one-dimensional wave equation.

Q12.19. Describe the "leap-frog" scheme of carrying out the FD-TD solution to the transmission-line equations for the lossless case.

Q12.20. Discuss the agreement between the values of V computed by the FD-TD method in Example 12.6 with those provided by the exact analytical solution.

PROBLEMS

Section 12.1

P12.1. By discretizing the region between $x = 0$ and $x = 1$ into five equal segments spaced 0.2 apart and applying the finite-difference method, solve for the approximate values of f at the four interior grid points for the one-dimensional differential equation

$$\frac{d^2f(x)}{dx^2} + 4f(x) = 0$$

with the boundary conditions specified as $f(0) = 0$ and $f(1) = 1$. Compare your answers with the exact solution to the differential equation.

P12.2. The cross section of an infinitely long arrangement of conductors normal to the page and that repeats endlessly in the plane of the page is shown in Fig. 12.21. For the grid points shown, find the values of V_1, V_2, V_3, V_4, and V_5, by writing equations consistent with (12.9) and solving them. Then find the approximate magnitude of the field intensity at grid point 2 and the approximate value of the surface charge density at point P, assuming that the spacing between the grid points is d and the medium between the conductors is free space.

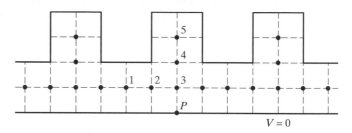

FIGURE 12.21   For Problem P12.2.

P12.3. The cross section of an arrangement of conductors, infinitely long normal to the page, is square, as shown in Fig. 12.22. Three sides are kept at 0 V and the fourth side is kept at 28 V. The region between the conductors is divided into a 4 × 4 grid of squares. Although there are nine grid points, there are only six unknown potentials V_A, V_B, ..., V_F, because of symmetry. **(a)** By writing equations consistent with (12.9) for these six potentials and solving the equations, find the values of the potentials. **(b)** Find the approximate magnitude of the electric field intensity at grid point B, assuming that the spacing between grid points is d. **(c)** Find the approximate surface charges per unit length of the arrangement on the 28-V conductor and the 0-V conductor.

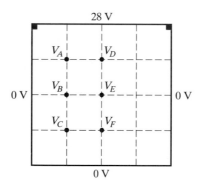

FIGURE 12.22   For Problem P12.3.

P12.4. In Fig. 12.2, assume that the region below the y-axis ($x < 0$) is a perfect dielectric of relative permittivity ε_r, whereas the region above the y-axis ($x > 0$) is free space. Show that the modified form of (12.9) is then given by

$$V_0 \approx \frac{V_1 + \varepsilon_r V_2}{2(1 + \varepsilon_r)} + \frac{V_3 + V_4}{4}$$

P12.5. For unequal spacings between grid points, as shown in Fig. 12.23, show that the generalization of (12.9) is given by

$$V_0 = \frac{V_1}{(1 + d_1/d_2)(1 + d_1 d_2/d_3 d_4)} + \frac{V_2}{(1 + d_2/d_1)(1 + d_2 d_1/d_4 d_3)}$$

$$+ \frac{V_3}{(1 + d_3/d_4)(1 + d_3 d_4/d_2 d_1)} + \frac{V_4}{(1 + d_4/d_3)(1 + d_4 d_3/d_1 d_2)}$$

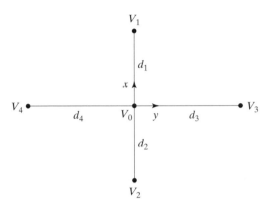

FIGURE 12.23   For Problem P12.5.

Section 12.2

P12.6. Consider a thin, straight cylindrical wire of length l and radius a ($\ll l$) bent in the middle to make a $90°$ angle and held at a potential of 1 V. By dividing the wire into four equal segments and assuming the charge density in each segment to be uniform, and using the method of moments, find the total charge on the wire if $l = 1$ m and $a = 1$ mm. To compute the potential at the center of a given segment due to the charge in another segment, assume the charge to be a point charge at the center of that segment.

P12.7. Consider a thin wire of radius 1 mm bent into the form of a square of sides 60 cm and held at a potential of 1 V. By dividing each side of the square into three equal segments and assuming the charge density in each segment to be uniform, and using the method of moments, find the total charge on the wire. To compute the potential at the center of a given segment due to the charge in another segment, assume the charge to be a point charge at the center of that segment.

P12.8. Consider two thin wires that are square-shaped as in Problem P12.7 and arranged such that the sides of one wire are directly above and parallel to the sides of the second wire at a spacing of 10 cm, so as to form a capacitor. Using the method of moments as in Problem P12.7, find the capacitance of the arrangement.

P12.9. A square-shaped conductor of area $3a \times 3a$, with a square-shaped hole of area $a \times a$ in the middle, as shown in Fig. 12.24, is held at a potential of 1 V. By dividing the conductor into eight squares, as shown in the figure, and using the method of moments, find the total charge on the conductor. To find the potential at the center point of a square due to the charge in another square, consider the charge in that square to be a point charge at the center of that square.

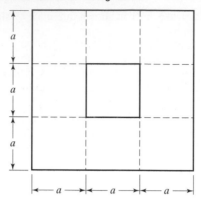

FIGURE 12.24   For Problem P12.9.

P12.10. Assume that a capacitor is made up of two parallel conductors, each having the shape shown in Fig. 12.24. If the spacing between the plates is a, find the capacitance of the arrangement by dividing each conductor into squares, as shown in Fig. 12.24, and applying the method of moments.

P12.11. The arrangement shown in Fig. 12.25 is that of a capacitor obtained by removing a square-shaped part of sides a from the center of a square-shaped conductor of sides $3a$ and displacing it by distance a directly above the hole. By dividing the lower plate as shown in the figure, find the capacitance of the arrangement.

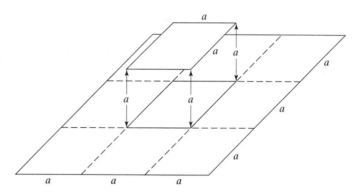

FIGURE 12.25   For Problem P12.11.

P12.12. A conductor having the shape of a cube of sides a is held at a potential of 1 V. By dividing each side into a 2×2 set of squares, and assuming the charge density in each square to be uniform, and using the method of moments, find the total surface charge on the conductor. To find the potential at the center of a square due to the charge in another square, consider the charge in that square to be a point charge at the center of the square.

Section 12.3

P12.13. For the parallel-strip line of Example 12.3, repeat the solution by considering the charges to be line charges along the centerlines of the substrips for writing the contributions to the potential difference between a given pair of substrips due to the charges in a different pair of substrips and using the formula given in Problem P4.29.

P12.14. Consider a parallel-strip line with unequal widths of the conductors, as shown in Fig. 12.26. Obtain the characteristic impedance of the line for the case of $k = 1$ by dividing the conductors into substrips as shown in the figure and using the method of moments. Note that from considerations of symmetry, there are only three unknown charge densities $\rho_{S1}, \rho_{S2},$ and ρ_{S3}. Write two equations by equating the expressions for the potential differences V_{12} and V_{13} to $(V_A - V_B)$ and the third equation from consideration of charge neutrality. Use the result of Problem P4.29 for writing the contributions to the potential differences in all cases.

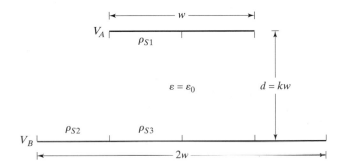

FIGURE 12.26   For Problem P12.14.

P12.15. Consider a transmission line having the cross-sectional view shown in Fig. 12.27. With the conductors of the line divided into substrips as shown in the figure, obtain the characteristic impedance by using the method of moments. Note that from considerations of symmetry, there are only three unknown charge densities $\rho_{S1}, \rho_{S2},$ and ρ_{S3}. Write two equations by equating the expressions for the potential differences V_{12} and V_{13} to $(V_A - V_B)$ and the third equation from consideration of charge neutrality. For writing the contribution to the potential difference between a given pair of substrips due to one of those substrips, use the result of Problem P4.29. But for writing the contribution to the potential difference between a given pair of substrips due to a third substrip, consider the charge in that substrip to be a line charge along the centerline of the substrip.

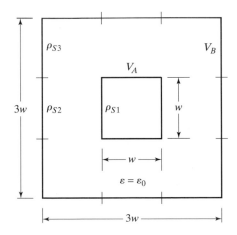

FIGURE 12.27   For Problem P12.15.

P12.16. For the enclosed-microstrip line of Fig. 12.11, repeat the computations of $\mathscr{C}$, $\mathscr{C}_0$, Z_0, and v_p, by finding the magnitude of the charge per unit length by considering the contour that passes through the center points of the squares adjacent to the center conductor, instead of the one shown in the figure.

Section 12.4

P12.17. Alternative to the representations (12.31a)–(12.31i), show that the elements of the element coefficient matrix $[C^{(e)}]$ in (12.30c) can be written as

$$[C^{(e)}] = \frac{1}{2}\begin{bmatrix} \cot\theta_2 + \cot\theta_3 & -\cot\theta_3 & -\cot\theta_2 \\ -\cot\theta_3 & \cot\theta_1 + \cot\theta_3 & -\cot\theta_1 \\ -\cot\theta_2 & -\cot\theta_1 & \cot\theta_1 + \cot\theta_2 \end{bmatrix}$$

where θ_1, θ_2, and θ_3 are the interior angles at the vertices 1, 2, and 3, respectively, of the triangular element, as shown in Fig. 12.28.

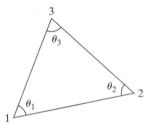

FIGURE 12.28   For Problem P12.17.

P12.18. Solve Example 12.5 by discretizing the region Fig. 12.16, as shown in Fig. 12.29.

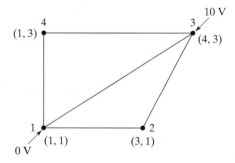

FIGURE 12.29   For Problem P12.18.

P12.19. By discretizing the region of Fig. 12.16 into three triangles, as shown in Fig. 12.30, solve for the potentials at global nodes 2, 4, and 5.

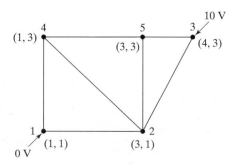

FIGURE 12.30   For Problem P12.19.

P12.20. Repeat Problem P12.29 for the region of Fig. 12.16 discretized into three triangles, as shown in Fig. 12.31.

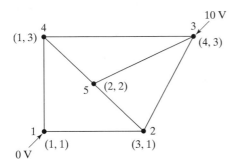

FIGURE 12.31   For Problem P12.20.

Section 12.5

P12.21. Show that the application of the central-difference approximation to the derivatives in (12.40) in conjunction with the space-time grid of Fig. 12.18 gives the result

$$E_x(i, j + 1) = \left(2 - 2\frac{\delta^2 c^2}{a^2}\right) E_x(i, j) + \frac{\delta^2 c^2}{a^2} [E_x(i + 1, j)$$
$$+ E_x(i - 1, j)] - E_x(i, j - 1)$$

where i and j refer to space (z) and time (t), respectively, and $c = 1/\sqrt{\mu_0 \varepsilon_0}$.

P12.22. In Example 12.6, assume that

$$V(z, 0) = 10 \sin \frac{\pi z}{6} \text{ V}$$

Using the same space-time grid as in Fig. 12.20, prepare a table similar to Table 12.1 for obtaining values of V at the grid points corresponding to $j = 4$.

P12.23. Repeat Problem 12.22 for

$$V(z, 0) = 10 \sin^3 \frac{\pi z}{12} \text{ V}$$

P12.24. Repeat Problem 12.22 for

$$V(z, 0) = \begin{cases} \frac{5}{3}z & \text{for } 0 \leq z \leq 6 \\ 20 - \frac{5}{3}z & \text{for } 6 \leq z \leq 12 \end{cases}$$

REVIEW PROBLEMS

R12.1. Consider two thin rectangular-shaped conductors of size $a \times 3a$ arranged as shown in Fig. 12.32. By dividing each conductor into three squares of sides a and applying the method of moments, find the capacitance of the arrangement. To find the potential at the center of a square due to the charge on another square, consider the charge on that square to be a point charge at the center of that square.

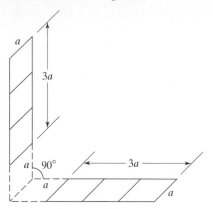

FIGURE 12.32 For Problem R12.1.

R12.2. Consider a transmission line having the cross section shown in Fig. 12.33. The inner conductor is a circle of radius a and the outer conductor is a square of sides $2a$. Using the grid points as shown in the figure and applying the finite-difference method, find the approximate value of the characteristic impedance of the line. (*Hint:* Use the result of Problem P12.5 for grid point 6.)

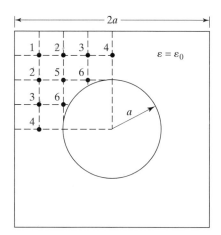

FIGURE 12.33 For Problem R12.2.

R12.3. Consider a square region divided into four right isosceles triangular finite elements, as shown in Fig. 12.34. Show that for linear variation of potential within each element, as represented by (12.18), the electric energy is proportional to the sum of the squares of the differences between the potential at global node 5 (the 90°-vertex) and the remaining two nodes (the 45°-vertices). For example, for element 1, it is proportional to $[(V_1 - V_5)^2 + (V_2 - V_5)^2]$. Further show that the finite-element method gives the same result for V_5 in terms of V_1, V_2, V_3, and V_4, as that given by the finite-difference method, that is,

$$V_5 = \tfrac{1}{4}(V_1 + V_2 + V_3 + V_4)$$

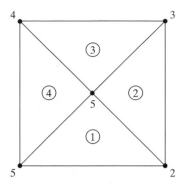

5   2   **FIGURE 12.34**   For Problem R12.3.

R12.4. Repeat the solution of Example 12.6 up to $t = 4 \times 10^{-8}$ s for each of the following two cases: **(a)** time step $= 2 \times 10^{-8}$ s and **(b)** time step $= 0.5 \times 10^{-8}$ s. For each case, compare the values of V obtained for $t = 4 \times 10^{-8}$ s with those from the exact analytical solution and comment on your results.

APPENDIX A

Curl, Divergence, Gradient, and Laplacian in Cylindrical and Spherical Coordinate Systems

In Chapter 4, we introduced the curl, divergence, gradient, and Laplacian and derived the expressions for them in the Cartesian coordinate system. In this appendix, we derive the corresponding expressions in the cylindrical and spherical coordinate systems. Considering first the cylindrical coordinate system, we recall from Section 1.3 that the infinitesimal box defined by the three orthogonal surfaces intersecting at point $P(r, \phi, z)$ and the three orthogonal surfaces intersecting at point $Q(r + dr, \phi + d\phi, z + dz)$ is as shown in Fig. A.1.

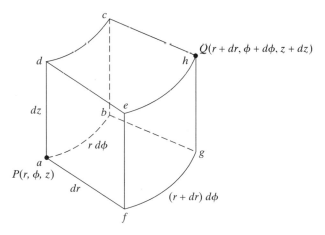

FIGURE A.1 Infinitesimal box formed by incrementing the coordinates in the cylindrical coordinate system.

From the basic definition of the curl of a vector introduced in Section 4.3 and given by

$$\nabla \times \mathbf{A} = \lim_{\Delta S \to 0} \left[\frac{\oint_C \mathbf{A} \cdot d\mathbf{l}}{\Delta S} \right]_{max} \mathbf{a}_n \tag{A.1}$$

we find the components of $\nabla \times \mathbf{A}$ as follows with the aid of Fig. A.1:

$$(\nabla \times \mathbf{A})_r = \lim_{\substack{d\phi \to 0 \\ dz \to 0}} \frac{\oint_{abcda} \mathbf{A} \cdot d\mathbf{l}}{\text{area } abcd}$$

$$= \lim_{\substack{d\phi \to 0 \\ dz \to 0}} \frac{\left\{ \begin{array}{l} [A_\phi]_{(r,z)} r\, d\phi + [A_z]_{(r,\phi + d\phi)}\, dz \\ - [A_\phi]_{(r,z+dz)} r\, d\phi - [A_z]_{(r,\phi)}\, dz \end{array} \right\}}{r\, d\phi\, dz} \tag{A.2a}$$

$$= \lim_{d\phi \to 0} \frac{[A_z]_{(r,\phi + d\phi)} - [A_z]_{(r,\phi)}}{r\, d\phi} + \lim_{dz \to 0} \frac{[A_\phi]_{(r,z)} - [A_\phi]_{(r,z+dz)}}{dz}$$

$$= \frac{1}{r} \frac{\partial A_z}{\partial \phi} - \frac{\partial A_\phi}{\partial z}$$

$$(\nabla \times \mathbf{A})_\phi = \lim_{\substack{dz \to 0 \\ dr \to 0}} \frac{\oint_{adefa} \mathbf{A} \cdot d\mathbf{l}}{\text{area } adef}$$

$$= \lim_{\substack{dz \to 0 \\ dr \to 0}} \frac{\left\{ \begin{array}{l} [A_z]_{(r,\phi)}\, dz + [A_r]_{(\phi,z+dz)}\, dr \\ - [A_z]_{(r+dr,\phi)}\, dz - [A_r]_{(\phi,z)}\, dr \end{array} \right\}}{dr\, dz} \tag{A.2b}$$

$$= \lim_{dz \to 0} \frac{[A_r]_{(\phi,z+dz)} - [A_r]_{(\phi,z)}}{dz} + \lim_{dr \to 0} \frac{[A_z]_{(r,\phi)} - [A_z]_{(r+dr,\phi)}}{dr}$$

$$= \frac{\partial A_r}{\partial z} - \frac{\partial A_z}{\partial r}$$

$$(\nabla \times \mathbf{A})_z = \lim_{\substack{dr \to 0 \\ d\phi \to 0}} \frac{\oint_{afgba} \mathbf{A} \cdot d\mathbf{l}}{\text{area } afgb}$$

$$= \lim_{\substack{dr \to 0 \\ d\phi \to 0}} \frac{\left\{ \begin{array}{l} [A_r]_{(\phi,z)}\, dr + [A_\phi]_{(r+dr,z)} (r + dr)\, d\phi \\ - [A_r]_{(\phi + d\phi,z)}\, dr - [A_\phi]_{(r,z)} r\, d\phi \end{array} \right\}}{r\, dr\, d\phi} \tag{A.2c}$$

$$= \lim_{dr \to 0} \frac{[rA_\phi]_{(r+dr,z)} - [rA_\phi]_{(r,z)}}{r\, dr} + \lim_{d\phi \to 0} \frac{[A_r]_{(\phi,z)} - [A_r]_{(\phi + d\phi,z)}}{r\, d\phi}$$

$$= \frac{1}{r} \frac{\partial}{\partial r} (rA_\phi) - \frac{1}{r} \frac{\partial A_r}{\partial \phi}$$

Combining (A.2a), (A.2b), and (A.2c), we obtain the expression for the curl of a vector in cylindrical coordinates as

$$\mathbf{\nabla} \times \mathbf{A} = \left[\frac{1}{r} \frac{\partial A_z}{\partial \phi} - \frac{\partial A_\phi}{\partial z} \right] \mathbf{a}_r + \left[\frac{\partial A_r}{\partial z} - \frac{\partial A_z}{\partial r} \right] \mathbf{a}_\phi$$

$$+ \frac{1}{r} \left[\frac{\partial}{\partial r} (rA_\phi) - \frac{\partial A_r}{\partial \phi} \right] \mathbf{a}_z$$

$$= \begin{vmatrix} \dfrac{\mathbf{a}_r}{r} & \mathbf{a}_\phi & \dfrac{\mathbf{a}_z}{r} \\[2mm] \dfrac{\partial}{\partial r} & \dfrac{\partial}{\partial \phi} & \dfrac{\partial}{\partial z} \\[2mm] A_r & rA_\phi & A_z \end{vmatrix}$$

(A.3)

To find the expression for the divergence, we make use of the basic definition of the divergence of a vector, introduced in Section 4.3 and given by

$$\mathbf{\nabla} \cdot \mathbf{A} = \lim_{\Delta v \to 0} \frac{\oint_S \mathbf{A} \cdot d\mathbf{S}}{\Delta v}$$

(A.4)

Evaluating the right side of (A.4) for the box of Fig. A.1, we obtain

$$\mathbf{\nabla} \cdot \mathbf{A} = \lim_{\substack{dr \to 0 \\ d\phi \to 0 \\ dz \to 0}} \frac{\left\{ \begin{array}{l} [A_r]_{r+dr}(r+dr)\, d\phi\, dz - [A_r]_r\, r\, d\phi\, dz + [A_\phi]_{\phi+d\phi}\, dr\, dz \\ - [A_\phi]_\phi\, dr\, dz + [A_z]_{z+dz}\, r\, dr\, d\phi - [A_z]_z\, r\, dr\, d\phi \end{array} \right\}}{r\, dr\, d\phi\, dz}$$

$$= \lim_{dr \to 0} \frac{[rA_r]_{r+dr} - [rA_r]_r}{r\, dr} + \lim_{d\phi \to 0} \frac{[A_\phi]_{\phi+d\phi} - [A_\phi]_\phi}{r\, d\phi}$$

(A.5)

$$+ \lim_{dz \to 0} \frac{[A_z]_{z+dz} - [A_z]_z}{dz}$$

$$= \frac{1}{r} \frac{\partial}{\partial r} (rA_r) + \frac{1}{r} \frac{\partial A_\phi}{\partial \phi} + \frac{\partial A_z}{\partial z}$$

To obtain the expression for the gradient of a scalar, we recall from Section 1.3 that in cylindrical coordinates,

$$d\mathbf{l} = dr\, \mathbf{a}_r + r\, d\phi\, \mathbf{a}_\phi + dz\, \mathbf{a}_z$$

(A.6)

Therefore,

$$d\Phi = \frac{\partial \Phi}{\partial r} dr + \frac{\partial \Phi}{\partial \phi} d\phi + \frac{\partial \Phi}{\partial z} dz$$

$$= \left(\frac{\partial \Phi}{\partial r} \mathbf{a}_r + \frac{1}{r} \frac{\partial \Phi}{\partial \phi} \mathbf{a}_\phi + \frac{\partial \Phi}{\partial z} \mathbf{a}_z \right) \cdot (dr\, \mathbf{a}_r + r\, d\phi\, \mathbf{a}_\phi + dz\, \mathbf{a}_z)$$

(A.7)

$$= \mathbf{\nabla}\Phi \cdot d\mathbf{l}$$

Thus,

$$\nabla\Phi = \frac{\partial\Phi}{\partial r}\,\mathbf{a}_r + \frac{1}{r}\frac{\partial\Phi}{\partial\phi}\,\mathbf{a}_\phi + \frac{\partial\Phi}{\partial z}\,\mathbf{a}_z \tag{A.8}$$

To derive the expression for the Laplacian of a scalar, we recall from Section 4.5 that

$$\nabla^2\Phi = \nabla\cdot\nabla\Phi \tag{A.9}$$

Then using (A.5) and (A.8), we obtain

$$\begin{aligned}
\nabla^2\Phi &= \frac{1}{r}\frac{\partial}{\partial r}\left(r\frac{\partial\Phi}{\partial r}\right) + \frac{1}{r}\frac{\partial}{\partial\phi}\left(\frac{1}{r}\frac{\partial\Phi}{\partial\phi}\right) + \frac{\partial}{\partial z}\left(\frac{\partial\Phi}{\partial z}\right)\\
&= \frac{1}{r}\frac{\partial}{\partial r}\left(r\frac{\partial\Phi}{\partial r}\right) + \frac{1}{r^2}\frac{\partial^2\Phi}{\partial\phi^2} + \frac{\partial^2\Phi}{\partial z^2}
\end{aligned} \tag{A.10}$$

Turning now to the spherical coordinate system, we recall from Section 1.3 that the infinitesimal box defined by the three orthogonal surfaces intersecting at $P(r, \theta, \phi)$ and the three orthogonal surfaces intersecting at $Q(r + dr, \theta + d\theta, \phi + d\phi)$ is as shown in Fig. A.2. From the basic definition of the curl of a vector given by (A.1), we then find the components of $\nabla \times \mathbf{A}$ as follows with the aid of Fig. A.2:

$$\begin{aligned}
(\nabla\times\mathbf{A})_r &= \lim_{\substack{d\theta\to0\\d\phi\to0}} \frac{\oint_{abcda}\mathbf{A}\cdot d\mathbf{l}}{\text{area } abcd}\\[2mm]
&= \lim_{\substack{d\theta\to0\\d\phi\to0}} \frac{\left\{\begin{array}{l}[A_\theta]_{(r,\phi)}r\,d\theta + [A_\phi]_{(r,\theta+d\theta)}\,r\sin(\theta+d\theta)\,d\phi\\ -\,[A_\theta]_{(r,\phi+d\phi)}r\,d\theta - [A_\phi]_{(r,\theta)}\,r\sin\theta\,d\phi\end{array}\right\}}{r^2\sin\theta\,d\theta\,d\phi}\\[2mm]
&= \lim_{d\theta\to0}\frac{[A_\phi\sin\theta]_{(r,\theta+d\theta)} - [A_\phi\sin\theta]_{(r,\theta)}}{r\sin\theta\,d\theta}\\[2mm]
&\quad + \lim_{d\phi\to0}\frac{[A_\theta]_{(r,\phi)} - [A_\theta]_{(r,\phi+d\phi)}}{r\sin\theta\,d\phi}\\[2mm]
&= \frac{1}{r\sin\theta}\frac{\partial}{\partial\theta}(A_\phi\sin\theta) - \frac{1}{r\sin\theta}\frac{\partial A_\theta}{\partial\phi}
\end{aligned} \tag{A.11a}$$

$$\begin{aligned}
(\nabla\times\mathbf{A})_\theta &= \lim_{\substack{d\phi\to0\\dr\to0}} \frac{\oint_{adefa}\mathbf{A}\cdot d\mathbf{l}}{\text{area } adef}\\[2mm]
&= \lim_{\substack{d\phi\to0\\dr\to0}} \frac{\left\{\begin{array}{l}[A_\phi]_{(r,\theta)}r\sin\theta\,d\phi + [A_r]_{(\theta,\phi+d\phi)}\,dr\\ -\,[A_\phi]_{(r+dr,\theta)}(r+dr)\sin\theta\,d\phi - [A_r]_{(\theta,\phi)}\,dr\end{array}\right\}}{r\sin\theta\,dr\,d\phi}
\end{aligned}$$

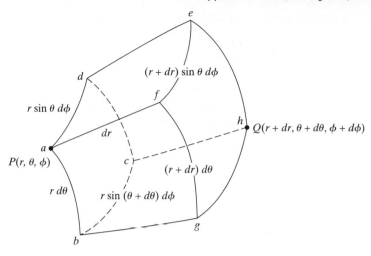

FIGURE A.2   Infinitesimal box formed by incrementing the coordinates in the spherical coordinate system.

$$= \lim_{d\phi \to 0} \frac{[A_r]_{(\theta, \phi + d\phi)} - [A_r]_{(\theta, \phi)}}{r \sin \theta \, d\phi}$$

$$+ \lim_{dr \to 0} \frac{[rA_\phi]_{(r, \theta)} - [rA_\phi]_{(r + dr, \theta)}}{r \, dr}$$

$$= \frac{1}{r \sin \theta} \frac{\partial A_r}{\partial \phi} - \frac{1}{r} \frac{\partial}{\partial r} (rA_\phi) \qquad\qquad (A.11b)$$

$$(\nabla \times \mathbf{A})_\phi = \lim_{\substack{dr \to 0 \\ d\theta \to 0}} \frac{\oint_{afgba} \mathbf{A} \cdot d\mathbf{l}}{\text{area } afgb}$$

$$- \lim_{\substack{dr \to 0 \\ d\theta \to 0}} \frac{\left\{ \begin{aligned} &[A_r]_{(\theta, \phi)} \, dr + [A_\theta]_{(r + dr, \phi)} (r + dr) \, d\theta \\ &- [A_r]_{(\theta + d\theta, \phi)} \, dr - [A_\theta]_{(r, \phi)} r \, d\theta \end{aligned} \right\}}{r \, dr \, d\theta}$$

$$= \lim_{dr \to 0} \frac{[rA_\theta]_{(r + dr, \phi)} - [rA_\theta]_{(r, \phi)}}{r \, dr} \qquad\qquad (A.11c)$$

$$+ \lim_{d\theta \to 0} \frac{[A_r]_{(\theta, \phi)} \, dr - [A_r]_{(\theta + d\theta, \phi)} \, dr}{r \, d\theta}$$

$$= \frac{1}{r} \frac{\partial}{\partial r} (rA_\theta) - \frac{1}{r} \frac{\partial A_r}{\partial \theta}$$

Combining (A.11a), (A.11b), and (A.11c), we obtain the expression for the curl of a vector in spherical coordinates as

$$\nabla \times \mathbf{A} = \frac{1}{r \sin \theta} \left[\frac{\partial}{\partial \theta} (A_\phi \sin \theta) - \frac{\partial A_\theta}{\partial \phi} \right] \mathbf{a}_r$$

$$+ \frac{1}{r} \left[\frac{1}{\sin \theta} \frac{\partial A_r}{\partial \phi} - \frac{\partial}{\partial r} (rA_\phi) \right] \mathbf{a}_\theta$$

$$+ \frac{1}{r} \left[\frac{\partial}{\partial r} (rA_\theta) - \frac{\partial A_r}{\partial \theta} \right] \mathbf{a}_\phi \qquad (A.12)$$

$$= \begin{vmatrix} \dfrac{\mathbf{a}_r}{r^2 \sin \theta} & \dfrac{\mathbf{a}_\theta}{r \sin \theta} & \dfrac{\mathbf{a}_\phi}{r} \\[2mm] \dfrac{\partial}{\partial r} & \dfrac{\partial}{\partial \theta} & \dfrac{\partial}{\partial \phi} \\[2mm] A_r & rA_\theta & r \sin \theta \, A_\phi \end{vmatrix}$$

To find the expression for the divergence, we make use of the basic definition of the divergence of a vector given by (A.4) and by evaluating its right side for the box of Fig. A.2, we obtain

$$\nabla \cdot \mathbf{A} = \lim_{\substack{dr \to 0 \\ d\theta \to 0 \\ d\phi \to 0}} \frac{\left\{ \begin{array}{l} [A_r]_{r+dr}(r + dr)^2 \sin \theta \, d\theta \, d\phi - [A_r]_r r^2 \sin \theta \, d\theta \, d\phi \\ + [A_\theta]_{\theta+d\theta} r \sin (\theta + d\theta) \, dr \, d\phi - [A_\theta]_\theta r \sin \theta \, dr \, d\phi \\ + [A_\phi]_{\phi+d\phi} r \, dr \, d\theta - [A_\phi]_\phi r \, dr \, d\theta \end{array} \right\}}{r^2 \sin \theta \, dr \, d\theta \, d\phi}$$

$$= \lim_{dr \to 0} \frac{[r^2 A_r]_{r+dr} - [r^2 A_r]_r}{r^2 \, dr} + \lim_{d\theta \to 0} \frac{[A_\theta \sin \theta]_{\theta+d\theta} - [A_\theta \sin \theta]_\theta}{r \sin \theta \, d\theta} \qquad (A.13)$$

$$+ \lim_{d\phi \to 0} \frac{[A_\phi]_{\phi+d\phi} - [A_\phi]_\phi}{r \sin \theta \, d\phi}$$

$$= \frac{1}{r^2} \frac{\partial}{\partial r} (r^2 A_r) + \frac{1}{r \sin \theta} \frac{\partial}{\partial \theta} (A_\theta \sin \theta) + \frac{1}{r \sin \theta} \frac{\partial A_\phi}{\partial \phi}$$

To obtain the expression for the gradient of a scalar, we recall from Section 1.3 that in spherical coordinates,

$$d\mathbf{l} = dr \, \mathbf{a}_r + r \, d\theta \, \mathbf{a}_\theta + r \sin \theta \, d\phi \, \mathbf{a}_\phi \qquad (A.14)$$

Therefore,

$$d\Phi = \frac{\partial \Phi}{\partial r} dr + \frac{\partial \Phi}{\partial \theta} d\theta + \frac{\partial \Phi}{\partial \phi} d\phi$$

$$= \left(\frac{\partial \Phi}{\partial r} \mathbf{a}_r + \frac{1}{r} \frac{\partial \Phi}{\partial \theta} \mathbf{a}_\theta + \frac{1}{r \sin \theta} \frac{\partial \Phi}{\partial \phi} \mathbf{a}_\phi \right) \qquad (A.15)$$

$$\cdot (dr \, \mathbf{a}_r + r \, d\theta \, \mathbf{a}_\theta + r \sin \theta \, d\phi \, \mathbf{a}_\phi)$$

$$= \nabla \Phi \cdot d\mathbf{l}$$

Thus,

$$\nabla \Phi = \frac{\partial \Phi}{\partial r} \mathbf{a}_r + \frac{1}{r} \frac{\partial \Phi}{\partial \theta} \mathbf{a}_\theta + \frac{1}{r \sin \theta} \frac{\partial \Phi}{\partial \phi} \mathbf{a}_\phi \qquad \text{(A.16)}$$

To derive the expression for the Laplacian of a scalar, we make use of (A.9), in conjunction with (A.13) and (A.16). Thus, we obtain

$$
\begin{aligned}
\nabla^2 \Phi &= \frac{1}{r^2} \frac{\partial}{\partial r} \left(r^2 \frac{\partial \Phi}{\partial r} \right) + \frac{1}{r \sin \theta} \frac{\partial}{\partial \theta} \left(\frac{1}{r} \frac{\partial \Phi}{\partial \theta} \sin \theta \right) \\
&\quad + \frac{1}{r \sin \theta} \frac{\partial}{\partial \phi} \left(\frac{1}{r \sin \theta} \frac{\partial \Phi}{\partial \phi} \right) \\
&= \frac{1}{r^2} \frac{\partial}{\partial r} \left(r^2 \frac{\partial \Phi}{\partial r} \right) + \frac{1}{r^2 \sin \theta} \frac{\partial}{\partial \theta} \left(\sin \theta \frac{\partial \Phi}{\partial \theta} \right) \\
&\quad + \frac{1}{r^2 \sin^2 \theta} \frac{\partial^2 \Phi}{\partial \phi^2}
\end{aligned}
\qquad \text{(A.17)}
$$

APPENDIX B

Units and Dimensions

In 1960, the International System of Units was given official status at the Eleventh General Conference on Weights and Measures held in Paris. This system of units is an expanded version of the rationalized meter–kilogram–second–ampere (MKSA) system of units and is based on six fundamental, or basic, units. The six basic units are the units of length, mass, time, current, temperature, and luminous intensity.

The international unit of length is the meter. It is exactly 1,650,763.73 times the wavelength in vacuum of the radiation corresponding to the unperturbed transition between the levels $2p_{10}$ and $5d_5$ of the atom of krypton-86, the orange-red line. The international unit of mass is the kilogram. It is the mass of the International Prototype Kilogram which is a particular cylinder of platinum-iridium alloy preserved in a vault at Sèvres, France, by the International Bureau of Weights and Measures. The international unit of time is the second. It is equal to 9,192,631,770 times the period corresponding to the frequency of the transition between the hyperfine levels $F = 4$, $M = 0$ and $F = 3$, $M = 0$ of the fundamental state $^2S_{1/2}$ of the cesium-133 atom unperturbed by external fields.

To present the definition for the international unit of current, we first define the newton, which is the unit of force, derived from the fundamental units meter, kilogram, and second in the following manner. Since velocity is the rate of change of distance with time, its unit is meter per second. Since acceleration is the rate of change of velocity with time, its unit is meter per second per second, or meter per second squared. Since force is mass times acceleration, its unit is kilogram-meter per second squared, also known as the newton. Thus, the newton is that force which imparts an acceleration of 1 meter per second squared to a mass of 1 kilogram. The international unit of current, which is the ampere, can now be defined. It is the constant current that when maintained in two straight, infinitely long, parallel conductors of negligible cross section and placed 1 meter apart in vacuum produces a force of 2×10^{-7} newton per meter length of the conductors.

The international unit of temperature is the kelvin. It is based on the definition of the thermodynamic scale of temperature by means of the triple point of water as a fixed fundamental point to which a temperature of exactly 273.16 kelvin is attributed. The international unit of luminous intensity is the candela. It is defined such that the luminance of a blackbody radiator at the freezing temperature of platinum is 60 candelas per square centimeter.

We have just defined the six basic units of the International System of Units. Two supplementary units are the radian and the steradian for plane angle and solid angle, respectively. All other units are derived units. For example, the unit of charge which is the coulomb is the amount of charge transported in 1 second by a current of 1 ampere; the unit of energy which is the joule is the work done when the point of application of a force of 1 newton is displaced a distance of 1 meter in the direction of the force; the unit of power which is the watt is the power that gives rise to the production of energy at a rate of 1 joule per second; the unit of electric potential difference which is the volt is the difference of electric potential between two points of a conducting wire carrying constant current of 1 ampere when the power dissipated between these points is equal to 1 watt; and so on. The units for the various quantities used in this book are listed in Table B.1, together with the symbols of the quantities and their dimensions.

TABLE B.1   Symbols, Units, and Dimensions of Various Quantities

Quantity	Symbol	Unit (Symbol)	Dimensions
Admittance	Y	siemens (S)	$M^{-1}L^{-2}TQ^2$
Area	A	square meter (m²)	L^2
Attenuation constant	α	neper/meter (Np/m)	L^{-1}
Capacitance	C	farad (F)	$M^{-1}L^{-2}T^2Q^2$
Capacitance per unit length	$\mathscr{C}$	farad/meter (F/m)	$M^{-1}L^{-3}T^2Q^2$
Cartesian coordinates	x	meter (m)	L
	y	meter (m)	L
	z	meter (m)	L
Characteristic admittance	Y_0	siemens (S)	$M^{-1}L^{-2}TQ^2$
Characteristic impedance	Z_0	ohm (Ω)	$ML^2T^{-1}Q^{-2}$
Charge	Q, q	coulomb (C)	Q
Conductance	G	siemens (S)	$M^{-1}L^{-2}TQ^2$
Conductance per unit length	$\mathscr{G}$	siemens/meter (S/m)	$M^{-1}L^{-3}TQ^2$
Conduction current density	$\mathbf{J}_e$	ampere/square meter (A/m²)	$L^{-2}T^{-1}Q$
Conductivity	σ	siemens/meter (S/m)	$M^{-1}L^{-3}TQ^2$
Current	I	ampere (A)	$T^{-1}Q$
Cutoff frequency	f_c	hertz (Hz)	T^{-1}
Cutoff wavelength	λ_c	meter (m)	L
Cylindrical coordinates	r, r_c	meter (m)	L
	ϕ	radian	—
	z	meter (m)	L
Differential length element	$d\mathbf{l}$	meter (m)	L
Differential surface element	$d\mathbf{S}$	square meter (m²)	L^2
Differential volume element	dv	cubic meter (m³)	L^3

(*Continued*)

TABLE B.1   (*Continued*)

Quantity	Symbol	Unit (Symbol)	Dimensions
Directivity	D	—	—
Displacement flux density	**D**	coulomb/square meter (C/m²)	$L^{-2}Q$
Electric dipole moment	**p**	coulomb-meter (C-m)	LQ
Electric field intensity	**E**	volt/meter (V/m)	$MLT^{-2}Q^{-1}$
Electric potential	V	volt (V)	$ML^2T^{-2}Q^{-1}$
Electric susceptibility	χ_e	—	—
Electron density	N_e	(meter)$^{-3}$ (m^{-3})	L^{-3}
Electronic charge	e	coulomb (C)	Q
Energy	W	joule (J)	ML^2T^{-2}
Energy density	w	joule/cubic meter (J/m³)	$ML^{-1}T^{-2}$
Force	**F**	newton (N)	MLT^{-2}
Frequency	f	hertz (Hz)	T^{-1}
Group velocity	v_g	meter/second (m/s)	LT^{-1}
Guide characteristic impedance	η_g	ohm (Ω)	$ML^2T^{-1}Q^{-2}$
Guide wavelength	λ_g	meter (m)	L
Impedance	Z	ohm (Ω)	$ML^2T^{-1}Q^{-2}$
Inductance	L	henry (H)	ML^2Q^{-2}
Inductance per unit length	$\mathscr{L}$	henry/meter (H/m)	MLQ^{-2}
Intensity	I	watt/square meter (W/m²)	MT^{-3}
Intrinsic impedance	η	ohm (Ω)	$ML^2T^{-1}Q^{-2}$
Length	l	meter (m)	L
Line charge density	ρ_L	coulomb/meter (C/m)	$L^{-1}Q$
Magnetic dipole moment	**m**	ampere-square meter (A-m²)	$L^2T^{-1}Q$
Magnetic field intensity	**H**	ampere/meter (A/m)	$L^{-1}T^{-1}Q$
Magnetic flux	ψ	weber (Wb)	$ML^2T^{-1}Q^{-1}$
Magnetic flux density	**B**	tesla or weber/square meter (T or Wb/m²)	$MT^{-1}Q^{-1}$
Magnetic susceptibility	χ_m	—	—
Magnetic vector potential	**A**	weber/meter (Wb/m)	$MLT^{-1}Q^{-1}$
Magnetization surface current density	**J**$_{mS}$	ampere/meter (A/m)	$L^{-1}T^{-1}Q$
Magnetization vector	**M**	ampere/meter (A/m)	$L^{-1}T^{-1}Q$
Mass	m	kilogram (kg)	M
Mobility	μ	square meter/volt-second (m²/V-s)	$M^{-1}TQ$
Permeability	μ	henry/meter (H/m)	MLQ^{-2}
Permeability of free space	μ_0	henry/meter (H/m)	MLQ^{-2}
Permittivity	ε	farad/meter (F/m)	$M^{-1}L^{-3}T^2Q^2$
Permittivity of free space	ε_0	farad/meter (F/m)	$M^{-1}L^{-3}T^2Q^2$
Phase constant	β	radian/meter (rad/m)	L^{-1}
Phase velocity	v_p	meter/second (m/s)	LT^{-1}
Polarization surface charge density	ρ_{PS}	coulomb/square meter (C/m²)	$L^{-2}Q$
Polarization vector	**P**	coulomb/square meter (C/m²)	$L^{-2}Q$
Power	P	watt (W)	ML^2T^{-3}
Power density	p	watt/square meter (W/m²)	MT^{-3}
Poynting vector	**P**	watt/square meter (W/m²)	MT^{-3}
Propagation constant	$\bar{\gamma}$	(meter)$^{-1}$ (m^{-1})	L^{-1}
Propagation vector	β	radian/meter (rad/m)	L^{-1}
Q factor	Q	—	—

(*Continued*)

TABLE B.1   (*Continued*)

Quantity	Symbol	Unit (Symbol)	Dimensions
Radian frequency	ω	radian/second (rad/s)	T^{-1}
Radiation resistance	R_{rad}	ohm (Ω)	$ML^2T^{-1}Q^{-2}$
Reactance	X	ohm (Ω)	$ML^2T^{-1}Q^{-2}$
Reflection coefficient	Γ	—	—
Refractive index	n	—	—
Relative permeability	μ_r	—	—
Relative permittivity	ε_r	—	—
Reluctance	$\mathcal{R}$	ampere (turn)/weber (A-t/Wb)	$M^{-1}L^{-2}Q^2$
Resistance	R	ohm (Ω)	$ML^2T^{-1}Q^{-2}$
Skin depth	δ	meter (m)	L
Spherical coordinates	$\begin{cases} r, r_s \\ \theta \\ \phi \end{cases}$	meter (m) radian radian	L — —
Standing wave ratio	SWR	—	—
Surface charge density	ρ_S	coulomb/square meter (C/m^2)	$L^{-2}Q$
Surface current density	$\mathbf{J}_y$	ampere/meter (A/m)	$L^{-1}T^{-1}Q$
Susceptance	B	siemens (S)	$M^{-1}L^{-2}TQ^2$
Time	t	second (s)	T
Transmission coefficient	τ	—	—
Unit normal vector	$\mathbf{a}_n$	—	—
Velocity	v	meter/second (m/s)	LT^{-1}
Velocity of light in free space	c	meter/second (m/s)	LT^{-1}
Voltage	V	volt (V)	$ML^2T^{-2}Q^{-1}$
Volume	V	cubic meter (m^3)	L^3
Volume charge density	ρ	coulomb/cubic meter (C/m^3)	$L^{-3}Q$
Volume current density	$\mathbf{J}$	ampere/square meter (A/m^2)	$L^{-2}T^{-1}Q$
Wavelength	λ	meter (m)	L
Work	W	joule (J)	ML^2T^{-2}

Dimensions are a convenient means of checking the possible validity of a derived equation. The dimension of a given quantity can be expressed as some combination of a set of fundamental dimensions. These fundamental dimensions are mass (M), length (L), and time (T). In electromagnetics, it is the usual practice to consider the charge (Q), instead of the current, as the additional fundamental dimension. For the quantities listed in Table B.1, these four dimensions are sufficient. Thus, for example, the dimension of velocity is length (L) divided by time (T), that is, LT^{-1}; the dimension of acceleration is length (L) divided by time squared (T^2), that is, LT^{-2}; the dimension of force is mass (M) times acceleration (LT^{-2}), that is, MLT^{-2}; the dimension of ampere is charge (Q) divided by time (T), that is, QT^{-1}; and so on.

To illustrate the application of dimensions for checking the possible validity of a derived equation, let us consider the equation for the phase velocity of an electromagnetic wave in free space, given by

$$v_p = \frac{1}{\sqrt{\mu_0 \varepsilon_0}}$$

We know that the dimension of v_p is LT^{-1}. Hence, we have to show that the dimension of $1/\sqrt{\mu_0\varepsilon_0}$ is also LT^{-1}. To do this, we note from Coulomb's law that

$$\varepsilon_0 = \frac{Q_1 Q_2}{4\pi FR^2}$$

Hence, the dimension of ε_0 is $Q^2/[MLT^{-2})(L^2)]$, or $M^{-1}L^{-3}T^2Q^2$. We note from Ampère's law of force applied to two infinitesimal current elements parallel to each other and normal to the line joining them that

$$\mu_0 = \frac{4\pi FR^2}{(I_1\,dl_1)(I_2\,dl_2)}$$

Hence the dimension of μ_0 is $[(MLT^{-2})(L^2)]/(QT^{-1}L)^2]$, or MLQ^{-2}. Now we obtain the dimension of $1/\sqrt{\mu_0\varepsilon_0}$ as $1/\sqrt{(M^{-1}L^{-3}T^2Q^2)(MLQ^{-2})}$, or LT^{-1}, which is the same as the dimension of v_p. It should be noted, however, that the test for the equality of the dimensions of the two sides of a derived equation is not a sufficient test to establish the equality of the two sides since any dimensionless constants associated with the equation may be in error.

It is not always necessary to refer to the table of dimensions for checking the possible validity of a derived equation. For example, let us assume that we have derived the expression for the characteristic impedance of a transmission line (i.e., $\sqrt{\mathcal{L}/\mathcal{C}}$) and we wish to verify that $\sqrt{\mathcal{L}/\mathcal{C}}$ does indeed have the dimension of impedance. To do this, we write

$$\sqrt{\frac{\mathcal{L}}{\mathcal{C}}} = \sqrt{\frac{\omega\mathcal{L}l}{\omega\mathcal{C}l}} = \sqrt{\frac{\omega L}{\omega C}} = \sqrt{(\omega L)\left(\frac{1}{\omega C}\right)}$$

We now recognize from our knowledge of circuit theory that both ωL and $1/\omega C$, being the reactances of L and C, respectively, have the dimension of impedance. Hence, we conclude that $\sqrt{\mathcal{L}/\mathcal{C}}$ has the dimension of $\sqrt{(\text{impedance})^2}$ or impedance.

Suggested Collateral and Further Reading

ADLER, R. B., L. J. CHU, and R. M. FANO, *Electromagnetic Energy Transmission and Radiation*. New York: John Wiley & Sons, 1960.

BOOTON, R. C., JR., *Computational Methods for Electromagnetics and Microwaves*. New York: John Wiley & Sons, 1992.

DAVIDSON, C. W., *Transmission Lines for Communications,* 2nd ed. New York: John Wiley & Sons, 1989.

DIAMENT, P., *Wave Transmission and Fiber Optics*. New York: Macmillan, 1990.

HAYT, W. H., JR., *Engineering Electromagnetics,* 5th ed. New York: McGraw-Hill, 1989.

JORDAN, E. C., and K. G. BALMAIN, *Electromagnetic Waves and Radiating Systems,* 2nd ed. Englewood Cliffs, NJ: Prentice Hall, 1968.

KRAUS, J. D., *Electromagnetics*, 4th ed. New York: McGraw-Hill, 1992.

LEE, D. L., *Electromagnetic Principles of Integrated Optics*. New York: John Wiley & Sons, 1986.

RAMO, S., J. R. WHINNERY, and T. VAN DUZER, *Fields and Waves in Communication Electronics,* 3rd ed. New York: John Wiley & Sons, 1994.

RAO, N. N., *Basic Electromagnetics with Applications*. Englewood Cliffs, NJ: Prentice Hall, 1972.

SADIKU, M. N. O., *Numerical Techniques in Electromagnetics*. Boca Raton, FL: CRC Press, 1992.

SALEH, B. E. A., and M. C. TEICH, *Fundamentals of Photonics*. New York: John Wiley & Sons, 1991.

SESHADRI, S. R., *Fundamentals of Transmission Lines and Electromagnetic Fields*. Reading, MA: Addison-Wesley, 1971.

SHEN, L. C., and J. A. KONG, *Applied Electromagnetism,* 3rd ed. Boston: PWS Publishers, 1995.

Answers to Selected Problems

CHAPTER 1

P1.1. (a) $1/3$; (b) 0.75; (c) $1/\sqrt{3}$

P1.4. (a) -2; (b) -1; (c) $-1, -5$; (d) 1

P1.6. Yes

P1.8. (a) 25; (b) 16; (c) $12\sqrt{2}$

P1.10. (b) $x + y + z = 4$

P1.13. $\frac{1}{\sqrt{6}}(\mathbf{a}_x + \mathbf{a}_y + \mathbf{a}_z)$

P1.16. $\sqrt{10}$

P1.19. $2\mathbf{a}_\theta$

P1.21. $\pm(\cos 2\phi\, \mathbf{a}_r - \sin 2\phi\, \mathbf{a}_\phi)$; (a) $\pm\mathbf{a}_y$; (b) $\pm(\frac{\sqrt{3}}{2}\mathbf{a}_r - \frac{1}{2}\mathbf{a}_\phi)$

P1.24. (a) $-mMG\dfrac{(x\mathbf{a}_x + y\mathbf{a}_y + z\mathbf{a}_z)}{(x^2 + y^2 + z^2)^{3/2}}$; (b) $-mMG\dfrac{(r_c\mathbf{a}_{rc} + z\mathbf{a}_z)}{(r_c^2 + z^2)^{3/2}}$; (c) $-\dfrac{mMG}{r_s^2}\mathbf{a}_{rs}$

P1.26. (a) $x^2 + 2y^2 = 9, z = 3$; (b) $6x = 3y = 2z$

P1.28. $r\sin^2\theta = 1, \phi = \pi/6$

P1.30. $f(z, t)$ represents a standing wave

P1.32. (a) Elliptical; (b) -2

P1.35. $\pm(2.5419 + j5.1593)$

P1.37. $0.8585\underline{/66.188°}$

P1.40. $1.5\sin t + 0.1\cos(3t + 143.13°)$

CHAPTER 2

P2.1. $\dfrac{0.1949Q^2}{\varepsilon_0 L^2}$ away from the center of the tetrahedron

P2.3. (a) No solutions; (b) $-108\pi\varepsilon_0$ C at $(3, 3, 2)$

P2.5. $\dfrac{d^2z}{dt^2} + \dfrac{3Qz}{4\pi\varepsilon_0 ma^3} = 0;\qquad \dfrac{1}{2\pi}\sqrt{\dfrac{3Q}{4\pi\varepsilon_0 ma^3}}$

P2.8. $\dfrac{2\pi a}{z\sqrt{r^2 + z^2}}\,\mathbf{a}_z$

P2.11. $0.046\mu_0(I\,dz)^2$ toward the origin

P2.14. **(a)** Same;   **(b)** $\sqrt{\dfrac{\pi d^2 w}{4\mu_0 lL}}$

P2.17. **(a)** $0.45\mu_0\mathbf{a}_z$;   **(b)** $-0.057\mu_0\mathbf{a}_z$

P2.19. (a) $-3B_0\mathbf{a}_x$;   **(b)** $B_0(\mathbf{a}_x + 2\mathbf{a}_y)$;   **(c)** $B_0(-3\mathbf{a}_x + 2\mathbf{a}_y + 2\mathbf{a}_z)$

P2.22. $\dfrac{E_0}{3B_0}(2\mathbf{a}_x - \mathbf{a}_y - 2\mathbf{a}_z)$

P2.25. **(a)** $-\dfrac{qE_0}{2}\,\mathbf{a}_z$;   **(b)** $\dfrac{qE_0}{2}\,\mathbf{a}_z$;   **(c)** 0

P2.28. **(b)** 0.431×10^{13} Hz; $1/\sqrt{2}$

P2.30. $\dfrac{\rho_{L0}}{2\pi\varepsilon_0}\left\{\dfrac{(x - d)\mathbf{a}_x + y\mathbf{a}_y}{(x - d)^2 + y^2} - \dfrac{(x + d)\mathbf{a}_x + y\mathbf{a}_y}{(x + d)^2 + y^2}\right\}$

P2.33. **(a)** $2\varepsilon_0 E_0(2\mathbf{a}_x + 2\mathbf{a}_y + \mathbf{a}_z)$;   **(b)** $2\varepsilon_0\mathbf{E}$;   **(c)** $\dfrac{D}{2\varepsilon_0}$

P2.35. $\dfrac{\omega Q a^2}{4}\,\mathbf{a}_z$

P2.38. $3k\mu_0$

CHAPTER 3

P3.1. **(a)** $1/2$;   **(b)** 0

P3.4. 0.5708

P3.7. 0

P3.9. $2\pi/3$

P3.12. **(a)** $-2B_0 v_0 \cos \pi(x - v_0 t)$;   **(b)** 0

P3.15. **(a)** $B_0 hb\omega \sin \omega t$;   **(b)** 0

P3.18. **(a)** 128 A;   **(b)** 8π A

P3.21. $0.0255 \cos\left(2\pi \times 10^6 t + \dfrac{\pi}{4}\right)$ A

P3.23. $\dfrac{B_0\pi^2}{2}$ Wb

P3.25. $\frac{1}{3}I$

P3.27. **(a)** $\frac{8}{3}$C;   **(b)** $\frac{32}{81}$C

P3.30. 0 for $r < a$, $\dfrac{\rho_0(r^3 - a^3)}{3r^2}\,\mathbf{a}_r$ for $a < r < 2a$; $\dfrac{7a^3\rho_0}{3r^2}\,\mathbf{a}_r$ for $r > 2a$

P3.32. $\dfrac{J_0 r^2}{3a}\,\mathbf{a}_\phi$ for $r < a$, $\dfrac{J_0 a^2}{3r}\,\mathbf{a}_\phi$ for $r > a$

P3.35. **(b)** $3\varepsilon_0 E_0 \cos\theta$;

      (c) $E_0(\cos\theta\,\mathbf{a}_r - \sin\theta\,\mathbf{a}_\theta), \dfrac{E_0 a^3}{r^3}(2\cos\theta\,\mathbf{a}_r + \sin\theta\,\mathbf{a}_\theta)$

P3.38. $2\varepsilon_0$

P3.40. $98\mu_0$

CHAPTER 4

P4.2. **(a)** $\dfrac{E_0}{3\times10^8}\sin 3\pi z \sin 9\pi \times 10^8 t\,\mathbf{a}_y$;

      (b) $\dfrac{E_0(0.6\mathbf{a}_x - 0.8\mathbf{a}_y)}{3\times10^8}\cos[3\pi\times10^8 t + 0.2\pi(4x + 3z)]$

P4.4. $\mathbf{B} = \dfrac{\alpha E_0}{\omega}e^{-\alpha z}\sin\omega t\,\mathbf{a}_z$; no

P4.6. $\pm 5/\sqrt{3}$

P4.9. **(a)** $\dfrac{\rho_0}{2a}(x^2 - a^2)\mathbf{a}_x$ for $-a < x < a, \mathbf{0}$ otherwise;

      (b) $\dfrac{\rho_0 r^2}{3a}\mathbf{a}_r$ for $0 < r < a, \dfrac{\rho_0 a^2}{3r}\mathbf{a}_r$ for $r > a$

P4.11. **(a)** Yes; **(b)** yes; **(c)** no

P4.14. $\nabla \times \mathbf{v} = -2\omega\mathbf{a}_z$

P4.17. **(a)** Both sides of divergence theorem are equal to $\frac{3}{2}$; **(b)** both sides of divergence theorem are equal to zero

P4.20. **(a)** $e^{-y}\sin x$; **(b)** $r\cos\phi$; **(c)** $-\dfrac{\cos\theta}{r^2}$

P4.22. $16x + 32y + z = 24$

P4.24. Direction lines are $(x^2 - y^2) = $ constant

P4.26. $V = \dfrac{\rho_{L0}}{4\pi\varepsilon}\ln\dfrac{\sqrt{r^2 + (z + a)^2} + (z + a)}{\sqrt{r^2 + (z - a)^2} + (z - a)}$; equipotential surfaces are

      $\dfrac{(c - 1)^2}{4c}\left(\dfrac{r}{a}\right)^2 + \left(\dfrac{c - 1}{c + 1}\right)^2\left(\dfrac{z}{a}\right)^2 = 1$, where c is a constant

P4.32. $120°, 101.6284°, 137.6641°$

P4.34. **(a)** $0.0233\dfrac{\rho_0^2 a^5}{\varepsilon_0}$;

      (b) $1.08a, 0.4763\rho_0$

P4.36. $0.0651\mu_0 I_0^2$

CHAPTER 5

P5.1. $\dfrac{\partial E_z}{\partial y} = -\dfrac{\partial B_x}{\partial t}, \dfrac{\partial H_x}{\partial y} = -J_z - \dfrac{\partial D_z}{\partial t}$

P5.4. **(c)**

P5.7. **(a)**

(d)

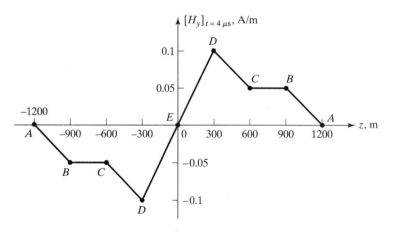

P5.10. $\mathbf{E} = 1.25E_0\left[\cos\left(2\pi \times 10^8 t - \dfrac{2\pi}{3}z + 0.2048\pi\right)\mathbf{a}_x\right.$

$\left. + \sin\left(2\pi \times 10^8 t - \dfrac{2\pi}{3}z + 0.2048\pi\right)\mathbf{a}_y\right]$

$\mathbf{H} = \dfrac{E_0}{96\pi}\left[-\sin\left(2\pi \times 10^8 t - \dfrac{2\pi}{3}z + 0.2048\pi\right)\mathbf{a}_x\right.$

$\left. + \cos\left(2\pi \times 10^8 t - \dfrac{2\pi}{3}z + 0.2048\pi\right)\mathbf{a}_y\right]$

P5.13. **(a)** Right circular; **(b)** left circular; **(c)** left elliptical; **(d)** right elliptical

P5.16. $\dfrac{|1 - k|}{|1 - 3k|}$; **(a)** $\dfrac{1}{2}$; **(b)** 1; **(c)** 0; **(a)** $\dfrac{2}{3}$; **(b)** $\dfrac{2}{5}$ or $\dfrac{1}{4}$

P5.18. $-\dfrac{fv_0}{c} \dfrac{4z^3 + z - 1}{\sqrt{(8z^2 + 1)^2(2z^4 + z^2 - 2z + 1)}}$; $\left(\dfrac{1}{4}, \dfrac{1}{4}, \dfrac{1}{2}\right)$

P5.20. **(a)** 21.224 m, 3.10;

(b) $91.82e^{-0.0533z} \cos(2\pi \times 10^6 t - 0.074z + 0.1988\pi)\, \mathbf{a}_x$ V/m

P5.23. $\mathbf{E} = \begin{cases} 14.493 \cos\left(2\pi \times 10^6 t - \dfrac{\pi}{150} z + 0.163\pi\right) \mathbf{a}_x & \text{for } z > 0 \\ 14.493 e^{0.05334z} \cos(2\pi \times 10^6 t + 0.07401z + 0.163\pi)\, \mathbf{a}_x & \text{for } z < 0 \end{cases}$

$\mathbf{H} = \begin{cases} 0.0384 \cos\left(2\pi \times 10^6 t - \dfrac{\pi}{150} z + 0.163\pi\right) \mathbf{a}_y & \text{for } z > 0 \\ 0.1675 e^{0.05334z} \cos(2\pi \times 10^6 t + 0.07401z + 0.964\pi)\, \mathbf{a}_y & \text{for } z < 0 \end{cases}$

P5.26. $2.25\mu_0, 16\varepsilon_0$

P5.28. $1.0883 \times 10^{-4}\,\mathrm{m^{-1}}$; 0.036276 rad/m; 1.732×10^8 m/s; 173.21 m; $(217.66 + j0.653)\,\Omega$; 9.189 km

P5.31. **(a)** 1 Hz; **(b)** 2 Hz

P5.33. **(a)** $\dfrac{V_0 I_0}{2\pi r^2 \ln(b/a)} \cos^2 \omega(t - \sqrt{\mu_0 \varepsilon_0} z)\, \mathbf{a}_z$, $\dfrac{V_0 I_0}{4\pi r^2 \ln(b/a)}\, \mathbf{a}_z$; **(b)** $\dfrac{1}{2} V_0 I_0$

P5.35. **(a)** $1.5791\, H_0^2\, e^{-2z}$ W/m²;

(b) $1.3654\, H_0^2$ W

P5.37. $\mathbf{E}_r - [0.4744 E_0 \cos(3\pi \times 10^5 t + 10^{-2}\pi z + 0.8976\pi)\, \mathbf{a}_x$
    $+ 0.1161 E_0 \cos(9\pi \times 10^5 t + 3 \times 10^{-3}\,\pi z + 0.9043\pi)\, \mathbf{a}_x]$ V/m
$\mathbf{E}_t = [0.3354 E_0 e^{-6.283 \times 10^{-3}z} \cos(3\pi \times 10^5 t - 9.425 \times 10^{-3}\pi z + 0.1476\pi)\, \mathbf{a}_x$
    $+ 0.1433 E_0 e^{-7.894 \times 10^{-3}z} \cos(9\pi \times 10^5 t - 22.504 \times 10^{-3}\pi z$
    $+ 0.0772\pi)\, \mathbf{a}_x]$ V/m

P5.40. $\mathbf{E}_r = -E_0 \cos(\omega t + \beta z)\, \mathbf{a}_x$; $\mathbf{E} = 2E_0 \sin \omega t \sin \beta z\, \mathbf{a}_x$

$\mathbf{H}_r - \dfrac{E_0}{\eta} \cos(\omega t + \beta z)\, \mathbf{a}_y$; $\mathbf{H} = \dfrac{2E_0}{\eta} \cos \omega t \cos \beta z\, \mathbf{a}_y$

$[\mathbf{J}_S]_{z=0} = \dfrac{2E_0}{\eta} \cos \omega t\, \mathbf{a}_x$

CHAPTER 6

P6.1. $-\dfrac{\rho_0 d^2}{6\varepsilon}$ for $x < -d$; $\dfrac{\rho_0}{2\varepsilon}\left(dx + x^2 + \dfrac{x^3}{3d}\right)$ for $-d < x < 0$;

$\dfrac{\rho_0}{2\varepsilon}\left(dx - x^2 + \dfrac{x^3}{3d}\right)$ for $0 < x < d$; $\dfrac{\rho_0 d^2}{6\varepsilon}$ for $x > d$

P6.4. **(a)** $\dfrac{\varepsilon_2 x}{\varepsilon_2 t + \varepsilon_1(d - t)} V_0$ for $0 < x < t$, $\dfrac{\varepsilon_2 t + \varepsilon_1(x - t)}{\varepsilon_2 t + \varepsilon_1(d - t)} V_0$ for $t < x < d$;

(b) $\dfrac{\varepsilon_1\varepsilon_2}{\varepsilon_2 t + \varepsilon_1(d - t)}$

P6.7. $\dfrac{V_0(r - b)}{a - b}$; $\dfrac{2\pi\varepsilon_0 b}{b - a}$

P6.8. 2.03365

P6.10. $2\pi\mu N^2 a^2 c \ln \dfrac{2a + b}{2a - b}$

P6.12. $\dfrac{\mu}{16\pi}$

P6.16. $\frac{1}{15}\mu_0$ H/m, $60\varepsilon_0$ F/m; $4\pi\ \Omega$

P6.19. $\frac{1}{9}\eta$

P6.22. 125 V, 25 Ω, 60 Ω, 2μs

P6.26.

(a)

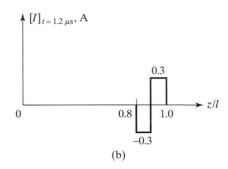

(b)

P6.29. $V^- = -0.2V^+, I^- = 0.002V^+; V^{++} = 0.2V^+, I^{++} = 0.004V^+$

P6.31. 150 m, $4\varepsilon_0$; $A = 8/15$

P6.33.

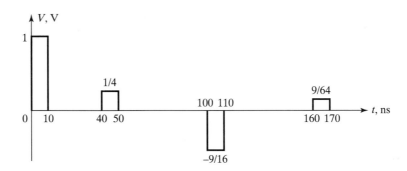

P6.35. **(a)** $\dfrac{dV^-}{dt} + \dfrac{1}{2CZ_0} V^- = \dfrac{V_0}{4CZ_0}$ for $t > T$;

(b) $\dfrac{V_0}{2} - \dfrac{V_0}{2} e^{-(1/2\ CZ_0)(t-T)}$ for $t > T$

P6.37 **(a)** L; **(b)** $\frac{1}{3}$

P6.40. **(a)**

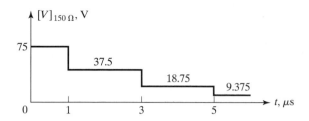

(b) 62.5×10^{-6} J

P6.43.

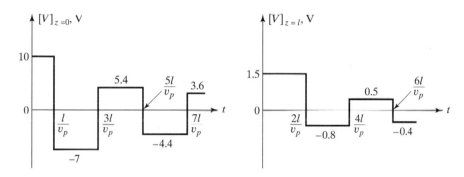

P6.45. **(a)** $38.4\,\Omega$; **(b)** $48.4\,\Omega$

P6.48. **(c)**

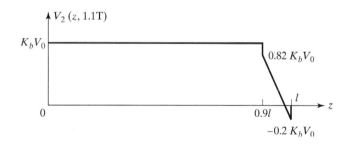

CHAPTER 7

P7.2.

d	V_{rms}, V	I_{rms}, A
0	$0.7906V_0$	0
$\frac{1}{3}l$	$0.6374V_0$	$0.0047V_0$
$\frac{1}{2}l$	$0.5V_0$	$0.0061V_0$
l	$0.3536V_0$	$0.0071V_0$

P7.5. $\dfrac{nv_p}{l}, n = 1, 2, 3, \ldots$

P7.7. 4.7746 cm

P7.9. 0.3229 GHz; 0.7920 GHz; 1.2698 GHz

P7.12. $(59.78 + j75.80)\,\Omega$

P7.14. 20 cm

P7.17. $(40 - j30)\,\Omega$; 30 W

P7.20. $30\,\Omega$; $75\,\Omega$

P7.22. 0.4119λ; 1.53

P7.26. **(a)** 1,562 MHz; **(b)** 4.0

P7.28. $(r - 2)^2 + x^2 = 1$

P7.30. **(a)** 2.0; **(b)** 1.56; **(c)** 2.24

P7.32. 0.126λ; 0.094λ

P7.34. 0.093λ to 0.159λ

P7.36. $(0.328\lambda, 0.458\lambda)$ or $(0.496\lambda, 0.042\lambda)$

P7.39. $(l_1 = 0.313\lambda, l_2 = 0.136\lambda)$
or $(l_1 = 0.168\lambda, l_2 = 0.364\lambda)$

P7.41. $(144.23 - j32.59)\,\Omega$; $(1.035 + j4.561) \times 10^{-4}\,\text{m}^{-1}$

P7.44. **(a)** 15.26 W; **(b)** 8.98 W; **(c)** 6.28 W

CHAPTER 8

P8.1. **(a)** $-0.0395 \sin 10^6 \pi t$ V;
(b) $-65.285 \sin 10^9 \pi t$ V

P8.3. **(a)** $\sigma \sqrt{\dfrac{\mu}{\varepsilon}}\, l = 3$; **(b)** $\sigma \sqrt{\dfrac{\mu}{\varepsilon}}\, l \ll 1$; **(c)** $\sqrt{\omega\mu\sigma}\, l \ll 1$ and $\dfrac{\sigma}{\omega\varepsilon} \gg 1$

P8.5. Inductor of value $L = \mu dl/w$ in parallel with a series combination of $\frac{1}{5}L$ and $\frac{1}{3}C$, where $C = \varepsilon w L/d$

P8.8. 534.881 A-t

P8.10. 8.4×10^{-4} Wb

P8.12. $\dfrac{V^2 w}{2d}\,(\varepsilon - \varepsilon_0)\mathbf{a}_x$

P8.15. **(a)** $\frac{1}{2}N^2 I^2 \pi a^2 (\mu - \mu_0)\mathbf{a}_x$ for $0 < x < b$, $-\frac{1}{2}N^2 I^2 \pi a^2 (\mu - \mu_0)\mathbf{a}_x$ for $l < x < (l + b)$, $\mathbf{0}$ otherwise; **(b)** $\frac{3}{4}N^2 I_0^2 \pi a^2 b(\mu - \mu_0)$ from mechanical to electrical form

CHAPTER 9

P9.2. **(a)** Yes; **(b)** $\dfrac{1}{12\pi}(-4\mathbf{a}_x + 5\mathbf{a}_y + 3\mathbf{a}_z) \cos [3\pi \times 10^7 t - 0.02\pi(3x + 4z)]$

P9.5. **(a)** 47.7465 MHz; **(b)** along $(-0.6\mathbf{a}_y + 0.8\mathbf{a}_z)$;
(c) $\dfrac{1}{12\pi}(-j0.5\mathbf{a}_x + 0.8\mathbf{a}_y + 0.6\mathbf{a}_z)e^{j(0.6y - 0.8z)}$;
(d) left-elliptical; **(e)** 0.1658 W

P9.7. $\dfrac{E_0}{8} \{3 \sin 20\pi x \, [\cos (10^{10}\pi t - 83.776z) + \cos (2 \times 10^{10}\pi t - 199.793z)]$

$\qquad\qquad - \sin 60\pi x \cos (2 \times 10^{10}\pi t - 91.293z)\} \, \mathbf{a}_y$

P9.9. **(a)** 0.25; **(b)** 0.4307; **(c)** 0.221

P9.11. **(a)** $\dfrac{0.866}{k}$; **(b)** $\dfrac{0.6378}{k}$; **(c)** $\dfrac{0.6495}{k}$

P9.13. **(a)** 1.7047×10^8 m/s; **(b)** 1.7321×10^8 m/s

P9.16. $\mathbf{E}_r = 0.0294E_0(\mathbf{a}_x + \mathbf{a}_z) \cos [6\pi \times 10^8 t + \sqrt{2}\pi(x - z)]$

$\qquad \mathbf{E}_t = 0.6863E_0(\mathbf{a}_x - \sqrt{2}\,\mathbf{a}_z) \cos [6\pi \times 10^8 t - \sqrt{2}\pi(\sqrt{2}x + z)]$

P9.18. **(a)** $3\varepsilon_0$;

$\qquad$**(b)** $\mathbf{E}_r = -0.5E_0\mathbf{a}_y \sin [6\pi \times 10^9 t + 10\pi(x - \sqrt{3}z)]$

$\qquad\qquad \mathbf{E}_t = E_0(0.2887\mathbf{a}_x - 0.5\mathbf{a}_z) \cos [6\pi \times 10^9 t - 10\pi(3x + \sqrt{3}z)]$

$\qquad\qquad\quad + 0.5E_0\mathbf{a}_y \sin [6\pi \times 10^9 t - 10\pi(3x + \sqrt{3}z)]$

P9.21. **(a)** 23; **(b)** $2/\sqrt{5}$

P9.24. 0.9152 (TE_0); 0.9130 (TE_1); 0.9090 (TE_2)

P9.26. $\tan \left(\dfrac{\pi d \sqrt{\varepsilon_{r1}}}{\lambda_0} \cos \theta_i \right) = -\dfrac{(\varepsilon_2/\varepsilon_1) \cos \theta_i}{\sqrt{\sin^2 \theta_i - (\varepsilon_2/\varepsilon_1)}}$

P9.30. $z = \dfrac{1}{\alpha} \ln \dfrac{\alpha x - 1 + \sqrt{\alpha^2 x^2 - 2\alpha x + \delta_0^2}}{\delta_0 - 1}; \dfrac{\delta_0^2}{2\alpha}, \dfrac{2\delta_0}{\alpha}$

P9.32. $x_a = \left[\dfrac{3(2m + 1)\lambda_0}{16n_0 \sqrt{\alpha}} \right]^{2/3}, m = 0, 1, 2, \dots$

CHAPTER 10

P10.2. 4.24 cm $\le a \le 4.47$ cm

P10.4. **(a)** 1.70; **(b)** 0.366 cm, $4.9\varepsilon_0$

P10.7. 4472.1 MHz (TE_{101}); 5385.2 MHz $(\text{TE}_{0,1,1})$; 5656.9 MHz $(\text{TE}_{1,0,2})$; 6403.1 MHz
$\qquad (\text{TE}_{0,1,2}, \text{TM}_{1,1,0})$; 6708.2 MHz $(\text{TE}_{1,1,1}, \text{TM}_{1,1,1})$

P10.10. 6.096 GHz $(\text{TE}_{0,1})$; 6.685 GHz $(\text{TE}_{3,1})$

P10.12. 0.4442 cm, 1.2726 cm

P10.14 **(a)** $\tan \dfrac{2\pi t}{\lambda_1} \sqrt{1 - \left(\dfrac{\lambda_1}{\lambda_c} \right)^2} \tan \dfrac{2\pi(d - t)}{\lambda_1} \sqrt{\dfrac{\varepsilon_2}{\varepsilon_1} - \left(\dfrac{\lambda_1}{\lambda_c} \right)^2}$

$\qquad\qquad = \dfrac{\sqrt{\varepsilon_1/\varepsilon_2} \sqrt{1 - (\lambda_1/\lambda_c)^2}}{\sqrt{(\varepsilon_2/\varepsilon_1) - (\lambda_1/\lambda_c)^2}};$

$\qquad$**(b)** 3.479 GHz

P10.16. 0.882×10^{-3} Np/m

P10.17. **(c)** 9276

P10.24. 0.7108

P10.25. $T_0 = \sqrt{2}\beta_z^{(2)}z; \sqrt{2}T_0$

P10.27. **(a)** $4.89\mu s$;

$\qquad$**(b)** 106 ps

P10.32. $101.5°$

P10.34. $\mathbf{E}_r = -\dfrac{E_0}{15}\,(7\mathbf{a}_x + \mathbf{a}_y)\cos(6\pi \times 10^9 t + 20\pi z)$

$\mathbf{E}_t = -\dfrac{E_0}{5}\,(2\mathbf{a}_x + \mathbf{a}_y)\cos(6\pi \times 10^9 t - 60\pi z)$

$\qquad + \dfrac{2E_0}{15}\,(\mathbf{a}_x - 2\mathbf{a}_y)\cos(6\pi \times 10^9 t - 40\pi z)$

$\mathbf{H}_r = -\dfrac{E_0}{15\eta_0}\,(\mathbf{a}_x - 7\mathbf{a}_y)\cos(6\pi \times 10^9 t + 20\pi z)$

$\mathbf{H}_t = -\dfrac{3E_0}{5\eta_0}\,(\mathbf{a}_x - 2\mathbf{a}_y)\cos(6\pi \times 10^9 t - 60\pi z)$

$\qquad + \dfrac{4E_0}{15\eta_0}\,(2\mathbf{a}_x + \mathbf{a}_y)\cos(6\pi \times 10^9 t - 40\pi z)$

P10.35. **(a)** 1.25 cm; **(b)** $-0.6E_0\cos(6\pi \times 10^9 t + 20\pi z)\,\mathbf{a}_x$; **(c)** 2.5 cm

CHAPTER 11

P11.4. 3.2038 V/m; 2.2503 V/m; 4.133×10^{-3} A/m

P11.6. 0.2024λ

P11.8. 6.075

P11.10. 2.1932 W

P11.12. 0.0167 A, 0.01 W

P11.14. **(a)** $\mathbf{E} = -\dfrac{\eta\beta I_0 L \sin\theta}{8\pi r}\sin(\omega t - \beta r)\,\mathbf{a}_\theta$

$\qquad \mathbf{H} = -\dfrac{\beta I_0 L \sin\theta}{8\pi r}\sin(\omega t - \beta r)\,\mathbf{a}_\phi$

$\qquad$ **(b)** $20\pi^2\left(\dfrac{L}{\lambda}\right)^2$, 1.5

P11.16. $\dfrac{\pi\eta}{6}\left(\dfrac{2\pi a}{\lambda}\right)^4$, 1.5

P11.18. **(a)** $|\cos\psi|\left|\sin\left(\dfrac{\pi}{2}\cos\psi\right)\right|$;

$\qquad$ **(b)** $|\cos\psi|\cos\left(\dfrac{\pi}{4}\cos\psi - \dfrac{\pi}{4}\right)$

P11.20. Five elements spaced 2λ apart, current amplitudes in the ratio 1:2:3:2:1, and progressive phase shift of 180°

P11.23. $|\sin\theta\,\sin(\beta d_1 \sin\theta)\cos(\beta d_2 \cos\theta)|$

P11.25. **(a)** 0.8284; **(b)** 4; **(c)** 0

P11.27. $\left(\dfrac{j\beta E_0 abe^{-j\beta r}}{4\pi r}\right)\left[\dfrac{\pi^2}{\pi^2 - \left(\dfrac{\beta a}{2}\sin\theta\cos\phi\right)^2}\right]$

$\qquad\times\left[\dfrac{\sin\left(\dfrac{\beta a}{2}\sin\theta\cos\phi\right)}{\dfrac{\beta a}{2}\sin\theta\cos\phi}\right]\left[\dfrac{\sin\left(\dfrac{\beta b}{2}\sin\theta\sin\phi\right)}{\dfrac{\beta b}{2}\sin\theta\sin\phi}\right]$;

(a) $\dfrac{4\lambda}{a}$; (b) $\dfrac{2\lambda}{b}$; (c) $1.44\dfrac{\lambda}{a}$; (d) $\dfrac{8\pi}{3}\dfrac{ab}{\lambda^2}$

P11.31. $12.543°$; $71.973°$; $91.368°$; $110.928°$; $133.672°$

CHAPTER 12

P12.3. (a) $V_A = 12$ V, $V_B = 5.25$ V, $V_C = 2$ V, $V_D = 14.75$ V, $V_E = 7$ V, $V_F = 2.75$ V;

(b) $|\mathbf{E}_B| = \dfrac{6.1033}{d}$ V/m; (c) $\pm61.25\varepsilon_0$ C/m

P12.6. $0.9242\varepsilon_0$ C

P12.9. $12.536\varepsilon_0 a$

P12.11. $3.873\varepsilon_0 a$

P12.14. $160.55\ \Omega$

P12.16. 227.8 pF/m; 38.2 pF/m; 35.7331 Ω; 1.2285×10^8 m/s

P12.19. $V_2 = 3.871$ V; $V_4 = 3.226$ V; $V_5 = 6.452$ V

P12.23. $V(4, 2) = V(4, 10) = 4.38$ V; $V(4, 4) = V(4, 8) = 3.245$ V; $V(4, 6) = 1.24$ V

Index